Table of Formulas

Perimeter and Area

Rectangle
Perimeter: $P = 2l + 2w$
Area: $A = lw$

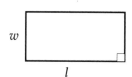

Parallelogram
Area: $A = bh$

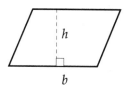

Square
Perimeter: $P = 4s$
Area: $A = s^2$

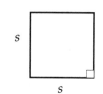

Trapezoid
Area: $A = \frac{1}{2}h(b_1 + b_2)$

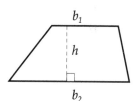

Triangle
Perimeter: P = the sum of the three sides
Area: $A = \frac{1}{2}bh$

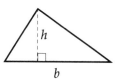

Circle
Circumference:
$C = 2\pi r$ or πd
Area: $A = \pi r^2$

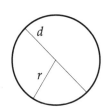

Right Triangle
Pythagorean Theorem:
$a^2 + b^2 = c^2$

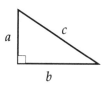

Volume and Surface Area

Rectangular Solid
Volume: $V = lwh$

Rectangular Prism
Surface Area:
$SA = 2(lh + wh + wl)$

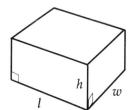

Cylinder
Volume: $V = \pi r^2 h$
Surface Area:
$SA = 2\pi r^2 + 2\pi rh$

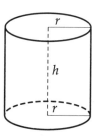

Cube
Volume : $V = s^3$
Surface Area:
$SA = 6s^2$

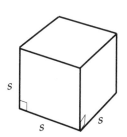

Sphere
Volume: $V = \frac{4}{3}\pi r^3$
Surface Area: $A = 4\pi r^2$

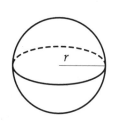

INTERMEDIATE ALGEBRA

INTERMEDIATE ALGEBRA

Sandra Pryor Clarkson
Hunter College

Barbara J. Barone
Hunter College

with **Mary Margaret Shoaf**   *Baylor University*

HOUGHTON MIFFLIN COMPANY   **Boston   New York**

*We dedicate this book
to our fathers for their complete faith in us,
to Shepherd G. Pryor III, and
in loving memory of Walter J. Primosch (1917–1995)
—S.P.C. and B.J.B.*

Senior sponsoring Editor: Maureen O'Connor
Associate Editor: Dawn Nuttall
Senior Project Editor: Cindy Harvey
Editorial Assistant: Michelle Francois
Senior Production/Design Coordinator: Carol Merrigan
Senior Manufacturing Coordinator: Marie Barnes
Marketing Manager: Sara Whittern, Ros Kane

Cover Design: Deborah Azerrad Savona
Cover Photo: Jim Schorer

Photo Credits: Chapter 1: AP/Wide World Photos; page 20, Roy Morsch/The Stock Market; page 54, David Young-Wolff/Photo Edit; Chapter 2: © 1992 O.G.S., Inc., Joseph H. Jacobson/The Picture Cube; Chapter 3: © 1990 Gabe Palmer/The Stock Market; Chapter 4: © David Ball/The Picture Cube; page 299, Photo Researcher; Chapter 5: Stephen Frisch/Stock Boston; Chapter 6: Carol Lee/The Picture Cube; page 444: The Chicago Historical Society; Chapter 7: Dave Bartruff/Stock Boston; Chapter 8: Kindra Clineff/The Picture Cube; Chapter 9: Tony Stone Images; Chapter 10: Frank Herholdt.

Printed in the U.S.A.

Library of Congress Number: 97-72454

ISBNs:
Text: 0-395-74425-3
Instructor's Annotated Edition: 0-395-87951-5

23456789-VH-01 00 99 98

CONTENTS

PREFACE

Intermediate Algebra serves as a bridge between the traditional and the reform text. It is a text that encourages students to explore mathematical ideas, formulate questions, and communicate mathematics using writing and graphing. This text embraces the best elements of the American Mathematical Association of Two-Year Colleges (AMATYC) standards and reform: motivating *applications* that drive the mathematics, including an abundance of *real-source data* graphs and tables; *technology* integrated where appropriate; *writing about mathematics*—in the exercises, as a new margin note feature, and as interactive chapter summaries; *group projects,* the majority of which involve real data; *variety in presentation* and illustration of mathematical concepts; and a *problem-solving approach* introduced early and integrated throughout.

Approach

Mastery   Many courses require students to demonstrate mastery of specific aspects of the material. Additionally, in this text, concepts and skills are integrated into a smoothly flowing text and not simply presented as isolated objectives. Section goals are identified, and the tests monitor achievement of those goals. Our exams give three questions for each objective. A student who answers two of the three questions correctly really knows the material.

Sensitivity to Learning Styles and Disabilities   Students have many opportunities to "see" mathematics in multiple formats—graphically, numerically, symbolically—and to use all these formats in solving problems. Additionally, Study Hints and Instructor Notes are given to aid students, including those with learning or perceptual difficulties. A section in the Instructor's Resource Manual is devoted to working with students with learning disabilities.

Motivation   We present problem situations from many areas of interest and concern that reflect the impact of mathematics in all aspects of modern life. Therefore, answers are not always artificially "nice" numbers but are often from real situations. This realism helps students develop genuine number sense.

Real-Life Applications

With respect to motivation and real data, newspapers, magazines, professional journals, and current and historical books were used as source material for the applications found in the following features.

Chapter and Section Applications   Each chapter has a *Chapter Lead-In* and a *Chapter Look-Back* and each section has a *Section Lead-In* and a *Section Follow-Up* that extends knowledge of mathematics in a real-life situation or in a mathematical exploration. These Lead-Ins relate peripherally to the section and chapter and help to interest students in the material to come.

Exercises and Examples   Examples and exercises involve data gathered from contemporary business, sports, current events, the sciences, and virtually all real-life situations. They cover a diverse selection of topics, including biology, physics, astronomy, real estate, architecture, engineering, and even elements of popular culture, such as movies. They appear early in the text and throughout all chapters. The wide variety and realism of the applications will appeal to students. The areas of application are identified.

Excursions   Excursions are extended project exercises that follow each section's exercise set. These problems can be assigned to individuals or groups. These exercises contain a wealth of real data and a written answer or justification is often required. Many are also open-ended and encourage creative thinking and problem solving.

- *Class Act*   These problems are especially designed to stimulate discussion and cooperation as student groups formulate the best solution. These can be assigned to individuals but may require more time than for groups.
- *Data Analysis*   These problems encourage graphical and/or statistical analysis. They are appropriate for both groups and individuals.
- *Posing Problems*   Numerical information is given in table, graph, or paragraph format. Students are encouraged to pose their own problems and solve them and share with others. These activities often stimulate lively classroom discussion.
- *Exploring . . .*   These problems often require some informed trial-and-error procedures to find a solution. Students will explore problems involving patterns, numbers, calculators, problem solving, and geometry.

Technology

Calculator Corners   All technology material has been specifically developed and written for this text. The Calculator Corners are integrated at appropriate places, providing detailed directions for how a graphing calculator can be useful—both in graphing and non-graphing topics. The Calculator Corners not only illustrate how a graphing calculator can make mathematical calculations easier, they also can take a mathematical idea one step further or present an alternate way to illustrate a mathematical concept. For classes not using calculators, the Calculator Corners are easy to identify and skip. Screen displays and directions are given for the Texas Instruments *TI 82/83* series, but the features discussed are found on all graphing calculator models.

The Calculator Corner examples and exercises were contributed by Dr. Mary Margaret Shoaf, who has given numerous workshops on the use of calculators in teaching mathematics. Additionally, her research and writing has focused on the use of calculators in enhancing student understanding. We are pleased to have her as a member of the writing team.

Writing/Group

Besides the Excursions and the section exercises identified with ✍, there are additional features which encourage student writing.

Margin Notes   In other texts, often student margin notes contain information for students to read only. We offer two types of student margin notes that ask students to give a written response, allowing students to interact immediately with the material.

- *Error Alert* These can be assigned individually or used as a catalyst for group discussion. A problem is presented along with an incorrect or incomplete solution. Students are asked to identify and correct the errors. Common student errors are presented in this feature. We ultimately want students to be able to identify and correct the errors they might make in their own work. No answers are given in the text in order to encourage students to rely on themselves for these answers.
- *Writer's Block* These questions require students to write about mathematics. Students explain, define, clarify, and interpret mathematical ideas, terms, and procedures.

Student Journal The student text comes with a separate Student Journal booklet. This journal contains interactive chapter summaries in which students must write their own summaries as prompted by questions. Each chapter summary covers all the definitions and rules in that chapter and asks students to provide definitions, complete statements and diagrams, fill in blanks, and write explanations in their own words. Students are encouraged to give an example of each definition and rule and to make notes to help them remember the material. Students are also encouraged to keep a homework journal, with pages numbered, in order to note the page numbers of problems that they have worked that best "model" the chapter's main ideas. Overall, the Student Journal is designed to help students make the most efficient use of their time while studying the material.

Chapter Pedagogy

Section Goals On the first page of each section, Section Goals list the terminal objectives for the section. These goals will be taught and practiced in that section, together with necessary "pre-skills."

Definitions and Rules Key words are defined and rules (or procedures) are clearly delineated in boxes set apart from the rest of the material. These features build vocabulary and aid students in communicating mathematics. Each mathematical term appears in bold type when it first occurs in the text.

Study Hints These are student margin notes that aid students in learning the material.

Warm-Ups Students are directed to a Warm-Up after each worked example. The Warm-Up parallels the example, reinforcing the concept just taught and building student confidence while also allowing students to interact immediately with the material. These special exercises appear right before the section exercises, and answers to all Warm-Ups are in the back of the text.

Connections to . . .

- *Probability and Statistics* Topics in probability and statistics are introduced throughout the text and reinforced periodically. Topics include mean, median, mode, standard deviation, range, line graphs, histograms, pie charts, and naming probabilities.

▪ *Geometry and Measurement* Topics in geometry and measurement are introduced throughout the text and reinforced periodically. Topics include area and perimeter of plane figures, surface area and volume of rectangular solids, cylinders, spheres and pyramids, similar triangles, parallel lines cut by a transversal, changing units, and scientific notation.

Problem-Solving Preparation Problem solving is emphasized throughout the text. A useful set of problem-solving strategies is developed in the text so that students learn to approach problems in an organized, efficient manner. Students are taught to reason mathematically using a four step procedure modeled on Polya's problem-solving work.

Estimation Throughout the text, estimation is taught and reinforced, where appropriate, to promote number sense.

Functions and Set Notation Functions and set notation are introduced early and reinforced throughout the text. Students completing this text will be well prepared for college algebra.

Assessment

We have incorporated a variety of assessment features into this text. Instructors might want to use a portfolio of these (and possibly also Warm-Ups, Excursions, Writer's Block, and Error Alert features) in evaluating student knowledge.

Skills Check The Skills Check sections determine whether students have the prerequisite skills necessary for that chapter. Passing this test does not allow a student to "skip" the chapter; instead, it lets them know if they are ready to begin. Students who miss problems are referred to appropriate sections for review.

Section Exercise Sets These exercises provide review and practice for all skills taught in the section. Exercises are arranged in pairs; answers to odd-numbered problems are at the back of the text.

Mixed Practice In all sections except the first section in a chapter, there are mixed practice problems that review and reinforce skills previously taught in that chapter. Answers to the odd-numbered problems are at the back of the text.

Chapter Review These exercises are divided into two sets. The first group of exercises provides a section-by-section review of the chapter's work. The second set is a mixed review that is not section referenced. All answers appear in the back of the text.

Chapter Test This test follows the chapter review material. It is mastery-based and contains three questions for each section goal in a chapter. Questions are organized by section goal and answers do *not* appear in the back of the text (solutions can be found in the Solutions Manual). A student who answers two questions correctly out of each group of three has most likely mastered the objective. The Chapter Test takes approximately one hour. An additional Chapter Test and suggestions for Assessment Alternatives—in keeping with the standards recommended by AMATYC and NCTM—appear in the Instructor's Resource Manual.

Cumulative Review   Beginning with Chapter 2, a Cumulative Review section appears at the end of each chapter. This review reinforces skills taught in all previous chapters including the current one. The exercises are mixed. Answers to all these problems are at the back of the text.

Supplements

FOR INSTRUCTOR USE

Instructor's Annotated Edition   The Instructor's Annotated Edition (IAE) is an exact replica of the student text and includes answers to all the exercises. All answers are given as red teacher annotations with the exception of those graphing answers that appear in the student answer sections at the back of the text. The IAE also contains helpful *Instructor Notes* in the margins. These Instructor Notes provide hints for teaching students, including those with learning or perceptual difficulties, and cautions about possible student misunderstandings.

Instructor's Resource Manual with Test Bank   The *Instructor's Resource Manual* contains information about how to organize a laboratory course using this text. Suggestions are also given about the use of cooperative learning groups and for working with students with certain types of perceptual or learning disabilities. There is also a section on how students can gain math confidence as well as a section listing the AMATYC standards. Finally, there is an additional Chapter Test for each chapter in the text. The questions are in random order and are slightly more challenging than the Chapter Tests in the text.

The *Test Bank* contains about 2000 test items. Items are grouped by section and goal and are also available in the Computerized Test Generator.

Computerized Test Generator   The Computerized Test Generator is the electronic version of the printed Test Bank. This user-friendly software permits an instructor to construct an unlimited number of customized tests from the 2000 test items offered. **On-line testing** and **gradebook** functions are also provided. It is available in Windows for the IBM PC and compatible computers.

Solutions Manual   The Solutions Manual contains full, worked-out solutions to all exercises in the text.

FOR STUDENT USE

Student Journal   This journal, which comes with the student edition, contains interactive chapter summaries in which students must write their own summaries as prompted by questions. Each chapter summary covers all the definitions and rules in that chapter and asks students to provide definitions, complete statements and diagrams, fill in the blanks, and write explanations in their own words.

Computer Tutor   The Computer Tutor is a text-specific, networkable, interactive, algorithmically driven software package. This powerful ancillary features full-color graphics, algorithmic exercises with extensive hints, animated solution steps, and a comprehensive classroom management system. It is available for the IBM PC and compatible computers and the Macintosh. A computer disk icon [CT] appears in the Section Goals box as a reminder that each section is covered in the tutor.

Videos Within each section of the text, a videotape icon ▭ appears in the Section Goals box. The icon contains the reference number of the appropriate video, making it easy for students to find the extra help they may need. Each video opens with a relevant application which is then solved at the end of the lesson.

Student Solutions Manual The Student Solutions Manual contains full, worked-out solutions to all the exercises whose answers are at the back of the text, namely all the Warm-Ups, the odd-numbered section exercises, all the Chapter Review exercises, and all the Cumulative Review exercises.

Acknowledgments

Many people have helped us directly with this project. In particular, we wish to thank Cindy Harvey, Kathy Deselle, George McLean, Dawn Nuttall, and Maureen O'Connor.

A special thanks must go to JoAnne Kennedy at LaGuardia Community College for authoring the *Test Bank, Solutions Manual,* and *Student Solutions Manual,* and to Mary Margaret Shoaf at Baylor University for authoring the Calculator Corners.

Through the years, each of us has people that inspire us, offer us opportunities, and teach us important lessons. The following have contributed to our growth in various ways: Mary Ellen Bohan; the late Ann Braddy; Mrs. Breiner; Mac Callaham; the late Tom Clarke; Tom Davis; the late Mary P. Dolciani Halloran and her husband James Halloran; Henry Edwards, Jr.; George Grossman; Eleanore Kantowski; Mrs. Leonardi; the late Sarah Anna Mathis; Ed Millman; the late Len Pikaart; Henry Pollak; Donna Shalala; June Smith; Andre Thibodeau; the late Mr. Towson; Zalman Usiskin; Bill Williams; Jim Wilson; and Gloria Wolinsky.

We would also like to thank the following reviewers for their suggestions:

Paul Allen, *University of Alabama;* Joseph Altinger, *Youngstown State University,* OH; Kathleen Burk, *Pensacola Junior College,* FL; Paul Dirks, *Miami-Dade Community College,* FL; Irene Doo, *Austin Community College,* TX; Jeff Koleno, *Lorain County Community College,* OH; Richard Leedy, *Polk Community College,* FL; William Livingston, *Missouri Southern State College;* Yvonne Lord, *University College of the Caribou,* British Columbia, Canada; Maria Maspons, *Miami-Dade Community College,* FL; Lamar Middleton, *Polk Community College,* FL; John Searcy; Samuel Self, *El Paso Community College,* TX; and John Shannon, *Northeastern University,* MA.

And, of course, our families contributed most of all!

S.P.C. and B.J.B.

INDEX OF APPLICATIONS

THE REAL NUMBER SYSTEM

"*S*he won 64% of the precinct." You may have heard election results reported this way. Percentages can also be expressed using other real numbers. In this case, 64% is the fraction $\frac{64}{100}$ (64 per 100). In other words, 64 voters out of every 100 voted for the candidate. This fraction can also be written as the decimal 0.64, or simplified to $\frac{16}{25}$. If you knew that there were 75,000 voters in the precinct, you could determine that 48,000 of the 75,000—64%—voted for the candidate.

■ *Find situations in the newspaper that are presented as percents.*
In what ways can percents make number information clear?

SKILLS CHECK

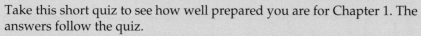

Take this short quiz to see how well prepared you are for Chapter 1. The answers follow the quiz.

1. Simplify: $18 + 3\frac{3}{5} + 2\frac{1}{2}$

2. Simplify: $6 \div 1\frac{7}{8}$

3. Simplify: $126 \div 12$

4. Simplify: $14\frac{7}{10} - 8\frac{3}{5}$

5. Simplify: $\left(2\frac{1}{2}\right)\left(3\frac{2}{3}\right)$

6. Simplify: $32.5 + 3.25$

7. Simplify: 0.34×4.9

8. Simplify: $1.2\overline{)40.08}$

9. Simplify: $4.7 - 0.976$

10. Simplify: $9\frac{1}{3} \div 1\frac{1}{6}$

ANSWERS: **1.** $24\frac{1}{10}$ **2.** $3\frac{1}{5}$ **3.** 10.5 **4.** $6\frac{1}{10}$ **5.** $9\frac{1}{6}$ **6.** 35.75 **7.** 1.666 **8.** 33.4 **9.** 3.724 **10.** 8

(These items are basic computations with whole numbers, fractions, and decimals that we will review in Chapter 1.)

CHAPTER LEAD-IN

We use numbers to represent amounts. Often, the numbers we use have to be interpreted in order to have meaning.

For example, if we say that 75% of the children on a bus have been sneezing, we have a sense of what proportion of the children have sneezed, but we have to know the number of students on the bus in order to know how many children have sneezed.

- Find examples in the newspaper where we know the proportion but not the number.

1.1 Introduction to the Real Numbers

SECTION LEAD-IN

You are working for 20 days and receive pay as follows:

Day 1	$1
Day 2	$2
Day 3	$4
Day 4	$8
.	.
.	.
.	.
Day 20	$524,288

Using this information, ask and answer at least four questions.

Real Numbers

Like arithmetic, algebra is concerned mainly with numbers. The numbers of algebra are the **real numbers.**

The real numbers include several groups of numbers, as shown below. We shall work almost exclusively with real numbers in this text.

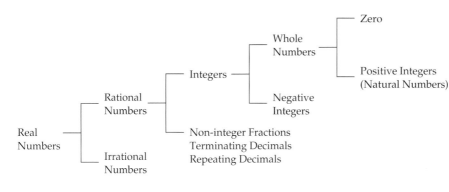

Opposites of Real Numbers

All the non-zero real numbers have opposites.

Opposite of a Real Number

The opposite of any non-zero real number n is the real number $-n$.

Also, for any non-zero real number n,

$$-(-n) = n$$

INSTRUCTOR NOTE

Remind students that n can be positive or negative. This appears to be a difficult concept for some.

ERROR ALERT

In these boxes you will identify the error and give a correct answer.

$-(-5) = -5$

Calculator Corner

You can use your graphing calculator's Home Screen to find the opposite of a number. Consider the following example.

Find the opposite of −2/3. (Note: the $\boxed{(-)}$ key means negative; the $\boxed{-}$ key means the operation of subtraction.)

```
-(-2/3)
```

```
-(-2/3)
           .6666666667
```

Press **ENTER.**

You can also obtain a fractional answer by telling the calculator to give the answer in fractional form.

```
-(-2/3)
```

```
MATH  NUM  HYP  PRB
1:▶Frac
2:▶Dec
3:³
4:³√
5:ˣ√
6:fMin(
7↓fMax(
```

```
-(-2/3)▶Frac
```

Press **ENTER.**

```
-(-2/3)▶Frac
              2/3
```

Press **ENTER.**

Some calculators will evaluate a "double negative sign" without the use of parentheses.

```
--2/3▶Frac
```

```
--2/3▶Frac
              2/3
```

Press **ENTER.**

Graphing Real Numbers

All real numbers can be graphed on a number line.

> **To graph a real number that is not an integer**
>
> 1. Determine which two integers it lies between.
> 2. Approximate its position between those integers on a number line.

▪▪▪

E<small>XAMPLE</small> **1**

Graph $-\frac{4}{5}$, 4.6, π, and the opposite of -0.8686 on one number line.

INSTRUCTOR NOTE

In this text, we use a "raised" minus sign to indicate a negative fraction.

S<small>OLUTION</small>

We identify the integers that each of those numbers lies between.

$-\frac{4}{5}$ lies between 0 and -1.

4.6 lies between 4 and 5.

π, which is approximately equal to 3.14159, lies between 3 and 4.

The opposite of -0.8686 is 0.8686, which lies between 0 and 1.

We draw a number line and mark off the integers from, say, -2 to 9. That range will include all four points.

To graph $-\frac{4}{5}$, we place a dot about four-fifths of the way from 0 to -1; to graph 4.6, we place a dot slightly more than halfway from 4 to 5. We graph π and 0.8686 similarly, obtaining the following graph.

▶ CHECK **Warm-Up 1**

Ordering Real Numbers

The number line shows that the real numbers have a certain **order.** For example, the figure above shows that -2 is always to the left of -1. We say that -2 is *less than* -1.

> A real number a is **less than** a real number b if a is to the left of b on the number line.

We write $a < b$ to mean a is less than b.

The figure above also shows that 1 is always to the right of -1. We say that 1 is *greater than* -1.

! ! !

ERROR ALERT

Identify the error and give a correct answer.

Compare -37.4 and -36.5.

Incorrect Solution:

Because 37.4 is larger than 36.5, -37.4 is larger than -36.5.

INSTRUCTOR NOTE

Students who spend some time graphing real numbers have an easier time dealing with the order of negative numbers.

INSTRUCTOR NOTE

There are various ways for students to remember which way to write the symbols $<$ and $>$. Remind them that the symbol points toward the smaller number.

A real number a is **greater than** a real number b if a is to the right of b on the number line.

We write $a > b$ to mean a is greater than b.

We know that

There is no greatest positive real number and there is no least negative real number.

• • •

EXAMPLE 2

Order 0.8, $-5\frac{3}{5}$, -820, and $6.2\overline{22}$ from least to greatest, using the symbol $<$.

SOLUTION

We identify the integers that each number lies between.

0.8 lies between 0 and 1.
$-5\frac{3}{5}$ lies between -5 and -6.
-820 lies at -820.
$6.2\overline{22}$ lies between 6 and 7.

Placing these numbers in order, left to right, as they would appear on a number line, we get

$$-820 < -5\frac{3}{5} < 0.8 < 6.2\overline{22}$$

If we ordered these numbers from greatest to least, we'd get

$$6.2\overline{22} > 0.8 > -5\frac{3}{5} > -820$$

▶ CHECK **Warm-Up 2**

STUDY HINT

To compare numbers, try using graph paper and writing the digits one to a box.

1	1	2	4	6	3	7	5
1	1	3	3	7	4	6	9
		1	1	2	3	7	8

Make sure you align ones digits when ordering whole numbers; align decimal points when ordering decimal numbers.

INSTRUCTOR NOTE

This study hint is particularly relevant for some students with learning disabilities.

Calculator Corner

You can use your graphing calculator's Home Screen to test the relationship between two numbers. That is, the calculator can help you determine if one number is greater or smaller than another number. As an example, consider the calculator work for the following example. The inequality signs can be found under the **2nd TEST** menu on many graphing calculators. You may need to consult the manual of your particular graphing calculator for directions.

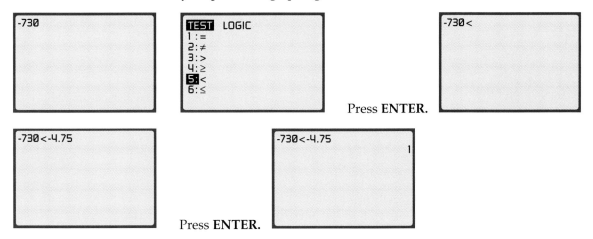

Press **ENTER.**

Press **ENTER.**

The calculator returns a response of 1 if the statement is true and a response of 0 if the statement is false. Notice the difference between the last calculator screen shown above and the one at the left below. Continuing to list the given numbers from least to greatest results in the screen at the right below.

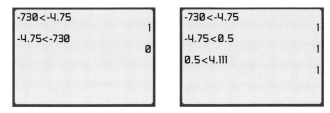

Now you should be able to write all four numbers from least to greatest.

Try the following problem on your own with the help of your graphing calculator.

Which of the following real numbers are greater than -3?

$$-2.999, -3.02, -17, -0.03, -2.00999$$

Definition

For a and b, both real numbers, $a < b$ if and only if there is a positive real number k that exists such that $a + k = b$.

For example, $6 < 19.5$, because $6 + 13.5 = 19.5$. Here, k is the number 13.5, a positive real number.

Absolute Value

In Example 1 we graphed the opposite of -0.8686 on the number line. The numbers -0.8686 and 0.8686 are the same *distance* from zero but in opposite directions. We say that they have the same *absolute value*.

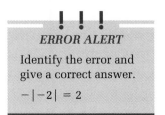

ERROR ALERT

Identify the error and give a correct answer.

$-|-2| = 2$

> **Absolute Value**
>
> The **absolute value** of a real number n, written $|n|$, is the distance of n from zero on a number line.

For example, $|4| = 4$ because 4 is 4 units from zero. And $|-4| = 4$ because -4 is 4 units from zero.

Because distances are never negative, absolute values are never negative. A number and its opposite have the same absolute value: $|n| = |-n|$.

Calculator Corner

You can use your graphing calculator's Home Screen to evaluate the absolute value of a number. On most graphing calculators, the letters **ABS** refer to the Absolute Value function. For example, you can find the absolute value of -7.5 as follows:

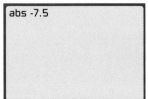

Press **ENTER.**

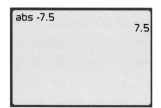

Explain why the following calculator result is different from the first example above. Why is the second answer a negative number while the first answer was positive?

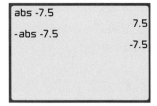

▪▪▪

EXAMPLE 3

Evaluate each of the following:

a. $-|7.24|$ **b.** the absolute value of the opposite of $4\frac{1}{2}$

SOLUTION

a. The absolute value of 7.24 is 7.24, and the opposite of that is -7.24. So $-|7.24| = -7.24$.

b. The opposite of $4\frac{1}{2}$ is $-4\frac{1}{2}$, and the absolute value of $-4\frac{1}{2}$ is $4\frac{1}{2}$. So $\left|-4\frac{1}{2}\right| = 4\frac{1}{2}$.

▶ *CHECK* **Warm-Up 3**

Rounding

Often numbers are **rounded** before being used. Population size, the national debt, and many other measures are often reported in rounded form.

To round a decimal number to a given place value

1. Locate and underline the digit *to the right* of the place you are rounding to.
2. If that digit is less than 5, change it and all digits to its right to zero.
3. If that digit is greater than or equal to 5, add 1 to the digit to *its left*. Then change the underlined digit and all digits to its right to zero.
4. If you are rounding to a decimal place, drop the underlined number and all digits to its right.

▪▪▪

EXAMPLE 4

Round 69.4545 to

a. the nearest tenth
b. two decimal places
c. the nearest thousandth
d. the nearest whole number
e. the nearest hundred

SOLUTION

a. Write 69.4<u>5</u>45 and round to 69.5.
b. To two decimal places means the same as to the nearest hundredth: Write 69.45<u>4</u>5 and round to 69.45.
c. Write 69.454<u>5</u> and round to 69.455.
d. Write 69.<u>4</u>545 and round to 69.
e. Write <u>6</u>9.4545 and round to 100.

▶ *CHECK* **Warm-Up 4**

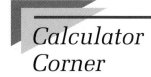

Calculator Corner

Most graphing calculators have the capability of rounding a result to a specified number of decimal places. The calculator can be instructed to give answers in tenths as shown in the following example. Use the **MODE** screen to set the number of decimal places.

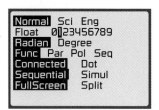  

Press **ENTER.**

You can also use the **MATH/NUMBER** utility of your graphing calculator to help you round decimal numbers to different decimal places. Round 69.4545 first to two decimal places and then to three decimal places.

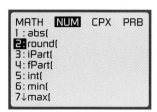 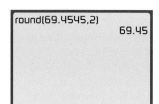 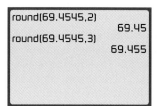

Rounding Fractions

It is often possible to estimate the results of computations with fractions. But to do so, we need to be able to round fractions and mixed numbers to the nearest whole number.

> **To round a fraction to the nearest whole number**
>
> If a proper fraction is equal to or greater than $\frac{1}{2}$, round it up to 1; otherwise, round it to zero.

■ ■ ■

EXAMPLE 5

Round $23\frac{15}{28}$ to the nearest whole number.

SOLUTION

We need to compare $\frac{15}{28}$ to $\frac{1}{2}$. To compare these fractions, we must write them with the same denominator. We use their **least common denominator** (LCD). To find the LCD, we factor both denominators into a product of primes:

$$28 = 7 \times 2^2 \times 1 \qquad 2 = 2 \times 1$$

Then we write all the prime factors with their greatest exponents:

$$7 \times 2^2 \times 1 = 28$$

The LCD of $\frac{15}{28}$ and $\frac{1}{2}$ is 28. So we have

$$\frac{15}{28} \quad \text{and} \quad \frac{1 \times 14}{2 \times 14} = \frac{14}{28}$$

Because $15 > 14$, we know that $\frac{15}{28} > \frac{1}{2}$. So we round the fraction up to 1. Then $23\frac{15}{28}$ rounded to the nearest whole number is $23 + 1 = 24$.

▶ *CHECK* **Warm-Up 5**

Writing Numbers in Scientific Notation

Very large and very small numbers often occur in scientific and engineering work. **Scientific notation,** a way of writing numbers, was developed to simplify calculations that involve such numbers.

A number written in scientific notation consists of two parts connected by a multiplication sign.

1. a decimal number greater than or equal to 1 and less than 10
2. a power of 10

To write a decimal number in scientific notation

1. Move the decimal point right or left to obtain a number n such that $1 \leq n < 10$. (We read this as "one is less than or equal to n and n is less than 10.")
2. Count the number of places p that the decimal point has been moved.
3. Multiply n by 10^p if the decimal point was moved to the left. Multiply by 10^{-p} if it was moved to the right. Eliminate any meaningless zeros.

WRITER'S BLOCK

Hiro read $1 \leq n < 10$ as "n is greater than or equal to 1 and less than 10." Explain how his reading can also be correct.

EXAMPLE 6

Write in scientific notation:

a. 4,060,000,000   **b.** −0.0000089

SOLUTION

a. We need to move the decimal point *to the left* 9 places to get a number n such that $1 \leq n < 10$: $4.060000000.$

So we multiply n by 10^9: 4.060000000×10^9.

The zeros to the right of the 6 are meaningless, so we eliminate them, getting

$$4.06 \times 10^9$$

b. Ignoring the minus sign, we move the decimal point *to the right* 6 places to obtain a number n such that $1 \leq n < 10$: $-0.0\ 0\ 0\ 0\ 0\ 8.9$

So we multiply the result by 10^{-6}. Eliminating the meaningless zeros on the left yields: $-0000008.9 \times 10^{-6} = -8.9 \times 10^{-6}$.

INSTRUCTOR NOTE

Suggest that students multiply to check that their rewriting is correct until they are comfortable with the process.

We reverse the process when we are given a number in scientific notation and must write it in **standard notation.**

▶ CHECK **Warm-Up 6**

To write a scientific number in standard notation

1. Move the decimal point a number of places equal to the exponent of 10. Move it to the right if the exponent is positive; move it to the left if the exponent is negative. (Add zeros as necessary.)
2. Eliminate the multiplication sign and power of 10.

▪▪▪

EXAMPLE 7

Write in standard notation:

a. 3.417×10^5 **b.** 8.03×10^{-12}

SOLUTION

a. Because the exponent is 5, we move the decimal point 5 places *to the right.*

$$3.417 \times 10^5 = 3.4\ 1\ 7\ 0\ 0.$$

So $3.417 \times 10^5 = 341{,}700$.

b. Because the exponent is -12, we move the decimal point 12 places *to the left.*

$$8.03 \times 10^{-12} = 0.0\ 0\ 0\ 0\ 0\ 0\ 0\ 0\ 0\ 0\ 8.0\ 3$$

So $8.03 \times 10^{-12} = 0.00000000000803$.

▶ CHECK **Warm-Up 7**

Practice what you learned.

SECTION FOLLOW-UP*

Day	1	2	3	4	· · ·	20
Earned	$1	$2	$4	$8	· · ·	$524,288

*Each section's Follow-Up refers to topics or questions presented in the section's Lead-In.

These are the questions we asked.

1. Using the base 2, write the amounts earned as powers of 2.

ANSWER:

Day	1	2	3	4	$\cdots$	20
Earned	2^0	2^1	2^2	2^3	$\cdots$	2^{19}

2. How did we find that $524,288 = 2^{19}$?

ANSWER: We used a *pattern*. We noticed that days 1, 2, 3, and 4 worked easily into a pattern showing that the amount earned was 2^{n-1}, where n was the number of the work day. We guessed that the pattern would work for day 20. That is, we decided that when the day was 20, the amount earned was 2^{20-1} or 2^{19}. We checked using a calculator, and we were correct.

3. Using the number $\frac{1}{2}$, describe the relationship between any two consecutive days.

ANSWER: Each day's earnings is $\frac{1}{2}$ what will be earned the next day.

4. Using the phrase "twice as much as," describe the relationship between any two consecutive days.

ANSWER: Each day's earnings is twice as much as was earned the day before.

1.1 WARM-UPS

Work these problems before you attempt the exercises.

1. Graph $\frac{-2}{3}$, 2.17, $-5.111\ldots$, and -3 on one number line.

2. Order from greatest to least using $>$:
$-3.927, -195.62, -503$, and $2\frac{1}{2}$ $2\frac{1}{2} > -3.927 > -195.62 > -503$

3. Evaluate the opposite of the absolute value of 17 and the absolute value of the opposite of -29. $-17; 29$

4. Round 173.98546 to the nearest ten, tenth, hundred, and thousandth. $170; 174.0; 200; 173.985$

5. Round to the nearest whole number: $2\frac{7}{12}$ 3

6. Write 70,540,000 in scientific notation. 7.054×10^7

7. Write 7.19×10^{-7} in standard notation. 0.000000719

1.1 EXERCISES

Note: Use your graphing calculator to check your results whenever possible.

In Exercises 1 through 8, find the opposite of the given number.

1. $-2\frac{1}{2}$ $2\frac{1}{2}$

2. -13 13

3. $|-1|$ -1

4. $-|-9|$ 9

5. $-|-22.1|$ 22.1

6. $\left|7\frac{1}{3}\right|$ $-7\frac{1}{3}$

7. $\left|-|-69|\right|$ -69

8. $-|-21.4|$ 21.4

In Exercises 9 through 12, find the number that is described.

9. the opposite of the opposite of -38 -38

10. the absolute value of the opposite of $5\frac{1}{2}$ $5\frac{1}{2}$

11. the opposite of the absolute value of a number x $-x$ if $x > 0$, or x if $x \leq 0$

12. the absolute value of the opposite of a number n n if $n \geq 0$, or $-n$ if $n < 0$

In Exercises 13 through 16, graph each group of numbers in order on the number line.

13. $-8\frac{2}{3}$, $-200\frac{2}{5}$, and -87.98

14. -64.5, $-\left(-96\frac{1}{2}\right)$, and $-\left(-48\frac{2}{3}\right)$

15. $2\frac{1}{3}$, $-5.\overline{666}$, and 14.7

16. $-29\frac{3}{4}$, 18.82, and 400.4

In Exercises 17 through 20, arrange each set of rational numbers in order from least to greatest. Use the symbol $<$.

17. $-|-2981|$, $-|-|264||$, $-|7|$
$-|-2981| < -|-|264|| < -|7|$

18. $-64\frac{2}{4}$, $|92|$, $27\frac{1}{3}$ $-64\frac{2}{4} < 27\frac{1}{3} < |92|$

19. 28, -58, 6.4, $-\left|48\frac{1}{5}\right|$ $-58 < -\left|48\frac{1}{5}\right| < 6.4 < 28$

20. $|-|-46||$, -20, 25, -0.05
$-20 < -0.05 < 25 < \left|-|-46|\right|$

In Exercises 21 through 28, use the symbol $>$, $<$, or $=$ to indicate the relationship between each pair of numbers.

21. -27 __$<$__ $-(-27)$

22. $-(-45)$ __$>$__ -45

23. $-|-54|$ __$=$__ $-|-54|$

24. $-|-67|$ __$=$__ $-|67|$

25. $-(-31)$ __$>$__ $-|-31|$

26. $-|-13|$ __$<$__ $-(-13)$

27. $-[-(-2)]$ __$<$__ $-(-|-2|)$

28. -2.5 __$=$__ $-[-(-2.5)]$

In Exercises 29 through 36, round each number to the nearest one, tenth, and thousandth.

29. 245.9081
246; 245.9; 245.908

30. 77.76619 78; 77.8; 77.766

31. 7.72536 8; 7.7; 7.725

32. 0.9891041
1; 1.0; 0.989

33. 1.617725 2; 1.6; 1.618

34. 39.8259 40; 39.8; 39.826

35. 0.1740132 0; 0.2; 0.174

36. 1021.8899
1022; 1021.9; 1021.890

In Exercises 37 through 40, determine whether the numbers are greater than, equal to, or less than $\frac{1}{2}$.

37. $\frac{5}{8}$ greater than $\frac{1}{2}$

38. $\frac{126}{218}$ greater than $\frac{1}{2}$

39. $\frac{6}{13}$ less than $\frac{1}{2}$

40. $\frac{132}{316}$ less than $\frac{1}{2}$

In Exercises 41 through 44, round each mixed number to the nearest whole number.

41. $15\frac{3}{8}$ 15 **42.** $29\frac{4}{9}$ 29 **43.** $31\frac{17}{21}$ 32 **44.** $26\frac{5}{21}$ 26

Work the following application problems.

45. Here are the densities of some substances, in grams per cubic centimeter (g/cm^3). Arrange them in order from greatest to least. rubber, petroleum, alcohol, beechwood

Alcohol	0.789332 g/cm^3
Rubber	0.9301152 g/cm^3
Petroleum	0.87865 g/cm^3
Beechwood	0.72 g/cm^3

46. There are three different ways of measuring a year. Each results in a measurement of approximately 365 days, but they all differ in the decimal approximations. Arrange these three years in order from greatest number of days to least. anomalistic year, sidereal year, trophical year

Sidereal year	365.2563656 days
Anomalistic year	365.2596 days
Trophical year	365.242198781 days

47. The first eight places of the decimal equivalents for $\frac{1}{12}, \frac{1}{13}, \frac{1}{14},$ and $\frac{1}{15}$ (not necessarily in order) are

0.07142857
0.07692307
0.06666666
0.08333333

0.08333333 > 0.07692307 > 0.07142857 > 0.06666666

Arrange these decimal numbers in order from greatest to least.

48. The following list gives the comparative masses of the planets (with Earth given mass 1). Put them in order from least mass to greatest.

Venus	0.81	Neptune	17.23
Mercury	0.06	Uranus	14.54
Pluto	0.17	Saturn	95.15
Mars	0.11	Jupiter	317.83
Earth	1		

Mercury, Mars, Pluto, Venus, Earth, Uranus, Neptune, Saturn, Jupiter

49. The number on the left has been rounded. Of the other three numbers, circle those that it could have been before rounding.

a. 6.9 (6.85) (6.94999) (6.9049)

b. 7.23 7.239 (7.23) (7.2349875)

c. 2.0 2.093 (1.99) (1.957867)

d. 3.04 (3.0399876) (3.04499) 3.045

e. 2.19 (2.1949) (2.18997) (2.1854)

f. 3.611 (3.61107) 3.61195 (3.610997)

50. Use the following chart to estimate the answer to parts (a) through (d). This chart gives the rotational velocity, at the equator, of each of the planets.

Venus	4.05 miles per hour
Mercury	6.73 miles per hour
Pluto	76.56 miles per hour
Mars	538 miles per hour
Earth	1040 miles per hour
Neptune	6039 miles per hour
Uranus	9193 miles per hour
Saturn	22,892 miles per hour
Jupiter	28,325 miles per hour

a. About how many times faster is Pluto's rotation than that of Venus? Is it closer to:

20 times?   2 times?   200 times?   0.2 times?   20 times

b. About how many times faster is Earth's rotation than that of Pluto? Is it closer to:

13 times?   1.3 times?   130 times?   0.13 times?   13 times

c. About how many times faster is Jupiter's rotation than that of Mercury? Is it closer to:

400 times?   4000 times?   4 times?   0.4 times?   4000 times

d. About how many times faster is Neptune's rotation than that of Mars? Is it closer to:

1.2 times?   12 times?   120 times?   0.12 times?   12 times

51. In 1987 total motor vehicle registrations in British Columbia, Québec, and Ontario were 2,175,032, 2,974,099, and 5,179,918, respectively. Estimate how many million vehicles were registered.   10,000,000 vehicles

52. The 1987 estimated revenue for the Republic of China was $16,349,000,000. Estimated expenses were $16,329,200,000. Estimate the difference to the nearest ten million. $20,000,000

53. There were 953,000 college graduates in the academic year 1981–1982 and 930,684 in the academic year 1971–1972. Estimate the change in the number of college graduates to the nearest ten thousand. Use your calculator to check your answer.   20,000 graduates

54. The greatest official altitude reached by an occupied balloon is 113,740 feet, whereas the greatest unofficial altitude reached by an occupied balloon is 123,800 feet. How much higher is the unofficial record, estimated to the nearest thousand? Use your calculator to check your answer.   10,000 feet

55. *Visible Light*   Use the information given in the Visible Light table to determine the color we would see for each of the wavelengths of light given in parts (a) through (h).

Visible Light

Color	Wavelength of light, angstroms (Å)	Color	Wavelength of light, angstroms (Å)
violet	3900 to 4550	yellow	5770 to 5970
blue	4550 to 4920	orange	5970 to 6220
green	4920 to 5770	red	6220 to 7700

a. 3928 Å violet **b.** 5775 Å yellow **c.** 7629 Å red **d.** 3996 Å violet

e. 4639 Å blue **f.** 6205 Å orange **g.** 5927 Å yellow **h.** 6277 Å red

56. *Wind Speed* The Beaufort Wind Scale table gives standard word descriptions for wind speeds. Use it to classify the wind speeds given in parts (a) through (h).

Beaufort Wind Scale

Description	Wind speed, km/hr	Description	Wind speed, km/hr
calm	less than 1	moderate gale	51 to 61
light air	1 to 5	fresh gale	62 to 74
light breeze	6 to 12	strong gale	75 to 87
gentle breeze	13 to 20	whole gale	88 to 102
moderate breeze	21 to 29	storm	103 to 120
fresh breeze	30 to 39	hurricane	greater than 120
strong breeze	40 to 50		

a. 55 km/hr moderate gale **b.** 42 km/hr strong breeze **c.** 38 km/hr fresh breeze

d. 19 km/hr gentle breeze **e.** 102 km/hr whole gale **f.** 120 km/hr storm

g. 83 km/hr strong gale **h.** 99 km/hr whole gale

i. ✏ Compare and contrast the use of numbers in the Beaufort Wind Scale table and the Visible Light table (see Exercise 55). How would you edit the tables to make them more useful?

A number such as 0.0368598 can be rounded to two significant digits as 0.037. In Exercises 57 through 64, round the number to two significant digits and then write it in scientific notation.

57. The density of Mercury at 0°C is 13.59509 grams per cubic centimeter.
1.4×10^1 grams per cubic centimeter

58. The mean wavelength of sodium light is 0.00005893 centimeter.
5.9×10^{-5} centimeter

59. The maximum density of water, which occurs at 3.98°C, is 0.999973 gram per cubic centimeter. 1.0×10^0 gram per cubic centimeter

60. The density of dry air at 0°C and 760 mm of mercury is 0.001293 gram per cubic centimeter. 1.3×10^{-3} gram per cubic centimeter

61. Number of telephone calls per day: 800,000,000 8.0×10^{8} calls

62. Size of DNA molecule: 0.00000217 mm 2.2×10^{-6} mm

63. Energy required to use a 10-watt flashlight for 1 month: 260,000,000,000,000 ergs 2.6×10^{14} ergs

64. Number of gallons of ice cream eaten in one month: 90,000,000 9.0×10^{7} gallons

In Exercises 65 through 72, rewrite each number as a decimal number.

65. The constant of gravitation K is 6.670×10^{-8} cm/sec/sec. 0.0000000667 centimeter per second per second

66. The approximate mass of a hydrogen atom is 1.67339×10^{-24} gram. 0.00000000000000000000000167339 gram

67. An acre is 4.3560×10^{4} square feet. 43,560 square feet

68. A square foot is 9.3×10^{-2} meter. 0.093 meter

69. Size of a hemoglobin molecule: 6.8×10^{-6} mm 0.0000068 mm

70. Number of pounds of advertising mail received by Americans in eight years: 2.92×10^{10} lbs 29,200,000,000 lbs

71. Money spent on lottery tickets per year: $\$5.2 \times 10^{9}$ $5,200,000,000

72. Amount of buffalo meat eaten each day in the United States: 1.8×10^{4} lbs 18,000 lbs

In Exercises 73 through 80, rewrite each number in scientific notation.

73. Number of pounds of advertising mail received by Americans in one year: 3,650,000,000 pounds 3.65×10^{9} pounds

74. Number of gallons of ice cream eaten each day: 3,000,000 3.0×10^{6} gallons

75. Number of telephone calls made per year: 292,000,000,000 2.92×10^{11} calls

76. Time for the solar system to complete one orbit around our galaxy: 7,080,000,000,000,000 seconds 7.08×10^{15} seconds

77. Number of years that there has been life on Earth: 3,000,000,000 years 3.0×10^{9} years

78. Time needed to compress a deuterium pellet by laser light: 0.000000001 second 1×10^{-9} second

79. Energy required to use a 10-watt flashlight for 1 minute: 6,000,000,000 ergs 6.0×10^{9} ergs

80. Energy needed to launch an Atlas rocket: 9,000,000,000,000,000,000 ergs 9.0×10^{18} ergs

In Exercises 81 through 86, rewrite each number in standard notation.

81. Number of pounds of chocolates eaten per day: 5.8×10^6 pounds
    5,800,000 pounds

82. Number of items bought per year that are shaped like Mickey Mouse or have a picture of Mickey Mouse on them: 1.825×10^9 1,825,000,000

83. Number of gallons of water used by Americans daily: 4.5×10^{11} gallons
    450,000,000,000 gallons

84. Energy given off by a hurricane: 5.0×10^{22} ergs   50,000,000,000,000,000,000,000 ergs

85. Number of seconds in the month of January: 2.6784×10^6 seconds
    2,678,400 seconds

86. Number of millimeters in a kilometer: 1.0×10^6 1,000,000 mm

87. *Research*   Find a very large number. Describe it (as we have). Now write it in scientific notation.   Answers may vary.

88. *Research*   Find a very small number. Describe it (as we have). Now write it in scientific notation.   Answers may vary.

▬▬▬▬▬▬▬▬▬▬▬▬▬▬▬▬▬▬▬▬▬▬▬▬▬▬▬▬▬▬

EXCURSIONS

Data Analysis

1. The Blue Trees Furniture Mart employs 200 workers. They are eligible for health care benefits after 3 months of full-time employment. The health benefits are then obtained for a "20% of the premium" payment. The following statements are true.

   ▪ 85% (170/200) of the workers are currently eligible for health care benefits.
   ▪ 70% (140/200) of the workers are currently covered by their employer's health care benefits.
   ▪ 82% (140/170) of the workers currently eligible for health care benefits are actually covered by those benefits.
   ▪ 94% (170/180) of the workers offered health care benefits are currently eligible for those benefits.
   ▪ 78% (140/180) of the workers offered health care benefits are currently covered by those benefits.
   ▪ 100% (200/200) of the workers are employed in an establishment that offers health care benefits to at least some of its workers.
   ▪ 90% (180/200) of the workers are offered health care benefits.

   Construct a table representing this information. Use this data to ask and answer four questions.

Posing Problems

2. Use the following Territorial Expansion table to ask and answer four questions.

Territorial Expansion

Accession	Date	Area[1]
United States	—	3,536,278
Territory in 1790	—	891,364
Louisiana Purchase	1803	831,321
Florida	1819	69,866
Texas	1845	384,958
Oregon	1846	283,439
Mexican Cession	1848	530,706
Gadsden Purchase	1853	29,640
Alaska	1867	591,004
Hawaii	1898	6,471
Other territory	—	4,664
Philippines	1898	115,600[2]
Puerto Rico	1899	3,426
Guam	1899	209
American Samoa	1900	77
Canal Zone[3]	1904	553
Virgin Islands of U.S.	1917	134
Trust Territory of Pacific Islands	1947	177[4]
All other	—	14
Total, 1990	—	**3,540,315**

1. Total land and water area in square miles. 2. Became independent in 1946. 3. Reverted to Panama. 4. Land area only; Palau only Trust Territory remaining. *Source:* Department of Commerce, Bureau of the Census, as cited in the *1996 Information Please® Almanac* (©1995 Houghton Mifflin Co.), p. 830. All rights reserved. Used with permission by Information Please LLC.

Class Act

3. Give three strategies for estimating the number of corn kernels in this photograph.

1.2 The Operations of Real Numbers

SECTION LEAD-IN

The mathematician Karl Friedrich Gauss (1777–1855) was rumored to be a disruptive student. His teacher once assigned him the task of finding the sum of the first 100 counting numbers

$$1 + 2 + \cdots + 100$$

He was 10 years old, and his teacher thought this task would keep him busy for quite some time. It did not. Gauss came back with a correct result quickly. Can you get the correct result before you finish this section?

Adding Real Numbers

In this section, we review the procedures for adding real numbers. In the addition problem $a + b = c$, a and b are called **addends,** and c is the **sum.**

To add two real numbers $(a + b)$

1. Note the signs of the two numbers.
2. If the signs of the numbers are the same, add their absolute values.
3. If the signs of the numbers are different, subtract the smaller absolute value from the larger.
4. Give the sum the same sign as the addend with the greater absolute value.

▪▪▪

EXAMPLE 1

a. Add: $\left(-3\frac{3}{4}\right) + \left(-2\frac{2}{5}\right)$ **b.** Add: $(-7.4) + (6.51)$

SOLUTION

a. Because the signs are the same, we add the absolute values.

$$\left|-3\frac{3}{4}\right| + \left|-2\frac{2}{5}\right| = 3\frac{3}{4} + 2\frac{2}{5}$$

We can add or subtract fractions only if they have the same denominator, so we must rewrite the two fractions as equivalent fractions with a common denominator. The least common denominator (LCD) is $2^2 \cdot 5 = 20$. We rewrite each fraction as an equivalent fraction with 20 as denominator.

$$3\frac{3}{4} = 3\frac{15}{20} \qquad \left[\frac{(3 \times 5)}{(4 \times 5)} = \frac{15}{20}\right]$$
$$2\frac{2}{5} = 2\frac{8}{20} \qquad \left[\frac{(2 \times 4)}{(5 \times 4)} = \frac{8}{20}\right]$$

Adding the integer parts and then the fractional parts gives us

INSTRUCTOR NOTE

Some students will need a review in finding the least common denominator and reducing to lowest terms.

ERROR ALERT

Identify the error and give a correct answer.

$$\frac{1}{3} + \frac{3}{8} = \frac{1}{\cancel{3}} + \frac{\cancel{3}}{8}$$
$$= \frac{1}{8}$$

$$3\tfrac{3}{4} + 2\tfrac{2}{5} = 3\tfrac{15}{20} + 2\tfrac{8}{20} = 5\tfrac{23}{20}$$

Because $\tfrac{23}{20}$ is an improper fraction, we simplify it by rewriting it as $1\tfrac{3}{20}$ and then adding this result to 5.

$$5\tfrac{23}{20} = 5 + 1\tfrac{3}{20} = 6\tfrac{3}{20}$$

We still have to attach a sign to $6\tfrac{3}{20}$. Because the addend with the greater absolute value, $-3\tfrac{3}{4}$, is negative, the sum is negative.

So $\left(-3\tfrac{3}{4}\right) + \left(-2\tfrac{2}{5}\right) = -6\tfrac{3}{20}$.

b. Because the signs are different, we subtract absolute values.

$$|-7.4| - |6.51| = 7.4 - 6.51$$

$$
\begin{array}{r}
7.40 \\
-6.51 \\
\hline
0.89
\end{array}
$$

7.40   Lining up decimal points

0.89   Subtracting and aligning the decimal point in the result

Because the number with the greater absolute value, -7.4, is negative, the sum is negative:

$$(-7.4) + (6.51) = -0.89$$

▶ *CHECK* **Warm-Up 1**

Calculator Corner

You can use your graphing calculator's Home Screen to help you add and subtract real numbers. Here are two examples, one involving decimals.

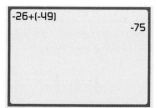

Some graphing calculators have the capability to perform operations using mixed fractions. Check the manual for your particular model to see if you can use mixed fractions on your calculator. If your calculator cannot use mixed fractions, you must first change a mixed fraction to its decimal representation or to an improper fraction in order to check your work.

Subtracting Real Numbers

Before we can develop a method for subtracting two real numbers, we need to define *additive inverses.*

Additive Inverses

Two numbers are **additive inverses** of each other if their sum is zero. The additive inverse of n is $-n$ because

$$n + (-n) = 0$$

Similarly, the additive inverse of $-n$ is n because

$$-n + n = 0$$

The additive inverse of a number is also its *opposite. Zero is its own additive inverse (or opposite).* Now we can subtract.

Definition of Subtraction

For any real numbers a and b,

$$a - b = a + (-b)$$

More simply, we can subtract one real number, b, from another, a, by *adding the inverse* of b to a. The number we begin with, a, is the **minuend;** the number being subtracted, b, is the **subtrahend;** and the result, c, is called the **difference.**

To subtract two real numbers $(a - b)$

1. Replace the subtrahend with its additive inverse, $-b$.
2. Rewrite the operation as addition by changing the operation sign.
3. Add $a + (-b)$.

! ! !

ERROR ALERT

Identify the error and give a correct answer.

Subtract: $(-29) - 32$

Incorrect Solution:

$(-29) + (32) = 3$

▪▪▪

EXAMPLE 2

a. Subtract: $-12\frac{17}{20} - \left(-8\frac{3}{20}\right)$ **b.** Subtract: $-4.5 - 6.89$

SOLUTION

a. To subtract, we replace $\left(-8\frac{3}{20}\right)$ with its additive inverse, $8\frac{3}{20}$, and rewrite the operation as addition.

$$-12\frac{17}{20} - \left(-8\frac{3}{20}\right) = -12\frac{17}{20} + \left(+8\frac{3}{20}\right)$$

Now we must perform the addition. The signs are different, so we subtract the absolute values.

$$12\frac{17}{20} - 8\frac{3}{20} = 4\frac{14}{20}$$

Because the number with the greater absolute value, $-12\frac{17}{20}$, is negative, the sum is negative. So $-12\frac{17}{20} - \left(-8\frac{3}{20}\right) = -4\frac{7}{10}$.

b. To subtract, we replace 6.89 with its additive inverse, -6.89, and rewrite the operation as addition.

$$-4.5 - 6.89 = -4.5 + (-6.89)$$

The signs are the same, so we add the absolute values.

$$-4.5 + (-6.89) = |-4.5| + |-6.89|$$
$$= 4.5 + 6.89 = 11.39$$

Because the number with the greater absolute value in the addition, -6.89, is negative, the sum is -11.39. And thus the result of the original subtraction is negative:

$$-4.5 - 6.89 = -11.39$$

▶ *CHECK* **Warm-Up 2**

Finally, let's combine addition and subtraction in one example.

▪▪▪
EXAMPLE 3

Simplify: $13\frac{7}{20} + (-6) - 4\frac{1}{2}$

SOLUTION

Working from left to right, we add two numbers at a time. We first add $13\frac{7}{20}$ and -6. The signs are different, so we subtract the absolute values.

$$13\frac{7}{20} + \left(-6\right) = \left|13\frac{7}{20}\right| - \left|-6\right| = 7\frac{7}{20}$$

The number in this pair with the greater absolute value, $13\frac{7}{20}$, is positive, so the result is positive. Replacing the first part of our problem with this result gives us

$$7\frac{7}{20} + \left(-4\frac{1}{2}\right)$$

The signs of these addends are different, so we have to subtract the lesser absolute value from the greater. First we write

$$7\frac{7}{20} - 4\frac{1}{2} = 7\frac{7}{20} - 4\frac{10}{20}$$

Because we cannot subtract $\frac{10}{20}$ from $\frac{7}{20}$, we must regroup from the 7. Because $1 = \frac{20}{20}$, we rewrite $7\frac{7}{20}$ as $6\frac{27}{20}$.

Then we subtract.

$$6\frac{27}{20} - 4\frac{10}{20} = 2\frac{17}{20}$$

The number in this addition with the greater absolute value, $7\frac{7}{20}$, is positive, so the result is positive:

$$13\frac{7}{20} + \left(-6\right) - 4\frac{1}{2} = 2\frac{17}{20}$$

▶ *CHECK* **Warm-Up 3**

Multiplying Real Numbers

In the multiplication $ab = c$, a and b are called **factors,** and c is the **product.**

To multiply two real numbers $(a \cdot b)$

1. Multiply the absolute values of the factors.
2. If the factors have the same sign, the product is positive.
3. If the factors have different signs, the product is negative.

■ ■ ■
EXAMPLE 4

a. Multiply: $(-0.6)(3.3)$ **b.** Multiply: $\left(-2\frac{1}{4}\right)\left(-6\frac{3}{8}\right)\left(\frac{1}{17}\right)$

SOLUTION

a. We first multiply their absolute values.

$$|-0.6||3.3| = 0.6(3.3) = 1.98$$

The factors together have two decimal places, so the product has two decimal places. Because the original factors have different signs, the product is negative.

So $(-0.6)(3.3) = -1.98$.

b. We first multiply their absolute values.

$$\left|-2\frac{1}{4}\right|\left|-6\frac{3}{8}\right|\left|\frac{1}{17}\right| = \left(2\frac{1}{4}\right)\left(6\frac{3}{8}\right)\left(\frac{1}{17}\right)$$

To multiply, we must convert the mixed numbers to improper fractions.

$$2\frac{1}{4} = \frac{[2(4) + 1]}{4} = \frac{9}{4}$$

$$6\frac{3}{8} = \frac{[6(8) + 3]}{8} = \frac{51}{8}$$

Now, to multiply the fractions, we multiply their numerators and their denominators.

$$\left(\frac{9}{4}\right)\left(\frac{51}{8}\right)\left(\frac{1}{17}\right) = \frac{(9 \cdot 51)}{(4 \cdot 8 \cdot 17)} = \frac{9 \cdot \overset{3}{\cancel{51}}}{4 \cdot 8 \cdot \underset{1}{\cancel{17}}} = \frac{27}{32}$$

Because the signs of the factors in the original question are the same, the product is positive. So $\left(-2\frac{1}{4}\right)\left(-6\frac{3}{8}\right)\left(\frac{1}{17}\right) = \frac{27}{32}$.

▶ *CHECK* **Warm-Up 4**

■■■
WRITER'S BLOCK

Using the numbers 2, 3, 5, and 6 and the words *sum* and *product*, write two different sentences describing a true relationship.

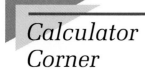

Calculator Corner

Multiplication on your graphing calculator's Home Screen can be done using parentheses or the multiplication sign. The screen at the right shows two ways to multiply.

```
(-93)(43)
              -3999
-93*43
              -3999
```

Dividing Real Numbers

In the division $a \div b = c$, we call a the **dividend**, b the **divisor**, and c the **quotient**.

> **To divide two real numbers ($a \div b$, where b is not zero)**
>
> 1. Divide the absolute value of a by the absolute value of b.
> 2. If a and b have the same sign, the quotient is positive.
> 3. If a and b have different signs, the quotient is negative.

▪▪▪

EXAMPLE 5

a. Divide: $(-2.89) \div 34$ **b.** Divide: $0.564 \div (-0.08)$

SOLUTION

a. First we must divide the absolute values of the two numbers.

$$|-2.89| \div |34| = 2.89 \div 34$$

The divisor has no decimal part, so we can go ahead and divide. Using long division or a calculator, we obtain 0.085. Because the signs of the original numbers are different, the quotient is negative.

So $(-2.89) \div 34 = -0.085$.

b. We begin by dividing the absolute values.

$$|0.564| \div |-0.08| = 0.564 \div 0.08$$

Before we can divide, we must remove the decimal point from the divisor. Multiplying both dividend and divisor by 100 gives us

$$0.564 \div 0.08 = \frac{0.564}{0.08} = \frac{(0.564)(100)}{(0.08)(100)} = \frac{56.4}{8}$$

Then, by either long division or calculator, we obtain

$$\frac{56.4}{8} = 7.05$$

Because the signs of the original numbers are different, the result is negative.

$$0.564 \div (-0.08) = -7.05$$

▶ *CHECK* **Warm-Up 5**

Multiplicative Inverse

The product of any real number and its *multiplicative inverse* is 1.

> **Multiplicative Inverse**
>
> The non-zero real numbers $\frac{a}{b}$ and $\frac{b}{a}$ are **multiplicative inverses,** and their product is 1. That is,
>
> $$\left(\frac{a}{b}\right)\left(\frac{b}{a}\right) = 1 \quad \text{and} \quad \left(\frac{b}{a}\right)\left(\frac{a}{b}\right) = 1$$

The multiplicative inverse is also called the **reciprocal.** *Zero does not have a reciprocal.*

▪▪▪
EXAMPLE 6

Divide: $\left(\frac{-5}{6}\right) \div \left(2\frac{1}{7}\right)$

SOLUTION

We must divide the absolute values of these numbers. So we have

$$\left|\frac{-5}{6}\right| \div \left|2\frac{1}{7}\right| = \frac{5}{6} \div \frac{15}{7}$$

To divide one fraction by another, we multiply by the reciprocal of the divisor and then simplify. The reciprocal of $\frac{15}{7}$ is $\frac{7}{15}$, so

$$\frac{5}{6} \div \frac{15}{7} = \frac{5}{6} \cdot \frac{7}{15}$$

$$= \frac{\overset{1}{\cancel{5}} \cdot 7}{6 \cdot \underset{3}{\cancel{15}}}$$

$$= \frac{7}{18}$$

Because the signs of the original numbers are different, the quotient is negative. So $\left(\frac{-5}{6}\right) \div \left(2\frac{1}{7}\right) = \frac{-7}{18}$.

▶ *CHECK* **Warm-Up 6**

INSTRUCTOR NOTE

Students often think that they can obtain the multiplicative inverse by inverting the fraction. They tend to ignore the integer part of a mixed number and/or the sign of the number. You may need to give them several examples that include these situations.

▪▪▪
> **WRITER'S BLOCK**
>
> A student made the observation that for real numbers a and b not zero,
>
> $$\frac{a}{b} = a \cdot \frac{1}{b} = \frac{1}{b} \cdot a$$
>
> Is this true or not? Justify your answer.

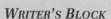
Simplifying Complex Fractions

> **Complex Fraction**
>
> A **complex fraction** is a fraction that has a fraction in either the numerator or the denominator or in both.

INSTRUCTOR NOTE

Students with perceptual difficulties
may benefit from using a different
color to highlight the fraction bar of
the *main* fraction in a complex form.

■ ■ ■

EXAMPLE 7

Simplify the complex fraction: $\dfrac{2\frac{2}{3}}{\frac{-5}{6}}$

SOLUTION

We rewrite the complex fraction as a division problem and then divide.

$$\frac{2\frac{2}{3}}{\frac{-5}{6}} = 2\frac{2}{3} \div \left(\frac{-5}{6}\right) \quad \text{Rewriting}$$

$$\left|2\frac{2}{3}\right| \div \left|\frac{-5}{6}\right| = \frac{8}{3} \div \frac{5}{6} \quad \begin{array}{l}\text{Changing to improper fractions}\\ \text{and dividing absolute values}\end{array}$$

$$= \left(\frac{8}{3}\right)\left(\frac{6}{5}\right) \quad \text{Writing as multiplication}$$

$$= \frac{8 \cdot \overset{2}{\cancel{6}}}{\underset{1}{\cancel{3}} \cdot 5} \quad \text{Multiplying and cancelling}$$

$$= \frac{16}{5} \quad \text{Simplifying}$$

$$= 3\frac{1}{5} \quad \text{Writing as a mixed number}$$

Because the signs in the original problem are different, the quotient is negative.

So $\dfrac{2\frac{2}{3}}{\frac{-5}{6}} = -3\frac{1}{5}$.

▶ CHECK **Warm-Up 7**

Calculator Corner

Your graphing calculator can be used for division. (Note: See the Calculator Corner on page 4 to review how to get an answer in fractional form.)

   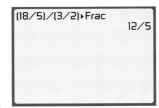

Press **ENTER.**

Question: Does it really matter that the parentheses are used in this problem? What would happen if the parentheses were left out?

   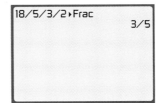

Press **ENTER.**

How did the calculator arrive at this answer? Write out the sequence of steps that the calculator performed to arrive at the answer 3/5.

Check Example 7 using your graphing calculator.

Practice what you learned.

SECTION FOLLOW-UP

According to some accounts, Gauss quickly solved the problem in the following manner:

```
                    100      100
        1    +       99      100
        2    +       98      100
        3    +       97      100
        4    +       96      100
        .            .        .
        .            .        .
        .            .        .
       49    +       51      100
       50                     50
                            ─────
                            5050
```

There are 49 pairs that sum to 100 plus the number 100 plus 50 more. The sum is 5050.

▪ Find the sum of the first 1000 counting numbers.

1.2 WARM-UPS

Work these problems before you attempt the exercises.

1. **a.** Find the sum of $-162\frac{3}{4}$ and $\left(-17\frac{3}{4}\right)$. $-180\frac{1}{2}$

   b. Add: -160 and 2.028 -157.972

2. **a.** Subtract $8\frac{1}{6}$ from $-6\frac{2}{3}$. $-14\frac{5}{6}$

   b. How much greater than -3.5 is 9.42? 12.92

3. Simplify: $8.3 - 17 - (-12.43)$ 3.73

4. **a.** Multiply: $(-1.03)(-0.76)$ 0.7828

   b. Multiply: $\left(-3\frac{1}{2}\right)\left(-\frac{1}{7}\right)\left(\frac{4}{5}\right)$ $\frac{2}{5}$

5. **a.** Divide: $(-2.24) \div (-28)$ 0.08

   b. Divide: $(-1.2) \div 2.5$ -0.48

6. Divide: $\left(-2\frac{3}{5}\right) \div \left(-1\frac{1}{10}\right)$ $2\frac{4}{11}$

7. Simplify: $\dfrac{-2\frac{1}{3}}{-\frac{5}{6}}$ $2\frac{4}{5}$

1.2 EXERCISES

Note: Use your graphing calculator to check your results whenever possible.

In Exercises 1 through 36, add or subtract as indicated.

1. $-459 + (-28)$ -487

2. $-388 + (-48)$ -436

3. $40.3 + 3.5$ 43.8

4. $-7.2 + 6.35$ -0.85

5. $-48\frac{5}{8} + 5\frac{3}{8}$ $-43\frac{1}{4}$

6. $5\frac{4}{5} + \left(-2\frac{3}{5}\right)$ $3\frac{1}{5}$

7. $(4.9) + (-3.88)$ 1.02

8. $(-252) + (-423)$ -675

9. $(-28) + (-35)$ -63

10. $-9.4 + (-5.8)$ -15.2

11. $-8.3 + (-49.3)$ -57.6

12. $2.2 + 8.42$ 10.62

13. $42 - 2.3$ 39.7

14. $6.23 - 3.42$ 2.81

15. $2.3 - 4.52$ -2.22

16. $-28.32 - 8.3$ -36.62

17. $\left(-6\frac{1}{4}\right) - \left(-23\frac{1}{8}\right)$ $16\frac{7}{8}$

18. $-8\frac{4}{11} - 5\frac{1}{11}$ $-13\frac{5}{11}$

19. $-3\frac{5}{8} - 6\frac{5}{8}$ $-10\frac{1}{4}$

20. $\left(-13\frac{1}{2}\right) - \left(-28\frac{1}{6}\right)$ $14\frac{2}{3}$

21. $\left(-8\frac{4}{5}\right) + \left(5\frac{9}{10}\right)$ $-2\frac{9}{10}$

22. $\left(9\frac{3}{8}\right) + \left(-2\frac{5}{8}\right)$ $6\frac{3}{4}$

23. $3\frac{2}{3} + \left(-25\frac{1}{3}\right)$ $-21\frac{2}{3}$

24. $\left(-8\frac{2}{3}\right) - \left(-2\frac{2}{5}\right)$ $-6\frac{4}{15}$

25. $\left(-3\frac{3}{7}\right) - \left(-4\frac{2}{3}\right)$ $1\frac{5}{21}$ **26.** $(-24.3) + (-8.95)$ -33.25 **27.** $45.3 + (-8.2)$ 37.1 **28.** $22\frac{3}{5} - 8\frac{2}{5}$ $14\frac{1}{5}$

29. How much greater is $29\frac{5}{12}$ than $-14\frac{5}{8}$? greater by $44\frac{1}{24}$ **30.** How much less than $45\frac{2}{5}$ is $-14\frac{7}{10}$? less by $60\frac{1}{10}$

31. How much less than -2.3 is -4.56? less by 2.26 **32.** How much greater than -16.3 is -9.28? greater by 7.02

33. What is the sum of $-27\frac{2}{9}$ and the opposite of 8? $-35\frac{2}{9}$ **34.** Find the sum of $6\frac{2}{3}$ and the opposite of $-4\frac{1}{2}$. $11\frac{1}{6}$

35. What is the total of $-5\frac{5}{6}$ and $\frac{1}{24}$ and $-18\frac{3}{8}$? $-24\frac{1}{6}$ **36.** Find the total of $-9\frac{5}{8}$ and $3\frac{5}{6}$ and $-12\frac{1}{8}$. $-17\frac{11}{12}$

In Exercises 37 through 40, use what you know about opposites to add and subtract mentally.

37. Simplify: $1\frac{1}{3} + \left(-2\frac{5}{8}\right) - 3\frac{1}{6} - \left(-2\frac{5}{8}\right) + 3 - 1\frac{1}{3}$ $-\frac{1}{6}$ **38.** Simplify: $-6\frac{1}{12} + 3\frac{5}{7} - \left(-6\frac{1}{12}\right) - 16 - 3\frac{5}{7}$ -16

39. Simplify:
$18.3 - 12.7 + 16.9 + (-18.3) - 16.9 + 12.7$ 0 **40.** Simplify:
$127.2 - 18.3 + 16.5 - (-18.3) + (-127.2)$ 16.5

In Exercises 41 through 72, multiply or divide as indicated.

41. $(-3)(18)$ -54 **42.** $(-6)(-11)$ 66 **43.** $(-7)\left(-4\frac{1}{3}\right)$ $30\frac{1}{3}$ **44.** $(-6)\left(-3\frac{2}{5}\right)$ $20\frac{2}{5}$

45. $(-0.42)(-0.02)$ 0.0084 **46.** $(-82)(0.04)$ -3.28 **47.** $(-560) \div (-8)$ 70 **48.** $105 \div (-7)$ -15

49. $(-98) \div (0.05)$ -1960 **50.** $(-74) \div (0.08)$ -925 **51.** $\left(-1\frac{5}{8}\right) \div \left(1\frac{3}{10}\right)$ $-1\frac{1}{4}$ **52.** $\left(-1\frac{8}{9}\right) \div \left(2\frac{4}{15}\right)$ $-\frac{5}{6}$

53. $(-0.16) \div (-0.2)$ 0.8 **54.** $(-0.38) \div (0.019)$ -20 **55.** $(-7)\left(2\frac{9}{14}\right)$ $-18\frac{1}{2}$ **56.** $\left(-2\frac{11}{12}\right)\left(-2\frac{2}{3}\right)$ $7\frac{7}{9}$

57. $(-0.16)(-0.25)$ 0.04 **58.** $(-0.35)(0.29)$ -0.1015 **59.** $\left(-8\frac{2}{5}\right) \div \left(-1\frac{4}{5}\right)$ $4\frac{2}{3}$ **60.** $\left(-6\frac{1}{3}\right) \div \left(-1\frac{5}{9}\right)$ $4\frac{1}{14}$

61. $\left(\frac{-1}{2}\right)\left(\frac{-5}{3}\right)\left(\frac{-6}{10}\right)$ $-\frac{1}{2}$ **62.** $\left(\frac{-5}{3}\right)\left(\frac{2}{10}\right)\left(\frac{-3}{5}\right)$ $\frac{1}{5}$ **63.** $\left(-1\frac{2}{3}\right)\left(-2\frac{1}{5}\right)\left(1\frac{1}{2}\right)$ $5\frac{1}{2}$ **64.** $\left(-1\frac{1}{8}\right)\left(3\frac{1}{3}\right)\left(-5\frac{1}{3}\right)$ 20

65. $(-1.1)(-2.1)(-5)$ -11.55 **66.** $(2.3)(-3.2)(-1.5)$ 11.04 **67.** $\left(-6\frac{1}{2}\right) \div \left(5\frac{1}{5}\right)$ $-1\frac{1}{4}$ **68.** $\left(3\frac{3}{5}\right) \div \left(-2\frac{7}{10}\right)$ $-1\frac{1}{3}$

69. $\dfrac{1\frac{1}{2}}{-6\frac{1}{4}}$ $-\frac{6}{25}$ **70.** $\dfrac{-\frac{54}{7}}{\frac{6}{14}}$ -18 **71.** $\dfrac{\frac{-3.232}{3.2}}{\frac{-121.2}{-2.4}}$ -0.02 **72.** $\dfrac{\frac{-226.6}{-0.05}}{\frac{-2.75}{0.0025}}$ -4.12

In Exercises 73 through 86, perform the indicated operations.

73. $-3\frac{5}{8} + 9\frac{5}{6} - 5\frac{1}{12}$ $1\frac{1}{8}$ **74.** $(-12)(-5)(4)$ 240 **75.** $(6.2)(-3.1)(-1.3)$ 24.986

76. $-\left|-18\frac{1}{2}\right| + 23\frac{7}{12} - 5\frac{1}{3}$ $-\frac{1}{4}$ **77.** $(-53)(2)(-8)$ 848 **78.** $2\frac{1}{6} \times \left(-3\frac{2}{5}\right) \times \frac{15}{19}$ $-5\frac{31}{38}$

79. $-12\frac{1}{3} + \left|-9\frac{2}{5}\right| - \left(-3\frac{7}{30}\right)$ $\frac{3}{10}$ **80.** $(-1.03)(-2.1)(19)$ 41.097 **81.** $18.15 - |19.2| - (-2.5)$ 1.45

82. $-32.8 + |-17.6| - 14.23$ -29.43 **83.** $(-15.8)(-2.5)(-1.2)$ -47.4 **84.** $(-3.15)(26)(-2.7)$ 221.13

85. $-26\frac{1}{3} - \left(-25\frac{3}{8}\right) - \left|\frac{5}{12}\right|$ $-1\frac{3}{8}$ **86.** $-15\frac{1}{8} + \left|-16\frac{2}{3}\right| - 2\frac{5}{12}$ $-\frac{7}{8}$

MIXED PRACTICE

By doing these exercises, you will practice the topics up to this point in the chapter.

87. Which is larger, -4.05 or $\left|-4\frac{3}{7}\right|$? $\quad \left|-4\frac{3}{7}\right|$

88. Subtract: $\left(-4\frac{1}{12}\right) - \frac{3}{8} \quad -4\frac{11}{24}$

89. Divide: $7\frac{4}{5} \div \left(-6\frac{3}{8}\right) \quad -1\frac{19}{85}$

90. Arrange 3.042, 3.04, and 3.4 in order from smallest to largest. $\quad 3.04 < 3.042 < 3.4$

91. Add: $-6 + 5\frac{1}{8} - 8\frac{3}{4} \quad -9\frac{5}{8}$

92. Simplify: $\dfrac{-6\frac{4}{5}}{8\frac{1}{2}} \quad -\frac{4}{5}$

93. Simplify: $-|-3.6| \quad -3.6$

94. Simplify: $0.34 - (-2.7) \quad 3.04$

95. Simplify: $(-20.5)(-8)(-2) \quad -328$

96. What is the opposite of 5? $\quad -5$

97. Simplify: $\left|-\left|-2\frac{1}{4}\right|\right| \quad 2\frac{1}{4}$

98. True or false: All integers can be written as rational numbers. $\quad$ true

99. Divide: $26 \div 18\frac{5}{11} \quad 1\frac{83}{203}$

100. Arrange -9.008, $-|-9.08|$, and -9 in order from greatest to least. $\quad -9 > -9.008 > -|-9.08|$

EXCURSIONS

Exploring Patterns

1. ✏ Do the first few of the following multiplications. Try to find a pattern and predict the answers to the last few. Use your calculator to find other patterns. Write a description of the pattern.

   a. $7 \times 11 \times 13$ **b.** $7 \times 11 \times 13 \times 2$ **c.** $7 \times 11 \times 13 \times 3$

   d. $7 \times 11 \times 13 \times 4$ **e.** $7 \times 11 \times 13 \times 7$ **f.** $7 \times 11 \times 13 \times 9$

2. ✏ Problems (a) through (d) exhibit a new pattern. What is it? Predict the results of (e) and (f). Justify your predictions and verify the results.

   a. 25×11 **b.** 25×111 **c.** 25×1111

   d. $25 \times 1,111,111,111$ **e.** $25 \times 111,111,110$ **f.** $111,111 \times 25$

3. ✏ Use your calculator to multiply each pair of numbers in turn. Stop multiplying when you discover a pattern to the answers. Indicate where you stopped multiplying, give the remaining answers, and describe in your own words the pattern you see.

a. 1.01×11	**b.** 9.9×123	**c.** 1.23×555	**d.** 1.05×7777
1.01×111	9.9×1234	1.23×5555	1.05×77777
1.01×1111	9.9×12345	1.23×55555	1.05×777777
1.01×11111	9.9×123456	1.23×555555	1.05×7777777
1.01×111111	9.9×1234567	1.23×5555555	1.05×77777777
1.01×1111111	9.9×12345678	1.23×55555555	1.05×777777777
		1.23×555555555	1.05×7777777777

CONNECTIONS TO *STATISTICS*

Finding the Range

In statistics we use numbers to describe the characteristics of sets of data. One of the measures that we use is that of "spread." That is, how far apart is the data? One basic measure of spread is the *range*. We use subtraction to find the range.

Range

The **range** of a set of data is the difference between the largest value and the smallest value in the set. It is given as a number with its unit, if it has a unit. A range is always positive or zero.

▪ ▪ ▪

EXAMPLE

Hourly readings on a pressure gauge were −1.80, −2.75, −2.25, and −1.25. The units are mb, millibars of mercury. What is the range of these readings?

SOLUTION

Our definition says that we must subtract the least value from the greatest. Because the least value is −2.75 and the greatest is −1.25, our problem becomes

$$-1.25 - (-2.75) = 1.5$$

1.5 mb is the range. Therefore, readings from −2.75 to −1.25 have a range of 1.5 millibars of mercury.

PRACTICE

1. The markings on a safe go from 31 to 158. What is their range?   127

2. The following numbers of feet were recorded: −3.4, 8.9, 32.18, 45, −17.3, and −0.002. What is the range of these numbers?   62.3 feet

3. Markings from −17 to 101 are indicated along a number line. What is the range of these markings?   118

4. What is the range of these test scores: 43, 25, 100, 18, 42, and 32?   82

5. What is the range of these fractional measures: $2\frac{1}{3}$, $5\frac{4}{7}$, $8\frac{3}{8}$, and $1\frac{1}{5}$? $7\frac{7}{40}$

6. A program randomly lists the numbers −102, 47, $188\frac{1}{2}$, 4.28, and −3.18. What is the range of these numbers? $290\frac{1}{2}$

7. *Flying and Diving* In 1961 people flew as high as 203.2 miles and dived as deep as 728 feet below sea level, using a gas mixture for breathing. What is the range of these record achievements? 931.2 miles

8. *Record Highs and Lows* Asia has a high point of 29,028 feet and a low point of 1302 feet below sea level; North America has a high point of 20,320 feet and a low point of 282 feet below sea level. Which continent has the greater range in elevation? What is its range? Asia; 30,330 feet

9. *Height and Depth Records* The following heights above and below sea level were recorded by a scientific team: 35 feet below, 46.3 feet above, 19 feet below, 25.17 feet above, 0.3 feet above, 135.4 feet below, 35 feet below, and 6.24 feet above. What is the range of these readings? 181.7 feet

10. *Height and Depth Records* In 1954 a man flew 93,000 feet high in a rocket plane. Another dived 13,287 feet below sea level in a bathyscaphe. What is the range of these records? 106,287 feet

SECTION GOALS

- *To identify and use the commutative, associative, and distributive properties*

- *To identify the identity elements for addition and multiplication*

- *To apply the multiplication properties of zero, one, and −1*

- *To raise real numbers to powers*

- *To find square roots and cube roots of real numbers*

- *To use the standard order of operations to simplify mathematical statements*

- *To estimate the results of calculations with real numbers*

1.3 A Closer Look at the Real Numbers

SECTION LEAD-IN

- Using the numbers 1, 2, 3, and 4 and the operations of +, −, ×, or ÷, represent all the numbers you can from 0 to 10. You may use grouping symbols.

Properties of Real Numbers

When real numbers are added and multiplied, certain properties apply. These properties are very useful in algebra, as you will see shortly.

Commutative, Associative, and Distributive Properties

These three properties enable us to rewrite algebraic or numerical statements to simplify computation.

Commutative Property of Addition

If a and b are real numbers, then

$$a + b = b + a$$

Commutative Property of Multiplication

If a and b are real numbers, then

$$ab = ba$$

Associative Property of Addition

If a, b, and c are real numbers, then

$$(a + b) + c = a + (b + c)$$

Associative Property of Multiplication

If a, b, and c are real numbers, then

$$(ab)c = a(bc)$$

These properties *do not* hold for subtraction or for division.

Distributive Property of Multiplication over Addition

If a, b, and c are real numbers, then

$$a(b + c) = ab + ac$$
$$(b + c)a = ba + ca$$

■■■
WRITER'S BLOCK

Does multiplication distribute over subtraction? Explain.

•••
EXAMPLE 1

Determine which property is illustrated in each of the following statements. Verify statements (a) and (b).

a. $[6 + (-8)] + (-3) = 6 + [(-8) + (-3)]$

b. $\left(4\frac{1}{2}\right)(-9 \cdot 8) = \left(4\frac{1}{2}\right)[8 \cdot (-9)]$

c. $(3x \cdot 10)2 = 2(3x \cdot 10)$

d. $-2 \cdot 3(-5y + 0.4) = (-2 \cdot 3)(-5y) + (-2 \cdot 3)(0.4)$

SOLUTION

a. The order of the addends remained the same; the grouping changed: associative property of addition. We verify as follows:

$$[6 + (-8)] + (-3) = 6 + [(-8) + (-3)]$$

$$(-2) + (-3) = 6 + (-11) \qquad \text{Adding within brackets}$$

$$-5 = -5 \quad \text{True} \qquad \text{Adding on each side}$$

b. The grouping remained the same; the order of the factors changed: commutative property of multiplication. To verify:

$$\left(4\frac{1}{2}\right)(-9 \cdot 8) = \left(4\frac{1}{2}\right)[8 \cdot (-9)]$$

$$\frac{9}{2}(-72) = \frac{9}{2}(-72) \quad \text{True}$$

c. The grouping remained the same; the order changed: commutative property of multiplication.

d. Here the factor $(-2 \cdot 3)$ has been distributed across the sum $(-5y + 0.4)$: distributive property.

▶ *CHECK* **Warm-Up 1**

Multiplication Properties of Zero, One, and −1; Additive and Multiplicative Identities

Addition Property of Zero

If a is any real number, then

$$a + 0 = 0 + a = a$$

Multiplication Property of Zero

If a is any real number, then

$$a \cdot 0 = 0 \cdot a = 0$$

Multiplication Property of One

If a is any real number, then

$$a \cdot 1 = 1 \cdot a = a$$

Multiplication Property of −1

If a is any real number, then

$$-1(a) = -a$$
$$-1(-a) = a$$

Because of the first and third properties, 0 is called the **additive identity**, and 1 is called the **multiplicative identity**.

▪▪▪

EXAMPLE 2

Identify the property used in each line of this simplification:

$$2 \cdot 3 \cdot 0 + 18(-3 + 2) = (2 \cdot 3) \cdot 0 + 18(-3 + 2) \quad (1)$$
$$6 \cdot 0 + 18(-3 + 2) = 0 + 18(-3 + 2) \quad (2)$$
$$18(-3 + 2) = 18(-1) \quad (3)$$
$$18(-1) = -18 \quad (4)$$

SOLUTION

Line (1) involves grouping in multiplication (associative property of multiplication).

$$2 \cdot 3 \cdot 0 + 18(-3 + 2) = (2 \cdot 3) \cdot 0 + 18(-3 + 2)$$

Line (2) first shows the multiplication $2 \cdot 3 = 6$. Then it gives the result of multiplying 6 by 0 (multiplication property of zero).

$$6 \cdot 0 + 18(-3 + 2) = 0 + 18(-3 + 2)$$

Line (3) shows first the result of adding zero (addition property of zero). Then adding within the parentheses yields

$$18(-3 + 2) = 18(-1)$$

Line (4) uses the multiplication property of -1.

$$18(-1) = -18$$

▶ *CHECK* **Warm-Up 2**

Zero as Divisor or Dividend

You must be careful when zero appears as a divisor or as a dividend. In division, $a \div b = c$ if and only if $bc = a$. Also $bc = a$ must be unique; that is, it must be true only for the specific numbers a, b, and c.

Now we can say that $0 \div b = 0$, because $b \cdot 0 = 0$ is true and because 0 is the only number for which it is true.

We cannot say $a \div 0 = m$ for some real number m, because $0 \cdot m = a$ is *not* true unless $a = 0$.

What about $0 \div 0$? Suppose that $0 \div 0$ were equal to n. Then we would have $0 \cdot n = 0$. This is true, but it is true for any real number n:

$$0 \cdot 1 = 0$$
$$0 \cdot 2 = 0$$
$$0 \cdot 3 = 0 \quad \text{And so on}$$

So n could be *any* real number, and $0 \cdot n = 0$ is not unique.

This tells us that

$$0 \div 0 \text{ has no solution}$$

and we saw that

$$a \div 0 \text{ has no solution}$$

We say that these divisions are **undefined.**

Properties of Division with Zero

$0 \div a = 0$ for all non-zero real numbers a.
$a \div 0$ is undefined for all real numbers.

There is no real number solution for $a \div 0$ or $0 \div 0$.

Using Other Operations on Real Numbers

Positive Exponents

In the exponential notation a^n, a is called the **base** and n is called the **exponent.**

The exponent indicates how many times the base is used as a factor.

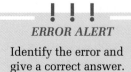

ERROR ALERT

Identify the error and give a correct answer.

Simplify: -2^3

Incorrect Solution:

$-2^3 = -2(3) = -6$

To raise a real number to a power

For a real number a and a positive integer n,

$$a^n \text{ means } \underbrace{a \cdot a \cdot \ldots \cdot a}_{n \text{ factors}}$$

Also, $a^1 = a$ and $a^0 = 1$. 0^0 is not defined.

The second power of a number is called its **square.** The third power of a number is called its **cube.**

▪▪▪

EXAMPLE 3

Simplify:

a. $(-3)^2$

b. $\left(\frac{2}{3}\right)^3$

c. $-(3.5)^2$

d. $(101)^0$

e. 2^1

SOLUTION

Using the definition of an exponent, we have

a. $(-3)^2 = (-3)(-3) = 9$

b. $\left(\frac{2}{3}\right)^3 = \left(\frac{2}{3}\right)\left(\frac{2}{3}\right)\left(\frac{2}{3}\right) = \frac{8}{27}$

ERROR ALERT

Identify the error and give a correct answer.

$5^0 = 0$

c. The base is 3.5, not -3.5. So

$$-(3.5)^2 = -[(3.5)(3.5)]$$
$$= -(12.25) = -12.25$$

d. $(101)^0 = 1$

e. $2^1 = 2$

▶ CHECK **Warm-Up 3**

Calculator Corner

Raise $-3/5$ to the fourth power.

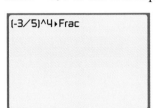

Press **ENTER.**

To raise a real number to a negative power, we use a definition.

> For all real numbers a not equal to zero, and all integers n,
> $$a^{-n} = \frac{1}{a^n}$$

This definition holds for both positive and negative n, as well as for $n = 0$.

■ ■ ■

EXAMPLE 4

Rewrite 9^{-2}, -4^{-3}, and $\frac{1}{3^{-3}}$ with positive exponents.

SOLUTION

According to our definition,

$$9^{-2} = \frac{1}{(9)^2} \qquad -4^{-3} = -(4^{-3}) = -\left(\frac{1}{4^3}\right) \qquad \frac{1}{3^{-3}} = \frac{1}{\frac{1}{3^3}} = 3^3$$

 CHECK **Warm-Up 4**

The reciprocal of a non-zero fraction $\frac{a}{b}$ is the fraction $\frac{b}{a}$. So for fractions,

> **Negative Powers of Fractions**
> For all integers a, b, and n, with a and b not equal to zero,
> $$\left(\frac{a}{b}\right)^{-n} = \left(\frac{b}{a}\right)^n$$

ERROR ALERT

Identify the error and give a correct answer.

$3^{-2} = -9$

▪▪▪

Example 5

Rewrite $\left(\frac{8}{x}\right)^{-2}$ with a positive exponent.

Solution

Because the reciprocal of $\frac{8}{x}$ is $\frac{x}{8}$,

$$\left(\frac{8}{x}\right)^{-2} = \left(\frac{x}{8}\right)^{2}$$

▶ CHECK **Warm-Up 5**

▪▪▪

WRITER'S BLOCK

Explain the difference between 2^{-3} and 2^{3}. Use the term *reciprocal*.

▪▪▪

Example 6

Rewrite $\left(\frac{-4}{7}\right)^{-2}$ without an exponent.

Solution

The reciprocal of $\left(\frac{-4}{7}\right)$ is $\left(\frac{-7}{4}\right)$, so

$$\left(\frac{-4}{7}\right)^{-2} = \left(\frac{-7}{4}\right)^{2} = \left(\frac{-7}{4}\right)\left(\frac{-7}{4}\right) = \frac{49}{16}$$

▶ CHECK **Warm-Up 6**

Roots

The opposite process of raising a number to the second power, or squaring it, is finding the *square root* of a number.

> **Square Root**
>
> A **square root** of a positive real number a is another real number b that, when squared, results in a.

Every positive real number has a positive square root and a negative square root. The positive square root of a real number a is written $\sqrt{a}$ and is called the **principal square root.** Thus $\sqrt{100} = 10$ because $10^2 = 100$. In words, the positive square root of 100 is 10 because the square of 10 is 100.

In a similar way, we say that a real number b is the **cube root** of a real number a if $b^3 = a$. As an example, we know that $2^3 = 2 \cdot 2 \cdot 2 = 8$, so 2 is the cube root of 8. We denote the cube root of a real number a by $\sqrt[3]{a}$

ERROR ALERT

Identify the error and give a correct answer.

$2^3 = 6$

▪▪▪

Example 7

Find:

a. the square of 9

b. the positive square root of 64

c. the cube of -1

d. the cube root of -125

Solution

a. $9^2 = 9 \cdot 9 = 81$

b. Because $8 \cdot 8 = 64$, we can write $\sqrt{64} = 8$.

c. $(-1)^3 = (-1)(-1)(-1) = -1$

d. Because $(-5)(-5)(-5) = -125$, we can write $\sqrt[3]{-125} = -5$.

▶ *CHECK* **Warm-Up 7**

Calculator Corner

a. Find the square of 9.

   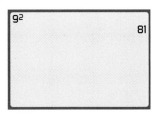

Press **ENTER.**

b. Find the positive square root of 64.

Press **ENTER.**

c. Find the cube of -1.

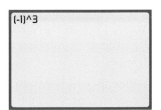   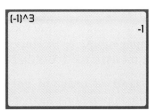

Press **ENTER.**

d. Find the cube root of -125.

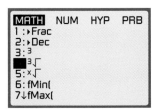   

Press **ENTER.**

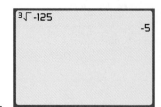

Press **ENTER.**

A **perfect square** is the square of a whole number.

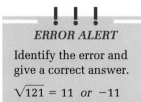
> **To find the square root of a perfect square**
>
> **1.** Find which two "tens" the square root lies between.
> **2.** Find the possible last digits of the square root.
> **3.** Test the possibilities.

▪▪▪

EXAMPLE 8

Find $\sqrt{7569}$.

SOLUTION

The square root of 7569 is smaller than 90 because $90^2 = 8100$. It is larger than 80 because $80^2 = 6400$. Thus we know that $80 < \sqrt{7569} < 90$. Therefore, the first digit of the square root is 8.

The last digit of 7569 is 9, so the last digit of its square root must be 3 or 7. Why? Because $3 \times 3 = 9$ and $7 \times 7 = 49$, and no other digit has a square ending in 9. Try both.

$$83 \times 83 = 6889? \quad \text{No}$$
$$87 \times 87 = 7569? \quad \text{Yes}$$

Because $87 \times 87 = 7569$, the square root of 7569 is 87.

▶ *CHECK* **Warm-Up 8**

The Standard Order of Operations

The **standard order of operations** specifies the order in which to perform indicated operations, so that every mathematical expression represents the same number to everyone.

> **Standard Order of Operations**
>
> **1.** Simplify within grouping symbols, working from the innermost out.
> **2.** Remove exponents.
> **3.** Multiply and divide from left to right.
> **4.** Add and subtract from left to right.

▪▪▪

EXAMPLE 9

Simplify: $15\left(\frac{-1}{5}\right)^{-2} + 3(1 + 4)^0 - \left(\frac{3}{4}\right)^{-1}$

SOLUTION

We begin by working inside the middle parentheses, retaining the zero exponent.

$$15\left(\frac{-1}{5}\right)^{-2} + 3(1 + 4)^0 - \left(\frac{3}{4}\right)^{-1}$$
$$= 15\left(\frac{-1}{5}\right)^{-2} + 3(5)^0 - \left(\frac{3}{4}\right)^{-1}$$

Next we simplify the exponents:

$$15\left(\frac{-1}{5}\right)^{-2} + 3(5)^0 - \left(\frac{3}{4}\right)^{-1}$$
$$= 15\left(\frac{-5}{1}\right)^{2} + 3(1) - \left(\frac{4}{3}\right)^{1} \quad \text{Writing reciprocals}$$
$$= 15(25) + 3(1) - \left(\frac{4}{3}\right) \quad \text{Simplifying}$$

We do multiplications and divisions in the order in which they occur from left to right, and then we do additions and subtractions in the same order.

$$15(25) + 3(1) - \left(\frac{4}{3}\right)$$
$$= 375 + 3 - \left(\frac{4}{3}\right) \quad \text{Multiplying}$$
$$= 378 - 1\tfrac{1}{3} \quad \begin{array}{l}\text{Adding; writing improper}\\\text{fraction as mixed number}\end{array}$$
$$= 376\tfrac{2}{3} \quad \text{Simplifying}$$

▶ *CHECK* **Warm-Up 9**

▪▪▪

EXAMPLE 10

Simplify: $5\left[\frac{1}{2} - \frac{2}{3}(9 - 1.8)\right] + 2^6\left[\frac{1}{4} + \left(\frac{-3}{8}\right)\right]^2$

SOLUTION

We begin by working within the innermost grouping symbols.

$$5\left[\frac{1}{2} - \frac{2}{3}(9 - 1.8)\right] + 2^6\left[\frac{1}{4} + \left(\frac{-3}{8}\right)\right]^2$$
$$= 5\left[\frac{1}{2} - \left(\frac{2}{3}\right)(7.2)\right] + 2^6\left[\frac{1}{4} + \left(\frac{-3}{8}\right)\right]^2$$
$$= 5\left[\frac{1}{2} - (4.8)\right] + 2^6\left[\frac{-1}{8}\right]^2$$
$$= 5[-4.3] + 2^6\left[\frac{-1}{8}\right]^2$$

Then we proceed as follows:

$$= 5[-4.3] + 64\left(\frac{1}{64}\right) \quad \text{Simplifying the exponents}$$
$$= (-21.5) + 1 \quad \text{Multiplying}$$
$$= -20.5 \quad \text{Adding}$$

▶ *CHECK* **Warm-Up 10**

! ! !
ERROR ALERT

Identify the error and give a correct answer.

Evaluate: $(-3^3) + 4^{-1}$

Incorrect Solution:
$(-3^3) + 4^{-1}$
$= -27 + (-4)$
$= -31$

! ! !
ERROR ALERT

Identify the error and give a correct answer.

Evaluate: $(-2)^{-2} + 5$

Incorrect Solution:
$(-2)^{-2} + 5$
$= \frac{-1}{4} + 5 = 4\tfrac{3}{4}$

Estimating Answers

In some calculations, an approximate result is enough.

Front-end estimation is an estimation technique that is useful when the numbers in a calculation include multidigit whole numbers or decimal numbers with whole-number (or integer) parts.

INSTRUCTOR NOTE

Remind students that there are many procedures for estimating. This is a very rough method.

> **To use front-end estimation**
> 1. For each number, identify the place value of the first non-zero digit.
> 2. Round the number to that place value.
> 3. Use these rounded numbers in computation.

▪ ▪ ▪

EXAMPLE 11

a. Use front-end estimation to determine an estimate for the computation $(1947)^2 + (3{,}287{,}210 \times 475)$.

b. Estimate the result of the computation $(0.00495)(4.3786)(0.00000999)$.

SOLUTION

a. We round 1947 to the nearest thousand: 2000. We round 3,287,210 to the nearest million: 3,000,000. And we round 475 to the nearest hundred: 500.

Our estimated computation becomes

$$(2000)^2 + (3{,}000{,}000 \times 500)$$
$$= 4{,}000{,}000 + 1{,}500{,}000{,}000$$
$$= 1{,}504{,}000{,}000$$

b. When some of the numbers in a calculation have only decimal parts, or a decimal number with a whole-number part less than 10, we use the same estimation procedure as before.

(0.00495) is rounded to 0.005.
(4.3786) is rounded to 4.
(0.00000999) is rounded to 0.00001.

Then the computation is estimated as $(0.005)(4)(0.00001) = 0.0000002$. This is the approximate result of the multiplication.

▶ *CHECK* **Warm-Up 11**

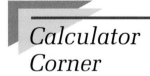

Calculator Corner

Enter the following problem onto your graphing calculator's Home Screen exactly as it is written. Then enter the same problem onto the Home Screen, except this time enter it using parentheses to enclose the entire denominator. How do the answers compare? Use estimation to decide which answer is correct.

$$\frac{(0.00000056)(24{,}340{,}000)}{(134{,}000)(0.0000078)}$$

```
(0.00000056)(24340000)/
(134000)(0.0000078)
              7.93411343E-10
```

```
(0.00000056)(24340000)/
((134000)(0.0000078))
              13.0409491
```

Practice what you learned.

SECTION FOLLOW-UP

Here are our solutions. Yours may differ.

$$(2 + 3) - (1 + 4) = 0$$
$$(2 + 3) \div (1 + 4) = 1$$
$$(2 + 4) \div (1 \times 3) = 2$$
$$1 \times (2 + 4) - 3\ \ = 3$$
$$(4 + 3) - (2 + 1) = 4$$
$$(4 + 3) - (2 \times 1) = 5$$
$$(4 + 3) - (2 - 1) = 6$$
$$(4 + 3) \times (2 - 1) = 7$$
$$(4 + 3) + (2 - 1) = 8$$
$$(4 + 3) + (2 \times 1) = 9$$
$$(2 + 3) + (1 + 4) = 10$$

▪ Now find what numbers you can from 11 to 20.

1.3 WARM-UPS

Work these problems before you attempt the exercises.

1. Determine whether each statement is true, and give the property illustrated by each true statement.

   a. $\left(5\frac{1}{3}\right)\left[\left(-2\frac{1}{2}\right)(-4)\right] = \left[\left(5\frac{1}{3}\right)\left(-2\frac{1}{2}\right)\right](-4)$ true; associative property of multiplication

   b. $3.2 + [6.8 + (-7.1)] = [6.8 + (-7.1)] + 3.2$ true; commutative property of addition

2. Fill in each blank with the name of the property used to reach the given step.

$$-(7 - 3) = -1(7 - 3)$$ multiplication property of −1

$$= -1[7 + (-3)]$$ Definition of subtraction of real numbers

$$= (-1)(7) + (-1)(-3)$$ distributive property

$$= -7 + 3$$ multiplication property of −1

$$= -4$$ Addition

3. Simplify: $\left(-\frac{1}{2}\right)^3$ and $-(2)^4$ $-\frac{1}{8}; -16$

4. Rewrite -2^{-2} and $-(-3)^{-3}$ with positive exponents. $-\frac{1}{2^2}; \frac{1}{3^3}$

5. Rewrite $\left(\frac{2}{3}\right)^{-1}$ and $\left(-1\frac{1}{2}\right)^{-2}$ with positive exponents. $\left(\frac{3}{2}\right)^1; \left(\frac{2}{3}\right)^2$

6. Rewrite $(-0.5)^{-2}$ and $\left(-1\frac{1}{2}\right)^{-3}$ without exponents. $4; -\frac{8}{27}$

7. Find the square of 16 and the principal square root of 81.   256; 9

8. Find: $\sqrt{8464}$ 92

9. Simplify: $[(8 - 0.4)^2 + 2(6 + 5 \div 2)] - 10^{-2}$ 74.75

10. Simplify: $5 - \left[\frac{1}{3}\left(6 \div \frac{1}{2}\right)\right]^2 + 2 \cdot \frac{1}{12}$ $-10\frac{5}{6}$

11. a. Use front-end estimation to determine a reasonable estimate of

$$\frac{(76985)(58996)(8989)}{(42930000000)}$$ 1080

   b. Estimate: $(0.0009768)(0.9368)(0.00982634)$ 0.000009

1.3 EXERCISES

Note: Use your graphing calculator to check your results whenever possible.

In Exercises 1 through 14, identify the property that is illustrated and then verify each statement by computation. For verifications, see the Solutions Manual.

1. $336 + [24 + (-5)] = [24 + (-5)] + 336$
   commutative property of addition

2. $3.2 + [6.8 + (-7.1)] = [6.8 + (-7.1)] + 3.2$
   commutative property of addition

3. $6(5 - 7) - 9 = (30 - 42) - 9$ distributive property

4. $2(3 + 5) + 7 = (6 + 10) + 7$ distributive property

5. $25 = -1(-25)$ multiplication property of −1

6. $-19 = -1 \cdot 19$ multiplication property of −1

7. $5 + (3 + 2) = (5 + 3) + 2$ associative property of addition

8. $16 + (5 - 8) = 16 + 5 + (-8)$
associative property of addition

9. $\left(5\frac{1}{3}\right)\left[\left(-2\frac{1}{2}\right)(-4)\right] = \left[\left(5\frac{1}{3}\right)\left(-2\frac{1}{2}\right)\right](-4)$
associative property of multiplication

10. $9(8 \cdot 10) = (9 \cdot 8) \cdot 10$ associative property of multiplication

11. $10 \cdot 9 = 2 \cdot 45$ associative property of multiplication

12. $10 \cdot 3 \cdot 5 = 3 \cdot 10 \cdot 5$ commutative property of multiplication

13. $8(6 + 16) = (8)(6) + (8)(16)$ distributive property

14. $15\frac{2}{3}(3 - 6) = 15\frac{2}{3}(3) - 15\frac{2}{3}(6)$ distributive property

In Exercises 15 through 28, replace the boxes (■) with numbers that make the resulting statements true. Identify the property illustrated by each statement. (Each box can represent a different number.)

15. $■[16 + (715 + a)] = 5[(■ + a) + 16]$ $5; 715$
commutative property of addition

16. $[(-9) + ■] + (-13) = (■) + [(-9) + 15]$ $15; -13$
commutative property of addition

17. $53 + [■ + 6.2] = [■ + (-2\frac{1}{2})] + 6.2$ $-2\frac{1}{2}; 53$
associative property of addition

18. $[(■) + 5] + (-3) = (-4) + [■ + (-3)]$ $-4; 5$
associative property of addition

19. $■ \cdot (a + 18^2) = (a + ■) \cdot 16$ $16; 18^2$
commutative property of multiplication

20. $3(■ + 5b) = (a + 5b)■$ $a; 3$
commutative property of multiplication

21. $■ + (-18 + 6a) = (-18 + ■) + 27$ $27; 6a$
commutative property of addition

22. $■(12 + 4y) = a(4y + ■)$ $a; 12$
commutative property of addition

23. $■ + \{3 \cdot [4 - ■(9 + 5)]\} = 2 + \{[4 - 8(9 + ■)] \cdot ■\}$
$2; 8; 5; 3$; commutative property of multiplication

24. $■ \cdot 87(b + 5) = ■(b + ■) \cdot 770$ $770; 87; 5$
commutative property of multiplication

25. $3 + 5(132 + ■) = ■(132 + 321) + ■$ $321; 5; 3$
commutative property of addition

26. $3(■ + 4) = 3(■ + x)$ $x; 4$
commutative property of addition

27. $[-16 \cdot ■] \cdot 5\frac{1}{2} = -16[256 \cdot ■]$ $256; 5\frac{1}{2}$
associative property of multiplication

28. $■ \cdot [75 \cdot (-1183)] = [425 \cdot ■](■)$ $425; 75; -1183$
associative property of multiplication

In Exercises 29 through 36, name the property used to obtain each line of the solution.

29. $19 \cdot 16 = [20 + (-1)] \cdot 16$ Writing 19 as $20 + (-1)$
$\quad = (20)(16) + (-1)(16)$ distributive property
$\quad = 320 + (-16)$ multiplication and multiplication property of -1
$\quad = 304$ Addition

30. $49 \cdot 41 = [50 + (-1)] \cdot 41$ Writing 49 as $50 + (-1)$
$\quad = 50 \cdot 41 + (-1)(41)$ distributive property
$\quad = 2050 + (-41)$ multiplication and multiplication property of -1
$\quad = 2009$ Addition

31. $45 \cdot 19 = 45[20 + (-1)]$ Writing 19 as $20 + (-1)$
$\quad = 45 \cdot 20 + 45 \cdot (-1)$ distributive property
$\quad = 900 + (-45)$ multiplication and multiplication property of -1
$\quad = 855$ Addition

32. $48 \cdot 99 = 48[100 + (-1)]$ Writing 99 as $100 + (-1)$
$\quad = 48 \cdot 100 + 48(-1)$ distributive property
$\quad = 4800 + (-48)$ multiplication and multiplication property of -1
$\quad = 4752$ Addition

33. $87 + (49 + 13) = 87 + (13 + 49)$ <u>commutative property of addition</u>

             $= (87 + 13) + 49$ <u>associative property of addition</u>

             $= 100 + 49$ Addition

             $= 149$ Addition

34. $124 + (47 + 16) = 124 + (16 + 47)$ <u>commutative property of addition</u>

             $= (124 + 16) + 47$ <u>associative property of addition</u>

             $= 140 + 47$ Addition

             $= 187$ Addition

35. $(4 \cdot 15) \cdot 5 = 4 \cdot (15 \cdot 5)$ <u>associative property of multiplication</u>

             $= 4 \cdot (5 \cdot 15)$ <u>commutative property of multiplication</u>

             $= (4 \cdot 5) \cdot 15$ <u>associative property of multiplication</u>

             $= 20 \cdot 15$ Multiplication

             $= 300$ Multiplication

36. $(6 \cdot 17) \cdot 5 = 6 \cdot (17 \cdot 5)$ <u>associative property of multiplication</u>

             $= 6 \cdot (5 \cdot 17)$ <u>commutative property of multiplication</u>

             $= (6 \cdot 5) \cdot 17$ <u>associative property of multiplication</u>

             $= 30 \cdot 17$ Multiplication

             $= 510$ Multiplication

In Exercises 37 through 52, determine whether each statement is true or false. If the statement is false, give one numerical example that illustrates this fact. Here, $a, b,$ and c are real numbers. Numerical examples may differ.

37. $a - 0 = 0 - a$ false

38. $g + 0 = 0 - g$ false

39. $1 \div (-c) = (-c) \div 1$ false

40. $a \div 1 = 1 \div a$ false

41. $a \div b = b \div a$ false

42. $(-a) \div b = b \div (-a)$ false

43. $-a - b = -b - a$ true

44. $a - b = b - a$ false

45. $a \div (b + c) = (a \div b) + (a \div c)$ false

46. $-a \div (b + c) = (-a \div b) + (-a \div c)$ false

47. $-a - b = -(a - b)$ false

48. $-(a + b) = -a + b$ false

49. $-(a - b) = a + b$ false

50. $-(a + b) = -a - b$ true

51. $-(a + b - c) = c - a - b$ true

52. $-(a - b - c) = c - b - a$ false

In Exercises 53 through 56, rewrite without an exponent.

53. $(1.5)^0$ 1

54. $(-3.8)^0$ 1

55. $-(0.37)^0$ -1

56. $\left(\frac{-5}{6}\right)^0$ 1

In Exercises 57 through 60, rewrite with a positive exponent.

57. $\left(\frac{5}{6}\right)^{-3}$ $\left(\frac{6}{5}\right)^3$

58. $\left(\frac{5}{6}\right)^{-4}$ $\left(\frac{6}{5}\right)^4$

59. $\left(-1\frac{1}{3}\right)^{-2}$ $\left(-\frac{3}{4}\right)^2$

60. $\left(-2\frac{1}{8}\right)^{-5}$ $\left(-\frac{8}{17}\right)^5$

In Exercises 61 through 64, find the requested power or root.

61. the positive square root of 9   3

62. the square of 25   625

63. the cube of 10   1000

64. the cube root of 64   4

In Exercises 65 through 80, simplify the expressions.

65. $(-32) + 5^{-1} \cdot 50$ -22

66. $-36 \cdot (12 \div 10)^{-2}$ -25

67. $-24 \div \left(8 \cdot \frac{1}{3}\right)^{-1}$ -64

68. $\left(12 \cdot \frac{1}{3^3}\right)^{-1} \div 3 - 1$ $\frac{-1}{4}$

69. $(2.1 + 4.3)^2 \div 6.4 - 30 \cdot 10^{-1}$ 3.4

70. $\left(\frac{2}{3}\right)^{-3} \cdot 8 - 3^{-4} \div 9^{-2}$ 26

71. $9\left(8 - 2\frac{1}{3}\right) - 4 \cdot 5\frac{1}{2}$ 29

72. $2.6(5 \div 0.1) + 19 \cdot 2$ 168

73. $-5(10)^{-3} + \{4.5 \div [(-2.4) + (1.5)(-0.28)]^3\}^0$ 0.995

74. $16 - \left\{3\left[2 - \left(\frac{5}{9} \div 2\frac{1}{4}\right)\right] \cdot 3^4 + 18\right\}^0$ 15

75. $3 + 2\{4 - [(6 - 7)^{-3} + 2]^2\}$ 9

76. $8 - 6\{5 - [(8 - 9)^{-3} + 4]^2\}$ 32

77. $(-1\{1[-1(-9 + 10)^{-2}]^{-3}\}^{-2})^{-3}$ -1

78. $(-1\{-1[1(8 - 9)^{-3}]^{-2}\}^{-3})^{-2}$ 1

79. $(3 - 2\{1 - [2(-1)^2]^2\}^2) \cdot 2 - 8$ -38

80. $\{4 - 3(2 - [3(-1)^3]^2)^2\} \cdot 3 - 5$ -434

In Exercises 81 through 84, estimate the answers by using front-end estimation.

81. $17263 + 3546 + 366 + 4$ $24{,}404$

82. $1432 + 6273 + 9007 + 42$ $16{,}040$

83. $(9776)^3 \cdot (-5738) - (-6288) \div (-987)$
$-6{,}000{,}000{,}000{,}000{,}000$

84. $(7826)^2 \div (-38) + 9681 \times 74$ $-900{,}000$

In Exercises 85 through 94, round to the place value of the first non-zero digit and estimate the result.

85. $\frac{(299)(2687)}{(-2513)(-33)}$ 10

86. $\frac{(27)(-9684)}{(1687)(15)}$ -7.5

87. $(213225)(5.542)(326)(4)$ $1{,}440{,}000{,}000$

88. $(2452)(221.33)(9007)(0.00042)$ $1{,}440{,}000$

89. $\frac{(92.136)(-5753)}{(-2.238)(-9813)}$ -27

90. $\frac{(13326)(-6.8)}{(9882)(-0.000074)}$ $100{,}000$

91. $\frac{(99)(5.6813)}{(-2523)(-53)}$ 0.004

92. $\frac{(213)(-9284)}{(2687)(2.5)}$ -200

93. $(-8255)(-13.2)(-2882)(-22)$ $4{,}800{,}000{,}000$

94. $(-298)(-2.13)(8572)(-2.625)$ $-16{,}200{,}000$

MIXED PRACTICE

By doing these exercises, you will practice the topics up to this point in the chapter.

95. Arrange $-7\frac{2}{5}, 7\frac{1}{4}$, and 7.3 in order from smallest to largest. $-7\frac{2}{5} < 7\frac{1}{4} < 7.3$

96. What property is illustrated by the statement $(-1)(18)(32) = (-18)(32)$?   multiplication property of -1

97. Simplify: $[13 - 6(2)]^{-3} + \sqrt{9} \div 12$ $1\frac{1}{4}$

98. Simplify: $\frac{-16\frac{2}{3}}{3\frac{1}{3}}$ -5

99. What property is illustrated by $18 + (17 + 15) = (17 + 15) + 18$?   commutative property of addition

100. Simplify: $(0 - 2)^{-4}(4^2 - 4)^2$ 9

EXCURSIONS

Class Act

1. In parts (a) through (d), each of the problems has missing digits. Find them.

a.
```
      2 _ _
  _ 7)74 _ _
     = =
     _ _
     00
      0
```

b.
```
      _ _ _
  4 _)92 _ _
      92
      _ _
      00
       0
```

c.
```
      _ 0 _
  7 _)829 _
      _ 9
      _ _ 5
      3 _ _
        0
```

d.
```
      _ 0 _
  8 _)92 _ 5
      _ 5
      _ 6 _
      7 _ _
        0
```

e. Make up four missing digit problems. Swap them with another group.

f. ✏ Explain to a friend how to solve the missing digit problems (a) to (d).

2. The following table lists the average weight of various animals.

a. What is the range of the weights?

b. What is the total weight of the three lightest animals? the three heaviest?

Animal	Weight
hedgehog	1.88 pounds
chinchilla	1.5 pounds
ferret	2.04 pounds
beaver	58.5 pounds
guinea pig	1.54 pounds
wild boar	302.5 pounds

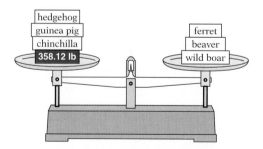

The balance scale above shows that the difference in weight between the three heaviest animals and the three lightest is 358.12 pounds. Rearrange the animals in groups so that the difference is

c. 354.04 pounds

d. 364.2 pounds

e. Arrange four animals so that the difference is 0.12 pounds.

f. Arrange three animals so that the difference is 1 pound.

g. ✏ Make up four other problems using this table. Solve them or swap with a classmate.

3. a. Take your pulse.

b. Estimate how long it would take for your pulse to beat a million times.

c. Write four more questions that involve estimation.

d. Have one person in the group look at the second hand on a watch. See if you can estimate when exactly one minute has passed.

e. ✏ Write a paragraph describing your estimation method. Did it work well? How would you alter your procedure next time? In estimation do you believe that "practice makes perfect"?

Exploring with Calculators

4. a. Input each of the following expressions into your graphing calculator Home Screen and obtain the result. Then place parentheses around the operation that should be performed first.

$$2 + 3 - 1 \qquad 2 \times 3 - 1 \qquad 2 + 3 \times 4$$
$$2 \times 3 + 4 \qquad 2 - 3 \times 4 \qquad 2 + 3 - 4$$
$$5 \times 6 + 3 \times 2 \qquad 5 + 6 + 3 \times 2 \qquad 5 + 6 \times 3 + 2$$
$$5 + 6 \times 3 \times 2 \qquad 5 + 6 - 3 + 2$$

b. Insert +, −, or × one time each in each of the following sequences of numbers in order to obtain the given result.

$$4 \quad 5 \quad 3 \quad 2 = 21 \qquad 4 \quad 5 \quad 3 \quad 2 = 17 \qquad 4 \quad 5 \quad 3 \quad 2 = 19$$
$$4 \quad 5 \quad 3 \quad 2 = 3 \qquad 4 \quad 5 \quad 3 \quad 2 = -9 \qquad 4 \quad 5 \quad 3 \quad 2 = 5$$

CONNECTIONS TO *GEOMETRY*

Introduction to Geometric Figures

In this text we will be investigating the uses of real numbers and algebra. Many real-life applications occur in geometry. But first we need some definitions before we can explore those applications.

Our approach to geometry and measurement is informal. This is a review of basic ideas and terminology. Some basic ideas—point, line, angle, and degree—are left undefined.

A **polygon** is a closed figure made up of three or more line **segments,** or parts of lines. (See the figure on the following page.) The segments are called the **sides** of the polygon. The point where two sides meet is a **vertex** and the sides form an **angle** at the vertex.

INSTRUCTOR NOTE

Because we did not introduce the notion of "closed curve," we did not use it in our definition. That formality can be left until later.

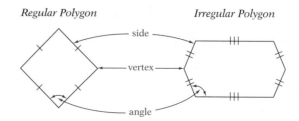

The measure of a side is called its **length.** If the sides of a polygon all have the same length and its angles all have the same measure, the polygon is called a **regular polygon.** If the lengths or angles differ, the polygon is **irregular.** Some polygons that you are probably familiar with, and some new ones, are shown here.

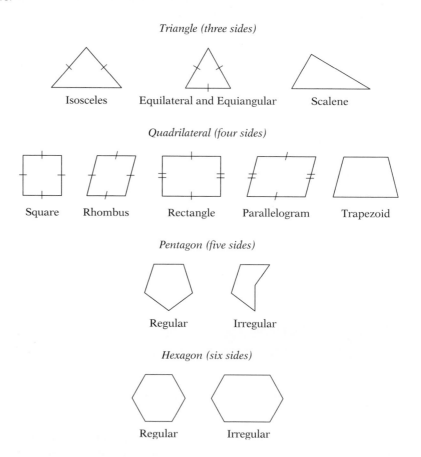

Triangles that have *two* sides of equal length are called **isosceles triangles.** Triangles that have *three* sides of equal length are called **equilateral triangles;** their angles are all of equal measure too. Those that have *no* sides equal in length are called **scalene triangles.**

Quadrilaterals are a little more difficult to classify. **Rectangles** are quadrilaterals that have two pairs of opposite sides of equal length and all four angles of equal measure; the angles are called **right angles,** or 90° angles. The sides that form right angles are said to be **perpendicular** to each other. **Squares** are quadrilaterals that have all four sides equal in length; their angles too are all right angles. If the angles are not right angles, the figure is called a **rhombus.** **Parallelograms** are quadrilaterals that have opposite sides equal in length and parallel. **Parallel** means that the perpendicular distance between the two lines is the same no matter where on the line the comparison is made. Parallel lines never intersect. (See the accompanying figure.)

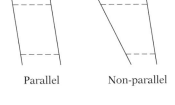

Parallel   Non-parallel

Trapezoids are quadrilaterals that have one pair of opposite sides parallel.

All the common polygons shown in these figures are **plane figures;** they exist only on a flat surface. Note that we use tick marks to show equal measure. Lines with the same number of marks have equal lengths.

The "angle" measure of a **circle** is 360°. A **right angle** has a measure of 90° (that is, $\frac{360}{4} = 90°$).

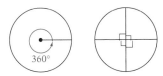

Right angles are found in squares, rectangles, and some triangles (called **right triangles**).

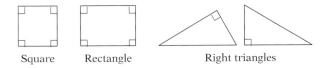

Square   Rectangle   Right triangles

We indicate right angles with little boxes. We usually leave those boxes out when the figure is clearly a square or a rectangle.

An **acute angle** has a measure of less than 90°. An **acute triangle** has all acute angles. An **obtuse angle** has a measure of more than 90°. An **obtuse triangle** has one obtuse angle.

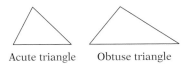

Acute triangle   Obtuse triangle

You will use many of these definitions and ideas in the geometry sections in this text.

1.4 The Language of Sets and Problem Solving

SECTION LEAD-IN

▪ Discuss strategies that can be used for estimating the number of people in a crowd. List three.
▪ Find references to numbers of people that attended various events. Which of these numbers do you believe were estimated?

An Introduction to the Language of Sets

A **set** is a collection of objects. The objects are called **members,** or **elements,** of the set. We name sets with capital letters, and we identify the members of a set by enclosing them in braces { }. Thus the statement $A = \{a, *, 1, 2\}$ tells us that A is the set that contains the elements a, $*$, 1, and 2.

The set that contains no elements is called the **empty set** and is denoted by either empty braces or the symbol $\emptyset$ (but not both). For example, we say the set E of whole numbers between 1 and 2 is an empty set, because there are no whole numbers between 1 and 2. We write $E = \{ \ \}$ or $E = \emptyset$.

Describing Sets

A set is correctly described when we can tell, from the description, whether any given object is a member of the set. Two methods are most often used. In **roster notation,** the elements of the set are listed. Set A above is given in roster form, and so is $N = \{1, 2, 3, 4, \ldots\}$. You should recognize N as the set of natural numbers. Set A is a finite set (its members can be counted), whereas set N is an infinite set (its members cannot be counted). We designate an infinite set by including the ellipsis symbol $(\ldots)$.

In **set-builder notation,** a typical element is named (usually with a letter) and carefully described. Thus

$$N = \{x \mid x \text{ is a natural number}\}$$

is a set-builder description of the set of natural numbers. The thin vertical line is read as "such that," so the entire set-builder statement is read, "N is the set of all elements x such that x is a natural number."

▪▪▪

EXAMPLE 1

a. Write the following sets in roster notation:

$$T = \{t \mid t \text{ is an even whole number less than 18}\}$$
$$X = \{x \mid x \text{ is an integer greater than } -6 \text{ and less than 4}\}$$

b. Write the following sets in set-builder notation:

$$Y = \{1, 3, 5, 7, \ldots\} \qquad Z = \{-2, -1, 0, 1\}$$

WRITER'S BLOCK

How do we read the set-builder notation for T and X in Example 1?

SOLUTION

a. Set T contains all the even whole numbers less than 18. We list them within braces.

$$T = \{0, 2, 4, 6, 8, \ldots, 16\}$$

And the description of set X tells us that

$$X = \{-5, -4, -3, -2, -1, 0, 1, 2, 3\}$$

b. The elements of Y are all the odd whole numbers, so

$$Y = \{y \mid y \text{ is an odd whole number}\}$$

The elements of Z are the integers greater than -3 and less than 2, so

$$Z = \{a \mid a \text{ is an integer } and \ a \text{ is greater than } -3 \text{ and less than 2}\}$$

▶ CHECK **Warm-Up 1**

Subsets

Set A is said to be a **subset** of set B when every element of set A is also an element of set B. The empty set is a subset of every set.

Set Operations

Two operations on sets, *union* and *intersection*, produce new sets.

Set Union

The **union** of two sets A and B is the set of all elements contained in either A or B or in both.

The union of A and B is written $A \cup B$.

> **Set Intersection**
> The **intersection** of two sets A and B is the set of all elements that are contained in both A and B.

The intersection of sets A and B is written $A \cap B$.

▪ ▪ ▪
EXAMPLE 2

a. Let $A = \{1, 2, 5\}$ and $B = \{2, 3, 4, 5\}$. Find $A \cup B$.

b. Given $E = \{x \mid x \text{ is a whole number}\}$ and $F = \{x \mid x \text{ is an integer less than 3}\}$, find $E \cap F$.

SOLUTION

a. $A \cup B$ contains the elements that are in either A or B.

We start by listing the elements in A:

$$\{1, 2, 5 \qquad \}$$

Then we include those additional elements from set B:

$$\{1, 2, 5, 3, 4\}$$

We usually give the answer as a set with the elements arranged in order.

$$A \cup B = \{1, 2, 3, 4, 5\}$$

b. We are looking for elements that these two infinite sets have in common. We can see those elements more clearly if we write the sets in roster form, one above the other. We get

$$E = \{0, 1, 2, 3, 4, \ldots\}$$
$$F = \{\ldots, -2, -1, 0, 1, 2\}$$

Now we can see that E and F have only the elements 0, 1, and 2 in common. So $E \cap F = \{0, 1, 2\}$.

▶ *CHECK* **Warm-Up 2**

STUDY HINT
Each element in a set is listed only once.

Set Complements

WRITER'S BLOCK
Distinguish between
{ } and {0}.

> Two sets A and A' are **complements** of each other when the intersection of the two sets is $\varnothing$ and their union is U, the universal set.

▪ ▪ ▪
EXAMPLE 3

Let $U = \{1, 3, 5, 7, 9\}$ and $A = \{1, 5\}$. Find A'.

SOLUTION

Start with the universal set {1, 3, 5, 7, 9}. The set A' will have all the same elements except for 1 and 5, the members of set A:

$$\{\cancel{1} \quad 3 \quad \cancel{5} \quad 7 \quad 9\}$$

So $A' = \{3, 7, 9\}$.

▶ *CHECK* **Warm-Up 3**

Complements have certain properties. If A and A' are complements, then

$$A \cap A' = \varnothing$$
$$A \cup A' = U$$

An Introduction to Problem Solving

We will be applying algebra to solve a variety of problems—especially word problems—in this text. The best way to attack such problems is with a systematic method, step by step. Here's one that works for many people.

Interpret the problem. You must read and understand the problem. Determine what information is given, what you are actually asked to find, and what information you need to supply.

Decide on a method for solving the problem. Do you have to add, subtract, multiply, or divide? There are sometimes several different ways to approach a problem. Decide which way will work best. Can you solve the problem directly, or must you solve for a piece of information first and then use it to find the answer?

Apply the method. Perform the actual computations in the proper order to obtain an answer.

Look back and reason. Once you have an answer, go back and check two things:

1. Check your calculations to make sure that you have not made any computational errors.
2. Reread the original question to make sure your answer is reasonable and valid. (An answer of $2\frac{1}{2}$ might be computationally accurate, but if you were asked for the number of children in a family, it would not be a reasonable answer.)

All together, these four steps are:

> **WRITER'S BLOCK**
>
> Explain when two sets will have the following property:
> $$A \cup B = A$$
> and
> $$A \cap B = \varnothing$$

Four Steps for Problem Solving

1. **INTERPRET** the problem.
2. **DECIDE** on a method for solving the problem.
3. **APPLY** the method.
4. Look back and **REASON**.

▪ ▪ ▪

EXAMPLE 4

Blood flows through the kidneys at a rate of 1.3 liters per minute. How much blood flows through the kidneys in 1 hour?

SOLUTION

INTERPRET We are given the amount of blood that flows each minute (1.3 liters). We are asked how much blood flows in 1 hour. We know there are 60 minutes in an hour.

DECIDE To find the amount of blood flowing in 1 hour, we must multiply the amount of blood that flows in 1 minute by 60 minutes.

APPLY Multiplying yields

$$\frac{1.3 \text{ liters}}{\cancel{\text{minute}}} \times \frac{60 \cancel{\text{ minutes}}}{\text{hour}} = 78 \frac{\text{liters}}{\text{hour}}$$

REASON Rereading the problem, we see that the blood flows at slightly more than 1 liter per minute, and our result shows the amount per hour to be slightly more than 60 liters per hour (60 minutes). The answer is reasonable. We can check the result by dividing 78 by 60. The answer checks.

So 78 liters of blood flow through the kidneys each hour.

▶ CHECK **Warm-Up 4**

! ! !
ERROR ALERT

Identify the error and give a correct answer.

$6000 divided between 2 people:

$$\frac{\$6000}{2} = \$3000$$

This answer is incomplete.

▪ ▪ ▪

EXAMPLE 5

A certain roller coaster ride lasts $1\frac{5}{6}$ minutes. If the entire ride is 3250 feet long, what is the average speed in feet per second?

SOLUTION

INTERPRET The ride is 3250 feet long and takes $1\frac{5}{6}$ minutes to complete. We want the speed in feet traveled per second. We know there are 60 seconds in a minute. (From here on, we will not indicate the units throughout the computations; however, we will always include them in our answer.)

DECIDE To find the time the ride takes in seconds, we must multiply $1\frac{5}{6}$ minutes times 60 seconds per minute. Then we can divide the distance traveled (3250 feet) by the time traveled in seconds. We will do the computations and then indicate the answer in feet per second.

APPLY First we find the time in seconds.

$$\left(1\frac{5}{6}\right) \times 60 = \left(\frac{11}{6}\right)(60) = 110 \text{ seconds}$$

Then we divide the length of the ride (3250 feet) by the time needed to travel it (110 seconds).

$$\frac{3250}{110} = 29\frac{60}{110} = 29\frac{6}{11}$$

The roller coaster travels at $29\frac{6}{11}$ feet per second.

REASON According to the original problem statement, the roller coaster travels about 3000 feet in about 100 seconds. Our result, then, should have been about 30 feet per second. The answer, $29\frac{6}{11}$ feet per second, is reasonable.

▶ *CHECK* **Warm-Up 5**

In the next example, you will need to know that the **markup rate** is the percent by which a dealer increases the cost to obtain the selling price. Note that the **discount rate** is the percent by which the original cost is reduced. The **percent of increase or decrease** is the ratio of the amount of change to the original amount.

▪▪▪

EXAMPLE 6

A coat costs retailers $180. It is sold for $230. What is the markup rate?

SOLUTION

INTERPRET We are being asked, "What percent of the cost is the given markup amount?" Because the amount the coat has been marked up is $230 − $180, or $50, we must find

"What percent of $180 is $50?"

Here, the *percent* is the missing value, the *base* is $180, and the *amount* is $50.

DECIDE We substitute the known values into the **percent equation.**

$$\text{Percent (as a decimal)} \times \text{base} = \text{amount}$$
$$n \times 180 = 50$$

A good estimate of the answer is 50 ÷ 200, or 0.25. So 25% is the estimated markup rate.

APPLY We solve the equation by dividing both sides by 180.

$$\frac{n \times 180}{180} = \frac{50}{180}$$
$$n = 0.27\overline{77} \quad \text{Dividing}$$

Rounded to the ten-thousandths place, n is 0.2778.

REASON The solution thus far is a decimal, but we are asked for a *percent*. So we must move the decimal point two places to the right and add a percent sign.

Thus the markup rate is 27.78%.

▶ *CHECK* **Warm-Up 6**

In the next example, we have to find a missing piece of information first and then use it to solve the problem.

INSTRUCTOR NOTE

Word problems always present a problem for students. The more that students practice, the easier word problems will become.

▪▪▪
WRITER'S BLOCK

Why did we get an estimated answer first? How should we use this estimate?

▪▪▪

EXAMPLE 7

An inventory of spare parts showed that 168 had rusted. This represented 56% of the total number of parts. How many parts had not rusted?

SOLUTION

DECIDE   Solving this problem requires two steps. First, we must find the total number of parts. Then we can subtract the number of rusted parts from that to find the number that did not rust. We know that 168 is 56% of the total number of parts.

The *percent* is 56%, which gives us the ratio $\frac{\text{part}}{\text{whole}} = \frac{56}{100}$. The *base* (or *whole*) is the missing value, and the *amount* (or *part*) is 168; they give us the ratio $\frac{168}{n}$.

APPLY   These fractions are equal. We set up a **proportion** and solve.

$$\frac{56}{100} = \frac{168}{n}$$ The proportion (equal fractions)

$56 \times n = 100 \times 168$ Cross-multiplying (If $\frac{a}{b} = \frac{c}{d}$, then $ad = bc$.)

$56 \times n = 16{,}800$ Multiplying out

$\frac{56 \times n}{56} = \frac{16{,}800}{56}$ Dividing by 56

$n = \frac{16{,}800}{56} = 300$ Simplifying

REASON   The total number of parts is 300.

DECIDE   Now we must subtract the number of rusted parts from the total.

APPLY $300 - 168 = 132$

REASON   So 132 parts were not rusted.

▶ CHECK **Warm-Up 7**

Practice what you learned.

◢ SECTION FOLLOW-UP

- How are operations of real numbers used in your crowd estimation?
- ***Research***   Find some numbers that are estimates in newspapers or magazines. What procedures do you think were used to obtain the estimates?

1.4 WARM-UPS

Work these problems before you attempt the exercises.

1. Describe the set of integers greater than -2 and less than 5, using both roster and set-builder notation.
   $\{-1, 0, 1, 2, 3, 4\}$; $\{x \mid x$ is an integer greater than -2 and less than 5$\}$

2. Let $A = \{1, 3, 5, 7\}$
   $B = \{2, 3, 4, 5\}$
   $C = \{1, 4, 5\}$

   a. Find $A \cup B$.   **b.** Find $(A \cup B) \cup C$.
   $\{1, 2, 3, 4, 5, 7\}$ $\{1, 2, 3, 4, 5, 7\}$

3. Let $A = \{1, 2, 3, 4, 5\}$
   $B = \{1, 3, 5\}$
   $U = \{1, 2, \ldots, 10\}$

   a. Find A'.   **b.** Find B'.   **c.** Find $(A \cap B)'$.
   $\{6, 7, 8, 9, 10\}$ $\{2, 4, 6, 7, 8, 9, 10\}$ B'

4. An astronaut can circle Earth in a space shuttle in 1.5 hours. How many complete times can the astronaut circle Earth in exactly 1 day?   16 times

5. If the roller coaster from Example 5 traveled a mile at its average speed, how long would it take to complete the trip? (Give your answer in minutes.)   about 3 minutes

6. A price ticket on a suit said "PRICE REDUCED $75." If the original price was $250, what is the discount rate (based on original price)?   30%

7. On a certain day, 22% of the workers called in sick. If the company employs 150 people, how many people did not call in sick?   117 people

1.4 EXERCISES

Note: Use your graphing calculator to check your results whenever possible.

In Exercises 1 through 4, determine whether the sets are empty.

1. $\{0\}$
   Not empty. 0 is an element.

2. $\{y \mid y$ is an even whole number between 3 and 6$\}$ Not empty. 4 is an element.

3. $\{x \mid x$ is even and odd$\}$
   Empty. No number is both even and odd.

4. $\{y \mid y$ is a natural number less than 1$\}$
   Empty. No natural number is less than 1.

In Exercises 5 through 12, write the sets in roster form.

5. $\{x \mid x$ is an even whole number$\}$
    $\{0, 2, 4, 6, \ldots\}$

6. $\{y \mid y$ is an integer greater than -5 and less than 3$\}$ $\{-4, -3, -2, -1, 0, 1, 2\}$

7. $\{t \mid t$ is a prime number (a number with exactly two factors—itself and one) less than 30$\}$
    $\{2, 3, 5, 7, 11, 13, 17, 19, 23, 29\}$

8. $\{$the odd numbers between 4 and 6$\}$
    $\{5\}$

9. $\{v \mid v$ is a positive-integer factor of 12$\}$
    $\{1, 2, 3, 4, 6, 12\}$

10. $\{$positive even-number factors of 50$\}$
    $\{2, 10, 50\}$

11. $\{y \mid y$ is an integer greater than -5 and less than 10$\}$
    $\{-4, -3, -2, -1, 0, 1, 2, 3, 4, 5, 6, 7, 8, 9\}$

12. $\{w \mid w$ is a whole number between 17 and 21$\}$
    $\{18, 19, 20\}$

In Exercises 13 through 16, answer true or false.

13. The set of integers is a subset of the set of real numbers.   true

14. The set of real numbers has the set of rational numbers as a subset.   true

15. The set of natural numbers is a subset of the set of integers.   true

16. The empty set is a subset of the irrational numbers.   true

In Exercises 17 through 20, find the indicated result when $A = \{0, 2, 3, 4, 5\}$; $B = \{0, 3, 5, 9\}$; and $C = \{0\}$.

17. $(A \cup B) \cup C$ $\{0, 2, 3, 4, 5, 9\}$

18. $(A \cap B) \cap C$ $\{0\}$

19. $(A \cup B) \cap C$ $\{0\}$

20. $C \cap (A \cup B)$ $\{0\}$

In Exercises 21 through 24, find $A \cup B$ and $A \cap B$.

21. $A = \{$whole numbers$\}$ $A \cup B = \{$integers$\}$ or $A \cup B = B$
    $B = \{$integers$\}$ $A \cap B = \{$whole numbers$\}$ or $A \cap B = A$

22. $A = \{$rational numbers$\}$ $A \cup B = \{$rational numbers$\}$ or $A \cup B = A$
    $B = \{$integers$\}$ $A \cap B = \{$integers$\}$ or $A \cap B = B$

23. $A = \{x \mid x$ is an even whole number$\}$ $A \cup B = \{n \mid n$ is a whole number$\}$
    $B = \{y \mid y$ is an odd whole number$\}$ $A \cap B = \varnothing$

24. $A = \{t \mid t$ is an integer multiple of 3$\}$ $A \cup B = \{t \mid t$ is an integer multiple of 3$\}$ or $A \cup B = A$
    $B = \{z \mid z$ is an integer multiple of 6$\}$ $A \cap B = \{z \mid z$ is an integer multiple of 6$\}$ or $A \cap B = B$

25. *High Jump*   In the 1988 summer Olympics in Seoul, Korea, the American who won the heptathlon, Jackie Joyner-Kersee, had come to the high jump with a previous personal-best jump of 6 feet 4 inches. In the Olympic event, however, she jumped 6 feet $1\frac{1}{4}$ inches. By how much did she miss her personal best? $2\frac{3}{4}$ inches

26. *High Jump*   In the 1996 summer Olympics in Atlanta, the winner of the heptathlon, Ghada Shouaa from Syria, jumped 6 feet 1.23 inches. By how much did she miss Jackie Joyner-Kersee's 1988 Olympic jump? her personal best jump? (Refer to Exercise 25.)   0.02 inches; 2.77 inches

27. *Caves*   The Cuyaguatega cave system in Cuba is $32\frac{7}{10}$ miles long. The Flintridge cave system in the United States is $148\frac{7}{10}$ miles longer. How long is the Flintridge system? $181\frac{2}{5}$ miles

28. *Glaciers*   The Antarctic glacier flows $84\frac{3}{5}$ yards in a week; the Greenland glacier flows $236\frac{9}{10}$ yards in the same time. How much farther does the Greenland glacier flow in a week? $152\frac{3}{10}$ yards

29. *Baseball*   The distance from the pitcher's mound to the home plate on a baseball diamond is $60\frac{1}{2}$ feet. If a player's stride is $2\frac{1}{2}$ feet long, how many steps would he take to walk from the mound to home plate?   25 steps

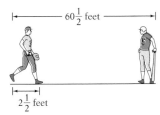

30. *Travel*   The road distance (in miles) between Memphis and Pittsburgh is approximately $1\frac{1}{4}$ times the air distance (in statute miles) between Pittsburgh and Memphis. If the air distance is approximately six hundred sixty statute miles, how many miles is the road distance?   825 miles

31. *Counterfeit Money*   During the Civil War, it was estimated that nearly one-third of all the currency in circulation was counterfeit. Approximately 1600 state banks were designing and printing their own money, and there were about 7000 different varieties of genuine bills. If there were 2 million bills in circulation, about how many were counterfeit? (*Hint:* Beware of unneeded information.) Use 0.333 as an estimate for $\frac{1}{3}$.   666,000 bills

32. *Classic Cars*   Only six 1930 Bugati Royale open cars were produced. If the total sales from these cars was $270,000 and they all sold for the same amount, about how much did each car sell for? Round your answer to the nearest whole number. $45,000

33. *Football*   The most points ever scored by one team over another in a football game was scored by Georgia Tech in 1916. The total number of points scored was 222. If a touchdown gives a team 6 points, what is the greatest number of touchdowns that could have been scored by Tech?   37 touchdowns

34. *Weight Lifting*   The heaviest weight ever lifted by a woman was lifted by Josephine Blatt in 1895. She lifted the equivalent of 26 women of average weight. If a woman of average weight weighs 135 pounds, how much weight did Blatt lift?   3510 pounds

35. *Frog Long Jump*   A trained frog can jump 17 feet $6\frac{3}{4}$ inches. If a man can jump about $\frac{3}{4}$ that distance, how far can he jump?   13 feet $2\frac{1}{16}$ inches

36. *Horse High Jump*   A horse has jumped 8 feet $1\frac{3}{4}$ inches high. The 1994 pole vault record is twice that plus $46\frac{1}{6}$ inches. About how high is the 1994 pole vault record jump?   20 feet $1\frac{2}{3}$ inches

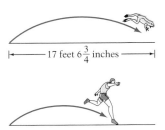

37. *Ocean Liner*   The Verrazano Narrows Bridge is how many times longer than the width of the Queen Elizabeth ocean liner if the bridge is 4260 feet long and the ocean liner is 119 feet wide? Round your answer to the nearest whole number.   36 times longer

38. *Roller Coaster Ride*   Six Flags Over Georgia, an amusement park, boasts one of the world's largest and tallest wooden roller coasters, the Scream Machine. It is 105 feet tall, and the cars travel 57 miles per hour (0.95 miles per minute). It takes about 1.5 minutes to complete one trip. How many miles long is the trip?   1.425 miles

39. *Mortgage Rate*   A mortgage company advertises that its current home mortgage rate is 11.75%, which represents a drop of 0.6% from the previous rate. What was the old rate?   12.35%

40. *Stock Prices*   A certain stock has fluctuated as follows: up $\frac{4}{8}$, down $\frac{3}{8}$, up $\frac{2}{8}$, up $\frac{1}{8}$. If the opening price was $3\frac{5}{8}$, what is the current price? $4\frac{1}{8}$

41. *Heart Rate*   Your heart rate changes with the amount of energy that you exert. If your heart rate was 70 beats per minute (bpm) before you exercised, increased 40 bpm during exercise, and then dropped 10 bpm after exercise, what was your heart rate after exercise?   100 bpm

42. *Sea Level*   An airplane is at an altitude of 29,000 feet above sea level. It flies over a valley that has a low point 25 feet below sea level. How far above the valley "floor" is the airplane?   29,025 feet

43. *Temperature Range*   The temperatures in a certain part of the world have been known to range from $-23°F$ to $102°F$. What is the range of these temperatures?   125° F

44. *Plant Growth*   A certain plant can grow at any altitude from 35 feet below sea level to 72 feet above sea level. What is the range of its "growth zone"?   107 feet

45. *Stock Prices*   The stock for a certain company closed yesterday at $31\frac{1}{8}$. Today it opened $3\frac{3}{8}$ points lower than yesterday's closing, and then it went to a level $6\frac{2}{8}$ points higher than yesterday's closing price. What is the range of these fluctuations? $9\frac{5}{8}$

46. *Sea Level*   If Mt. Everest's base were at the bottom of the Marianas Trench, deep under the Pacific Ocean, its peak would be 7169 feet below sea level. Mt. Everest is 29,028 feet tall. How deep is the Marianas Trench?   36,197 feet

47. *Scale Readings*   The changes in a reading on a scale were -2, $3\frac{1}{2}$, 10.5, -16, -15.5, and -4.5. What does the scale read now in relation to its starting point?   24 less than at the start

48. ***Tsunami*** The highest recorded tsunami (tidal wave) was 220 feet high. The Statue of Liberty is 305 feet tall with its pedestal. How much taller is the statue than this wave?   85 feet

49. ***Hammer Throw*** In the 1996 summer Olympics, Lance Deal, the silver medal winner from the United States, threw the hammer 76.94 meters, 75.62 meters, 77.26 meters, and 81.12 meters. By how much did his longest throw exceed his shortest throw?   5.5 meters

50. ***Ocean Size*** There are 139,500,000 square miles of salt water oceans in the world. This is 139,170,000 more than the number of square miles of fresh water lakes. How much area is covered by fresh water lakes?
330,000 square miles

51. ***Bowling*** In 1984 there were 8,401,000 members in the American Bowling Congress. This was 868,000 fewer than in 1983. What was the enrollment in 1983?   9,269,000 members

52. ***Rent Increases*** Your rent increases 15% after the first year and 10% more the second year. If your rent before any increases was $200, what is your rent after the two increases? $253

53. ***Sales Tax*** Find the sales tax on a dress that costs $124.80 if the tax rate is 5.5%. $6.87 (*Note:* We will always round money up.)

54. ***Discount Prices*** A coat is marked "20% off." How much money will you save if the coat was originally priced at $175? $35

55. ***Price Reduction*** A bicycle originally priced at $220 was reduced by $52.80. What was the percent reduction?   24%

56. ***Land Area*** Ice covers 10% of the land area of Earth. If Earth's land area is approximately 58,433,000 square miles, how much is covered by ice?
5,843,300 square miles

57. ***Carbon Weight*** Anthracite is 95% carbon by weight. If a specimen weighs 24 ounces, how much of it is carbon?   22.8 ounces

58. ***Fish Harvest*** 95% of the 75 million tons of fish harvested each year come from the oceans. How many millions of tons of fish come from other water sources?   3,750,000 tons

59. ***Ocean Area*** The surface area of Earth is approximately 196,949,970 square miles. Approximately 147,712,470 square miles are covered by oceans. Round these numbers to the nearest million, and determine what

percent of Earth is covered by oceans. Round to the nearest whole percent. What percent is land?   75%; 25%

60. a. *Grade Point Average* Su's freshman GPA was 3.15. It went up by 0.34, down by 0.15, and up by 0.08 in her next three years. What was her final GPA?   3.42

b. ✎ How is GPA calculated at your school?   Answers may vary.

61. *Sea Level* In Africa the highest point of elevation is Mt. Kilimanjaro at 19,340 feet above sea level. The lowest point is a spot in Egypt that is 436 feet below sea level. A man is standing at the lowest spot. He is 6 feet $3\frac{1}{3}$ inches tall. Another man, 5 feet $8\frac{1}{4}$ inches tall, is standing on top of Mt. Kilimanjaro. Find the difference in the elevations of the tops of these men's heads. Draw a picture showing this problem situation. $19,775\frac{59}{144}$ feet

62. *Meter Reader* A meter is checked and is found to read over or under the actual level by these amounts: $+0.004$, -0.36, -0.004, $+0.5$, and -0.003. What is the range of these readings?   0.86

63. a. *Tip Size* A bill for a meal in a restaurant is $56. If you decide to leave a tip of 18%, what is the total cost of the meal? $66.08

b. ✎ In some locations, people "double the tax" to determine the size of a tip. Determine the tax on food in your location. Would you consider a "tax-doubled" tip sufficient? Explain.   Answers may vary.

64. a. *Realtor Commission* How much of a commission will a Realtor in a large city receive on a sale of a $246,000 house if the rate of commission is 12%? $29,520

b. ✎ Write a plan that a Realtor can use to find what price house she would need to sell to earn a given commission.   Answers may vary.

65. a. *Hockey* When David Williams was one of the San José Sharks, he scored his first game-winning NHL goal on February 6, 1992, against the Chicago Blackhawks. The Sharks eventually won 5 to 2. What fraction of the Sharks' goals did Williams score? What fraction of the total goals did he score? What fraction of the total goals did the Blackhawks score? Write these fractions in order from smallest to largest.

$\frac{1}{5}, \frac{1}{7}, \frac{2}{7}$
$\frac{1}{7}, \frac{1}{5}, \frac{2}{7}$

b. *Research* What does the phrase "game-winning goal" mean in hockey?   Answers may vary.

66. *Cross-Country Runners* According to *Runner's World* magazine (March 1992), the first successful cross-country run (coast to coast) was in 1890. It took John Ennis 80 days and 5 hours to complete the run. The fastest crossing, in 46 days, was accomplished 90 years later.

a. In what year was the fastest crossing?   1980

b. *Research* Find the distance of a "cross-country" run.   Answers may vary.

67. **Trip Reimbursement** Dr. Callaham drives from Dahlonega to Athens for a class (70 miles), then to Atlanta for a meeting (65 miles), and then back through Athens to Dahlonega. If he gets $0.23 reinbursement per mile driven for travel expenses when he drives to meetings and classes, how much should he be paid for this trip? $62.10

68. **a. Crossword Puzzles** A crossword puzzle appeared for the first time in the *New York Times* magazine section in 1942. If a puzzle appeared each Sunday after that, how many puzzles appeared in the next 51 years? (Disregard leap years and their effect on your solution.) 2652 puzzles

   b. ✏ Explain to a friend how to find the results if leap years are included. Answers may vary.

69. **Motorcycles** Motorcycles often have a two-cylinder engine. To determine the engine size, multiply the displacement of the cylinder by the number of cylinders. The displacement (D) of one cylinder is found by using the formula

$$\text{displacement} = \pi \times \left(\frac{\text{bore}}{2}\right)^2 \times \text{stroke}$$

   where

$$\pi \approx 3.14$$
$$\text{bore} = \text{diameter of the cylinder in centimeters}$$
$$\text{stroke} = \text{distance a piston moves in centimeters}$$
$$\text{displacement} = \text{volume of a cylinder in cubic centimeters}$$

   Find the displacement for a two-cylinder Harley Davidson® motorcycle with a bore of 88.8 millimeters and a stroke of 108.0 millimeters.
   1337 cubic centimeters

70. **a. Pottery Sale** Peggy and Michael have a pottery sale the first Sunday in December each year. They feed breakfast to everyone who comes to the sale. It costs an average of $1.39 for each person, and they spent $300 on breakfast this year. How many people attended? 216 people

   b. ✏ Plan a breakfast for a family or a group of friends. How much will it cost per person? Describe how you found your estimated cost per person. Answers may vary.

Use the following chart to answer Exercises 71 through 74.

Menu Planning

Serving size	Calories	Serving size	Calories
beef, rib, lean roasted (4 oz)	273	rice, boiled (4 oz)	400
chicken, mixed meat, roasted (4 oz)	206	potatoes, baked (4 oz)	115
carrots, cooked (4 oz)	20	apple pie (1 slice)	350
cabbage, cooked (4 oz)	16	chocolate chip cookie (1)	50

71. Plan a meal that consists of beef or chicken, a vegetable, potato or rice, and a dessert and that has fewer than 390 calories. Answers may vary.

72. Plan a meal that consists of beef or chicken, a vegetable, potato or rice, and a dessert and that has more than 1000 calories. Answers may vary.

73. Plan a meal that consists of beef or chicken, a vegetable, potato or rice, and a dessert and that has fewer than 1000 calories and more than 950 calories. Answers may vary.

74. ✏ Write your own question that can be answered using the Menu Planning table. Answer your question. Answers may vary.

75. *Travel Time* Len Pikaart used to drive his Corvette all over the state of Georgia to give math workshops for teachers. He usually averaged 695 miles a week during the school year, and each semester was 15 weeks long. How many semesters would it take him to travel 150,000 miles? 15 semesters

76. *Text Author* Professor Nakahara writes mathematics books for his students. He can write 7 problems every 10 minutes. How many hours will he have to work to write 6000 math problems? 143 hours

77. *Elevator Operator* Loraine runs an elevator at college. The elevator holds 23 people. How many trips must Loraine make to transport 123 people from floor three to floor twelve? 6 trips

78. *Runner* Ruth runs around the lower loop in the park. The distance is 1.7 miles. How many times must she run the loop in order to have run a total distance of at least 26 miles? 16 loops

79. *Newspaper Prices* Each week, Babatunde buys the *New York Times* 5 times, the *New York Post* 6 times, the *Wall Street Journal* twice, and *USA Today* once. In addition, he buys the Sunday *New York Times*. The cost of each paper is shown in the following table. What does Babatunde spend in a year for papers? $533

	Daily	Sunday
New York Times	$0.60	$2.00
New York Post	$0.50	$1.00
Wall Street Journal	$0.75	—
USA Today	$0.75	—

80. *Long-Distance Charges* One long-distance phone company advertises that, on the weekend, it costs 11 cents a minute to call anywhere in the United States. Stacie and A.J. speak long-distance from Twentynine Palms to Sacramento for 1 hour and 28 minutes one weekend, using this phone plan. How much does this call cost? $9.68

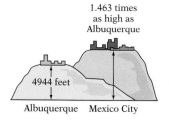
1.463 times as high as Albuquerque

4944 feet

Albuquerque   Mexico City

81. *Sea Level* Albuquerque is 4944 feet above sea level. Mexico City is about 1.463 times as high. How much higher is Mexico City? Give your answer to the nearest 10 feet. 2290 feet higher

82. *Creek Widths* Black Creek in Mississippi ranges in width from 6.67 yards to 33.33 yards. What is the difference between these two widths? A duck in the center of this creek is how far from the shore at the widest part? 26.66 yards; 16.67 yards

83. *Kitty Treats* Each year, on my cat Rail's birthday, I feed her 10 treats for each year old she is. When she is 10 years old, how many birthday treats will she have eaten altogether? 550

84. *Landscaping* A woman wants to arrange 36 stones into the shape of a square with the same number on each side to surround a flower bed. How many stones will there be on each side? 10

MIXED PRACTICE

By doing these exercises, you will practice the topics up to this point in the chapter.

85. Subtract: $-16\frac{1}{3} - \left(-8\frac{7}{8}\right)$ $-7\frac{11}{24}$

86. Supply the missing numbers to make the statement true. Name the property illustrated.
■ + (6 + ■) = (4 + ■) + 2 4; 2; 6
associative property of addition

87. Simplify: $\dfrac{\frac{5}{9}}{\frac{16}{45}}$ $1\frac{9}{16}$

88. Simplify: $[(2^2 - 2^{-3} + 2^3) \div 2^{-3}]^2$ 9025

89. $A = \{0, 2, 4\}$ and $B = \{1, 3, 5\}$. Find $A \cap B$. $\varnothing$

90. What is 16% of 0.03125? 0.005

91. What property is illustrated by the statement $\left(\frac{1}{4}\right)(4 + 8) = 1 + 2$? distributive property

92. Is $\dfrac{(3890)(1651)(694)}{(39500)(18116)}$ closest to 7, to 70, or to 700? closest to 7

93. Round 63.98545 to the nearest tenth, hundredth, and thousandth. 64.0; 63.99; 63.985

94. Multiply: $\left(\frac{-1}{3}\right)\left(-2\frac{1}{7}\right)(5)\left(-\frac{1}{5}\right)$ $-\frac{5}{7}$

95. Simplify: $[1 - 16 \div (2 + 4) \div 2]^2$ $\frac{1}{9}$

96. *Number of Apes* King Kong was 50 feet tall. The Empire State Building is 1472 feet tall. How many apes of this size, standing feet to head, would it take to reach the height of the Empire State Building? 30 apes

97. *Earth's Land Area* Meadows and pastures cover approximately 11,686,600 square miles. This is approximately 20% of the estimated land area of Earth. Find the estimated land area of Earth. 58,433,000 square miles

98. *Depth of a Cave* The world's deepest cave is Pierre St-Martin, located in the Pyrenees. It is 4370 feet deep. That is 1906 feet deeper than Ghar Parau in Iran. How deep is the Ghar Parau cave? 2464 feet

EXCURSIONS

Class Act

1. **a.** Estimate the number of tickets available for each section of this football stadium. Answers may vary.

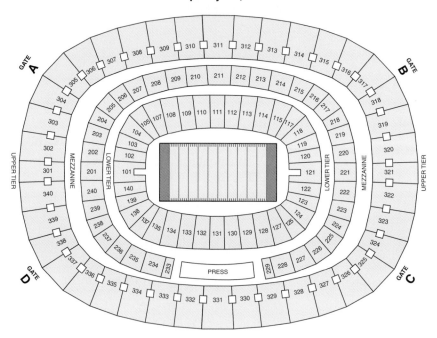

Meadowlands Sports Complex
Capacity: 77,716

Reprinted with permission by Giants Stadium.

    b. Write a paragraph describing your estimation procedure. Answers may vary.

CHAPTER LOOK-BACK

- "Three out of four doctors recommend. . . ."
- "Ninety percent of all cars we manufactured since 1980 are on the road today."
- "The average family size this year is 2.8."

None of these numbers is meaningful as such. However, with a bit more information or additional insight, we can understand and use this information. This chapter, especially the emphasis on problem solving in Section 1.4, has prepared you well for applying the mathematics you will learn in this text.

CHAPTER **1**
REVIEW PROBLEMS

The following exercises will give you a good review of the material presented in this chapter.

SECTION 1.1

1. True or false: The rational numbers include the integers.   true

2. What is the opposite of $-(-6)$? -6

3. Arrange -8.932, -8.7, and -8.75 in order from least to greatest. $-8.932 < -8.75 < -8.7$

4. Arrange $\left|-\left|\frac{9}{5}\right|\right|$, $-\left|\frac{11}{3}\right|$, and 2.4 in order rom greatest to least. $2.4 > \left|-\left|\frac{9}{5}\right|\right| > -\left|\frac{11}{3}\right|$

5. Round 86.2935 to the nearest tenth, hundredth, and thousandth.   86.3; 86.29; 86.294

6. Estimate the result of this computation. (Do not calculate.)
$(79375)(1875) \div 7787635$ 20

7. Estimate the result of this computation. (Do not calculate.)
$\dfrac{(35{,}987{,}335)(68{,}765)}{(68{,}999{,}585)}$ 40,00

8. Estimate (do not calculate): $(18987)(7358) \div 18688$ 7000

SECTION 1.2

9. Subtract: $\left(-13\frac{2}{5}\right) - 6\frac{3}{4}$ $-20\frac{3}{20}$

10. Add: $-17\frac{1}{2} + 8\frac{3}{4} + 11\frac{7}{12}$ $2\frac{5}{6}$

11. Add: $\left(13\frac{2}{7}\right) + \left(-14\frac{3}{8}\right)$ $-1\frac{5}{56}$

12. Subtract: $0.009 - 2.398$ -2.389

13. Add: $(-3.7) + (-6.5) + 3.9$ -6.3

14. Subtract: $35.4 - (-18.6)$ 54

15. Divide: $-3069 \div (-15)$ 204.6

16. Multiply: $\left(-8\frac{2}{5}\right)\left(-2\frac{1}{7}\right)$ 18

17. Simplify: $\dfrac{8\frac{2}{3}}{-2\frac{3}{5}}$ $-3\frac{1}{3}$

18. Multiply: $(-2.5)(-0.08)(-1.2)$ -0.24

19. Divide: $-8\frac{1}{2} \div 17$ $-\frac{1}{2}$

20. Simplify: $\dfrac{\frac{-18}{21}}{\frac{9}{7}}$ $-\frac{2}{3}$

SECTION 1.3

In Exercises 21 through 24, fill in the blanks with numbers to make true statements. What property is illustrated in each exercise?

21. $\blacksquare(3x + 2) = (\blacksquare + 2)5$ $5; 3x$
commutative property of multiplication

22. $\blacksquare(\blacksquare + 16) = 7[\blacksquare + (-63)]$ $7; -63; 16$
commutative property of addition

23. $\blacksquare + [17 + (-3)] = (27 + \blacksquare) + (-3)$ $27; 17$
associative property of addition

24. $(2x + \blacksquare)9 = \blacksquare \cdot 9 + 15 \cdot \blacksquare$ $15; 2x; 9$
distributive property

25. What is the square of 12? 144

26. Write $\dfrac{1}{3^{-2}}$ as an integer. 9

Simplify each of the following expressions.

27. $2^2 + 4(2 - 5)^2$ 40

28. $\left(-3\frac{3}{5}\right) + \left(2\frac{2}{5}\right)\left(-1\frac{1}{4}\right) - \left(1\frac{1}{2}\right)^2$ $-8\frac{17}{20}$

29. $(-12.5) + (4.3)^2 - 3(10)^{-2} - 4$ 1.96

30. $2 \div 3 + 7 \div 8 - 11 \div 12$ $\frac{5}{8}$

31. $4^2 - 2 \cdot 3 \cdot 5 + 3^2$ -5

32. $[(2^2 + 1)^2 + 1]^2$ 676

SECTION 1.4

Use the following information to answer Exercises 33 through 35.

$$A = \{5, 6, 7\}, B = \{6, 13\}, \text{ and } C = \{0, 5, 6\}$$

33. Find $(A \cup B) \cap C$ $\{5, 6\}$ **34.** Find $A \cap C$. $\{5, 6\}$ **35.** Find $B \cup C$. $\{0, 5, 6, 13\}$

36. What is the new price of an item that was discounted 20% if the price was originally $2500? $2000

37. Describe the set of even integers between and including -6 and 4, using both roster and set-builder notation. $\{x$ is an even integer and $-6 \le x \le 4\}; \{-6, -4, -2, 0, 2, 4\}$

38. Given $A = \{$integers less than 3 and greater than $-3\}$, $B = \{$whole numbers less than 5$\}$, and $C = \{$real numbers$\}$

   a. Find $B \cap C$. $\{0, 1, 2, 3, 4\}$ **b.** Find $A \cap (B \cap C)$. $\{0, 1, 2\}$

39. The cars on a roller coaster can each carry a dozen people. Two cars are used at a time, with one running while the other one is loading and unloading. The roller coaster operates 10 hours per day. The ride takes 4 minutes. Loading and unloading takes 4 minutes also. How many people can ride during the day?   1800 people

40. Dresses on 24 mannequins have buttons. If 40% of the dresses on mannequins do not have buttons, how many mannequins are there? *40 mannequins*

MIXED REVIEW

41. Find the cube of 3. *27.*

42. Determine which is larger, $-\left|\frac{23}{4}\right|$ or $-\left|-\frac{23}{5}\right|$. $-\left|-\frac{23}{5}\right|$ *is larger*

43. Round 956.9812 to the nearest tenth, hundredth, and thousandth. *957.0; 956.98; 956.981*

44. Add: $-7.4 + 83.9$ *76.5*

45. Multiply: $\left(10\frac{2}{7}\right)\left(-3\frac{8}{9}\right)$ *−40*

46. Simplify: $\{[(2^2 - 1)^2 - 1]^2 - 1\}^2$ *3969*

CHAPTER 1 TEST

This exam tests your knowledge of the material in Chapter 1.

1. **a.** Arrange $|-4.6|$, $-|-8.4|$, and 6.2 in order from largest to smallest. $6.2 > |-4.6| > -|-8.4|$

   b. Round 67.95003 to the nearest hundredth and thousandth. *67.95; 67.950*

   c. Write $\left(\frac{1}{5}\right)^{-3}$ as an integer. *125*

2. **a.** Add: $-85 + (-39)$ *−124*

   b. Subtract: $\left(-3\frac{3}{7}\right) - \left(-6\frac{1}{2}\right)$ $3\frac{1}{14}$

   c. Add: $-3.7 + 9.4 + (-8.12)$ *−2.42*

3. **a.** Multiply: $(-6.12)(0.03)$ *−0.1836*

   b. Divide: $(-384) \div (-3)$ *128*

   c. Divide: $\dfrac{\frac{-45}{7}}{\frac{36}{5}}$ $-\frac{25}{28}$

4. **a.** Fill in the boxes with numbers that make the statement true. What property is illustrated?

   $\blacksquare(4x + (-6)) = \blacksquare \cdot \blacksquare + 5(-6)$ *5; 5; 4x; distributive property*

   b. Simplify: $[18 - 2(1 - 4)^3] \div 3 + 7^2$ *73*

   c. Simplify: $-2^2 + 4(2 - 5)^2 - \left(3 \cdot \frac{1}{5}\right)^0$ *31*

5. **a.** The Des Allemands Catfish Festival takes place each July in Louisiana. During the festival weekend, some 300 people are needed to run the festival, cook, and sell 15,900 pounds of catfish. If each person sells the same amount, how many pounds of catfish are sold per person?   53 pounds

   b. In the 1988 New York Marathon, 6866 runners were between the ages of 40 and 49. If this represented approximately 29% of the runners, how many runners were there in all?   23,676 runners

   c. The world's deepest mine is Western Deep in South Africa. It reaches 12,600 feet deep. The world's deepest drilling site, the Kola peninsula in the former Soviet Union, is 31,911 feet deep. How much deeper is the Kola site?   19,311 feet

INTRODUCTION TO ALGEBRA

*H*ave you ever heard a weather forecast that called for variable winds? You knew that the speed of the wind was expected to vary, or change, during the day. Likewise, if you have ever performed or designed a scientific experiment, you know that the variable is that part of the experiment that changed. The term *variable* is also used in algebra, where it has a specific, yet similar, meaning. As with meteorology and other sciences, algebra requires a specific vocabulary.

■ *List five terms whose meaning depends on the context or discipline in which they are used.*

SKILLS CHECK

Take this short quiz to see how well prepared you are for Chapter 2. The answers follow the quiz.

1. Simplify: $7 + 5 - 9 - 6 + 3$

2. Simplify $-8(10 - 6)$ using the distributive property.

3. Add: $-8 + \left(-10\frac{1}{2}\right)$

4. What is the multiplicative inverse of -1?

5. Subtract: $-62\frac{1}{4} - \left(-59\frac{4}{9}\right)$

6. Simplify: $9 + 5 \div 3 - 8$

7. Find the additive and multiplicative inverses of -5.

8. Use the associative property to supply the missing numbers in this true statement:
$2 + (\blacksquare + 4) = (\blacksquare + 3) + 4$

9. Multiply: $\left(-5\right)\left(-6\frac{3}{4}\right)\left(-8\frac{1}{2}\right)$

10. Divide: $(-2.6) \div (0.8)$

ANSWERS: **1.** 0 [Section 1.3]   **2.** -32 [Section 1.3]   **3.** $-18\frac{1}{2}$ [Section 1.2]   **4.** -1 [Section 1.3]
5. $-2\frac{29}{36}$ [Section 1.2]   **6.** $2\frac{2}{3}$ [Section 1.3]   **7.** additive inverse: $+5$; multiplicative inverse: $-\frac{1}{5}$
[Section 1.3]   **8.** $2 + (\underline{3} + 4) = (\underline{2} + 3) + 4$ [Section 1.3]   **9.** $-286\frac{7}{8}$ [Section 1.2]   **10.** -3.25
[Section 1.2]

CHAPTER LEAD-IN

Many problems have to be translated into language that we understand before we can solve them.

- About how many miles have you walked in the last year?
- About how many feet high is your school building?
- About how long does your hair grow in a minute?
- About how many times does your heart beat in 10 years?

Often the language we use includes a linear equation.

2.1 Introduction to Algebraic Terms and Expressions

SECTION LEAD-IN

Before you work a word problem, you must read and interpret the given information and determine what question is being asked.

"There are 3 times as many students as there are professors. . . ." So starts a famous problem that many students find difficult to express using algebraic notation.

1. Please fill in the following blanks:

   a. There are 90 students, so there are ___30___ professors.

   b. There are 18 professors, so there are ___54___ students.

2. Complete these sentences:

   a. The number of students is _three times_ the number of professors.

   b. The number of professors is _one-third_ the number of students.

The language of a problem situation is extremely important.

Algebraic Terminology

In arithmetic, we use the digits 0 through 9 to write the symbols for numbers. Each of these symbols (such as 64 and 159.5) always represents the same number, so they are called *constants.*

> **Constant**
> A **constant** is a symbol that stands for a certain number. It has a fixed value.

In algebra, we also use symbols to represent unknown numbers. We call them *variables.*

> **Variable**
> A **variable** is a symbol that stands for a number that is yet to be determined. Its value can vary.

Almost any written symbol can be used to represent a variable, but we generally use letters. Because a variable stands for a number, we can multiply it by itself, obtaining such products as x^2 or m^5 or w^n. We sometimes also use letters (usually from the beginning of the alphabet) to represent constants. Then we must make clear whether these letters stand for variables or constants.

> **Algebraic Expression**
>
> An **algebraic expression** consists of numbers and variables linked together by any operation.

Thus $y + 5$, $x^2 - 8x$, and $4t^2y - 8t + 7$ are all algebraic expressions. Expressions may contain one or more terms.

> **Term**
>
> A **term** consists of variables and/or constants connected by multiplication or division.

Thus $\frac{1}{5}n$, 3, x, $3.6x^3y^2$, and $4n$ are terms. We put terms together to make expressions.

The sign that precedes each term in an expression is considered part of that term. We usually write each term with the variables in alphabetical order, so we would write xy instead of yx.

A term has two parts—a *variable part* and a *coefficient*.

> **Variable Part**
>
> The **variable part** of a term consists of all the variables, together with their exponents.

> **Coefficient**
>
> The **coefficient** of a term is the number, along with the sign, that multiplies the variable part. A term with no number has the coefficient 1 or −1, depending on its sign.

If a term consists only of a constant, it is sometimes called a **constant term.**

In the algebraic expression $4t^2 + 3rt - 7r^2 + 2$, the terms are $4t^2$, $3rt$, $-7r^2$, and 2. The coefficients are, in order, 4, 3, −7, and 2. The variable parts of the first three terms are, in order, t^2, rt, and r^2. The fourth term has no variable part.

Evaluating Algebraic Expressions

STUDY HINT

When substituting the values for variables, make sure you always use parentheses. This is especially important when substituting negative values, as it will keep you from losing a minus sign.

To "evaluate an algebraic expression" means to find its numerical value. We can do this only when we know the values represented by its variables.

> **To evaluate a given algebraic expression**
>
> 1. Substitute the numerical value for each variable.
> 2. Use the standard order of operations to simplify the expression.

...

EXAMPLE 1

Evaluate $-3(7r^3 - 4t^2 \cdot 9y) - 5rt$ when $r = 2$, $t = 0$, and $y = 3$.

SOLUTION

First we substitute the values for the variables.

$$-3(7r^3 - 4t^2 \cdot 9y) - 5rt$$
$$= -3[7(2)^3 - 4(0)^2 \cdot 9(3)] - 5(2)(0)$$

Then we use the order of operations to simplify, beginning inside the brackets.

$-3[7(2)^3 - 4(0)^2 \cdot 9(3)] - 5(2)(0)$

$= -3[7(8) - 4(0) \cdot 9(3)] - 5(2)(0)$ Removing exponents

$= -3[56 - 0 \cdot 27] - 10(0)$ Multiplying

$= -3[56 - 0] - 0$ Multiplying

$= -3[56] - 0$ Subtracting

$= -168 - 0$ Multiplying

$= -168$ Simplifying

▶ CHECK **Warm-Up 1**

Calculator Corner

The graphing calculator can evaluate any expression containing one or more variables. The variables *do not* always have to be the commonly used x and y. As long as you have **STO**red a numerical value for each variable that is used in the expression, the calculator will find the result. This is an excellent way to check your own work.

Evaluate $x - a^2$ when $x = 20$ and $a = -4$ using the Home Screen and the **STO**re features of your graphing calculator. On Texas Instruments Graphing calculators, press the $\boxed{\text{STO}\blacktriangleright}$ key to produce the arrow you see on screen. The calculator "**STO**res" the value 20 for x and -4 for a and then uses them to evaluate $x - a^2$, as shown in the accompanying figure.

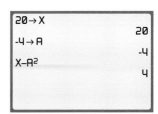

...

EXAMPLE 2

Evaluate

$$\frac{-3n(0.5y^3 - t)^2}{-3nt} + 6n^{-2}$$

when $n = \frac{2}{3}$, $y = -4$, and $t = -2$.

SOLUTION

We begin by substituting the known values and then simplifying.

$$\frac{-3\left(\frac{2}{3}\right)[0.5(-4)^3 - (-2)]^2}{-3\left(\frac{2}{3}\right)(-2)} + 6\left(\frac{2}{3}\right)^{-2}$$

$$= \frac{-3\left(\frac{2}{3}\right)[0.5(-64) - (-2)]^2}{-3\left(\frac{2}{3}\right)(-2)} + 6\left(\frac{2}{3}\right)^{-2} \qquad \begin{array}{l}\text{Removing exponent}\\\text{within parentheses}\end{array}$$

$$= \frac{-3\left(\frac{2}{3}\right)[(-32) - (-2)]^2}{-3\left(\frac{2}{3}\right)(-2)} + 6\left(\frac{2}{3}\right)^{-2} \qquad \begin{array}{l}\text{Multiplying within}\\\text{parentheses}\end{array}$$

$$= \frac{-3\left(\frac{2}{3}\right)(-30)^2}{-3\left(\frac{2}{3}\right)(-2)} + 6\left(\frac{2}{3}\right)^{-2} \qquad \begin{array}{l}\text{Simplifying inside}\\\text{parentheses}\end{array}$$

$$= \frac{-3\left(\frac{2}{3}\right)(900)}{-3\left(\frac{2}{3}\right)(-2)} + 6\left(\frac{9}{4}\right) \qquad \begin{array}{l}\text{Removing the remaining}\\\text{exponents}\end{array}$$

$$= \frac{-3\left(\frac{2}{3}\right)(900)}{-3\left(\frac{2}{3}\right)(-2)} + 6\left(\frac{9}{4}\right) \qquad \text{Cancelling}$$

$$= \frac{900}{(-2)} + \frac{54}{4} \qquad \text{Multiplying}$$

$$= -450 + 13\frac{1}{2} \qquad \text{Simplifying}$$

$$= -436\frac{1}{2} \qquad \text{Adding}$$

▶ *CHECK* **Warm-Up 2**

! ! !
ERROR ALERT

Identify the error and give a correct answer.

Evaluate $-a^2 + b$ when $a = -3$ and $b = 4$.

Incorrect Solution:

$-a^2 + b$
$(-3^2) + 4$
$9 + 4$
13

■ ■ ■

EXAMPLE 3

Evaluate $-4t^{-3}\left[2x^2y^3 - 2\left(\frac{1}{3}xyt^{-2}\right)\right]$ when $x = 6$, $y = 1$, and $t = 2$.

SOLUTION

First we substitute the values for each variable, being careful to insert parentheses where they are needed.

$$-4(2)^{-3}\left\{2(6)^2(1)^3 - 2\left[\left(\frac{1}{3}\right)(6)(1)(2)^{-2}\right]\right\}$$

Next we remove exponents.

$$-4\left(\frac{1}{8}\right)\left\{2(36)(1) - 2\left[\left(\frac{1}{3}\right)(6)(1)\left(\frac{1}{4}\right)\right]\right\}$$

Then we simplify within the braces.

$$-4\left(\tfrac{1}{8}\right)\left\{72 - 2\left[\left(\tfrac{1}{3}\right)(6)\left(\tfrac{1}{4}\right)\right]\right\}$$
$$= -4\left(\tfrac{1}{8}\right)\left[72 - 2\left(\tfrac{1}{2}\right)\right]$$
$$= -4\left(\tfrac{1}{8}\right)[72 - 1]$$
$$= -4\left(\tfrac{1}{8}\right)[71]$$

Finally, we perform the multiplications.

$$\left(\tfrac{-1}{2}\right)(71) = -35\tfrac{1}{2}$$

▶ *CHECK* **Warm-Up 3**

Calculator Corner

You can use your graphing calculator to decide if both sides of an equation are equal. You do this by "asking" the calculator if a statement is true or false. Study the calculator steps for $(a + b)^2 = (a + b)(a + b)$. We can check this equation using different values for a and b. Let $a = 3$, and let $b = 5$. Then

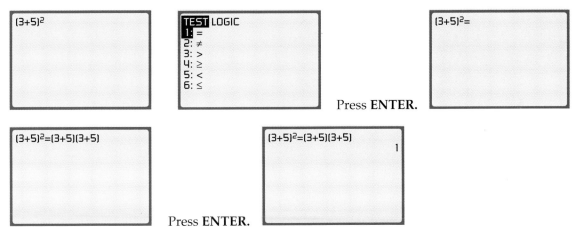

Press **ENTER.**

Press **ENTER.**

The graphing calculator "returns" an answer of 1 when the statement is true. However, if the statement is false, the calculator returns an answer of 0, as in the following example.

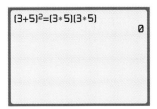

Use this procedure to check some of your real-number properties like $a(b + c) = ab + ac$.

Sometimes, we are asked to evaluate expressions and are *not* given replacements for all the variables. We then say that we have evaluated "in terms of" a particular variable.

▪ ▪ ▪

EXAMPLE 4

Evaluate $3x^2 - 8y^2 + 3 - 2xyt$ in terms of t when $x = 3$ and $y = 4$.

SOLUTION

Substituting the given values for x and y yields

$$3x^2 - 8y^2 + 3 - 2xyt = 3(3)^2 - 8(4)^2 + 3 - 2(3)(4)t$$

Using the order of operations, we then have

$$
\begin{aligned}
3(3)^2 &- 8(4)^2 + 3 - 2(3)(4)t \\
&= 3(9) - 8(16) + 3 - 2(3)(4)t \quad &&\text{Removing exponents} \\
&= 27 - 128 + 3 - 24t \quad &&\text{Multiplying} \\
&= -98 - 24t \quad &&\text{Simplifying}
\end{aligned}
$$

So $3x^2 - 8y^2 + 3 - 2xyt$, evaluated as required, is $-98 - 24t$.

▶ CHECK **Warm-Up 4**

Combining Like Terms

In algebra, we are often concerned with terms that are alike in an important way. We call them *like terms.*

> **Like Terms**
>
> **Like terms** are terms that have the same variable part.

$9y$ and $2y$ are like terms.
$6xy$ and $-8yx$ are like terms.
$3r$ and $4r^2$ are *not* like terms—the variable is the same, but the exponents differ.
$5t^2$ and $5x^2$ are *not* like terms—the variables differ.

One way to simplify certain algebraic expressions is to combine like terms. For that, we can use the properties of real numbers. As an example, suppose you were asked to combine $4x + 3x$. You could apply the distributive property, obtaining

$$4x + 3x = (4 + 3)x = 7x$$

Here is another example:

$$
\begin{aligned}
-9rt &+ 16 - 8rt \\
&= (-9rt - 8rt) + 16 \quad &&\text{Commutative and associative properties} \\
&= (-9 - 8)rt + 16 \quad &&\text{Distributive property} \\
&= -17rt + 16
\end{aligned}
$$

Here we first used the associative and commutative properties to group like terms. Then we used the distributive property to combine their coefficients.

To combine like terms

Group like terms, and then combine their coefficients.

INSTRUCTOR NOTE

Remind students that like terms must have identical variables with identical exponents.

▪ ▪ ▪

EXAMPLE 5

Combine like terms:

$$-4x^2 + 4x^2 + 6x + 5x - 2x^2 + 8$$

SOLUTION

First we must identify the like terms. There are three terms containing x^2 and two terms containing x. To simplify, we use the commutative and associative properties to rearrange the terms. We then combine the coefficients of like terms.

$$-4x^2 + 4x^2 + 6x + 5x - 2x^2 + 8$$
$$= (-4x^2 + 4x^2 - 2x^2) + (6x + 5x) + 8$$
$$= (-4 + 4 - 2)x^2 + (6 + 5)x + 8$$
$$= -2x^2 + 11x + 8$$

Thus $-4x^2 + 4x^2 + 6x + 5x - 2x^2 + 8 = -2x^2 + 11x + 8$

▶ *CHECK* **Warm-Up 5**

! ! !
ERROR ALERT

Identify the error and give a correct answer.

$4x + 7x = 11x^2$

In this next example, each term has more than one variable.

▪ ▪ ▪

EXAMPLE 6

Combine like terms:

$$-3r^2t^2 - 9rt^2 - 7r^2t - 4rt^2 + 6r^2t^2$$

SOLUTION

Inspection of this expression shows that there are only two sets of like terms.

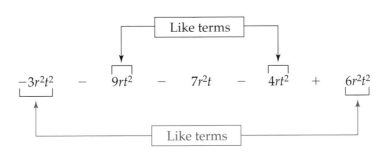

! ! !
ERROR ALERT

Identify the error and give a correct answer.

Simplify:
$3x - 4x + 2x^2 - 5y$

Incorrect Solution:

$3x - 4x + 2x^2 - 5y = x^2 - 5y$

Rearranging and combining coefficients, we get

$$-3r^2t^2 - 9rt^2 - 7r^2t - 4rt^2 + 6r^2t^2 \qquad \text{Original expression}$$

$$= (6r^2t^2 - 3r^2t^2) + (-9rt^2 - 4rt^2) - 7r^2t \qquad \text{Commutative and associative properties}$$

$$= (6 - 3)r^2t^2 + (-9 - 4)rt^2 - 7r^2t \qquad \text{Distributive property}$$

$$= 3r^2t^2 - 13rt^2 - 7r^2t \qquad \text{Simplifying}$$

So $-3r^2t^2 - 9rt^2 - 7r^2t - 4rt^2 + 6r^2t^2$ simplifies to $3r^2t^2 - 13rt^2 - 7r^2t$.

▶ *CHECK* **Warm-Up 6**

In the last example, we have parentheses within brackets. We work within the innermost grouping symbols first.

▪ ▪ ▪

EXAMPLE 7

Simplify: $-6(3y^2 + 5) + 2(y^2 - 3y) - 6[4(y - 2)]$

SOLUTION

Clearing the parentheses that occur within the brackets, we have

$$-6(3y^2 + 5) + 2(y^2 - 3y) - 6[4(y - 2)]$$
$$= -6(3y^2 + 5) + 2(y^2 - 3y) - 6(4y - 8)$$

Then, using the distributive property, we get

$$= -18y^2 - 30 + 2y^2 - 6y - 24y + 48$$

Rearranging by using the commutative and associative properties gives us

$$-18y^2 - 30 + 2y^2 - 6y - 24y + 48$$
$$= (-18 + 2)y^2 + (-6 - 24)y + (-30 + 48)$$
$$= -16y^2 - 30y + 18$$

Thus $-6(3y^2 + 5) + 2(y^2 - 3y) - 6[4(y - 2)] = -16y^2 - 30y + 18$

▶ *CHECK* **Warm-Up 7**

Practice what you learned.

SECTION FOLLOW-UP

Let x be the number of professors and y be the number of students. Then ordered pairs (x, y) can be written that satisfy the equation

Number of students = 3 times number of professors

$(30, 90)$ is one ordered pair that shows this relationship.

1. Write three ordered pairs that fit this relationship.

2. Fill in the blanks to make ordered pairs that fit this relationship.

$$\{(35, \underline{105}), (100, \underline{300}), (\underline{144}, 432), (\underline{367}, 1101)\}$$

3. Finish this statement using the variables as given in part 2:

$$x = \underline{\frac{1}{3}}\ y$$

4. Finish this statement using the variables as given in part 2:

$$y = \underline{3}\ x$$

2.1 WARM-UPS

Work these problems before you attempt the exercises.

1. Evaluate $-m(r^3 - r \div t) + 3rt$ when $m = 3, r = 4$, and $t = -2$.
   -222

2. Evaluate
   $$\frac{2r\left(\frac{1}{2}t^2 - v\right)^2}{tv^3} - 7v^{-3}$$
   when $r = 6, t = 3$, and $v = -1$. -114

3. Evaluate $8x^3y^2 - [-3(x^3 - y^2) + y^{-2} \cdot x^2]$ when $x = -1$ and $y = -3$. $-102\frac{1}{9}$

4. Evaluate $9y^2 - 2y + 7xy - 6xt$ in terms of t when $x = -2$ and $y = 3$. $33 + 12t$

5. Combine like terms: $5r + 8r^2 - 6r^2 - 8r$ $-3r + 2r^2$

6. Combine like terms: $9y^2x - 6y^2x + 5yx^2 - 5x^2y + 3x^2y$ $3xy^2 + 3x^2y$

7. Simplify: $-5(2x^2 + 4) + 3(3x^2 - 5) - 3[-2(x^2 + 5)]$ $5x^2 - 5$

2.1 EXERCISES

Note: Use your graphing calculator to check your results whenever possible.

1. List the terms and coefficients in $-11r^2t + 5xt^3 - 8rt^3 + 7rx$.   terms: $-11r^2t, 5xt^3, -8rt^3, 7rx$; coefficients: $-11, 5, -8, 7$

2. List the terms and variable parts in $\frac{1}{8}xy^2 - 16x^2y + 8x - 6y$.   terms: $\frac{1}{8}xy^2, -16x^2y, 8x, -6y$; variable parts: xy^2, x^2y, x, y

3. List the terms and coefficients in $xtz - 3xz^2 + tz$.   terms: $xtz, -3xz^2, tz$; coefficients: $1, -3, 1$

4. List the terms, coefficients, and variable parts in $-3rt + rt - 4 + r^2t^3$.   terms: $-3rt, rt, -4, r^2t^3$; coefficients: $-3, 1, -4, 1$; variable parts: rt, rt, none, r^2t^3

In Exercises 5 through 16, evaluate the algebraic expressions when

$$n = 3 \qquad x = -1 \qquad r = 0 \qquad t = 4 \qquad y = -2$$

5. nx^3 -3

6. n^2r 0

7. $t^3x^2y^3$ -512

8. $-4n^3r$ 0

9. $6t^2x^2y^3$ -768

10. $-2n^2x$ 18

11. $-8t^{-2}x^4y^{-3}$ $\frac{1}{16}$

12. $-4n^{-2}rt^{-3}$ 0

13. $6t^3y - 3ty \div n$ -760

14. $-n + nt \div x$ -15

15. $8r + 6t - 3n \div y$ $28\frac{1}{2}$

16. $-12ny - tn^3$ -36

In Exercises 17 through 28, evaluate each algebraic expression when

$$r = 2 \qquad y = -1 \qquad n = -3 \qquad x = 4$$

17. $-2n(n^3 + rx - y)$ -108

18. $-5(x \div x + nx)$ 55

19. $-(xy^2 + x - y^2)$ -7

20. $(x^2y - n^3)8$ 88

21. $-2(nx + r) - 5rx$ -20

22. $3(nx - y) + 2ny$ -27

23. $-6xy + 5(nx + x)$ -16

24. $-5n^2y - (ny - x)$ 46

25. $\dfrac{x}{(-r + xy)}$ $-\frac{2}{3}$

26. $\dfrac{5nx}{9(ry - x)}$ $1\frac{1}{9}$

27. $\dfrac{6(-5n + x)n}{3r}$ -57

28. $\dfrac{2x(r - 7x)}{-xy}$ -52

In Exercises 29 through 40, evaluate each algebraic expression when

$$n = -3 \qquad r = \frac{1}{2} \qquad t = 1 \qquad x = -4 \qquad y = -5$$

Be sure that you follow the standard order of operations.

29. $5xy - 4nx^3 + (ny - 3t)$ -656

30. $2nt - 2t^2y - 3t(ny + 7x)$ 43

31. $7nx^3 - 3x^3 - 6r - 4(n^3r^2)$ 1560

32. $6rx^3 - 5(n^2 + 2n^2x^3) + 4x$ 5507

33. $2n^3(y - x)^{-2}$ -54

34. $2ty^{-3}(4x - 6y)$ $-\frac{28}{125}$

35. $6nx^{-3}(rx - n)$ $\frac{9}{32}$

36. $4x(2x^{-3} - x) + n$ $-66\frac{1}{2}$

37. $n^{-3}(r + x^2) + 2x$ $-8\frac{11}{18}$

38. $-4n^{-2}(8r - n^3) - 3x$ $-1\frac{7}{9}$

39. $\dfrac{6n \div 9ny}{n - y} + 3n^{-2}$ $-14\frac{2}{3}$

40. $\dfrac{-4tx \div 8y}{x} + (2t)^{-2}$ $2\frac{3}{4}$

41. Evaluate $4xy^2 + 6xt^2 - 4n + 9xt$ in terms of n when $x = -3, y = -6$, and $t = \frac{1}{2}$. $-450 - 4n$

42. Evaluate $-3yt^3 + 6ny - 18y - 3x$ in terms of x when $n = \frac{1}{4}, t = \frac{1}{2}$, and $y = -6$. $101\frac{1}{4} - 3x$

43. Evaluate $-\frac{1}{5}x^2y + 3ry - 7r^3x - 2t$ in terms of t when $r = -2, x = 5$, and $y = -\frac{1}{3}$. $283\frac{2}{3} - 2t$

44. Evaluate $4m^3n^2 - mr^2 + 7 + 3m^2t$ in terms of t when $m = \frac{1}{4}, n = \frac{1}{2}$, and $r = \frac{1}{8}$. $7\frac{3}{256} + \frac{3}{16}t$

In Exercises 45 through 48, identify the different variable parts in each algebraic expression, and then combine like terms.

45. $3ny^2 - 7ny^3 - 5ny^2 - 12ny + 8n^2y - 7ny^2 + 6ny^3 - 3$ variable parts: $ny^2, ny^3, ny, n^2y; -9ny^2 - ny^3 - 12ny + 8n^2y - 3$

46. $9yt - 8ty + 3y^2t - 6t^2y + 7t^2y + 11y^2ts + 8$ variable parts: $ty, ty^2, t^2y, sty^2; ty + 3ty^2 + t^2y + 11sty^2 + 8$

47. $yr + 2rt^2 + 3rt + 12rt^2 - 9 - 3rt + 4tr + 6$ variable parts: $ry, rt^2, rt; ry + 14rt^2 + 4rt - 3$

48. $3yt^2 - 4t^2y - 8t^2 + 4yt - 9yt - 8t^2y - 12$ variable parts: $t^2y, t^2, ty; -9t^2y - 8t^2 - 5ty - 12$

In Exercises 49 through 72, combine like terms.

49. $n^3 + 4n^2 - n + n^2 + n^3 + n^2 - 3n^2 - 18 + 24$
$2n^3 + 3n^2 - n + 6$

50. $6r - 7r^2 + 9r^2 + 4r - 9r - 6r + 3r - 2r^2 + 8$
$-2r + 8$

51. $11t^2 - 6t - 3t^2 + 2t + 12t + 12t - 23t + 4t^2$
$12t^2 - 3t$

52. $5xy + 6yx - 1 + 8xy - 4 + 5xy - 9xy +$
$12yx$ $27xy - 5$

53. $3(2t - 5) + 4(3t + 2)$ $18t - 7$

54. $4(3mt + 2) - 3(2mt - 5)$
$6mt + 23$

55. $-6(kt - 8) - 6kt + 8$
$-12kt + 56$

56. $-10rm - 6 - 6(rm - 8)$
$-16rm + 42$

57. $-2(-8 + k^3n) - 9(-k^3n - 5)$
$7k^3n + 61$

58. $8(8t^2m - 8) - 2(t^2m - 6)$
$62t^2m - 52$

59. $-2(-3 + k) - 9(-k - 5)$
$7k + 51$

60. $2(t - 3) + 3(n - 3) - 4t + 3n$
$-2t + 6n - 15$

61. $-5(y - 3) - 6(m + 8) +$
$5m - 3y$ $-m - 8y - 33$

62. $7(3n - 6) + 4n + 5 + 6(n + 2)$ $31n - 25$

63. $6r^2t + 5 + 2(r^2t + 2) - 6(8rt - 2)$
$8r^2t - 48rt + 21$

64. $-9(4n - 6) + 3(x - 8) + 3n + (-5n)$
$-38n + 3x + 30$

65. $2(4 - x) - 7x + 2n + 3(n + 8)$ $5n - 9x + 32$

66. $-\frac{2}{3}\left(3y - \frac{1}{5}\right) - 6\left(y - \frac{1}{2}\right)$ $-8y + 3\frac{2}{15}$

67. $-1.2(4y - 3) + 2y - 6 + 6(y - 7)$ $3.2y - 44.4$

68. $8(4x - 6) + 9x + 3 + 2(x + 2)$ $43x - 41$

69. $0.2(n + x) - 1.3(n + x) + 4.5x$ $-1.1n + 3.4x$

70. $\frac{3}{4}x^2 - \frac{2}{5}(x^2 - y) + 1.3(y - x^2)$ $-0.95x^2 + 1.7y$

71. $-6[2(n^2y + 5n^2y) - 9(n^2y - 3n^2y)] -$
$15ny^2 - 9$ $-180n^2y - 15ny^2 - 9$

72. $4.2(n + t) + (-8.5)(t + n) - 2.3n$ $-6.6n - 4.3t$

EXCURSIONS

Exploring with Calculators

1. The surface area of a cylinder is calculated using the formula $SA = 2\pi r^2 + 2\pi rh$, where h is the height of the cylinder and r is the radius of the circular base of the cylinder. Use your graphing calculator to find the surface area of the following four cylinders. Remember that your calculator has π already stored in memory. (Note: See the Calculator Corner on page 79 to review how to store values for variables.)

```
9→H
                    9
4→R
                    4
2πR²+2πRH
             326.725636
```

a. $h = 9$ inches $r = 4$ inches   (See the accompanying figure.)

The **2nd ENTRY** feature of your graphing calculator will "recall" the formula for the surface area so that you do not need to re-enter it for each new problem!

b. $h = 11.9$ inches $r = 1.6$ inches

c. $h = 23.06$ meters $r = 16.73$ meters

d. $h = 579.001$ feet $r = 801.9003$ feet

2. Suppose you receive 1 cent on Monday, 2 cents on Tuesday, 3 cents on Wednesday, 4 cents on Thursday, and so on. How much money would you have at the end of twenty days?

SECTION GOALS

▪ *To solve linear equations using the addition principle, the multiplication principle, or both*

▪ *To recognize and classify linear equations*

▪ *To solve equations that require simplification*

▪ *To solve equations that involve absolute value*

2.2 Solving Linear Equations Using the Addition and Multiplication Principles

SECTION LEAD-IN

In each of the following problems, ask and answer a question that can be answered using the given data.

a. In 1987, the U.S. Post Office sold $31,028,300,000 worth of stamps.

b. In 1997, the cost of first class postage was $0.32 for each letter up to one ounce and then 20 cents for each additional ounce.

c. Fresh tuna steak costs $4.49 a pound.

d. Two hundred twenty people paid 75 cents each to buy a *Wall Street Journal*.

e. One-half cup of three-bean salad has 120 calories.

f. One hundred percent cotton flannel costs five-and-a-half dollars a yard.

g. One-fourth of a 9-inch pie is 405 calories.

h. One-half cup of strawberry ice cream has 232 calories and 0.75 cup of strawberry frozen nonfat yogurt has 135 calories.

An **equation** is a mathematical statement in which two expressions are set equal to each other.

Linear Equation

A **linear equation in one variable** is an equation that contains only constants and a single variable that has an exponent of 1. It is also called a **first-degree equation.**

A **solution,** or **root,** of an equation is a number that makes the equation true when it replaces the variable. The set containing all the solutions of an equation is called the **solution set** of that equation.

The Addition Principle of Equality

Consider the equation

$$x + 4 = 9$$

You can probably see that the solution is 5, because $5 + 4 = 9$. However, when the solution is not obvious, we can find it by applying the **addition principle.**

Addition Principle of Equality

For real numbers a, b, and c,

$$\text{if } a = b, \text{ then } a + c = b + c$$

This principle tells us that the equations

$$x = b \quad \text{and} \quad x + c = b + c$$

are **equivalent equations,** or equations with the same solution set. To use the addition principle, we add the same number to both sides of an equation in order to isolate the variable (to get it alone on one side of the equation).

• • •

Example 1

Find the solution set for $y + 8 = 3$

Solution

To isolate y, we need to end up with $y + 0$ on the left. To get this, we add -8 (the additive inverse of 8) to both sides of the equation and then simplify.

$$y + 8 + (-8) = 3 + (-8) \quad \text{Adding } -8 \text{ to both sides}$$
$$y + 0 = 3 + (-8) \quad \text{Additive inverse property}$$
$$y = -5 \quad \text{Simplifying}$$

The solution set is $\{-5\}$.

To check that -5 is a solution, we substitute it for y in the original equation.

▶ *CHECK* **Warm-Up 1**

Calculator Corner

You can easily check your solution to an equation involving one operation using your graphing calculator. For instance, let $x - 4 = 10$.

Let Y_1 equal the left side of this equation and let Y_2 be the right side of the equation. Then use the **TRACE** feature of your graphing calculator to find the point where Y_1 and Y_2 intersect. This is the point where the two sides of the equation are equal. We simply traced the graph of the slanted line to where $y = 10$ to find the value of x that makes this true.

 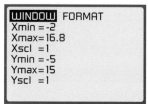 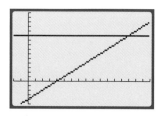 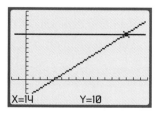

The point of intersection is (14, 10). The result we obtain through algebra is $x = 14$. These two lines intersect at $x = 14$, therefore, confirming the answer that we find using algebra. This simple method of graphing both sides of an equation and finding the point where the two equations intersect can be used to check and confirm your paper-and-pencil work. We will learn more about graphing in Chapter 3.

The Multiplication Principle of Equality

To solve an equation such as $3x = 12$, we must transform it so that the coefficient of x is 1. To do so, we use the **multiplication principle.**

> **Multiplication Principle of Equality**
>
> For real numbers a, b, and c, where c is not equal to zero,
>
> $$\text{if } a = b, \text{ then } ca = cb$$

In other words, the equations

$$x = b \quad \text{and} \quad cx = cb$$

are equivalent equations. To use the multiplication principle, we multiply both sides of an equation by the same number—the reciprocal of the coefficient of x. That gives x the new coefficient 1.

∎∎∎

EXAMPLE 2

Find the solution set for $9x = 117$

SOLUTION

We multiply both sides of the equation by $\frac{1}{9}$ (the reciprocal, or multiplicative inverse, of 9) and then simplify.

$$9x = 117 \qquad \text{Original equation}$$

$$\left(\tfrac{1}{9}\right)(9x) = \left(\tfrac{1}{9}\right)(117) \qquad \text{Multiplying by } \tfrac{1}{9} \text{ on both sides}$$

$$1x = \tfrac{117}{9} \qquad \text{Multiplicative inverse property}$$

$$x = 13 \qquad \text{Simplifying}$$

The solution set is {13}. Check the answer in the original equation.

▶ CHECK **Warm-Up 2**

■ ■ ■

EXAMPLE 3

Solve for t: $20 = \frac{-1}{4}t$

SOLUTION

We multiply by the reciprocal of $\frac{-1}{4}$.

$$\left(\frac{-4}{1}\right)20 = \left(\frac{-4}{1}\right)\left(\frac{-1}{4}\right)t \quad \text{Multiplying by reciprocal}$$

$$-80 = \frac{4}{4}t \qquad\qquad \text{Simplifying}$$

$$-80 = t$$

The check is left for you to do.

▶ CHECK **Warm-Up 3**

Calculator Corner

You must be careful when entering equations into your calculator when parentheses are needed in the equation. For instance, how would you enter the equation

$$\left(\frac{2x}{3}\right) + 2 = 9$$

using two operations (division and addition) into your graphing calculator?

 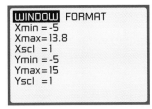 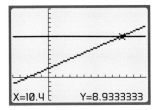 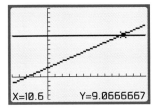

What do you notice happens when you **TRACE** on this graph? The **TRACE** feature does not land on $x = 10.5$, the algebraic solution to the problem (use algebra to check that this answer is correct). It goes from $x = 10.4$ to $x = 10.6$. So how can you confirm your answer obtained using algebra when **TRACE** only *suggests* the correct answer? Some graphing calculators are able to calculate points of intersection using a **CALC**ulate feature.

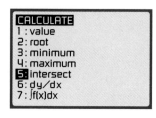 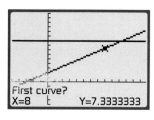 Press **ENTER.**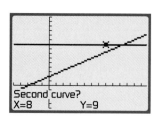

Press **ENTER.** 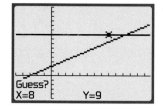 Press **ENTER.**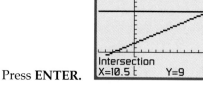

The **INTERSECT** utility of your graphing calculator *confirms* your answer $x = 10.5$

.

Solving Linear Equations Using Both Principles

We can solve a linear equation of the type $x + b = c$ with the addition principle; and we can solve an equation of the type $ax = c$ with the multiplication principle. To solve an equation of the type

$$ax + b = c$$

we must use *both* the addition principle and the multiplication principle.

To solve any linear equation in one variable

1. Use the properties of real numbers to simplify both sides of the equation.
2. Use the addition principle to isolate the term that contains the variable.
3. Use the multiplication principle to give the variable a coefficient of 1.

∎ ∎ ∎

EXAMPLE 4

Solve for x: $4x + 16 = 12x$

SOLUTION

We use the addition principle to remove the variable term from the left side of the equation.

$$4x + 16 = 12x \qquad \text{Original equation}$$
$$4x + 16 + (-4x) = 12x + (-4x) \qquad \text{Adding } -4x \text{ to both sides}$$
$$16 = 12x - 4x \qquad \text{Additive inverse property}$$

Next we use the distributive property to combine like terms.

$$16 = 12x - 4x$$
$$16 = (12 - 4)x \qquad \text{Distributive property}$$
$$16 = 8x \qquad \text{Simplifying}$$

Finally, we apply the multiplication principle.

$$\left(\tfrac{1}{8}\right)16 = \left(\tfrac{1}{8}\right)8x \quad \text{Multiplying by } \tfrac{1}{8}$$

$$2 = x \qquad \text{Simplifying}$$

Check by substituting 2 for x in the original equation.

▶ *CHECK* **Warm-Up 4**

▪ ▪ ▪
EXAMPLE 5

Solve: $-4.8 = 0.95r + 0.9$

SOLUTION

To isolate the variable term $0.95r$, we add -0.9 to both sides of the equation.

$$-4.8 + (-0.9) = 0.95r + 0.9 + (-0.9)$$
$$-5.7 = (0.95)r \qquad \text{Simplifying}$$

Then, to obtain the coefficient 1 for r, we multiply both sides of the equation by $\frac{1}{0.95}$ and simplify.

$$\left(\tfrac{1}{0.95}\right)(-5.7) = \left(\tfrac{1}{0.95}\right)(0.95)r$$
$$-6 = r$$

Check your answer in the original equation.

▶ *CHECK* **Warm-Up 5**

Equations that Require Simplification

One-variable linear equations are the most basic equations, but even they can be complicated. Where necessary, simplify each side of an equation first, before you apply the addition and/or multiplication principles.

▪ ▪ ▪
EXAMPLE 6

Solve: $6 = \frac{5n}{3} - 2n - 8$

SOLUTION

We rewrite the equation as $6 = \frac{5}{3}n - 2n - 8$. Then, simplifying on the right side and applying the addition principle, we get

$$6 = \left(\tfrac{5}{3} - 2\right)n - 8 \qquad \text{Distributive property}$$

$$6 = \left(\tfrac{-1}{3}\right)n - 8 \qquad \text{Simplifying}$$

$$6 + 8 = \left(\tfrac{-1}{3}\right)n - 8 + 8 \qquad \text{Addition principle}$$

$$14 = \left(\tfrac{-1}{3}\right)n \qquad \text{Simplifying}$$

▪▪▪
WRITER'S BLOCK

Explain to a friend how to solve an equation like
$16 = 10 + x.$

▪▪▪
WRITER'S BLOCK

List the steps to solve an equation like
$120 = 4x.$

(We could have moved 6, but that would have resulted in 0 on one side of the equation. This would not have helped us. In fact, it would have made solving much more complicated!)

Because the coefficient of n is $\frac{-1}{3}$, we multiply both sides of the equation by $\frac{-3}{1}$, the multiplicative inverse of $\frac{-1}{3}$.

$$14 = \left(\frac{-1}{3}\right)n$$
$$(-3)(14) = (-3)\left(\frac{-1}{3}\right)n$$
$$-42 = n$$

Check your answer in the original equation.

▶ *CHECK* **Warm-Up 6**

ERROR ALERT

Identify the error and give a correct answer.

$6 + 5\,(2x - 1) = -4$
$11\,(2x - 1) = -4$
$22x - 11 = -4$
$22x = -4 + 11$
$22x = 7$
$x = \frac{7}{22}$

■■■

EXAMPLE 7

Solve for x: $-11(5x - 4) = 8(x + 12) - (3x + 4)$

SOLUTION

Applying the distributive property first and continuing, we have

$-11(5x - 4) = 8(x + 12) - (3x + 4)$	Original equation
$-55x + 44 = 8x + 96 - 3x - 4$	Distributive property
$-55x + 44 = 5x + 92$	Simplifying
$-55x + 44 + (-5x) = 5x + 92 + (-5x)$	Adding $-5x$
$-60x + 44 = 92$	Simplifying
$-60x + 44 + (-44) = 92 + (-44)$	Adding -44
$-60x = 48$	
$\left(\frac{-1}{60}\right)(-60x) = (48)\left(\frac{-1}{60}\right)$	Multiplying by $\frac{-1}{60}$
$x = \frac{-48}{60}$	Simplifying
$x = \frac{-4}{5}$	Writing in simplest form

Remember to check your answer in the original equation.

▶ *CHECK* **Warm-Up 7**

Calculator Corner

Your graphing calculator can also help you to check your answers for more complex linear equations, such as $3(x - 3) - 4(x + 1) = 5x - 6$. If you use the **TRACE** feature of your calculator, you may not land on the *exact* point of intersection. If your calculator has a **CALC**ulate **INTERSECT** feature, use it to confirm your result. (Note: See the Calculator Corner on pages 91 and 92 to review how to find the intersection points of two graphs.)

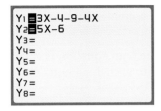
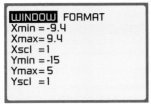
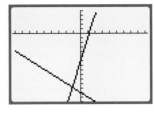
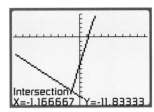

The calculator gives the answer in decimal form. On many graphing calculators, you can use the Home Screen and easily convert this decimal result into a fraction. Go to your Home Screen by pressing **2nd QUIT**; then press the x-variable key followed by the **MATH** key and choose the **FRAC** option. Now you can see the answer as a fraction. Check this answer in the original equation.

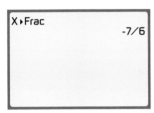

▪▪▪
EXAMPLE 8

Find the solution set of $n + 5 = 5(0.5n + 4) - 0.5n$.

SOLUTION

We begin by using the distributive property to remove the parentheses. Then we simplify each side of the equation.

$$n + 5 = 5(0.5n + 4) - 0.5n \qquad \text{Original equation}$$
$$n + 5 = 2.5n + 20 - 0.5n \qquad \text{Distributive property}$$
$$n + 5 = 2n + 20 \qquad \text{Simplifying}$$

Next we use the addition principle twice.

$$n + 5 + (-20) = 2n + 20 + (-20) \qquad \text{Adding } -20$$
$$n - 15 = 2n \qquad \text{Simplifying}$$
$$n - 15 + (-n) = 2n + (-n) \qquad \text{Adding } -n$$
$$-15 = n \qquad \text{Simplifying}$$

So the solution set of $n + 5 = 5(0.5n + 4) - 0.5n$ is $\{-15\}$.

▶ CHECK **Warm-Up 8**

Conditional, Identity, and Contradictory Equations

The equations we have solved so far have had only one solution; later we shall solve equations that have two or more solutions. An equation that has one, two, or any countable number of solutions is called a **conditional equation.** There are two other possibilities:

An equation can have all real numbers as solutions; such an equation is called an **identity equation.** An equation can have no solutions; such an equation is called a **contradictory equation.**

▪ ▪ ▪

EXAMPLE 9

Solve each equation, if possible.

a. $2x + 1 = 5x - (3x - 1)$ **b.** $5x + 9 = 3x + 2x - 7$

SOLUTION

a. We remove the parentheses and then solve, obtaining

$$2x + 1 = 5x - (3x - 1) \quad \text{Original equation}$$
$$2x + 1 = 5x - 3x + 1 \quad \text{Removing parentheses}$$
$$2x + 1 = 2x + 1 \quad \text{Combining on right}$$

The two sides of the equation are identical, so every real number is a solution. If we continue solving, we eventually get to

$$x = x$$

Because every real number equals itself, every real number is a solution to this equation and to the original equation. Thus the original equation is an identity.

b. Simplifying each side first yields

$$5x + 9 = 3x + 2x - 7$$
$$5x + 9 = 5x - 7$$

Adding $-5x$ to both sides (to eliminate x from the left side), we get

$$5x + (-5x) + 9 = 5x + (-5x) - 7$$
$$9 \neq -7$$

This is false, so the original equation has no solutions. That equation is contradictory, and its solution set is $\varnothing$. We could also have tried to solve this equation by adding -9 to both sides. We would have obtained.

$$5x + 9 + (-9) = 5x - 7 + (-9)$$
$$5x \neq 5x - 16 \quad \text{Simplifying}$$

This is impossible; a number cannot equal 16 less than itself. Again, this means that there are no solutions, and the equation is contradictory.

▶ *CHECK* **Warm-Up 9**

Equations that Involve Absolute Value

Let $|x| = 3$, where x may be any algebraic expression. We know that $|x|$ is positive, but we don't know whether x represents a positive real number or a negative real number. If it is positive, then $x = 3$; if it is negative, then $x = -3$. So the equation $|x| = 3$ really represents the two equations.

$$x = 3 \quad \text{and} \quad x = -3$$

More generally, the equation

$$|x| = b, \text{ where } b \geq 0$$

represents the two equations

$$x = b \quad \text{and} \quad x = -b$$

We must solve both of these equations to find the solution of $|x| = b$. Of course, if b is negative, there is no solution, because an absolute value cannot be negative.

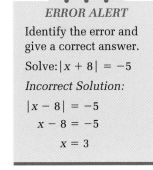

▪▪▪

EXAMPLE 10

a. Find the solution set for $|x - 4| = 8$.　**b.** Solve: $|x - 2| = -5$

SOLUTION

a. The given absolute value equation represents the equations $x - 4 = 8$ and $x - 4 = -8$. We must solve both to find the solution set.

$x - 4 = 8$	$x - 4 = -8$	The equations
$x - 4 + 4 = 8 + 4$	$x - 4 + 4 = -8 + 4$	Adding 4 to both sides
$x = 12$	$x = -4$	Simplifying

We check both solutions in the original equation.

$|x - 4| = 8$	$|x - 4| = 8$
$|12 - 4| = 8$	$|-4 - 4| = 8$
$|8| = 8$	$|-8| = 8$
$8 = 8$　True	$8 = 8$　True

So $|x - 4| = 8$ has the solution set $\{12, -4\}$. Most conditional linear equations in one variable that involve absolute values have two solutions.

b. The given equation is $|x - 2| = -5$. This cannot possibly be true: An absolute value can never be negative.

So the equation is contradictory; it has no solutions.

▶ *CHECK* **Warm-Up 10**

Calculator Corner

a. Find the solution to $|x - 2| = 3$. (Note: On the *TI-83*, press MATH NUM 1 to get the absolute value function.)

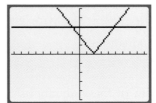

$x = -1$ and $x = 5$.

b. Find the solution to $|3x + 2| = |x + 2|$.

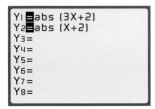

$x = 0$ and $x = -1$.

An equivalent way to look at this would be to graph $|3x + 2| - |x + 2| = 0$.

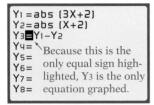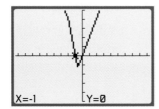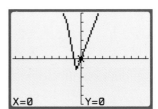

▪ ▪ ▪

EXAMPLE 11

Find the solution set for $|4x + 9| - 5 = 12$.

SOLUTION

We can isolate the absolute value by adding 5 to each side.

$$|4x + 9| - 5 + 5 = 12 + 5$$
$$|4x + 9| = 17 \qquad \text{Simplifying}$$

Now we solve the two equations that this equation represents.

$4x + 9 = 17$	$4x + 9 = -17$	Adding -9
$4x + 9 + (-9) = 17 + (-9)$	$4x + 9 + (-9) = -17 + (-9)$	to both sides
$4x = 8$	$4x = -26$	Simplifying
$\left(\frac{1}{4}\right)(4x) = \left(\frac{1}{4}\right)(8)$	$\left(\frac{1}{4}\right)(4x) = \left(\frac{1}{4}\right)(-26)$	Multiplying
$x = 2$	$x = -6\frac{1}{2}$	Simplifying

The solution set is $\left\{2, -6\frac{1}{2}\right\}$. The check is left for you to do.

▶ CHECK **Warm-Up 11**

Practice what you learned.

SECTION FOLLOW-UP

Now go back and ask another question for each problem situation that uses a different operation.

2.2 WARM-UPS

Work these problems before you attempt the exercises.

1. Solve for n: $-4 = n + 9$
   $n = -13$

2. Solve for n: $-32n = 288$
   $n = -9$

3. Solve for n: $\frac{-8}{9}n = 408$
   $n = -459$

4. Solve for x: $2.45x = 2.5x + 5$
   $x = -100$

5. Solve for y: $7 - 4y = 12$
   $y = -1\frac{1}{4}$

6. Solve for t: $-19 + \frac{8t}{5} - 7 = 6$
   $t = 20$

7. Simplify: $2x - 5(2x - 6) = 4(x - 5)$ $x = 4\frac{1}{6}$

8. Solve for n: $6(2n + 8) - n = n + 8$ $n = -4$

9. a. Solve for y: $2y - 4 = 6y - (4 + 4y)$
      all real numbers

   b. Find the solution set: $4x - 5 = -30 + 24x - 20x$
      no solution, or $\varnothing$

10. Solve for t: $-8 = |t - 3|$
    no solution

11. Solve for w: $|w - 5| + 6 = 7$
    $w = 6; w = 4$

2.2 EXERCISES

Note: Use your graphing calculator to check your results whenever possible.

In Exercises 1 through 8, use either the addition principle or the multiplication principle to solve the equation.

1. $10 = x + 3$ $x = 7$ **2.** $n - 27 = -34$ $n = -7$ **3.** $-\frac{1}{2}y = 25$ $y = -50$ **4.** $-\frac{1}{5}x = 15$
$x = -75$

5. $\frac{3}{4} + y = \frac{1}{8}$ $y = -\frac{5}{8}$ **6.** $-27n = -243$ $n = 9$ **7.** $n - 1.9 = -2.6$ **8.** $96.4 = -0.2y$
$n = -0.7$ $y = -482$

In Exercises 9 through 46, solve the equation.

9. $18 = 3x + 6$ $x = 4$ **10.** $-5 - n = 22$ $n = -27$ **11.** $-6 - n = 34$ $n = -40$ **12.** $4 - 2r = 3$ $r = \frac{1}{2}$

13. $-n - 1.8 = 1.2$ **14.** $-t + 0.12 = 16$ **15.** $5t + 1.9 = 1.6$ **16.** $7r - 1.4 = 3.5$
$n = -3$ $t = -15.88$ $t = -0.06$ $r = 0.7$

17. $-3 - 6x = -0.15$ **18.** $24y - 12 = 3y$ $y = \frac{4}{7}$ **19.** $6n + 2.1 = 1.2$ **20.** $5.1r + 1.3 = 6.4$
$x = -0.475$ $n = -0.15$ $r = 1$

21. $-1.4 - 9r = -5.0$ **22.** $2x - 1.2 = 2.8$ $x = 2$ **23.** $\frac{n}{6} + 7 = -8$ $n = -90$ **24.** $6 = 8 - \frac{r}{4}$
$r = 0.4$ $r = 8$

25. $-27 + \frac{5}{9}x = -34$ **26.** $-35 + \frac{3}{5}y = -62$ **27.** $\frac{5}{8}n - 6 = 29$ $n = 56$ **28.** $\frac{4}{9}r - 12 = 32$
$x = -12\frac{3}{5}$ $y = -45$ $r = 99$

29. $\frac{x}{4} + 7 = 18$ $x = 44$ **30.** $\frac{t}{4} + 21 = 36$ $t = 60$ **31.** $\frac{1}{4}n + 8 = 16$ $n = 32$ **32.** $27 = \frac{4}{5}r - 9$ $r = 45$

33. $-29 = \frac{x}{7} + 11$ **34.** $\frac{t}{3} + 9 = -16$ $t = -75$ **35.** $-4 + 2.4r = 12$ **36.** $0.6n + 18 = 30$
$x = -280$ $r = 6\frac{2}{3}$ $n = 20$

37. $-3x + \frac{9}{10} = 15$ **38.** $\frac{n}{6} + 9 = 26$ $n = 102$ **39.** $\frac{r}{5} + 8 = 32$ $r = 120$ **40.** $-57 = \frac{y}{8} - 41$
$x = -4\frac{7}{10}$ $y = -128$

41. $5x + 10 = 15x + 35$ **42.** $6x - 6 = 3x + 8$ **43.** $11y + 8 = 15y - 9 + 2y$
$x = -2\frac{1}{2}$ $x = 4\frac{2}{3}$ $y = 2\frac{5}{6}$

44. $21t - 18 = 12 - 6t$ **45.** $15n + 12 = 16n + 3$ **46.** $27 = 20y - 5 + 12y$ $y = 1$
$t = 1\frac{1}{9}$ $n = 9$

In Exercises 47 through 50, determine whether each of the following equations is an identity, a contradiction, or a conditional equation.

47. $x + 3 = 3$ **48.** $x + 2 = 2x + 4$ **49.** $x + 5 = x + 6$ **50.** $2(y - 8) = 2y - 16$
conditional   conditional   contradiction   identity

In Exercises 51 through 99, solve for the indicated variable, if possible.

51. $3v + 9 = 11v - 15$ $v = 3$ **52.** $27x + 12 = 13x + 5$ $x = -\frac{1}{2}$

53. $11y - 12 + 3y = 7y + 2$ $y = 2$ **54.** $26t - 19 + 5t = -3t - 7$ $t = \frac{6}{17}$

55. $7y - 3 + 12 - 3y = 6y - 8$ $y = 8\frac{1}{2}$ **56.** $2t + 3 - 8 - 5t = 3t + 13$ $t = -3$

57. $\frac{2}{3}n - 8 + 12 = 16 + \frac{4}{3}n - 8$ $n = -6$ **58.** $\frac{10}{8}y + 18 - \frac{7}{8}y = \frac{3}{8}y + 25$ contradiction; no solution

59. $7(2 + 4x) = 11 - 4x$ $x = -\frac{3}{32}$ **60.** $8(4 + 3x) = 27 + 3x$ $x = -\frac{5}{21}$

61. $14.42 = 0.7(1.3x - 8)$ $x = 22$ **62.** $0.4(2x - 1.2) = 8.6 - 0.2x$ $x = 9.08$

63. $2(7x + 4) - 8 = x - 6$ $x = -\frac{6}{13}$

64. $(12x - 5)8 = 4 - x$ $x = \frac{44}{97}$

65. $3r + 4r - (6r + 12) = 18 - 12r$ $r = 2\frac{4}{13}$

66. $26 - 3t - (4t - 7) = 8t - 5$ $t = 2\frac{8}{15}$

67. $23 = 0.5(1.2x - 2) - 2.4x$ $x = -13\frac{1}{3}$

68. $0.4(2x + 1.2) = 8.52 + 0.2x$ $x = 13.4$

69. $17 = -5(8x - 4) + 3(2x - 4)$ $x = -\frac{9}{34}$

70. $-106 = 5(-4y - 6) + 6(3y + 2)$ $y = 44$

71. $(8.2 + t)0.3 + 5 = 2(2.03 + t)$ $t = 2$

72. $7(3x - 2) - 9 = -3(8x - 5)$ $x = \frac{38}{45}$

73. $-12(5x - 3) + 6 = 11(2x + 4)$ $x = -\frac{1}{41}$

74. $(7x - 10)5 = 6(-x + 19)$ $x = 4$

75. $\frac{2}{5}(x + 4) = \frac{3}{4}(2x - 5)$ $x = 4\frac{19}{22}$

76. $\frac{1}{2}(8 - x) = \frac{2}{3}(x + 9)$ $x = -1\frac{5}{7}$

77. $\frac{1}{8}(2x - 8) = \frac{1}{4}(x - 4)$ identity; all real numbers

78. $6 - 3(x - 4) + 2x = 2[3(2x - 5)]$ $x = 3\frac{9}{13}$

79. $4 + 3(t - 7) + 4t = 5[-2(2t + 4)]$ $t = -\frac{23}{27}$

80. $-2[-1(x - 30)] + 4(7 - 3x) = 8(4x - 5) - 6$
$x = \frac{1}{3}$

81. $|p| = 4$
$p = 4; p = -4$

82. $|m| = 6$
$m = 6; m = -6$

83. $|x - 4| = -5$
no solution

84. $|y + 3| = -8$
no solution

85. $26 = |4x - 1|$ $x = 6\frac{3}{4}; x = -6\frac{1}{4}$ **86.** $12 = |9x - 3|$ $x = 1\frac{2}{3}; x = -1$ **87.** $|12x - 4| = 8$ $x = 1; x = -\frac{1}{3}$

88. $|3x + 2| = 7$ $x = 1\frac{2}{3}; x = -3$ **89.** $6 = |2x + 5|$ $x = \frac{1}{2}; x = -5\frac{1}{2}$ **90.** $|6x - 9| = 5$ $x = 2\frac{1}{3}; x = \frac{2}{3}$

91. $|3y - 4| = -7$ no solution   **92.** $|4x - 8| = 6$ $x = 3\frac{1}{2}; x = \frac{1}{2}$ **93.** $|-3x - 5| + 4 = 8$
$x = -3; x = -\frac{1}{3}$

94. $|4x + 3| + 8 = 12$
$x = \frac{1}{4}; x = -1\frac{3}{4}$ **95.** $|8 - x| + 2 = 7$ $x = 3; x = 13$ **96.** $|10 + 2x| - 5 = 9$
$x = 2; x = -12$

97. $|2x - 3x + 5| + 9 = 8$ no solution   **98.** $7 - 3y = |3y + 4y - 6|$
$y = 1\frac{3}{10}; y = -\frac{1}{4}$ **99.** $|11 - 2t| = 6t + 5 - 11t$
$t = -2$

MIXED PRACTICE

By doing these exercises, you will practice the topics up to this point in the
chapter.

100. Evaluate $-13ab^2$ when $a = -1$ and $b = -2$.
52

101. Solve: $26 - x = -86$ $x = 112$

102. How many terms are there in
$-3 + 6x - 2x^2$? What are they?
three; $-3, 6x, -2x^2$

103. Identify the variable part, the exponent,
and the coefficient of $-7a^3$.   variable part: a^3;
exponent: 3; coefficient: -7

104. Combine like terms: $8a^2 - 3a + 2a^2$ $10a^2 - 3a$

105. Solve: $\frac{2t}{3} - 5 = -12$ $t = -10\frac{1}{2}$

106. Evaluate: $3xy - 2x^2$ when $x = -10$ and $y = 5$.
-350

107. Solve: $9 - 8y = 3$ $y = \frac{3}{4}$

108. Solve: $\frac{x}{8} + 5 = -23$ $x = -224$

109. Solve: $2(y - 6) + 7 = 9 + 3y$ $y = -14$

110. Name the terms in $16n + 24$. Which is the
constant term?   terms: $16n, 24$; constant term: 24

111. Simplify: $-4(3m + 5) - 32m + 10$
$-44m - 10$

112. Simplify: $3x - 4x + 8 - 3x - 9 - 2x$ $-6x - 1$ **113.** Solve: $|3x - 5| = 9$ $x = 4\frac{2}{3}; x = -1\frac{1}{3}$

EXCURSIONS

Class Act

1. Place $+$, $-$, $\times$, and $\div$ signs between the numbers in each row to give the indicated answers. Use a calculator. Compare your results with other groups, and find each result as many ways as possible. Be sure to use ( ) and [ ] where needed.

a. 0 1 2 3 4 5 6 7 8 9 = 0

b. 0 1 2 3 4 5 6 7 8 9 = 1

c. 10 9 8 7 6 5 4 3 2 1 = 0

d. 0 1 2 3 4 5 6 7 8 9 = 2

e. 0 1 2 3 4 5 6 7 8 9 = −3

f. 0 1 2 3 4 5 6 7 8 9 = −4

g. 0 1 2 3 4 5 6 7 8 9 = 10

h. 10 9 8 7 6 5 4 3 2 1 = 1

i. 0 1 2 3 4 5 6 7 8 9 = 9

j. 0 1 2 3 4 5 6 7 8 9 = 100

Exploring Numbers

2. a. Let $A = B$
Let $B = 3$
Then $A = ?$

b. Let $A = B$
Let $B = C + 1$
Let $C = D \div 3$
Let $D = 27$
Find A, B, and C

c. Let $A = B - 6$
Let $B = C + 8$
Let $C = D \div 4$
Let $D = E + 10$
Let $E = F$
Let $F = 6$
Find A, B, C, D, and E

d. Let $C = 5$
Let $A = B + 7$
Let $B = 24 \div D$
Let $D = C - 2$
Find A, B, C, and D

e. Write four number rules. Share them with another group.

Exploring with Calculators

3. a. Find the solution to $|x - 2| = 2x + 1$. Two solutions are found algebraically: -3 and $\frac{1}{3}$. But this does not occur when the graphical solution is obtained. Why not?

   b. Find the solution to $|x + 2| = |x - 1|$. Notice that these two absolute values are "parallel." How can you modify this problem so that there are no solutions?

4. Use your graphing calculator to help you solve the following equation: $5(3x + 4) - x = -2(x + 5)$. Write out calculator steps (or sketch the screens) to solve this equation. How can you check your answer?

CONNECTIONS TO *PROBABILITY*

Introduction to Probability

The **probability** of an event is the numerical likelihood that the event will occur.

In an **experiment,** we gather data. We can list all possible outcomes in a universal set called a **sample space, S.** Any subset of this sample space is called an **event.**

We can describe any event by using a rule (such as set-builder notation) or by listing the outcomes (such as roster notation). Each outcome in a sample space is a **simple event.**

When we toss two coins, we have four possible **outcomes.** Let H stand for heads and T for tails, then the possible outcomes can be written as HH, HT, TH, TT. The sample space of the experiment is $\{HH, HT, TH, TT\}$. Experience shows us that each of these events is **equally likely** and has a **probability** of

$$p = \frac{1}{\text{total number of outcomes in sample space}} \quad \text{or} \quad \frac{1}{n(S)}$$

So the probability of HH, written $p(HH)$, is $\frac{1}{4}$; $p(HT) = \frac{1}{4}$; $p(TH) = \frac{1}{4}$, and $p(TT) = \frac{1}{4}$.

The sum of the probabilities of the simple events in a sample space is 1.

To give the probability of any event, we use the rule:

$$\text{Probability of event E} \longrightarrow p(E) = \frac{n(E) \longleftarrow \text{Number of outcomes in } E}{n(S) \longleftarrow \text{Number of outcomes in } S}$$

▪▪▪

EXAMPLE

Toss two coins. Find

a. the probability of getting exactly one head.

b. the probability of getting at least one tail.

c. the probability of getting two heads or two tails.

SOLUTION

a. The event of getting exactly one head is $\{HT, TH\}$.

$$\text{Let } A = \{HT, TH\}$$

The probability is

$$p(A) = \frac{n(A)}{n(S)} = \frac{2}{4}$$

> **▪▪▪**
> **WRITER'S BLOCK**
> Explain why *HT* and *TH* are both listed.

b. $B = \{HT, TH, TT\}$, so

$$p(B) = \frac{n(B)}{n(S)} = \frac{3}{4}$$

c. $C = \{HH, TT\}$, so

$$p(C) = \frac{n(C)}{n(S)} = \frac{2}{4}$$

We may reduce the fractions or leave them in this form.

◢

Just as with sets, two events A and B can be complementary, if $A \cap B = \emptyset$ and $A \cup B = S$.

▪▪▪

EXAMPLE

Let $S = \{HH, HT, TH, TT\}$. Let event A be no more than one head. Let event B be no tails.

a. List A and B.

b. Find the probability of each event.

c. Use the rules $A \cap A' = \emptyset$ and $A \cup A' = U$, and show that A and B are complements.

SOLUTION

a. $A = \{HT, TH, TT\}$

$B = \{HH\}$

b. $p(A) = \frac{n(A)}{n(S)} = \frac{3}{4}$

$p(B) = \frac{n(B)}{n(S)} = \frac{1}{4}$

c. $A \cap B = \{HT, TH, TT\} \cap \{HH\} = \emptyset$

$A \cup B = \{HT, TH, TT\} \cup \{HH\} = \{HH, HT, TH, TT\}$

Since $\{HH, HT, TH, TT\} = U$, we can say

$$A \cup B = U$$

◢

Note that in part (b), $p(A) + p(B) = 1$. This property is true for all complementary sets.

If A and A' are complements of each other,

$$p(A) + p(A') = 1$$
$$1 - p(A) = p(A')$$
$$1 - p(A') = p(A)$$

▪▪▪

EXAMPLE

Ten balls with numbers from 0 to 9 are placed in a container. The chance, or probability, of choosing the ball with the number 6 on it, if you draw just one ball, is 1 out of 10, or $\frac{1}{10}$. Under these same conditions, what are the chances of *not* drawing the ball with the number 6 on it?

SOLUTION

The number required by the problem is the chance that the number 6 will *not* be drawn. This number is equal to 1 minus the chance that it will be drawn. So we write 1 as $\frac{10}{10}$ and subtract $\frac{1}{10}$.

$$\frac{10}{10} - \frac{1}{10} = \frac{9}{10}$$

Thus there is a $\frac{9}{10}$ probability of *not* drawing the ball with a 6 on it.

◢

PRACTICE

1. Toss one die. Find the probability of getting a 3. Find the probability of getting an even number. $\frac{1}{6}; \frac{3}{6}$

2. Toss two coins. Find the probability of getting at least one head. Find the probability of getting three heads. $\frac{3}{4}; 0$

3. Show that the events {2, 4, 6} and {1, 3, 5} are complements when tossing one die is the experiment. {2, 4, 6} ∪ {1, 3, 5} = {1, 2, 3, 4, 5, 6}
{2, 4, 6} ∩ {1, 3, 5} = ∅

4. Flip three coins. Are the events {all heads} and {no heads} complements of each other? Explain. No. There are 6 possible outcomes in addition to *HHH* and *TTT*.

5. The probability that it will rain tonight is $\frac{1}{3}$. The probability that it will not rain is what? $\frac{2}{3}$

6. The probability that I will get tickets for the Collective Soul concert is $\frac{3}{7}$. The probability that I will not get tickets is what? $\frac{4}{7}$

In the experiment of tossing three coins, let $S = \{HHH, HHT, HTH, THH, HTT, THT, TTH, TTT\}$.

Event *A* is the event of exactly one tail.
Event *B* is the event of two or more heads.
Event *C* is the event of exactly two heads or exactly two tails.
Event *D* is no more than one head.
Event *E* is no less than two heads.

7. **a.** Find $p(A)$ $\frac{3}{8}$ **b.** Find $p(B)$ $\frac{4}{8}$

8. **a.** Find $p(C')$ $\frac{2}{8}$ **b.** Find $p(E')$ $\frac{4}{8}$

9. **a.** Give an event F different from A, B, C, D, or E. List the elements and give the probability of F.   Answers may vary.

   b. Make up a different sample space and define four events of that sample space. List the elements of S and your four events.   Answers may vary.

10. **a.** You have probably heard the expression, "You're one in a million." What is the probability that you are *not* one in a million? $\frac{999,999}{1,000,000}$

   b. The chances of winning a certain scratch-off game are 3 out of 35. What is the probability of losing? $\frac{32}{35}$

2.3 Translating Between English and Algebra

SECTION GOALS

- *To translate English phrases into algebra*
- *To translate algebraic expressions into words*
- *To solve literal equations*
- *To solve and evaluate formulas*

SECTION LEAD-IN

The number of calories in a cup of homemade popcorn depends on how the corn is popped and whether you add butter. The following numbers are approximately correct:

air popped, butter and salt	45 cal/cup
oil popped, with salt	35 cal/cup
oil popped, no salt	35 cal/cup
oil popped, with butter and salt	55 cal/cup
air popped, no salt	25 cal/cup

a. How many calories per cup does butter add to air-popped popcorn?   20 calories

b. How many calories would be in $8\frac{1}{2}$ cups of oil-popped popcorn that has butter and salt?   467.5 calories

c. You eat buttered, salted, oil-popped popcorn that has $192\frac{1}{2}$ calories. How much popcorn did you eat?   3.5 cups

d. ✏ Write the plan you used to answer questions (a) through (c).

e. Ask and answer four questions using this data.

Translating from English to Algebra

We can use algebra to solve many types of real problems. To do so, though, we must be able to translate English descriptions into algebraic terms.

To translate an English phrase into algebra

1. Identify the quantities involved and express them as symbols.
2. Identify the operation.
3. Connect the symbols with the operation sign.

Here are examples of English phrases that can be translated into algebraic expressions.

Operation	English Phrase	Algebraic Translation
Addition	*the sum of* 6 and a number	$6 + x$
	3 *more than* yesterday's amount	$x + 3$
	7 *increased by* a number	$7 + x$
	4 *plus* a certain amount	$4 + x$
	3 *added to* a quantity	$x + 3$
	3 years *older than* José	$J + 3$
Subtraction	the *difference of* 6 and x	$6 - x$
	4 *less than* Mary's age	$M - 4$
	6 *minus* an amount	$6 - x$
	4 *subtracted from* a number	$n - 4$
	7 dollars *decreased by* a number	$7 - n$
	2 *fewer than* Stephen's	$S - 2$
	4 *less* the number of peas	$4 - n$
Multiplication	the *product of* x and y	xy
	2 *times* the length x	$2x$
	3 *multiplied by* a number	$3n$
	one-half of a recipe	$\frac{1}{2}r$
Division	the *quotient of* a number and 5	$\frac{x}{5}$
	3 *divided by* x	$3 \div x$
	the *ratio of* length to width	$\frac{L}{W}$
Exponentiation	*square of* the hypotenuse x	x^2
	an edge *cubed*	e^3

•••

EXAMPLE 1

Rewrite each expression in algebraic notation.

a. one and five-tenths less than a number   **b.** the quotient of $\frac{1}{4}$ and x

c. the product of 7, and the difference of a number and 5

INSTRUCTOR NOTE

Remind students that subtraction and division are not commutative. The order that the symbols are written in does matter.

ERROR ALERT

Identify the error and give a correct answer.

Translate into algebra: 5 less than q

Incorrect Solution:

$5 - q$

SOLUTION

a. The quantities are "one and five-tenths" and "a number" and are expressed as the symbols 1.5 and **x.** "Less than" calls for subtraction, so the expression can be translated as $x - 1.5$.

b. The quantities are "$\frac{1}{4}$" and "a number." "Quotient" means division, so the expression is written $\frac{1}{4} \div x$.

c. We want to find the product of two quantities: "7" and "the difference of a number and 5." We translate the second quantity as $x - 5$, so the entire expression becomes

$$7(x - 5)$$

Note: Parentheses must be used here; $7x - 5$ is *not* the same as $7(x - 5)$.

▶ CHECK **Warm-Up 1**

Sometimes you may wish to symbolize a statement that requires additional information. Look at this example.

...

EXAMPLE 2

Write an algebraic expression for "the value of the coins in a stack of nickels."

ERROR ALERT

Identify the error and give a correct answer.

Translate into algebra: the quotient of w and 7

Incorrect Solution:

$7 \div w$

SOLUTION

We can let n be the number of nickels in the stack. We write it down for later reference:

$$n = \text{number of nickels}$$

Now we want their value. To find it, we need the additional information that each coin (a nickel) has a value of 5 cents (0.05).

Then the total value of the stack is found by multiplication. The value is $(0.05)(n)$, or $0.05n$.

▶ CHECK **Warm-Up 2**

Translating from Algebra to English

There may be more than one way to express a given algebraic phrase in English. This method can help:

> **To translate an algebraic phrase into English**
>
> 1. Determine the operation, and choose an appropriate word that indicates it.
> 2. Identify quantities connected by this operation.
> 3. Write the phrase.

▪▪▪

EXAMPLE 3

Rewrite $6.4n - 15$ in words.

SOLUTION

The operation subtraction connects the two quantities $6.4n$ (6.4 times n) and 15. This can be expressed as

The difference of 6.4 times n, and 15

Another way to say it is "six and four-tenths times a number, minus fifteen."

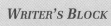 CHECK **Warm-Up 3**

Solving Literal Equations and Formulas

A **literal equation** is an equation that contains several letters, which may represent variables or constants. To *solve* a literal equation means to find an algebraic expression for one of the variables—you will always be told which one. You solve as for any equation, by using algebraic principles and properties to isolate the variable of interest.

▪▪▪

> ▪▪▪
> **WRITER'S BLOCK**
> What is the difference between a linear equation and a literal equation?

EXAMPLE 4

Solve for w: $2w + e = 3w + y$

SOLUTION

We need to apply the addition principle twice to isolate w on the right side.

$2w + e = 3w + y$	Original equation
$2w + e + (-2w) = 3w + y + (-2w)$	Adding $-2w$
$e = w + y$	Simplifying
$e + (-y) = w + y + (-y)$	Adding $-y$
$e - y = w$	Simplifying

▶ CHECK **Warm-Up 4**

> ▪▪▪
> **WRITER'S BLOCK**
> What is meant by *solving* a literal equation?

▪▪▪

EXAMPLE 5

Solve for r: $\frac{3r}{8} = h - 6$

SOLUTION

We first rewrite this equation as $\frac{3}{8}r = h - 6$. Then we multiply both sides by the reciprocal of $\frac{3}{8}$, namely $\frac{8}{3}$. We insert parentheses around $h - 6$ to make sure that the whole quantity is multiplied by $\frac{8}{3}$.

INSTRUCTOR NOTE

When using the multiplication principle, students should remember to apply the distributive property, multiplying *every* term on both sides of the equation.

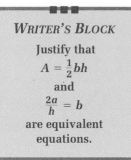

$$\left(\frac{8}{3}\right)\frac{3}{8}r = \left(\frac{8}{3}\right)(h - 6) \qquad \text{Multiplying by } \frac{8}{3}$$

$$r = \left(\frac{8}{3}\right)(h - 6) \qquad \text{Simplifying}$$

$$= \left(\frac{8}{3}\right)h - \left(\frac{8}{3}\right)6 \qquad \text{Distributive property}$$

$$= \frac{8}{3}h - 16 \qquad \text{Simplifying}$$

▶ CHECK **Warm-Up 5**

Practical Applications

Formulas, such as the area formula $A = \left(\frac{1}{2}\right)bh$, are also literal equations.

Often we must evaluate a particular variable in such a formula. This can be done in two different ways. The first method involves substituting all known values and then solving for the unknown variable.

• • •
EXAMPLE 6

The height of a projectile in feet at a time t seconds after launch is given by the formula $h = -16t^2 + vt$, where v is the initial velocity of the projectile. Find the height of a projectile launched at an initial velocity of 40 feet per second after it has traveled for 2 seconds.

SOLUTION

We begin by substituting all known values. Then we solve for the unknown variable.

$$h = -16t^2 + vt \qquad \text{Original equation}$$

$$= -16(2)^2 + 40(2) \qquad \text{Substituting}$$

$$= -16(4) + 40(2) \qquad \text{Simplifying exponents}$$

$$= -64 + 80 \qquad \text{Multiplying}$$

$$= 16 \qquad \text{Adding}$$

So the height after 2 seconds is 16 feet.

▶ CHECK **Warm-Up 6**

In the second method of evaluating a variable in a formula, we solve for the unknown variable first. Then we substitute the given numerical values in the resulting equation. Both methods yield the same result. Choose whichever works best for you.

• • •
EXAMPLE 7

Find Q in the total-cost formula

$$T = F + QV$$

when T is \$78, F is \$60, and V is 4 cubic feet. In this formula, T is the total cost, F is the wholesale price, Q is the storage cost per cubic foot, and V is the stored volume.

SOLUTION

Our immediate goal is to solve for Q on the right side of the equation.

$$T = F + QV \qquad \text{Original equation}$$
$$T + (-F) = F + QV + (-F) \quad \text{Adding } -F$$
$$T - F = QV \qquad \text{Simplifying}$$
$$\left(\tfrac{1}{V}\right)(T - F) = \left(\tfrac{1}{V}\right)(QV) \qquad \text{Multiplying by } \tfrac{1}{V}$$
$$\left(\tfrac{1}{V}\right)(T - F) = Q \qquad \text{Simplifying}$$

We write this as

$$Q = \tfrac{1}{V}(T - F)$$

Now we substitute the given numerical values and simplify.

$$Q = \left(\tfrac{1}{V}\right)(T - F)$$
$$= \left(\tfrac{1}{4}\right)(78 - 60) \quad \text{Substituting}$$
$$= \left(\tfrac{1}{4}\right)(18) \qquad \text{Simplifying}$$
$$= 4.5$$

So Q is \$4.50 when $T = \$78$, $F = \$60$, and $V = 4$ cubic feet.

You should check such a calculation by substituting all known values in the original formula and simplifying. If the result is a true statement, the solution is correct.

▶ *CHECK* **Warm-Up 7**

Practice what you learned.

SECTION FOLLOW-UP

Using the popcorn data, ask and answer a question that can be solved by the following:

a. addition

b. subtraction

c. multiplication

d. division

c. any two operations

f. any three operations

2.3 WARM-UPS

Work these problems before you attempt the exercises.

1. Write in algebraic notation the sum of three times a number and sixty-two. $3n + 62$

2. Write in algebraic notation the value in cents of the coins in a stack of dimes.
   number of dimes in a stack = d; $10d$ cents

3. Write $27(c + 9)$ in words.
   Answers may vary. The sum of c and 9, multiplied by twenty-seven.

4. Solve for r: $5r - 8n = 2r + 6n$
   $r = \frac{14n}{3}$

5. Solve for y: $\frac{4y}{5} = x - 2$ $y = \frac{5}{4}x - \frac{5}{2}$

6. The number of trees T destroyed to print B copies of an average-sized book is given by the formula $T = \frac{B}{147}$. How many trees would be saved if 40,000 copies of a book were printed on recycled paper?   approximately 272 trees

7. The formula for the distance s that a baseball travels, in feet, when the ball is hit n miles per hour is

$$\frac{s}{5.195} = n$$

Find the speed at which you must hit the ball for it to travel 400 feet for a home run.   77 miles per hour

2.3 EXERCISES

Note: Use your graphing calculator to check your results whenever possible.

In Exercises 1 through 36, write each verbal expression as an algebraic expression. Define variables as needed.

1. four less than a number $n - 4$

2. x decreased by 8 $x - 8$

3. the square of t t^2

4. 5 fewer than m $m - 5$

5. the product of five hundred twenty-eight and a number $528 \cdot n$

6. the quotient of fifty-two and a number $52 \div n$

7. the sum of eighty-five hundredths and a number $0.85 + n$

8. the difference of a number and eleven and five-tenths $n - 11.5$

9. \$1.94 less than the price of butter
   price of butter = B; $B - \$1.94$

10. eight days after David's birthday
    David's birthday = D; $D + 8$

11. one and one-fourth times as expensive as leather
    cost of leather = C; $1\frac{1}{4}C$

12. three and three-quarter pounds more than her sister's birth weight   sister's birth weight = w;
    $w + 3\frac{3}{4}$

13. one monthly installment on an annual insurance bill   annual insurance bill = I; $\frac{1}{12} \cdot I$

14. the annual rent budgeted in weekly payments   annual rent = R; $\frac{1}{52} \cdot R$

15. the cost of 15 items at n cents each   cost of one item = n; $15n$

16. half the sum of r and 1.5   sum of r and 1.5 = $r + 1.5$; $\frac{1}{2}(r + 1.5)$

17. twice the difference of 8 and x difference of 8 and x = $8 - x$; $2(8 - x)$

18. the number of cents in n dimes and t nickels   number of dimes = n; number of nickels = t; $10n + 5t$

19. the number of coins in n dimes and y nickels   number of dimes = n, number of nickels = y; $n + y$

20. the sum of Maxwell's age and Anne's age if Maxwell is half Anne's age   Anne's age = A, Maxwell's age = $\frac{1}{2}A$; $A + \frac{1}{2}A$

21. twice Catherine's age three years from now   Catherine's age now = C; $2(C + 3)$

22. the difference of one-quarter of a number and twelve   a number = n; $\frac{1}{4}n - 12$

23. the sum of the product of 8 and x, and 6   product of 8 and x = $8x$; $8x + 6$

24. the difference of the product of 4 and y, and 9   product of 4 and y = $4y$; $4y - 9$

25. the sum of a number and the number decreased by one-fifth   a number = n; $n + \left(n - \frac{1}{5}\right)$

26. the sum of Marie's age and Jason's age if Marie is 2 years older than Jason.   Jason's age = J, Marie's age = $J + 2$; $J + (J + 2)$

27. the square of a number plus the number increased by two   a number = n; $n^2 + (n + 2)$

28. the cube of the product of a number and 7   a number = n; $(7n)^3$

29. the quotient of the square of 4 times a number, and 23   a number = n; $(4n)^2 \div 23$

30. one-fourth the product of a number increased by five and that number   a number = n; $\frac{1}{4}(n + 5)n$

31. the difference between a number and six more than that number   a number = n; $n - (n + 6)$

32. one-third the product of a number increased by four and that number   a number = n; $\frac{1}{3}(n + 4)n$

33. the sum of four-sevenths less than a number, and that number   a number = n; $\left(n - \frac{4}{7}\right) + n$

34. the quotient of a number and that number increased by 4   a number = n; $n \div (n + 4)$

35. the difference between a number and three-tenths more than that number   a number = n; $n - (n + 0.3)$

36. the product of a number and fifty-two thousandths more than that number   a number = n; $n(n + 0.052)$

In Exercises 37 through 56, write each algebraic phrase in words.

37. $15 - y$ the difference of 15 and y, or y less than 15

38. $\left(2\frac{1}{2}\right)y$ the product of $2\frac{1}{2}$ and y

39. $\frac{19.6}{x}$ the quotient of 19.6 and x

40. $y + 3\frac{1}{2}$ the sum of y and $3\frac{1}{2}$, or $3\frac{1}{2}$ more than y

41. $3.4k + 6$ the sum of the product of 3.4 and k, and 6

42. $1.7x - 9x$ the product of 1.7 and x minus the product of 9 and x.

43. $6 + x^2$ the sum of 6 and the square of x

44. $(20y)(-11y)$ the product of 20 and y times the product of -11 and y

45. $2.5(y - 3)$ the product of 2.5 and the difference of y and 3

46. $4\frac{1}{2}(k + 5)$ the product of $4\frac{1}{2}$ and the sum of k and 5

47. $(5.2x) \div (18x)$ the product of 5.2 and x divided by the product of 18 and x

48. $(6x + 26)^2$ the square of the sum of the product of 6 and x and 26

49. $(n + 6)c$ the sum of n and 6 times c

50. $(6 + n) \div 3$ the sum of 6 and n divided by 3

51. $(n \div 7) + 2$ the quotient of n and 7 plus 2

52. $(18 \div n) - 5$ the quotient of 18 and n minus 5

53. $6n \div c$ the product of 6 and n divided by c

54. $nt - b$ the product of n and t minus b

55. $(n - 10) + dc$ the difference of n and 10 plus the product of d and c

56. $(n - b)(c + 5)$ the difference of n and b times the sum of c and 5

In Exercises 57 through 72, use the given variable and represent the English expression in symbols.

57. When *r* rooms are rented at $155 per room, the amount collected is
_____ 155*r* dollars _____.

58. When 180 rooms are rented at *s* per room, the amount collected is
_____ 180*s* dollars _____.

59. You had *d* dollars and you spent $4.15. How much do you have left?
_____ *d* − 4.15 dollars _____

60. When you had a $20 bill and you spent *d* dollars, the amount you received in change was _____ 20 − *d* dollars _____.

61. The cost of *n* bagels at 50 cents and 6 coffees at *p* cents is
_____ 50*n* + 6*p* cents _____.

62. The cost of 16 books at *p* dollars each and *c* pencils at 25 cents each is
_____ 16*p* + 0.25*c* dollars _____.

63. Each of *n* people have *h* hamburgers. How many hamburgers are there?
_____ *nh* hamburgers _____

64. My age is *a* years. Last year my age was _____ *a* − 1 years _____.

65. I am mailing *c* cards with one stamp on each. How many stamps do I need? _____ *c* stamps _____

66. When there is 6% tax, the cost of *n* items at $1 each is *n* +
_____ 0.06 _____ *n* dollars.

67. What is the number of people when *n* rooms have *p* people in each room?
_____ *np* people _____

68. What is the total number of people on a bus that is half full if the bus holds *b* people? _____ *b*/2 people _____

69. When *n* tickets are sold at $6 each and 80 tickets are sold at *d* dollars each, the total collected is _____ 6*n* + 80*d* dollars _____.

70. When *x* people go to Atlanta by train and *m* people travel by air, how many more travel by air than by train? _____ *m* − *x* people _____

m passengers

71. I am *n* years old and my daughter is *d* years younger. Our total age in two years is (*n* + 2) + [(*n* − *d*) + 2] years.

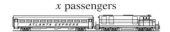

x passengers

72. I earn *d* dollars and my sister earns *s* dollars. How much less than I does she earn? _____ *d* − *s* dollars _____

In Exercises 73 through 92, solve for the indicated variable.

73. Solve for f: $y + f = e$ $f = e - y$

74. Solve for t: $x - t = m$ $t = -m + x$

75. Solve for w: $j - w = r$ $w = -r + j$

76. Solve for y: $y - t = f$ $y = f + t$

77. Solve for w: $\frac{w}{j} = r$ $w = jr$

78. Solve for f: $\frac{f}{y} = e$ $f = ye$

79. Solve for x: $\frac{x}{h} = \frac{e}{t}$ $x = \frac{he}{t}$

80. Solve for j: $\frac{j}{w} = \frac{r}{e}$ $j = \frac{wr}{e}$

81. Solve for j: $4j + w = r$ $j = \frac{1}{4}(r - w)$, or $\frac{r}{4} - \frac{w}{4}$

82. Solve for j: $7 + 9j = 10c + j$ $j = \frac{10c}{8} - \frac{7}{8}$

83. Solve for m: $m + \left(\frac{x}{y}\right) = r$ $m = r - \frac{x}{y}$

84. Solve for t: $t - \left(\frac{y}{x}\right) = h$ $t = h + \frac{y}{x}$

85. Solve for x: $-t - \left(\frac{x}{z}\right) = y$ $x = -yz - tz$

86. Solve for y: $-t + \left(\frac{y}{z}\right) = r$ $y = rz + tz$

87. Solve for m: $4y - \left(\frac{m}{x}\right) = r$ $m = -xr + 4xy$

88. Solve for y: $2y - \left(\frac{t}{x}\right) = y$ $y = \frac{t}{x}$

89. Solve for x: $\frac{5x}{4} = y + 3$ $x = \frac{4}{5}y + \frac{12}{5}$

90. Solve for t: $\frac{8t}{12} = w - 10$ $t = \frac{3}{2}w - 15$

91. Solve for y: $6y(t + x) = z$ $y = \frac{z}{6(t + x)}$

92. Solve for t: $3t(x - m) = r$ $t = \frac{r}{3(x - m)}$

In Exercises 93 through 109, find the information requested.

93. *Business Interest*   Given that $I = PRT$, where I is the interest, P is the amount of money deposited, R is the percent, and T is the time in years, find P in terms of I, R, and T. $P = \frac{I}{RT}$

94. *Automobile Weight*   The weight of a car W is related to the tire pressure in pounds per square inch. The area of each tire's contact with the ground A is given by the formula

$$\frac{W}{4A} = P$$

   a. Solve for A in terms of W and P. $A = \frac{W}{4P}$

   b. Find the value of A when P is 28 pounds per square inch and W is 2708 pounds. $A \approx 24 \text{ in.}^2$

95. *Shoe Size and Foot Length*   Suppose shoe size and foot length in inches are related by the formula

$$S = 3.1F - 23.9$$

   a. Solve for F in terms of S.   b.   Find F when S is size 9.   *about 10.6 inches*
   $F = \frac{1}{3.1}(S + 23.9)$

96. *Business Markup*   Given that $S = C + Cm$, where S is the selling price, C is the cost to the retailer, and m is the mark-up rate, find m in terms of S and C. $m = \frac{S - C}{C}$

97. *Geometry*   Given that $P = 2L + 2W$ where P is the perimeter of a rectangle, W is the width, and L is the length:

   a. Find W in terms of P and L.   b.   Find L in terms of P and W.
   $W = \frac{P}{2} - L$ $L = \frac{P}{2} - W$

98. **Geometry** Given that $A = 0.5h(b_1 + b_2)$ where A is the area of a trapezoid, h is the height, and b_1 and b_2 are the bases:

    a. Find b_2 in terms of A, h, and b_1.
       $$b_2 = \frac{A}{0.5h} - b_1$$

    b. Find h in terms of A, b_1, and b_2.
       $$h = \frac{A}{0.5b_1 + 0.5b_2}$$

99. **Wildlife Conservation** In wildlife conservation efforts, a number of animals M are often trapped temporarily, marked in some way, and returned to their habitats. At some later time, after the marked animals have mixed in with the others, a random sample S is trapped and the number m of marked animals is counted and used to estimate the size of the population P. It is assumed that the proportion of marked animals in the entire population will be the same as the proportion of marked animals in the sample. The formula is

    $$\frac{M}{P} = \frac{m}{S}$$

    Solve for P in terms of M, m, and S. $P = \frac{MS}{m}$

100. **Geometry** The formula for the point-slope form of the equation of a line is

    $$y - y_1 = m(x - x_1)$$

    Solve for x in terms of the other variables. $x = \frac{y - y_1}{m} + x_1$

101. **Mathematics Formula** In long division, a dividend D divided by a divisor V gives a quotient Q plus a remainder r. We write this as

    $$\frac{D}{V} = Q + \frac{r}{V}$$

    Solve for V in terms of the other variables. $V = \frac{D - r}{Q}$

102. **Deflection of a Beam** In engineering mechanics, the deflection d of a beam of length l is

    $$d^2 = \frac{l^2 \cdot a}{2l + a}$$

    where a is a constant determined by the material the beam is made of. Solve for a in terms of the other variables. $a = \frac{2d^2 l}{l^2 - d^2}$

103. **Blimps** Given $L = V(D_a - D_h)$, where L is the lifting capacity (in pounds) of a blimp, V is the volume of the blimp, D_a is the density of air, and D_h is the density of helium. Solve for D_a in terms of the other variables.
    $$D_a = \frac{L}{V} + D_h$$

104. **Building a Highway** A formula for determining how much expansion is necessary to allow for in building a highway is

    $$I = kl(T - t)$$

    where I is the expansion of a highway of length l at temperature T when it was constructed at temperature t. k is the expansion coefficient.

    a. Solve the formula for t in terms of the other variables. $t = T - \frac{I}{kl}$

    b. Find the construction temperature if a two-mile stretch of highway expands 4.8 feet on a hot day (95°F). (Use 1 mile = 5280 feet.) $k = 1.2 \times 10^{-5}$.   about 57°F

105. *Deposit Time*   Given $T_1 = PRT + P$, where T_1 is the total amount of money in the bank, P is the principal (the originally deposited money), R is the rate expressed in decimal form, and T is the time in years, find the amount of time for which $1500 was deposited in a bank at an interest rate of 0.05 if the total amount in the bank is now $1725. $T = 3$ years

106. *Sound*   If v is the velocity of sound in air and g is the acceleration of gravity, then the time t required for a person to hear the sound of a dropped object hitting a surface d units below is

$$t = \frac{d}{v} + \sqrt{\frac{2d}{g}}$$

Solve for v in terms of the other variables. $v = \dfrac{d}{t - \sqrt{\frac{2d}{g}}}$

107. *Pendulum*   Given that T is the time in seconds for one complete swing of a pendulum x feet long, then

$$\left(\frac{T}{2\pi}\right)^2 = \frac{x}{32}$$

Find the length of a pendulum that takes 2 seconds to complete one swing. (Use 3.14 as an approximation for π).   about 3.25 feet

108. *Depreciation*   A formula for depreciating certain business property is

$$V = C\left(1 - \frac{n}{N}\right)$$

where V is its value at the end of n years, C is its original cost, and the depreciation is taken over N years. Solve this formula for N in terms of V, C, and n. $N = \dfrac{nC}{C - V}$

109. *Cross-Training*   Runners who cross-train on treadmills can convert their pace on an inclined treadmill to track speed with the formula

$$m(1 - 3r) = t$$

where m is the time per mile on an inclined treadmill, r is the percent treadmill grade or incline (written as a decimal), and t is the time per mile on the track. A runner who runs a 10-minute mile on a treadmill inclined $5\frac{1}{2}\%$ will average what speed on the track?   about 8 minutes and 21 seconds, or 8.35 minutes, per mile

MIXED PRACTICE

By doing these exercises, you will practice the topics up to this point in the chapter.

110.   Solve: $-9|18 + x| = -90$ $x = -8; x = -28$

111.   Simplify: $6x - 9x + 56y + 16x - 35y$
$13x + 21y$

112.   Solve for c: $ce - 4 = f$ $c = \dfrac{f + 4}{e}$

113.   Evaluate $-18y + 4x^3 - 12x$ when $x = -1$ and $y = -3$.   62

114.   Solve: $\frac{w}{3} + 276 = 376$ $w = 300$

115.   Find the solution set for $|-x + 36| = -184$.
no solution, or $\varnothing$

116.   Simplify: $12r - 42r - (11r + 14r) + r(-12)$ $-67r$ **117.**   Solve for c: $\frac{c}{h} - w = y$ $c = h(y + w)$

118. Solve: $5|x + 5| = 180$ $x = 31; x = -41$

119. Solve: $2x + 9 = 7x - 3$ $x = 2\frac{2}{5}$

120. Solve for x: $5x + y = -3x + v$ $x = \frac{v - y}{8}$

121. Solve for w: $P = 2(l + w)$ $w = \frac{P}{2} - l$

EXCURSIONS

Class Act

1. **a.** The tallest skyscrapers in the United States are: in Chicago—the Sears tower (1454 feet tall, 110 stories), the Standard Oil Building (1136 feet tall, 80 stories), and the John Hancock Center (1127 feet tall, 100 stories); and in New York—The World Trade Center (1350 feet tall, 110 stories) and the Empire State Building (1250 feet tall, 102 stories).

   Ask four questions that can be answered using the skyscraper data. Answer those questions.

   b. Find some data that you can use to make up some word problems. Swap your word problems with another group.

2. Each person in your group should toss a coin 20 times and keep a record of heads and tails. Put all your data together and answer the following questions.

   a. What is the group total of heads and of tails?

   b. With a fair coin, the probability of a head is $\frac{1}{2}$ and the probability of a tail is $\frac{1}{2}$. Discuss your results. Is your coin fair?

   c. Would it be possible for someone tossing a fair coin to get 10 heads in a row? Justify your answer.

   d. What do you think is meant by a "fair" coin?

CONNECTIONS TO *STATISTICS*

Reading a Table

You may sometimes have to refer to a complex table or chart to obtain information you need. Table 1 is such a table, and it can be confusing. The color notations have been added to help you interpret the information in the table, above and below it, and at the left. Note, especially, the *"total"* and *"average"* rows, which can get in the way of reading the table.

1　TABLE 1

2　U.S. Travel to Foreign Countries—Travelers and Expenditures: 1975 to 1985

3　Travelers in thousands; expenditures in millions of dollars, except as indicated. Covers residents of United States and Puerto Rico.

Item and Area	1975	1979	1980	1983	1984	1985	
Total overseas travelers	**6,354**	**7,835**	**8,163**	**9,628**	**11,252**	**12,309**	4
Region of destination:							
Europe and Mediterranean	3,185	4,068	3,934	4,780	5,760	6,457	7
Caribbean and Central America	2,065	2,533	2,624	2,989	3,313	3,497	
South America	447	434	594	535	557	553	
Other	657	800	1,011	1,324	1,622	1,802	
Total Expenditures abroad	**6,417**	**9,413**	**10,397**	**13,556**	**15,449**	**16,482**	
Canada	1,306	1,599	1,817	2,160	2,416	2,694	
Mexico	1,637	2,460	2,564	3,618	3,599	3,531	
Total overseas areas	3,474	5,354	6,016	7,778	9,434	10,257	
Europe and Mediterranean	1,918	3,185	3,412	4,201	5,171	5,857	8
Average per trip (dollars)*	602	783	867	882	897	(NA)	
Caribbean and Central America	787	1,019	1,134	1,428	1,786	1,830	
South America	242	288	392	408	357	365	
Japan	131	142	185	276	400	458	
Other	396	720	893	1,465	1,720	1,747	

9　NA　Not available in dollars.
10　Adapted from *Statistical Abstracts of the United States, 1988.*

1　table number
2　title: describes the information in the table
3　units
4　years for which data was compiled
5　places visited
6　places visited
7　numbers of travelers (in thousands)
8　travel costs (in millions of dollars); *Average per trip is in dollars
9　explanatory note
10　source of data

▪▪▪

EXAMPLE

a. Using Table 1, determine the total amount spent by people traveling from the United States to Canada during 1983.

b. How many people traveled to South America from the United States in 1979?

SOLUTION

a. Look across the *row* that starts "Canada" to where it *intersects* the *column* headed "1983." The number in that intersection is 2160. This is the number of millions of dollars spent, so the answer is 2160 million dollars, or $2,160,000,000.

b. In the section that refers to travelers, look across the row that starts "South America" until you are under the column headed "1979." The number there is 434, so the answer is 434 thousand (or 434,000) people.

◢

Travelers from the United States who visit other countries may need to determine the money exchange rates to budget for their trips. The accompanying table shows various exchange rates—that is, how much $1 was worth in foreign currencies—for late September 1989.

▪▪▪

EXAMPLE

A Mexican hotel advertises rooms at 225,000 pesos per night. How much is this in U.S. dollars?

SOLUTION

Using Table 2 on page 121, we first find the exchange rate for Mexico: 2387 pesos per $1. We then set up a proportion with the information we have and solve for the unknown value.

$$\frac{2387 \text{ pesos}}{1 \text{ dollar}} = \frac{225{,}000 \text{ pesos}}{n \text{ dollar}}$$

$$\frac{2387}{1} = \frac{225{,}000}{n} \qquad \text{Dropping units}$$

$$2387 \times n = 225{,}000 \times 1 \qquad \text{Cross-multiplying}$$

$$\frac{2387 \times n}{2387} = \frac{225{,}000}{2387} \qquad \text{Dividing by 2387}$$

$$n = 94.26 \qquad \text{Simplifying and rounding}$$

So 225,000 pesos is $94.26.

◢

> ▪▪▪
>
> **WRITER'S BLOCK**
>
> *Research* Find up-to-date exchange rates and compare them to those shown in Table 2. Which have gone up? Which have gone down? What reasons can you give for these changes?

PRACTICE

Use Table 1 to answer Questions 1 through 5.

1. How much more was spent for U.S. travel to Japan in 1985 than was spent 10 years before that? $327,000,000

2. How many people traveled from the United States to the Caribbean and Central America in 1985?   3,497,000 people

3. What were the expenditures by travelers in Japan during 1984?
   $400,000,000

TABLE 2 Foreign Exchange Rates

Country	Foreign exchange for $1	Country	Foreign exchange for $1	Country	Foreign exchange for $1
Argentina (austral)	550.96	Holland (guilder)	2.00	Singapore (dollar)	1.82
Australia (dollar)	1.20	Hong Kong (dollar)	7.37	South Africa (rand)	2.53
Austria (schilling)	12.50	India (rupee)	14.97	South Korea (won)	602.05
Belgium (franc)	37.17	Indonesia (rupiah)	1,515	Spain (peseta)	109.91
Brazil (cruzado)	4.05	Ireland (pound)	0.67	Sweden (kroner)	6.07
Britain (pound)	0.59	Israel (shekel)	1.65	Switzerland (franc)	1.55
Canada (dollar)	1.13	Italy (lira)	1,287	Tahiti (franc)	111.37
Chile (peso)	259.81	Japan (yen)	133.98	Taiwan (dollar)	23.65
Colombia (peso)	359.32	Jordan (dinar)	0.60	Thailand (baht)	23.65
Denmark (krone)	6.89	Mexico (peso)	2,387	Turkey (lira)	1,634
Ecuador (sucre)	328.08	New Zealand (dollar)	1.58	Venezuela (bolivar)	33.06
Egypt (pound)	2.14	Norway (kroner)	6.52	W. Germany (mark)	1.78
Finland (mark)	4.03	Philippines (peso)	18.74	Yugoslavia (dinar)	18519
France (franc)	6.09	Portugal (escudo)	145.77		
Greece (drachma)	145.65	Saudi Arabia (riyal)	2.39		

4. What was the average amount spent per trip in Europe and the Mediterranean during 1984? $897

5. An average vacation to Europe and the Mediterranean would have cost how much more in 1983 than in 1975? $280

Use Table 2 to answer Questions 6 through 10.

6. In which countries was the dollar worth less than one of the foreign currency units? Britain, Ireland, and Jordan

7. A tourist brought back 9000 Japanese yen from a trip. How much is this in U.S. currency? $67.17

8. A visitor to London, England, bought jewelry for 250 pounds. How much did she pay in U.S. currency? $423.73

9. A trip to Norway resulted in the purchase of an embroidered blouse for 260.8 kroner. That same blouse was available in New York for $92. Was it more or less expensive in New York? more expensive

10. Money that we had invested in Canada was exchanged for 118 U.S. dollars. About how many Canadian dollars is this? $133.34

2.4 Solving Linear Inequalities

SECTION LEAD-IN

Let $S = NV + X$, where

    S is the score on an exam
    N is the number of questions correct
    V is the value for each question
    X is the number of extra credit points

Choose numbers to fill in the blanks in the following statements to give four different problems. Use the given formula to solve each problem.

a. Find the score on an exam when 27 questions are answered correctly at _____ points each, and an additional 5 points are given for extra credit.

b. Find the number of extra credit points when there are 30 questions, each question counts _____ points, there are _____ questions correct, and the total score is 178.

c. Find the value for each question when 10 points are given for extra credit, 98 is the total score, and there are _____ questions on the exam.

d. Find the number of questions on the exam when there were _____ points counted per question, the total score was _____, and there were _____ points given for extra credit.

SECTION GOALS

- To solve and graph linear inequalities in one variable

- To solve and graph compound inequalities in one variable

- To solve absolute value inequalities

WRITER'S BLOCK

Explain the procedures you used to choose numbers for the problems in the Section Lead-In.

> **Inequalities**
>
> An **inequality** is a mathematical statement in which two expressions are related by one of the following symbols: $<$ ("is less than"), $>$ ("is greater than"), $\leq$ ("is less than or equal to"), or $\geq$ ("is greater than or equal to").

An inequality that contains only constants and a single variable with an exponent of 1 is a **linear inequality** in one variable.

A **solution** of an inequality is a number that makes the inequality true when it replaces the variable. Most inequalities have many solutions.

We solve linear inequalities much as we solve linear equations. We apply two principles that enable us to rewrite inequalities in the very simple form $x < $ a number or $x > $ a number. (In what follows, we can always substitute the symbol $\leq$ for $<$, and the symbol $\geq$ for $>$.)

The Addition Principle of Inequality

Addition Principle of Inequality

For real numbers a, b, and c,

$$\text{if } a < b, \text{ then } a + c < b + c$$
$$\text{if } a > b, \text{ then } a + c > b + c$$

In words, if we add the same real number to both sides of an inequality, the solution set of the inequality remains unchanged. We use this principle to isolate the variable term on one side of the inequality.

▪▪▪

EXAMPLE 1

Find the solution set for the inequality $x + 9 > 21$. Check your solution.

SOLUTION

To isolate x, we add -9 to both sides of the inequality and simplify.

$x + 9 > 21$	Original equation
$x + 9 + (-9) > 21 + (-9)$	Adding -9 to both sides
$x > 12$	Simplifying

The solution set we obtain is {all real numbers greater than 12}.

To check the solution, we substitute 13, which is in the proposed solution set, and 11, which is not. We get

$x + 9 > 21$		$x + 9 > 21$	
$13 + 9 > 21$		$11 + 9 > 21$	
$22 > 21$	True	$20 > 21$	False

Thus we have verified the solution set for $x + 9 > 21$, which we can write as $\{x \mid x > 12\}$. We read it as "the set of all real numbers x such that x is greater than 12.

▶ *CHECK* **Warm-Up 1**

STUDY HINT

To check a solution to an inequality, we substitute, into the original inequality, a number that is in the proposed solution set; it must give a true statement when the inequality is then simplified. Then we substitute a number that is not in the solution set; it must give a false *statement.*

Calculator Corner

The inequality $-3 + x + 5 < 7$ can be worked on your graphing calculator in two ways. One way would be to have the calculator give the solution on a number line. You might need to consult your graphing calculator manual to learn how to insert inequality signs.

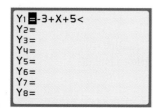

Press **2nd MATH.**

Press **ENTER.**

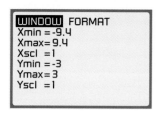

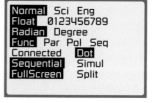

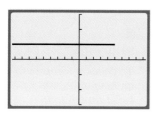

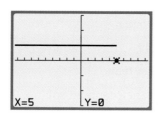

You may want to change the calculator to **DOT MODE.**

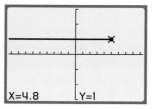

Press **GRAPH.**   Use **TRACE.**

For all x-values shown, when $y = 1$, the statement is true; when $y = 0$, the statement is false. The calculator uses the horizontal line $y = 1$ to show that for all real numbers less than 5 the inequality is true. However, for all real numbers greater than or equal to 5 the inequality is false.

A second way is to examine the inequality as two separate equations.

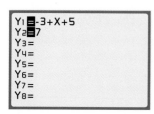

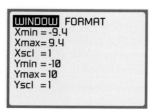

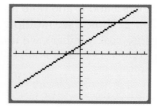

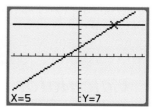

The graph now shows where the line $y = -3 + x + 5$ (the slanted line) is less than (*in this case below*) the line $y = 7$.

The Multiplication Principle of Inequality

As you might have guessed, there is also a multiplication principle for inequalities, and we use it to remove coefficients from unknown terms. It has two separate parts.

Multiplication Principle of Inequality, Positive Multipliers

For all real numbers a, b, and c, where $c > 0$,

$$\text{if } a < b, \text{ then } ac < bc$$
$$\text{if } a > b, \text{ then } ac > bc$$

In words, if we multiply both sides of an inequality by the same *positive* real number, the solution set of the inequality is not changed. We use this principle to give the variable the coefficient 1.

▪ ▪ ▪

EXAMPLE 2

Find the solution set for the inequality $2.4y \geq -6$.

SOLUTION

Using the multiplication principle, we multiply both sides of the inequality by $\frac{1}{2.4}$.

$$2.4y \geq -6 \qquad \text{Original equation}$$
$$\left(\frac{1}{2.4}\right)2.4y \geq \left(\frac{1}{2.4}\right)(-6) \qquad \text{Multiplying by } \frac{1}{2.4}$$
$$y \geq -2.5 \qquad \text{Simplifying}$$

So the solution set of $2.4y \geq -6$ is $\{y \mid y \geq -2.5\}$. The check is done as before. We leave this for you to do.

▶ *CHECK* **Warm-Up 2**

Now consider two real numbers a and b, and note that if $a > b$, then $-a < -b$. For example, $7 > -2$, so $-7 < 2$. Thus, if we multiply both sides of an inequality by -1 (or any other *negative* real number), we must reverse the inequality sign. Otherwise, the resulting inequality will not be true.

Multiplication Principle of Inequality, Negative Multipliers

For all real numbers a, b, and c, where $c < 0$,

$$\text{if } a < b, \text{ then } ac > bc$$
$$\text{if } a > b, \text{ then } ac < bc$$

▪▪▪
EXAMPLE 3

Find the solution set for the inequality $-\frac{5}{6}t \geq 25$.

SOLUTION

To obtain a coefficient of 1 for t, we multiply both sides of the inequality by $-\frac{6}{5}$. Because $-\frac{6}{5}$ is negative, we reverse the inequality as well.

$$\left(-\frac{5}{6}\right)t \geq 25 \qquad \text{Original equation}$$

$$\left(-\frac{6}{5}\right)\left(-\frac{5}{6}\right)t \leq \left(-\frac{6}{5}\right)25 \qquad \text{Multiplying by } -\frac{6}{5} \text{ (and so reversing the inequality)}$$

Then, we simplify.

$$t \leq -30$$

Check the solution set $\{t \mid t \leq -30\}$ with a real number less than -30 and one greater than -30.

▶ CHECK **Warm-Up 3**

Calculator Corner

Solve the inequality $-\frac{3}{4}x \geq 15$. One way to solve the inequality graphically would be to graph each side of the inequality separately.

 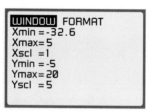 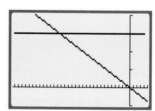 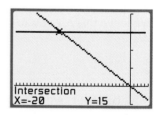

Remember that the inequality is asking, "Where is the line $y = -\frac{3}{4}x$ greater than or equal to the horizontal line $y = 15$." The fourth screen shows that the point of intersection for the two lines is at $x = -20$. So the answer is all real numbers less than or equal to -20.

Using Both Principles

There are times when we must apply both the addition principle and the multiplication principle to solve an inequality. We always apply the addition principle first, to isolate the variable term. Then we use the multiplication principle to give the variable the coefficient of 1. However, before doing any of this, we must simplify each side of the inequality, if possible.

To solve a linear inequality in one variable

1. Use the properties of real numbers to simplify both sides of the inequality.
2. Use the addition principle to isolate the term that contains the variable.
3. Use the multiplication principle to give the variable a coefficient of 1.

▪▪▪

EXAMPLE 4

Find the solution set for $-8.3 - 0.5r < 3.6 - 8.5r$.

SOLUTION

To isolate the terms that contain r on one side of the inequality, we must use the addition principle twice.

$$-8.3 - 0.5r + 8.3 < 3.6 - 8.5r + 8.3 \qquad \text{Adding } 8.3$$
$$-0.5r < 11.9 - 8.5r \qquad \text{Simplifying}$$
$$-0.5r + 8.5r < 11.9 - 8.5r + 8.5r \qquad \text{Adding } 8.5r$$
$$8r < 11.9 \qquad \text{Simplifying}$$

Then we apply the multiplication principle to give the variable a coefficient of 1.

$$\left(\tfrac{1}{8}\right)(8r) < \left(\tfrac{1}{8}\right)(11.9) \qquad \text{Multiplying by } \left(\tfrac{1}{8}\right)$$
$$r < 1.4875 \qquad \text{Simplifying}$$

You should check that the solution set is $\{r \mid r < 1.4875\}$.

▶ CHECK **Warm-Up 4**

▪▪▪
WRITER'S BLOCK

How are equations and inequalities alike? Compare
$3 + x > 5$ and
$3 + x = 5$.

▪▪▪
WRITER'S BLOCK

How do equations and inequalities differ? Contrast
$3x > 15$ and $3x = 15$.

Calculator Corner

Consider graphing an inequality involving two lines. This type of inequality can also be solved easily using your graphing calculator. Put your calculator in DOT mode. Graph $-14x - 3(x + 9) > 7 + 2x$.

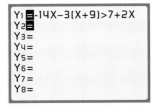

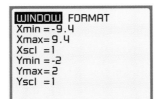

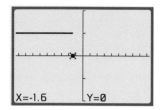 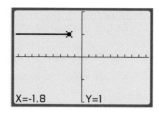

By using **TRACE** the calculator indicates that the inequality is *true* for x's less than -1.8. If you converted the algebraic answer $x < \frac{-34}{19}$ to a decimal answer, the result would be -1.789473684, which would round off to $x < -1.8$.

The same result can also be found by graphing the two sides of the inequality separately. You are then asking where is $Y_1 > Y_2$. This would be the same as asking where Y_1 is above Y_2. The point of intersection of the two lines is -1.789474, which is -1.8 if rounded to one decimal position.

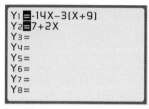 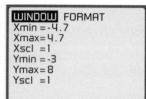 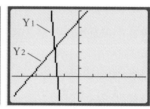 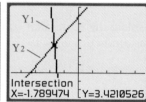

Graphing Solutions to Inequalities

We graph a real number on a number line by placing a filled-in circle at the point that represents that number. For example, the following figure shows the graph of the real number 12. The solution of a linear equation is a real number, so we can also graph such a solution on a number line. The figure also shows the graph of the solution to the equation $x + 9 = 21$, which is $\{12\}$.

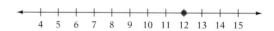

The solution of an inequality usually is an uncountable (or infinite) set of real numbers. For example, the solution set for $x + 9 > 21$ is $\{x \mid x > 12\}$. On the graph, we place an open circle at the point 12 to show that the solution does *not* include the real number 12; then we draw a thick arrow to the right along the number line to indicate that all real numbers greater than 12 are included.

To graph an inequality that contains $\geq$ or $\leq$, we fill in the circle to show that the point *is included*.

In Example 3, the solution set was $\{t \mid t \leq -30\}$. We first sketch a part of the number line that includes -30, as shown in the following figure; then we draw a filled-in circle at -30, because the inequality includes -30, and a thick arrow to the left of -30 to indicate the inclusion of all numbers less than 30.

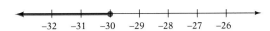

In Example 4, the solution set was $\{r \mid r < 1.4875\}$. We sketch the part of the number line from -2 through 3, as shown below. Then, estimating that 1.4875 is close to 1.5, we draw an open circle at 1.5, because that value is not included, and a thick arrow to the left to indicate the inclusion of all real numbers less than 1.4875.

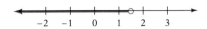

In the next example, we must simplify the inequality as a first step.

▪▪▪

EXAMPLE 5

Find the solution set for $7(-2n + 4) - 2n \geq 6n - 5$. Graph the results.

SOLUTION

First using the distributive property to remove the parentheses, we have

$$7(-2n + 4) - 2n \geq 6n - 5 \quad \text{Original equation}$$
$$-14n + 28 - 2n \geq 6n - 5 \quad \text{Distributive property}$$
$$-16n + 28 \geq 6n - 5 \quad \text{Combining like terms}$$

Now we can apply the addition principle (twice) and then the multiplication principle.

$$-16n + 28 + (-28) \geq 6n - 5 + (-28) \quad \text{Adding } -28$$
$$-16n \geq 6n - 33 \quad \text{Simplifying}$$
$$-16n + (-6n) \geq 6n - 33 + (-6n) \quad \text{Adding } -6n$$
$$-22n \geq -33 \quad \text{Simplifying}$$
$$\left(-\frac{1}{22}\right)(-22n) \leq \left(-\frac{1}{22}\right)(-33) \quad \text{Multiplying by } -\frac{1}{22} \text{ (and so reversing the inequality)}$$
$$n \leq 1\frac{1}{2} \quad \text{Simplifying}$$

Checking will show that the solution set is $\left\{n \mid n \leq 1\frac{1}{2}\right\}$. The graph is shown below.

ERROR ALERT

Identify the error and give a correct answer.

Solve: $-5x + 9 > 12$

Incorrect Solution:

$$-5x + 9 + (-9)$$
$$> 12 + (-9)$$
$$-5x > 3$$
$$\left(-\frac{1}{5}\right)(-5x) > 3\left(-\frac{1}{5}\right)$$
$$x > \frac{-3}{5}$$

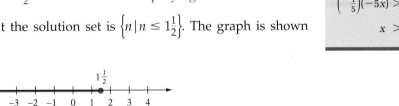

▶ *CHECK* **Warm-Up 5**

Compound Inequalities

A **compound inequality** may consist of either the intersection of two (or more) inequalities or the union of two (or more) inequalities. A good way to see the meaning of a compound inequality, and to solve it, is to

1. Solve the individual inequalities.

2. Graph the solutions.
3. Find and graph the intersection or union (whichever is required) of the individual solutions.

Intersection of Inequalities

Recall from Section 1.4 that ∩ and *and* are both used as mathematical symbols for "intersection," so the **intersection,** or **conjunction,** of two inequalities is written as

$$A < B \quad and \quad C > D$$

or as

$$A < B \quad \cap \quad C > D$$

Any of the four inequality signs ($<$, $>$, $\geq$, or $\leq$) may appear in such statements; A, B, C, and D here represent algebraic expressions or numbers.

Recall that the intersection of two sets is composed of all elements that are members of both sets. Thus the intersection of two inequalities is composed of all real numbers that satisfy both inequalities. These are the real numbers that appear in the solution sets of both inequalities.

▪▪▪

EXAMPLE 6

Solve the compound inequality $x + 4 > 7$ *and* $x - 5 < 2$.

SOLUTION

ERROR ALERT

Identify the error and give a correct answer.

Solve: $2x + 6 < 2(x + 5)$

Incorrect Solution:

$2x + 6 < 2x + 10$

$2x < 2x + 4$

$0 < 4$

No solutions because variable was eliminated.

We first solve each inequality separately.

$$
\begin{array}{c|c}
x + 4 > 7 & x - 5 < 2 \\
x + 4 + (-4) > 7 + (-4) & x - 5 + 5 < 2 + 5 \\
x > 3 & x < 7
\end{array}
$$

These solutions are graphed in figures (a) and (b). From the figures, we can see that their intersection consists of all real numbers that are both greater than 3 and less than 7, as sketched in figure (c). The numbers 3 and 7 are *not* elements of the intersection; these numbers do not appear in both solutions.

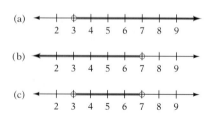

We then may write the solution to the original compound inequality as the intersection of the two solutions, either as

$$\{x \mid x > 3 \ and \ x < 7\}$$

or as

$$\{x \mid x > 3 \cap x < 7\}.$$

▶ *CHECK* **Warm-Up 6**

When a conjunction of inequalities is given in condensed form, it can be solved as a "three-expression inequality," as the next example shows.

▪▪▪
EXAMPLE 7

Solve and graph the compound inequality $-9 < 4 - 2x < 8$.

SOLUTION

As always, to solve this inequality, we must isolate the variable and transform its coefficient to 1. We proceed as follows:

$-9 < 4 - 2x < 8$	Original inequality
$-9 + (-4) < 4 - 2x + (-4) < 8 + (-4)$	Adding -4 to all members
$-13 < -2x < 4$	Simplifying
$\left(\frac{-1}{2}\right)(-13) > \left(\frac{-1}{2}\right)(-2x) > \left(\frac{-1}{2}\right)4$	Multiplying by $\frac{-1}{2}$ (and so reversing the inequalities)
$6.5 > x > -2$	Simplifying
$-2 < x < 6.5$	Rewriting

The solution is graphed in the figure below. You should check that the graph is the intersection of $x > -2$ and $x < 6.5$.

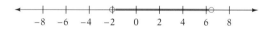

▶ *CHECK* **Warm-Up 7**

A compound inequality that is given in condensed form can be separated into its component inequalities before it is solved. Thus the inequality of Example 7 could have been written as

$$-9 < 4 - 2x \ and \ 4 - 2x < 8$$

and then solved by the method of Example 6.

Union of Inequalities

The **union,** or **disjunction,** of two inequalities is written as

$$A < B \ or \ C > D$$

or

$$A < B \cup C > D$$

where both *or* and $\cup$ are symbols for "union." Again, any of the four inequality symbols may appear in the inequalities.

▪▪▪

EXAMPLE 8

Solve and graph the compound inequality $x - 5 < -8 \cup 4x + 3 \geq 7$.

SOLUTION

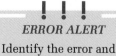
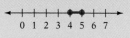
We solve each inequality separately.

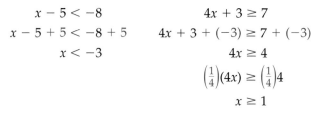

These solutions are graphed in figures (a) and (b). Their union, which consists of every point that is in at least one of these solutions, is shown in figure (c).

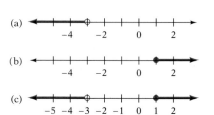

The union of the individual algebraic solutions is $\{x \mid x < -3 \text{ or } x \geq 1\}$. The inequalities cannot be written in condensed form.

▶ CHECK **Warm-Up 8**

Absolute Value Inequalities

Like equations, inequalities can contain absolute values. Absolute value inequalities can be translated into compound inequalities and then solved.

The "less than" inequality $|ax + b| < c$ means

$$-c < ax + b < c$$

Similarly, $|ax + b| \leq c$ means

$$-c \leq ax + b \leq c$$

To see this, first note that $|ax + b| < c$ means that the real numbers represented by $ax + b$ are within a distance of c units from zero on a number line. So the graph of this inequality looks like the following figure. The same graph is obtained by plotting the compound inequality $-c < ax + b \cap ax + b < c$.

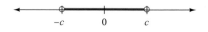

▪ ▪ ▪

EXAMPLE 9

Find and graph the solution set for $|3n - 4| \leq 5$.

SOLUTION

The equation can be translated into

$$-5 \leq 3n - 4 \leq 5$$

which we can solve in its condensed form.

$$-5 \leq 3n - 4 \leq 5$$
$$-5 + 4 \leq 3n - 4 + 4 \leq 5 + 4 \quad \text{Adding } 4$$
$$-1 \leq 3n \leq 9 \quad \text{Simplifying}$$
$$\left(\tfrac{1}{3}\right)(-1) \leq \left(\tfrac{1}{3}\right)(3n) \leq \left(\tfrac{1}{3}\right)(9) \quad \text{Multiplying by } \left(\tfrac{1}{3}\right)$$
$$-\tfrac{1}{3} \leq n \leq 3 \quad \text{Simplifying}$$

This is the solution of the original absolute value inequality. It is graphed below. It should be checked with a number less than $-\tfrac{1}{3}$, with a number between $-\tfrac{1}{3}$ and 3, and with a number greater than 3.

▶ *CHECK* **Warm-Up 9**

The "greater than" inequality $|ax + b| > c$ means
$$ax + b > c \quad \text{or} \quad ax + b < -c$$
Similarly, $|ax + b| \geq c$ means
$$ax + b \geq c \quad \text{or} \quad ax + b \leq -c$$

The inequality $|ax + b| > c$ means that on a number line, the numbers $ax + b$ are farther than c units from zero, as shown below. You should be able to see that the union of the inequalities given in the foregoing box means the same thing and has the same graph.

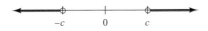

▪ ▪ ▪

EXAMPLE 10

Find and graph the solution set for $|-x + 3| > -2$.

SOLUTION

We translate this inequality into

$$-x + 3 > -2 \cup -x + 3 < 2$$

and solve the inequalities separately.

$$-x + 3 > -2 \qquad\qquad -x + 3 < 2$$
$$-x + 3 + (-3) > -2 + (-3) \qquad -x + 3 + (-3) < 2 + (-3)$$
$$-x > -5 \qquad\qquad -x < -1$$
$$x < 5 \qquad\qquad x > 1$$

The solution set is $\{x \mid x < 5 \cup x > 1\}$, or, more accurately, all real numbers. It is graphed in the following figure and requires three points for a check.

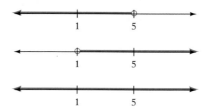

The solution of a problem of this type will always be all real numbers. The absolute value of any non-zero real number is always positive, and a positive number is always greater than a negative number.

▶ *CHECK* **Warm-Up 10**

Practice what you learned.

SECTION FOLLOW-UP

- Write a problem similar to the one presented in this section's Lead-In. The formula you use must differ from that in the Lead-In. Swap with a classmate.

2.4 WARM-UPS

Work these problems before you attempt the exercises.

Check: Answers may vary.
$y = -2 \qquad\qquad y = 0$
$0.3 \geq -0.7$ True $0.3 \geq 1.3$ False

1. Find the solution set for $0.3 \geq y + 1.3$ and then check your answer. $\{y \mid y \leq -1\}$

2. Find the solution set: $27 \leq 3r$ $\{r \mid r \geq 9\}$

3. Find the solution set: $\left(-\frac{7}{8}\right)r < 18$ $\left\{r \mid r > -20\frac{4}{7}\right\}$

4. Find the solution set: $-8.6r + 6 < -3.6r - 1.25$ $r > 1.45$

5. Find the solution set: $-5m - 8 > 4(m - 5) - 3m$ $\{m \mid m < 2\}$

6. Solve: $x + 9 < 12$ *and* $x - 9 < 7$ $\{x \mid x < 3\}$

7. Solve and graph: $10 \geq x - 4 > -15$ $\{x \mid -11 < x \leq 14\}$

8. Solve and graph the inequality: $x - 6 > 5 \cup x - 3 \geq 7$ $\{x \mid x \geq 10\}$

9. Find and graph the solution set: $|2n - 12| \leq 10$ $\{n \mid 1 \leq n \leq 11\}$

10. Find and graph the solution set: $|4x + 5| > 11$ $\left\{x \mid x < -4 \cup x > 1\frac{1}{2}\right\}$

2.4 EXERCISES

Note: Use your graphing calculator to check your results whenever possible.

In Exercises 1 through 8, determine whether the given real number is an element of the solution set for the given inequality.

1. $-3: m + 4 > 7$ no
2. $3: 6 + t \leq 9$ yes
3. $-4: -12 < y - 8$ no
4. $-11: 17 \geq -v + 5$ yes

5. $-1: -6m \geq 8$ no
6. $3: 45 > 15m$ no
7. $26: -125 < 5m$ yes
8. $-7: 15y \leq 120$ yes

In Exercises 9 through 16, find the solution set for the inequality.

9. $-24 + t > 9$ $\{t \mid t > 33\}$
10. $0.5m \geq 175$ $\{m \mid m \geq 350\}$
11. $1.5y - 3 \geq 6$ $\{y \mid y \geq 6\}$
12. $2.4 < 6x - 1.8$ $\{x \mid x > 0.7\}$

13. $7.7 > 1.1y - 4.4$ $\{y \mid y < 11\}$
14. $0.5v - 7.1 \leq 9.3$ $\{v \mid v \leq 32.8\}$
15. $-16 < 3.2t$ $\{t \mid t > -5\}$
16. $\left(\frac{4}{5}\right)x > 36$ $\{x \mid x > 45\}$

In Exercises 17 through 40, find and graph the solution set.

17. $\frac{1}{2} + m \leq \frac{5}{6}$ $\left\{m \mid m \leq \frac{1}{3}\right\}$

18. $7.8 \leq t - 9.3$ $\{t \mid t \geq 17.1\}$

19. $\frac{-1}{6} - \left(\frac{2}{3}\right)y \leq \frac{1}{2}$ $\{y \mid y \geq -1\}$

20. $\frac{7}{12} \geq \left(-\frac{1}{2}x\right) - \frac{1}{3}$ $\left\{x \mid x \geq -1\frac{5}{6}\right\}$

21. $\frac{4}{15} < \left(-\frac{3}{5}y\right) - \frac{2}{3}$ $\left\{y \mid y < -1\frac{5}{9}\right\}$

22. $\left(-3\frac{3}{4}\right)v - \frac{3}{4} > \frac{1}{8}$ $\left\{v \mid v < \frac{-7}{30}\right\}$

23. $45 \geq -5y + 4y$ $\{y \mid y \geq -45\}$

24. $-8v + 6v < 42$ $\{v \mid v > -21\}$

25. $\frac{-3}{8} < \frac{-3}{4} + m$ $\left\{m \mid m > \frac{3}{8}\right\}$

26. $-6.5 < 0.5m$ $\{m \mid m > -13\}$

27. $-2y + 3y \leq 9 - 2y$ $\{y \mid y \leq 3\}$

28. $12 - 8x > -11x + 18x$ $\left\{x \mid x < \frac{4}{5}\right\}$

29. $-3y + 7 \le -4y + 5y$ $\left\{ y \mid y \ge \frac{7}{4} \right\}$

30. $-27v + 11v < 15 - 3v$ $\left\{ v \mid v > \frac{-15}{13} \right\}$

$\xleftarrow{\hspace{2cm}}$
$\frac{-16}{13}\ \frac{-15}{13}\ \frac{-14}{13}\ \frac{-13}{13}$

31. $12y + 6 < 9y + 8$ $\left\{ y \mid y < \frac{2}{3} \right\}$

32. $12x + 3 \ge 10x + 5$ $\{ x \mid x \ge 1 \}$
$\xrightarrow{\hspace{2cm}}$
$0\quad 1\quad 2\quad 3\quad 4$

33. $23y + 9 > 14y + 11$ $\left\{ y \mid y > \frac{2}{9} \right\}$

34. $8 + 21r \le 11r + 5$ $\left\{ r \mid r \le -\frac{3}{10} \right\}$
$\xleftarrow{\hspace{2cm}}$
$\frac{-5}{10}\quad \frac{-3}{10}\quad \frac{-1}{10}\ 0$

35. $-64 \ge \frac{-2}{3}m$ $\{ m \mid m \ge 96 \}$

36. $0.02r \le 0.008$ $\{ r \mid r \le 0.4 \}$
$\xleftarrow{\hspace{2cm}}$
$0.3\quad 0.4\quad 0.5\quad 0.6$

37. $(4y + 5)2 \ge 6y$ $\{ y \mid y \ge -5 \}$

38. $3(x + 2) \le 8x$ $\left\{ x \mid x \ge \frac{6}{5} \right\}$
$\xrightarrow{\hspace{2cm}}$
$\frac{5}{5}\quad \frac{6}{5}\quad \frac{7}{5}\quad \frac{8}{5}\quad \frac{9}{5}$

39. $7.4 < 2.8 - m$ $\{ m \mid m < -4.6 \}$

40. $r - \frac{2}{5} > \frac{-2}{3}$ $\left\{ r \mid r > \frac{-4}{15} \right\}$
$\xleftarrow{\hspace{2cm}}$
$\frac{-6}{15}\ \frac{-5}{15}\ \frac{-4}{15}\ \frac{-3}{15}\ \frac{-2}{15}$

In Exercises 41 through 52, graph the solution set for each compound inequality.

41. $x \ge -5$ or $x < 3$

42. $x \le -6$ or $x \le -9$
$\xleftarrow{\hspace{2cm}}$
$-9\quad -8\quad -7\quad -6$

43. $y < 4$ and $y \ge 8$ No solution

44. $t \le 10$ or $t > 5$ All real numbers
$\xleftrightarrow{\hspace{2cm}}$

45. $y \le 3 \cap y \ge 5$ No solution

46. $t \ge 2 \cup t > 5$
$\xleftarrow{\hspace{2cm}}$
$0\quad 2\quad 4\quad 6$

47. $r \ge 6 \cup r > 4$

48. $n > 5 \cap n < 8$
$\xleftrightarrow{\hspace{2cm}}$
$2\quad 4\quad 6\quad 8$

49. $-5 \le x < 3$

50. $6 > y \ge 2$
$\xleftrightarrow{\hspace{2cm}}$
$0\quad 2\quad 4\quad 6\quad 8$

51. $-5 < t \le 1$

52. $-4 \le x \le 2$
$\xleftrightarrow{\hspace{2cm}}$
$-4\quad -2\quad 0\quad 2$

In Exercises 53 through 72, solve and graph the solution set of each compound inequality.

53. $6 < x - 3 < 12$ $\{ x \mid 9 < x < 15 \}$

54. $6 > 2x - 1 > -2$ $\left\{ x \mid \frac{-1}{2} < x < 3\frac{1}{2} \right\}$
$\xleftrightarrow{\hspace{2cm}}$
$0\quad 1\quad 2\quad 3\quad 4$

55. $-4 < -2x + 3 < 22$ $\left\{ x \mid -9\frac{1}{2} < x < 3\frac{1}{2} \right\}$

56. $15 \ge x - 4 \ge -18$ $\{ x \mid -14 \le x \le 19 \}$
$\xleftrightarrow{\hspace{2cm}}$
$-14\ -7\quad 0\quad 7\quad 14\ 21$

57. $-6 > x + 2 > -9$ $\{ x \mid -11 < x < -8 \}$

58. $-13 \le 3x + 5 < 18$ $\left\{ x \mid -6 \le x < 4\frac{1}{3} \right\}$
$\xleftrightarrow{\hspace{2cm}}$
$-7\ -5\ -3\ -1\quad 1\quad 3\quad 5$

59. $26 > -4x - 2 \ge 10$ $\{ x \mid -7 < x \le -3 \}$

60. $4.5 \le -5x - 3 \le 7.5$
$\xleftarrow{\hspace{2cm}}$
$-2.5\ -2\ -1.5\ -1\ -0.5$
$\{ x \mid -2.1 \le x \le -1.5 \}$

61. $x < 1$ or $x > 4$ $\{ x \mid x < 1 \cup x > 4 \}$

62. $x > -3$ or $x < 5$ All real numbers
$\{ x \mid x$ is any real number $\}$
$\xleftarrow{\hspace{2cm}}$

63. $x + 7 > 9$ or $3x - 5 \ge 19$ $\{ x \mid x > 2 \}$

64. $2x < 6$ or $x + 3 > 1$ All real numbers
$\{ x \mid x$ is any real number $\}$

65. $3x < 3$ and $x > 2$ $\varnothing$

66. $-6x > 12$ and $x + 1 < 6$
$\xleftarrow{\hspace{2cm}}$
$-7\ -6\ -5\ -4\ -3\ -2$
$\{ x \mid x < -2 \}$

67. $x + 5 \le 9$ and $x - 3 \ge -4$ $\{ x \mid -1 \le x \le 4 \}$

68. $x + 3 \ge 5$ and $2x \le 6$
$\xleftrightarrow{\hspace{2cm}}$
$0\quad 1\quad 2\quad 3\quad 4$
$\{ x \mid 2 \le x \le 3 \}$

69. $4y < 8$ or $y - 8 > 2$ $\{ y \mid y < 2 \cup y > 10 \}$

70. $3x - 11 < 4$ or $4x + 9 \ge 1$
$\{ x \mid x$ is any real number $\}$
$\xleftarrow{\hspace{2cm}}$

71. $4x + 9 < 13$ or $3x + 2 < 5$ $\{ x \mid x < 1 \}$

72. $-3x + 7 < 22$ or $2x - 8 < 12$ All real numbers
$\{ x \mid x$ is any real number $\}$
$\xleftarrow{\hspace{2cm}}$
All real numbers

In Exercises 73 through 80, find and graph the solution set for each absolute value inequality.

73. $|m| < 3$ $\{m \,|\, -3 < m < 3\}$

74. $|x| \le 1$ $\{x \,|\, -1 \le x \le 1\}$

75. $|x + 1| \ge 9$ $\{x \,|\, x \ge 8 \cup x \le -10\}$

76. $|n - 5| > 7$ $\{n \,|\, n > 12 \cup n < -2\}$

77. $|3y + 4| > 10$ $\left\{y \,\middle|\, y > 2 \text{ or } y < \dfrac{-14}{3}\right\}$

78. $|2x - 1| < 5$ $\{x \,|\, -2 < x < 3\}$

79. $|2r - 4| \ge 8$ $\{r \,|\, r \ge 6 \cup r \le -2\}$

80. $|5r - 3| \le 6$ $\left\{r \,\middle|\, -\dfrac{3}{5} \le r \le \dfrac{9}{5}\right\}$

In Exercises 81 through 88, graph each inequality.

81. $9(y - 5) \ge (3y - 2)8$ $\left\{y \,\middle|\, y \le \dfrac{-29}{15}\right\}$

82. $9(4x - 1) < 3(2x + 2)$ $\left\{x \,\middle|\, x < \dfrac{1}{2}\right\}$

83. $2(y - 15.3) \le 2(12y + 10)$ $\{y \,|\, y \ge -2.3\}$

84. $0.7t - 8 > 0.3(4t + 8)$ $\{t \,|\, t < -20.8\}$

85. $0.5(6y + 8) > 0.8y - 7$ $\{y \,|\, y > -5\}$

86. $4(2x + 3.35) \ge -2.2 + 2x$ $\{x \,|\, x \ge -2.6\}$

87. $11(7y - 5) > 2(y - 3)$ $\left\{y \,\middle|\, y > \dfrac{49}{75}\right\}$

88. $16(r - 4) \le 2(r - 7)$ $\left\{r \,\middle|\, r \le \dfrac{25}{7}\right\}$

MIXED PRACTICE

By doing these exercises, you will practice the topics up to this point in the chapter.

89. Solve: $\dfrac{x}{9} = 24$ $x = 216$

90. Solve: $|3 - 2x| = 9$ $x = -3; x = 6$

91. Find the solution set: $-x + 8 \ge 15$ $\{x \,|\, x \le -7\}$

92. Solve for v: $5v - x = -9v + x$ $v = \dfrac{1}{7}x$

93. Find the solution set: $-3x + 4 < 7x + 5$ $\left\{x \,\middle|\, x > -\dfrac{1}{10}\right\}$

94. *Consecutive Integers*   The sum of three consecutive odd integers is -99. Find the second integer. -33

95. Solve for x: $rx + e = f$ $x = \dfrac{f - e}{r}$

96. Solve: $|3x + 5| = 3$ $x = -\dfrac{2}{3}; x = -2\dfrac{2}{3}$

97. Write an algebraic expression for one-third of the product of four and a number.   a number $= n$, product of 4 and $n = 4n$; $\dfrac{1}{3}(4n)$

98. *Geometry*   Given that $A = \dfrac{1}{2}bh$, where A is the area of a triangle, b is the base, and h is the height, find b in terms of A and h. $b = \dfrac{2A}{h}$

99. *Geometry*   Given that $V = LWH$, where V is the volume of a rectangular solid, L is the length, W is the width, and H is the height, find W in terms of V, L, and H. $W = \dfrac{V}{LH}$

100. *Geometry*   Given that $C = 2\pi r$, where C is the circumference of a circle and r is the radius, find r in terms of C and π. $r = \dfrac{C}{2\pi}$

EXCURSIONS

Posing Problems

1. What information would you like that you can find in this table? Ask and
    answer four questions using these data.

Resident Population, by Age Group, Race, and Hispanic Origin, 1995[1] (in thousands)

Age	White	Black	Hispanic origin[2]	American Indian, Eskimo & Aleut	Asian & Pacific Islanders	All persons
Under 5	12,524	2,932	3,191	176	770	19,593
5–9	12,630	2,665	2,658	191	670	18,814
10–14	12,641	2,739	2,424	198	711	18,713
15–19	12,291	2,689	2,280	171	636	18,067
20–24	12,184	2,507	2,334	158	690	17,873
25–29	13,126	2,447	2,488	150	752	18,963
30–34	15,686	2,679	2,533	157	821	21,876
35–39	16,508	2,661	2,158	152	797	22,276
40–44	15,370	2,288	1,729	136	713	20,236
45–49	13,666	1,780	1,313	111	585	17,455
50–54	10,850	1,326	962	84	413	13,635
55–59	8,844	1,095	782	65	320	11,102
60–64	8,141	954	636	53	266	10,050
65–69	8,220	889	543	42	228	9,922
70–74	7,544	677	405	34	174	8,834
75–79	5,802	498	258	23	104	6,685
80–84	3,911	311	189	15	58	4,484
85–89	2,046	160	64	8	22	2,300
90–94	864	80	38	4	8	994
95–99	229	22	9	1	3	264
100 and over	42	7	2	1	1	53
All ages	193,519	31,606	26,978	1,931	8,743	262,777
16 and over	152,755	22,496	18,237	1,328	6,454	201,270
18 and over	147,835	21,422	17,332	1,258	5,199	193,046
65 and over	28,658	2,644	1,508	128	598	33,536

1. July 1, 1995. 2. Persons of Hispanic origin may be of any race. The information on the total and Hispanic population shown in this table was collected in the 50 states and the District of Columbia, and, therefore, does not include residents of Puerto Rico. *Source:* U.S. Bureau of the Census, as cited in the *1996 Information Please® Almanac* (©1995 Houghton Mifflin Co.), p. 830. All rights reserved. Used with permission by Information Please LLC.

2. Ask and answer four questions using the data in the following table.
    Share your questions with a classmate.

Currency Exchange Rate per New York Dollar			
	Today	6 mos. ago	1 yr. ago
Australian dollar	1.2870	1.3200	1.3446
Austrian schilling	**10.440**	**10.388**	**9.900**
Belgian franc	30.58	30.38	28.99
British pound	**.6481**	**.6503**	**.6246**
Canadian dollar	1.3720	1.3713	1.3558
French franc	**5.0605**	**5.0725**	**4.8560**
German mark	1.4848	1.4755	1.4062
Greek drachma	**237.10**	**243.55**	**226.64**
Hong Kong dollar	7.7340	7.7309	7.7377
Irish punt	**.6224**	**.6326**	**.6105**
Israeli shekel	3.1495	3.1412	3.0210
Italian lira	**1518.00**	**1572.50**	**1587.60**
Japanese yen	107.90	106.95	91.60
Mexican peso	**7.5050**	**7.4550**	**6.1350**
Dutch guilder	1.6640	1.6548	1.5795
Norwegian krone	**6.4095**	**6.4485**	**6.2140**

Exploring Problem Solving

3. The senior class officers at the College of Wakefield have arranged for class members to attend a performance of the *Nutcracker Ballet* in New York City in December. The bus has a capacity of 75 people and will cost $725 plus $40 for each passenger. Each ticket for the performance costs $50. The class officers have decided to charge each participant $125 for the trip and ballet performance. How many people must go on the trip so that the college does not lose money? Use your graphing calculator to help you construct a table that shows the costs for 20 to 75 students in increments of 5.

2.5 Solving Word Problems that Involve Linear Equations and Inequalities

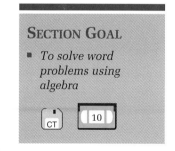

SECTION GOAL

▪ *To solve word problems using algebra*

SECTION LEAD-IN

Ask a question, answer it, and tell how you can check your answer.

a. In 1987, 1.2 million high-school seniors spent $14,549,475 to take the SAT exams.

b. There is a sound-activated burglar alarm that mimics the barking of a dog. You can buy it for $69.95.

c. It costs about ten thousand dollars to freeze Wollman rink in Central Park at the beginning of each winter.

d. It costs $1.5 billion dollars a year to replace stolen road signs.

e. P. T. Barnum's famous six and one-half ton elephant, Jumbo, ate two hundred pounds of food daily.

f. On a standard badminton shuttlecock, there are 14 to 16 feathers, each $2\frac{1}{2}$ to $2\frac{3}{4}$ inches long.

g. The Suez canal is 105 miles long.

h. The leaves in Central Park were raked by volunteers last year. It took 4546 people-hours to rake all these leaves.

We use algebra to solve problems from many other disciplines—the sciences, business, and social studies—as well as other branches of mathematics. Most problems are described in English, so the first step is to translate the problem into an equation.

STUDY HINT

Before you work a problem, take a few seconds to write the problem in your notebook and record its page number. If you do this, you can always find the section in the text to go back to for review.

To solve a word problem by translating it into an equation

1. Translate the English phrases into algebraic phrases.
2. Equate them as indicated by the word or phrase that implies "equals."
3. Solve the resulting equation.
4. Check your result by ensuring that it is reasonable in terms of the original problem.

···

EXAMPLE 1

On September 17, Stacie always receives a birthday check from her aunt for $5 plus $2 for each year of her age. On what birthday did she receive $37?

SOLUTION

We let y stand for Stacie's age in years; then on each birthday, she receives $5 + $2y$. When she receives $37, we have the equation

$$\$37 = \$5 + 2y$$

or, without the dollars signs,

$$37 = 5 + 2y$$

Solving, we get

$$37 + (-5) = 5 + 2y + (-5) \quad \text{Addition property}$$
$$32 = 2y \quad \text{Simplifying}$$
$$\left(\tfrac{1}{2}\right)(32) = \left(\tfrac{1}{2}\right)(2y) \quad \text{Multiplication property}$$
$$16 = y \quad \text{Simplifying}$$

Stacie received $37 on her sixteenth birthday. We check by returning to the problem statement. On her sixteenth birthday, she would have received

$$\$5 + (\$2)(16) = 5 + 32 = 37 \quad \text{True}$$

▶ *CHECK* **Warm-Up 1**

"Age" problems seem artificial, but they offer a good test of your ability to solve word problems. We translate the information into an equation and solve.

▪▪▪

EXAMPLE 2

Your age is twice your sister's age, minus 18. You are 50. How old is your sister?

SOLUTION

We want your sister's age. Call it x. We know your age, both as "twice your sister's age, minus 18" and as "50." They are equal, so

<u>Twice your sister's age, minus 18, equals 50.</u>

$$2x \quad - \quad 18 \quad = \quad 50$$

Solving, we find that

$$2x - 18 + 18 = 50 + 18 \quad \text{Addition principle}$$
$$2x = 68 \quad \text{Simplifying}$$
$$\left(\tfrac{1}{2}\right)(2x) = \left(\tfrac{1}{2}\right)(68) \quad \text{Multiplication principle}$$
$$x = 34 \quad \text{Simplifying}$$

So your sister is 34 years old. The check is left for you to do.

▶ *CHECK* **Warm-Up 2**

Consecutive integers are integers that follow each other in ascending order. The numbers 1, 2, and 3, for example, are consecutive integers. So are -210 and -209. If n represents the first of a group of consecutive integers, the next integer is $n + 1$, then $n + 2$, then $n + 3$, and so on.

The numbers, 1, 3, 5, . . . are **consecutive odd integers.** If x represents an odd integer, the next consecutive odd integers are $x + 2, x + 4$, and so on. The same is true for **consecutive even integers:** If y represents an even integer, the next consecutive even integers are $y + 2, y + 4$, and so on.

▪▪▪

EXAMPLE 3

The sum of three consecutive odd integers is 45. Find the integers.

SOLUTION

Let the first integer be x. Then, because they are consecutive odd integers, the second integer is $x + 2$, and the third integer is $x + 4$. We then translate and solve.

<u>The sum of three odd consecutive integers is 45.</u>

$$x + (x + 2) + (x + 4) = 45$$
$$3x + 6 = 45 \quad \text{Combining like terms}$$
$$3x = 39 \quad \text{Adding } -6 \text{ to both sides}$$
$$x = 13 \quad \text{Simplifying}$$

The first integer (x) is 13; the second integer is $x + 2$, or 15; and the third integer is $x + 4$, or 17. We leave the check for you to do.

▶ *CHECK* **Warm-Up 3**

STUDY HINT

Often the solution to an equation is not the answer to the word problem it represents. After solving the equation, always go back and answer the question asked by the problem.

STUDY HINT

Check your results every time.

STUDY HINT

When you draw a diagram, indicate all the relevant information, label the diagram carefully, and indicate any missing information.

Often, we use formulas to solve problems. The distance formula

$$\text{Rate} \times \text{time} = \text{distance} \quad \text{or} \quad r \cdot t = D$$

relates the speed (or rate r) at which one travels, the time (t) of travel, and the distance (D) traveled. If you know any two of these quantities, you can use the formula to find the third—provided that the units are compatible.

In the next example, we will begin by drawing a diagram. Sketches can be drawn for almost all geometry problems and for any problems that involve measurement.

▪▪▪

EXAMPLE 4

Two trains leave the same location at the same time, heading in different directions. One travels at a speed of 95 miles per hour, and the second at 80 miles per hour. The first stops traveling after 4 hours. After how many hours will the second train have covered the same distance as the first?

SOLUTION

We are given information about two trains; we diagram the trains' travel in the following figure. Now we can apply the distance formula to each.

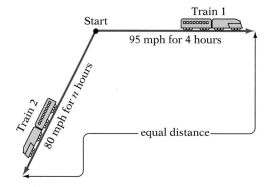

We let t represent the time traveled by the second train. Then

$$\begin{array}{ccccc} r & \cdot & t & = & D \end{array}$$

Train 1   95 mph · 4 hours = 380 miles

Train 2   80 mph · t hours = 80t

The two trains travel the same distance, so we can equate the two distances.

$$380 = 80t$$

Multiplying both sides by $\frac{1}{80}$ yields the solution.

$$t = \frac{380}{80} = \frac{19}{4} = 4\frac{3}{4}$$

So the second train must travel for $4\frac{3}{4}$ hours. The check is left for you to do.

▶ *CHECK* **Warm-Up 4**

The procedures for solving word problems involving inequalities can easily be adapted from equations and applied to inequalities.

▪ ▪ ▪

EXAMPLE 5

Thirty-two minus twice a number is greater than or equal to eighty-six and less than 108. Find the solution set.

SOLUTION

After locating the inequality relation, we symbolize the phrases to the left and right of it, and connect them with the appropriate inequality sign.

$$\text{thirty-two minus twice a number} \geq \text{eighty-six}$$
$$32 - 2x \qquad\qquad \geq \qquad 86$$

and

$$\text{thirty-two minus twice a number} < 108$$
$$32 - 2x \qquad\qquad < 108$$

Then, we solve the inequalities.

$$32 - 2x \geq 86$$

$32 - 2x + (-32) \geq 86 + (-32)$	Adding -32
$-2x \geq 54$	Simplifying
$\left(-\frac{1}{2}\right)(-2x) \leq \left(-\frac{1}{2}\right)(54)$	Multiplying by $\left(-\frac{1}{2}\right)$
$x \leq -27$	Simplifying

and

$32 - 2x < 108$	
$32 - 2x + (-32) < 108 + (-32)$	Adding -32
$-2x < 76$	Simplifying
$\left(-\frac{1}{2}\right)(-2x) > \left(-\frac{1}{2}\right)(76)$	Multiplying by $\left(-\frac{1}{2}\right)$
$x > -38$	Simplifying

So, the solution set is $\{x \mid -38 < x \leq 27\}$.

▶ *CHECK* **Warm-Up 5**

▪ ▪ ▪

EXAMPLE 6

The triangle inequality states that the sum of the lengths of any two sides a and b of a triangle must be greater than the length of the third side c. If side a is 18 inches and side c is 39 inches, what is the smallest whole number length that side b can be? What is the greatest whole number length that side b can be?

SOLUTION

Translating the statement into algebra, we have:

side a + side b is greater than side c

$$a + b > c$$

Evaluating the expression and substituting the values for a and c yields

$$18 + b > 39$$

We then solve the inequality.

$$18 + b > 39$$
$$18 + b + (-18) > 39 + (-18) \quad \text{Adding } -18$$
$$b > 21 \quad\quad\quad\quad \text{Simplifying}$$

Side b must be greater than 21 inches. And also, we know that b is limited because

$$\text{side } c + \text{side } a > \text{side } b$$
$$39 + 18 > b$$
$$57 > b$$

So

$$b < 57$$

Side b must be less than 57 inches. So, the smallest that side b can be is 22 inches; the greatest that side b can be is 56 inches.

▶ *CHECK* **Warm-Up 6**

Practice what you learned.

SECTION FOLLOW-UP

What operations did you use in asking and answering questions at the beginning of this section?

Go back and ask a question that involves

1. addition

2. subtraction

3. multiplication

4. division

5. 2 or more operations

2.5 WARM-UPS

Work these problems before you attempt the exercises.

1. Charlie and Viretta's business requires that they be away from the office 16 days a year, plus 12 days for every big account they manage. They are away from the office 184 days a year. How many big accounts do they manage?   14 accounts

2. Anita is 10 years younger than her boyfriend. Half of his age plus 19.5 is her age. How old are they?   The boyfriend is 59; Anita is 49.

3. Twice the sum of three consecutive odd integers, minus 15, is 111. Find the integers.   19, 21, 23

4. Two trains leave the same location at the same time running on parallel tracks. One travels at 60 miles per hour for 5 hours and then stops on a siding. The second travels at 45 miles per hour. How many hours later will the second train overtake the first? $6\frac{2}{3}$ hours

5. During the day, the pulse rate of an individual can vary from 46 to 130 beats per minute. How can that individual's pulse rate vary in beats per 10 minutes?   from 460 to 1300 beats per 10 minutes

6. Josie wants to purchase 18 sodas at a small delicatessen near work. Cold ones sell for 79 cents per can; warm ones are as low as $3.99 for a 6-pack. There is no tax on sodas. How much money will she need for the purchase?   from $11.97 to $14.22

2.5 EXERCISES

Note: Use your graphing calculator to check your results whenever possible.

Solve each word problem by translating it into an equation where necessary.

1. *Manufacturing Expenses*   A company's expenses include fixed costs of 2.5 thousand dollars plus $800 per unit manufactured. If its expenses in a week total $460,100, how many units were manufactured?   572 units

2. *Taxi Fare*   A taxi ride costs $1.35, plus 18 cents per $\frac{1}{5}$ mile. A man travels home and the ride costs $6.03. How far did he travel?   5.2 miles

3. *Snakes*   For a certain snake species, the female's total length (x) and tail length (y) in millimeters are related in the following way (approximately):

$$y = 0.13x - 1.2$$

A snake has a tail length of 154.8 millimeters. What is its total length?   1200 millimeters

4. ***Renting an Apartment*** In the mid-1990s in some large cities, the relationship between the amount spent on a finder's fee (x) and the amount of monthly rent (y) one had to pay for an apartment in some desirable areas could be represented by the equation

$$y = -0.12x + \$1000$$

A person who rented through this procedure pays \$520 a month rent. How much did he spend to find the apartment? \$4000

5. ***Number Problem*** The difference between two numbers is 48. Their sum is -598. What are the two numbers? -323 and -275

6. ***Number Problem*** The difference of one-fourth of a number and 36 is -99. Find the number. -252

7. ***Towing Charges*** An automobile club advertises that its towing price in Chicago is the first 5 miles free and then \$2.50 per mile; in New Jersey, the price is \$5 for the first mile and \$1.50 for each additional mile; in Atlanta, the first 10 miles are free, and then the price is \$2 per mile. What is the price in each area for towing 25 miles?
The price for towing 25 miles in Chicago is \$50; in New Jersey, \$41; in Atlanta, \$30.

8. ***Land Acquisition*** In 1968 Mexico gained some land from the United States because the Rio Grande river was diverted into a new course. The amount of land, in acres, that Mexico gained is the quotient of the year in which this happened and 4, less 54.82. How much land did Mexico gain?
437.18 acres

9. ***Age Problem*** Joella is 4 years older than her husband. The sum of their ages is 118. How old is each? Joella is 61; her husband is 57.

10. ***Immigration*** According to a business weekly, the total number of immigrants to the United States in a certain year was 1,041,000. Of these, about 50% were close relatives of U.S. citizens, 13.4% were skilled workers, and about 19% were undocumented. What percent have not been accounted for? Explain. What are the *numbers* of immigrants in the three listed categories? We do not know the percent that have not been accounted for. Some close relatives could also be skilled workers, for example. 520,500 were close relatives; 139,494 were skilled workers; and 197,790 were undocumented.

11. ***Age Problem*** The sum of the ages of Jorge's three children is 10 years less than his age. If one child is 16, another is 10, and Jorge is 56, find the age of the third child. 20 years old

12. ***Age Problem*** Three sisters' ages total 29 years. If the first sister is 5 years older than the second, and the second is one-fourth of the age of the third, find the ages of the three sisters. 9, 4, and 16 years

13. ***Number Problem*** Find three consecutive odd integers such that their sum is 43 more than -1000. $-321, -319,$ and -317

14. ***Number Problem*** Twice the sum of three consecutive odd integers is -894. Find the integers. $-151, -149,$ and -147

15. *Number Problem*   The sum of 4 consecutive integers is -2. What are the integers? $-2, -1, 0,$ and 1

16. *Number Problem*   One-third of the sum of three consecutive integers is zero. What are the integers? $-1, 0,$ and 1

17. *Triathlon*   According to the *New York Times,* Dan O'Brien had a world record triathlon score that was 44 points higher than the record set by Daley Thompson in 1984. The total of their scores is 17,738. What was Dan O'Brien's score?   8891

18. *Shot-Put Throw*   O'Brien's shot-put distance was greater than Thompson's by 3 feet $2\frac{1}{2}$ inches. Twice O'Brien's distance, minus Thompson's distance, is 58 feet. What are the two distances?   Thompson's distance is 51 feet 7 inches. O'Brien's distance is 54 feet $9\frac{1}{2}$ inches.

19. *Hurdles*   O'Brien's time for the 110-meter hurdles was 0.35 seconds less than Thompson's time. Twice the sum of their times is 56.62 seconds. How fast was each athlete?   Thompson's time was 14.33 seconds. O'Brien's time was 13.98 seconds.

20. *Javelin Throw*   Thompson's javelin throw exceeded that of O'Brien by a certain number. Six times that number is 46 feet. The sum of their distances is $420\frac{1}{3}$ feet. What was O'Brien's javelin distance? $206\frac{1}{3}$ feet

21. *Travel Time*   At 3 P.M. two cars start traveling in opposite directions from the same point. One travels north at 40 miles per hour, and the other travels south at 50 miles per hour. After how many hours are they 315 miles apart?   after 3.5 hours

22. *Travel Time*   Car A travels 55 miles per hour for 6 hours. Car B travels the same distance at 40 miles per hour. How many hours does car B travel?   8.25 hours

23. *Travel Time*   Stanley rides his bicycle at 15 miles per hour for 50 miles. Laurel rides her bike for the same amount of time and travels 10 miles further. At what speed did she travel?   18 miles per hour

24. *Travel Time*   A car leaves a train depot at 1:00 P.M. and travels at 60 miles per hour. A train leaves the same station $1\frac{1}{2}$ hours later and travels at 80 miles per hour in the same direction as the car. At what time will the train overtake the car?   7 P.M.

25. *Borrowing Videos*   Educational videos can be borrowed from the school library for $1.90 for the first day and $0.70 for each additional day. When I returned the video I borrowed, it cost me $10.30. How long did I keep it?   13 days

26. *Exchange Rate in Kenya*   In August 1992 in Kenya, the exchange rate for one dollar in shillings was 3.99 shillings more than in August 1991. The sum of the two rates is 47.45.

    a.   What were the rates in August of each of those years?   The exchange rate in August 1991 was 21.73 shillings/dollar; in August 1992 it was 25.72 shillings/dollar.

    b.   *Research*   What is the current exchange rate in Kenya?

27. *Postage Costs* In December 1977, first-class postage was 13 cents for the first ounce and 11 cents for each additional ounce or fraction of an ounce.

  a. If the cost of sending a package first class was $4.86, how much did the package weigh?   44 ounces

  b. *Research* Using the current postage rates, what weight package could you mail first class for $4.86?   Answers may vary.

28. *Long-Distance Charges* From a hotel in a big city, a long-distance telephone call costs $0.75 for the call plus a charge of 58 cents for the first minute and 23 cents for each additional minute. Upon checking out, I had to pay $6.62 for a call to Blytheville, Arkansas. How many minutes did I talk?   24 minutes

29. *Age Problem* Marisol's age is between 35 and 50. Sansi's age is 3 more than half Marisol's age. If the sum of their ages is 60 years, find their ages.
Marisol is 38; Sansi is 22.

30. *Ice Hockey* David and Martin are brothers who play hockey. In 1997, the sum of two times David's jersey number and 6 times Martin's number was their father's age. The difference between David's number and Martin's number was −7. If their father was 66 that year, what were their hockey numbers?   Martin's number was 10. David's number was 3.

31. *Coffee Exports* Coffee exports in 1992 were bountiful, running into the millions of bags. The sum of the number of millions of bags exported that year and the weight per bag is 204. The difference between half the weight per bag and the number of millions of bags exported is −6. How many millions of bags were exported? How much did each bag weigh?
72,000,000 bags were exported. Each bag weighed 132 pounds.

32. *Classrooms* Professor Churchill teaches MATH 150, and Professor Williams teaches STAT 311 in classrooms on different floors in the college. Professor Churchill's room number is 205 greater than Professor Williams'. Three times the sum of their room numbers is 4281. In what rooms do these professors teach?   Professor Churchill's room is 816. Professor Williams' room is 611.

33. *Running Shoes* Running shoes sell for $69 to $135 per pair depending on the brand, the specific features, and the place the shoes were purchased. A marathon runner may need from three to seven pairs of shoes per year. How much should she budget for shoes?   from $207 to $945

34. *Commuting Costs* A commuter spends from $6.75 to $12.50 round trip depending on his method of transportation. How much can he spend for 25 round trips?   from $168.75 to $312.50

35. *Health Food* A box of cereal containing oat bran may vary in weight from 13.5 ounces to 14.49 ounces and still be listed as weighing 14 ounces. A shipment that contains 120 such boxes, then, also can vary in weight. What is the least and most such a shipment can weigh?   least: 1620 ounces; most: 1738.8 ounces

36. *Number Problem* I'm thinking of a whole number. Five times this number is greater than 360, and one-half of this number is less than thirty-seven. What is the number?   73

37. *Calorie Content* The calorie content for baked potatoes varies from 90 to 200 calories depending on the size. Over a two-week period, a dieter has a plain baked potato each night. How few and how many calories might he have consumed in potatoes during this period? from 1260 to 2800 calories

38. *Price Reduction* A designer suit may vary in price from $199 to $275 depending on the store. A series of sales has reduced these prices to 65% and 75% of the original prices. What are the smallest and largest sale prices on this suit? smallest: $129.35; largest: $206.25

39. *Box Sizes* The measurements for a certain box-making machine can vary plus or minus 0.02 inches. A packing box is constructed that measures $8\frac{1}{4}$ inches by $19\frac{1}{2}$ inches by 27 inches according to this machine. What is the volume of the smallest and largest boxes possible because of the error? smallest: about 4325 cubic inches; largest: about 4362 cubic inches

40. *Homework Time* It takes 20 to 30 minutes to work a complicated statistics problem and write it up for submission to Georgianna's teacher. If she works 18 problems, how many hours will she work? 6 to 9 hours

MIXED PRACTICE

By doing these exercises, you will practice the topics up to this point in the chapter.

41. Find the solution set: $-9 < x + 4 < -6$
$\{x \mid -13 < x < -10\}$

42. Solve: $2(-9x + 6) = 24$
$x = \frac{-2}{3}$

43. Solve for x: $ax + b = y$ $\quad x = \frac{y - b}{a}$

44. Find the solution set: $-5 < 3x - 5 < 12$
$\{x \mid 0 < x < 5\frac{2}{3}\}$

45. Solve: $15x + 0.2 = 0.8$ $\quad x = 0.04$

46. Solve for y: $2yx = 3 + a$
$y = \frac{3 + a}{2x}$

47. Find the solution set: $1.1 - 3x \le 1.2 + x$
$\{x \mid x \ge -0.025\}$

48. Solve: $-9(12x - 5) = 9$ $\quad x = \frac{1}{3}$

49. One-third of a number minus that number is 64. Find the unknown number. -96

50. *Travel Time* At 3 P.M. two cars started traveling in opposite directions from the same point. One traveled north at 45 miles per hour, and the other traveled south at 50 miles per hour. After how many hours were they 500 miles apart? after $5\frac{5}{19}$ hours

51. Write as an algebraic expression the sum of a number and the quotient of that number and ten. $n + (n \div 10)$

52. *Geometry* The second angle of a triangle measures 58° more than the first, and the first measures 14° more than the third. Find the measures of the three angles. (Use the fact that the sum of the measures of the three angles is 180°.) first angle $= 45\frac{1}{3}°$; second angle $= 103\frac{1}{3}°$; third angle $= 31\frac{1}{3}°$

53. Write $4x - 7x$ as an English phrase. the difference between the product of 4 and a number and the product of 7 and that number

EXCURSIONS

Posing Problems

1. How are the following formulas alike? How are they different? Ask and answer one question for each formula.

   a. Let $Q = L \div M$, where
   Q is the cost per unit
   L is the price of the item
   M is the number of units in the item

   b. Let $d = m \div v$, where
   d is the density of an object in grams per cubic centimeter
   m is the mass of the object in grams
   v is the volume of the object in cubic centimeters

   c. Let $C = A \div S$, where
   A is the total number of calories
   S is the number of servings
   C is the number of calories per serving

   d. Let $F = D \div U$, where
   F is the fuel in gallons used
   D is the distance traveled
   U is the rate of usage in miles per gallon

2. How are the following formulas alike? How are these formulas different? Ask and answer at least one question for each formula.

   a. Let $P = 20s$, where P is the total number of pecks a woodpecker makes in s seconds.

   b. Let $rt = d$, where
   d is the distance in miles
   r is the rate or speed in miles per hour
   t is the time in hours

   c. Let $P = I^2R$, where
   P is the power loss in watts
   I is the current in amperes
   R is the resistance in ohms

   d. Let $V = at$, where
   V is the velocity of an object
   a is the rate of acceleration
   t is the elapsed time in seconds

   e. Let $F = ma$, where
   F is the force on an object
   m is the mass of the object
   a is the rate of acceleration of the object

CHAPTER LOOK-BACK

Most word problems encountered in your mathematics class are presented in a form so that your work is straightforward. Real-world problems, however, are not always so neatly presented.

Problems such as

About how many miles have you walked in the last year?

must be approached first by formulating the problem in a solvable way. That is, we must INTERPRET this problem—what does it mean? We must make a plan and DECIDE on a strategy. We APPLY this strategy, and then, after obtaining an answer, we check to see that the answer makes sense. We check it, and then we look back. We use REASON to determine the validity of our solution.

Explain how you approached each problem at the beginning of Chapter 2.

CHAPTER 2
REVIEW PROBLEMS

The following exercises will give you a good review of the material presented in this chapter.

SECTION 2.1

1. Give the variable part and the coefficient of $-6x^4$.
   variable part: x^4; coefficient: -6

2. Name all the terms in $-2 + 6x - 5x^2$.
   $-2, 6x, -5x^2$

3. What is the exponent in $-7x^{-2}$? -2

4. Evaluate $x^2 - 3x + 2x^3$ when $x = -1$. 2

5. Evaluate $xy^2 - x^2y$ when $x = -2$ and $y = 3$. -30

6. Evaluate $3x^5 - 2x^4 - x$ when $x = -1$. -4

In Exercises 7 through 12, simplify by combining like terms.

7. $3x^2 - 2x + 4x^2 - 7x^2 + 2x$ $\quad 0$

8. $rt^2 - tr + 2r^2t - 3rt^2 - 5rt + 6rt^2$
   $4rt^2 - 6rt + 2r^2t$

9. $8ab - 2a + 5b - 7ab + 6b + 8a$ $\quad ab + 6a + 11b$

10. $9(a - 6) + 8(9 - a)$ $\quad a + 18$

11. $4(2 + x) - 8(x - 1)$ $\quad 16 - 4x$

12. $3y - 2(x + y) + 7(y - x)$ $\quad 8y - 9x$

SECTION 2.2

13. Solve: $-26 + x = -84$ $\quad x = -58$

14. Solve: $\frac{2t}{3} + 26 = -37$ $\quad t = \frac{-189}{2}$

15. Solve: $\frac{1}{2}x = 44$ $\quad x = 88$

16. Solve: $-x - 4 = 12$ $\quad x = -16$

17. Solve: $3y - 4 = 7$ $\quad y = \frac{11}{3}$

18. Solve: $\frac{-r}{84} = 26$ $\quad r = -2184$

19. Solve: $-9(2x - 4) = 18$ $\quad x = 1$

20. Solve: $|y - 12| = |8|$
    $y = 20; y = 4$

21. Solve: $4(2x + 5) = 21$
    $x = \frac{1}{8}$

22. Solve: $-7x + 3 = -9x + 14$
    $x = \frac{11}{2}$

23. Solve: $|y| - 28 = 178$
    $y = -206; y = 206$

24. Solve: $2m + 5 = 8m - 2$
    $m = \frac{7}{6}$

SECTION 2.3

25. Solve for e: $4e + 3f = 9e - 8f$ $\quad e = \frac{11}{5}f$

26. Solve for x: $\frac{x}{a} + f = c$ $\quad x = a(c - f)$

27. Solve for s: $ts - 4w = 11r$ $\quad s = \frac{11r + 4w}{t}$

28. Solve for w: $mw + mr = xt$ $\quad w = \frac{xt - mr}{m}$

SECTION 2.4

29. Find the solution set: $-3x \leq 24.6$ $\quad \{x \mid x \geq -8.2\}$

30. Find the solution set: $-3(4 - 2x) < 19$ $\quad \left\{x \mid x < 5\frac{1}{6}\right\}$

31. Solve: $|x - 11| < -97$ $\quad$ no solution

32. Find the solution set: $\frac{3}{4}c < 16$ $\quad \left\{c \mid c < \frac{64}{3}\right\}$

33. Find and graph the solution set:
    $-x - 20 < -14 \text{ or } 2x > 18$ $\quad \{x \mid x > -6\}$

34. Find and graph the solution set:
    $|x - 5| > 15$ $\quad \{x \mid x > 20 \cup x < -10\}$

35. Find and graph the solution set:
    $7 - 2x > 5x + 3 \text{ or } x \geq 5$ $\quad \left\{x \mid x < \frac{4}{7} \text{ or } x \geq 5\right\}$

36. Find and graph the solution set:
    $3x - 2(x + 5) < 6 \text{ and } x \leq -1$ $\quad \{x \mid x \leq -1\}$

SECTION 2.5

37. Write in words: $x - 7$ *the difference between x and 7*

38. Translate into algebra: two and two-thirds the size of a walnut   *size of a walnut: w;* $2\frac{2}{3}w$

39. Let x be an integer. What is the opposite of this integer? $-x$

40. Let $x - 2$ be an odd integer. What is the next odd integer? x

41. Four-fifths of the sum of a number and 2 is twelve. Find the number.   *13*

42. My age is equal to twice your age plus 15. If I am 49, how old are you?   *17 years*

43. You need two pieces of wood, one three times longer than the other. The total length of the two pieces is 20.8 feet. How long are the two pieces?   *15.6 feet and 5.2 feet*

44. n less than 18 is -84. What is n? $n = 102$

MIXED REVIEW

45. Solve: $-5(x - 6) = 8 - 6(x - 3x)$ $x = \frac{22}{17}$

46. Solve: $4y - 7(5y + 6) = -2(y - 6)$ $y = \frac{-54}{29}$

47. Evaluate $x^2y^4 + 3xy^3$ when x is 2 and y is -4.   *640*

48. Combine like terms: $5r^2t + 6rt^2 - 7rt^2 + 2r^2t - 9$ $7r^2t - rt^2 - 9$

49. Find and graph the solution set:
$-3y - 6 < 4y + 12$ $\{y \mid y > \frac{-18}{7}\}$

50. Solve for t: $te + 3 = rw + e$ $t = \frac{rw + e - 3}{e}$

51. Evaluate $-5(x^2 - 3t)^{-2}$ when x is 5 and t is -2. $-\frac{5}{961}$

52. Find and graph the solution set:
$-1 < \frac{1}{8} - y < \frac{-3}{4}$ $\{y \mid \frac{7}{8} < y < \frac{9}{8}\}$

53. Find and graph the solution set:
$9 - |y + 3| < 25$ $\{y \mid y \text{ is a real number}\}$

54. Find and graph the solution set:
$-7 \leq x - 6 \leq 12$ $\{x \mid -1 \leq x \leq 18\}$

CHAPTER 2 TEST

This exam tests your knowledge of the material in Chapter 2.

1. **a.** Evaluate $x^2 - 2xy + y^2$ when $x = -2$ and $y = 3$.   *25*

   b. Name the terms and identify the coefficients and variable parts in $2xy - 3x^2$.
   terms: 2xy, $-3x^2$; coefficients: 2, -3; variable parts: xy, x^2

   c. Simplify: $20t - 36c - 5(t + 2c) - 8(7c + t)$ $7t - 102c$

2. Solve:

   a. $\frac{-t}{15} = 126$ $t = -1890$ **b.** $|-84 + c| = 62$ $c = 146; c = 22$ **c.** $-810 = -9x + 85 - 3x$ $x = 74\frac{7}{12}$

3. For parts (a) and (b), solve for the indicated variable.

   a. e: $re + f = c$ $e = \frac{c - f}{r}$ **b.** t: $\frac{t}{x} - h = r$ $t = x(r + h)$ **c.** Write "three times the sum of a number and 12" in algebra.
   $3(n + 12)$

4. Find the solution set for the inequality. In part (c), graph the solution set as well.

   a. $t - 7 < 6$ {$t \mid t < 13$}

   b. $-x + 100 \leq -431$ {$x \mid x \geq 531$}

   c. $x - 3 < 12$ and $3x + 5 < 7$
   {$x \mid x < \frac{2}{3}$}

5. Solve:

   a. **Consecutive Integer Problem** The sum of two consecutive even integers is -86. Find the integers. -44 and -42

   b. **Woodworking** You are making shelves for your living room to display statues. Out of a piece of wood 10 feet long, you want to cut two boards. One board should be two and one-third times the length of the other board. How long are the two boards? 7 feet and 3 feet

   c. **Renting Videocameras** Videocameras can be rented at the store for \$36.85 for the first three hours plus \$0.45 for each additional hour. How much does it cost you to rent the camera for 7 hours? \$38.65

CUMULATIVE REVIEW

CHAPTERS 1–2

The following exercises will help you maintain the skills you have learned in this and the previous chapter.

1. Add: $\left(-6\frac{3}{8}\right) + 2\frac{1}{4} + \left(-29\frac{1}{2}\right)$ $-33\frac{5}{8}$

2. Solve: $-3.6(6.3x + 45) = 405$ $x = -25$

3. Given $A = \{1, 8\}$ and $B = \{0, 1, 2, 3\}$, find the union of A and B. $A \cup B = \{0, 1, 2, 3, 8\}$

4. Find the solution set:
   $-4 < -3x + 5 < 9$ {$x \mid \frac{-4}{3} < x < 3$}

5. Solve for y: $2x + 3y = 8$ $y = \frac{8}{3} - \frac{2}{3}x$

6. Round 82.9765 to the nearest ten, one, hundredth, and thousandth.   80; 83; 82.98; 82.977

7. Use front-end estimation to find the value of
   $$\frac{(879.67)(2.398995)}{(59.87668)}$$ 30

8. Solve for m: $y = mx + b$ $m = \frac{y - b}{x}$

9. Arrange $\left|\frac{-14}{6}\right|$, $\left|\frac{-18}{3}\right|$, and $-\left|\frac{-12}{5}\right|$ in order from largest to smallest. $\left|\frac{-18}{3}\right| > \left|\frac{-14}{6}\right| > -\left|\frac{-12}{5}\right|$

10. Divide -4.602 by -0.0002.   23,010

11. Simplify: $2^0 - 2^{-2} + 2^2 \cdot 3^2 - 5^2$ $\frac{47}{4}$

12. Simplify: $(-0.006) + 2 - (-2.309)$ 4.303

13. Solve: $6 - 2t \geq 9 + t$ $t \leq -1$

14. Evaluate $c^2e^2 - 4(c - et^2)$ when $c = 4, e = -5$, and $t = -2$.   304

15. Solve: $\frac{x}{9} + 3 = 5$ $x = 18$

16. Write algebraically: the product of the sum of x and 6, and y $(x + 6) \cdot y$

17. Multiply: $(-0.2)(-0.001)(3.5)(60)$ 0.042

18. Divide $-3\frac{1}{9}$ by $2\frac{1}{3}$. $\frac{-4}{3}$

19. Simplify $-3(7x - 2) + 5(7x - 2)$ by multiplying and combining like terms. $14x - 4$

20. *Buying Books*   Dr. Lefkarites wants to purchase a book but she has exactly $20 with her. The price of the book plus $8\frac{1}{4}\%$ tax must be less than or equal to $20. What is the largest amount, before tax, that Dr. Lefkarites can pay for the book? $18.48

3

LINEAR EQUATIONS AND INEQUALITIES IN TWO VARIABLES

*T*hink of the last graph you saw. It may have been a line graph, showing one or more quadrants of a coordinate system. For example, a graph of the United States' national debt could use quadrant IV with the *y*-axis representing dollar amounts (+ represents surplus and − represents debt) and the *x*-axis representing time in years. So, in an ordered pair (*x*, *y*), *x* would represent the year and *y,* the debt. If the nation's revenue ever exceeded its liability, quadrant I could be included.

■ *Locate a graph from today's newspaper and describe, in words, the information on that graph. Which method is more efficient in conveying information?*

Skills Check

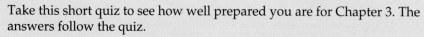

Take this short quiz to see how well prepared you are for Chapter 3. The answers follow the quiz.

1. Add: $-18 + 46$

2. Solve for r: $2r = 3(r - 5)$

3. Subtract: $23 - (-38)$

4. Simplify: $\frac{-1}{2}(8 - 4.5)$

5. Solve for x: $|x - 7| = 4$

6. Solve for t: $-3t - 7 > 10$

7. Simplify: $-8y + 2y - (-14y)$

8. What is the multiplicative inverse of $-4\frac{1}{2}$?

9. Graph $x < 9$ on the number line.

10. Solve for y: $x = my + b$

ANSWERS: 1. 28 [Section 1.2] 2. $r = 15$ [Section 2.2] 3. 61 [Section 1.2] 4. -1.75 [Section 1.3]
5. $x = 11, 3$ [Section 2.2] 6. $t < -5\frac{2}{3}$ [Section 2.4] 7. $8y$ [Section 2.1] 8. $\frac{-2}{9}$ [Section 1.2]
9. [number line with open circle at 9, marks 6 7 8 9 10 11] [Section 2.4] 10. $y = \frac{x - b}{m}$ [Section 2.3]

CHAPTER LEAD-IN

The Wind Chill Factors table shows how cold it feels when the wind is blowing at a given speed. For example, at 10°F, with the wind blowing 10 miles per hour, the temperature feels (and chills) as though it were −9°F.

Wind Chill Factors

Wind speed (mph)	Thermometer reading (degrees Fahrenheit)																
	35	30	25	20	15	10	5	0	−5	−10	−15	−20	−25	−30	−35	−40	−45
5	33	27	21	19	12	7	0	−5	−10	−15	−21	−26	−31	−36	−42	−47	−52
10	22	16	10	3	−3	−9	−15	−22	−27	−34	−40	−46	−52	−58	−64	−71	−77
15	16	9	2	−5	−11	−18	−25	−31	−38	−45	−51	−58	−65	−72	−78	−85	−92
20	12	4	−3	−10	−17	−24	−31	−39	−46	−53	−60	−67	−74	−81	−88	−95	−103
25	8	1	−7	−15	−22	−29	−36	−44	−51	−59	−66	−74	−81	−88	−96	−103	−110
30	6	−2	−10	−18	−25	−33	−41	−49	−56	−64	−71	−79	−86	−93	−101	−109	−116
35	4	−4	−12	−20	−27	−35	−43	−52	−58	−67	−74	−82	−89	−97	−105	−113	−120
40	3	−5	−13	−21	−29	−37	−45	−53	−60	−69	−76	−84	−92	−100	−107	−115	−123
45	2	−6	−14	−22	−30	−38	−46	−54	−62	−70	−78	−85	−93	−102	−109	−117	−125

Notes: This chart gives equivalent temperatures for combinations of wind speed and temperatures. Wind speeds of higher than 45 mph have little additional cooling effect. *Source: 1996 Information Please® Almanac* (©1995 Houghton Mifflin Co.), p. 392. All rights reserved. Used with permission by Information Please LLC.

- Ask and answer four questions using this information.

3.1 Linear Equations in Two Variables: Introduction

SECTION LEAD-IN

We can use algebra to model the relationship between two measures.* Then when we know the value of one of these measures, we can find the value of the other.

Let $c + 40 = F$, where

> F is the temperature Fahrenheit
> c is the number of cricket chirps in 15 seconds

- What is the temperature when the crickets are chirping 55 times each 15 seconds?

SECTION GOALS

- *To verify solutions of linear equations in two variables*
- *To find the missing coordinate in an ordered pair*
- *To graph ordered pairs*
- *To make a table of values and graph solutions of linear equations*

Graphing Solutions

> **Linear Equation in Two Variables**
>
> A linear, or first-degree, equation in two variables is an equation that can be put in the *standard form* $Ax + By = C$. Here A, B, and C are constants, and A and B are not both equal to zero.

In a linear equation, the two variables x and y must appear in separate terms and must have an exponent of one.

Finding Solutions to Linear Equations

When an equation is in the form

$$y = ax + b, a \neq 0$$

x is called the **independent variable** and y is called the **dependent variable.** That is, the value of y depends on the value assigned to x. The equation tells us how to find the value of y when we know x. We could solve this equation for x. Then y would be the independent variable and x the dependent variable.

Verifying Solutions

A **solution** of an equation in two variables x and y is a pair of real numbers that makes the equation true when substituted for the variables. The pair of numbers is usually written in parentheses as (x-value, y-value), or simply (x, y), and is then called an **ordered pair** of numbers. The pair is *ordered* because the value of the independent variable is always the first one listed.

> **WRITER'S BLOCK**
>
> In your own words, tell how you can find the value of x when you know the value of y and $y = 3x - 2$.

> **WRITER'S BLOCK**
>
> State a rule in your own words for finding the value of y when you know the value of x and $2y = 6x + 8$.

---

*The other words that we might have used here are *variables, varying amounts,* or *related measurements.*

An equation in two variables can have many solutions. The **solution set** for a linear equation in two variables consists of all the ordered pairs that are solutions to the equation.

To determine whether a particular ordered pair is a solution of an equation, we substitute the values into the equation. If the result is a true statement, the ordered pair is a solution.

▪ ▪ ▪

EXAMPLE 1

Which of the following ordered pairs is a solution for the equation $-3x - 2y = 9$?

a. $(-3, 0)$ **b.** $(5, 3)$

SOLUTION

We first substitute each ordered pair into the equation $-3x - 2y = 9$ and then evaluate the result

a. To test $(-3, 0)$, we substitute -3 for x and 0 for y.

$$-3x - 2y = 9$$
$$-3(-3) - 2(0) = 9$$
$$9 - 0 = 9 \quad \text{True}$$

So $(-3, 0)$ is a solution of the equation $-3x - 2y = 9$.

WRITER'S BLOCK

When is an ordered pair of real numbers a solution to a linear equation?

b. We substitute similarly to evaluate $(5, 3)$.

$$-3x - 2y = 9$$
$$-3(5) - 2(3) = 9$$
$$-15 - 6 = 9$$
$$-21 = 9 \quad \text{False}$$

So $(5, 3)$ is not a solution of the equation $-3x - 2y = 9$.

▶ CHECK **Warm-Up 1**

Calculator Corner

Test whether the ordered pair $(1, 0)$ is a solution of the equation $2x + 4y = 8$. The result is 2, not 8. Therefore, the ordered pair $(1, 0)$ is not a solution to the equation.

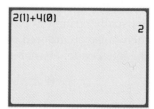

Some graphing calculators have the capability to recall the previous line so that you can substitute new numbers for x and y. For instance, pressing the **2nd ENTRY** keys on the *TI-82* graphing calculator would result in the screen at the left below. Replace the ordered pair $(1, 0)$ with the ordered pair $(4, 0)$ and press **ENTER** to get the screen at the right below. The result is 8, so the ordered pair $(4, 0)$ is a solution to the equation.

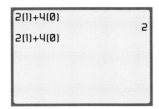

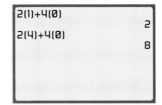

Now try the ordered pair $(-2, 3)$ on your own.

If we know one value in an ordered pair that is a solution to an equation, we can find the other value. To do so, we substitute the known value in the equation and solve for the other variable.

▪▪▪

EXAMPLE 2

Find a solution to the equation $-2x + y = 4$ when $x = 5$.

SOLUTION

To find the ordered pair $(5, y)$, we substitute 5, the value for x, in the equation and solve for y.

$$
\begin{array}{ll}
-2x + y = 4 & \text{Original equation} \\
-2(5) + y = 4 & \text{Substituting for } x \\
-10 + y = 4 & \text{Simplifying} \\
-10 + y + 10 = 4 + 10 & \text{Addition principle} \\
y = 14 & \text{Simplifying}
\end{array}
$$

You can check that $(5, 14)$ is a solution by substituting both values in the equation.

▶ *CHECK* **Warm-Up 2**

Graphing Points

We graph an ordered pair of numbers on a **Cartesian** (or rectangular) **coordinate system,** which is made up of a horizontal number line and a vertical number line that intersect (cross) at the zero points. When we are graphing ordered pairs (x, y), the number lines are called the x-axis and the y-axis, as shown in the following figure. (The independent variable is usually assigned the horizontal axis.)

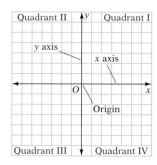

The axes divide the coordinate system (or coordinate **plane**) into four regions called **quadrants.** These are numbered quadrants I, II, III, and IV, counterclockwise, starting at the upper right. The point O where the axes intersect is called the **origin.**

INSTRUCTOR NOTE

Visually impaired students may need help in locating points. Grid paper with bold lines every inch and raised line drawing kits are available and are very useful.

To graph an ordered pair of numbers (x, y)

1. Starting from the origin, move horizontally right (for positive x) or left (for negative x) along the x-axis to x.
2. Then move vertically up (for positive y) or down (for negative y) exactly y units.
3. Mark a dot right there on the coordinate plane, and label it (x, y). The dot represents the ordered pair (x, y).

INSTRUCTOR NOTE

Students sometimes confuse the x- and y-coordinates when graphing. For this reason, you may want to avoid assigning points such as $(-2, -2)$ when assessing a student's understanding.

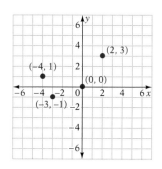

■ ■ ■

EXAMPLE 3

Graph the following ordered pairs.

a. $(2, 3)$ **b.** $(0, 0)$

c. $(-4, 1)$ **d.** $(-3, -1)$

SOLUTION

a. The point is $(2, 3)$. The first coordinate, 2, tells us to move 2 units to the right from the origin along the x-axis. The second coordinate, 3, tells us to move 3 units up parallel to the y-axis.

b. The point $(0, 0)$ is the origin.

c. The point $(-4, 1)$ is 4 units left and 1 unit up from the origin.

d. The point $(-3, -1)$ is 3 units left and 1 unit down.

The accompanying figure shows all four points.

▶ *CHECK* **Warm-Up 3**

■ ■ ■

EXAMPLE 4

Find the coordinates of points A, B, and C in the accompanying figure.

SOLUTION

a. Point A is 2 units to the right of the origin (in the positive x direction), so its x-coordinate is 2. It is 3 units above the x-axis (in the positive y direction), so its y-coordinate is 3. Its coordinates are then (2, 3).

b. Point B has an x-coordinate of -4 and a y-coordinate of -2, so its coordinates are $(-4, -2)$.

c. Point C has coordinates $(0, -1)$.

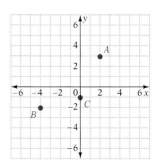

▶ *CHECK* **Warm-Up 4**

Graphing Solutions of Two-Variable Equations

Solutions of linear two-variable equations are ordered pairs of real numbers, so they can be graphed on the coordinate plane. To help us organize our results, we construct a table of values for the equation. Such a table lists the values for several solution pairs.

...

EXAMPLE 5

Find and graph three solutions of the equation $x = 6 + 2y$.

SOLUTION

Because this equation is already solved for x, we choose several values for the other variable (here, y) and show them in a **table of values.** It is a good idea to choose both positive and negative values, but any values will do.

x	y
	-1
	0
	1

Next we substitute the chosen y-values into the equation solved for x, one at a time, to find the corresponding x-values. For $y = -1$, we get

$$x = 6 + 2y = 6 + 2(-1) = 6 - 2 = 4$$

For $y = 0$,

$$x = 6 + 2y = 6 + 2(0) = 6$$

For $y = 1$,

$$x = 6 + 2y = 6 + 2(1) = 8$$

Finally, we place these values on the table next to the proper values of y.

x	y
4	-1
6	0
8	1

INSTRUCTOR NOTE

Although it is not necessary that we rewrite equations to solve for a given variable first, this makes calculations easier.

■ ■ ■
WRITER'S BLOCK
Two points determine a line. Explain.

This is a table of values for the equation $x = 6 + 2y$. Its ordered pairs are graphed in the following figure. All three pairs are part of the solution set for $x = 6 + 2y$.

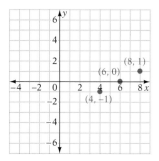

CHECK **Warm-Up 5**

WRITER'S BLOCK

What is the difference between *linear* and *collinear*? When might you use each word?

It would be a good idea now to find several more solutions (x, y) of the equation $x = 6 + 2y$ and to graph them on the same graph. When you do, you will see that all the plotted points lie in a straight line. The graph of all the solutions of a first-degree equation in two variables is a line. That is why it is called a *linear* equation.

Calculator Corner

A table of values can also be constructed on some graphing calculators using the **2nd TblSet** and the **2nd TABLE** utilities. For example, using the equation $y = -12 - 3x$, first enter the equation into the calculator. Remember to use the negative key $\boxed{(-)}$ for -12 and the subtract key $\boxed{-}$ $-3x$.

Press the **2nd TblSet** keys and let TblMin equal -1 and the increment ΔTbl equal 1, as shown at the left below. The increment is how much the x-value in the table will change. (Note: The *TI-83* graphing calculator uses TblStart instead of TblMin.) Press the **2nd TABLE** key to obtain the table shown at the right below. Check to see that you get the same values for y by constructing the table using the paper-and-pencil method.

```
Y₁ ◼-12-3X
Y₂=
Y₃=
Y₄=
Y₅=
Y₆=
Y₇=
Y₈=
```

```
TABLE SETUP
 TblMin=-1
 △Tbl=1
Indpnt:  Auto   Ask
Depend:  Auto   Ask
```

X	Y₁
-1	-9
0	-12
1	-15
2	-18
3	-21
4	-24
5	-27

X=-1

In the next section, we will learn more about linear equations.

Practice what you learned.

SECTION FOLLOW-UP

We know that $F = c + 40$. When $c = 55$, we can evaluate

$$F = (55) + 40$$
$$F = 95°$$

So when the crickets are chirping 55 times each 15 seconds, it is very hot—95° Fahrenheit.

3.1 WARM-UPS

Work these problems before you attempt the exercises.

1. Is $(2, -7)$ a solution of the equation $5x - y = 17$?   yes

2. For what value of A in $Ax + 2y = -3$ is $(6, -3)$ a solution to that equation? $\frac{1}{2}$

3. Graph and label the ordered pairs:

   a. $(-3, 4)$ **b.** $(3, -4)$ **c.** $(3, 4)$ **d.** $(-3, -4)$

4. One coordinate of each plotted point is given below. Give the other coordinate, name the point, and name the quadrant in which it appears.

   $(\underline{\quad}, 4), (\underline{\quad}, -5), (\underline{\quad}, -7), (-5, \underline{\quad})$

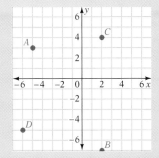

Point	Quadrant
$(2, 4) = C$	I
$(-6, -5) = D$	III
$(2, -7) = B$	IV
$(-5, 3) = A$	II

5. Make a table and graph three solutions of $4x - 2y = 6$.
   Tables may vary.

5.

x	y
0	−3
1	−1
2	1

3.1 EXERCISES

Note: Use your graphing calculator to check your results whenever possible.

In Exercises 1 through 8, determine whether each equation is linear or nonlinear.

1. $\frac{4}{x} = y$ nonlinear

2. $3x = 7$ linear

3. $3x + 4 = 6$ linear

4. $-6x + 5g = 11$
linear

5. $x^2 - 4y = 6$ nonlinear

6. $x^2 - x = -17$ nonlinear

7. $3t - 4 = 18kr$ nonlinear

8. $-10 + 2mt = n$
nonlinear

In Exercises 9 through 16, solve the equation first for x and then for y.

9. $25y + 5 = -91x$ $x = \frac{-25}{91}y - \frac{5}{91}; y = \frac{-91}{25}x - \frac{1}{5}$

10. $-3 + 21y = 9x$ $x = \frac{7}{3}y - \frac{1}{3}; y = \frac{3}{7}x + \frac{1}{7}$

11. $-9 - x = 36y$ $x = -36y - 9; y = -\frac{1}{36}x - \frac{1}{4}$

12. $-92x + 9 = 7y$ $x = \frac{-7}{92}y + \frac{9}{92}; y = \frac{-92}{7}x + \frac{9}{7}$

13. $-78x + 19 = 17y$ $x = \frac{-17}{78}y + \frac{19}{78}; y = \frac{-78}{17}x + \frac{19}{17}$

14. $8 - y = 15x$ $x = -\frac{1}{15}y + \frac{8}{15}; y = -15x + 8$

15. $27x - 6 = 4y$ $x = \frac{4}{27}y + \frac{2}{9}; y = \frac{27}{4}x - \frac{3}{2}$

16. $96y - 15 = 2x$ $x = 48y - \frac{15}{2}; y = \frac{1}{48}x + \frac{5}{32}$

In Exercises 17 through 20, state a rule, in words, to describe how to find the value of the dependent variable when given the value of the independent variable. Verify your rule by using the given value.

17. $y = 17x - 21; x = -1$ To find y: Multiply 17 times -1, obtaining -17. Subtract 21 (or add negative 21), obtaining -38. So y is equal to -38.

18. $2y + 12 = x; y = -6$ To find x: Multiply 2 times -6, obtaining -12. Add -12 and 12, obtaining zero. So x is equal to zero.

19. $0.3y - 17 = x; y = 50$ To find x: Multiply 0.3 times 50, obtaining 15. Compute 15 plus -17, obtaining -2. So x is equal to -2.

20. $y = \frac{1}{2}x - 12; x = 20$ To find y: Multiply $\frac{1}{2}$ times 20, obtaining 10. Add 10 and -12, obtaining -2. So y is equal to -2.

In Exercises 21 through 28, determine whether each ordered pair is a solution of the given equation.

21. $(-4, 9)$ and $\left(-\frac{3}{5}, 5\frac{3}{5}\right); x + y = 5$ yes; yes

22. $(-7, -5)$ and $\left(-\frac{1}{2}, 0\right); y - 2x = 1$ no; yes

23. $(2, -8)$ and $(-5, -11); x = y + 6$ no; yes

24. $(-6, -14)$ and $(-5, 8); 3x - 2y = -31$ no; yes

25. $(-4, 3)$ and $(-9999, 3); y = 3$ yes; yes

26. $(0, 9)$ and $(9, 876); x = 9$ no; yes

27. $\left(2, \frac{1}{2}\right)$ and $\left(-3, -1\frac{1}{2}\right); 2x - 4y = 0$ no; yes

28. $(0, 4.5)$ and $(2, 6.5); x - y = -4.5$ yes; yes

In Exercises 29 through 32, determine the missing value in ordered pairs in (a) through (d) that makes each ordered pair a solution of the given equation.

a. $(5, y)$ **b.** $(-3, y)$ **c.** $(x, 4)$ **d.** $(x, -6)$

29. $5x - y = 8$
   a. $(5, 17)$ **b.** $(-3, -23)$ **c.** $\left(2\frac{2}{5}, 4\right)$ **d.** $\left(\frac{2}{5}, -6\right)$

30. $6x - 2y = 5$
   a. $\left(5, 12\frac{1}{2}\right)$ **b.** $\left(-3, -11\frac{1}{2}\right)$ **c.** $\left(2\frac{1}{6}, 4\right)$ **d.** $\left(-1\frac{1}{6}, -6\right)$

31. $7x = y - 11$
   a. $(5, 46)$ **b.** $(-3, -10)$ **c.** $(-1, 4)$ **d.** $\left(-2\frac{3}{7}, -6\right)$

32. $-8x + 3y = 9$
   a. $\left(5, 16\frac{1}{3}\right)$ **b.** $(-3, -5)$ **c.** $\left(\frac{3}{8}, 4\right)$ **d.** $\left(-3\frac{3}{8}, -6\right)$

In Exercises 33 through 40, find a value for A such that the given ordered pair is a solution of $Ax + 2y = 1$.

33. $(5, -1)$ $\frac{3}{5}$

34. $(2, -3)$ $3\frac{1}{2}$

35. $(3, -1)$ 1

36. $(-6, 2)$ $\frac{1}{2}$

37. $(3.5, -1)$ $\frac{6}{7}$

38. $(5, -6)$ $2\frac{3}{5}$

39. $(-2, -1)$ $-1\frac{1}{2}$

40. $(-4, -3)$ $-1\frac{3}{4}$

In Exercises 41 through 44, complete the table of values to determine three pairs that are solutions of the given equation.

41. $y = 2.5x - 5$

x	y
6	10
2	0
-2	-10

42. $y = 0.5x - 1.5$

x	y
2	-0.5
-6	-4.5
-2	-2.5

43. $y = 5x$

x	y
2	10
-2	-10
6	30

44. $x = 2y$

x	y
-18	-9
-24	-12
0	0

In Exercises 45 through 48, give the coordinates of each point and identify the quadrant in which it appears.

45.

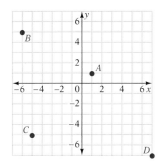

$A = (1, 1);$ QI
$B = (-6, 5);$ QII
$C = (-5, -5);$ QIII
$D = (7, -7);$ QIV

46.

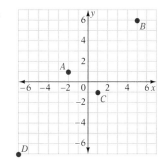

$A = (-2, 1);$ QII
$B = (5, 6);$ QI
$C = (1, -1);$ QIV
$D = (-7, -7);$ QIII

47.

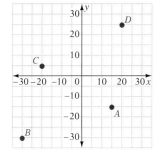

$A = (15, -15);$ QIV
$B = (-30, -30);$ QIII
$C = (-20, 5);$ QII
$D = (20, 25);$ QI

48.

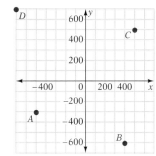

$A = (-500, -300);$ QIII
$B = (400, -600);$ QIV
$C = (500, 500);$ QI
$D = (-700, 700);$ QII

Calculator Corner

Exercises 49 through 60 can be checked with your graphing calculator **TABLE** utility by using **ASK** on the **TblSet** menu as follows. Note that it is then possible to input various values of x, but it is not possible to input values for y (the calculator automatically computes y in this setting). Consider the following for Exercise 49.

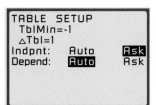
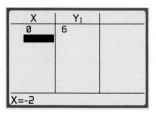
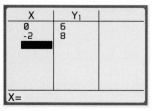

Continue in this manner until you have all the points you want.

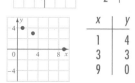

In Exercises 49 through 60, set up a table of values containing three pairs (x, y) that are solutions of the given equation.　Tables may vary.

49. $x + y = 6$

x	y
0	6
−2	8
−3	9

50. $-2x + y = 15$

x	y
0	15
1	17
−1	13

51. $-y - 4x = 40$

x	y
−10	0
0	−40
−9	−4

52. $-y + x = -20$

x	y
−20	0
−10	10
0	20

53. $-5x = 8 - y$

x	y
2	18
−2	−2
0	8

54. $-2y + 9 = x$

x	y
1	4
−1	5
3	3

55. $-2y = 4x$

x	y
0	0
2	−4
−2	4

56. $y = x - 16$

x	y
16	0
8	−8
0	−16

57. $2x - 2 = y$

x	y
1	0
0	−2
−2	−6

58. $-3y + 2 = 4x$

x	y
2	−2
−4	6
8	−10

59. $5x - 5y = 40$

x	y
8	0
0	−8
16	8

60. $12x + y = -5$

x	y
0	−5
−1	7
1	−17

In Exercises 61 through 68, make a table of values and graph three solutions for each equation.　Tables may vary.

61. $y - 3x = 10$

x	y
0	10
−3	1
−5	−5

62. $-2y + 3x = 2$

x	y
0	−1
2	2
−2	−4

63. $0.5x = 3 - 5y$

x	y
−4	1
−14	2
6	0

64. $-2y + 9 = x$

x	y
1	4
3	3
9	0

65. $-2y = x$

x	y
0	0
2	−1
4	−2

66. $-y = x + 8$

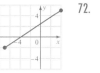

x	y
0	−8
−8	0
−10	2

67. $12x + 2 = 4y$

x	y
0	$\frac{1}{2}$
2	$6\frac{1}{2}$
−2	$-5\frac{1}{2}$

68. $y + 5 = 4x$

x	y
0	−5
1	−1
2	3

In Exercises 69 through 72, graph the points using your graphing calculator and draw a straight line connecting them.

69. $(-2, 3)$ and $(5, -8)$

70. $(-7, -2)$ and $(4, 5)$

71. $(-6, 4)$ and $(3, 4)$

72. $(-2, 6)$ and $(-2, -5)$

70.

72.

EXCURSIONS

Posing Problems

1. Ask and answer four questions about the data in the following table.

National Recreation Areas

Name and location	Total acreage	Name and location	Total acreage
Amistad (Tex.)	58,500.00	Gauley River (W. Va.)	10,300.00
Bighorn Canyon (Wyo.-Mont.)	120,296.22	Glen Canyon (Ariz.-Utah)	1,236,880.00
Chattahoochee River (Ga.)	9,259.91	Golden Gate (Calif.)	73,179.90
Chickasaw (Okla.)	9,930.95	Lake Chelan (Wash.)	61,886.98
Coulee Dam (Wash.)	100,390.31	Lake Mead (Ariz.-Nev.)	1,495,665.52
Curecanti (Colo.)	42,114.47	Lake Meredith (Tex.)	44,977.63
Cuyahoga Valley (Ohio)	32,524.76	Ross Lake (Wash.)	117,574.59
Delaware Water Gap (Pa.-N.J.)	67,204.92	Santa Monica Mountains (Calif.)	150,050.00
Gateway (N.Y.-N.J.)	26,310.93	Whiskeytown-Shasta-Trinity (Calif.)	42,503.46

Source: 1996 Information Please® Almanac (©1995 Houghton Mifflin Co.), p. 578. All rights reserved. Used with permission by Information Please LLC.

2. The following table lists the salaries of the New York Rangers hockey team during the 1994 season when they won the Presidents' Trophy and the Stanley Cup. Ask and answer four questions based on this data.

Rangers' Salaries*

Player	Salary	Player	Salary	Player	Salary
Mark Messier	$2,533,000	Eddie Olczyk	850,000	Stephane Matteau	425,000
Brian Leetch	1,805,000	Jeff Beukeboom	725,000	Brian Noonan	400,000
Glenn Anderson	1,250,000	Doug Lidster	700,000	Jay Wells	400,000
Adam Graves	1,150,000	Sergei Nemchinov	600,000	Mike Hudson	375,000
Steve Larmer	1,100,000	Craig MacTavish	550,000	Mike Hartman	310,000
Mike Richter	1,000,000	Nick Kypreos	525,000	Alexander Karpovtsev	275,000
Esa Tikkanen	979,000	Alexei Kovalev	450,000	Sergei Zubov	250,000
Kevin Lowe	950,000	Greg Gilbert	425,000		
Glenn Healy	850,000	Joe Kocur	425,000		

*Excluding bonuses other than for signing or reporting

CONNECTIONS TO *MEASUREMENT*

Working with Temperature

In this Connection, we discuss the measurement of temperatures. We will convert temperature measurements among the three most common scales.

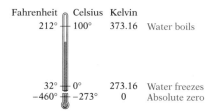

In the United States, we measure temperature on the Fahrenheit scale, developed in 1714 by a German physicist named Gabriel Fahrenheit. He chose to label the freezing point of water thirty-two degrees (32°F) and the boiling point of water 212 degrees (212°F).

In other countries, temperature is measured on the Celsius scale. This scale was developed by a Swedish astronomer, Anders Celsius, in 1742. On this scale, zero degrees (0°C) is the freezing point of water, and 100 degrees (100°C) is the boiling point of water.

The Kelvin scale is used mainly in the sciences. It was developed in the late 1800s. Absolute zero (0 K) on the Kelvin scale is theoretically the lowest possible temperature—the one at which all molecular movement ceases. The freezing point of water is 273.16 K, and its boiling point is 373.16 K. (*Note:* The degree sign is *not* used with Kelvin temperatures.) Temperatures on the Fahrenheit and Celsius scales can be negative numbers (below zero). Most scientists use Celsius or Kelvin scales, not Fahrenheit.

To change a temperature from degrees Celsius (C) to degrees Fahrenheit (F)

Use either of the following formulas:

$$F = \frac{9}{5}C + 32$$

or

$$F = (C \times 1.8) + 32$$

To change a temperature from degrees Fahrenheit (*F*) to degrees Celsius (*C*).

Use either of the following formulas:

$$C = \frac{5}{9}(F - 32)$$

or

$$C = (F - 32) \div 1.8$$

We use one of these formulas in the first example.

▪▪▪
EXAMPLE

The lowest temperature ever recorded in New Hampshire was −46 degrees Fahrenheit in 1925. Express this temperature to the nearest degree Celsius.

SOLUTION

We substitute the Fahrenheit temperature into the formula and simplify.

$$
\begin{aligned}
C &= (F - 32) \div 1.8 \\
&= (-46 - 32) \div 1.8 \quad \text{Substituting} \\
&= (-78) \div 1.8 \quad\quad \text{Simplifying} \\
&= -43.3\overline{3} \quad\quad\quad \text{Dividing}
\end{aligned}
$$

The temperature was approximately −43°C.

◢

In our last example, we will change a Celsius temperature to kelvins.

To change a temperature from degrees Celsius (C) to kelvins (K)

Use the formula

$$K = C + 273.16$$

▪▪▪
EXAMPLE

A temperature of −100°C is equivalent to what temperature on the Kelvin scale?

SOLUTION

We substitute the Celsius temperature into the proper formula and simplify.

$$K = C + 273.16$$
$$= -100 + 273.16$$
$$= 173.16$$

So $-100°C$ is equivalent to 173.16 K.

◤

PRACTICE

Work the following word problems. Round answers to the nearest hundredth of a degree.

1. Dalol, Ethiopia, the hottest spot in the world, has an average yearly temperature of 94 degrees Fahrenheit. A place that is 110 degrees colder is what temperature in degrees Celsius? $-26.67°C$

2. The coldest temperature ever recorded was $-128.5°F$ in Vostok, Antarctica. Change this temperature to degrees Celsius. $-89.17°C$

3. The hottest temperature on record in the inhabitable world was $136°F$, recorded in El Azizia, Libya, in 1892. Change this temperature to degrees Celsius. $57.78°C$

4. Mid-latitude areas are described as "warm climate" areas if they have an average cold temperature of not less than $-3°C$ and not more than $18°C$. Find the difference between these two extremes in degrees Fahrenheit. $37.8°F$

5. The tropical rainy climate has average monthly temperatures that never fall below $18°C$. Express this temperature in degrees Fahrenheit. $64.4°F$

6. Wood's metal has a melting point of $150°F$. This is such a low melting point that a spoon made of it will melt in a cup of hot tea. Express this melting point in kelvins. $338.72 K$

7. The ideal temperature of a fluid for quick absorption is $40°F$. If a fluid has a temperature of 300 K, is it ideal, too hot, or too cold?   too hot

8. **a.** The element calcium melts at $810°C$. What is this temperature in degrees Fahrenheit?

   b. The melting point of aluminum is $660.2°C$. What is this temperature in kelvins?   a. $1490°F$; b. $933.36 K$

9. Corals live at $23°C$. Express this temperature in degrees Fahrenheit. $73.4°F$

10. The longest hot spell on record was 162 consecutive days of $100°F$ in Marble Bar, Australia. Change this temperature to degrees Celsius. $37.78°C$

3.2 Graphing Linear Equations I

SECTION LEAD-IN

Given

$$F = c + 40,$$

where F is the Fahrenheit temperature and c is the number of cricket chirps in fifteen seconds, answer the following questions.

1. As the temperature gets colder, what happens to the number of cricket chirps?

2. As the number of chirps goes up, how has the temperature changed?

3. What do you think happens to crickets in the winter?

4. Think of another real-life situation that can be described by an equation such as

$$x = a + b$$

Finding the Intercepts and Slope of a Line

If we graph all the ordered pairs that are solutions to a linear equation, we obtain a straight line. We say that the line is the graph of the equation. (See the following figure.) However, it is not necessary to graph many ordered pairs, because any two points on a line determine the location of that line.

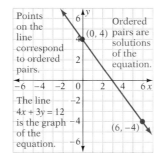

Points on the line correspond to ordered pairs.

Ordered pairs are solutions of the equation.

The line $4x + 3y = 12$ is the graph of the equation.

To graph a linear equation in two variables

1. Find two or more solutions of the equation.
2. Graph the solutions (ordered pairs) as points.
3. Draw a straight line through these points.

• • •

EXAMPLE 1

Graph: $4x - 2y = 12$

SOLUTION

We begin by solving the equation for y.

$4x - 2y = 12$	Original equation
$4x - 2y + (-4x) = 12 + (-4x)$	Addition principle
$-2y = 12 - 4x$	Simplifying
$\left(-\frac{1}{2}\right)(-2y) = \left(-\frac{1}{2}\right)(12 - 4x)$	Multiplication principle
$y = -6 + 2x$	Simplifying

Now we pick any three values for x. We arbitrarily choose $x = 1$, 0, and -1. Then we have to complete the ordered pairs $(1, y)$, $(0, y)$, and $(-1, y)$. We do this by substituting for x.

For $x = 1$: $y = -6 + 2x = -6 + 2(1) = -4$
For $x = 0$: $y = -6 + 2x = -6 + 2(0) = -6$
For $x = -1$: $y = -6 + 2x = -6 + 2(-1) = -8$

Thus three solutions to the given equation are $(1, -4)$, $(0, -6)$, and $(-1, -8)$. To organize this information, we list our computed solutions in a table of values. Here, we have

x	y	
1	-4	the ordered pair $(1, -4)$
0	-6	
-1	-8	

Now we graph the points, draw a line through them, and label the line as shown in the following figure.

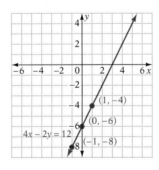

Because all three points lie on the line, we know the line is the correct graph. All points on the line represent solutions of the equation $4x - 2y = 12$.

▶ *CHECK* **Warm-Up 1**

▪▪▪

EXAMPLE 2

Graph: $-3x - y = 7$

SOLUTION

First we solve the equation for y.

$$-3x - y = 7 \qquad \text{Original equation}$$
$$-y = 3x + 7 \qquad \text{Adding } 3x \text{ to both sides}$$
$$y = -3x - 7 \qquad \text{Multiplying by } -1$$

Now we pick any three values for x and complete the ordered pairs $(1, y)$, $(0, y)$, and $(-1, y)$ by substituting for x.

$$\text{For } x = 1: \qquad y = -3x - 7 = -3 - 7 = -10$$
$$\text{For } x = 0: \qquad y = -3x - 7 = 0 - 7 = -7$$
$$\text{For } x = -1: \quad y = -3x - 7 = 3 - 7 = -4$$

Thus three solutions to the given equation are $(1, -10)$, $(0, -7)$, and $(-1, -4)$.

x	y
1	-10
0	-7
-1	-4

We then graph and label the points as shown in the following figure.

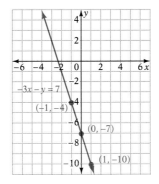

All points on the line represent solutions of the equation $-3x - y = 7$, and the solution set includes $\{(1, -10), (0, -7), (-1\ -4)\}$.

▶ *CHECK* **Warm-Up 2**

The x- and y-Intercepts of a Graph

A graph intersects the x- and y-axes at points called **intercepts**.

> The graph of a linear equation intersects the x-axis at a point $(a, 0)$ called the **x-intercept** of the graph. It intersects the y-axis at a point $(0, b)$ called the **y-intercept**.

We have a way to find the intercepts algebraically, without having to graph the equation.

> **To find the x- and y-intercepts of the graph of an equation**
>
> 1. Set y equal to 0 and solve the equation for x. The x-intercept will be $(x, 0)$.
> 2. Set x equal to 0 and solve the equation for y. The y-intercept will be $(0, y)$.

···

EXAMPLE 3

Find the x- and y-intercepts of the graph of the equation $-3x + 2y = 7$ and use them to graph the equation.

SOLUTION

To find the x- and y-intercepts, we substitute 0 for y and x, respectively, and solve.

x-intercept: Let $y = 0$.	*y-intercept:* Let $x = 0$.
$-3x + 2y = 7$	$-3x + 2y = 7$
$-3x + 2(0) = 7$	$-3(0) + 2y = 7$
$-3x = 7$	$2y = 7$
$x = -2\frac{1}{3}$	$y = 3\frac{1}{2}$

So the x- and y-intercepts are $\left(-2\frac{1}{3}, 0\right)$ and $\left(0, 3\frac{1}{2}\right)$, respectively.

To graph the equation, we simply graph the two intercepts, draw a line through them, and label the line as shown in the accompanying figure.

To find a third point on the line to check our work, we can substitute any value for x or y. We let $x = -5$ and get

$$-3(-5) + 2y = 7$$
$$15 + 2y = 7$$
$$2y = -8$$
$$y = -4$$

The point $(-5, -4)$ is on the line.

▶ CHECK **Warm-Up 3**

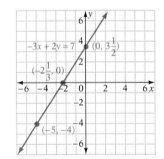

Calculator Corner

Many graphing calculators have a special feature that enables you to have **DECIMAL** numbers when you graph a function and then want to **TRACE** on that function. Consult your calculator's manual to see how to do this on your calculator. On the *TI-82* the steps are as follows.

 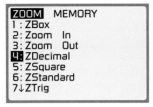 

Now **TRACE** on the graph to find the *x*- and *y*-intercepts. The next two screens show that the *y*-intercept is (0, 2) and the *x*-intercept is (−1, 0).

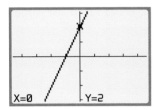 

You can use the Home Screen to check to see if your answers are correct. Press **2nd QUIT** to go from the **GRAPH** to the **Home Screen.**

First, let $x = -1$ and see if $y = 0$.   Now let $x = 0$ and see if $y = 2$.

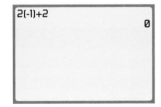 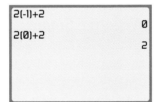

The Slope of a Line

This figure shows a line and two points, (0, 0) and (3, 2), on the line. To move from (0, 0) to (3, 2), we must move vertically up 2 units on the graph and then horizontally to the right 3 units. We say that the **slope** of this line is

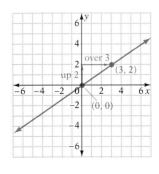

$$\text{Slope} = \frac{\text{change in } y}{\text{change in } x} = \frac{2 \text{ units (up)}}{3 \text{ units (right)}} = \frac{2}{3}$$

The slope of a line is a measure of its "slantedness."

This figure shows the same line with another point, (6, 4), marked on it. To move from (0, 0) to (6, 4), we would have to move up 4 units and to the right 6 units. The slope of the line now is

$$\text{Slope} = \frac{\text{change in } y}{\text{change in } x} = \frac{4 \text{ units (up)}}{6 \text{ units (right)}} = \frac{4}{6} = \frac{2}{3}$$

which is the same slope we calculated before. A line has only one slope, and *any* two points on the line can be used to find it.

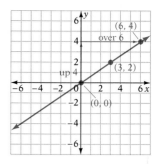

The Slope Formula

If we know the coordinates of any two points on a line, we can find the slope of the line with the following formula.

> **Slope of a Line**
>
> Let (x_1, y_1) and (x_2, y_2) be two distinct points on a line. Then the slope *m* of the line is
>
> $$m = \frac{\text{change in } y}{\text{change in } x} = \frac{y_2 - y_1}{x_2 - x_1}$$
>
> provided that x_2 is not equal to x_1.

Positive and Negative Slopes

•••
EXAMPLE 4

Find the slope of the line that passes through the points (5, 3) and (4, −2).

SOLUTION

The points and the line are sketched in the following figure.

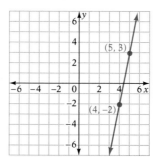

We let (5, 3) be the point (x_1, y_1), and we let (4, −2) be the point (x_2, y_2).

Then

$$m = \frac{y_2 - y_1}{x_2 - x_1} = \frac{(-2) - 3}{4 - 5} = \frac{-5}{-1} = 5$$

▶ CHECK **Warm-Up 4**

Note that the line in Example 4 slants up from left to right and therefore has a positive slope.

Calculator Corner

Use the Home Screen to confirm the numerical slope of the line containing the points (5, 3) and (4, −2) from Example 4. See the screen at the right.

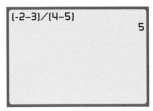

It is also possible to enter a formula on the Home Screen and use it repeatedly without having to re-enter it each time. For example, rewrite the formula for slope as $m = \dfrac{D - B}{C - A}$ and the two points as $(A, B) = (5, 3)$ and $(C, D) = (4, -2)$.

STORE 5 into A, 3 into B, 4 into C, and -2 into D. (Note: See the Calculator Corner on page 79 to review how to store values for variables.)

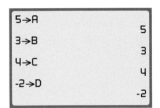

Then type in the new equation for slope: $\dfrac{D - B}{C - A}$.

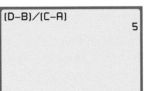

> Lines that slant *upward* from left to right have **positive slopes**.

If you graph the line in Warm-Up 4, you will see that it slants downward from left to right. It has a negative slope.

> Lines that slant *downward* from left to right have **negative slopes**.

Horizontal and Vertical Lines

> Let a and b be constants. Then a vertical line has the equation $x = a$, the x-intercept $(a, 0)$, and an undefined (or infinite) slope. A horizontal line has the equation $y = b$, the y-intercept $(0, b)$, and zero slope.

WRITER'S BLOCK

A line with a positive slope has both an x- and a y-intercept. Explain.

ERROR ALERT

Identify the error and give a correct answer.

$(x_2, y_2) = (3, -4)$
$(x_1, y_1) = (2, 1)$

$$\frac{x_2 - x_1}{y_2 - y_1} = \frac{3 - (-2)}{-4 - 1}$$

$$= \frac{5}{-5}$$

$$= -1$$

▪ ▪ ▪

EXAMPLE 5

Find the slope and intercepts of the line that passes through the points $(-1, 3)$ and $(2, 3)$.

SOLUTION

We graph the points and the line that passes through them, as in the figure on the left following Example 6. There we see that the line is horizontal and lies 3 units above the x-axis. Its y-intercept is $(0, 3)$, and it has no x-intercept. The slope is zero.

▶ CHECK **Warm-Up 5**

▪ ▪ ▪

EXAMPLE 6

Find the slope and intercepts of the line that passes through the points $(7, 1)$ and $(7, -3)$.

SOLUTION

The points and the line are graphed in the figure on the right following Example 6. The line is vertical and lies 7 units to the right of the y-axis. Its x-intercept is $(7, 0)$, and it has no y-intercept. The slope of this line is undefined.

▶ CHECK **Warm-Up 6**

The graphs for Example 5 and Example 6 are below. Use the slope formula to verify the slope of each line.

WRITER'S BLOCK
What are some real-life examples of vertical and horizontal lines?

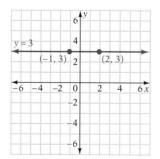

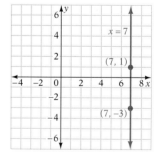

Calculator Corner

GRAPH each of the following equations on your graphing calculator. What do these four lines have in common?

GRAPH each of the following equations on your graphing calculator. What do these four lines have in common?

State a conjecture about how the value of *m* affects the graph of a straight line.

GRAPH each of the following groups of equations on your graphing calculator.

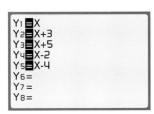 

State a conjecture about how the value of the constant *b* affects the graph of a straight line.

Practical Applications

Linear equations and their graphs are very useful for describing real situations and relationships. The last example illustrates four different ways of expressing a relationship between real variables.

■ ■ ■
EXAMPLE 7

The amount of uranium-235 that can be extracted from its ore (pitchblende) is given by the equation $U + 0.5 = 14P$, where U is the amount of uranium extracted, in pounds, and P is the available amount of pitchblende, in tons.

a. Graph this equation, taking into consideration the limits of the actual situation.

b. Use your graph to determine how much uranium-235 can be extracted from one-half ton of ore.

c. In this situation, what is the meaning of the ordered pair (2.5, 34.5)?

INSTRUCTOR NOTE

We suggest encouraging students to check answers by substituting the result into the original verbal problem and not the equation.

SOLUTION

a. The amount of uranium that can be extracted depends on the amount of ore that is available for the extraction process. Accordingly, P must be the independent variable and U the dependent variable. To make sure our equation indicates this, we first rewrite it as

$$U = 14P - 0.5$$

Then, choosing the values $-1, 0$, and 1 for P, we obtain the ordered pairs and the graph shown in the following figure on the left. Note that the scale on the horizontal (independent variable) axis has been "stretched" to show the graph clearly.

P	U
-1	-14.5
0	-0.5
1	13.5

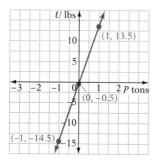 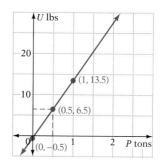

Because there is no such thing as a negative amount of pitchblende or uranium, only the part of the graph shown in the figure above on the right (the first quadrant) applies to the actual situation.

b. We find $\frac{1}{2}$ on the P-axis, move up from it to the line, and then move horizontally across to the U-axis, where we find that $\frac{1}{2}$ ton of pitchblende yields 6.5 pounds of uranium-235.

c. Because P is the independent variable, it is the first element in the ordered pair (P, U). Thus the pair $(2.5, 34.5)$ means that 2.5 tons of pitchblende yields 34.5 pounds of uranium-235.

▶ CHECK **Warm-Up 7**

In Example 7, we expressed the relationship between the independent variable P and the dependent variable U (1) as an equation, (2) as a graph, (3) in ordered pairs, and (4) in a table.

Practice what you learned.

SECTION FOLLOW-UP

1. As the temperature gets colder, the number of cricket chirps modeled by

$$F = c + 40°$$

gets smaller.

2. As the number of chirps goes up, we would find that the temperature has risen.

3. We can infer from this equation that below 40° Fahrenheit, crickets do not chirp. Research to find out what happens to them.

3.2 WARM-UPS

Work these problems before you attempt the exercises.

1. Graph $2x - y = 6$ using a table of values with $x = 1, 0,$ and 3.   Tables may vary.

   1.
x	y
1	−4
0	−6
3	0

2. Graph: $7y - 5x = 8$

3. Find the x- and y-intercepts of:

   a. $y = 3x$ x- and y-intercepts: (0, 0)

   b. $y = 5x + 2$ y-intercept: (0, 2); x-intercept: $\left(-\frac{2}{5}, 0\right)$

   c. $x = -2y - 3$ y-intercept: $\left(0, -1\frac{1}{2}\right)$; x-intercept: (−3, 0)

4. Find the slope of the line containing the points $(-4, -2)$ and $(5, -4)$. $-\frac{2}{9}$

5. a. Find the slope of the line that passes through the points $(-5, 8)$ and $(-17, 8)$.   m = 0

   b. What are the x- and y-intercepts of the line that passes through the points $(9, 0)$ and $(-20, 0)$?   The x-intercept is every point on the x-axis. The y-intercept is the origin, the point (0, 0).

6. What are the slopes and the x- and y-intercepts of the lines that make up the polygon in the accompanying graph?

7. My age (y) can be represented as an equation expressed in terms of my daughter's age (x). The equation is $y = x + 27$. Set up a table of values showing our ages when my daughter was 21, when I was 30, and when I was 40. Graph this equation.

   7.
x	y
21	48
3	30
13	40

Note: $\overline{AB}$ indicates the line segment between A and B.
slope of $\overline{AB} = 0$
slope of $\overline{BC}$ is undefined
slope of $\overline{CD} = 0$
slope of $\overline{AD}$ is undefined
x-intercepts: (2, 0) and (−4, 0)
y-intercepts: (0, −5) and (0, 4)

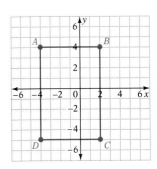

3.2 EXERCISES

Note: Use your graphing calculator to check your results whenever possible.

In Exercises 1 through 20, graph the equation using a table of values. *Tables may vary.*

1. $5x = 10 - 5y$

2. $2y - x = 5$

3. $5y = 3 + x$

4. $2x - 5 = y$

5. $-2y - 2 = x$

6. $-3x - 2 = 4y$

7. $y + 5x = 4$

8. $6y - 2x = 4$

9. $-x + 3y = 1$

10. $-2x - y = 2$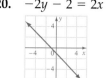

11. $-5y = 4 + x$

12. $-2x - 4 = 10y$

13. $2y - 2 = 4x$

14. $2x = y$

15. $2y = 6x$

16. $3y - 6 = 9x$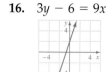

17. $4x - 8 = 4y$

18. $x + 4y = 3$

19. $7x - 7y = 0$

20. $-2y - 2 = 2x$

In Exercises 21 through 28, solve each problem and then graph each equation to verify that your solution is reasonable.

21. *Cable TV Bill*   Your cable TV bill includes a fixed $23.95 monthly charge plus $3.95 for each pay-per-view movie (x) you order. This relationship is represented by $y = 3.95x + 23.95$. If you ordered one movie per day for 20 days, what did you pay?   *$102.95*

22. *Bob Dylan Concert*   You ordered tickets for the "Music of Bob Dylan" concert at the Garden. The price (y) was $80 per ticket ($x$) plus a $6.25 service charge per ticket plus a $10 charge for the order. The equation that expresses this relationship is $y = 80x + 6.25x + 10$. You paid $355. How many tickets did you buy?   *4 tickets*

22.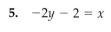

23. *Hockey Points System*   In hockey, a point system determines what teams make the playoffs. Teams receive 2 points for a win (w), 1 point for a tie (t), and none for a loss. A team that earns 62 points can have a combination of wins and ties. The equation is $62 = 2w + t$. Find three combinations of wins and ties that solve this equation.   *Answers may vary.*
20 wins/22 ties; 25 wins/12 ties; 29 wins/4 ties

24. **Sales Tax** In a certain city, the sales tax is 8.25%. You have $75 to spend. The equation that describes the cost (c) of an item as the sum of the sale price (p) plus 8.25% of the sale price is $c = p + 0.0825p$. What is the largest sale price an item can have so that you pay exactly $75, including the sales tax? $69.28

24.

25. **Overnight Mail** The cost of an overnight mailing service is represented by $y = 0.50x + 3.95$. Each package costs $3.95 plus 50 cents per pound.

a. Find the cost of mailing a package that weighs 10 pounds. $8.95

b. You pay $16.45 to mail a package. How much did the package weigh? 25 pounds

c. You mail three packages and pay $49.35. How much did the three packages weigh? 75 pounds

26. **Cost of Rental Car** The cost of a rental car is $39 and an additional 18 cents per mile. The equation representing this is $y = 0.18x + 39$.

26.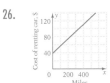

a. Find the cost of renting a car and driving 150 miles. $66

b. You only have $100. How far can you drive? 338 miles

c. You pay $48.36. How many miles did you drive? 52 miles

27. **Travel Time** The relationship between how far a certain car travels (y) and how far a bus travels (x) is $y = \left(1\frac{1}{2}\right)x - 10$.

a. How far has the bus traveled when the car has gone 110 miles? 80 miles

b. The bus has traveled 40 miles. How far has the car traveled? 50 miles

c. ✏ Graph the equation and explain what the point $(30, 35)$ means.

   The point $(30, 35)$ means that when the bus has traveled 30 miles, the car has traveled 35 miles.

28. **Travel Time** At 60 miles per hour, the relationship between the distance traveled (d) and the time traveled (t) is $d = 60t$.

a. Find d when the time is 3 hours. 180 miles

b. Find d when the time is $8\frac{1}{2}$ hours. 510 miles

c. Find t when the distance is 1820 miles. $30\frac{1}{3}$ hours

28.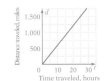

In Exercises 29 through 32, graph the equation and use the graph to find the x- and y-intercepts. Check by substituting into the original equation.

29. $y = x + 2$
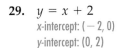
x-intercept: $(-2, 0)$
y-intercept: $(0, 2)$

30. $y = x - 3$
x-intercept: $(3, 0)$
y-intercept: $(0, -3)$
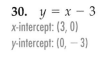

31. $y = -x - 3$
x-intercept: $(-3, 0)$
y-intercept: $(0, -3)$

32. $y = -x + 2$
x-intercept: $(2, 0)$
y-intercept: $(0, 2)$

In Exercises 33 through 40, find the *x*- and *y*-intercepts of the graph of the equation algebraically.

33. $3x - 2 = y$

x-intercept: $\left(\frac{2}{3}, 0\right)$

y-intercept: $(0, -2)$

34. $3 + 2x = y$

x-intercept: $\left(-\frac{3}{2}, 0\right)$

y-intercept: $(0, 3)$

35. $5x = y + 2$

x-intercept: $\left(\frac{2}{5}, 0\right)$

y-intercept: $(0, -2)$

36. $6x = 3 + y$

x-intercept: $\left(\frac{1}{2}, 0\right)$

y-intercept: $(0, -3)$

37. $3x + 2y = 8$

x-intercept: $\left(2\frac{2}{3}, 0\right)$

y-intercept: $(0, 4)$

38. $9x + 7y = 2$

x-intercept: $\left(\frac{2}{9}, 0\right)$

y-intercept: $\left(0, \frac{2}{7}\right)$

39. $5y - 3x + 2 = 0$

x-intercept: $\left(\frac{2}{3}, 0\right)$

y-intercept: $\left(0, -\frac{2}{5}\right)$

40. $18x + 2y - 7 = 0$

x-intercept: $\left(\frac{7}{18}, 0\right)$

y-intercept: $\left(0, 3\frac{1}{2}\right)$

In Exercises 41 through 49, find the *x*- and *y*-intercepts algebraically, and use them to sketch the graph of the equation.

41. $y = 2x + 3$

x-intercept: $\left(-\frac{3}{2}, 0\right)$

y-intercept: $(0, 3)$

42. $y = 2x - 4$

x-intercept: $(2, 0)$

y-intercept: $(0, -4)$

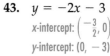

43. $y = -2x - 3$

x-intercept: $\left(-\frac{3}{2}, 0\right)$

y-intercept: $(0, -3)$

44. $y = -2x + 4$

x-intercept: $(2, 0)$

y-intercept: $(0, 4)$

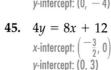

45. $4y = 8x + 12$

x-intercept: $\left(-\frac{3}{2}, 0\right)$

y-intercept: $(0, 3)$

46. $2y = 6x - 10$

x-intercept: $\left(\frac{5}{3}, 0\right)$

y-intercept: $(0, -5)$

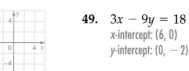

47. $y = 2$

no x-intercept

y-intercept: $(0, 2)$

48. $x = -3$

x-intercept: $(-3, 0)$

no y-intercept

49. $3x - 9y = 18$

x-intercept: $(6, 0)$

y-intercept: $(0, -2)$

In Exercises 50 through 64, find the slope of the line that passes through the indicated points.

50. $(-2, 1)$ and $(2, 5)$

$m = 1$

51. $(-1, 5)$ and $(-6, 3)$

$m = \frac{2}{5}$

52. $(6, -2)$ and $(16, -2)$

$m = 0$

53. $(-1, 1)$ and $(0, 4)$

$m = 3$

54. $(5, 1)$ and $(6, -4)$

$m = -5$

55. $(-8, -3)$ and $(-7, 2)$

$m = 5$

56. $(3, 5)$ and $(2, -1)$

$m = 6$

57. $(7, -1)$ and $(-6, 2)$

$m = -\frac{3}{13}$

58. $(-3, 5)$ and $(5, 1)$

$m = -\frac{1}{2}$

59. $(3, -2)$ and $(-3, -2)$

$m = 0$

60. $(6, 3)$ and $(1, 2)$

$m = \frac{1}{5}$

61. $(-8, 6)$ and $(-3, 2)$

$m = -\frac{4}{5}$

62. $(-1, -1)$ and $(0, 0)$

$m = 1$

63. $(1, -3)$ and $(-2, -1)$

$m = -\frac{2}{3}$

64. $(3, -2)$ and $(4, 5)$

$m = 7$

In Exercises 65 through 68, find the slope of each side of the given polygon from the graph.   *Note:* $\overline{AB}$ indicates the line segment between *A* and *B*.

65.

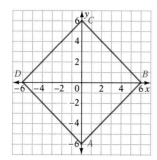

Slope of $\overline{AB} = 1$

Slope of $\overline{BC} = -1$

Slope of $\overline{CD} = 1$

Slope of $\overline{AD} = -1$

66.

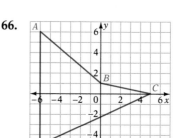

Slope of $\overline{AB} = -\frac{5}{6}$

Slope of $\overline{BC} = -\frac{1}{5}$

Slope of $\overline{CD} = \frac{5}{11}$

Slope of $\overline{AD}$ is undefined.

67.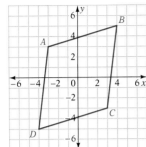

Slope of $\overline{AB} = \frac{2}{7}$
Slope of $\overline{BC} = 8$
Slope of $\overline{CD} = \frac{2}{7}$
Slope of $\overline{AD} = 8$

68.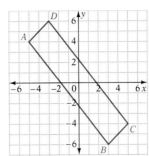

Slope of $\overline{AB} = -1\frac{1}{4}$
Slope of $\overline{BC} = 1$
Slope of $\overline{CD} = -1\frac{1}{4}$
Slope of $\overline{AD} = 1$

In Exercises 69 through 72, you are given the slope of a line and a point on the line. Find a second point on the line.   Answers may vary.

69. slope: $\frac{2}{3}$; point: $(-2, 1)$ (1, 3)

70. slope: -4; point: $(-5, 4)$ $(-4, 0)$

71. slope: -2; point: $(6, -3)$ $(7, -5)$

72. slope: $\frac{-3}{4}$; point: $(7, -2)$ $(11, -5)$

In Exercises 73 through 76, find the slope of the line that is described.

73. The y-axis   undefined

74. The slope of the x-axis $m = 0$

75. The line that passes through the origin and through the point $(-3, 1)$
$m = \frac{-1}{3}$

76. The line that passes through the point $(-2, 5)$ and through the origin
$m = \frac{-5}{2}$

MIXED PRACTICE

By doing these exercises, you will practice the topics up to this point in the chapter.

77. Find the missing coordinate of the point $(x, -3)$ that lies on the graph of $2y = 3x - 4$. $x = \frac{-2}{3}$

78. Graph $2x = y + 3$ using a table of values.   Tables may vary.

78.

79. Verify whether $(-2, -1)$ is a solution of the equation $y = \left(\frac{1}{2}\right)x + 1$.   no

80. Find the missing coordinate of the point $(-9, y)$ that lies on the graph of $y = 2x$. $y = -18$

81. Verify whether $(0, 4)$ is a solution to $y = 4x + 4$.   yes

82. Solve $7 + 3y = 2x$ for x and for y. $x = \frac{3}{2}y + \frac{7}{2}; y = \frac{2}{3}x - \frac{7}{3}$

83. Is $(6, -3)$ a solution of the equation $2y - 8x = -54$?   yes

84. Graph $9x - 4y = 7$ using a table of values.   Tables may vary.

84.

85. Solve $x - 8y = 9$ for x and for y. $x = 8y + 9; y = \frac{1}{8}x - \frac{9}{8}$

86. Find the slope of the line that passes through the origin and through the point $(-4, -6)$. $m = \frac{3}{2}$

87. Graph and label the points $A = (3, 4)$, $B = (-2, -1)$, $C = (-3, -4)$, and $D = (-2, 1)$.

88. Solve $11x - 7 = 2y$ for x and for y. $x = \frac{2}{11}y + \frac{7}{11}; y = \frac{11}{2}x - \frac{7}{2}$

89. Find the x- and y-intercepts of the graph of the equation $y - 2x = 7$.   x-intercept: $\left(-\frac{7}{2}, 0\right)$; y-intercept: $(0, 7)$

90. Graph $y = 2x + 5$ using a table of values.   Tables may vary.

90.

91. Graph $-2y + 1 = x$ using a table of values.   Tables may vary.

92. Graph the equation $x = -4$. What is its slope?   the slope is undefined

92.

93. Is $(-4, 2)$ on the graph of the equation $4y = -2x$?   yes

94. Identify three points that lie on the line whose equation is $y = x + 5$. $(0, 5); (-1, 4); (-2, 3)$ Answers may vary.

EXCURSIONS

Data Analysis

1. ✏ Using the following data, analyze the nutrients and calories per ounce for energy bars A through G. Use your analysis to recommend an energy bar for your health-conscious friends. Justify your recommendation.

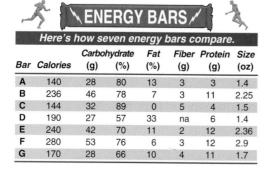

ENERGY BARS

Here's how seven energy bars compare.

Bar	Calories	Carbohydrate (g)	Carbohydrate (%)	Fat (%)	Fiber (g)	Protein (g)	Size (oz)
A	140	28	80	13	3	3	1.4
B	236	46	78	7	3	11	2.25
C	144	32	89	0	5	4	1.5
D	190	27	57	33	na	6	1.4
E	240	42	70	11	2	12	2.36
F	280	53	76	6	3	12	2.9
G	170	28	66	10	4	11	1.7

CONNECTIONS TO *GEOMETRY*

Parallel Lines

A straight angle has a measure of 180°. It is called a **line.** Two angles are complements of each other, or **complementary,** if their measures total 90°. Two angles are supplements of each other, or **supplementary,** if their measures total 180°.

Two distinct lines either **intersect** or do not. When they intersect, they intersect in just one point. The angles formed opposite each other are called **vertical angles.** Vertical angles have equal measure. Angles formed next to each other are **adjacent angles.** Adjacent angles formed by two intersecting lines are supplementary.

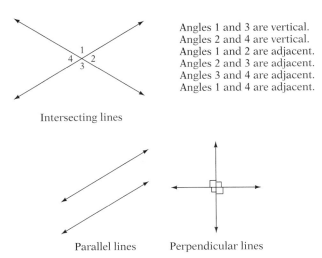

Angles 1 and 3 are vertical.
Angles 2 and 4 are vertical.
Angles 1 and 2 are adjacent.
Angles 2 and 3 are adjacent.
Angles 3 and 4 are adjacent.
Angles 1 and 4 are adjacent.

Intersecting lines

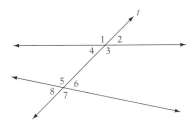

Parallel lines   Perpendicular lines

Lines in a plane that do not intersect are called **parallel lines.** Lines that intersect in right angles are called **perpendicular lines.**

When a line intersects two or more lines in different points, it is called a **transversal.**

We give names to the angles:

Exterior angles	1, 2, 7, 8
Interior angles	3, 4, 5, 6
Corresponding angles	1 and 5, 2 and 6, 4 and 8, 3 and 7
Alternate exterior angles	1 and 7, 2 and 8
Alternate interior angles	3 and 5, 4 and 6

When two parallel lines are cut by a transversal,

1. corresponding angles are equal in measure.

2. alternate exterior angles are equal in measure.

3. alternate interior angles are equal in measure.

···

EXAMPLE

Lines *a* and *b* are parallel and angle 2 is 60°. Find the measure of all other angles.

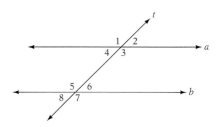

SOLUTION

We use ∠ to indicate angle.

∠2 = ∠4 = 60°	Vertical angles
∠1 = 120°	Adjacent to ∠2 and supplementary: 180° − 60° = 120°
∠1 = ∠3 = 120°	Vertical angles
∠2 = ∠8 = 60°	Alternate exterior angles
∠8 = ∠6 = 60°	Vertical angles
∠3 = ∠5 = 120°	Alternate interior angles
∠5 = ∠7 = 120°	Vertical angles

So ∠2 = ∠4 = ∠6 = ∠8 = 60°
   ∠1 = ∠3 = ∠5 = ∠7 = 120°

PRACTICE

1. Redo the example without using the idea of equality of vertical angles to justify a result.   Answers may vary.

2. Redo the example using only vertical angles, supplementary angles, or alternate interior angles to justify each result.   Answers may vary.

For Questions 3 through 8, use the following figure.

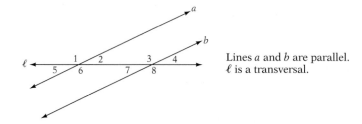

Lines *a* and *b* are parallel.
ℓ is a transversal.

Find the measure of each angle when

3. ∠6 = 150° ∠1 = ∠3 = ∠6 = ∠8 = 150°
              ∠2 = ∠4 = ∠5 = ∠7 = 30°

4. ∠8 = 125° ∠1 = ∠3 = ∠6 = ∠8 = 125°
              ∠2 = ∠4 = ∠5 = ∠7 = 55°

5. ∠5 = 32°   ∠1 = ∠3 = ∠6 = ∠8 = 148°
              ∠2 = ∠4 = ∠5 = ∠7 = 32°

6. ∠7 = 20°   ∠1 = ∠3 = ∠6 = ∠8 = 160°
              ∠2 = ∠4 = ∠5 = ∠7 = 20°

7. ✏ Show how you know that ∠5 + ∠3 = 180°.   *See the Solutions Manual.*

8. ✏ Show how you know that ∠1 + ∠4 = 180°.   *See the Solutions Manual.*

3.3 Graphing Linear Equations II

SECTION LEAD-IN

Here is a model for the relationship between Fahrenheit and Celsius temperatures.

$$C = \frac{5}{9}(F - 32)$$

where C is the temperature in degrees Celsius and F is the temperature in degrees Fahrenheit. Using this model, answer the following questions.

1. Solve for F.

2. Determine which scale has the greatest numerical value at temperatures above 100? below zero?

3. Find out at what temperature C and F have the same numerical value.

Using the Slope-Intercept Form of a Linear Equation to Graph the Equation

From a linear equation, we can find the slope and y-intercept of its graph. We then can use that information to graph the equation.

Slope-Intercept Form

The **slope-intercept form** of a linear equation is

$$y = mx + b$$

In that form, m is the slope of the graph of the equation, and $(0, b)$ is the y-intercept.

SECTION GOALS

- *To rewrite linear equations in slope-intercept form and to identify the slope and y-intercept*

- *To graph equations using the slope and y-intercept*

- *To determine when two lines are parallel or perpendicular*

- *To write the equation of a line using the slope-intercept form*

- *To write the equation of a line using the point-slope form*

- *To write the equation of a line given the slope and/or sets of points*

- *To write the equation of a line that passes through a point and is perpendicular (or parallel) to another line*

- *To solve verbal applications of slope problems*

Finding the Slope and y-Intercept

Any linear equation in two variables can be rewritten in slope-intercept form by solving the equation for the dependent variable y. Then the slope-intercept form of the equation immediately tells us the slope and y-intercept of its graph.

ERROR ALERT

Identify the error and give a correct answer.

Find the slope of the line whose equation is $2y = x + 4$.

Incorrect Solution:

The coefficient of x is 1, so the slope is 1.

•••

EXAMPLE 1

a. Find the slope and y-intercept of the graph of $2x + y = 6$.

b. Then graph the equation.

SOLUTION

a. We put the equation in slope-intercept form by solving for y.

$$2x + y = 6 \quad \text{becomes} \quad y = -2x + 6$$

By comparing this result with the general slope-intercept form, we easily read off m and b.

$$y = -2x + 6$$
$$y = mx + b$$

The graph has slope -2 and y-intercept $(0, 6)$.

ERROR ALERT

Identify the error and give a correct answer.

Find the slope and y-intercept of

$2y = 3x + 6$.

Incorrect Solution:

Slope: 3

y-intercept: 6

b. The y-intercept is a point on the line, so we graph it first as shown below.

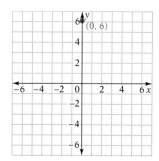

The slope is -2; we first write it as a fraction, $\frac{-2}{1}$, so it is in the form

$$m = \frac{\text{change in } y}{\text{change in } x} = \frac{-2}{1}$$

This now tells us that for each change in x of 1, the change in y is -2. So to find a second point on the line, we start at $(0, 6)$ and move

2 units down (in the *negative y* direction), and

1 unit right (in the *positive x* direction)

This gives us another point on the line—the point $(1, 4)$ in the following figure. We draw the line that passes through the two points and label it as the graph of $2x + y = 6$.

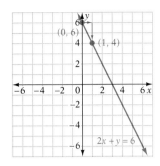

▶ CHECK **Warm-Up 1**

You now have two ways to graph a linear equation: (1) finding and plotting two or three solutions and drawing the line through them, and (2) finding the slope and *y*-intercept and using them to draw the line. Unless you are instructed otherwise, use whichever method you prefer or whichever seems best in a particular problem.

Parallel and Perpendicular Lines

The figure below on the left shows the graphs of the equations

$$y = 2x - 7 \quad \text{and} \quad y = 2x - 3$$

These two lines have the same slope, 2, as we can see from both the figure and their equations.

> Two lines are **parallel lines** if their slopes are equal. All vertical lines are parallel, although they have undefined slopes.

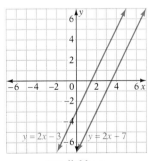

Parallel lines

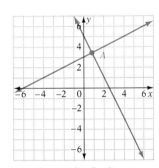

Perpendicular lines

The figure above on the right shows the graphs of the equations

$$y = \frac{1}{2}x + 3 \quad \text{and} \quad y = -2x + 5.$$

These lines intersect, or cross each other, at point *A* in a 90° angle, or right angle, and are said to be *perpendicular*. Their slopes $\left(\frac{1}{2} \text{ and } -2\right)$ are negative reciprocals.

Two lines are **perpendicular lines** if their slopes are negative reciprocals—that is, if the product of their slopes is -1. All vertical lines are perpendicular to all horizontal lines.

Calculator Corner

GRAPH each of the following equations on your graphing calculator.

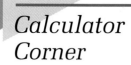

What do these lines have in common? State a conjecture about lines having the same slope.

Study the graphs of the following two equations on grids that have two different scales.

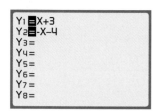 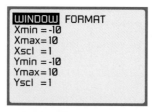 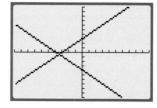

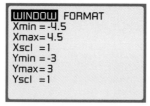 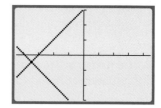

Study the graphs of the following two equations on grids that have two different scales.

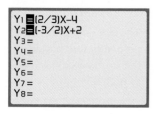 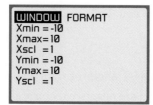 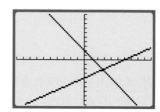

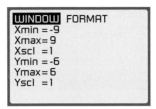
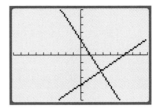

The second graph is the most "visually correct" representation of these perpendicular lines. Why is this true? State a conjecture about the slope of perpendicular lines.

Give the equations of two perpendicular lines and graph them on a grid that shows them to be perpendicular.

...

EXAMPLE 2

Which pairs of the following equations have graphs that are parallel to each other? Which are perpendicular to each other? Use the equations to answer.

(a) $4y = 2x + 32$ (b) $2y - x = 10$ (c) $y = -\frac{1}{2}x + 4$ (d) $y = -2x + 4$

SOLUTION

We first rewrite equations (a) and (b) in slope-intercept form.

(a) $4y = 2x + 32$ becomes $y = \frac{1}{2}x + 8$

(b) $2y - x = 10$ becomes $y = \frac{1}{2}x + 5$

(c) $y = \left(\frac{-1}{2}\right)x + 4$

(d) $y = -2x + 4$

Equations (a) and (b) have the same x-coefficient, so their graphs are parallel; both have slope $\frac{1}{2}$.

Equations (a) and (d) and equations (b) and (d) have x-coefficients that are negative reciprocals. Thus equations (a) and (d) have perpendicular graphs, as do equations (b) and (d).

You should check your results by graphing.

▶ *CHECK* **Warm-Up 2**

> **WRITER'S BLOCK**
>
> Explain how you can tell when two equations have graphs that are parallel.

> **WRITER'S BLOCK**
>
> Give examples of things in your environment that have parallel lines; perpendicular lines.

Writing the Equation of a Line

"The equation of a line" means "the equation whose graph is that line." We can write the equation of any line if we have certain information about the line. For example, if we know the slope and y-intercept of a line, we can use the slope-intercept form to write its equation.

Slope-Intercept Form

Given the slope and the y-intercept:

> **To write an equation of a line given the slope (m) and the y-intercept (0, b)**
>
> 1. Substitute the values for m and b into the slope-intercept form, $y = mx + b$.
> 2. Simplify the resulting equation.

▪▪▪

EXAMPLE 3

Write the equation of the line that has slope -5 and y-intercept $(0, -2)$.

SOLUTION

We have $m = -5$ and $b = -2$. Substituting into the slope-intercept form and simplifying gives us

$$y = mx + b$$
$$y = (-5)x + (-2)$$
$$y = -5x - 2$$

So the equation of the line is $y = -5x - 2$.

▶ CHECK **Warm-Up 3**

❗❗❗

ERROR ALERT

Identify the error and give a correct answer.

Find the equation of the line that has slope 4 and passes through the point (2, 3).

Incorrect Solution:

$y = mx + b$
$3 = 4(2) + b$
$3 = 8 + b$
$-5 = b$

Equation: $3 = 4(2) - 5$

Given the slope and any point:

> **To write an equation of a line given the slope (m) and a point (x_1, y_1)**
>
> 1. Substitute the value for m and the coordinates of (x_1, y_1) into the slope-intercept form, $y = mx + b$.
> 2. Solve for b.
> 3. Substitute the values for m and b in the slope-intercept form.
> 4. Simplify the resulting equation.

▪▪▪

EXAMPLE 4

Find the equation of the line that has slope 6 and passes through the point $(-1, -2)$.

SOLUTION

We have $m = 6$ and $(x, y) = (-1, -2)$. Substituting these values into the slope-intercept form and solving for b, we obtain

$$y = mx + b$$
$$-2 = 6(-1) + b$$
$$-2 = -6 + b \quad \text{Simplifying}$$
$$4 = b \quad \text{Adding 6 to both sides}$$

Now we know that the slope is 6 and the y-intercept is $(0, 4)$. Substituting m and b into the slope-intercept form one more time, we get

$$y = mx + b$$
$$y = (6)x + 4 \quad \text{Substituting}$$
$$y = 6x + 4 \quad \text{Simplifying}$$

So the equation of the line is $y = 6x + 4$.

▶ *CHECK* **Warm-Up 4**

Given two points on a line:

> **To write an equation of a line given two points (x_1, y_1) and (x_2, y_2) on the line**
> **1.** Find the slope $m = \frac{y_2 - y_1}{x_2 - x_1}$.
> **2.** Substitute the slope and the coordinates of either point into the slope-intercept form, $y = mx + b$.
> **3.** Solve for b.
> **4.** Substitute the values for m and b into the slope-intercept form.
> **5.** Simplify the resulting equation.
>
> If $x_1 = x_2$, the slope is undefined, the line is vertical, and its equation is $x = x_1$.
>
> If $y_1 = y_2$, the slope is zero, the line is horizontal, and its equation is $y = y_1$.

▪ ▪ ▪
EXAMPLE 5

Find the equation of the line that passes through the points $(5, 1)$ and $(-3, -7)$.

SOLUTION

Because we do not have the slope, we must calculate it. We let $(5, 1)$ be the point (x_1, y_1), and we let $(-3, -7)$ be (x_2, y_2). Then,

$$m = \frac{y_2 - y_1}{x_2 - x_1} = \frac{(-7) - 1}{(-3) - 5} = \frac{-8}{-8} = 1$$

Now we need to find b. To do so, we substitute $m = 1$ and the coordinates of *either* point into the slope-intercept form. We use the second point, $(-3, -7)$, obtaining

$$y = mx + b \qquad \text{Slope-intercept form}$$
$$-7 = (1)(-3) + b \quad \text{Substituting}$$
$$-7 = -3 + b \qquad \text{Simplifying}$$
$$-4 = b \qquad\qquad \text{Adding 3 to both sides}$$

So the slope m is 1 and the y-intercept b is -4. We substitute these into $y = mx + b$ once more.

$$y = (1)x + (-4) \quad \text{or} \quad y = x - 4$$

▶ *CHECK* **Warm-Up 5**

Using the Point-Slope Form

A linear equation in two variables can also be written in the point-slope form,

$$y - y_1 = m(x - x_1)$$

Again, m is the slope of the graph of the equation, and (x_1, y_1) is a point on the graph. This form gives us a simpler way to write the equation of a line when we know the slope and the coordinates of one point.

> **To write an equation for a line, given the slope m and one point (x_1, y_1), using the point-slope form**
>
> 1. Substitute the value of m and the coordinates (x_1, y_1) in the point-slope form, $y - y_1 = m(x - x_1)$.
> 2. Simplify the resulting equation.

■■■

EXAMPLE 6

Use the point-slope form to write the equation of the line that passes through the point $(-3, 2)$ and has slope -5.

SOLUTION

We have $m = -5$, $x_1 = -3$, and $y_1 = 2$. Substituting these values in the point-slope form and simplifying, we get

$y - y_1 = m(x - x_1)$	Point-slope form
$y - 2 = (-5)[x - (-3)]$	Substituting
$y - 2 = -5(x + 3)$	Simplifying
$y - 2 = -5x - 15$	Distributive property
$y = -5x - 13$	Adding 2 to both sides

So the equation is $y = -5x - 13$.

▶ CHECK **Warm-Up 6**

■■■

EXAMPLE 7

STUDY HINT

Some problems that involve slopes and points can be solved in more than one way. Unless you are told otherwise, the choice of method is yours. Look for the one that you find easiest to use.

Write the equation of the line that passes through the point $(1, 3)$ and is perpendicular to the graph of $2x + y = 5$.

SOLUTION

We first put the equation $2x + y = 5$ in slope-intercept form so that we can easily determine the slope.

$$2x + y = 5 \quad \text{becomes} \quad y = -2x + 5$$

The slope of this line is -2, so the slope of the line perpendicular to it is the negative reciprocal of -2, or

$$-\left(\frac{-1}{2}\right) = \frac{1}{2}$$

We now use the point-slope form to find the equation of a line that has slope $\frac{1}{2}$ and passes through the point $(1, 3)$.

$$
\begin{align}
(y - y_1) &= m(x - x_1) && \text{Point-slope form} \\
y - 3 &= \left(\frac{1}{2}\right)(x - 1) && \text{Substituting slope and point} \\
y - 3 &= \left(\frac{1}{2}\right)x - \frac{1}{2} && \text{Distributive property} \\
y &= \left(\frac{1}{2}\right)x + 2\frac{1}{2} && \text{Adding 3 to both sides}
\end{align}
$$

Substitute $(1, 3)$ for x and y to check whether this line actually does pass through the point $(1, 3)$.

▶ *CHECK* **Warm-Up 7**

■ ■ ■

EXAMPLE 8

An experimental car gets 35 miles per gallon on the highway and 30 miles per gallon in town. The gas tank holds 20 gallons. The equation that represents the number of miles y that the car can travel when x of the 20 gallons are used on the highway and $(20 - x)$ gallons are used in town is

$$y = 35x + 30(20 - x)$$

How many miles can the car travel on one tank of gas? Graph this equation and find three solutions.

SOLUTION

We set up a table of values and graph the equation as shown in the following figure.

x	y
0	600
15	675
20	700

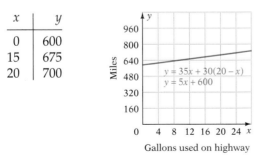

Gallons used on highway

Because the tank holds 20 gallons, the only solutions that are meaningful are those that have x-values equal to numbers from zero to 20. We substitute to find solutions.

$$\text{For } x = 1: \quad y = 35(1) + 30(20 - 1)$$
$$= 35 + 570$$
$$= 605$$

$$\text{For } x = 5: \quad y = 35(5) + 30(20 - 5)$$
$$= 175 + 450$$
$$= 625$$

$$\text{For } x = 10: \quad y = 35(10) + 30(20 - 10)$$
$$= 350 + 300$$
$$= 650$$

So the car can travel 605 miles if 1 gallon is used on the highway and 19 are used in town. In that case, $35 \cdot 1 = 35$ miles represent highway driving and $30(20 - 1) = 30 \cdot 19 = 570$ miles are driven in town.

In like manner, when $x = 5$, then 175 miles are highway miles and 450 miles are driven in town, for a total of 625 miles. And when $x = 10$, then 350 miles are highway miles and 300 are driven in town.

▶ *CHECK* **Warm-Up 8**

Practice what you learned.

SECTION FOLLOW-UP

1.
$$C = \frac{5}{9}(F - 32)$$
$$\frac{9}{5}C = F - 32$$
$$\frac{9}{5}C + 32 = F$$

2. When the temperature is greater than 100°, in either C or F, the Fahrenheit number is greater. When the temperature is less than 0°, in either C or F, the Fahrenheit number is greater. (There are exceptions. See the answer to part 3.)

3. When C and F both equal $-40°$, the measures are equivalent. (When the temperature is below $-40°$, the Celsius numerical value is greater.)

3.3 WARM-UPS

Work these problems before you attempt the exercises.

1. Rewrite $2y + 6x = 12$ in slope-intercept form. Identify the slope and y-intercept. $y = -3x + 6; m = -3,$ y-intercept: $(0, 6)$

2. Rewrite each equation in slope-intercept form. Then identify the pairs of equations that have graphs that are parallel to each other, those that are perpendicular to each other, and those that intersect each other.

   (a) $3y - 6x + 6 = 0$ (b) $2y - 12 = 4x$

   (c) $2y + x = 10$ (d) $4y - 2x + 8 = 0$

2. (a) $y = 2x - 2$
   (b) $y = 2x + 6$
   (c) $y = \frac{-1}{2}x + 5$
   (d) $y = \frac{1}{2}x - 2$

   The graphs of equations (a) and (b) are parallel. The graphs of equations (a) and (c) and those of equations (b) and (c) are perpendicular and intersect. The graphs of equations (a) and (d), (b) and (d), and (c) and (d) intersect only.

3. Find the equation of the line that has slope $-\frac{3}{8}$ and passes through the origin. $y = \frac{-3}{8}x$

4. Find the equation of the line that has slope 2.6 and passes through the point $(-2, 3.5)$. $y = 2.6x + 8.7$

5. Find the equation of the line that passes through the points $(-2, -3)$ and $(2, -5)$. $y = \frac{-1}{2}x - 4$

6. Use the point-slope form to write the equation of the line that passes through the point $(0, -3)$ and has slope -2. $y = -2x - 3$

7. Write an equation of the line that passes through the origin and is parallel to the graph of $y = 2x - 798$. $y = 2x$

8. The sale price of an article is given by the equation

$$y = 0.65p$$

where p is the price before the sale. An article that originally sold for \$194 goes on sale for what price?   \$126.10

3.3 EXERCISES

Note: Use your graphing calculator to check your results whenever possible.

In Exercises 1 through 8, rewrite each equation in slope-intercept form to help you find the slope and y-intercept of its graph. Do not graph.

1. $y + 2x = 4$
   $m = -2;$ y-intercept: $(0, 4)$

2. $3x + y = 7$
   $m = -3;$ y-intercept: $(0, 7)$

3. $3x - y = 5$
   $m = 3;$ y-intercept: $(0, -5)$

4. $-5x + y = 2$
   $m = 5;$ y-intercept: $(0, 2)$

5. $6y - 3x = 9$
   $m = \frac{1}{2};$ y-intercept: $\left(0, \frac{3}{2}\right)$

6. $2x + 6y = 8$
   $m = \frac{-1}{3};$ y-intercept: $\left(0, \frac{4}{3}\right)$

7. $5y - 10x = 15$
   $m = 2;$ y-intercept: $(0, 3)$

8. $12x + 15y = -18$
   $m = \frac{-4}{5};$ y-intercept: $\left(0, \frac{-6}{5}\right)$

In Exercises 9 through 16, use the slope and y-intercept to graph the equation.

9. $y - 2x = 6$ **10.** $y + 4x = 5$ **11.** $5(x + y) = 10$ **12.** $-3(x + y) = 9$

13. $-y = 2x + 3$ **14.** $3x + y = 4$ **15.** $-2y + 4x = 8$ **16.** $-3y - 6x = -15$

In Exercises 17 through 36, write both equations in slope-intercept form, give the slopes of the lines, and determine whether their graphs are parallel, perpendicular, or intersecting. Do not graph.

17. $y = -2x + 1$
$y = \frac{-1}{2}x + \frac{5}{2}$
slopes: -2 and $-\frac{1}{2}$; intersecting

18. $y = 2x + 5$
$15 + 8y = x \quad y = \frac{1}{8}x - \frac{15}{8}$
slopes: 2 and $\frac{1}{8}$; intersecting

19. $y = x + 7$
$y + x = 7 \quad y = -x + 7$
slopes: 1 and -1; perpendicular and intersecting

20. $y = \frac{1}{4}x - 6$
$4x + 3y = 9 \quad y = \frac{-4}{3}x + 3$
slopes: $\frac{1}{4}$ and $\frac{-4}{3}$; intersecting

21. $y = -x + 6$
$x - y = -3 \quad y = x + 3$
slopes: -1 and 1; perpendicular and intersecting

22. $y = \frac{1}{2}x - 5$
$-2x - y = 5 \quad y = -2x - 5$
slopes: $\frac{1}{2}$ and -2; perpendicular and intersecting

23. $4y = 2x + 4 \quad y = \frac{1}{2}x + 1$
$2x - 4 = 4y \quad y = \frac{1}{2}x - 1$
slopes: $\frac{1}{2}$ and $\frac{1}{2}$; parallel

24. $5y = -2x + 1 \quad y = \frac{-2}{5}x + \frac{1}{5}$
$2x - 7 = 5y \quad y = \frac{2}{5}x - \frac{7}{5}$
slopes: $\frac{-2}{5}$ and $\frac{2}{5}$; intersecting

25. $y = 8x + 3$
$\frac{-1}{8}x + y = 7 \quad y = \frac{1}{8}x + 7$
slopes: 8 and $\frac{1}{8}$; intersecting

26. $6y = 3x - 4 \quad y = \frac{1}{2}x - \frac{2}{3}$
$-3y = 6x - 5 \quad y = -2x + \frac{5}{3}$
slopes: $\frac{1}{2}$ and -2; perpendicular and intersecting

27. $y = 8x - 5$
$8x - y = 6 \quad y = 8x - 6$
slopes: 8 and 8; parallel

28. $y = x - 1$
$-y = x + 4 \quad y = -x - 4$
slopes: 1 and -1; perpendicular and intersecting

29. $6y = -3x + 4 \quad y = \frac{-1}{2}x + \frac{2}{3}$
$-3y = 6x - 5 \quad y = -2x + \frac{5}{3}$
slopes: $\frac{-1}{2}$ and -2; intersecting

30. $y = 2x - 7$
$8y + 4x = 16 \quad y = \frac{-1}{2}x + 2$
slopes: 2 and $\frac{-1}{2}$; perpendicular and intersecting

31. $6y = 2x - 6 \quad y = \frac{1}{3}x - 1$
$6x - 2y = 1 \quad y = 3x - \frac{1}{2}$
slopes: $\frac{1}{3}$ and 3; intersecting

32. $y = 3x - 2$
$y - 3x = 6 \quad y = 3x + 6$
slopes: 3 and 3; parallel

33. $y = -2x + 4$

$\frac{1}{2}x + y = 11$ $y = -\frac{1}{2}x + 11$

slopes: -2 and $-\frac{1}{2}$; intersecting

34. $y = 2x + 5$

$2x - y = 8$ $y = 2x - 8$

slopes: 2 and 2; parallel

35. $y = \frac{1}{4}x + 3$

$4x + y = 7$ $y = -4x + 7$

slopes: $\frac{1}{4}$ and -4; perpendicular and intersecting

36. $y = \frac{2}{3}x - 7$

$3y - 2x = 11$ $y = \frac{2}{3}x + \frac{11}{3}$

slopes: $\frac{2}{3}$ and $\frac{2}{3}$; parallel

In Exercises 37 through 44, graph the line that is described.

37. The line that is parallel to the graph of $y = 3x - 2$ and has the y-intercept $(0, 3)$

38. The line that is parallel to the graph of $y = -2x + 1$ and has the y-intercept $(0, -3)$

38.

39. The line that is parallel to the graph of $y = x - 5$ and has the y-intercept $(0, 2)$

40. The line that is parallel to the graph of $y = -x + 5$ and has the y-intercept $(0, -3)$

40.

41. The line that is perpendicular to the graph of $y = 5$ and has the x-intercept $(2, 0)$

42. The line that is perpendicular to the graph of $y = -3x$ and has the y-intercept $(0, -2)$

42.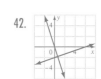

43. The line that is parallel to the graph of $y = 3$ and passes through the origin

44. The line that is perpendicular to the graph of $x = -2$ and passes through the origin

44.

In Exercises 45 through 50, write the equation of the line that has the given slope and y-intercept. Write your answers in slope-intercept form.

45. slope: -2

y-intercept: $\left(0, \frac{1}{3}\right)$

$y = -2x + \frac{1}{3}$

46. slope: $\frac{3}{5}$

y-intercept: $(0, 5)$

$y = \frac{3}{5}x + 5$

47. slope: $\frac{-1}{2}$

y-intercept: $(0, 8)$

$y = \frac{-1}{2}x + 8$

48. slope: -4

y-intercept: $\left(0, \frac{2}{5}\right)$

$y = -4x + \frac{2}{5}$

49. slope: $\frac{3}{5}$

y-intercept: $(0, -6)$

$y = \frac{3}{5}x - 6$

50. slope: -3

y-intercept: $\left(0, \frac{1}{4}\right)$

$y = -3x + \frac{1}{4}$

In Exercises 51 through 58, write the equation of the line that has the given slope and passes through the given point. Write your answers in slope-intercept form.

51. slope: -4

point: $(-9, -2)$
$y = -4x - 38$

52. slope: 3

point: $(2, 3)$
$y = 3x - 3$

53. slope: 2

point: $(-4, 1)$
$y = 2x + 9$

54. slope: 1

point: $(-3, 5)$
$y = x + 8$

55. slope: $\frac{1}{2}$

point: $(5, -2)$
$y = \frac{1}{2}x - 4\frac{1}{2}$

56. slope: $\frac{-3}{5}$

point: $(-2, 6)$
$y = -\frac{3}{5}x + 4\frac{4}{5}$

57. slope: $-\frac{1}{4}$

point: $(1, -1)$
$y = -\frac{1}{4}x - \frac{3}{4}$

58. slope: $\frac{1}{3}$

point: $(2, -2)$
$y = \frac{1}{3}x - 2\frac{2}{3}$

In Exercises 59 through 66, write the equation of the line that passes through the given points. Write your answers in slope-intercept form.

59. $(2, 9)$ and $(-1, 1)$

$y = 2\frac{2}{3}x + 3\frac{2}{3}$

60. $(1, 2)$ and $\left(0, 2\frac{1}{4}\right)$

$y = -\frac{1}{4}x + 2\frac{1}{4}$

61. $(2, 2)$ and $(4, -5)$

$y = -3\frac{1}{2}x + 9$

62. $(3, 5)$ and $(5, 7)$

$y = x + 2$

63. $(-1, 6)$ and $(-5, 2)$

$y = x + 7$

64. $(2, -3)$ and $(10, -9)$

$y = -\frac{3}{4}x - 1\frac{1}{2}$

65. $(-15, 3)$ and $(-5, 1)$

$y = -\frac{1}{5}x$

66. $(3, 5)$ and $(-3, -5)$

$y = 1\frac{2}{3}x$

In Exercises 67 through 74, write the equation of the line that has the given slope and passes through the given point. Write your answers in slope-intercept form.

67. slope: -2

point: $(-1, 2)$
$y = -2x$

68. slope: -6

point: $(3, -4)$
$y = -6x + 14$

69. slope: -3

point: $(-2, -2)$
$y = -3x - 8$

70. slope: -9

point: $(0, 0)$
$y = -9x$

71. slope: 5

point: $(5, 3)$
$y = 5x - 22$

72. slope: 8

point: $(4, 6)$
$y = 8x - 26$

73. slope: $\frac{1}{3}$

point: $(1, -3)$
$y = \frac{1}{3}x - 3\frac{1}{3}$

74. slope: -3

point: $(2, 4)$
$y = -3x + 10$

In Exercises 75 through 82, write the equation of the line that passes through the given points. Simplify all equations and write them in terms of y.

75. $(-1, 1)$ and $(3, 4)$

$y = \frac{3}{4}x + 1\frac{3}{4}$

76. $(-7, -4)$ and $(2, -6)$

$y = -\frac{2}{9}x - 5\frac{5}{9}$

77. $(4, -2)$ and $(1, -2)$

$y = -2$

78. $(12, 3)$ and $(-1, -1)$

$y = \frac{4}{13}x - \frac{9}{13}$

79. $(2, -3)$ and $(5, -6)$

$y = -x - 1$

80. $(3, 5)$ and $(2, -1)$

$y = 6x - 13$

81. $(-4, 2)$ and $(10, 3)$

$y = \frac{1}{14}x + 2\frac{2}{7}$

82. $(-6, 1)$ and $(5, -4)$

$y = -\frac{5}{11}x - 1\frac{8}{11}$

In Exercises 83 through 88, write the equation of the line that passes through the given point and is parallel to the graph of the given line. Simplify all equations and write them in terms of y.

83. point: $(-1, 1)$

line: $y - 2x = 5$
$y = 2x + 3$

84. point: $(0, 0)$

line: $x + y = 9$
$y = -x$

85. point: $(-7, 4)$

line: $y = 3x + 1$
$y = 3x + 25$

86. point: $(2, -3)$

line: $x - y = 7$
$y = x - 5$

87. point: $(12, 3)$

line: $3y - 2x = 6$
$y = \frac{2}{3}x - 5$

88. point: $(-2, 0)$

line: $8x - 7y = 7$
$y = \frac{8}{7}x + 2\frac{2}{7}$

In Exercises 89 through 94, write the equation of the line that passes through the given point and is perpendicular to the graph of the given line. Use any method. Simplify all equations and write them in terms of y.

89. point: $(-3, 6)$

line: $y + 3x = 7$
$y = \frac{1}{3}x + 7$

90. point: $(-3, -2)$

line: $2x - 4y = 9$
$y = -2x - 8$

91. point: $(-1, 4)$

line: $-2x - 8y = 1$
$y = 4x + 8$

92. point $(1, 2)$

line: $y = 2x + 5$
$y = -\frac{1}{2}x + 2\frac{1}{2}$

93. point: $(5, -2)$

line: $y + 3x = 4$
$y = \frac{1}{3}x - 3\frac{2}{3}$

94. point: $(-6, -2)$

line: $5x - y = 9$
$y = -\frac{1}{5}x - 3\frac{1}{5}$

95. An equation for converting a temperature from Fahrenheit *(F)* to Celsius *(C)* is: $F = \left(\frac{9}{5}\right)C + 32$. What does the point $(0, 32)$ mean? What does the slope of this relationship mean in real life?   *When the temperature is 0°C, the equivalent temperature is 32°F. For every 5° that the temperature changes on the Celsius scale, it changes 9°F.*

96. **a.** Write the equation for changing a temperature from degrees Celsius to kelvins (units on the Kelvin temperature scale) if one point that satisfies the equation is $(C, K) = (250, 523)$ and another is $(C, K) = (0, 273)$.   *K = C + 273*

   b. The *K*-intercept is the Kelvin temperature at which water freezes $(0°C)$. Find the *K*-intercept. What does it mean in real life?   *K-intercept: (0, 273); water freezes at 273 K*

   c. Find the Kelvin temperature at which water boils $(100°C)$.   *Water boils at 373 K.*

97. *Election Day Temperatures* In a certain city, the relationship between the temperature on Election Day *(T)* and the number of people voting *(V)* can be represented as a line that passes through the points $(50, 10{,}000)$ and $(90, 5000)$.

   a. Identify the slope.   *slope: −125*

   b. Find the equation of this line.   *V = −125T + 16,250*

   c. What are the *x*- *(T-)* and *y*- *(V-)* intercepts? What meaning do they have in real life (if any)?   *T-intercept: (130, 0); V-intercept (0, 16,250). The T-intercept means that when it is 130°, no one votes. The V-intercept means that when it is 0°, 16,250 people will vote.*

98. *Calories and Exercises* Calories burned off in exercise *(c)* and minutes exercised *(m)* exhibit a linear relationship when ordered as *(m, c)*.

   a. For a particular runner, two days of exercise yields $(60, 552)$ and $(82, 754.4)$. Write an equation that represents this linear relationship.   *c = 9.2m*

   b. What is the *c*-intercept of the graph of this relationship? What does it mean in real life?   *(0, 0). When you don't run, you don't burn off calories. This is technically* not *correct because you are always burning off calories. Therefore, this point has no meaning in real life.*

   c. How many calories does this runner burn every 10 minutes?   *92 calories*

99. *Taxi Fares* The rate charged by a cab company is represented by $y = 29x + 1.90$, where *x* is the number of tenths of a mile traveled.

   a. Write the equation of the new rate charged by this company when it keeps the same rate per mile but raises the initial charge 10 cents.   *y = 29x + 2.00*

   b. Can *x* ever be negative? Why or why not?   *No. Because x represents distance, and distance is* not *negative.*

   c. Do *x*- and *y*-intercepts have any real-life meaning for the graphs of this equation? Why or why not?   *No. (0, 2) would mean that you rode no miles (or no tenths of a mile) and paid $2.00. Hardly likely. The y-intercept would require x to be negative. That is not possible here.*

100. *Restaurant Charges* The cost for a buffet special at the Skylite Diner is a "plate fee" plus an amount per ounce. The cashier writes down the number of ounces *(z)* and the price owed *(p)* as an ordered pair. She writes $(10, 4.85)$ and $(22, 8.33)$ for two orders.

   a. Write the equation that represents this situation.   *p = 0.29z + 1.95*

   b. If you buy 15 ounces, what will you pay?   *$6.30*

c. Your plate of food costs $4.27. How many ounces of food did you select?   8 ounces

d. What does the slope mean in real life?   For every ounce of food you put on your plate, your cost will increase by 29¢.

101. *Cab Fares*   Three men take cabs from their homes to the airport. The miles they traveled (x) and the amount they paid (y) are as follows:

$$(x, y) = (2.4, \$4.20)$$
$$(x, y) = (8, \$11)$$
$$(x, y) = (6, \$8.70)$$

Did their cab companies all charge the same rate? (*Hint:* Are these points collinear? That is, do they all lie along the same line?)   no

102. *Health Club Fees*   A health club charges a fixed fee plus an amount per service used (aerobics class, free weights, racketball court, and so on). Three patrons are given the following bills:

Patron 1   four services, $15
Patron 2   twelve services, $21
Patron 3   seven services, $18

The third person was overcharged. Write the equation that represents the correct pricing information, and correct the charges for patron 3.
$y = \frac{3}{4}x + 12$; Patron 3 should have been charged $17.25 instead of $18.00.

MIXED PRACTICE

By doing these exercises, you will practice the topics up to this point in the chapter.

103. Determine whether the line $-3x + y = 7$ contains the point $(0, 6)$.   no

104. Find the equation of the line that passes through the points $(8, 0)$ and $(-3, -8)$.
$y = \frac{8}{11}x - \frac{64}{11}$

105. Solve $-y = x + 7$ for x and for y.
$x = -y - 7$; $y = -x - 7$

106. Find A when $Ax + 4y = 12$ and one solution is $\left(-1, -4\frac{1}{2}\right)$. $A = -30$

107. Find the x- and y-intercepts of the graph of $2y + 7x = -3$.
x-intercept: $\left(-\frac{3}{7}, 0\right)$; y-intercept: $\left(0, -\frac{3}{2}\right)$

108. Write the equation of the line that is parallel to the graph of $y = 3x - 2$ and passes through the origin. $y = 3x$

109. Find the slope and the y-intercept of the line whose equation is $-2y + 4x = 9$.
slope: 2; y-intercept: $\left(0, \frac{-9}{2}\right)$

110. Graph: $10x - 5y = 20$

EXCURSIONS

Data Analysis

1. ✏   Compare and contrast the two telephone companies in the following table. Which company seems to offer the best value to you? Justify your answer.

Long-Distance Calls

City to City	Minutes	Telephone Company		City to City	Minutes	Telephone Company	
		A	B			A	B
New York City to Boston	17	$3.62	$1.86	Rockville Centre to Kansas City	5	1.34	0.68
Great Neck to Chicago	2	0.57	0.27	New Canaan to Cincinnati	26	6.19	3.32
Greenwich to Atlanta	20	4.79	2.66	Flushing to Milwaukee	3	0.81	0.40
White Plains to Philadelphia	6	1.26	0.63	Tarrytown to San Diego	15	4.01	2.21
Newark to Denver	14	3.56	1.96	New York City to Oklahoma City	59	14.68	8.13
Darien to New Orleans	42	10.48	5.74	Edison to Colorado Springs	23	5.79	3.23
Hackensack to Houston	21	5.29	2.91	Elmsford to Tucson	12	3.23	1.74
Jersey City to San Francisco	2	0.63	0.30	The Oranges to Baton Rouge	36	9.00	4.92
Forest Hills to Washington, D.C.	10	2.30	1.10	Stamford to Indianapolis	18	4.32	2.36
Garden City to Minneapolis	19	4.80	2.57	Farmingdale to Louisville	7	1.74	0.91
Elmsford to Salt Lake City	48	12.59	6.88	New York City to St. Louis	14	3.38	1.88
Paramus to Detroit	9	2.21	1.10	Brooklyn to Pittsburgh	5	1.26	0.58
New York City to Los Angeles	16	4.27	2.36				

2. Choose the car of your dreams and research the price. Using the information in the following table, finance your car and give the amount you would owe down and your cost per month (don't forget the interest rate). How would you find your *real* total cost by the end of your payments? Justify your bank choice.

CAR LOAN RATES ----------------

Bank	New Car			Used Car		
	Term (months)	Interest rate (%)	% down	Term (months)	Interest rate (%)	% down
A	48	6.00	15	36	12.50	15
B	48	10.90	20	48	13.40	20
C	48	13.00	20	36	14.75	20
D	60	10.50	10	48	13.40	10
E	60	11.00	10	60	13.00	10
F	48	12.50	0	48	13.50	0
G	48	11.90	20	60	12.90	20
H	60	10.95	10	60	13.25	10
I	24	6.75	25	36	14.00	25
J	48	8.25	20	36	9.50	20
K	36	8.25	25	48	14.25	30

CONNECTIONS TO *MEASUREMENT*

Applying Ratio and Proportion

Sometimes information is given in one form but we want it in another form. We use measurements to describe length, area, volume, capacity, mass, time, and temperature. Two systems of measurement are used in the United States, the U.S. Customary system and the metric system. The metric system is used almost exclusively throughout the rest of the world.

In this Connection, we compare the two systems informally and **convert,** or rename, measurements within the systems and from system to system. The ideas of ratio and proportion are helpful in doing this.

The first measure we discuss is **length.** The most common units of length are shown in Table 1. (Note: All tables referred to in this Connection appear on the inside covers of this text.) Here are some examples of length:

> The wire in a paper clip is usually about 1 millimeter thick.
> An average-size man's long step is about 1 meter.
>   (A meter is a little more than a yard.)

•••

EXAMPLE

3245 centimeters is how many millimeters? (3245 cm = ? mm)

SOLUTION

To answer this question, we will set up a proportion involving ratios of centimeters to millimeters. We will use the fact that 10 millimeters is the same as 1 centimeter to help us convert this measurement. Then 3245 centimeters is to n millimeters as 1 centimeter is to 10 millimeters.

$$\frac{3245 \text{ cm}}{n \text{ mm}} = \frac{1 \text{ cm}}{10 \text{ mm}}$$

Once we check that the units are the same on both sides, we can drop them.

$$\frac{3245}{n} = \frac{1}{10}$$

$$3245 \times 10 = 1 \times n \quad \text{Cross-multiply}$$

$$32,450 = n$$

So 3245 centimeters is the same as 32,450 millimeters.

In the metric system, spaces are sometimes used instead of commas to separate the periods (groups of three digits). Then 32,450 would be written 32 450.

Units of **area** are shown in Table 2. Area is a measure of the extent of a surface. One square unit of area can be thought of as the area of a square whose side is 1 unit long.

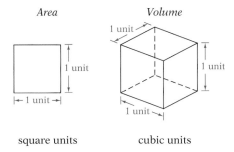

Area   Volume

square units   cubic units

Volume is the measure of the three-dimensional space in a container. Volume units are given in Table 3. The measure of liquid volume is usually called **capacity** (Table 4).

One milliliter is the same as 1 cubic centimeter. A teaspoon holds about 5 mL. A liter is just a bit larger than a quart.

▪▪▪
EXAMPLE

1950 ounces is equal to how many quarts?

SOLUTION

We first must look up the relationship between ounces and quarts. Table 4 shows no direct relationship. However, it does show that 16 ounces is 1 pint and that 2 pints is 1 quart. We shall need two proportions.

Our first proportion is

$$\frac{1950 \text{ oz}}{n \text{ pt}} = \frac{16 \text{ oz}}{1 \text{ pt}} \quad \text{or} \quad \frac{1950}{n} = \frac{16}{1}$$

Then we get

$$1950 \times 1 = 16 \times n \quad \text{Cross-multiplying}$$
$$\frac{975}{8} = n \qquad\qquad \text{Simplifying}$$

So 1950 ounces is the same as $\frac{975}{8}$ pints.

Now we want to determine the number of quarts equivalent to $\frac{975}{8}$ pints. We set up another proportion:

$$\frac{\frac{975}{8} \text{ pt}}{n \text{ qt}} = \frac{2 \text{ pt}}{1 \text{ qt}} \quad \text{or} \quad \frac{121.875}{n} = \frac{2}{1}$$

Then we have

$$121.875 \times 1 = 2 \times n \quad \text{Cross-multiplying}$$
$$60.9375 = n \qquad\qquad \text{Simplifying}$$

So $\frac{975}{8}$ pints is the same as 60.9375 quarts and is also equivalent to 60 quarts 1 pint 14 ounces.

◢

Table 5 lists units of another measurement—**mass,** or the amount of matter in an object. Although many people think that weight and mass mean the same thing, in science, mass and weight are different concepts. Mass is the amount of matter in an object. Weight is a measure of the force of gravity exerted on an object. However, it is common to use the metric units for mass to describe the metric weight of an object.

> The mass of one paper clip is about 1 gram.
> A kilogram is slightly more than 2 pounds.
> The mass of a sub-compact car is about 1 ton.

Units of **time** are listed in Table 6. They are used in both the metric and the U.S. customary systems.

In the next conversion example, we will use ratios to change two units at the same time.

▪ ▪ ▪

EXAMPLE

Professor Bill runs a 9-minute mile (9 minutes per mile). How many feet per second does he run?

SOLUTION

He runs $\frac{9 \text{ minutes}}{\text{mile}}$. We want to write this as $\frac{? \text{ feet}}{\text{second}}$. We need to use these relationships:

$$1 \text{ minute} = 60 \text{ seconds} \longrightarrow \frac{60 \text{ sec}}{1 \text{ min}} \quad \text{or} \quad \frac{1 \text{ min}}{60 \text{ sec}}$$

$$1 \text{ mile} = 5280 \text{ feet} \longrightarrow \frac{5280 \text{ ft}}{1 \text{ mi}} \quad \text{or} \quad \frac{1 \text{ mi}}{5280 \text{ ft}}$$

We have to chose the units to rewrite

$$\frac{9 \text{ min}}{\text{mi}} \cdot \qquad = \frac{? \text{ ft}}{\text{sec}}$$

We choose

$$\frac{9 \text{ min}}{\text{mi}} \left(\frac{60 \text{ sec}}{1 \text{ min}} \right) \left(\frac{1 \text{ mi}}{5280 \text{ ft}} \right) = \frac{5410 \text{ sec}}{5280 \text{ ft}}$$

Now we have the right units. However, they are in the wrong order. We invert the ratio to

$$\frac{5280 \text{ ft}}{540 \text{ sec}}$$

And then rewrite as a unit ratio.

$$9.8 \text{ ft/sec}$$

◢

PRACTICE

The heights of four South American volcanoes are given in Exercises 1 through 4. Use a calculator to convert them as requested. In all cases, round to the nearest hundredth of a unit.

1. Aconcagua, 22,834 feet. Convert to miles.
4.32 miles

2. Chimborazo, 20,560 feet. Convert to yards.
6853.33 yards

3. Antisana, 18,713 feet. Convert to inches.
224,556 inches

4. Cotopaxi, 19,344 feet. Convert to miles.
3.66 miles

The jumping heights actually recorded for certain animals are given in Exercises 5 through 8. Convert these heights into the given units.

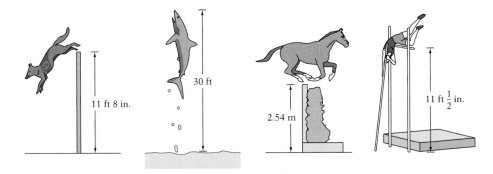

11 ft 8 in.

30 ft

2.54 m

11 ft $\frac{1}{2}$ in.

5. A German Shepherd dog in K-9 training jumped 11 feet 8 inches. Convert to yards. Express as a fraction. $3\frac{8}{9}$ yards

6. A Mako shark can jump 30 feet into the air. Convert to inches.   360 inches

7. An Australian horse has been recorded jumping 2.54 meters. Convert to centimeters.   254 centimeters

8. The first recorded pole vault over 11 feet was 11 feet $\frac{1}{2}$ inch. Convert this to inches. $132\frac{1}{2}$ inches

Some animal long-jump records are given in Exercises 9 through 12. Represent them in the requested units.

9. Frog, 17 feet $6\frac{3}{4}$ inches. Represent as inches.
$210\frac{3}{4}$ inches

10. Snow leopard, 15.24 meters. Represent this as centimeters.   1524 centimeters

11. Greyhound, 9.14 meters. Represent this as millimeters.   9140 millimeters

12. Horse, 32 feet 10 inches. Represent as yards expressed as a fraction. $10\frac{17}{18}$ yards

Animals can travel at very high speeds. In Exercises 13 through 16, convert each of the given speeds as requested.

13. Red kangaroo, 45 miles per hour. Rewrite as kilometers per hour.   72 kilometers per hour

14. Cheetah, 98 kilometers per hour. Rewrite as miles per hour.   61.25 miles per hour

15. Gazelle, 68 kilometers per hour. Rewrite as miles per hour. 42.5 miles per hour

16. Ostrich, 50 miles per hour. Rewrite as kilometers per hour. 80 kilometers per hour

The weights of various playing balls are given in Exercises 17 through 20. Rewrite them in the requested units.

17. Table tennis ball, 0.09 ounce. Rewrite in grams. 2.52 grams

18. Jai alai ball, 127 grams. Rewrite in ounces expressed to the nearest ounce. 5 ounces

19. Badminton "bird," 5.5 grams. Rewrite in kilograms. 0.0055 kilogram

20. Racquetball, 1.4 ounces. Rewrite in pounds. 0.0875 pound

3.4 Graphing Linear Inequalities

SECTION GOALS

- *To graph linear inequalities in one or two variables*
- *To graph inequalities that contain absolute value expressions*

SECTION LEAD-IN

In meteorology, the wind chill, H (heat loss), is related to the wind speed, V; the neutral skin temperature, S; the air temperature, T; and the constants A, B, and C. The formula is

$$H = (A + B\sqrt{V} - CV)(S - T)$$

where $S > T$.

a. Using what you have learned about equations and variables, discuss what happens to the value of H if all other variables remain constant but S becomes smaller.

b. What if only T changes and it gets larger?

c. Solve this equation for T.

Linear Inequalities in Two Variables

> **Linear Inequalities in Two Variables**
>
> A **linear inequality in two variables** is an inequality that can be placed in the standard form $ax + by < c$, where a, b, and c are constants and a and b are not both equal to zero. (If a or b is zero, it is an inequality in *one variable*.)

The symbols $>$, $\geq$, and $\leq$ can replace $<$ in this definition.

The graph of a linear inequality in two variables is a portion, or **region,** of the coordinate plane. The region is bounded by the line whose equation is formed when we substitute an equal sign for the inequality symbol. The line is part of the graph for $\geq$ and $\leq$ inequalities, but it is not part of the graph for $<$ and $>$ inequalities.

To graph a linear inequality in two variables

1. Replace the inequality sign with an equal sign and graph the resulting equation (a line).
2. Draw the line of step 1 solid for a $\leq$ or $\geq$ inequality; draw it dashed for a $<$ or $>$ inequality.
3. Substitute the coordinates of any point into the original inequality, and simplify.
4. If the statement resulting from step 3 is true, shade the region of the plane that contains the point. If it is false, shade the region that does not contain the point.

The graph of the inequality is the shaded region plus the solid line (if any).

The graph of an inequality is the graph of its solution set. In step 3 of the graphing process, we are actually testing to see whether the point is in that solution set.

▪▪▪
EXAMPLE 1

Graph: $x + 2y < 8$

SOLUTION

Substituting an equal sign for the "less than" sign gives us the equation $x + 2y = 8$. To obtain two points on this line for graphing, we set x and y, in turn, equal to zero. This gives us the intercepts $(0, 4)$ and $(8, 0)$.

$$x + 2y = 8 \qquad x + 2y = 8$$
$$0 + 2y = 8 \qquad x + 2\,(0) = 8$$
$$y = 4 \qquad x = 8$$

We graph these points and draw the line through them, as shown in the figure on the left on page 212. The line is drawn dashed because the graph of this "less than" inequality does not include points that lie on the line.

A good "test point" for determining which region to shade is the origin, $(0, 0)$. It makes the next step very easy. Substituting $x = 0$ and $y = 0$ in the original equality yields

$$x + 2y < 8$$
$$0 + 2\,(0) < 8$$
$$0 < 8 \quad \text{True}$$

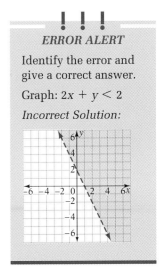

WRITER'S BLOCK

Write directions for a friend to use when graphing a linear inequality.

STUDY HINT

It is a good idea to test an inequality by using two points—one on either side of the line. Only one of the two will give a true statement, and the other will serve as a check.

Therefore, we shade the region that contains the origin (the lower region). The figure on the right below is the completed graph.

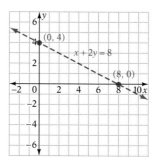 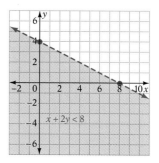

▶ *CHECK* **Warm-Up 1**

In the figure on the right above, the point $(2, 5)$ is in the upper region of the plane. If we substitute that point into the original inequality, we get

$$x + 2y < 8$$
$$2 + 2(5) < 8$$
$$12 < 8 \quad \text{False}$$

This confirms that the *lower* region should be shaded.

ERROR ALERT

Identify the error and give a correct answer.

Graph: $x + y < 5$

Incorrect Solution:

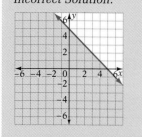

INSTRUCTOR NOTE

The most common error in graphing inequalities is shading. Remind students that picking a point on the line will not help to determine the shading. We try to use (0, 0) when practical so that substitutions will be easier, but this won't work if the y-intercept is zero.

■■■
EXAMPLE 2

Graph $16x - 4y \geq 9$ and identify three points in the solution set.

SOLUTION

The associated equation is $16x - 4y = 9$. In slope-intercept form, this equation becomes

$$y = 4x - \frac{9}{4}$$

so its graph has a slope of 4 and the y-intercept $\left(0, -\frac{9}{4}\right)$. It is drawn solid in the figure on the left on page 213 to show that it is part of the graph of the inequality.

The point $(0, 0)$ is to the left of the line. It yields, in the inequality,

$$16x - 4y \geq 9$$
$$16(0) - 4(0) \geq 9$$
$$0 \geq 9 \quad \text{False}$$

The point $(2, 0)$ is to the right of the line. It yields

$$16(2) - 4(0) \geq 9$$
$$32 \geq 9 \quad \text{True}$$

Thus we shade the region to the right of the line, obtaining the figure on the right on page 213. This is the graph of $16x - 4y \geq 9$. Any point in the shaded area or on the line $16x - 4y \geq 9$ is part of the solution. So, in particular, point $A\ (2, -6)$, point $B\ (6, 6)$, and point $C\ (6, -5)$ are points in the solution set.

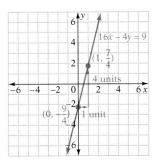 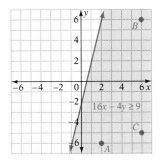

► CHECK **Warm-Up 2**

Every point of the graph of an inequality is a solution of that inequality. For example, every point in the shaded area in the figure on the right above, and every point on the line shown there, satisfies $16x - 4y \geq 9$ when its coordinates are substituted into that inequality. This solution set may be written

$$\{(x, y) \mid 16x - 4y \geq 9\}$$

and the figure shows the graph of this set.

Linear Inequalities in One Variable

The equations $x = a$ and $y = b$ have graphs that are vertical and horizontal lines, respectively. We use them to graph linear inequalities *in one variable* on the plane.

INSTRUCTOR NOTE

Students tend to misgraph horizontal and vertical lines. Practice will help avoid this problem.

▪ ▪ ▪

EXAMPLE 3

Graph on the coordinate plane: **a.** $x \leq 6$ **b.** $y > 0$

SOLUTION

a. The line $x = 6$ is a vertical line 6 units to the right of the y-axis. Numbers less than 6 are to the left of this line, so the graph is that shown in the figure below on the left.

b. The line $y = 0$ is the x-axis. Numbers greater than 0 are above this axis, so the graph is that shown in the figure below on the right. Note that the x-axis is dashed here.

WRITER'S BLOCK

When do we use a solid line and when a dashed line?

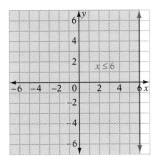 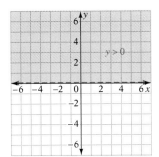

► CHECK **Warm-Up 3**

Inequalities that Contain Absolute Value

For absolute value expressions,

$$|a| < b \quad \text{means} \quad -b < a < b$$

and

$$|a| > b \quad \text{means} \quad a < -b \text{ or } a > b$$

These "translations" hold even when a is an expression involving two variables. We use them to solve and graph absolute value inequalities.

▪▪▪

EXAMPLE 4

Graph the solution set for $|x + y| > 4$.

SOLUTION

We rewrite this inequality as $x + y > 4$ *or* $x + y < -4$.
Finding the x- and y-intercepts of the associated equations, we have

	$x + y = 4$			$x + y = -4$	
	For $x = 0$:	For $y = 0$:		For $x = 0$:	For $y = 0$:
	$0 + y = 4$	$x + 0 = 4$		$0 + y = -4$	$x + 0 = -4$
	$y = 4$	$x = 4$		$y = -4$	$x = -4$
Intercepts:	$(0, 4)$	$(4, 0)$		$(0, -4)$	$(-4, 0)$

We use these intercepts to graph the two associated equations as dashed lines, as is done (separately) in figures (a) and (b) below. Next we must decide which regions to shade. We test the point $(0, 0)$ in each case.

For $x + y > 4$:
$$0 + 0 < -4$$
$$0 > 4 \quad \text{False}$$

For $x + y < -4$:
$$0 + 0 < -4$$
$$0 < -4 \quad \text{False}$$

So $(0, 0)$ is not a solution of either inequality, and we shade as shown in figures (a) and (b). The union of these inequalities is the solution set for the given absolute value inequality. Its graph is the union of the graphs in figures (a) and (b), shown in figure (c).

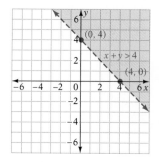
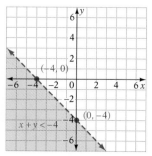
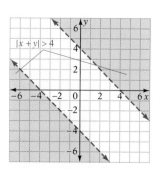

▶ CHECK Warm-Up 4

...

EXAMPLE 5

Graph the solutions of $|2x + 3y| \leq 6$.

SOLUTION

This inequality can be written in condensed form as $-6 \leq 2x + 3y \leq 6$ or as the conjunction $-6 \leq 2x + 3y$ *and* $2x + 3y \leq 6$.

The graphs of the latter inequalities are drawn in figures (a) and (b) below, respectively. Their intersection, the required solution, is shown in figure (c).

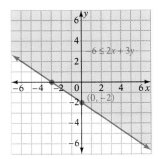 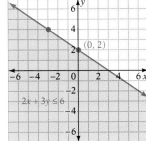 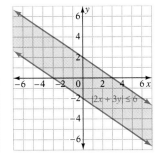

▶ CHECK **Warm-Up 5**

Practice what you learned.

SECTION FOLLOW-UP

a. As S becomes smaller, H becomes smaller. Explain.

b. As T becomes larger, H becomes smaller. Explain.

c. $T = S - \dfrac{H}{A + B\sqrt{V} - CV}$

3.4 WARM-UPS

Work these problems before you attempt the exercises.

1. Graph: $x + y < 5$

2. Graph: $-9x + 3y \leq -3$

3. Graph: $-y + 2 > 0$

4. Graph: $|2x + 2y| > 6$

5. Graph: $|4x - 8| \leq -2y$

3.4 EXERCISES

Note: Use your graphing calculator to check your results whenever possible.

In Exercises 1 through 24, graph the inequality.

1. $x - 3 \geq y$

2. $y \geq x + 5$

3. $y \leq 4x - 7$

4. $y < 2x + 5$

5. $9x - 2 > y$

6. $4x - 5 \geq y$

7. $x + y \geq 7$

8. $4x + 7 > y$

9. $x - 2y \geq 4$

10. $x + y < 1$

11. $y \leq 2x$

12. $y + x \geq 3$

13. $x \leq 7 + y$

14. $4x + y \geq -3$

15. $x - 2y \leq 8$

16. $\frac{x}{3} + y > 7$

17. $x + \frac{y}{4} \geq 1$

18. $\frac{y}{5} - x \geq -1$

19. $9(x + y) \geq -27$

20. $2(x + y) < 4$

21. $2(y - 3) < x$

22. $5(x + y) > 10$

23. $2(x + y) \geq -6$

24. $-3(y - x) \leq 12$

In Exercises 25 through 28, graph the parts of each compound inequality in one color, and then graph the union or intersection, as required, in a second color.

25. $x > 5$ or $x \leq -2$ **26.** $y \geq 2$ or $y \leq -1$ **27.** $x \geq -2$ and $x < 5$ **28.** $y \geq -3$ and $y < 4$

In Exercises 29 through 34, graph the compound inequalities.

29. $-2 < x < 5$ **30.** $-4 < x < 0$ **31.** $-6 \leq x < 1$

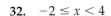

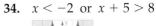

32. $-2 \leq x < 4$ **33.** $x + 3 \leq 7$ or $x > 5$ **34.** $x < -2$ or $x + 5 > 8$

In Exercises 35 through 38, match each pair of inequalities to the correct graphs.

A. $x \geq 0$ and $y \geq 0$ **B.** $x \leq 0$ and $y \leq 0$

C. $x \geq 0$ and $y \leq 0$ **D.** $x \leq 0$ and $y \geq 0$

35. 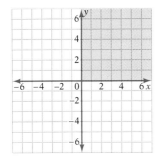   **36.**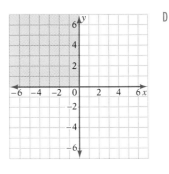

37. 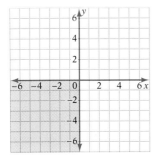   **38.**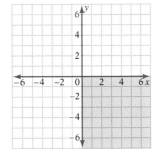

In Exercises 39 through 42, match each pair of inequalities to the correct graphs.

A. $y \leq -x$ and $x \leq 0$ **B.** $y \geq -x$ and $(x \leq 0$ or $y \leq 0)$

C. $x \geq 0$ and $y \leq -x$ **D.** $y \geq -x$ or $x \geq 0$

39. 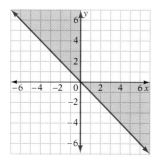 B

40. 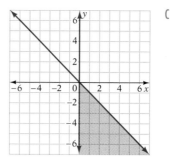 C

41. 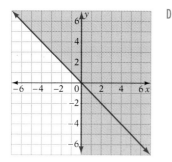 D

42. 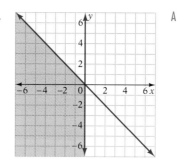 A

In Exercises 43 through 48, graph the absolute value inequalities.

43. $|x + 3| \leq 2y$

44. $|2x - 4| \leq 2y$

45. $|3x - 3y| \leq 9$

46. $|8x - 2y| \leq 6$

47. $|2y - 6x| > 2$

48. $|3x - 6| > y$

In Exercises 49 through 54, graph the compound inequalities.

49. $|y| \leq 2$ and $|x| \leq 3$

50. $|y| < 3$ and $|x| > -2$

51. $|2x - y| > 3$ and $y < 3$

52. $|x + 3y| > 2$ and $y < 1$

53. $|2y| > 6$ and $|x + 3| \leq 1$

54. $|3y| > 6$ and $|x - 2| \leq 3$

MIXED PRACTICE

By doing these exercises, you will practice the topics up to this point in the chapter.

55. Find three points that lie in the solution set for $-3x + y \geq -9$. (0, 9), (0, 10),
(0, 11). Answers may vary.

56. Graph the inequality: $-9x - 3y \leq -6$

56.

57. Write the equation of the line that passes through the points (6, 9) and
$(-3, -5)$. $y = \frac{14}{9}x - \frac{1}{3}$

58. Find the slope and y-intercept of the line whose equation is $-3y + 2x = 12$.
slope: $\frac{2}{3}$; y-intercept: $(0, -4)$

59. Find the equation of the line that has slope -6 and y-intercept $(0, -4)$.
$y = -6x - 4$

60. Graph: $2x - y = 6$

60.

61. Graph the inequality: $|3x + y| < 4$

62. An equation representing the cost of a rental car is $y = 0.59x + 25$, where
y is the total cost and x is the number of miles traveled in excess of 150
(the first 150 miles are free). A person who paid $39.75 traveled how far?
175 miles

EXCURSIONS

Class Act

In forensic science, it is sometimes important to be able to tell something about
a person from investigating a bone. The height of a person can be calculated
from knowing the lengths of certain major bones: the femur (F), the tibia (T),
the humerus (H), and the radius (R). When the length of one of these bones is
known, one of the following formulas is used to determine the height. (All
measurements are in centimeters.)

Male	Female
$h = 69.09 + 2.24F$	$h = 61.41 + 2.32F$
$h = 81.69 + 2.39T$	$h = 72.57 + 2.53T$
$h = 73.57 + 2.97H$	$h = 64.98 + 3.14H$
$h = 80.41 + 3.65R$	$h = 73.50 + 3.88R$

After the age of thirty, the height of a person begins to decrease at the rate of
approximately 0.06 centimeter per year.

Using these linear equations, ask and answer four questions. Compare your
questions with your classmates.

3.5 Sets, Relations, Functions, and Graphs

SECTION LEAD-IN

The formula for the temperature T at height h (in feet) when the ground temperature is t is

$$T = t - \frac{h}{100}$$

Use this formula to ask and answer 2 questions about temperatures.

Introduction

The graphs that we sketched in previous sections of this chapter serve as examples of the graphs of *relations* and *functions*. "Relation" and "function" are more basic concepts than, say, "ordered pair" and "equation in two variables." They form a common bond among many areas of mathematics.

Let A and B be sets. Then the set $A \times B$ consists of all ordered pairs (a, b) such that a is an element of A and b is an element of B. That is,

$$A \times B = \{(a, b) \mid a \text{ is in } A \text{ and } b \text{ is in } B\}$$

We will confine our discussions to sets containing real numbers from here on. For example, if A and B are the set of all real numbers, then $A \times B$ (real numbers × real numbers) consists of all ordered pairs of real numbers. It has the entire coordinate plane as its graph.

Relations and Graphs of Relations

A relation is a subset of a set of ordered pairs.

> **Domain and Range of a Relation**
>
> The set of all first elements of a relation is called the **domain** of the relation. The set of all second elements is called the **range**.

▪▪▪

EXAMPLE 1

Let $A = \{1, 2, 3\}$ and $B = \{3, 4\}$.

a. Write $A \times B$ as a set of ordered pairs.

b. Find three relations on $A \times B$, and give the domain and range of each.

SOLUTION

a. We pair each element of A with an element of B, obtaining

$$A \times B = \{(1, 3), (2, 3), (3, 3), (1, 4), (2, 4), (3, 4)\}$$

b. For the relation "the second number in the pair is even," the pairs are

$$R = \{(1, 4), (2, 4), (3, 4)\}$$

The domain of R is $\{1, 2, 3\}$, and the range is $\{4\}$.

For the relation "the numbers in the pair are integers," the pairs are

$$S = \{(1, 3), (2, 3), (3, 3), (1, 4), (2, 4), (3, 4)\}$$

As you can see, $S = A \times B$. Its domain is $\{1, 2, 3\}$, and its range is $\{3, 4\}$.

For the relation "the second number is one greater than the first," the ordered pairs are

$$I = \{(2, 3), (3, 4)\}$$

The domain is $\{2, 3\}$, and the range is $\{3, 4\}$.

▶ *CHECK* **Warm-Up 1**

The relations that are of most interest to us (and to most mathematicians) are those that actually exhibit some sort of relationship between the first and second elements of each ordered pair.

For example, the subset $\{(3, 3)\}$ of $A \times B$ in Example 1 is the set of all pairs that exhibit the relationship "second element equals first element." The subset

$$U = \{(1, 3), (2, 3), (1, 4), (2, 4), (3, 4)\}$$

contains all pairs of $A \times B$ that are related by "second element is greater than first element." Thus we could also write the relation U as

$$U = \{(a, b) \mid (a, b) \text{ is in } A \times B \text{ and } b > a\}$$

The graph of a relation is the graph of its ordered pairs. For example, the graph of U as we have defined it consists of the five points shown in the following figure.

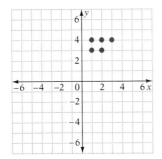

Functions and Graphs of Functions

Let X and Y be sets. Then a **function** f from X to Y is a rule or other means for assigning to each element of X *exactly one element of Y*.

The set X is called the domain of the function, and the set Y is called its range.

A function f is actually an *instruction* for making a set of ordered pairs (x, y) such that each x-value is used with only one y-value. The result of applying

▪▪▪
WRITER'S BLOCK
Explain why all functions are also relations.

that instruction is a relation—a set of ordered pairs. The domain and range of the relation are the domain X and range Y of the function. The x- and y-values are elements of X and Y, respectively. The **graph of a function** is the graph of its ordered pairs.

▪ ▪ ▪

EXAMPLE 2

Let $f =$ "add 3 to the first element to get the second element," and let $X = \{1, 2, 3, 5\}$.

a. Find the ordered pairs produced by the function.

b. Find its range.

SOLUTION

a. The function assigns to each element x of the domain X the number $x + 3$. Taking the elements of the domain in order, we obtain the set of ordered pairs.

$$R = \{(1, 4), (2, 5), (3, 6), (5, 8)\}$$

b. The range of relation R is $\{4, 5, 6, 8\}$, so that is the range of f.

▶ CHECK **Warm-Up 2**

Writing out function rules, as in Example 2, can be awkward, so several other ways of stating function rules have been adopted. One is simply to list the pairs that the function produces. Thus, for the function f in Example 2, we could write

$$f = \{(1, 4), (2, 5), (3, 6), (5, 8)\}$$

WRITER'S BLOCK

Define the domain and range of a function.

Functional Notation

The most useful functions have all the real numbers as the domain, then, however, lists of ordered pairs become awkward as means for stating functions. Instead, mathematicians use what is called **functional notation.** In this notation, the function rule is written as a mathematical expression that includes a variable. The variable (usually x) represents domain elements. The corresponding range elements are represented by the symbol $f(x)$, which is read "f of x." The two are equated, to show that each range value depends on the corresponding domain value.

As an example, the function "add 3 to the first element to get the second element" is written

$$f(x) = x + 3$$

Suppose the domain of f is all real numbers. Then this notation tells us that

when x is 2, then $f(2) = 2 + 3 = 5$

This gives us the pair $(2, 5)$. Similarly,

when $x = -1.7$, then $f(-1.7) = -1.7 + 3 = 1.3$

This gives us the pair $(-1.7, 1.3)$

More generally, a function f generates the pairs $(x, f(x))$ for all x in the domain of f. Finding the value of $f(x)$ for a given value of x is called "evaluating the function at x."

▪ ▪ ▪

EXAMPLE 3

The function $f(x) = 2x - 2$ has the domain $X = \{x \mid x \text{ is a real number}\}$. Evaluate the function at $x = 0$, $x = 1$, and $x = 2$. Then graph the function.

SOLUTION

Substituting 0, 1, and then 2 for x yields

$$f(0) = 2(0) - 2 = -2$$

$$f(1) = 2(1) - 2 = 0$$

$$f(2) = 2(2) - 2 = 2$$

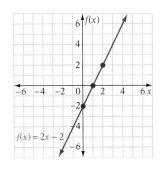

To graph the function, first note that $f(0)$, $f(1)$, and $f(2)$ give us the function pairs $(0, -2)$, $(1, 0)$, and $(2, 2)$. Now look at the function $f(x) = 2x - 2$.

It is not only a function; it is also a linear equation with independent variable x, dependent variable $f(x)$, and solutions $(0, -2)$, $(1, 0)$, and $(2, 2)$. The graph of this equation is a straight line through the points that correspond to those solutions, as sketched in the accompanying figure. Moreover, every solution of the equation $f(x) = 2x - 2$ is an ordered pair of the function $f(x) = 2x - 2$. Thus the graph in the accompanying figure is also the graph of the function.

▶ *CHECK* **Warm-Up 3**

> A function of the form $f(x) = ax + b$ is called a **linear function.** Its graph is a straight line with slope a and vertical intercept $(0, b)$.

We often extend functional notation to equations in two variables. For example, we might write

$$y = f(x)$$

to emphasize that, in a certain equation, y is the dependent variable whose value depends on the value of x. If the equation were, say,

$$y = x^2 + 2$$

we might write $f(0) = 2$, $f(5) = 27$, and $f(t) = t^2 + 2$, meaning that the ordered pairs $(0, 2)$, $(5, 27)$, and $(t, t^2 + 2)$ are solutions of the equation. Strictly speaking, though, functional notation is for functions.

The Vertical-Line Test for Functions

We can determine whether any graph is the graph of a function by applying the following test.

> **Vertical-Line Test**
>
> A graph is the graph of a function if no vertical line intersects it at more than one point.

The idea here is that all points on a vertical line have the same second coordinate. If two or more points on a graph are intersected by a vertical line, then they too have the same second coordinate. Thus the corresponding ordered pairs have the same second element, and the graph is not the graph of a function.

▪ ▪ ▪

EXAMPLE 4

Determine whether each graph is the graph of a function.

a.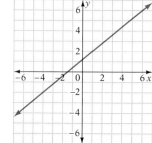

b.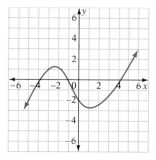

c.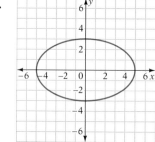

d.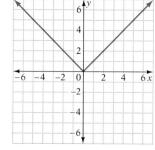

SOLUTION

a. This is the graph of a function. At no place along the graph does a vertical line cross more than one point. Every line except a vertical line is the graph of a function.

b. This is also the graph of a function, even though it is curved.

c. This is not the graph of a function, because some vertical lines intersect it in two points. It is actually an ellipse, a curve that we shall discuss in Chapter 9. Note that it has two y-intercepts.

d. This too is the graph of a function. It is actually the graph of $f(x) = |x|$.

▶ CHECK **Warm-Up 4**

Practice what you learned.

◣ **SECTION FOLLOW-UP**

We asked and answered these questions.

1. Solve for h in terms of T and t. $h = 100(t - T)$
2. Assume that this formula is written for T and t in degrees Celsius. Find the height necessary for the temperature to drop from 29°C to 13°C.   1600 feet

3.5 WARM-UPS

Work these problems before you attempt the exercises.

1. Let R = "the sum of the numbers in the ordered pair is even." For $A \times B$ as defined in Example 1, give the pairs in R and its domain and range. $R = \{(1, 3), (2, 4), (3, 3)\}$; Domain: $\{1, 2, 3\} = A$; Range: $\{3, 4\} = B$

2. Let f = "twice the first element is the second element," and let $X = \left\{\frac{1}{2}, 1, 1\frac{1}{2}\right\}$. Find the ordered pairs produced by the function and find the range. $\left\{\left(\frac{1}{2}, 1\right), (1, 2), \left(1\frac{1}{2}, 3\right)\right\}$; Range: $\{1, 2, 3\}$

3. Evaluate $f(0)$, $f(2)$, and $f(a)$ when $f(x) = 3x^2 - 2$. -2; 10; $3a^2 - 2$

4. Use a vertical-line test to determine which are functions.

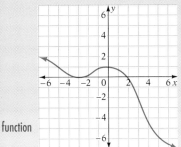

function

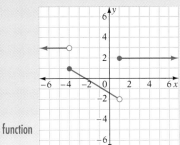

function

3.5 EXERCISES

Note: Use your graphing calculator to check your results whenever possible.

In Exercises 1 through 6, give the domain and range of each relation, and then graph it.

1. $R = \{(x, y) \mid x \text{ is an even whole number and } y \text{ is an odd whole number}\}$
   Domain: {even whole numbers}; Range: {odd whole numbers}

2. $R = \{(x, y) \mid x \text{ is greater than 3 and } y < -6\}$
   Domain: $\{x \mid x > 3\}$; Range: $\{y \mid y < -6\}$

2.

3. $R = \{(x, y) \mid x$ is 3 or x is 4 and y is any real number$\}$
   Domain: {3, 4}; Range: {all real numbers}

4. $R = \{(x, y) \mid x$ is 1 or 2 or 3 and y is a whole number and $y < x\}$
   Domain: {1, 2, 3}; Range: {0, 1, 2}

5. $R = \{(x, y) \mid x$ is a whole number and $y = 5\}$
   Domain: {whole numbers}; Range: {5}

6. $R = \{(x, y) \mid x > y$ and $y = 10\}$ Domain: $\{x \mid x > 10\}$; Range: {10}

4.

6.

In Exercises 7 through 14, the functions are defined on the real numbers. Rewrite using functional notation, and list four ordered pairs in the function.
Answers may vary.

7. $f =$ "the second element is 3 times the first element"   {(0, 0), (1, 3), (2, 6), (3, 9)}

8. $f =$ "twice the first element plus two is the second element"   {(0, 2), (1, 4), (2, 6), (3, 8)}

9. $f =$ "the first element minus three is the second element"   {(0, −3), (1, −2), (2, −1), (3, 0)}

10. $f =$ "the first element is three times the second element"   {(0, 0), (3, 1), (6, 2), (9, 3)}

11. $f =$ "the second element is four less than twice the first element"   {(0, −4), (1, −2), (2, 0), (3, 2)}

12. $f =$ "one half the first element minus 5 is the second element"
    {(0, −5), (2, −4), (4, −3), (6, −2)}

13. $f =$ "the second element is the product of the first element and the first element minus one"   {(0, 0), (1, 0), (2, 2), (3, 6)}

14. $f =$ "the second element is the product of the first element and 6"
    {(0, 0), (1, 6), (2, 12), (3, 18)}

In Exercises 15 through 22, evaluate $f(x)$ at the given values.

15. $f(x) = 2x + 7$; evaluate at $x = 0, 2, −1$
    7; 11; 5

16. $f(x) = 3x − 8$; evaluate at $x = 5, 0, −3$
    7; −8; −17

17. $f(x) = x^2 − 2$; evaluate at $x = 0, 4, −3$
    −2; 14; 7

18. $f(x) = 2x^2$; evaluate at $x = 0, 3, −2$
    0; 18; 8

19. $f(x) = −2x + 1$; find $f(0), f(a), f(x + 2)$
    (*Hint:* Treat "$x + 2$" as just another number.)
    1; −2a + 1; −2x − 3

20. $f(x) = x − 2$; find $f(0), f(a), f(x + 3)$
    −2; a − 2; x + 1

21. $f(x) = 3 − 2x$; find $f(0), f(a), f(x − 1)$
    3; 3 − 2a; 5 − 2x

22. $f(x) = 2 − 3x$; find $f(0), f(a), f(x − 2)$
    2; 2 − 3a; 8 − 3x

In Exercises 23 through 28, use the vertical-line test to determine which graphs are the graphs of functions.

23.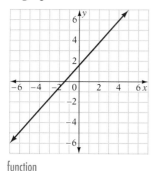

function

24.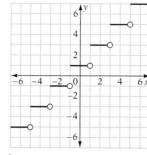

function

25.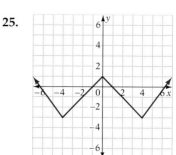

function

26.

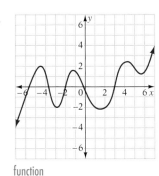

function

27.

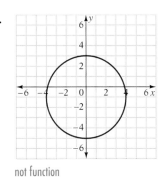

not function

28.

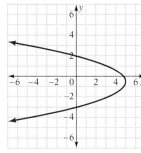

not function

MIXED PRACTICE

By doing these exercises, you will practice the topics up to this point in the chapter.

29. Graph: $-y < 3x + 3$

30. Find the slope and y-intercept of $-9(x - 3) = y + 2x$.
slope: -11; y-intercept: $(0, 27)$

31. Evaluate $f(x) = x^2 + 2x - 7$ for $x = -1$. -8

32. Find the x- and y-intercepts of $3y = 2x + 12$. x-intercept: $(-6, 0)$; y-intercept: $(0, 4)$

33. Find three members of the solution set of $3x \geq -y$.   Answers may vary. $(0, 1)$; $(0, 0)$; $(0, 2)$

34. Write the equation of the line that is perpendicular to $y = -2x - 5$ and passes through the origin. $y = \frac{1}{2}x$

35. Give the range and domain of $R = \{(0, 3), (6, 2), (5, -3)\}$.   Range: $\{-3, 2, 3\}$ Domain: $\{0, 5, 6\}$

36. Show that $(-3, 5)$ is part of the solution set of $|x - 2| \leq y$.
See the Solutions Manual.

37. $(x, 2)$ is a solution of $-5x + 2y = 10$. Find x. $\frac{-6}{5}$

38. Write an equation of the line that has slope 3 and an x-intercept of $(-3, 0)$.
$y = 3x + 9$

39. Write the equation that represents the cost (P) of an article that is determined by a tax of 8.25% of the ticket price (T) to the ticket price.
$P = 1.0825T$

40. Graph $f(x) = 2x - 1$ where x is a real number.

EXCURSIONS

Class Act

Use the formula for relative humidity and the table to answer parts (a) and (b).

$$\text{Relative humidity} = \frac{\text{partial pressure of } H_2O}{\text{saturated vapor pressure of } H_2O} \times 100$$

The units of partial pressure are millimeters of Mercury, or "torr."

Saturated Vapor Pressure of Water

Temperature (°C)	torr (= mmHg)	Temperature (°C)	torr (= mmHg)	Temperature (°C)	torr (= mmHg)
−50	0.030	20	17.5	70	234
−10	1.95	25	23.8	80	355
0	4.58	30	31.8	90	526
5	6.54	40	55.3	100	760
10	9.21	50	92.5	120	1489
15	12.8	60	149	150	3570

a. On one very hot day, the temperature is 25°C and the partial pressure of water vapor in the air is 21.0 torr. What is the relative humidity?

b. When the temperature was 10°C, the relative humidity was 65%. Find the partial pressure of water vapor in the air.

c. Ask and answer two additional questions using this data.

CHAPTER LOOK-BACK

Here are questions we asked and answered.

1. How cold will it feel with a temperature of 15°F and a 45-mile-per-hour wind?   −30°F

2. How cold will it feel with a temperature of 20 degrees below zero Fahrenheit and a 5-mile-per-hour wind?   −26°F

3. A temperature of −5°F and a 5-mile-per-hour wind will feel how cold?   −10°F

4. The temperature is 10 degrees below zero Fahrenheit, with a wind of 15 miles per hour. What difference in temperature will be felt if the wind increases to 40 miles per hour?
    It will feel 24°F colder (−69°F).

▭   Write a paragraph describing what effect wind speed has on how cold you feel. Use the information from the Wind Chill Factors table in the Chapter Lead-In.

CHAPTER 3
REVIEW PROBLEMS

The following exercises will give you a good review of the material presented in this chapter.

SECTION 3.1

1. Find a solution of $3y = 20x + 2$ with $x = 3$.
$y = 20\frac{2}{3}$

2. Is $(-9, 5)$ a solution of $9x + y = 86$?
no

3. Find A in $Ax + 2y = -5$ with $(2, 4)$ as a solution of the equation.
$A = -6\frac{1}{2}$

4. Identify the points A, B, C, and D, and name the quadrant each lies in.

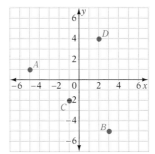

$A = (-5, 1)$, quadrant II
$B = (3, -5)$, quadrant IV
$C = (-1, -2)$, quadrant III
$D = (2, 4)$, quadrant I

5. Find three solutions for $2y - 2x = 3$.
Answers may vary. $\left(0, \frac{3}{2}\right); \left(1, \frac{5}{2}\right); \left(-1, \frac{1}{2}\right)$

6. Find a solution of $2x - 8y = 5$ with $y = 1$.
$\left(\frac{13}{2}, 1\right)$

SECTION 3.2

7. Graph: $x - y = 2$

8. Graph: $y - x = 3$

9. Graph: $x + 2y = 6$

10. Graph: $3x - y = 8$

In Exercises 11 and 12, assume that your score S on a test was determined by the equation

$$S = 8(12 - m) + 4$$

where m is the number of problems missed out of 12.

11. **a.** A score of 100 indicates that you got 12 right. Verify this with the ordered pair (0, 100).
See the Solutions Manual.
b. The ordered pair (4, 68) means what in terms of this test? A score of 8 correct (4 missed) gives a 68 on the test.

12. You got 10 right. What is your score? 84

13. Find the slope of the line that passes through the points $(0, 0)$ and $(-4, -5)$. $m = \frac{5}{4}$

14. Find the y- and x-intercepts of the graph of $-3y - 7x = 66$. y-intercept: $(0, -22)$; x-intercept: $\left(-9\frac{3}{7}, 0\right)$

15. Find the slope of the line that passes through the points $(6, 0)$ and $(-5, 4)$. $m = \frac{-4}{11}$

16. Find the slope of the line that passes through the points $(-9, 3)$ and $\left(-\frac{1}{2}, 4\right)$. $m = \frac{2}{17}$

SECTION 3.3

17. Write $x - y = 8$ in slope-intercept form.
    $y = x - 8$

18. Determine whether $\frac{1}{2}x - 4 = y$ and
    $x + 9 = 2y$ are parallel, perpendicular, or
    intersecting only. parallel

19. Rewrite $8x + 9 = 6y$ in slope-intercept form.
    $y = \frac{4}{3}x + \frac{3}{2}$

20. Determine whether $-3y - 7 = x$ and
    $x = 3y + 2$ are parallel, perpendicular, or
    intersecting only. intersecting only

21. Write $x - y = 1$ in slope-intercept form.
    $y = x - 1$

22. Determine whether $2y - 4 = x$ and
    $y = 5x + 12$ are parallel, perpendicular, or
    intersecting only. intersecting only

23. Write the equation of the line that passes
    through the point $(-8, -6)$ and has slope -2.
    $y = -2x - 22$

24. Find the equation of the line that passes
    through the points $(-1, -1)$ and $(-3, -2)$.
    $y = \frac{1}{2}x - \frac{1}{2}$

25. Write the equation of the line that has slope
    -6 and passes through the point $(0, -3)$.
    $y = -6x - 3$

26. Find the equation of the line that has slope
    5 and y-intercept $(0, -12)$.
    $y = 5x - 12$

27. Find (in slope-intercept form) the equation
    of the line that passes through the points
    $(-3.5, 6)$ and $(-1, 1)$. $y = -2x - 1$

28. Write an equation for the line that has slope
    -1 and passes through the point $(6, 0)$.
    $y = -x + 6$

SECTION 3.4

29. Graph: $-2y < 4x + 6$

30. Find two points that do not lie in the solution
    set for $8x - 3y > 15$.   Answers may vary. (0, 0) and (1, 0)

31. Graph: $3y - 5 \leq 2x$

32. Graph: $|8x - y| \geq 5$

SECTION 3.5

33. Find $f(0)$ and $f(-1)$ when $f(x) = 2x - 3$.
    $-3; -5$

34. Find $f(0)$ and $f(-2)$ when $f(x) = 3x + 5$.
    $5; -1$

35. Find the domain and range of
    $S = \{(0, 3), (1, 5), (4, 6), (8, 6)\}$
    Domain: {0, 1, 4, 8}; Range: {3, 5, 6}

36. Find four ordered pairs of the relation
    $R = \{$the second number in the ordered pair
    is half the first number$\}$ Answers may vary.
    (2, 1), (6, 3), (4, 2), (10, 5)

MIXED REVIEW

37. Rewrite $x - 5 = 6y$ in slope-intercept form.
    $y = \frac{1}{6}x - \frac{5}{6}$

38. Graph: $4y - 3x = 2$

39. Find the equation of the line that has slope -1 and y-intercept $(0, 3)$.
    $y = -x + 3$

40. Is $(3, -5)$ a solution of $x + y = 2$?   no

CHAPTER 3 TEST

This exam tests your knowledge of the material in Chapter 3.

1. **a.** Identify which of these two equations is a linear equation, and solve it for x and for y.

$$x^2 + 7y = 8 \quad \text{or} \quad 2y = 3x + 12$$

The second is the linear equation. $\frac{2}{3}y - 4 = x; y = \frac{3}{2}x + 6$

b. Is $(1, 2)$ a solution of $3x + 3y = 9$?   yes

c. Find B in $4x + By = -8$ if one solution of the equation is $(-1, 2)$.
$B = -2$

2. In parts (a) and (b), graph each equation using a table of values.

2a.

a. $-2y - 1 = x$ **b.** $3x - 2y = 5$

c. A child-care worker charges $5 plus $2 per child per hour. The hourly earnings are thus given by $y = 2x + 5$. Graph this equation and determine what the point $(5, 15)$ means.   If the child-care worker has 5 children, he or she earns $15 an hour.

b.

3. **a.** Find the x- and y-intercepts of $5y - 6 = 3x$.   x-intercept: $(-2, 0)$; y-intercept: $\left(0, \frac{6}{5}\right)$

b. Find the slope of the line that passes through the points $(-1, 3)$ and $(5, 7)$. $m = \frac{2}{3}$

c.

c. A line has x- and y-intercepts $(2, 0)$ and $(0, -5)$, respectively. What is its slope? $m = \frac{5}{2}$

4. **a.** Write $5x - 3y + 8 = 0$ in slope-intercept form, and identify the slope and y-intercept. $y = \frac{5}{3}x + \frac{8}{3}$; slope: $\frac{5}{3}$; y-intercept: $\left(0, \frac{8}{3}\right)$

b. Are $2y + 5 = x$ and $x + 6 = 2y$ parallel, perpendicular, or intersecting?
parallel

c. Are $x = 5$ and $y - 4 = 0$ parallel, perpendicular, or neither?   perpendicular

5. **a.** Write the equation of a line that has slope 7 and y-intercept $\left(0, \frac{-1}{2}\right)$. $y = 7x - \frac{1}{2}$

b. Write the equation of the line that passes through $(-3, 2)$ and has a slope of -1. $y = -x - 1$

c. Find the equation of the line that passes through the points $(6, 2)$ and $(-4, -5)$. $y = \frac{7}{10}x - 2\frac{1}{5}$

6. Graph each inequality

    a. $3y - 2x > 5$ **b.** $x \le 6$ and $y > -2$ **c.** $|2x - 3y| \le 5$

7. **a.** Find $f(-1)$ when $f(x) = 2x^2 + 3x$. -1

    b. Let $R = \{(2, -3), (5, -2), (-1, -1), (6, -3)\}$, and find its domain and range. Is R a function? Domain: $\{-1, 2, 5, 6\}$, Range: $\{-1, -2, -3\}$; yes

    c. Graph the function $\{(x, y) \mid x$ is real and $y = 3x - 2\}$.

6a.

b.

c.

CUMULATIVE REVIEW

CHAPTERS 1–3

The following exercises will help you maintain the skills you have learned in this and previous chapters.

1. Simplify: $(0.05 - 0.02)^2 - 3(1.2 - 2.3)$
   3.3009

2. Rewrite -5^{-2} without exponents.
   $-\frac{1}{25}$

3. Divide $15\frac{2}{5}$ by $5\frac{1}{2}$. $2\frac{4}{5}$

4. Solve for x: $\frac{x}{a} + b = c$

   $x = a(c - b)$ or $x = ac - ab$

5. Susan has $n + 6$ quarters. How many cents does she have? She has $25(n + 6)$ cents.

6. Solve: $|x + 7| = -4$

   no solution

7. Evaluate $y - 5 - (5 - y)$ when $y = 10$. 10

8. Round 89.39499 to the nearest ten, tenth, hundredth, and thousandth.
   90; 89.4; 89.39; 89.395

9. Fill in the missing values and tell what property is illustrated. distributive property
   $(6 + \underline{\;\;7\;\;})3 = 6 \cdot \underline{\;\;3\;\;} + 7 \cdot \underline{\;\;3\;\;}$

10. Let $A = \{1, 3, 5\}$, and let $B = \{0, 1, 2, 3, 5\}$.
   a. Find the intersection of A and B. $A \cap B = \{1, 3, 5\}$
   b. Find the union of A and B.
      $A \cup B = \{0, 1, 2, 3, 5\} = B$

11. Solve: $2.4 - 3x \ge 6 + x$ $x \le -0.9$

12. Graph: $12x - 3y = 9$

13. Find the equation of the line that passes through the points $(-6, -2)$ and $(6, 5)$.
   $y = \frac{7}{12}x + 1\frac{1}{2}$

14. Subtract: $\left(-6\frac{2}{3}\right) - \left(-9\frac{1}{2}\right)$
   $2\frac{5}{6}$

15. Solve: $-2x + 7 = x - 3$
   $x = 3\frac{1}{3}$

16. Evaluate $xy - xy + xy^2 - x^2y$ when $x = -1$ and $y = -8$. -56

17. Graph the inequality $4x - 3y \ge 8$.

18. Use front-end estimation to approximate the value of $\dfrac{(38.925)(7.1111799)}{(69.9908295)}$. 4

19. Solve: $-3y + 7 > 6$ $y < \frac{1}{3}$

20. Find the x- and y-intercepts of the graph of $3x - 2y = 17$. x-intercept: $\left(5\frac{2}{3}, 0\right)$; y-intercept: $\left(0, -8\frac{1}{2}\right)$

SYSTEMS OF EQUATIONS AND INEQUALITIES

*E*rosion and explosive volcanic eruptions can significantly reduce the shape and height of a mountain peak. You could estimate the mountain's original shape by graphing systems of linear equations. Imagine *x*- and *y*-axes superimposed on the mountain with the *y*-axis bisecting the mountain. The left and right sides of the mountain represent two linear equations. Their point of intersection is the solution to the two equations and a rough estimate of the mountain's original shape and height. In this situation, the two lines formed a "model" of the real situation.

■ *Find a situation where two straight lines can be used to predict a result. What does the point of intersection represent?*

SKILLS CHECK

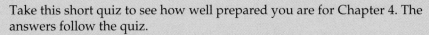

Take this short quiz to see how well prepared you are for Chapter 4. The answers follow the quiz.

1. Solve: $3(9x - 7) = 24$

2. Solve: $|x + 3| > 5$

3. Combine: $9x + 5x - 7x - 6x$

4. Find the slope and y-intercept of the graph of $y + x = 4$.

5. Solve: $\frac{3}{8} + \frac{x}{6} = -5$

6. Graph: $2y + 3x = 7$

7. Solve: $-11y - 7 < 8$

8. Solve for t: $w = xt + k$

ANSWERS: **1.** $x = 1\frac{2}{3}$ [Section 2.2] **2.** $x > 2$ or $x < -8$ [Section 2.4] **3.** x [Section 2.1]
4. slope: -1; y-intercept: $(0, 4)$ [Section 3.2] **5.** $x = -32\frac{1}{4}$ [Section 2.2] **6.**
[Section 3.2] **7.** $y > -1\frac{4}{11}$ [Section 2.4] **8.** $t = \frac{w - k}{x}$ [Section 2.3]

CHAPTER LEAD-IN

Almost every decision we make in our personal or professional lives is made under some sort of limiting conditions: We have only so much money to spend, only so much time available, or only so much raw material and machinery to use in producing goods. We can use a mathematical technique called **linear programming** in such decision-making situations.

Using linear programming, we can solve problems such as the following:

A baker makes both pan and shortbread cookies. Each batch of pan cookies requires 4 cups of sugar and 2 cups of butter. Each batch of shortbread cookies requires 3 cups of sugar and 3 cups of butter. The baker makes a $2.50 profit on each batch of pan cookies and a $2.75 profit on each batch of shortbread cookies. He has 76 cups of sugar, 56 cups of butter, and as much of the other ingredients as is needed. How many batches of each type of cookie should he make to maximize his profits?

4.1 Solving Systems of Linear Equations

SECTION LEAD-IN

An experimental car gets 35 miles per gallon on the highway and 30 miles per gallon in town. The gas tank holds 20 gallons. The equation that represents the number of miles (y) that the car can travel when x of the 20 gallons are used on the highway and $(20 - x)$ gallons are used in town is

$$y = 35x + 30(20 - x)$$

- How many miles can the car travel on one tank of gas?
- What are the limiting factors in solving this equation?
- Write all inequalities that this problem implies.

SECTION GOALS

- To identify a system of linear equations as consistent, inconsistent, or dependent

- To determine whether a given ordered pair is the solution of a system of linear equations

- To solve systems of linear equations by graphing

- To solve systems of linear equations by addition

- To solve systems of linear equations by substitution

Introduction

Recall that a linear equation is an equation in which each term has at most one variable, and each variable appears only to the first power.

> **System of Linear Equations**
>
> A **system of linear equations** is two or more linear equations that contain the same variables.

Systems of linear equations can be solved using algebra. We can sometimes also find the solutions to two-variable systems by graphing.

Graphs of Systems of Linear Equations

We can graph a system of two equations by drawing the graphs of both equations on the same coordinate axes. The point at which the graphs intersect is on *both* graphs, so the coordinates of this point make both equations true.

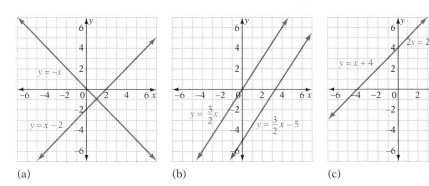

(a)   (b)   (c)

The lines in figure (a) have different slopes. As a result, they intersect in one point, and the system they represent has *one solution*. That solution is given by the coordinates of the point of intersection.

> A system that has at least one solution is called a **consistent system.**

The lines in figure (b) have the same slope, $\frac{3}{2}$, but different y-intercepts. They do not intersect but instead are parallel. Therefore, the system they represent has *no solution*.

> A system that has no solution is called an **inconsistent system.**

The lines in figure (c) have the same slope and the same y-intercept. They are really the same line. They intersect in an infinite number of points, so the system they represent has an *infinite number of solutions*. Any ordered pair that satisfies one equation also satisfies the other; these equations are *dependent*.

> A system that has infinitely many solutions is called a **dependent system.**

Note: Dependent systems are also consistent systems because they have a solution.

Solving Linear Systems by Graphing

We can solve a system of linear equations by graphing the equations. If the graphs intersect, the coordinates of the point of intersection provide the solution. This graphical solution must always be checked by substituting the coordinates of the point of intersection in the original equations.

STUDY HINT

When checking a solution of a system of equations, always check it in both equations. An ordered pair must make both equations true to be a solution.

> **To solve a system of linear equations by graphing**
> 1. Graph and label both equations.
> 2. Estimate the coordinates of the point of intersection of the two lines, if it exists. This is the solution.
> 3. Check by substituting the ordered pair in the original equations.

■ ■ ■

EXAMPLE 1

Solve this system of equations by graphing:
$$y = 3x + 2$$
$$3y = -x + 6$$

SOLUTION

First we solve the second equation for y to put it in slope-intercept form.

$$y = 3x + 2 \qquad 3y = -x + 6$$
$$y = -\frac{1}{3}x + 2$$

These two equations are not identical, and their graphs do not have equal slopes. In fact, the slopes are negative reciprocals. This means that the graphs of the lines are perpendicular; they will intersect in exactly one point.

We can use the slopes and y-intercepts to graph the equations. The graph of the first equation has slope 3 and y-intercept $(0, 2)$. The graph of the second equation has slope $-\frac{1}{3}$ and y-intercept $(0, 2)$. The graphs of these equations are shown below. The lines appear to intersect at $(0, 2)$. We check this result by substituting in the two original equations.

$y = 3x + 2$	$3y = -x + 6$
$2 = 3(0) + 2$	$3(2) = -(0) + 6$
$2 = 2$ True	$6 = 6$ True

So $(0, 2)$ is the single solution of the system. Its solution set is written $\{(0, 2)\}$.

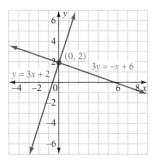

▶ CHECK **Warm-Up 1**

Calculator Corner

In order to graph two equations, you must first get them into the "$y = $" form. Then graph the two equations and **TRACE** to find the point of intersection. You could also find the intersection point by using the **CALC**ulation utility if your graphing calculator has that feature. Use $-2y = -6 - x$ and $y = 3 + 3x$.

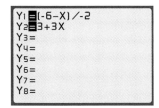

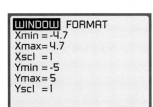

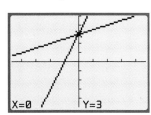

You can check your results by using your graphing calculator's Home Screen to evaluate each equation at the value $x = 0$. If the point $(0, 3)$ is indeed the point of intersection, then y should be equal to 3 in each equation.

 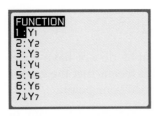 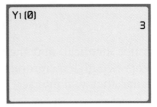

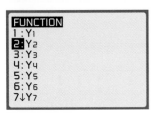 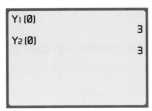

Because both equations give a result of 3 when $x = 0$, the point of intersection is $(0, 3)$.

···

EXAMPLE 2

Solve this system of equations by graphing: $-4x + y = 9$
$-8x + 2y = 6$

SOLUTION

Putting the equations in slope-intercept form yields

$$-4x + y = 9 \qquad -8x + 2y = 6$$
$$y = 4x + 9 \qquad y = 4x + 3$$

Because the equations have the same slope and different y-intercepts, the graphs are parallel. The system is inconsistent, and the solution set is $\varnothing$ (the empty set.) A graph of the two equations confirms this fact.

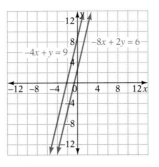

▶ CHECK **Warm-Up 2**

We use the graphing method on two-variable systems only. Graphing three-variable systems can be done on a computer, but we will show you how to use other procedures.

> **WRITER'S BLOCK**
>
> Explain how you find an intersection point of the graph of two linear equations by graphing.

> **STUDY HINT**
>
> *The graphical method of solving systems of equations has one serious drawback. Pinpointing the exact solution (the coordinates of a point of intersection on a graph) is often difficult. This is especially so when the coordinates are fractions or decimal numbers. So when an exact solution is required, you should use one of the algebraic methods presented in the following discussion.*

Solving Systems of Linear Equations Algebraically

The Addition (or Elimination) Method

The *same term* can be added to both sides of an equation without changing the solution set of the equation. Similarly, *equal expressions* can be added to the two sides of an equation without changing the solution set. That is,

> **Addition Principle of Equality (restated)**
> If $a = b$ and $c = d$, then $a + c = b + d$.

We can use this principle to eliminate one of the variables from a system of two equations in two unknowns. This leaves an equation that can easily be solved for the remaining variable.

▪▪▪
EXAMPLE 3

Solve this system of equations by addition: $6y + 2x = 9$ (1)
$$5y - 2x = 2 \quad (2)$$

(We have numbered the equations for ease in referring to them.)

SOLUTION

If we add the left sides and the right sides of these equations, the x-term drops out. We are left with an equation in y only.

$$
\begin{array}{ll}
6y + 2x = 9 & \text{Equation (1)} \\
\underline{5y - 2x = 2} & \text{Equation (2)} \\
11y + 0 = 11 & \text{Adding left and right sides} \\
y = 1 & \text{Multiplying both sides by } \frac{1}{11}
\end{array}
$$

We now know that $y = 1$ is part of the solution of the system. We substitute 1 for y in either equation and solve for x.

$$
\begin{array}{ll}
6y + 2x = 9 & \text{Equation (1)} \\
6(1) + 2x = 9 & \text{Substituting 1 for } y \\
6 + 2x = 9 & \text{Simplifying} \\
2x = 3 & \text{Adding } -6 \text{ to both sides} \\
x = 1.5 & \text{Multiplying by } \frac{1}{2}
\end{array}
$$

So $(1.5, 1)$ is the solution of this system. Checking both values in both equations gives us

$$
\begin{array}{ll}
6y + 2x = 9 & \qquad 5y - 2x = 2 \\
6(1) + 2(1.5) = 9 & \qquad 5(1) - 2(1.5) = 2 \\
6 + 3 = 9 & \qquad 5 - 3 = 2 \\
9 = 9 \quad \text{True} & \qquad 2 = 2 \quad \text{True}
\end{array}
$$

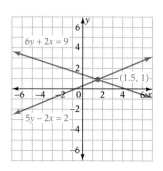

Indeed, $\{(1.5, 1)\}$ is the solution set for this system of equations. The accompanying figure shows that the graphs of the equations intersect at that point.

▶ CHECK **Warm-Up 3**

In the next example, we cannot just add left sides and right sides because that doesn't eliminate one variable. However, we can first multiply both sides of one equation by a non-zero number (without changing its solution set). Then we can add the result to the other equation to eliminate one variable.

▪▪▪

EXAMPLE 4

Solve by addition: $6y - 2x = 6$ (1)
$3y - 3x = 10$ (2)

SOLUTION

We will be able to eliminate the y-terms if we multiply Equation (2) by -2. That gives us the equivalent system

$$6y - 2x = 6$$
$$-6y + 6x = -20$$

Now we can add the left and right sides to eliminate y and solve for x.

$$
\begin{array}{ll}
6y - 2x = 6 & \text{Equation (1)} \\
\underline{-6y + 6x = -20} & \text{Equation (2)} \times -2 \\
0 + 4x = -14 & \text{Adding left and right sides} \\
x = \frac{-14}{4} & \text{Multiplying by } \frac{1}{4} \\
x = \frac{-7}{2} & \text{Simplifying}
\end{array}
$$

So $x = \frac{-7}{2}$ is part of the solution. Substituting $\frac{-7}{2}$ for x in the first equation yields

$$
\begin{array}{ll}
6y - 2x = 6 & \text{Equation (1)} \\
6y - 2\left(\frac{-7}{2}\right) = 6 & \text{Substituting for } x \\
6y + 7 = 6 & \text{Multiplying} \\
6y = -1 & \text{Adding } -7 \text{ to both sides} \\
y = \frac{-1}{6} & \text{Multiplying by } \frac{-1}{6}
\end{array}
$$

So $\left(-\frac{7}{2}, -\frac{1}{6}\right)$ is a tentative solution: Checking in both equations confirms it; $\left\{\left(-\frac{7}{2}, -\frac{1}{6}\right)\right\}$ is the solution set.

▶ *CHECK* **Warm-Up 4**

To solve the next system by addition, we must first multiply *both* equations by constants.

▪▪▪

EXAMPLE 5

Solve by addition: $11x + 2y = 13$ (1)
               $12 - 7y = 5x$ (2)

SOLUTION

First we rewrite the equations with the variables in the same order on the same side. That makes finding the solution easier.

$$11x + 2y = 13 \quad \text{Equation (1)}$$
$$-5x - 7y = -12 \quad \text{Equation (2) rewritten}$$

We want either the coefficients of x or those of y to be additive inverses so that one variable will be eliminated. The coefficients of y will be additive inverses if we multiply Equation (1) by 7 and Equation (2) by 2. Then we can add the left side and the right side, eliminating y, and solve for x.

$$77x + 14y = 91 \quad \text{Equation (1)} \times 7$$
$$\underline{-10x - 14y = -24} \quad \text{Equation (2) (rewritten)} \times 2$$
$$67x = 67 \quad \text{Adding left and right sides}$$
$$x = 1 \quad \text{Multiplying by } \tfrac{1}{67}$$

Thus $x = 1$ is part of the solution. We substitute 1 for x in Equation (1) and solve for y.

$$11(1) + 2y = 13 \quad \text{Substituting for } x$$
$$11 + 2y = 13 \quad \text{Simplifying}$$
$$2y = 2 \quad \text{Adding } -11 \text{ to both sides}$$
$$y = 1 \quad \text{Multiplying by } \tfrac{1}{2}$$

The solution set is $\{(1, 1)\}$. Check this in both equations.

▶ *CHECK* **Warm-Up 5**

STUDY HINT

In Example 5, we could have multiplied Equation (1) by 5 and Equation (2) by 11 instead. First decide which variable to remove. (The choice is yours.) Then select the proper multipliers.

We have seen that a system of linear equations in two variables is a set of two linear equations in two variables, usually x and y. Its solution is an ordered pair (x, y) that makes both equations true.

A **system of linear equations in three variables** is a set of three linear equations in three variables (usually x, y, and z). A solution of such a system is an ordered triple (x, y, z) that makes all three equations true. We will solve these three-variable systems algebraically.

A *consistent* system of linear three-variable equations has at least one ordered triple in its solution set. For most such systems, the most direct method of solution is the addition (or elimination) method. It must be applied twice to remove the same variable from two of the equations.

▪ ▪ ▪
Example 6

Solve this system of equations:
$$\begin{aligned} x + 3y + z &= 6 \quad (1) \\ -x + y + 3z &= -2 \quad (2) \\ 2x + 2y - 1z &= 1 \quad (3) \end{aligned}$$

Solution

If we add Equations (1) and (2), the x-terms drop out.

$$\begin{aligned} x + 3y + z &= 6 \quad &\text{Equation (1)} \\ -x + y + 3z &= -2 \quad &\text{Equation (2)} \\ \hline 4y + 4z &= 4 \quad &\text{Adding left and right sides; new Equation (4)} \end{aligned}$$

Now if we multiply Equation (2) by 2 and then add the result to Equation (3), the x-terms drop out here too.

$$\begin{aligned} -2x + 2y + 6z &= -4 \quad &\text{Equation (2)} \times 2 \\ 2x + 2y - 1z &= 1 \quad &\text{Equation (3)} \\ \hline 4y + 5z &= -3 \quad &\text{Adding left and right sides; new Equation (5)} \end{aligned}$$

Now we have eliminated the x-term from two equations and have thereby obtained a system of two linear equations in two unknowns.

$$\begin{aligned} 4y + 4z &= 4 \quad &\text{Equation (4)} \\ 4y + 5z &= -3 \quad &\text{Equation (5)} \end{aligned}$$

We can solve this system by addition.

Multiplying Equation (4) by -1 and adding yields

$$\begin{aligned} -4y - 4z &= -4 \quad &\text{Equation (4)} \times -1 \\ 4y + 5z &= -3 \quad &\text{Equation (5)} \\ \hline z &= -7 \quad &\text{Adding left and right sides} \end{aligned}$$

So $z = -7$ is one part of the solution. Substituting this value in Equation (4) gives

$$\begin{aligned} 4y + 4(-7) &= 4 \quad &\text{Substituting } z = -7 \text{ in Equation (4)} \\ 4y &= 32 \quad &\text{Simplifying} \\ y &= 8 \quad &\text{Multiplying by } \tfrac{1}{4} \end{aligned}$$

So $y = 8$ is a second part of the solution. Now we can substitute for y and z in any of our original equations and find the corresponding value of x. We choose Equation (1).

$$\begin{aligned} x + 3(8) + (-7) &= 6 \quad &\text{Substituting } y = 8, z = -7 \text{ in Equation (1)} \\ x + 17 &= 6 \quad &\text{Simplifying} \\ x &= -11 \quad &\text{Adding } -17 \end{aligned}$$

So the solution is the ordered triple $(-11, 8, -7)$. You should check it in original Equations 2 and 3.

$$-x + y + 3z = -2 \qquad\qquad 2x + 2y - 1z = 1$$
$$-(-11) + 8 + 3(-7) = -2 \qquad 2(-11) + 2(8) - (-7) = 1$$
$$19 - 21 = -2 \quad \text{True} \qquad\qquad -6 + 7 = 1 \quad \text{True}$$

The solution checks. It is the only solution, and the solution set is $\{(-11, 8, -7)\}$.

▶ CHECK **Warm-Up 6**

Just like systems of linear equations in two variables, systems of linear equations in three variables may be inconsistent or dependent.

The Substitution Method

In the substitution method, we use one equation to find an expression for one variable. We then substitute the expression into the second equation to eliminate that variable.

▪ ▪ ▪

EXAMPLE 7

Solve: $y = 3x - 7$ (1)
$\quad\quad 4y - 3x = 8$ (2)

SOLUTION

Equation (1) tells us that y is equal to $3x - 7$. Thus we can substitute "$3x - 7$" for y in Equation (2) and then solve.

$$4y - 3x = 8 \qquad \text{Equation (2)}$$
$$4(3x - 7) - 3x = 8 \qquad \text{Substituting for } y$$
$$12x - 28 - 3x = 8 \qquad \text{Multiplying out}$$
$$9x - 28 = 8 \qquad \text{Combining like terms}$$
$$9x = 36 \qquad \text{Adding 28 to both sides}$$
$$x = 4 \qquad \text{Multiplying by } \tfrac{1}{9}$$

So $x = 4$ is part of the solution. We can substitute 4 for x in one of the original equations and solve for the corresponding value of y. Equation (1) seems easier:

$$y = 3x - 7 \qquad \text{Equation (1)}$$
$$y = 3(4) - 7 \qquad \text{Substituting for } x$$
$$y = 5 \qquad \text{Simplifying}$$

The solution set is $\{(4, 5)\}$. You should always check *both* equations to verify the solution. Another way to check it is to graph the system.

▶ CHECK **Warm-Up 7**

❗❗❗
ERROR ALERT

Identify the error and give a correct answer.

Solve by substitution:
$$2x + 4y = 5$$
$$y = 3x - 5$$

Incorrect Solution:
$$2x + 4y = 5$$
$$2x + 4(3x - 5) = 5$$
$$2x + 12x - 5 = 5$$
$$14x - 5 = 5$$
$$14x = 10$$
$$x = \frac{10}{14} = \frac{5}{7}$$
$$y = \frac{-40}{14} = \frac{20}{7} = 2\tfrac{6}{7}$$

The substitution method can be used to solve a three-variable system when (1) one or more of the equations is missing a variable and (2) the missing variable is not the same in all equations. In the next example, Equation (2) is missing the z-term, and Equation (3) is missing the y-term.

▪▪▪
EXAMPLE 8

Solve the system:
$$\begin{aligned} x + y + z &= 11 \quad (1) \\ x - y &= 3 \quad (2) \\ x - z &= 7 \quad (3) \end{aligned}$$

SOLUTION

We can rewrite Equations (2) and (3) in terms of other variables and then substitute the results in Equation (1) where all the variables appear.

$$x - y = 3 \quad \text{is rewritten as} \quad y = x - 3$$
$$x - z = 7 \quad \text{is rewritten as} \quad z = x - 7$$

In this way, we have values for y and z expressed in terms of x. We can then substitute into Equation (1) and solve.

$$\begin{aligned} x + y + z &= 11 \qquad \text{Equation (1)} \\ x + (x - 3) + (x - 7) &= 11 \qquad \text{Substituting for } y \text{ and } z \text{ in Equation (1)} \\ 3x - 10 &= 11 \qquad \text{Simplifying} \\ 3x &= 21 \qquad \text{Adding 10 to both sides} \\ x &= 7 \qquad \text{Multiplying by } \tfrac{1}{3} \end{aligned}$$

Substituting $x = 7$ in Equations (2) and (3), we find the values for y and z.

$$\begin{array}{ll} x - y = 3 & \qquad x - z = 7 \\ 7 - y = 3 & \qquad 7 - z = 7 \\ -y = -4 & \qquad z = 0 \\ y = 4 & \end{array}$$

So the solution set for this system of equations is $\{(7, 4, 0)\}$.

▶ *CHECK* **Warm-Up 8**

Practice what you learned.

WRITER'S BLOCK

How should you decide whether to use *addition* or *substitution* to solve a system of equations?

SECTION FOLLOW-UP

The inequalities implied by the problem in the Section Lead-In are

Highway miles possible on one tank of gas ≤ 700 miles $[(35)(20) \text{ miles}]$

Town miles possible on one tank of gas ≤ 600 miles   [(30)(20) miles]

Gas left in tank ≤ 20 gallons

$600 \leq$ total miles ≤ 700

4.1 WARM-UPS

Work these problems before you attempt the exercises.

1. Solve by graphing:
$y = 2x - 5$
$-2x + 7 = y$ $\{(3, 1)\}$

2. Solve by graphing:
$y = -3x - 13$
$7y = 7x - 7$ $\{(-3, -4)\}$

3. Solve by addition:
$-2y - 5x = 11$
$2y + 3x = -5$ $\{(-3, 2)\}$

4. Solve by addition:
$3x - 4y = 7$
$2y + 3x = 20$ $\left\{\left(5\frac{2}{9}, 2\frac{1}{6}\right)\right\}$

5. Solve by addition:
$2x = 2y - 5$
$6y + 4x = 4$ $\{(-1.1, 1.4)\}$

6. Solve:
$3x - 2y + 5z = 13$
$5x + 2y + 7z = 15$
$6x - 2y - 7z = 7$ $\{(2, -1, 1)\}$

7. Solve by substitution:
$2y + 3x = 7$
$y - x = 1$ $\{(1, 2)\}$

8. Solve:
$3y - 4z = -26$
$-3x + z = -43$
$5x + 3y = 57$
$\{(15, -6, 2)\}$

4.1 EXERCISES

Note: Use your graphing calculator to check your results whenever possible.

In Exercises 1 through 6, determine whether the ordered pair shown is a solution to the given system of linear equations.

1. $(-2, -4)$: $6x = 4 + 4y$
$x - y = 2$ yes

2. $(7, -3)$: $4y = -2x + 2$
$y = -x + 4$ yes

3. $(-5, 12)$: $2y - x = 29$
$6x + 5y = 30$ yes

4. $(4, -3)$: $x - y = 7$
$2y + x = 2$ no

5. $(-7, -5)$: $x - y = -2$
$-x + 2y = -3$ yes

6. $(2, 0)$: $x - y = 2$
$y + 2x = 4$ yes

In Exercises 7 through 12, determine, without graphing, whether the given system is consistent, inconsistent, or dependent.

7. $y = -2x - 2$
$y - x = 1$ consistent

8. $y = 2x + 5$
$y = 2x - 6$ inconsistent

9. $y = 2x + 1$
$y = x + 2$ consistent

10. $y = x - 1$
$y - x = 1$ _inconsistent_

11. $3x = 2y + 5$
$6x = 4y + 10$ _dependent and consistent_

12. $y - 3x = 7$
$2y - 6x = 14$ _dependent and consistent_

In Exercises 13 through 18, use a table of values to graph each system, and determine the point of intersection of the graphs. _Tables may vary._

13. $2x + 5y = 12$
$y = 3x - 1$
$(1, 2)$

14. $6x - 8y = 34$
$y = 3x - 2$
$(-1, -5)$

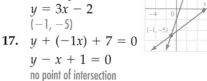

15. $3x + 2y = 20$
$3y = 2x + 4$
$(4, 4)$

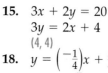

16. $4x + y = 7$
$4x + y = -3$
no point of intersection

17. $y + (-1x) + 7 = 0$
$y - x + 1 = 0$
no point of intersection

18. $y = \left(-\dfrac{1}{4}\right)x + \dfrac{5}{4}$
$y = 5x - 4$
$(1, 1)$

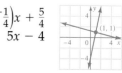

In Exercises 19 through 33, use the slope and y-intercept to graph each system, and then find the solution of the system.

19. $4y = -3x$
$5y = 3x$
$\{(0, 0)\}$

20. $y = 2x - 4$
$y = x - 1$
$\{(3, 2)\}$

21. $y = 2x + 1$
$y = -2x + 1$
$\{(0, 1)\}$

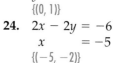

22. $3x = y - 1$
$y = 1 - 3x$
$\{(0, 1)\}$

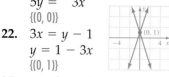

23. $x = 4$
$x + y = 5$
$\{(4, 1)\}$

24. $2x - 2y = -6$
$x = -5$
$\{(-5, -2)\}$

25. $y = 2x - 1$
$x = 3$
$\{(3, 5)\}$

26. $y + 2x = 4$
$x = 4$
$\{(4, -4)\}$

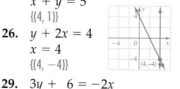

27. $x = 1$
$5x - y = 6$
$\{(1, -1)\}$

28. $y + 1 = 3x - 2$
$y = -2(x - 1)$
$\{(1, 0)\}$

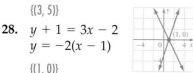

29. $3y + 6 = -2x$
$3y + 15 = x$
$\{(3, -4)\}$

30. $\dfrac{2}{3}y = -\dfrac{2}{3}x$
$\left(\dfrac{1}{3}\right)y = \left(\dfrac{5}{3}\right)x - 2$

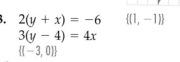

31. $3(y - 4) = -2x$
$3(y + 3) = 5x$
$\{(3, 2)\}$

32. $2y = -5x + 6$
$2y = 3x - 10$
$\{(2, -2)\}$

33. $2(y + x) = -6$
$3(y - 4) = 4x$
$\{(-3, 0)\}$ $\{(1, -1)\}$

In Exercises 34 through 45, use the addition method to solve each system of equations.

34. $3x + y = 7$
$4x - y = 7$
$\{(2, 1)\}$

35. $x + 5y = -18$
$-x + 3y = 2$
$\{(-8, -2)\}$

36. $5x - y = 20$
$6x + y = 46$
$\{(6, 10)\}$

37. $x + 5y = 30$
$-x + 2y = 19$
$\{(-5, 7)\}$

38. $2y + 5x = 15$
$2y - 3x = -1$
$\{(2, 2.5)\}$

39. $11x = -8y + 47$
$-3x = -8y + 5$
$\left\{\left(3, 1\dfrac{3}{4}\right)\right\}$

40. $-3y + 5x = 7$
$-18y + 30x = 42$
dependent; infinitely many solutions

41. $7y - 3x = 14$
$14y - 6x = 22$
inconsistent; no solution

42. $-7y - 6x = 5$
$11y + 8x = -10$
$\{(1.5, -2)\}$

43. $7y - 5x = 3$
$-2y + 3x = 7$
$\{(5, 4)\}$

44. $7y - 9x = 24$
$5y - 2x = 26$
$\{(2, 6)\}$

45. $4y + 3x = 9$
$6y + 4x = 18$
$\{(-9, 9)\}$

In Exercises 46 through 57, use the substitution method to solve each system of equations.

46. $y = -x + 9$
$6y - 5x = 10$
$\{(4, 5)\}$

47. $y = x + 15$
$3y + 5x = 21$
$\{(-3, 12)\}$

48. $x = 3y + 4$
$2x = -3y - 5$
$\left\{\left(-\dfrac{1}{3}, -1\dfrac{4}{9}\right)\right\}$

49. $y + x = 7$
$y = x + 4$
$\left\{\left(1\dfrac{1}{2}, 5\dfrac{1}{2}\right)\right\}$

50. $y - x = 6$
$4y + 3x = -4$
$\{(-4, 2)\}$

51. $-y + x = 8$
$2y + 3x = 7$
$\{(4.6, -3.4)\}$

52. $x - 6y = 7$
$3y - x = -4$
$\{(1, -1)\}$

53. $26y + 4x = 13$
$y + 4x = -12$
$\{(-3.25, 1)\}$

54. $6y = 6x - 6$
$2y = 6x$
$\{(-0.5, -1.5)\}$

55. $6y + 2 = 4x$
$4x = 3y$
$\left\{\left(-\frac{1}{2}, -\frac{2}{3}\right)\right\}$

56. $3y + 2x = 7$
$4y - 2x = 14$
$\{(-1, 3)\}$

57. $16y - 4x = -4$
$8x - 40y = -8$
$\{(9, 2)\}$

In Exercises 58 through 69, use any method to solve each system of equations.

58. $-2y - 6x = -98$
$9y - 2x = 6$
$\{(15, 4)\}$

59. $11y - 12x = 14$
$18y - 3x = -27$
$\{(-3, -2)\}$

60. $y - 2x = 6$
$-3y + 2x = 6$
$\{(-6, -6)\}$

61. $15y = x + 19$
$3y = -x + 17$
$\{(11, 2)\}$

62. $3y - 4x = 12$
$-5y + 2x = -27$
$\{(1.5, 6)\}$

63. $0.8x - 0.1y = 0.69$
$0.6x - 0.5y = 0.73$
$\{(0.8, -0.5)\}$

64. $0.9x + 0.5y = -0.21$
$0.2y - 0.7x = -0.19$
$\{(0.1, -0.6)\}$

65. $1.6x - 3.5y = 3.03$
$-2.8x - 1.3y = -1.59$
$\{(0.8, -0.5)\}$

66. $\frac{3}{10}x - \frac{1}{5}y = -\frac{2}{5}$
$\frac{1}{2}x - \frac{2}{5}y = -\frac{7}{10}$
$\left\{\left(-1, \frac{1}{2}\right)\right\}$

67. $\frac{1}{8}x - 1\frac{1}{2}y = -4\frac{3}{4}$
$-\frac{3}{8}x - \frac{1}{5}y = \frac{3}{20}$
$\{(-2, 3)\}$

68. $\frac{1}{4}x - \frac{1}{2}y = -2$
$\frac{1}{8}x + 2\frac{1}{2}y = 12\frac{3}{4}$
$\{(2, 5)\}$

69. $\frac{2}{5}x + 5y = -1\frac{4}{5}$
$-\frac{1}{10}x - 2y = \frac{3}{4}$
$\left\{\left(\frac{1}{2}, -\frac{2}{5}\right)\right\}$

In Exercises 70 through 81, solve each system of linear equations by using elimination.

70. $2x - 4y + 6z = 44$
$x - 3y - z = 14$
$4x - 7y + 5z = 63$
$\{(5, -4, 3)\}$

71. $2x + y - z = 5$
$-x + 3y + z = -2$
$-4x + 3y + z = -11$
$\{(3, 0, 1)\}$

72. $2x + 2y + 4z = -6$
$2x - y - z = 44$
$3x + y + 2z = 29$
$\{(16, -5, -7)\}$

73. $12x - y - 3z = 81$
$4x - 2y - 6z = 22$
$2x + 4y + 8z = 18$
$\{(7, -3, 2)\}$

74. $x + 2y + 3z = 17$
$-2x + 4y - 6z = -26$
$x - 3y + z = 4$
$\{(3, 1, 4)\}$

75. $x - y - 2z = 20$
$x - 3y + z = 31$
$-x - y - 2z = 26$
$\{(-3, -13, -5)\}$

76. $3x + 5y - 2z = 17$
$4x + 4y + 4z = 8$
$2x + y + 3z = 3$
$\{(7, -2, -3)\}$

77. $9x + 12y + 6z = -39$
$6x - 3y + 3z = -3$
$2x - 4y + 6z = 0$
$\{(-1, -2, -1)\}$

78. $-5x - 3y = 39$
$-4x + y + 4z = -16$
$-3x + 2y + 6z = -37$
$\{(-3, -8, -5)\}$

79. $x + y + 2z = -7$
$-x - 3y = 33$
$2x + 5y + 8z = -46$
$\{(3, -12, 1)\}$

80. $-3x - y - 4z = 17$
$-2x - y - z = 3$
$-x - 2y - 3z = 4$
$\{(-2, 5, -4)\}$

81. $-4x + 5y + 3z = 8$
$x + 2y + 3z = 2$
$-x + y + z = -1$
$\{(3, 7, -5)\}$

In Exercises 82 through 87, solve each system of equations by using substitution.

82. $5x - 2y + 3z = -4$
$4x - 7y + 5z = -16$
$x - y + z = 0$
$\{(-8, 12, 20)\}$

83. $x + 2y = 5$
$-2x - 6z = -30$
$-3y + z = 1$
$\{(3, 1, 4)\}$

84. $x - y = 10$
$x + z = -4$
$-y - 2z = 24$
$\{(6, -4, -10)\}$

85. $-y + 2z = -15$
$x + y = 3$
$3x + 3z = -6$
$\{(8, -5, -10)\}$

86. $-5x + y = 31$
$-3x - 2z = 25$
$2y - z = -6$
$\{(-7, -4, -2)\}$

87. $-x - y = -9$
$2x - z = -9$
$-3y - z = -38$
$\{(-2, 11, 5)\}$

EXCURSIONS

Class Act

1. In the total-cost formula

    $$T = F + QV$$

    T is total cost, F is flat fee, Q is amount per service, and V is number of services.

    a. If F remains constant and Q remains constant, how does V change as T gets larger? As T gets smaller?

    b. If F and T remain constant and V gets larger, what happens to Q?

    c. Think of another real-life situation that can be modeled by an equation in this form.

2. Answer parts (a) through (d) with one of the following:

    "remains true;" "is false;" or "cannot predict"

    What happens to the inequality

    $$ax \leq b$$

    if

    a. both sides of the inequality are multiplied by a positive integer?

    b. both sides of the inequality are multiplied by zero?

    c. both sides of the inequality are multiplied by a positive number between zero and one?

    d. a negative number is added to both sides of the inequality?

    Give a real-life situation that can be modeled by an inequality.

3. Have you ever wondered what day of the week a certain date fell on? or will fall on? There is an algebraic expression that you can use to find the day of the week for any date. The expression, taken from the *NCTM Sourcebook of Applications of School Mathematics,* is

    $$\frac{d + 2m + \left[\frac{3(m + 1)}{5}\right] + y + \left[\frac{y}{4}\right] - \left[\frac{y}{100}\right] + \left[\frac{y}{400}\right] + 2}{7}$$

    where d is the day of the month, m is the month, and y is the year. This expression (actually, this fraction) must be computed in a certain way.

    ▪ For each fraction enclosed in brackets [ ] in the numerator, use the integer part only. (After doing the indicated division, drop the remainder.)

    ▪ Consider January to be the thirteenth month of the preceding year, and February the fourteenth month of the preceding year (with January and

February, use $y - 1$ instead of y for the year). (Consider March through December to be, as always, the third through twelfth months.)

- Consider Sunday to be the first day of the week, Saturday is either the seventh or the "zeroth" day of the week.
- When you simplify the expression by dividing by the denominator 7, the remainder will be the desired day of the week.

Find the day of the week on which Columbus is said to have come to America (October 12, 1492).

Posing Problems

4. Use the data from this table to ask and answer four questions.

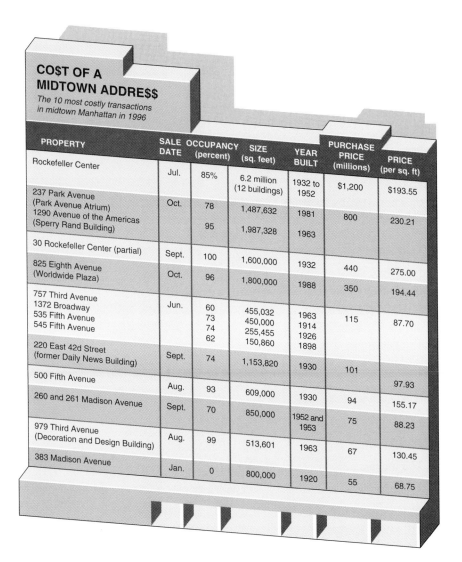

CO$T OF A MIDTOWN ADDRE$$
The 10 most costly transactions in midtown Manhattan in 1996

PROPERTY	SALE DATE	OCCUPANCY (percent)	SIZE (sq. feet)	YEAR BUILT	PURCHASE PRICE (millions)	PRICE (per sq. ft)
Rockefeller Center	Jul.	85%	6.2 million (12 buildings)	1932 to 1952	$1,200	$193.55
237 Park Avenue (Park Avenue Atrium)	Oct.	78	1,487,632	1981	800	230.21
1290 Avenue of the Americas (Sperry Rand Building)		95	1,987,328	1963		
30 Rockefeller Center (partial)	Sept.	100	1,600,000	1932	440	275.00
825 Eighth Avenue (Worldwide Plaza)	Oct.	96	1,800,000	1988	350	194.44
757 Third Avenue	Jun.	60	455,032	1963	115	87.70
1372 Broadway		73	450,000	1914		
535 Fifth Avenue		74	255,455	1926		
545 Fifth Avenue		62	150,860	1898		
220 East 42d Street (former Daily News Building)	Sept.	74	1,153,820	1930	101	97.93
500 Fifth Avenue	Aug.	93	609,000	1930	94	155.17
260 and 261 Madison Avenue	Sept.	70	850,000	1952 and 1953	75	88.23
979 Third Avenue (Decoration and Design Building)	Aug.	99	513,601	1963	67	130.45
383 Madison Avenue	Jan.	0	800,000	1920	55	68.75

Source: Cushman & Wakefield from *New York Times*, 12 January 1997.

CONNECTIONS TO *STATISTICS*

Mean, Median, and Mode

Finding the Mean

An important concept that is used in calculating grades, in controlling the quality of products, and in many other situations is the "average" of a set of measurements. The average of a set of numbers is one of three measures that indicate where the "center" of the set is. The **mean** is one of those measures. The others are *median* and *mode*. We will discuss all three in this Connection. The mean can be thought of as the balance point in a set of measurements.

To find the mean of a set of n numbers

1. Add the numbers together.
2. Divide by n.

Let's look at an example.

▪ ▪ ▪

EXAMPLE

A woman measures the heights of some shrubs in front of her house. They measure 33.4 inches, 35.2 inches, 42.5 inches, and 38.1 inches. What is the average height?

SOLUTION

Here $n = 4$, because there are four measurements. To find the mean, we add all the measurements and then divide that total by 4.

$$
\begin{array}{ll}
\text{Add:} \quad 33.4 & \text{Divide by 4:} \quad 4\overline{)149.2}^{\,37.3} \\
\qquad\quad\; 35.2 & \qquad\qquad\qquad\; \underline{12} \\
\qquad\quad\; 42.5 & \qquad\qquad\qquad\; 29 \\
\qquad\quad\; \underline{38.1} & \qquad\qquad\qquad\; \underline{28} \\
\qquad\quad\; 149.2 & \qquad\qquad\qquad\; 12 \\
& \qquad\qquad\qquad\; \underline{12} \\
& \qquad\qquad\qquad\; 0
\end{array}
$$

The average of the four heights is 37.3 inches.

Finding the Median

The **median** of a set of numbers is the "central number" of that set. It can be found using the following method.

> **To find the median of a set of *n* numbers**
>
> 1. Arrange the numbers in order.
> 2. Find the "central number" of those numbers. The "central number" is the number such that there are as many values greater than it as there are values less than it.

▪ ▪ ▪

EXAMPLE

Find the median of 43, 39, 47, 26, and 28.

SOLUTION

Arrange the numbers in order.

$$26 \quad 28 \quad 39 \quad 43 \quad 47$$

The central number is 39, because there are 2 numbers on either side of it. So 39 is the median.

◢

Finding the median involves more computation when we have an even number of values.

▪ ▪ ▪

EXAMPLE

A child cuts some string in the following lengths: 43.1 inches, 47.2 inches, 42 inches, and 48.5 inches. What is the median length?

SOLUTION

To find the median, we first write the numbers in order:

$$42 \quad 43.1 \quad 47.2 \quad 48.5$$

There is no "central number" because there are four numbers. To find the median in such a case, we must find the average of the two middle numbers—here 43.1 and 47.2. We add them and divide their sum by 2.

$$\frac{43.1 + 47.2}{2} = \frac{90.3}{2} = 45.15$$

So 45.15 inches is the median length.

◢

Finding the Mode

The **mode** is also relatively easy to find. It is the value that occurs most frequently in a set of numbers. A given set of numbers may have one mode, no mode, or many modes.

> **To find the mode of a set of *n* numbers**
> 1. Determine how many times each number appears in the set (by tallying).
> 2. The mode is the number (or numbers) that appears (or appear) most.

▪▪▪

EXAMPLE

Find the mode of the following numbers:

$$1, 3, 2, 4, 5, 3, 7, 2, 3, 7, 3$$

SOLUTION

First we write down all the numbers in order. Then we tally how many of each there are.

```
        1  /
        2  //
 ⟶      3  ////
        4  /
        5  /
        7  //
```

Because the number 3 occurs four times, and no other number occurs as many times, 3 is the mode of this set of numbers.

◀

Calculator Corner

You can use your graphing calculator to find the mean, median, and mode of a set of data. For the 1994 New York Marathon, the ages of the first 15 male finishers were

$$26, 33, 31, 30, 30, 25, 28, 33, 32, 25, 29, 28, 32, 24, 33$$

Use the **STAT**istics utility on your graphing calculator to help you answer the following questions about the data.

a. What is the mode? **b.** What is the median? **c.** What is the mean?

Press the **STAT** key on your graphing calculator and enter the data (you may need to consult your graphing calculator's manual for instructions).

To find the mode, have your graphing calculator **SORT** the data in either ascending or descending order. (Press [2nd] [1] to show "L1" on the screen.)

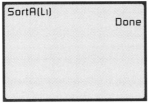

The **MODE** is 33.

Now do **1-Var Stats** to find the median and mean of the data by using the **STAT/CALC** utility. ($\overline{x}$ is the symbol for the mean.)

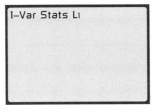

The mean is 29.26666667 and the median is 30.

PRACTICE

1. The lengths of the spans of the five longest steel arch bridges in the world are given in the following table.

New River Gorge	1699.58 feet
Bayonne	1625.4 feet
Sydney Harbour	1649.66 feet
Fremont	1254.9 feet
Port Mann	1199.95 feet

   What are the mean and median lengths for these five steel arch bridges?
   mean, 1485.898 feet; median, 1625.4 feet

2. The lengths of three of the longest canals in the world are as follows: White Sea, 141.3 miles; Suez, 100.25 miles; and Volga, 61.75 miles. What is the mean for these canal lengths?   101.10 miles

3. Find the mean land area of these five cities in the United States.

Columbus, Ohio	189.272 square miles
Honolulu, Hawaii	617 square miles
Washington, D.C.	68.25 square miles
Seattle, Washington	144.6 square miles
Cleveland, Ohio	79 square miles

219.6244 square miles

4. ✎ Ask and answer four questions using data from Exercises 1 through 3.

5. The last six Apollo missions had durations of $245\frac{1}{3}$ hours, 143 hours, $215\frac{3}{4}$ hours, $295\frac{1}{4}$ hours, $266\frac{1}{3}$ hours, and $301\frac{1}{2}$ hours. What was the median duration of these six missions? $255\frac{5}{6}$ hours

6. In a certain six-year period, the following numbers of people have finished the New York City marathon: 13,599; 14,546; 14,492; 15,887; 19,689; and 21,244. What was the median number of finishers during these six years? What was the mean?   median, 15,216.5; mean, 16,576 finishers

7. On six of the entries in the Tall Ship race in New York City on July 4, 1976, the numbers of crew members were as follows: 104, 189, 99, 162, 236, and 16. What was the median number of crew members on these six ships?
median: 133 crew members

8. Organize the following data in a table, and determine the mean number of people per year who immigrated to the United States from France in the period 1931–1980. The number of people who immigrated from France to the United States during the period 1931–1940 was 12,623. During 1941–1950, the number rose 26,186; during 1951–1960, the number increased 12,312. From 1961–1970, the number dropped by 5884, and during the period 1971–1980, the number dropped 20,168.   3457 immigrants per year

9. The following heights above and below sea level were recorded by a scientific team: 35 feet below, 46.3 feet above, 19 feet below, 25.17 feet above, 0.3 feet above, 135.4 feet below, 35 feet below, and 6.24 feet above. What is the range of these readings? What was the mean height recorded?
range, 181.7 feet; mean, −18.30 feet

10. The estimated numbers of people who consider themselves Muslim, Buddhist, Protestant, Catholic, or Jewish are as follows:

Muslim	935 million
Buddhist	303 million
Protestant	73.5 million
Catholic	50.5 million
Jewish	5.9 million

What is the mean number of people (to the nearest tenth of a billion) who consider themselves part of the five major religions?   0.3 billion

4.2 Applying Systems of Linear Equations

SECTION LEAD-IN

Living within a budget involves a number of limiting factors. Make a budget and write a series of equations and inequalities that describe the limits of your spending in categories such as

- food
- transportation
- savings

What determines these limits?

Systems of linear equations can be used to solve a greater variety of practical problems than can one-variable equations. They can also be used to solve familiar problems more easily. We first assign variables to the unknown quantities; then we set up the system of equations and solve.

STUDY HINT

When checking verbal problems, return to the problem statement instead of to the equations because you may have written incorrect equations!

▪▪▪
EXAMPLE 1

A boat travels 300 miles in an easterly direction with the current in 6 hours. Coming back against the current, the trip takes 8 hours. What are the rate of the current and the rate of the boat in still water?

With the current
300 mi, 6 hr

Current ⟶

Against the current
300 mi, 8 hr

Current ⟶

SOLUTION

We first assign variables to the unknown quantities.

> Let x be the rate of the boat in still water.
>
> Let y be the rate of the current.

Then

> Rate of the boat with the current $= x + y$
>
> Rate of the boat against the current $= x - y$

We know that the product of rate and time equals distance, so we can make a table.

	Rate	× Time	= Distance
With current	$(x + y)$ ×	6	300
Against current	$(x - y)$ ×	8	300

The table immediately gives us a pair of equations:

$$(x + y)6 = 300 \quad \text{or} \quad 6x + 6y = 300 \quad (1)$$
$$(x - y)8 = 300 \quad \text{or} \quad 8x - 8y = 300 \quad (2)$$

Solving this system by elimination yields

$$48x + 48y = 2400 \quad \text{Equation (1)} \times 8$$
$$48x - 48y = 1800 \quad \text{Equation (2)} \times 6$$
$$\overline{96x \qquad\quad = 4200} \quad \text{Adding left and right sides}$$
$$x = 43.75 \quad \text{Multiplying by } \tfrac{1}{96}$$

Substituting 43.75 for x in Equation (1) and solving for y, we have

$$6(43.75) + 6y = 300$$
$$262.50 + 6y = 300$$
$$6y = 37.50$$
$$y = 6.25 \text{ mph}$$

The rate of the boat in still water is 43.75 miles per hour, and the rate of the current is 6.25 miles per hour.

▶ *CHECK* **Warm-Up 1**

Recall that the place values of the digits in 265 are, from the left, hundreds, tens, and ones. So 265 is equivalent to $2 \cdot 100 + 6 \cdot 10 + 5 \cdot 1$. We use this concept to solve the next problem. This type of problem is often called a "number" problem.

▪ ▪ ▪
EXAMPLE 2

The sum of the digits in a two-digit number is 12. If the digits are reversed, the new number is 54 more than the original number. Find the original number.

SOLUTION

We assign the variables as follows:

$$\text{Let } x = \text{units, or ones, digit}$$
$$\text{Let } y = \text{tens digit}$$

Then

$$x = \text{value of ones digit}$$
$$10y = \text{value of tens digit}$$
$$10y + x = \text{value of original number}$$
$$10x + y = \text{value of number with digits reversed}$$

Now we can translate the first phrase into algebra.

The sum of the digits is 12.
$$x + y \quad = 12 \quad \text{Equation (1)}$$

Then we translate and simplify the second phrase:

The number with digits reversed is 54 more than the original number.
$$10x + y \qquad = \qquad 10y + x + 54$$

So

$$9x - 9y = 54 \quad \text{Equation (2)}$$

We then solve the two equations. We can multiply Equation 1 by 9 to eliminate the y-terms.

$$
\begin{array}{rl}
9x - 9y = & 54 \quad \text{Equation (2)} \\
\underline{9x + 9y = 108} & \quad \text{Equation (1)} \times 9 \\
18x = 162 & \\
x = \quad 9 &
\end{array}
$$

Substituting 9 for x in Equation (1) yields

$$
\begin{aligned}
x + y &= 12 \\
9 + y &= 12 \\
y &= 3
\end{aligned}
$$

The original number is $10y + x = 39$. The check is left for you to do.

▶ *CHECK* **Warm-Up 2**

Another type of verbal problem is the mixture problem. In one type of mixture problem, we are given information about two or more groups of items (coins, stamps, or the like) and are asked questions about a mixture of these groups. A table is useful in compiling the given information.

■ ■ ■

EXAMPLE 3

A bank containing dimes and quarters has been spilled on the floor. There is a total of $6.75, and there are twice as many dimes as quarters. How many of each coin is there?

SOLUTION

We set up a table of monetary values, letting x represent the number of dimes and y the number of quarters.

Item	Number	Monetary Value
Dimes	x	$0.10x$
Quarters	y	$0.25y$
Total		$6.75

The table gives us the first equation:

$$0.10x + 0.25y = \$6.75$$

From the problem statement, we also have a relationship between the numbers of dimes and quarters:

$$x = 2y$$

We now have a system of two equations. To solve it, we substitute $2y$ for x in the first equation and then solve.

$$
\begin{aligned}
0.10(2y) + 0.25y &= 6.75 \quad \text{Substituting } 2y \text{ for } x \\
0.45y &= 6.75 \quad \text{Multiplying and combining like terms} \\
y &= 15 \quad \text{Multiplying by } \tfrac{1}{0.45}
\end{aligned}
$$

Substituting 15 for y in the second equation yields

$$x = 2y = 2(15) = 30$$

So we should have 30 dimes and 15 quarters. We go back to the problem statement to check. Sure enough, 30 is twice 15, so we do have twice as many dimes as quarters. And

$$\$0.10(30) + \$0.25(15) = \$3.00 + \$3.75 = \$6.75$$

So we do have a total of $6.75. The solution checks. There are 30 dimes and 15 quarters.

▶ *CHECK* **Warm-Up 3**

Another type of mixture problem involves solutions (liquids) with different concentrations.

∎∎∎

EXAMPLE 4

How many ounces of a 6% acid solution must be added to how many ounces of a 12% acid solution to produce 75 ounces of a mixture that is 10% acid?

SOLUTION

We are asked for the amounts of two solutions to mix. We let x and y represent the numbers of ounces of the 6% and 12% solutions:

Let x = ounces of 6% solution needed

Let y = ounces of 10% solution needed

The amount of acid in the 6% solution plus the amount of acid in the 12% solution equals the amount of acid in the mixture. Now we can make a table showing what we know. The right column tells how much acid is in each of the solutions.

Item	Amount of solution	Amount of acid in solution
6% solution	x	$0.06x$
12% solution	y	$0.12y$
10% mixture	75	$(0.10)(75)$

We can write two equations from the information in the two right columns of the table:

1. The amounts mixed together must add up to the total number of ounces, so

$$x + y = 75$$

2. The amounts of acid in the solutions must add up to the amount in the mixture, so

$$0.06x + 0.12y = 0.10(75)$$

Simplifying, we have the system

$$
\begin{aligned}
x + \quad y &= 75 \quad (1) \\
0.06x + 0.12y &= 7.5 \quad (2)
\end{aligned}
$$

Solving by substitution, we get

$$y = 75 - x \qquad \text{Solving for } y \text{ using Equation (1)}$$
$$0.06x + 0.12(75 - x) = 7.5 \qquad \text{Substituting for } y \text{ in Equation (2)}$$
$$0.06x + 9 - 0.12x = 7.5 \qquad \text{Simplifying}$$
$$-0.06x + 9 = 7.5 \qquad \text{Combining like terms}$$
$$-0.06x = -1.5 \qquad \text{Adding } -9 \text{ to both sides}$$
$$x = 25 \qquad \text{Multiplying by } -\tfrac{1}{0.06}$$

Substituting 25 for x to solve for y, we get

$$x + y = 75$$
$$25 + y = 75$$
$$y = 50$$

Going back to the original problem to check, we see that we do have 75 ounces of mixture. The amount of acid in the mixture is

$$0.06(25) + 0.12(50) = 1.5 + 6 = 7.5 \text{ ounces}$$

The percent of acid in the mixture is

$$\tfrac{7.5}{75} \times 100\% = 10\%$$

▶ *CHECK* **Warm-Up 4**

Many real-life applications can be solved by using systems of equations.

▪ ▪ ▪
Example 5

Walter and Irene are buying bushes to landscape around their home. They can buy either 15 rose bushes and 5 azaleas for $975 or 10 rose bushes and 6 azaleas for $850. Find the cost of each type of bush.

Solution

We let

$$x = \text{cost of one rose bush}$$
$$y = \text{cost of one azalea bush}$$

We can represent the cost of 15 rose bushes by $15x$ and the cost of 5 azaleas by $5y$; this selection costs $975. Similarly, the selection of 10 rose bushes and 6 azaleas, $10x + 6y$, is $850. The equations are

$$15x + 5y = 975 \quad (1)$$
$$10x + 6y = 850 \quad (2)$$

We solve by elimination:

$$-30x - 10y = -1950 \qquad \text{Equation (1)} \times (-2)$$
$$\underline{30x + 18y = 2550} \qquad \text{Equation (2)} \times 3$$
$$8y = 600 \qquad \text{Adding left and right sides}$$
$$y = 75 \qquad \text{Multiplying by } \tfrac{1}{8}$$

Substituting 75 for y in Equation (1), we have

$$15x + 5(75) = 975$$
$$15x + 375 = 975$$
$$15x = 600$$
$$x = 40$$

Each rose bush costs $40 and each azalea plant costs $75. The check is left for you to do.

▶ CHECK **Warm-Up 5**

In the next example, we see an application involving investments and can use a system of linear equations with three variables to solve the problem.

▪▪▪

EXAMPLE 6

A total of $15,000 is invested in three different accounts that pay, respectively, 4%, 9%, and 11% interest per year. There is $1000 more invested at 11% than at 9%. And together, the accounts earn $1150 in interest per year. How much money is invested at each rate?

SOLUTION

This is really a mixture problem. We assign variables and then set up a table showing the given information.

Let $x =$ amount invested at 4%

Let $y =$ amount invested at 9%

Let $z =$ amount invested at 11%

Then the various amounts and the interest they earn are as follows:

	Amount	Rate	Interest
	x	4%	$0.04x$
	y	9%	$0.09y$
	z	11%	$0.11z$
Total	$15,000		1150

The table gives us only two equations. We look back into the problem statement for a third equation; and there we find a relation between y and z:

$$z = 1000 + y$$

Then the problem, as stated, is translated into a system of three equations:

$$x + y + z = 15000$$
$$0.04x + 0.09y + 0.11z = 1150$$
$$z = 1000 + y$$

The equations are easier to deal with when the variables are sorted and decimal numbers are eliminated:

$$x + y + z = 15000 \qquad \text{Equation (1)}$$
$$4x + 9y + 11z = 115000 \qquad \text{Equation (2) (original} \times 100)$$
$$-y + z = 1000 \qquad \text{Equation (3) (original rewritten)}$$

To solve, we multiply Equation (1) by -4 and then add the result to Equation (2).

$$
\begin{array}{ll}
4x + 9y + 11z = 115000 & \text{Equation (2)} \\
\underline{-4x - 4y - 4z = -60000} & \text{Equation (1)} \times (-4) \\
5z + 7z = 55000 & \text{Adding; new Equation (4)}
\end{array}
$$

We now add this equation in y and z to the result of multiplying Equation (3) by 5.

$$
\begin{array}{ll}
5y + 7z = 55000 & \text{Equation (4)} \\
\underline{-5y + 5z = 5000} & \text{Multiplying Equation (3) by 5} \\
12z = 60000 & \text{Adding left and right sides} \\
z = 5000 & \text{Multiplying by } \frac{1}{12}
\end{array}
$$

WRITER'S BLOCK

Explain to a friend why we chose to substitute in Equation (3).

We now have the value of z. Going back to Equation (3) and substituting 5000 for z gives us

$$
\begin{array}{ll}
-y + z = 1000 & \\
-y + 5000 = 1000 & \text{Substituting } z = 5000 \\
y = 4000 &
\end{array}
$$

Then, from Equation (1),

$$
\begin{array}{ll}
x + y + z = 15000 & \\
x + 4000 + 5000 = 15000 & \text{Substituting } y = 4000; z = 5000 \\
x = 6000 &
\end{array}
$$

So $6000 was invested at 4%, $4000 at 9%, and $5000 at 11%.

▶ *CHECK* **Warm-Up 6**

▪▪▪

EXAMPLE 7

Plastic knives, forks, and spoons are sold in prepackaged groups. One grouping contains 8 forks, 8 spoons, and 5 knives for $1.42. A second grouping contains 5 forks and 5 spoons but no knives and sells for $0.70. A third group contains 6 knives and 6 forks and sells for $0.78. Find the cost of each knife, fork, and spoon, assuming that they cost the same no matter what the grouping.

SOLUTION

We assign variables and represent each grouping as an equation.

$$\text{Let } x = \text{cost per fork}$$
$$\text{Let } y = \text{cost per spoon}$$
$$\text{Let } z = \text{cost per knife}$$

Then

$$8x + 8y + 5z = 1.42 \quad (1)$$
$$5x + 5y = 0.70 \quad (2)$$
$$6x + 6z = 0.78 \quad (3)$$

Because not all variables appear in each equation, we will solve the second and third equations for y and z, respectively, in terms of x, and then we will use substitution to solve for x.

$$5x + 5y = 0.70 \quad \text{becomes} \quad y = 0.14 - x$$
$$6x + 6z = 0.78 \quad \text{becomes} \quad z = 0.13 - x$$

Substituting the expressions for y and z in Equation (1) and then solving, we have

$$8x + 8y + 5z = 1.42$$
$$8x + 8(0.14 - x) + 5(0.13 - x) = 1.42$$
$$8x + 1.12 - 8x + 0.65 - 5x = 1.42$$
$$1.77 - 5x = 1.42$$
$$-5x = -0.35$$
$$x = 0.07$$

Substituting 0.07 for x in Equation (2) gives us

$$5(0.07) + 5y = 0.70$$

from which we find $y = 0.07$. Then substituting in Equation (1) gives

$$8(0.07) + 8(0.07) + 5z = 1.42$$

which yields $z = 0.06$.

So forks and spoons cost 7 cents each, and knives cost 6 cents each.

▶ CHECK Warm-Up 7

Practice what you learned.

SECTION FOLLOW-UP

Here is a sample budget based on an income of $1500 per month.

Rent (R)	$R = \$500$
Utilities (U)	$0 < U \leq \$50$
Food (F)	$\$150 < F \leq \$400$
Transportation (T)	$0 \leq T \leq \$90$
Savings (S)	$0 \leq S \leq \$300$
Miscellaneous (M)	$0 \leq M \leq \$200$
Entertainment (X)	$0 \leq X \leq \$200$

$$R + U + F + T + S + M + X \leq \$1500$$

4.2 WARM-UPS

Work these problems before you attempt the exercises.

1. *Travel Time*   An airplane travels 266 miles in two hours with the wind. It takes the same plane 1.5 hours to travel 162 miles against the wind. Find the speed of the wind.   12.5 mph

2. *Lunar Cycle*   The number of days in the lunar (moon) cycle is a two-digit number. Four times the sum of the two digits is five times the ones digit. The sum of the ones digit and 4 is six times the tens digit. How many days long is the lunar cycle?   28 days

3. *Movie Tickets*   A movie theater showing *Rocky Horror Picture Show* and *Scream* sells tickets at $6 for adults and $2 for students. A total of 275 tickets were sold, giving a revenue of $1050. How many of each type were sold?   adult tickets: 125; student tickets: 150

4. *Mixed Nuts*   Pointers Mixed Nuts has 10% pecans. Nestors Nuts has 18% pecans. How many ounces of each should be mixed to get 20 ounces of a nut mixture that is 15% pecans?   Pointers: 7.5 ounces; Nestors: 12.5 ounces

5. *Taxi Service*   San Berdoo car service charges $1.50 for the first mile and 15 cents for each additional fourth of a mile. Redlands car service charges $2.50 for the first mile and 10 cents for each additional fifth of a mile. At what point will the charges be identical for both car services for the same distance? (*Hint:* Write equations representing these two situations. Use x for miles traveled and y for total cost.)   At 11 miles the charges will be the same for both car services.

6. *Investing Money*   Pam and Bob have investments totaling $8000 in three accounts: a savings account paying 5% annual interest, a money market account paying 6%, and a checking account paying 4%. If the annual interest from the three accounts is $430, and the interest from the savings and money market accounts is $410, find how much was invested at each rate. $4000 at 5%; $3500 at 6%; $500 at 4%

7. *Elevator Weight Limits*   A small elevator has a weight limit of 200 pounds. Blake, Bobby, and Catherine want to use it. Blake and Catherine weigh 119 pounds together; Bobby and Catherine weigh 133 pounds, and Blake and Bobby weigh 142 pounds. Can these three people ride together?   Blake: 64 pounds; Catherine: 55 pounds; Bobby: 78 pounds; yes

4.2 EXERCISES

Note: Use your graphing calculator to check your results whenever possible.

1. **Basketball** In a basketball game, Blue Edwards of the Vancouver Grizzlies scored 19 points, consisting of both 2-point and 3-point baskets. Altogether he had 8 baskets. How many 2-point and how many 3-point baskets were there?   2-point baskets: 5; 3-point baskets: 3

2. **Geometry** Complementary angles are two angles whose measures total 90°. Of two complementary angles, one is five degrees more than one-fourth the size of the other. Find the measures of the two angles.   68°; 22°

3. **Travel Time** A small plane travels 3000 kilometers in 5 hours with the help of a tailwind. Returning takes 6 hours against the wind. Find the speed of the wind. Find the speed of the plane in still air.   wind: 50 km/hr; plane: 550 km/hr

4. **Hamburger Mix** How many ounces of ground beef that is 75% lean must be mixed with ground beef that is 90% lean to get 65 ounces of ground beef that is 85% lean?   75%; $21\frac{2}{3}$ ounces; 90%; $43\frac{1}{3}$ ounces

5. **Deli Costs** The Dixie Diner in Dahlonega, Georgia, is selling platters with a brand-name ham and a local butcher's brand mixed. Three pounds of the brand-name mixed with 5 pounds of the local ham costs $17.88. Four pounds of each costs $18.56. About how much of the brand-name ham can you buy for $25?   9.4 pounds

6. **Concert Tickets** For a Red Hot Chili Peppers concert, 7500 tickets were sold for $12.50 and $10. Altogether, promoters collected $78,125. How many of each ticket were sold? $10 tickets: 6,250; $12.50 tickets: 1,250

7. **Salaries** A cabinet maker earns $10 more per hour than his apprentice. For a 40-hour week, their combined earnings were $1280. How much does each earn?   cabinet maker: $21 per hour; apprentice: $11 per hour

8. **Island Hopping** A sight-seeing boat completes a round-trip to an offshore island in 5 hours. Its average rate going is 20 miles per hour and its average rate returning is 30 miles per hour. How far is the island from shore and what was the traveling time to the island?   60 miles; 3 hours

9. **Investing Money** I decided to invest in two accounts—savings, paying 3.5% interest and a market account, paying 6.5%. I invested amount A in savings and amount B in the market account. In one year, I earned $51.88 in interest. The next year, I invested A in the market account and B in savings earning $6.24 more than I did last year in interest. What are the two amounts A and B?   A: $654; B: $446

10. **Solutions** A cleaning solution contains 4% ammonia. A second solution contains 8% ammonia. How many ounces of each should be mixed to give 20 ounces of a 5% ammonia solution?   4% solution: 15 ounces; 8% solution: 5 ounces

11. ***Space Facilities*** The vehicle assembly building at Cape Canaveral, Florida, is one of the largest buildings in the world. It is 525 feet high. The difference between its width and length is 198 feet. Three-fourths of its width is 19 more than its length. What is its volume? 194,716,200 cubic feet

12. ***First Escalator*** The world's first escalator was installed in Coney Island, New York. The year it was installed has its last two digits equal to six times two less than the first two digits. The sum of the numbers formed by the first two digits and by the last two digits is 114. What year was the escalator installed? 1896

13. ***Fan Mail*** Rin-Tin-Tin was a famous canine actor. There were several Rin-Tin-Tins. The year that the original one died and the number of letters per week he was receiving from fans are related. The year is 68 less than the number of letters. Twice the year plus $\frac{1}{2}$ the number of letters is four thousand, eight hundred sixty-four. What year did Rin-Tin-Tin die? How many letters was he receiving per week at that time? 1932; 2000 letters

14. ***Coal and Gravel Mixture*** How many tons of a mixture of 25% coal and gravel must be mixed with a 45% mixture to produce 50 tons of a 35% coal and gravel mixture? 25%: 25 tons; 45%: 25 tons

15. ***Oat Bran*** Ochoa Oats has 25% oat bran. Octagon Oats has 65% oat bran. Three times as much Octagon Oats as Ochoa Oats were mixed to give 32 ounces of oats. What is the percent oat bran in this new mixture? 55%

16. ***Seed Spitting Contest*** According to *Sports Illustrated*, Ernest Corpus won the youth division of the World Championship Watermelon Seed Spitting contest in Luling, Texas. The world senior record is held by Lee Wheelis. The total distance spit by Ernest and Lee is 87 feet 5 inches. Lee spit 50 feet 1 inch farther than Ernest. What are the distances of the two winning watermelon spits? Lee: 68'9"; Ernest: 18'8"

17. ***Elevator Speed*** The world's fastest passenger elevator is in Tokyo. The difference between the number of stories in the building and the number of miles per hour its elevator travels is 38. Three times the speed is 6 more than the number of stories in the building. How tall is the building? How fast does its elevator travel? 60 stories; 22 mph

18. ***Heart Rate*** In the vocabulary of runners, cyclists, and swimmers, a target zone is the range of heart rate considered ideal for best aerobic conditioning. The lowest limit y is a function of age A and is given by

$$y = \frac{-2}{3}A + 150$$

For a particular runner, the difference between the lower limit and her age is 30. How old is the runner? What is the lowest limit for this runner?
The runner is 72 years old; the lower heart rate limit is 102 beats per minute.

19. ***Pickle Packing*** A pickle packer has two cauldrons of brine (salt water). The first cauldron contains one pound of salt per gallon. The second contains 2 pounds per gallon. How many gallons of each must

be mixed to give one hundred gallons of brine with 1.6 pounds per gallon of salt?   first cauldron: 40 gallons; second cauldron: 60 gallons

20. **Taxes**   In Fitzgerald and in Ocilla, two neighboring towns, the tax on small electric appliances differs by 0.4%. The cost of a radio, with tax, in Fitzgerald is $37.59. In Ocilla, the same radio is $37.45. How much is the radio without tax? $35

21. **Fence Construction**   A three-sided fence is constructed around a cabin on Green River and its rectangular lot, using the river's edge as one side. (See the figure.) This fence is 215 feet long. One side of the fence is 38 feet shorter than the side opposite the river. What is the area of the lot?
5723 square feet

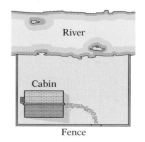

River

Cabin

Fence

22. **Flipping Coins**   Wilson was flipping dimes and quarters and got 13 heads and 14 tails. Altogether he had $4.35. How many of each coin does Wilson have?   dimes: 16; quarters: 11

23. **Recreation**   Two people, one on rollerblades and one race walking, are 18 miles apart. They will meet in 2 hours if they head toward each other. They will meet in 4 hours if they head in the same direction. Find the rate of each.   rollerblader: 6.75 mph; race walker: 2.25 mph

24. **Assembly Lines**   Three assembly lines together produce 7400 car parts per week. The first two assembly lines together produce 4700 car parts, and the last two together produce 5200 parts. How many parts does each line produce by itself?   line 1: 2200 parts; line 2: 2500 parts; line 3: 2700 parts

25. **Fruit Picking**   Gladys picked a total of 64 quarts of fruit over a period of three days. She picked 6 quarts more on the first day than on the third day, and she picked 2 quarts less on the second day than on the third. How many quarts of fruit did she pick each day?   day 1: 26 quarts; day 2: 18 quarts; day 3: 20 quarts

26. **Investing Money**   Marcel invests $10,000 in three different accounts. The first, a money market account, pays 9% annual interest; the second, a checking account, pays 4%; and a third, a savings account, pays 7%. He has $1000 less in the checking account than in the savings account, and he earned $761 interest altogether last year. How much does he have in each account?   money market account: $5600; checking account: $1700; savings account: $2700

27. **Investing Money**   Josefina divides $15,000 into three investments: a savings account paying 5% annual interest, a bond paying 7%, and a money market account paying 9%. The annual interest from the three accounts is $1070. There is $2000 more invested in the money market account than in the bond. How much money has Josefina invested at each rate? $5000 at 5%; $4000 at 7%; $6000 at 9%

28. **Counting Coins**   A purse filled with change is holding $5.45 and has 41 coins. There are nickels, dimes, and quarters. If we remove the nickels, the remaining coins have a value of $4.55. How many of each coin was there originally?   nickels: 18; dimes: 8; quarters: 15

29. *Number Puzzle*   I am thinking of a three-digit number. It is the length (in miles) of the longest stretch of straight rail track in the world, which is located in Australia. The sum of the hundreds and ones digits is equal to the tens digit. The quotient of the sum of the three digits and 3 is one less than the ones digit. Eight times the hundreds digit is equal to the sum of the other two digits. What is the number?   297

30. *Book Sellers*   Tom Clarke is selling statistics, algebra, and calculus textbooks. He sold 20 statistics and 32 calculus texts for $1987. He sold 16 calculus and 40 algebra texts for $2026. Then he sold 30 of each kind of text for $3195. How much does each kind of text sell for?   statistics: $29.75; algebra: $33.25; calculus: $43.50

31. *Basketball*   Reggie Lewis, who played basketball for the Boston Celtics, Tony Gwyn, a baseball player for the San Diego Padres, and Reggie Jackson, a baseball hall-of-famer had jersey numbers related in the following ways. Twice Gwyn's number minus twice Lewis' number plus 3 times Jackson's number is 100. Four times the sum of Lewis' and Jackson's number is 316. Twice Lewis' number minus the other two numbers is 7. What are the three athletes' jersey numbers?   Jackson: 44; Lewis: 35; Gwyn: 19

MIXED PRACTICE

By doing these exercises, you will practice the topics up to this point in the chapter.

32. Graph the equations $y = 2x - 1$ and $y + x = 5$ and find their point of intersection.   (2, 3)

32.

33. The sum of two numbers is 135. Twice the first minus the second is 99. What are the two numbers?   78 and 57

34. Solve: $2x - y = -64$
$$3y - z = 133$$
$$4x = -48$$
$$\{(-12, 40, -13)\}$$

35. Solve: $-8x + 11y = -311$
$$16x - y = 349$$
$$\{(21, -13)\}$$

36. *Day-Care Regulations*   A day-care professional always requires that the juice served to children contain at least 20% pure juice. To economize, he mixes a prepared juice drink that contains 10% juice with real orange juice to make an 18-ounce pitcher of a drink that contains 20% orange juice. How much juice and how much drink does he mix together for each pitcher?   16 ounces of drink; 2 ounces of juice

37. Solve by graphing: $y = \dfrac{-3}{2}x + 3$
$$y = \dfrac{1}{3}x - 4\dfrac{1}{3} \quad \{(4, -3)\}$$

38. Solve: $x - 4z = -57$
$$y = 21$$
$$x - 3y = -48 \quad \{(15, 21, 18)\}$$

39. Solve by substitution: $x\quad = -3y$

$$5x - 3 = \quad y$$

$\{(0.5625, -0.1875)\}$ or $\left\{\left(\frac{9}{16}, -\frac{3}{16}\right)\right\}$

40. Solve: $-2y + 3z = 14$

$$x + 5y\quad = 35$$

$$8z = 64 \quad \{(10, 5, 8)\}$$

41. Solve by elimination: $6y = \quad 3x - 9$

$$-5y = -3x + 8 \quad \{(1, -1)\}$$

EXCURSIONS

Data Analysis

1. With only three games remaining in the 1995 shortened hockey season, the 1994 Stanley Cup winning New York Rangers had 45 points. The Eastern Conference NHL standings at that time are given in the following table:

NHL Standings, 1995

NHL Team	Points	NHL Team	Points
Philadelphia	58	Tampa Bay	37
Ottawa	19	NY Islanders	35
Québec	61	Florida	41
Boston	51	Montréal	42
Washington	49	NJ Devils	50
Buffalo	46	Hartford	43
NY Rangers	45	Pittsburgh	61

    a. What is the mode?

    b. What is the median?

    c. What is the mean?

    d. Which of these measures, in your opinion, best describes the "average" of the data? Justify your answer.

2. The production of "trash" is a continuing problem for the world. The top twenty "trash" producers are listed in the table on page 269 (not in any particular order).

    a. What is the mode?

    b. What is the median?

    c. What is the mean?

    d. Which of these measures, in your opinion, best describes the "average" of the data? Justify your answer.

Country	Waste (lbs. per capita per day)	Country	Waste (lbs. per capita per day)
Australia	4.2	Denmark	2.6
Saudi Arabia	2.4	Israel	2.4
Switzerland	2.2	Qatar	2.4
United Arab Emirates	2.4	United Kingdom	2.2
New Zealand	4	Oman	2.4
Finland	2.4	Canada	3.7
Iraq	2.4	United States	3.3
Bahrain	2.4	Netherlands	2.6
France	4	Kuwait	2.4
Norway	2.9	Luxembourg	2.2

Source: 1993 Information Please® Environmental Almanac (©1995 Houghton Mifflin Co.), p. 339. All rights reserved. Used by permission from Information Please LLC.

3. How does the following table relate to the graphs? Tell what you have learned from the combination graphs and table. How does your city compare?

DOW JONES TRAVEL INDEX

TRENDS IN TRAVEL COSTS

Overall Index +8%*

Hotel Index +4%*

* Percent change from the first week of January 1996 to the first week of January 1997.

WHAT THE AVERAGE ROOM AND CAR WILL COST NEXT WEEK

CITY	DAILY ROOM RATE	DAILY CAR-RENTAL RATE	CITY	DAILY ROOM RATE	DAILY CAR-RENTAL RATE
Atlanta	$135	$48	Minneapolis	$119	$42
Boston	172	43	New Orleans	146	44
Chicago	151	45	New York	212	62
Cleveland	124	46	Orlando	109	35
Dallas	140	46	Phoenix	156	31
Denver	115	34	Pittsburgh	130	34
Detroit	116	51	St. Louis	112	51
Houston	118	38	San Francisco	173	45
Los Angeles	160	35	Seattle	133	36
Miami	140	34	Washington, D.C.	160	44

The Overall Index tracks the cost of business fares on 20 major routes plus hotel and car-rental rates in 20 cities. Hotel and car-rental data based on rates from 10 leading hotel chains and six car-rental companies. Prices surveyed Dec. 31 for Tuesday-Thursday travel during the week of Jan. 5.

Source: Reprinted by permission of The Wall Street Journal ©1997 Dow Jones & Company, Inc. All rights reserved worldwide.

Class Act

4. Write a paragraph that explains this data to a person who cannot see the graph. Check your work. What does this graph show visually that is the most difficult to explain verbally?

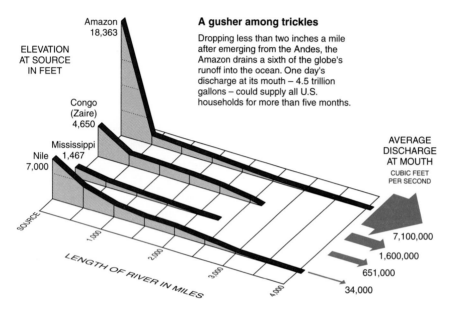

ELEVATION
AT SOURCE
IN FEET

Amazon
18,363

Congo
(Zaire)
4,650

Mississippi
Nile 1,467
7,000

SOURCE

LENGTH OF RIVER IN MILES

1,000   2,000   3,000   4,000

A gusher among trickles

Dropping less than two inches a mile after emerging from the Andes, the Amazon drains a sixth of the globe's runoff into the ocean. One day's discharge at its mouth – 4.5 trillion gallons – could supply all U.S. households for more than five months.

AVERAGE
DISCHARGE
AT MOUTH
CUBIC FEET
PER SECOND

7,100,000

1,600,000

651,000

34,000

Source: National Geographic Image Collection.

CONNECTIONS TO *STATISTICS*

Broken-Line Graphs

Reading Broken-Line Graphs

In some graphs, points are plotted and connected by lines to present information. The figure at the top of page 271 is such a **broken-line graph.** The position of each dot represents the average price of tea, in cents per pound, for the year shown directly beneath it. The dots are connected by lines so that readers can see more clearly the changes, up and down, in the data.

▪▪▪

EXAMPLE

In what year did tea first average more than 90 cents per pound?

SOLUTION

We locate 90 cents on the price scale at the left. A horizontal line at 90 cents crosses the graph just about at the 1983 mark. Between 1983 and 1984, the graph rises higher than 90 cents.

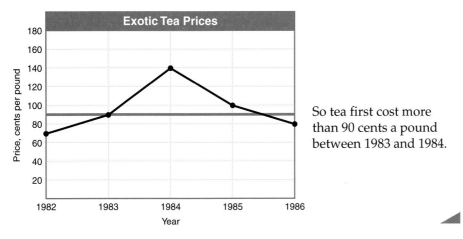

So tea first cost more than 90 cents a pound between 1983 and 1984.

We can sometimes use a broken-line graph to represent two or more sets of data. The next example shows such a use.

▪▪▪
EXAMPLE

This next graph shows the average prices in cents per pound of imported coffee, cocoa, and tea from 1982 to 1986.

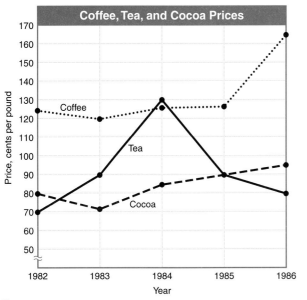

In what year did the price of tea exceed that of coffee and cocoa?

SOLUTION

As we look at this figure, we can see that the graph for tea is higher than that for coffee and cocoa at just one point. This point corresponds to the year 1984. Thus, the cost of tea exceeded that of coffee and that of cocoa in 1984.

Constructing Broken-Line Graphs

We construct a broken-line graph by plotting the points and joining them with line segments, or short pieces of a line.

PRACTICE

Use the Coffee, Tea, and Cocoa Prices graph to answer Exercises 1 through 6.

1. What is the difference in cost between the the most expensive and the least expensive of these three goods in 1986?   85¢

2. In what year were the costs of cocoa and tea the same?   1985

3. In 1984 what was the difference in the prices of coffee and cocoa?   40¢

4. In 1985 what would you have paid altogether for a pound of cocoa and a pound of tea?   about $1.80

5. If you had bought a pound each of coffee, cocoa, and tea in 1986, about what would you have paid altogether (to the nearest 10 cents)? $3.40

6. Ask and answer a question about this graph.   Answers may vary.

The following graph shows the winning times for the men's Olympic 800-meter run for the years 1896 through 1960. Use this graph to answer Exercises 7 through 10.

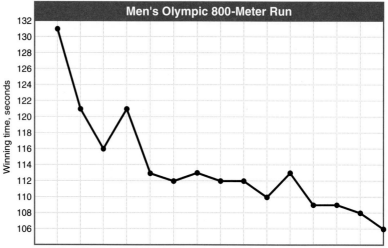

* Interim games in Athens but *not* official Olympics games.

7. What is the difference between the longest and the shortest record times shown?   25 seconds

8. During which two pairs of succeeding Olympics did the winning time remain the same?   1924 and 1928; 1948 and 1952

9. Make up three additional questions about the graph and answer them.   Answers may

10. ***Research***   Find the winning times for the men's Olympic 800-meter run for the years 1964–1996. Redraw the graph, adding those points.

4.3 Solving Systems of Linear Inequalities

SECTION LEAD-IN

In chemistry we may have limiting factors when we mix chemical elements and compounds to create another compound.

In such a case, only one of the reactants may be completely consumed. This reactant (or reagent as it is sometimes called) is the limiting reactant, or limiting reagent.

When zinc metal reacts with hydrochloric acid, zinc chloride and hydrogen are produced.

$$Zn + 2HCl \longrightarrow ZnCl_2 + H_2$$
$$\underline{1} \; Zn + \underline{2} \; HCl \longrightarrow \underline{1} \; ZnCl_2 + \underline{1} \; H_2$$

In this reaction 1 *mole* of Zn and 2 *moles* of HCl produce 1 *mole* of $ZnCl_2$ and 1 *mole* of H_2.

▪ What would be produced by 8 moles of Zn and 5 moles of HCl?
▪ Which reactant is the limiting reactant?

A system of linear inequalities consists of two or more linear inequalities that contain the same variables. The best way to solve such a system is by graphing, just as in solving simple inequalities. In fact, to solve a system, we actually graph each inequality individually. The solution of the system is the intersection of the solutions of all the individual inequalities.

▪▪▪
EXAMPLE 1

By graphing, find the solution set of this system of inequalities.

$$5x + y < 5$$
$$3x + 8y > 4$$

SOLUTION

We shall graph both inequalities and then find the portion of the plane where the two graphs overlap. That portion represents all ordered pairs whose coordinates satisfy both inequalities. It thus represents the solution set for the system.

To graph the inequalities, we first must graph the lines they suggest. Those lines have the equations

$$5x + y = 5 \quad \text{and} \quad 3x + 8y = 4$$

or, in slope-intercept form,

$$y = -5x + 5 \quad \text{and} \quad y = \frac{-3}{8}x + \frac{1}{2}$$

We graph the first of these as a broken line to show that the line is not part of the graph (because the inequality is of the "less than" type). To determine which region of the plane is part of the inequality, we substitute the test point $(0, 0)$ into the inequality. We find that the origin is part of the solution, so we shade the region containing the origin, as shown at right.

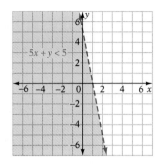

We then graph the inequality $3x + 8y > 4$ in exactly the same way, on the same set of axes. This gives us the graph on the left below. Points that are *doubly shaded* are part of the graphs of both inequalities and therefore form the graph of the solution of the system. All points of the graph on the right below represent the ordered pairs of the solution set of the system. No part of either line is in the solution set.

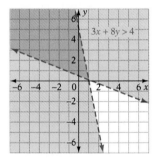
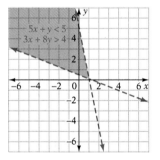

▶ *CHECK* **Warm-Up 1**

Calculator Corner

Graph: $3y \geq 2x + 6$
$y < 5x - 2$

If we solve both inequalities for y, the results will be

$$y \geq \frac{2}{3}x + 2 \quad \text{and} \quad y < 5x - 2$$

If we graph the y's as linear functions, the graphs look as follows.

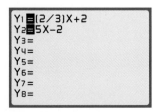
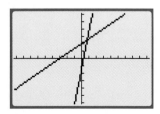

Now use your graphing calculator to **Shade** that area of the graph that we are investigating. The lower bound of the system is Y1 (we want to shade the area *greater* than this) and the upper bound of the system is Y2 (we want to shade the area *less* than this). The command for shading is found under the **DRAW** feature of the *TI-82*. The command on the Home Screen would be written as

Shade(lower bound,upper bound, resolution)   or

$$\text{Shade}\left(\tfrac{2}{3}X + 2, 5X + 2, 2\right) \text{ or the equivalent statement}$$

Shade(Y1, Y2, 2)

(**Resolution** simply describes the density of the shading. The shading capabilities of graphing calculators vary widely. You may want to consult your calculator manual for the **Shad**ing instructions for your particular model.)

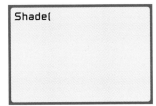
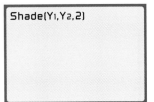

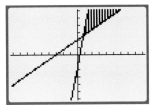

Press **ENTER.**

Now press any of the arrow keys on your calculator to activate the **screen cursor.** Move the cursor to the point (3, 5), which is in the shaded area of your graph.

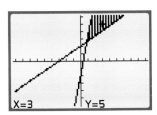

Points within the shaded area satisfy the conditions of both linear inequalities *at the same time.* To verify this, use the **STO**re function to substitute these values for *x* and *y* into the *original inequalities* as shown on the screen below. Nine is definitely greater than or equal to 6, so the point (3, 5) satisfies the first inequality. Ten is greater than 2, so the point (3, 5) also satisfies the second inequality.

```
3→X
               3
5→Y
               5
3Y–2X
               9
5X–Y
              10
```

Now move the **screen cursor** to a point that *is not* in the shaded area.

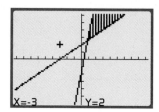

Evaluate each of the *original inequalities* for the point not in the shaded area, $(-3, 2)$ as shown on the screen below. Twelve is definitely greater than or equal to 6, so the point $(-3, 2)$ satisfies the first inequality. However, -17 *is not* greater than 2, so the point $(-3, 2)$ *does not* satisfy the second inequality. Therefore, the point $(-3, 2)$ does not belong in the solution set for this system of inequalities.

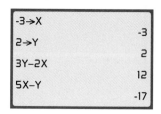

▪▪▪

EXAMPLE 2

Graph the solution of the system of inequalities $3x + y > 10$ and $6x - 3y \leq 9$.

SOLUTION

We write the associated equations in slope-intercept form, obtaining

$$y = -3x + 10 \quad \text{and} \quad y = 2x - 3$$

The two equations are graphed in the figure at the right. The line $y = 2x - 3$ is drawn as a solid line because its inequality ($\leq$) includes the equality; we graph $y = -3x + 10$ as a dashed line. *Note:* the point of intersection of these two lines is not a solution of the system. We could have used an open point to show this. Either form is acceptable.

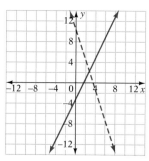

Then, using the test point $(0, 0)$, we shade the proper region of the plane for each inequality.

$$\begin{array}{ll} 3x + y > 10 & 6x - 3y \leq 9 \\ 3(0) + 0 > 10 & 6(0) - 3(0) \leq 9 \\ 0 > 10 \quad \text{False} & 0 \leq 9 \quad \text{True} \end{array}$$

The solution of the system is the doubly shaded region in the figure on the left below, including the part of the solid line above the intersection. It is redrawn in the figure on the right.

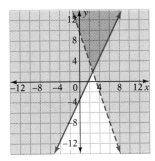   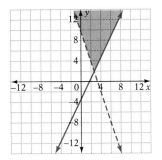

▶ *CHECK* **Warm-Up 2**

▪▪▪

EXAMPLE 3

Find the solution: $5x - 6y < 3$
$$y < \frac{5}{6}x - 7$$

SOLUTION

In slope-intercept form, the equations associated with this system are

$$y = \frac{5}{6}x - \frac{1}{2} \quad \text{and} \quad y = \frac{5}{6}x - 7$$

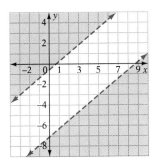

These lines have the same slope $\frac{5}{6}$ and so are parallel. The equations thus have no common solution, but the inequalities might. To find out, we graph the inequalities, obtaining the accompanying figure. The graph shows that the inequalities have no common points. The solution set is $\varnothing$.

▶ *CHECK* **Warm-Up 3**

Calculator Corner

Investigate the system of inequalities:

$$y > 0.5x + 2 \quad \text{and} \quad y > -3x - 4$$

Notice that both inequalities are *greater than* situations. How would you shade this on your graphing calculator?

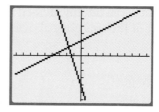

Remember that the **Shade** command is **Shade(lower bound, upper bound, resolution)**. Put this onto your Home Screen and use an arbitrary large number for the upper bound, say 15.

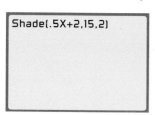

Press **ENTER.**

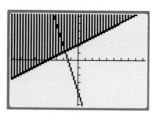

Press **ENTER.**

Notice that the area of the graph that satisfies *both inequalities* is the area that has been shaded twice—once with a resolution of 2 and then again with a resolution of 3.

In our last example, we solve a system that includes an absolute value inequality.

▪ ▪ ▪

EXAMPLE 4

Graph this system:

$$|y| < 3$$
$$2y + 8x > 4$$

SOLUTION

Recall that the inequality $|ax + b| < c$ represents an intersection of inequalities:

$$-c < ax + b \quad \text{and} \quad ax + b < c$$

Thus the inequality $|y| < 3$ means

$$-3 < y \quad \text{and} \quad y < 3$$

This intersection is graphed in the left figure below. When we superimpose the graph of the inequality $2y + 8x > 4$, we obtain the result shown in the middle figure below. The graph of the system (and of the solution to the system) is the region where the two graphs overlap. It is shown in the right figure below.

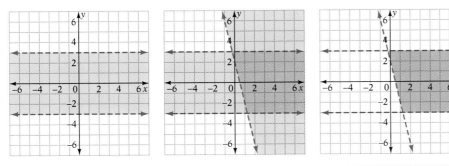

▶ CHECK **Warm-Up 4**

Practice what you learned.

SECTION FOLLOW-UP

The reaction we are considering is

$$Zn + 2HCl \longrightarrow ZnCl_2 + H_2$$

This chemical equation works in a fashion similar to other equations

$$1Zn + 2HCl \longrightarrow 1ZnCl_2 + 1H_2$$
$$2Zn + 4HCl \longrightarrow 2ZnCl_2 + 2H_2$$
$$3Zn + 6HCl \longrightarrow 3ZnCl_2 + 3H_2$$

The amount of $ZnCl_2$ and H_2 produced by 8 moles of Zn and 5 moles of HCl is limited by the HCl.

According to our chemical equations, the amount of H_2 and $ZnCl_2$ that can be produced by HCl is in the ratio of 1 to 2. That is, if we start with x moles of HCl, we can obtain only $\frac{x}{2}$ moles of H_2 and $\frac{x}{2}$ moles of $ZnCl_2$.

$$2 \text{ moles HCl} \longrightarrow \frac{1}{2} \cdot 2 = 1 \text{ mole } H_2 \text{ (and } ZnCl_2)$$

$$4 \text{ moles HCl} \longrightarrow \frac{1}{2} \cdot 4 = 2 \text{ moles } H_2 \text{ (and } ZnCl_2)$$

and

$$5 \text{ moles HCl} \longrightarrow \frac{1}{2} \cdot 5 = \frac{5}{2} \text{ moles } H_2 \text{ (and } ZnCl_2)$$

HCl is the limiting reactant in this equation. Only $\frac{5}{2}$ moles of Zn would be used for this reaction. To use all 8 moles of Zn, we would need 16 moles of HCl.

4.3 WARM-UPS

Work these problems before you attempt the exercises.

1. By graphing, find the solution set of the system

$$y - x \geq 3$$
$$2y - 5x \geq 8$$

2. Graph the solution set of the system

$$x < 3y + 4$$
$$y \leq 5$$

3. By graphing, find the solution set of the system

$$y - 2x < 6$$
$$y + 3 > 2x$$

4. Graph the solution set for the system $|y| < 3$ and $8x + 2y > 8$.

4.3 EXERCISES

Note: Use your graphing calculator to check your results whenever possible.

In Exercises 1 through 24, solve the system of inequalities by graphing.

1. $4y - 5 \geq x$
   $3x + 7y < 0$

2. $2y - 5 > x$
   $-3x - y < 1$

3. $y - x > \quad 6$
   $2y + 4x \geq -12$

4. $3y + 6x \leq 9$
   $4y + 2 > x$

5. $5y - 5x > \quad 10$
   $y + \quad x > -1$

6. $x - y > 4$
   $y + x \leq 2$

7. $y + 2x < 4$
   $2y - \quad x < 0$

8. $2y - x > \quad 6$
   $3y + x < -1$

9. $y + 3x \geq 5$
   $y + 6 > x$

10. $2y + 4 \leq x$
    $y + x < 4$

11. $2y + x < 3$
    $-y + 3 > x$

12. $x > 2y - 4$
    $y - x < 1$

13. $y \leq 2x - 4$
    $2x - 5 \leq y$

14. $x \geq y + 3$
    $3y + x \geq 9$

15. $y + x \geq 7$
    $x - y < 1$

16. $2y + 3 > x$
$x + y < 6$

17. $3y - x \leq 6$
$x \leq 2y$

18. $8y - 4 \geq 12x$
$-\frac{1}{3}x - y \geq -5$

19. $y \geq 7 - x$
$x - 3 \leq y$

20. $x > 3y + 15$
$3y - x < 1$

21. $2y + 2x < 6$
$|y| < 3$

22. $4y - 2x \geq 8$
$|y| < 6$

23. $7y + x < 4$
$|y + 1| < 6$

24. $x + 6 < y$
$|y + 1| > 4$

In Exercises 25 through 32, place the proper sign ($<, >, \geq,$ or $\leq$) in each system of inequalities such that the given graph represents the solution set for the system.

25. $x \underline{\ \geq\ } -3$
$y \underline{\ >\ } -3$
$x \underline{\ <\ } 4$
$y \underline{\ \leq\ } 3$

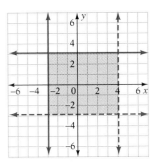

26. $x \underline{\ \leq\ } -3$
$y \underline{\ \leq\ } -3$
$x \underline{\ <\ } 4$
$y \underline{\ <\ } 3$

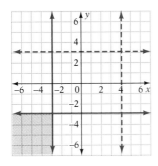

27. $|y| \underline{\ >\ } 3$
$|x - 0.5| \underline{\ \geq\ } 3.5$

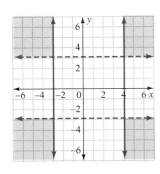

28. $|x - 1| \underline{\ \leq\ } 3$
$|y| \underline{\ >\ } 3$

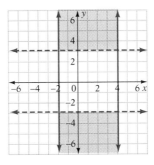

29. $y \underline{\ >\ } 3x + 4$
$y \underline{\ \leq\ } -2x + 4$
$y \underline{\ \leq\ } -\frac{1}{3}x - 4$

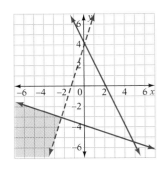

30. $y \underline{\ \leq\ } 3x + 4$
$y \underline{\ >\ } -2x + 4$
$y \underline{\ >\ } -\frac{1}{3}x - 4$

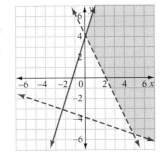

31. $y \underline{\quad >\quad} 3x + 4$
$y \underline{\quad \leq \quad} -2x + 4$
$y \underline{\quad \geq \quad} -\frac{1}{3}x - 4$

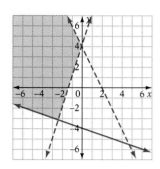

32. $y \underline{\quad \leq \quad} 3x + 4$
$y \underline{\quad \leq \quad} -2x + 4$
$y \underline{\quad \geq \quad} -\frac{1}{3}x - 4$

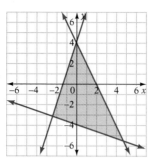

In Exercises 33 through 35, graph these systems of inequalities.

33. $y \leq 2x$
$y \leq -3x + 15$
$y \geq 0.5x - 7$

34. $y \leq -2x - 1$
$y \leq -2x + 7$
$y \geq \frac{1}{2}x - 4$
$y \leq 0.5x + 2$

35. $y \leq -x + 7$
$y \leq 2x + 7$
$y \geq 2x - 1$
$y \geq -x - 5$

MIXED PRACTICE

By doing these exercises, you will practice the topics up to this point in the chapter.

36. Graph the system: $y \leq \frac{3}{2}x$
$y \geq \frac{-7}{5}x$

37. Solve by addition: $45x + 27y = -22.5$
$15x + 90y = 195$
$\{(-2, 2.5)\}$

38. Without graphing, determine whether the graphs of these equations are parallel, perpendicular, or neither.
$2y + x = 9$
$3x + y = 4$ neither

39. Determine whether this system of equations is consistent, inconsistent, or dependent.
$x + y = 1$
$2x + 3y = 2$ consistent

40. Graph the solution set: $3y + 1 < x$
$|y + 1| > 2$

41. Solve: $21x - 10y = 111.5$
$2x + 18y = -141$ $\{(1.5, -8)\}$

42. Solve graphically: $y = \frac{1}{4}x - \frac{1}{4}$
$x = 1$
$\{(1, 0)\}$

43. Find the solution set: $5x - 2y = -41$
$-x + z = 3$
$-x + 6z = -7$
$\{(-5, 8, -2)\}$

44. A box has the following characteristics: The sum of its length, width, and depth is 23 inches. The perimeter of one side is 36 inches. The perimeter of another side is 26 inches. What is its volume?   400 cubic inches

EXCURSIONS

Posing Problems

1. Ask and answer four questions that can be answered with this data.

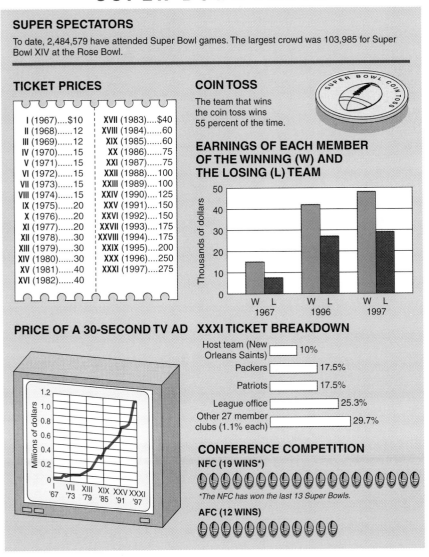

SUPER BOWL FACTS

SUPER SPECTATORS

To date, 2,484,579 have attended Super Bowl games. The largest crowd was 103,985 for Super Bowl XIV at the Rose Bowl.

TICKET PRICES

I (1967)....$10	XVII (1983)....$40
II (1968)......12	XVIII (1984)......60
III (1969)......12	XIX (1985)......60
IV (1970)......15	XX (1986)......75
V (1971)......15	XXI (1987)......75
VI (1972)......15	XXII (1988)....100
VII (1973)......15	XXIII (1989)....100
VIII (1974)......15	XXIV (1990)....125
IX (1975)......20	XXV (1991)....150
X (1976)......20	XXVI (1992)....150
XI (1977)......20	XXVII (1993)....175
XII (1978)......30	XXVIII (1994)....175
XIII (1979)......30	XXIX (1995)....200
XIV (1980)......30	XXX (1996)....250
XV (1981)......40	XXXI (1997)....275
XVI (1982)......40	

COIN TOSS

The team that wins the coin toss wins 55 percent of the time.

EARNINGS OF EACH MEMBER OF THE WINNING (W) AND THE LOSING (L) TEAM

PRICE OF A 30-SECOND TV AD

XXXI TICKET BREAKDOWN

Host team (New Orleans Saints) 10%
Packers 17.5%
Patriots 17.5%
League office 25.3%
Other 27 member clubs (1.1% each) 29.7%

CONFERENCE COMPETITION

NFC (19 WINS*)

*The NFC has won the last 13 Super Bowls.

AFC (12 WINS)

Source: National Football League.

Data Analysis

2. Analyze this data and write a paragraph about baseball salaries. Compare your work with that of another group.

1994–1996 BASEBALL PAYROLLS

TEAM	1996	1995	1994
Atlanta Braves	$53,422,000	$47,023,444	$44,100,972
Baltimore Orioles	55,127,855	48,739,636	38,711,487
Boston Red Sox	38,516,402	38,157,750	36,337,937
California Angels	25,140,142	34,702,577	24,528,385
Chicago Cubs	32,605,000	36,797,696	32,546,333
Chicago White Sox	44,827,833	40,750,782	40,144,836
Cincinnati Reds	43,676,946	47,739,109	41,458,052
Cleveland Indians	47,615,507	40,180,750	31,705,667
Colorado Rockies	40,958,990	38,039,871	23,654,508
Detroit Tigers	17,955,500	28,663,667	41,118,509
Florida Marlins	25,286,000	22,961,781	19,524,361
Houston Astros	29,613,000	33,614,668	33,092,500
Kansas City Royals	19,980,250	31,181,334	40,667,375
Los Angeles Dodgers	37,313,500	36,725,956	38,837,526
Milwaukee Brewers	11,701,000	17,407,384	24,786,857
Minnesota Twins	21,254,000	15,362,750	25,053,237
Montreal Expos	17,264,500	13,116,557	18,771,000
New York Mets	24,890,167	13,097,944	30,903,583
New York Yankees	61,511,870	58,165,252	47,512,342
Oakland Athletics	22,524,093	33,372,722	34,574,000
Philadelphia Phillies	30,403,458	30,333,350	31,143,000
Pittsburgh Pirates	16,994,180	17,665,833	21,503,250
St. Louis Cardinals	38,595,666	28,679,250	29,622,052
San Diego Padres	33,141,026	25,008,834	13,774,268
San Francisco Giants	34,646,793	33,738,683	42,260,538
Seattle Mariners	43,131,001	37,984,610	28,463,110
Texas Rangers	41,080,028	35,888,726	32,399,097
Toronto Blue Jays	28,728,577	42,233,500	42,265,168
Termination Pay	*25,008,776*	*25,515,176*	*30,783,861*
Total payrolls	**$937,905,284**	927,334,416	909,459,950
Average player salary	**$1,099,875**	1,094,440	1,154,486
Percent increase/decrease over previous year	8.6%	−5.2%	0.005%

Source: Major League Player Relations Committee.

3. Use the following graph to answer the questions.

   a. Calculate the slopes of the line segments for each year to find between what two years the income changed most rapidly.

   b. ✎ Explain why we can use the method in part (a).

   c. What information will you get if you calculate the change for every two years?

   d. Who might be interested in information from this graph? Tell them four noteworthy facts about this data.

Real Median Family Income After Taxes, 1981 to 1994 ($)

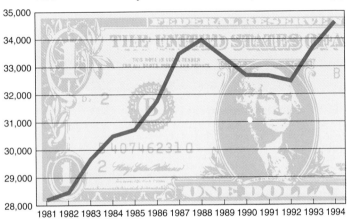

Source: 1996 Information Please® Business Almanac (©1995 Houghton Mifflin Co.), p. 102. All rights reserved. Used with permission by Information Please LLC.

Class Act

4. a. Write two different verbal situations that could be shown by the following graph.

   b. Make a list that gives the "points" on this graph.

   c. Write equations for the lines whose "pieces" make up this graph. (There are 5.)

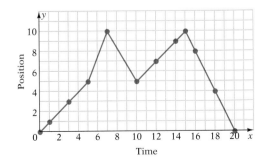

4.4 Optional Topic: Working with Determinants: Cramer's Rule

SECTION LEAD-IN

We name triangles by naming each angle. Triangle *ABC* is shown in color in the figure. The sum of the angles in each triangle is 180°. The following equations are true.

$$A + B + C = 180$$
$$D + E + F = 180$$
$$D + G + I = 180$$
$$C + H + I = 180$$
$$A + F + J = 180$$
$$B + E + H + J + G = 540$$

Use this information to find

$$A + D + C + F + I$$

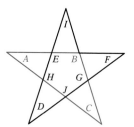

In this section, we introduce another method for solving systems of linear equations. But before we get to the actual method, we need to discuss some new ideas.

Matrices

We use the word **matrix** (plural **matrices**) to mean any rectangular array of numbers. The **rows** of a matrix are horizontal; the **columns** are vertical. Here are three matrices.

$$[3 \quad 2 \quad 1]$$

$$\begin{bmatrix} 3 & -2 \\ 5 & 7 \end{bmatrix}$$

$$\begin{bmatrix} 6 & 8 & -3 & 5 \\ 2 & -1 & 2 & -4 \end{bmatrix}$$

1×3	2×2	2×4
a "one-by-three" matrix	a "two-by-two" matrix	a "two-by-four" matrix

"One-by-three" indicates one row and three columns; "two-by-two" indicates two of each; "two-by-four" indicates two rows and four columns.

Determinants

A **determinant** is a square matrix.

$$\begin{vmatrix} 1 & 2 \\ -3 & 6 \end{vmatrix}$$

$$\begin{vmatrix} 18 & 1.1 & 1 \\ 3 & 7 & -4 \\ 0.9 & 1.2 & 3 \end{vmatrix}$$

Second order determinant   Third order determinant

Think of a determinant as a frame that holds the elements (the numbers) in specific positions. If we shift or change the numbers, we change the determinant.

Evaluating a 2 × 2 Determinant

A 2 × 2 ("two-by-two") determinant has two rows and two columns. It has a numerical value that is found as follows:

> The 2 × 2 determinant
>
> $$\begin{vmatrix} a_1 & b_1 \\ a_2 & b_2 \end{vmatrix}$$
>
> has the value $a_1 b_2 - a_2 b_1$.

In other words, the value of a 2 × 2 determinant is the difference of the diagonal products of its elements.

$$\begin{vmatrix} a_1 & b_1 \\ a_2 & b_2 \end{vmatrix} = a_1 b_2 - a_2 b_1$$

▪▪▪

EXAMPLE 1

Evaluate: **a.** $\begin{vmatrix} -2 & -4 \\ 0 & 5 \end{vmatrix}$ **b.** $\begin{vmatrix} 1 & 1 \\ 1 & 1 \end{vmatrix}$ **c.** $\begin{vmatrix} 0 & 6 \\ 0 & -3 \end{vmatrix}$

SOLUTION

We apply the "diagonal formula" in each part.

a. $(-2 \cdot 5) - [0 \cdot (-4)] = -10$

This determinant has the value -10.

b. $(1 \cdot 1) - (1 \cdot 1) = 0$

Any determinant whose entries are all equal has the value zero.

c. $[0 \cdot (-3)] - (0 \cdot 6) = 0$

When any column or row of a determinant contains only zeros, the determinant has the value zero.

▶ CHECK **Warm-Up 1**

Calculator Corner

A graphing calculator can also be used to work with matrices. The following is a Matrix example using the *TI-82* or *TI-83* graphing calculator. (You may need to consult your graphing calculator's manual for directions on how to use matrices with your model.)

When you press the ⎡**MATRX**⎤ key on the calculator, you will see the following:

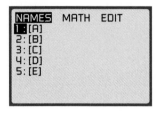 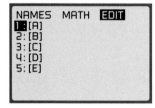

Press **ENTER** to define **MATRIX [A]**.

Enter 2 twice to define **MATRIX [A]** as a 2 × 2 Matrix, and then enter the following into **MATRIX [A]**.

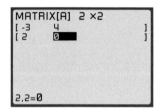

Now find the determinant of **MATRIX [A]**. Press **2**nd **QUIT** to get back to the Home Screen. Press the ⎡**MATRX**⎤ Key and choose **MATH**.

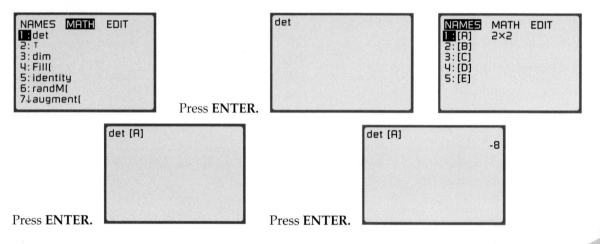

Press **ENTER.**

Press **ENTER.**   Press **ENTER.**

Evaluating a 3 × 3 Determinant

Evaluating a 3 × 3 determinant is a little more complicated. We shall use a method called *expanding by cofactors.*

Minors and Cofactors

The **minor** of any element of a 3 × 3 determinant is the 2 × 2 determinant that is formed by removing the row and column that contain that element. Thus, for example, the minor of the element 4 in the determinant

$$D = \begin{vmatrix} 3 & 4 & 1 \\ 6 & 2 & -1 \\ 0 & 7 & 8 \end{vmatrix}$$

is the determinant

$$\begin{vmatrix} 3 & 4 & 1 \\ 6 & 2 & -1 \\ 0 & 7 & 8 \end{vmatrix}$$

which we write as

$$\begin{vmatrix} 6 & -1 \\ 0 & 8 \end{vmatrix}$$

The **cofactor** of the element is its minor multiplied by either 1 or −1, according to the position of that element in the original 3 × 3 determinant and this **cofactor multiplier chart:**

$$\begin{vmatrix} 1 & -1 & 1 \\ -1 & 1 & -1 \\ 1 & -1 & 1 \end{vmatrix}$$

To obtain the cofactor of an element, multiply the minor by the value that occupies the same position as the element.

Note that −1 appears in the chart in the position of our element 4 (top row, middle column), so we multiply the minor of 4 by −1 to obtain its cofactor. The cofactor of the element 4 in D is

$$(-1)\begin{vmatrix} 6 & -1 \\ 0 & 8 \end{vmatrix} = (-1)[6 \cdot 8 - 0 \cdot (-1)] = (-1)(48) = -48$$

‼‼
ERROR ALERT

Identify the error and give a correct answer.

What is the cofactor of 5?

$$\begin{vmatrix} 3 & 4 & 6 \\ 2 & 0 & 5 \\ 7 & 1 & 2 \end{vmatrix}$$

Incorrect Solution:

$-1[(3)(4) - (7)(1)]$
$= -1[12 - 7]$
$= -1[5]$
$= -5$

▪▪▪

EXAMPLE 2

Find the cofactor of (a) −2 and (b) 5 in the determinant.

$$\begin{vmatrix} 1 & 0 & -4 \\ -2 & 5 & -7 \\ 3 & 2 & 6 \end{vmatrix}$$

SOLUTION

a. Eliminating the row and column that contain −2 yields

$$\begin{vmatrix} 1 & 0 & -4 \\ -2 & 5 & -7 \\ 3 & 2 & 6 \end{vmatrix} = \begin{vmatrix} 0 & -4 \\ 2 & 6 \end{vmatrix}$$

The position of −2 requires that we multiply by −1, so the cofactor is

$$(-1)\begin{vmatrix} 0 & -4 \\ 2 & 6 \end{vmatrix} = (-1)[0 \cdot 6 - 2 \cdot (-4)] = (-1)(8) = -8$$

b. The minor of 5 is

$$\begin{vmatrix} 1 & -4 \\ 3 & 6 \end{vmatrix}$$

Because 5 is located in a "+1" chart position, its cofactor is also

$$\begin{vmatrix} 1 & -4 \\ 3 & 6 \end{vmatrix} = (1)(6) - [3 \cdot (-4)] = 6 + 12 = 18$$

▶ *CHECK* **Warm-Up 2**

Expanding by Cofactors

To find the value of a 3 × 3 determinant, multiply each of the three elements in any column or row by the value of its cofactor and then add these products. This process is called **expanding by cofactors.**

The choice of which row or column to use is yours. The result will be the same no matter which you use. But choosing a row or column that contains a zero will lessen your work.

■ ■ ■
EXAMPLE 3

Find the value of the determinant: $\begin{vmatrix} -3 & -2 & 1 \\ 4 & 6 & 0 \\ 5 & 2 & 7 \end{vmatrix}$

SOLUTION

Let us expand on the middle row, which contains a zero.

The element 4 gives us the term

$$(4)(-1)\begin{vmatrix} -2 & 1 \\ 2 & 7 \end{vmatrix} = (-4)[-2 \cdot 7) - (2 \cdot 1)] = (-4)(-16) = 64$$

The minor of 4
From the chart
The element

The 6 gives us

$$(6)(1)\begin{vmatrix} -3 & 1 \\ 5 & 7 \end{vmatrix} = 6[(-3 \cdot 7) - (5 \cdot 1)] = 6(-26) = -156$$

As we expect, the 0 gives us

$$(0)(-1)\begin{vmatrix} -3 & -2 \\ 5 & 2 \end{vmatrix} = 0$$

So the value of the determinants is

$$64 + (-156) + 0 = -92$$

▶ *CHECK* **Warm-Up 3**

Here's another example, done more compactly.

■ ■ ■
EXAMPLE 4

Evaluate the determinant: $\begin{vmatrix} -3 & -2 & -5 \\ 0 & 6 & 3 \\ 8 & 7 & 4 \end{vmatrix}$

SOLUTION

We expand on the first column, obtaining

First   Second   Third
entry   entry   entry

$$(-3)(1)\begin{vmatrix} 6 & 3 \\ 7 & 4 \end{vmatrix} \quad + 0 + \quad (8)(1)\begin{vmatrix} -2 & -5 \\ 6 & 3 \end{vmatrix}$$

$$= (-3)[(6 \cdot 4) - (7 \cdot 3)] + 8\{(-2 \cdot 3) - [6 \cdot (-5)]\}$$

$$= (-3)(3) + 8(24) = 183$$

The value of the determinant is 183.

▶ CHECK **Warm-Up 4**

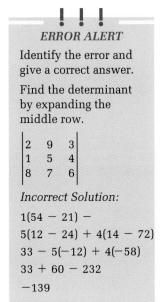

Calculator Corner

Find the determinant of the following 3 × 3 Matrix using the **MATRIX** utility on your graphing calculator. (Note: See the Calculator Corner on pages 287 and 288 to review how to define matrices and find determinants.)

$$\begin{vmatrix} -3 & 5 & 0 \\ 1 & -2 & 4 \\ 0.5 & -4 & 1 \end{vmatrix}$$

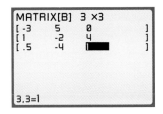

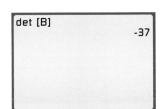

Solving Systems of Equations by Using Determinants

We can use determinants to solve systems of linear equations, applying a method called *Cramer's rule.* We shall apply it first to two-variable systems and then to three-variable systems.

Cramer's rule for two-variable systems: Cramer's rule may look difficult in general form, but it is easy to apply to specific systems.

Cramer's Rule for 2 × 2 Systems

The system of equations

$$a_1x + b_1y = d_1$$
$$a_2x + b_2y = d_2$$

has the solutions

$$x = \frac{D_x}{D} \quad \text{and} \quad y = \frac{D_y}{D}$$

where

$$D = \begin{vmatrix} a_1 & b_1 \\ a_2 & b_2 \end{vmatrix} \qquad D_x = \begin{vmatrix} d_1 & b_1 \\ d_2 & b_2 \end{vmatrix} \qquad D_y = \begin{vmatrix} a_1 & d_1 \\ a_2 & d_2 \end{vmatrix}$$

provided that D is not zero.

If the determinant D is zero, the system is either inconsistent or dependent. In either case, Cramer's rule cannot be used.

To use this method, first we find the determinant D, which is made up of the coefficients of the variables in the proper order. Then we find the "variable" determinants D_x and D_y by substituting the constants for the x- and y-coefficients, respectively. Last we find the quotients $\frac{D_x}{D}$ and $\frac{D_y}{D}$.

•••

EXAMPLE 5

Solve this system with Cramer's rule: $x + 3y = 12$
$2x - 2y = 14$

SOLUTION

The "coefficients" determinant is

$$D = \begin{vmatrix} 1 & 3 \\ 2 & -2 \end{vmatrix} = 1(-2) - 2(3) = -8$$

To find D_x, we substitute the constants (in order) for the x-coefficient column.

$$D_x = \begin{vmatrix} 12 & 3 \\ 14 & -2 \end{vmatrix} = 12(-2) - 14(3) = -66$$

To find D_y, we substitute the constants for the y-coefficient column.

$$D_y = \begin{vmatrix} 1 & 12 \\ 2 & 14 \end{vmatrix} = 1(14) - 2(12) = -10$$

Now

$$x = \frac{D_x}{D} = \frac{-66}{-8} = 8.25 \qquad y = \frac{D_y}{D} = \frac{-10}{-8} = 1.25$$

So the solution set is {(8.25, 1.25)}. You should check it by substituting in the original equations.

▶ CHECK **Warm-Up 5**

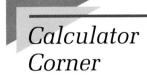

Calculator Corner

You can use your graphing calculator's **MATRIX** utility to solve systems of equations. The following illustrates this utility on the *TI-82* and *TI-83* graphing calculator. (You may need to consult your model's manual.)

$$x + 3y = 12$$
$$2x - 2y = 14$$

Enter the "coefficients" Matrix as **MATRIX [A]**, D_x as **MATRIX [B]**, and D_y as **MATRIX [C]**. (Note: See the Calculator Corner on pages 287 and 288 to review how to define matrices and find determinants.)

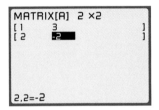
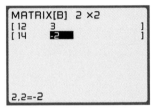
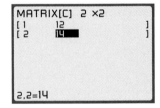

Now perform the operations on the matrices following Cramer's rule. (Note: See the Calculator Corner on page 79 to review how to store values for variables.)

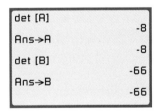
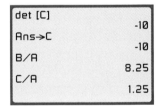

So the solution set is (8.25, 1.25). Now try the following system on your own for practice.

$$4x - 6y = 0$$
$$2x + 3y = 24 \quad \{(6, 4)\}$$

Cramer's Rule for Three-Variable Systems

Cramer's rule for three-variable systems is a direct extension of the rule for two-variable systems. And it is applied in exactly the same way.

Cramer's Rule for 3 × 3 Systems

The system of equations

$$a_1x + b_1y + c_1z = d_1$$
$$a_2x + b_2y + c_2z = d_2$$
$$a_3x + b_3y + c_3z = d_3$$

has the solution

$$x = \frac{D_x}{D} \quad \text{and} \quad y = \frac{D_y}{D} \quad \text{and} \quad z = \frac{D_z}{D}$$

where

$$D = \begin{vmatrix} a_1 & b_1 & c_1 \\ a_2 & b_2 & c_2 \\ a_3 & b_3 & c_3 \end{vmatrix} \quad D_x = \begin{vmatrix} d_1 & b_1 & c_1 \\ d_2 & b_2 & c_2 \\ d_3 & b_3 & c_3 \end{vmatrix}$$

$$D_y = \begin{vmatrix} a_1 & d_1 & c_1 \\ a_2 & d_2 & c_2 \\ a_3 & d_3 & c_3 \end{vmatrix} \quad D_z = \begin{vmatrix} a_1 & b_1 & d_1 \\ a_2 & b_2 & d_2 \\ a_3 & b_3 & d_3 \end{vmatrix}$$

provided D is not equal to 0.

▪▪▪

EXAMPLE 6

Solve with Cramer's rule:

$$\begin{array}{rcrcrcr} 2x & - & & & 4z & = & 20 \\ -2x & - & 2y & + & 3z & = & -22 \\ x & + & y & - & z & = & 9 \end{array}$$

SOLUTION

The determinant of coefficients is

$$D = \begin{vmatrix} 2 & 0 & -4 \\ -2 & -2 & 3 \\ 1 & 1 & -1 \end{vmatrix}$$

Note that we consider a missing variable (here, y in the first equation) to have a coefficient of zero. We evaluate this determinant by expanding on the first row.

$$D = (2)(1)\begin{vmatrix} -2 & 3 \\ 1 & -1 \end{vmatrix} + 0 + (-4)(1)\begin{vmatrix} -2 & -2 \\ 1 & 1 \end{vmatrix}$$
$$= 2[(-2)(-1) - (1)(3)] - 4[(-2)(1) - (1)(-2)]$$
$$= (2)(-1) - 4(0) = -2$$

We find D_x by replacing the x-coefficients column with the constants.

$$D_x = \begin{vmatrix} 20 & 0 & -4 \\ -22 & -2 & 3 \\ 9 & 1 & -1 \end{vmatrix}$$

Expanding on the top row, we obtain

$$D_x = (20)(1)\begin{vmatrix} -2 & 3 \\ 1 & -1 \end{vmatrix} + 0 + (-4)(1)\begin{vmatrix} -22 & -2 \\ 9 & 1 \end{vmatrix}$$

$$= 20[(-2)(-1) - (1)(3)] - 4[(-22)(1) - (9)(-2)]$$

$$= (20)(-1) - 4(-4) = -4$$

We find D_y by replacing the y-coefficients column with the constants.

$$D_y = \begin{vmatrix} 2 & 20 & -4 \\ -2 & -22 & 3 \\ 1 & 9 & -1 \end{vmatrix}$$

Expanding on the first column gives

$$D_y = 2(1)\begin{vmatrix} -22 & 3 \\ 9 & -1 \end{vmatrix} + (-2)(-1)\begin{vmatrix} 20 & -4 \\ 9 & -1 \end{vmatrix} + (1)(1)\begin{vmatrix} 20 & -4 \\ -22 & 3 \end{vmatrix}$$

$$= 2[(-22)(-1) - (9)(3)] + 2[(20)(-1) - (9)(-4)] + 1[(20)(3) - (-22)(-4)]$$

$$= 2(-5) + 2(16) + (-28)$$

$$= 10 + 32 + (-28) = -6$$

We find D_z by replacing the z-coefficients column with the constants.

$$D_z = \begin{vmatrix} 2 & 0 & 20 \\ -2 & -2 & -22 \\ 1 & 1 & 9 \end{vmatrix}$$

Expanding on the middle column gives

$$D_z = 0 + (-2)(1)\begin{vmatrix} 2 & 20 \\ 1 & 9 \end{vmatrix} + (1)(-1)\begin{vmatrix} 2 & 20 \\ -2 & -22 \end{vmatrix}$$

$$= 0 + (-2)[(2)(9) - (1)(20)] + (-1)[(2)(-22) - (-2)(20)]$$

$$= (-2)(-2) + (-1)(-4)$$

$$= 4 + 4 = 8$$

Last, we have

$$x = \frac{D_x}{D} \quad \text{and} \quad y = \frac{D_y}{D} \quad \text{and} \quad z = \frac{D_z}{D}$$

so

$$x = \frac{D_x}{D} = \frac{-4}{-2} = 2 \qquad y = \frac{D_y}{D} = \frac{-6}{-2} = 3 \qquad z = \frac{D_z}{D} = \frac{-8}{2} = -4$$

Thus the ordered triple $(2, 3, -4)$ is the solution to this system of equations. You should check each solution.

▶ *CHECK* **Warm-Up 6**

Practice what you learned.

SECTION FOLLOW-UP

We rearrange and add the first five equations:

$$
\begin{array}{r}
A + B + C = 180 \\
D + E + F = 180 \\
D + G + I = 180 \\
C + H + I = 180 \\
A + F + J = 180 \\
\hline
2A + B + 2C + 2D + E + 2F + G + H + 2I + J = 900
\end{array}
$$

Rearrange the resulting equation and then subtract the sixth equation.

$$
\begin{array}{r}
2A + 2C + 2D + 2F + 2I + B + E + G + H + J = 900 \\
- B - E - G - H - J = -540 \\
\hline
2A + 2C + 2D + 2F + 2I = 360 \\
2(A + C + D + F + I) = 360 \\
A + C + D + F + I = 180
\end{array}
$$

We say that the sum of the measures of the "exterior" angles of the star is 180°.

4.4 WARM-UPS

Work these problems before you attempt the exercises.

1. Find the value of the determinant:

$$\begin{vmatrix} 2 & 4 \\ 5 & -1 \end{vmatrix} \quad -22$$

2. Find the cofactor of 6:

$$\begin{vmatrix} -2 & 7 & 5 \\ 1 & 6 & 2 \\ 4 & 5 & -8 \end{vmatrix} \quad -4$$

3. Find the value of this determinant by expanding on a column.

$$\begin{vmatrix} 5 & 6 & -1 \\ 8 & 6 & 3 \\ 2 & 1 & -3 \end{vmatrix} \quad 79$$

4. Show that you get the same value of the determinant when you use two different columns.

$$\begin{vmatrix} -1 & -3 & -5 \\ 2 & 4 & 6 \\ 0 & 7 & -4 \end{vmatrix}$$

-36; Calculations may vary.

5. Solve by using Cramer's rule:

$$3x - 4y = -4$$
$$5x - 3y = -3$$

$\{(0, 1)\}$

6. Solve by using determinants:

$$x + y + z = -9$$
$$2 = x - y$$
$$x - z = 7$$

$\{(0, -2, -7)\}$

![decorative bar]

4.4 EXERCISES

Note: Use your graphing calculator to check your results whenever possible.

In Exercises 1 through 12, find the value of the determinant.

1. $\begin{vmatrix} 3 & 5 \\ 2 & 4 \end{vmatrix}$ 2

2. $\begin{vmatrix} 6 & 7 \\ 8 & 1 \end{vmatrix}$ -50

3. $\begin{vmatrix} 9 & -1 \\ 20 & -2 \end{vmatrix}$ 2

4. $\begin{vmatrix} -3 & -9 \\ -6 & -10 \end{vmatrix}$ -24

5. $\begin{vmatrix} -4 & -8 \\ -5 & -12 \end{vmatrix}$ 8

6. $\begin{vmatrix} 5 & -8 \\ -6 & -3 \end{vmatrix}$ -63

7. $\begin{vmatrix} -10 & 8 \\ 5 & 4 \end{vmatrix}$ -80

8. $\begin{vmatrix} 6 & 3 \\ -2 & -4 \end{vmatrix}$ -18

9. $\begin{vmatrix} 13 & 10 \\ 4 & 18 \end{vmatrix}$ 194

10. $\begin{vmatrix} -1 & -11 \\ 12 & 2 \end{vmatrix}$ 130

11. $\begin{vmatrix} 3 & 9 \\ -10 & -4 \end{vmatrix}$ 78

12. $\begin{vmatrix} -7 & -5 \\ 6 & 8 \end{vmatrix}$ -26

In Exercises 13 through 20, find the cofactors indicated.

13. cofactor of 3
cofactor of 0
$\begin{vmatrix} 1 & 2 & -5 \\ 1 & 2 & 7 \\ 0 & 1 & 3 \end{vmatrix}$ 0; 24

14. cofactor of 12
cofactor of 8
$\begin{vmatrix} 3 & 8 & 5 \\ 1 & 12 & 2 \\ 4 & -6 & 4 \end{vmatrix}$ $-8; 4$

15. cofactor of 10
cofactor of 3
$\begin{vmatrix} 5 & 6 & 10 \\ 3 & -6 & 2 \\ -7 & 4 & 6 \end{vmatrix}$ $-30; 4$

16. cofactor of -4
cofactor of -1
$\begin{vmatrix} 7 & 5 & -1 \\ 2 & 8 & 1 \\ -4 & 6 & 0 \end{vmatrix}$ 13; 44

17. cofactor of 9
cofactor of 10
$\begin{vmatrix} -7 & 4 & 9 \\ 2 & 2 & 4 \\ 10 & 3 & 3 \end{vmatrix}$ $-14; -2$

18. cofactor of 12
cofactor of 1
$\begin{vmatrix} 11 & -4 & 1 \\ 3 & 12 & 2 \\ -5 & 4 & 9 \end{vmatrix}$ 104; 72

19. cofactor of -18
cofactor of 6
$\begin{vmatrix} 6 & 0 & 5 \\ -18 & 1 & 3 \\ 0 & 2 & 0 \end{vmatrix}$ 10; -6

20. cofactor of 15
cofactor of 9
$\begin{vmatrix} 0 & 9 & 3 \\ 6 & 0 & 15 \\ 1 & 5 & 0 \end{vmatrix}$ 9; 15

In Exercises 21 through 28, find the value of the 3×3 determinant.

21. $\begin{vmatrix} 6 & 13 & -8 \\ 4 & 2 & 5 \\ -3 & 0 & 4 \end{vmatrix}$ -403

22. $\begin{vmatrix} 1 & 0 & -2 \\ 8 & 14 & 7 \\ 4 & 1 & -9 \end{vmatrix}$ -37

23. $\begin{vmatrix} 15 & 3 & -9 \\ -8 & 5 & -3 \\ 3 & 6 & 13 \end{vmatrix}$ 2097

24. $\begin{vmatrix} 16 & 2 & 0 \\ -3 & 7 & -1 \\ -2 & 1 & 4 \end{vmatrix}$ 492

25. $\begin{vmatrix} 0 & -4 & 1 \\ 15 & -2 & 6 \\ 7 & -1 & 8 \end{vmatrix}$ 311

26. $\begin{vmatrix} 14 & -2 & -10 \\ 8 & 2 & 3 \\ 1 & 4 & 11 \end{vmatrix}$ 10

27. $\begin{vmatrix} 1 & 2 & 0 \\ 0 & 3 & 5 \\ -2 & 0 & 6 \end{vmatrix}$ -2

28. $\begin{vmatrix} 8 & 5 & 0 \\ 10 & 9 & 0 \\ 0 & 3 & -16 \end{vmatrix}$ -352

In Exercises 29 through 48, find the solution set by using Cramer's rule.

29. $2x + 2y = -1$
$x - 2y = -2$ $\left\{\left(-1, \frac{1}{2}\right)\right\}$

30. $4x + 3y = 4$
$3x - 4y = 11.25$ $\{(1.99, -1.32)\}$

31. $5x - 4y = 1$
$-6x - 9y = -30$ $\left\{\left(1\frac{20}{23}, 2\frac{2}{23}\right)\right\}$

32. $-x + 4y = -6$
$2x - 8y = 9$
solution can't be found using Cramer's rule because $D = 0$

33. $5x - y = 3$
$10x + y = 0$ $\{(0.2, -2)\}$

34. $4x - 6y = 0$
$2x + 3y = 24$ $\{(6, 4)\}$

35. $y + x = 2$
$2y - 3x = 19$ $\{(-3, 5)\}$

36. $3y - 2x = 29$
$y + 5x = 18$ $\left\{\left(1\frac{8}{17}, 10\frac{11}{17}\right)\right\}$

37. $y + 12 = x$
$-3y - 20 = -x$ $\{(8, -4)\}$

38. $-3y + x = -5$
$-2y + x = -5$
$\{(-5, 0)\}$

39. $-4x + 6y = 0$
$2x + 3y = 36$
$\{(9, 6)\}$

40. $13y = 9x + 4$
$-3y = -4x + 7$
$\{(4.12, 3.16)\}$

41. $3y = 2x + 6$
$y = -x - 4$
$\{(-3.6, -0.4)\}$

42. $3y = 2x - 7$
$6y = 5x - 22$
$\{(8, 3)\}$

43. $6x - 12y = 18$
$-4x + 8y = -12$
solution can't be found by Cramer's rule because $D = 0$

44. $3y = \frac{15}{2}x - \frac{15}{2}$
$12x - 18y = 45$
$\{(0, -2.5)\}$

45. $6x - 4y + 10z = -20$
$4y + 2z = 14$
$2y - 8z = 16$
$\{(1, 4, -1)\}$

46. $-6x + 4y - 6z = 24$
$-4x + 2y - 6z = 14$
$-2x - 8y + 12z = -6$
$\left\{\left(-3, 2, \frac{1}{3}\right)\right\}$

47. $-3x - 3y - 3z = -15$
$3x - 6y - 6z = -21$
$-9x - 3y + 6z = 6$
$\{(1, 1, 3)\}$

48. $-2x + y - z = -2$
$-x - y + 3z = 21$
$-x - y = 3$ $\left\{\left(-\frac{7}{3}, -\frac{2}{3}, 6\right)\right\}$

In Exercises 49 through 52, solve the system of equations by using a calculator and Cramer's rule. (Check your owner's manual for operating instructions.)

49. $-4x - 4y - 4z = -24$
$-8x + 4y - 4z = 12$
$-4x + 8y - 12z = -24$ $\left\{\left(-2\frac{1}{3}, 3\frac{1}{3}, 5\right)\right\}$

50. $5x + 5y + 5z = 30$
$-5x + 5y + 5z = 10$
$5x + 10y + 5z = 40$
$\{(2, 2, 2)\}$

51. $-12x - 12y - 18z = -114$
$-6x - 6y - 12z = -60$
$-18x + 16y - 30z = -21$
$\{(3.5, 4.5, 1)\}$

52. $-3x - 3y - 6z = -24$
$-9x - 6y + 3z = -45$
$9x + 3y - 9z = 33$
$\{(4, 2, 1)\}$

MIXED PRACTICE

By doing these exercises, you will practice the topics up to this point in the chapter.

53. Find the solution set for $y = 3x + 8$ and $y = \frac{-5}{2}x - 3$ by graphing. $\{(-2, 2)\}$

54. Solve using elimination: $-10x = 8 - 5y$
$0.4y = 2x - 0.5$
$\{(0.95, 3.5)\}$

55. Find the cofactor of 8: $\begin{vmatrix} 2 & 9 & 3 \\ 5 & 8 & 0 \\ 0 & 6 & 4 \end{vmatrix}$
8

56. Solve using Cramer's rule: $3x - 5y + 4z = -45$
$2x - 7z = 6$
$10y + z = 48$
$\{(-4, 5, -2)\}$

57. Solve using substitution: $x - z = 3$
$y + 2z = 1$
$2x - y = 13$
$\{(5, -3, 2)\}$

58. Solve using Cramer's rule: $x + y + z = 50$
$x - z = -20$
$x = 10$
$\{(10, 10, 30)\}$

59. Evaluate the determinant: $\begin{vmatrix} 2 & 6 & 0 \\ 5 & 0 & 8 \\ 0 & -1 & 3 \end{vmatrix}$
-74

60. Solve: $5x - z = 23$
$4x - 3y = 23$
$x - y - z = 4$
$\{(5, -1, 2)\}$

61. A plane flying with the wind completes a 150-mile trip in 1 hour. Coming back, with the plane flying against the wind, the trip takes $1\frac{1}{2}$ hours. What is the speed of the plane and of the wind? plane: 125 mph; wind: 25 mph

62. An automobile radiator holds 20 liters. How much pure antifreeze will you need to add to a solution of water and antifreeze that is 4% antifreeze to obtain a coolant solution that is 25% antifreeze? 4.375 liters of pure antifreeze

63. The sum of my age and Bill's is 110. The difference between my age and Barbara's is 9 years. The sum of all three of our ages is 151. How old are we? I am 50, Bill is 60, and Barbara is 41.

EXCURSIONS

Posing Problems

1. Ask and answer four questions using this data.

WORLD WEATHER IN JULY			
City	**Average High/Low (°F)**	**City**	**Average High/Low (°F)**
Beijing	88/70	Madrid	87/63
Boston	80/63	Miami	88/76
Budapest	82/62	Moscow	73/55
Chicago	81/66	New York	82/66
Delhi	96/81	Paris	76/58
Dublin	67/52	Phoenix	104/77
Frankfurt	77/58	Rio de Janeiro	75/63
Geneva	77/58	Rome	87/67
Hong Kong	87/78	San Francisco	65/53
Houston	92/74	Stockholm	71/57
Jerusalem	87/63	Sydney	60/46
Johannesburg	63/39	Tokyo	83/70
London	71/56	Toronto	79/59
Los Angeles	81/60	Washington	87/68

Class Act

2. Estimate the size of the marked portion of the building in this photograph at the right. Justify your answer and describe the method(s) you used.

Data Analysis

3. Study the tables on the following two pages. Which team is the "biggest"? Answer this question in four different ways using the given data.

1996 N.F.C. Champions: Green Bay Packers

No.	Player	Position	Height	Weight	Years
4	Brett Favre	QB	6-2	225	6
9	Jim McMahon	QB	6-1	195	15
13	Chris Jacke	K	6-0	205	8
17	Craig Hentrich	P	6-3	200	3
18	Doug Pederson	QB	6-3	215	4
21	Craig Newsome	CB	5-11	190	2
25	Dorsey Levens	FB	6-1	235	3
27	Calvin Jones	RB	5-11	205	3
28	Roderick Mullen	CB	6-1	204	2
30	William Henderson	FB	6-1	248	2
32	Travis Jervey	RB	5-11	225	2
33	Doug Evans	CB	6-0	190	4
34	Edgar Bennett	RB	6-0	217	5
36	LeRoy Butler	S	6-0	200	7
37	Tyrone Williams	CB	5-11	195	R
39	Mike Prior	S	6-0	208	11
40	Chris Hayes	S	6-0	200	R
41	Eugene Robinson	S	6-0	195	12
46	Michael Robinson	CB	6-1	192	R
51	Brian Williams	LB	6-1	235	2
52	Frank Winters	C	6-3	295	10
53	George Koonce	LB	6-1	243	5
54	Ron Cox	LB	6-2	235	7
55	Bernardo Harris	LB	6-2	243	2
56	Lamont Hollingst	LB	6-3	243	3
59	Wayne Simmons	LB	6-2	248	4
62	Marco Rivera	G	6-4	295	R
63	Adam Timmerman	G	6-4	295	2
64	Bruce Wilkerson	T	6-5	305	10
65	Lindsay Knapp	G	6-6	300	4
67	Jeff Dellenbach	C	6-6	300	12
68	Gary Brown	T	6-4	315	3
71	Santana Dotson	DT	6-5	285	5
72	Earl Dotson	OT	6-3	315	4
73	Aaron Taylor	G	6-4	305	3
77	John Michels	T	6-7	290	R
80	Derrick Mayes	WR	6-0	200	R
81	Desmond Howard	WR	5-10	180	5
82	Don Beebe	WR	5-11	183	8
83	Jeff Thomason	TE	6-4	250	4
84	Andre Rison	WR	6-1	195	8
85	Terry Mickens	WR	6-0	198	3
86	Antonio Freeman	WR	6-0	190	2
88	Keith Jackson	TE	6-2	258	9
89	Mark Chmura	TE	6-5	250	5
90	Darius Holland	DT	6-4	310	2
91	Shannon Clavelle	DE	6-2	287	2
92	Reggie White	DE	6-5	300	12
93	Gilbert Brown	NT	6-2	325	4
94	Bob Kuberski	NT	6-4	295	2
95	Keith McKenzie	DE	6-2	242	R
96	Sean Jones	DE	6-7	283	13
98	Gabe Wilkins	DE	6-4	305	3

Source: National Football League.

1996 A.F.C. Champions: New England Patriots

No.	Player	Position	Height	Weight	Years
4	Adam Vinatieri	K	6-0	200	R
11	Drew Bledsoe	QB	6-5	233	4
15	Ray Lucas	WR	6-2	201	R
16	Scott Zolak	QB	6-5	235	6
19	Tom Tupa	P	6-4	220	8
21	Ricky Reynolds	CB	5-11	190	10
22	David Meggett	RB	5-7	195	8
23	Terry Ray	SS	6-1	205	5
24	Ty Law	CB	5-11	196	2
25	Larry Whigham	SS	6-2	205	3
26	Jerome Henderson	CB	5-10	188	6
27	Mike McGruder	CB	5-10	178	7
28	Curtis Martin	RB	5-11	203	2
30	Corwin Brown	FS	6-1	200	4
31	Jimmy Hitchcock	CB	5-10	188	2
32	Willie Clay	FS	5-10	198	5
35	Marrio Grier	FB	5-10	225	R
36	Lawyer Milloy	SS	6-0	208	R
41	Keith Byars	FB	6-1	255	11
45	Otis Smith	CB	5-11	190	7
48	Lovett Purnell	TE	6-2	250	R
52	Ted Johnson	LB	6-3	240	2
53	Chris Slade	LB	6-5	245	4
54	Tedy Bruschi	LB	6-1	245	R
55	Willie McGinest	LB	6-5	255	3
58	Marty Moore	LB	6-1	244	3
59	Todd Collins	LB	6-2	242	4
61	Bob Kratch	G	6-3	288	8
62	Dave Richards	T	6-5	315	9
63	Heath Irwin	G	6-4	273	R
64	Dave Wohlabugh	C	6-3	292	2
67	Mike Gisler	C	6-4	300	4
68	Max Lane	T	6-6	305	3
71	Todd Rucci	G	6-5	291	4
72	Devin Wyman	DT	6-7	307	R
74	Chris Sullivan	DE	6-4	279	R
75	Pio Sagapolutele	DT	6-6	297	6
76	William Roberts	G	6-5	298	13
78	Bruce Armstrong	T	6-4	295	10
80	Troy Brown	WR	5-9	190	4
81	Hason Graham	WR	5-10	176	2
82	Vincent Brisby	WR	6-2	188	4
83	Dietrich Jells	WR	5-10	186	R
84	Shawn Jefferson	WR	5-11	180	6
85	John Burke	TE	6-3	248	3
86	Mike Bartrum	TE	6-5	245	3
87	Ben Coates	TE	6-5	245	6
88	Terry Glenn	WR	5-10	184	R
90	Chad Eaton	WR	5-10	184	R
92	Ferric Collons	DE	6-6	285	3
95	Dwayne Sabb	LB	6-4	248	5
96	Mike Jones	DE	6-4	295	6
97	Mark Wheeler	DT	6-3	285	5

Source: National Football League.

CHAPTER LOOK-BACK

We first assign variables to the amounts of cookies baked:

Let x = number of batches of shortbread cookies baked

Let y = number of batches of pan cookies baked

We then translate the constraints into inequalities, reasoning that because we cannot have negative numbers of batches of cookies,

$$x \geq 0 \quad \text{and} \quad y \geq 0$$

To bake x batches of shortbread cookies, the baker needs $3x$ cups of sugar; to bake y batches of pan cookies requires $4y$ cups of sugar. The baker has 76 cups of sugar, so the sugar constraint is

$$3x + 4y \leq 76$$

Similarly, x batches of shortbread cookies require $3x$ cups of butter; y batches of pan cookies require $2y$ cups of butter. The baker has 56 cups of butter, so the butter constraint is

$$3x + 2y \leq 56$$

These are the only constraints.

For x batches of shortbread and y batches of pan cookies, the baker makes a profit of P.

$$P = 2.75x + 2.5y$$

We are asked to maximize this profit, so P is the objective function. The linear programming problem is then to maximize the objective function

$$P = 2.75x + 2.5y$$

subject to the constraints

$$3x + 4y \leq 76$$
$$3x + 2y \leq 56$$
$$x \geq 0$$
$$y \geq 0$$

We first graph the system of inequalities.

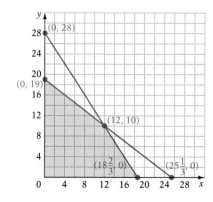

Every point within the shaded area (including the lines) is a solution of the system. However, only points at the intersections of the boundary lines can give a maximum or minimum of the objective function. (You can check this statement by substituting other values in the objective function.) We substitute the coordinates of each of these corner points in the objective function to see which produces a maximum.

$$P = 2.75x + 2.5y$$

For $(0, 0)$: $P = 0$

For $(0, 19)$: $P = 2.75(0) + 2.5(19) = \$47.50$

For $(12, 10)$: $P = 2.75(12) + 2.5(10) = \$58$

For $\left(18\frac{2}{3}, 0\right)$: $P = 2.75\left(18\frac{2}{3}\right) + 2.5(0) = \$51.34$

Because 58 is the largest value of P, the point $(12, 10)$ maximizes the objective function. The baker should make 12 batches of shortbread and 10 batches of pan cookies. His profit will be $58.

CHAPTER 4
REVIEW PROBLEMS

The following exercises will give you a good review of the material presented in this chapter.

SECTION 4.1

1. Solve by graphing: $y = \frac{-1}{4}x + 3$
$y = \frac{5}{4}x - 3$ $\{(4, 2)\}$

2. Solve by graphing: $4y = x - 16$
$y = \frac{3}{4}x - 2$ $\{(-4, -5)\}$

3. Solve by graphing: $y + x = 1$
$2y = -5x - 10$ $\{(-4, 5)\}$

4. Solve by graphing: $x = 2$
$y = -3$ $\{(2, -3)\}$

5. Solve by addition: $6x - 2y = 18$
$3x + 5y = 27$ $\{(4, 3)\}$

6. Solve by addition: $8x - 5y = -34$
$7x + 2y = -17$ $\{(-3, 2)\}$

7. Solve by substitution: $x - y = 10$
$5x + 3y = 26$ $\{(7, -3)\}$

8. Solve by substitution: $x - 13y = -3$
$2x + 5y = 25$ $\{(10, 1)\}$

9. Solve by addition: $2x + 2y - 3z = 22$
$4x + 4y + z = 30$
$x - y + 2z = -6$
$\{(3, 5, -2)\}$

10. Solve by substitution: $x + 2y = 3$
$y - 3z = -1$
$x - z = -2$
$\{(-1, 2, 1)\}$

SECTION 4.2

11. A stack of 20 quarters and dimes is worth $3.95. How many of each coin are in the stack?   13 quarters; 7 dimes

12. Two towns, Ashley and Blaine, are 190 miles apart. A train leaves Ashley at noon, traveling 28 miles per hour. At the same time, another train leaves Blaine traveling 48 miles per hour. The two trains are traveling toward each other. What time will they meet, and how far from Ashley will they be?   2:30 P.M.; 70 miles

13. Potato salad sells for $1.30 per pound; fruit salad sells for $2.50 per pound. For a party, we buy a total of 10 pounds of salad. We pay $18.40 altogether. How much of each salad do we buy?   5.5 pounds of potato salad; 4.5 pounds of fruit salad

14. How much water should be evaporated from 75 ounces of a 2% solution of salt to produce a 5% solution?   45 ounces of water

15. Some nickels, dimes, and quarters are in a stack. There are 10 coins in all. The nickels and dimes total 60 cents. The quarters and dimes total $1.25. What is the value of the stack? $1.35

16. Three runners ran a marathon relay totaling 26 miles. Anita and Ruth ran 20 miles altogether. Vincent and Ruth together ran 15 miles. How many miles did each run?   Ruth: 9 miles; Anita: 11 miles; Vincent: 6 miles

SECTION 4.3

17. Solve graphically: $x \leq 2$
$$5 > y$$

18. Solve by graphing: $x > -1$
$$y \leq 2$$

19. Solve by graphing: $y \leq \frac{-1}{5}x + 4$
$$y > \frac{4}{7}x + 4$$

20. Solve by graphing: $y > -x - 3$
$$y \leq 5x + 9$$

MIXED REVIEW

21. Solve: $-x + 15y = -123$ $\{(18, -7)\}$
$$21x - 5y = 413$$

22. Solve by graphing: $y = \frac{-7}{5}x - 5$ $\{(-5, 2)\}$
$$y = \frac{1}{5}x + 3$$

23. Solve: $x + y + 2z = 16$ $\{(-4, 8, 6)\}$
$$3x + 2y + 3z = 22$$
$$2x + 2y + 3z = 26$$

24. Solve: $7y - 2x = 9$ $\{(6, 3)\}$
$$3x - 7y = -3$$

25. Graph: $y \leq -6x + 5$
$$y \geq x - 2$$

26. Solve: $-5y + 4z = -33$ $\{(10, 5, -2)\}$
$$-x - 8z = 6$$
$$4x - 5y = 15$$

CHAPTER 4 TEST

This exam tests your knowledge of the material in Chapter 4.

1. Solve by graphing:
   a. $5y - 2x = 5$
   $6 + 3y = 3x$
   $\{(5, 3)\}$

   b. $y = 2x + 1$
   $y = \frac{-1}{2}x - 4$
   $\{(-2, -3)\}$

   c. $7y - 3x = 21$
   $x - 1 = y$
   $\{(7, 6)\}$

2. a. Solve by addition:

$2x + 5 = 7y$

$15 - 21y = 6x$

$\left\{\left(0, \frac{5}{7}\right)\right\}$

b. Solve by substitution:

$x - 6 = y$

$5x = y + 2$

$\{(-1, -7)\}$

c. Solve: $x + 2y - z = 3$

$4x - 3y + 2z = 13$

$2x + 4y + z = 12$

$\{(3, 1, 2)\}$

3. Solve:

a. A boat steams upstream, full throttle, at 20 miles per hour against the current; its trip upstream takes $2\frac{1}{2}$ hours. The return trip downstream takes $1\frac{1}{2}$ hours. How fast is the current, and what is the speed of the boat in still water?   current: $6\frac{2}{3}$ mph; speed in still water: $26\frac{2}{3}$ mph

b. How many ounces of a 10% cleaning solution must be mixed with 16 ounces of water to give an 8% cleaning solution?   64 ounces

c. Georgio, Jayson, and Rámon are having birthdays. The sum of Georgio's and Jayson's ages is 56. Jayson's and Rámon's ages total 47 years, and those of Georgio and Rámon add up to 55. How old is each man?   Georgio is 32; Jayson is 24; Rámon is 23.

4. Graph to find the solution set of each system of inequalities:

a. $x \geq 2$

$2x - y \geq 3$

b. $y - x \leq -5$

$y + x > 5$

c. $|x| \leq 3$

$3x + y \leq 4$

CUMULATIVE REVIEW

CHAPTERS 1–4

The following exercises will help you maintain the skills you have learned in this and previous chapters.

1. Solve: $3 < 2x + 5 < 9$

$-1 < x < 2$

2. Translate to algebra: "twice the difference of x and y" $2(x - y)$

3. Determine whether these equations represent lines that are parallel, perpendicular, or intersecting.

$3x = y + 4$

$y = -\frac{1}{3}x + 5$

perpendicular and intersecting

4. Replace the boxes with numbers that make the statement true. What property is illustrated? $\blacksquare(5 - 7) = \blacksquare(5) + 3(\blacksquare)$

3; 3; -7; distributive property

5. Solve: $-3x + 5 = -9(x - 7) - 6x$

$x = 4\frac{5}{6}$

6. Determine the x- and y-intercepts of the graph of $y = -x - 5$.   x-intercept: $(-5, 0)$;

y-intercept: $(0, -5)$

7. Evaluate $rt - r^2t + rt^2 + (r - t)^2$ when $t = -4$ and $r = 0$.   16

8. Simplify: $(2 - 3)^2 + 9(-6)^2 - 2^{-1}$ $324\frac{1}{2}$

9. Evaluate: $-|-6|$ -6

10. Combine like terms:

$7x^2y - 5xy^2 + 8x^2y - 3xy^2 + 11x^3 + 12y^2$

$15x^2y - 8xy^2 + 11x^3 + 12y^2$

11. Graph $y = 3x + 2$ by using the slope and y-intercept. •

12. Simplify: $3^{-2} + 1^5 - 6^0 + 2^2$ $4\frac{1}{9}$

13. Multiply: $(-3)(-2)(5)(-6)$ -180

14. If $f(x) = x^2 + 2x + 7$, find $f(-5)$. 22

15. Solve graphically: $x \le 5$
$$x - 2 \le y$$

16. Complete this table of values for $5y = -3x + 2$.

x	y
-1	1
$\frac{2}{3}$	0
4	-2

17. Solve by graphing: $y = \frac{1}{2}x$
$$y = -\frac{3}{2}x - 4$$
$\{(-2, -1)\}$

18. Find the equation of the line that passes through the points $(-3, -4)$ and $(-6, -5)$.
$y = \frac{1}{3}x - 3$

19. Find the additive inverse of $3x^2 - 2x + 7$.
$-3x^2 + 2x - 7$

20. Evaluate x^0 when x is not equal to zero.
1

5

MONOMIALS AND POLYNOMIALS

Suppose you were a chemist and needed to calculate the number of magnesium atoms in 0.0000011 gram. You would need to multiply that mass by Avogadro's number divided by magnesium's gram molecular weight—0.0000011 g × 602,300,000,000,000,000,-000,000 atoms/24.31 g. Fortunately, scientific notation and the product rules for exponents allow you to rewrite the equation as

$$\frac{(1.1 \times 10^{-6} \times 6.023 \times 10^{23})}{(2.431 \times 10^{1})}$$

and calculate the answer: 2.7×10^{16} atoms, written without scientific notation as 27,000,000,000,000,000.

■ *In what other fields or situations might the use of scientific notation simplify measurements or calculations?*

Take this short quiz to see how well prepared you are for Chapter 5. The answers follow the quiz.

1. Add: $\left(-11\frac{1}{2}\right) + \left(-13\frac{3}{7}\right)$

2. Divide: $24\overline{)7214448}$

3. Simplify: -6^{-3}

4. Subtract $-83 - 29$

5. Simplify: $x^2 + 6x - 3x + 2$

6. Simplify: $-3(y - 4)$

7. Multiply: $(-12.5)(2.4)(-6)$

8. Find the opposite of -6.4.

ANSWERS: **1.** $-24\frac{13}{14}$ [Section 1.2]   **2.** 300,602 [Section 1.2]   **3.** $-\frac{1}{216}$ [Section 1.3]
4. -112 [Section 1.2]   **5.** $x^2 + 3x + 2$ [Section 2.1]   **6.** $-3y + 12$ [Section 1.3]   **7.** 180 [Section 1.2]
8. 6.4 [Section 1.1]

CHAPTER LEAD-IN

Certain genes can be carried in the genetic code of an organism and not always be observable. One such gene determines attached earlobes; the gene for unattached, or hanging, earlobes is **dominant.***

A person with the F gene will have hanging earlobes.

F: hanging earlobes

f: attached earlobes

Each person carries two genes for this particular characteristic.

FF	Ff	fF	ff
Hanging	Hanging	Hanging	Attached

The Hardy-Weinberg equation is used to model the distribution of genes in the population for characteristics that are wholly controlled by two genes. It is

$$p^2 + 2pq + q^2 = 1$$

where

p = proportion of population with Gene I

q = proportion of population with Gene II

- For our example, if Gene I is F and Gene II is f, what does each term in the Hardy-Weinberg equation mean?

*Source: Gregory Fiore, "An Out of Math Experience: Quadratic Equations and Polynomial Multiplication as used in Genetics." *The AMATYC Review,* Volume 17, Number 1 (Fall 1995): pp. 20–27.

INSTRUCTOR NOTE

Students with dyslexia or perceptual difficulties have trouble differentiating between p and q.

5.1 Operations with Monomials

SECTION LEAD-IN

The probability of having children of a certain sex in a family that consists of exactly two children can be found by using the equation

$$p^2 + 2pq + q^2 = 1$$

(What is this equation called?)

The probability of a boy is $\frac{1}{2}$.

The probability of a girl is $\frac{1}{2}$.

▪ What does each term of the above equation mean in terms of boys and girls?

SECTION GOALS

- ▪ To multiply monomials
- ▪ To divide monomials
- ▪ To raise monomials to integer powers
- ▪ To multiply and divide using scientific notation.

Introduction

A **monomial** is the product of a constant and one or more variables raised to non-negative integer powers. Monomials are algebraic expressions and therefore represent real numbers. A monomial is the simplest possible polynomial. This chapter is about polynomials. In this section, we discuss several rules that involve exponents on real numbers, and we apply them to monomials. Recall that for real numbers a and positive integers n,

$$a^n = \underbrace{a \cdot a \cdot a \cdots \cdot a}_{n \text{ factors}}$$

$a^1 = a$

$a^0 = 1$ if a is not equal to 0

Multiplying Monomials

There is an easy way to multiply terms that contain the same base.

> **Product rule for exponents**
> For any non-zero real number a, and any integers m and n,
> $$a^m \cdot a^n = a^{m+n}$$

INSTRUCTOR NOTE

It is useful to give numerical examples that can be calculated to verify the rules for exponents.

That is, to multiply two numbers with the same base, add their exponents.

▪▪▪
EXAMPLE 1

Multiply: $(5x^4)(9xy^6)$

SOLUTION

Monomials are easier to multiply when we can see all the coefficients and the exponents. So we rewrite this multiplication as

$$(5x^4)(9x^1y^6)$$

We have to multiply the coefficients 5 and 9; we multiply the variables by adding the exponents of each like base. So

$$(5x^4)(9x^1y^6) = (5)(9)x^{4+1}y^6$$
$$= 45x^5y^6$$

▶ CHECK **Warm-Up 1**

▪ ▪ ▪

EXAMPLE 2

Multiply: $(2x^3y^4)(-4xy^2)(-3.5x^2y^6)$

SOLUTION

Rewriting each term to show the exponents, we have

$$(2x^3y^4)(-4x^1y^2)(-3.5x^2y^6)$$

Then, using the commutative property to regroup the coefficients and variable parts, we obtain

$$(2)(-4)(-3.5)(x^3)(x^1)(x^2)(y^4)(y^2)(y^6) \quad \text{Rearranging}$$
$$= 28x^{3+1+2}y^{4+2+6} \quad \text{Adding exponents}$$
$$= 28x^6y^{12} \quad \text{Simplifying}$$

▶ CHECK **Warm-Up 2**

Our rules for exponents enable us to treat terms with negative exponents as monomials. In this next example, two of the variables have negative exponents. We simply add them as we did positive exponents.

▪ ▪ ▪

EXAMPLE 3

Multiply: $(8m^2n^5t^3)(9m^{-4}n^{-6})$

SOLUTION

Regrouping coefficients and variables gives us

$$(8)(9)(m^2)(m^{-4})(n^5)(n^{-6})(t^3)$$

Then simplifying yields

$$72m^{2+(-4)}n^{5+(-6)}t^3 = 72m^{-2}n^{-1}t^3$$

▶ CHECK **Warm-Up 3**

Dividing Monomials

Suppose we are asked to find

$$x^7 \div x^5$$

We can write out

$$\frac{x^7}{x^5} = \frac{(x)(x)(x)(x)(x)(x)(x)}{(x)(x)(x)(x)(x)} = x^2$$

or we can use the quotient rule:

Quotient rule for exponents

For any non-zero real number a and any integers m and n,

$$a^m \div a^n = a^{m-n}$$

In words, to divide powers with the same base, subtract the exponents.

▪▪▪
EXAMPLE 4

Divide: $\dfrac{25m^4n^{-5}}{15m^{-3}n}$

SOLUTION

Using the quotient rule yields

$$\frac{25m^4n^{-5}}{15m^{-3}n} = \frac{\overset{5}{\cancel{25}}}{\underset{3}{\cancel{15}}}m^{4-(-3)}n^{-5-1}$$

$$= \frac{5}{3}m^7n^{-6}$$

Then we eliminate the negative exponent.

$$\frac{5}{3}m^7n^{-6} = \frac{5m^7}{3} \cdot \frac{1}{n^6} \quad \text{Reciprocal rule}$$

$$= \frac{5m^7}{3n^6} \qquad \text{Simplifying}$$

▶ *CHECK* **Warm-Up 4**

▪▪▪
EXAMPLE 5

Divide: $\dfrac{56r^5z^6}{4rt^8z^{-2}}$

SOLUTION

Again, we apply the quotient rule and simplify. But first we write the numerator and denominator with all the same variables.

$$\frac{56r^5z^6}{4rt^8z^{-2}} = \frac{56r^5t^0z^6}{4r^1t^8z^{-2}}$$

$$= \frac{\overset{14}{\cancel{56}}}{\underset{1}{\cancel{4}}}r^{5-1}t^{0-8}z^{6-(-2)} \quad \text{Quotient rule}$$

$$= 14r^4t^{-8}z^8 \qquad\qquad \text{Simplifying}$$

$$= \frac{14r^4z^8}{t^8} \qquad\qquad \text{Reciprocal rule}$$

▶ *CHECK* **Warm-Up 5**

Raising Powers to Powers

> **Power rule for exponents**
> For any real number a and integers m and n,
> $$(a^m)^n = a^{mn}$$

In other words, to raise a power to a power, multiply the exponents.

> **Power rule for products**
> For any real numbers a and b and integer m,
> $$(a \cdot b)^m = a^m b^m$$

We can combine our two power rules into still another (and generally more useful) rule:

> **Power rule for products with exponents**
> For any integers m, n, and t,
> $$(a^m b^n)^t = a^{mt} b^{nt}$$

In words, to raise a term containing exponents to a power, multiply each exponent by the power. Thus, for example,

$$(w^3xy)^{12} = w^{3 \cdot 12}x^{1 \cdot 12}y^{1 \cdot 12} = w^{36}x^{12}y^{12}$$

STUDY HINT

When a term that is raised to a power contains a coefficient, be sure to raise the coefficient to the given power also.

▪ ▪ ▪

EXAMPLE 6

Simplify: $(-12r^3t^4z^7)^3$

SOLUTION

To simplify this expression, we must raise (-12) to the third power *as well as* r^3, t^4, and z^7.

$$(-12r^3t^4z^7)^3 = (-12)^{1 \cdot 3}r^{3 \cdot 3}t^{4 \cdot 3}z^{7 \cdot 3} \quad \text{Power rule}$$
$$= (-12)^3 r^9 t^{12} z^{21} \quad \text{Simplifying}$$
$$= -1728 r^9 t^{12} z^{21} \quad \text{Simplifying}$$

▶ *CHECK* **Warm-Up 6**

Recall that for non-zero real numbers a and b and integer n,

$$a^{-n} = \frac{1}{a^n} \quad \text{and} \quad \left(\frac{a}{b}\right)^{-n} = \left(\frac{b}{a}\right)^n$$

In the next few examples, we will need to apply these rules to monomials.

▪ ▪ ▪

ERROR ALERT

Identify the error and give a correct answer.

Simplify: $(4a^9)^3$

Incorrect Solution:
$(4a^9)^3 = 12a^{27}$

EXAMPLE 7

Simplify

$$(8xy^5)^{-3}$$

and write the result using only positive exponents.

SOLUTION

Using the power rule, we have

$$(8xy^5)^{-3} = 8^{1(-3)}x^{1(-3)}y^{5(-3)} \quad \text{Power rule}$$
$$= 8^{-3}x^{-3}y^{-15} \quad \text{Product rule}$$

Because the problem asked us to use only positive exponents, we must rewrite $(8^{-3})x^{-3}y^{-15}$ as

$$\frac{1}{8^3} \cdot \frac{1}{x^3} \cdot \frac{1}{y^{15}} = \frac{1}{512x^3y^{15}}$$

An alternative method can also be used to solve this problem. We can eliminate the negative exponent first, obtaining

INSTRUCTOR NOTE

Remind students that a negative exponent does not influence the sign of its base.

$$(8xy^5)^{-3} = \frac{1}{(8xy^5)^3}$$
$$= \frac{1}{(8^{1 \cdot 3}x^{1 \cdot 3}y^{5 \cdot 3})} \quad \text{Power rule}$$
$$= \frac{1}{512x^3y^{15}} \quad \text{Simplifying}$$

The choice is yours.

▶ *CHECK* **Warm-Up 7**

ERROR ALERT

Identify the error and give a correct answer.

Simplify: $(2x)^5$

Incorrect Solution:
$(2x)^5 = 2x^5$

Calculator Corner

The graphing calculator can also be used to check your work with negative exponents and give the answer in fraction form. (Note: See the Calculator Corner on page 4 to review how to get an answer in fractional form.)

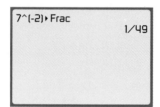

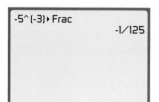

There is also a power rule for fractions (and quotients).

Power Rule for Fractions

For any real numbers a and b, where b is not equal to zero, and any integer m,

$$\left(\frac{a}{b}\right)^m = \frac{a^m}{b^m}$$

▪▪▪

EXAMPLE 8

Simplify: $\left(\dfrac{-3m^2n^4}{7v^3}\right)^2$

SOLUTION

When we simplify this expression, we must be careful to raise both numerator and denominator to the second power. You will remember to do so if you apply the fraction rule as the first step.

$$\left(\frac{-3m^2n^4}{7v^3}\right)^2 = \frac{(-3m^2n^4)^2}{(7v^3)^2} \qquad \text{Fraction rule}$$

Now, multiplying the exponents by the power 2 in both numerator and denominator gives us

$$\frac{(-3m^2n^4)^2}{(7v^3)^2} = \frac{(-3)^{1\cdot2}m^{2\cdot2}n^{4\cdot2}}{(7)^{1\cdot2}v^{3\cdot2}} \qquad \text{Power Rule}$$

$$= \frac{9m^4n^8}{49v^6} \qquad \text{Simplifying}$$

▶ *CHECK* **Warm-Up 8**

▪▪▪
EXAMPLE 9

Simplify: $\left(\frac{3x^2y^2}{18x^3y}\right)^{-4}$

SOLUTION

First we simplify the fraction within the parentheses.

$$\left(\frac{3x^2y^2}{18x^3y}\right)^{-4} = \left(\frac{y}{6x}\right)^{-4}$$

Then we eliminate the negative exponent

$$\left(\frac{y}{6x}\right)^{-4} = \left(\frac{6x}{y}\right)^{4}$$

Then we apply the fraction rule for powers.

$$\left(\frac{6x}{y}\right)^{4} = \frac{1296x^4}{y^4}$$

▶ CHECK **Warm-Up 9**

Operations in Scientific Notation

Before we begin Example 10, let's review the procedures for using scientific notation.

To write a decimal number in scientific notation

1. Move the decimal point right or left to obtain a number n such that $1 \le n < 10$.
2. Count the number of places, p, that the decimal point has been moved.
3. Multiply n by 10^p if the decimal point was moved to the left. Multiply by 10^{-p} if it was moved to the right. Eliminate any meaningless zeros.

To write a scientific number in standard notation

1. Move the decimal point a number of places equal to the exponent of 10. Move it to the right if the exponent is positive; move it to the left if the exponent is negative. (Add zeros as necessary.)
2. Eliminate the multiplication sign and power of 10.

Calculator Corner

Writing the number 10,300,000 in scientific notation would give the answer 1.03×10^7. Now check your answer with your calculator. If your answer is correct, the calculator will return a "1," meaning the statement is true.

If your answer is wrong, the calculator will return a "0," meaning the statement is false. (Note: See the Calculator Corner on pages 6 and 7 to review how to test a statement.)

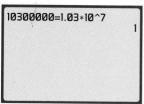

Notice the difference here:

Now try the following on your own.

1. 12,300,000 1.23×10^7

2. 123,456,000 1.23456×10^8

3. 222,800,000 2.228×10^8

4. 943,006,000 9.43006×10^8

5. 11,562,300,000 1.15623×10^{10}

6. 56,123,456,000 5.6123456×10^{10}

Numbers expressed in scientific notation can be multiplied and divided very easily, much like monomials.

■ ■ ■

EXAMPLE 10

a. Multiply: $(7.9 \times 10^{18})(2.5 \times 10^{-9})$

b. Estimate the result of 938,727,000 divided by 29,800.

SOLUTION

a. To multiply the two numbers in scientific notation, multiply the coefficients and then multiply the powers of 10.

$$(7.9 \times 10^{18})(2.5 \times 10^{-9}) = (7.9)(2.5) \times 10^{18 + (-9)}$$
$$= 19.75 \times 10^9$$

To write the number in scientific notation, we put the decimal number part in proper scientific notation and then simplify.

$$19.75 \times 10^9 = (1.975 \times 10^1) \times 10^9$$
$$= 1.975 \times 10^{10}$$

b. First we write the numbers in scientific notation. Then we round each coefficient to the nearest whole number and divide.

$$\frac{9.38727 \times 10^8}{2.98 \times 10^4} = \frac{9 \times 10^8}{3 \times 10^4}$$
$$= \frac{9 \times 10^{8-4}}{3}$$
$$= 3 \times 10^4$$

So 938,727,000 divided by 29,800 is about 3×10^4 or 30,000.

▶ *CHECK* **Warm-Up 10**

Practice what you learned.

SECTION FOLLOW-UP

In the equation

$$p^2 + 2pq + q^2 = 1$$

p^2 is the probability of exactly 2 boys.
q^2 is the probability of exactly 2 girls.
$2pq$ is the probability of exactly one boy and one girl.

Those probabilities are

$$p^2 = p \cdot p = \left(\tfrac{1}{2}\right)\left(\tfrac{1}{2}\right) = \tfrac{1}{4} \quad \text{Probability of 2 boys}$$

$$q^2 = q \cdot q = \left(\tfrac{1}{2}\right)\left(\tfrac{1}{2}\right) = \tfrac{1}{4} \quad \text{Probability of 2 girls}$$

$$2pq = 2(p)(q) = 2\left(\tfrac{1}{2}\right)\left(\tfrac{1}{2}\right) = \tfrac{1}{2} \quad \text{Probability of one boy and one girl}$$

▪ In a population of 1000 families that have exactly two children each, what would we expect to find?
250 families with 2 boys; 250 families with 2 girls; 500 families with 1 boy and 1 girl

5.1 *WARM-UPS*

Work these problems before you attempt the exercises.

1. Multiply: $(3n^5t)(-2n)$
   $-6n^6t$

2. Multiply:
   $(-2x^2y^3)(-4.6x^4y^5)(8x^5)$
   $73.6x^{11}y^8$

3. Multiply: $(-9r^2t^{-3})(-8t^5)$
   $72r^2t^2$

4. Divide: $\dfrac{28t^3}{7t^{-5}r^{-2}}$
   $4t^8r^2$

5. Divide: $\dfrac{3x^2y^3}{18x^4y^5z}$
   $\dfrac{1}{6x^2y^2z}$

6. Simplify: $(3ny^2)^4$
   $81n^4y^8$

7. Simplify: $(4mn^2)^{-3}$
   $\dfrac{1}{64m^3n^6}$

8. Simplify $\left(\dfrac{2rt^3}{x}\right)^{-8}$ and rewrite the result using only positive exponents. $\dfrac{x^8}{256r^8t^{24}}$

9. Divide: $\left(\dfrac{27r^3t^5}{9r^7t^9}\right)^{-4}$ $\dfrac{r^{16}t^{16}}{81}$

10. **a.** Multiply: $(9.1 \times 10^{17})(17.4 \times 10^{-9})$
    1.5834×10^{10}

    b. Divide: $\dfrac{3.2 \times 10^5}{8 \times 10^{16}}$
    4.0×10^{-12}

5.1 EXERCISES

Note: Use your graphing calculator to check your results whenever possible.

In Exercises 1 through 8, multiply the monomial:

1. $(-4x^2y)(3x^3y^2)$
$-12x^5y^3$

2. $(-14x^3y^3)(9x^5y)$
$-126x^8y^4$

3. $\left(\frac{5}{6}x^2y\right)(-3x^3y^6)$
$-\frac{5x^5y^7}{2}$

4. $(-7x^4y^9)\left(\frac{6}{7}y^5x^8\right)$
$-6x^{12}y^{14}$

5. $\left(\frac{1}{2}x^5y\right)\left(\frac{2}{5}x^4y\right)$
$\frac{x^9y^2}{5}$

6. $\left(\frac{5}{6}y^2x\right)\left(\frac{1}{4}x^2y\right)$
$\frac{5x^3y^3}{24}$

7. $(-4.3x^2y^3)(2.1x^3y^4)$
$-9.03x^5y^7$

8. $(-0.7x^5)(1.5xy^7)$
$-1.05x^6y^7$

In Exercises 9 through 16, rewrite each term without negative exponents:

9. m^{-4} $\frac{1}{m^4}$

10. n^{-3} $\frac{1}{n^3}$

11. $(-p)^{-2}$ $\frac{1}{p^2}$

12. $(-r)^{-4}$ $\frac{1}{r^4}$

13. $3x^{-2}y^{-1}$ $\frac{3}{x^2y}$

14. $4y^{-3}x^{-2}$ $\frac{4}{y^3x^2}$

15. $-6x^{-5}y^3$ $\frac{-6y^3}{x^5}$

16. $-3y^{-8}x^2$ $\frac{-3x^2}{y^8}$

In Exercises 17 through 24, multiply the terms.

17. $(12x)(-5x^{-9})$ $\frac{-60}{x^8}$

18. $(7x)(-18x^{-2})$ $\frac{-126}{x}$

19. $(4x^{-2}y)(3x^3)$ $12xy$

20. $(8x^3y)(4x^{-3}y)$ $32y^2$

21. $(7m^4t)(-12mt^{-6})$
$\frac{-84m^5}{t^5}$

22. $(-10tu^3)(5tu^{-2})$
$-50t^2u$

23. $(-17xy)(-2xwy^{-2})$
$\frac{34wx^2}{y}$

24. $(-24x^2y)(-3xy^{-2})$
$\frac{72x^3}{y}$

In Exercises 25 through 44, simplify each expression by raising the term to the indicated power. Rewrite all negative exponents as positive.

25. $(11x^2y^5)^2$ $121x^4y^{10}$

26. $(2w^2x^3)^2$ $4w^4x^6$

27. $(-3xy^2)^3$ $-27x^3y^6$

28. $(-8wx^3)^4$ $4096w^4x^{12}$

29. $\left(\frac{2w^3x^4}{3y^4}\right)^5$ $\frac{32w^{15}x^{20}}{243y^{20}}$

30. $\left(\frac{4w^3y^2}{7x^4}\right)^3$ $\frac{64w^9y^6}{343x^{12}}$

31. $(5x^{-3}y^2)^4$ $\frac{625y^8}{x^{12}}$

32. $(2w^2x^{-4})^2$ $\frac{4w^4}{x^8}$

33. $(-2x^{-5}y^3)^2$ $\frac{4y^6}{x^{10}}$

34. $(-4w^{-2}x^4)^3$ $\frac{-64x^{12}}{w^6}$

35. $\left(-\frac{8x^3}{9y^{-4}}\right)^2$ $\frac{64x^6y^8}{81}$

36. $\left(\frac{-w^3}{2x^2y^{-5}}\right)^3$ $\frac{-w^9y^{15}}{8x^6}$

37. $(w^2xy)^{-3}$ $\frac{1}{w^6x^3y^3}$

38. $(w^3xy)^{-2}$ $\frac{1}{w^6x^2y^2}$

39. $(2xy^{-3})^{-5}$ $\frac{y^{15}}{32x^5}$

40. $(5wx^{-3})^{-2}$ $\frac{x^6}{25w^2}$

41. $(-4w^2x^3)^{-3}$ $-\frac{1}{64w^6x^9}$

42. $(-3r^6t^4)^{-5}$ $-\frac{1}{243r^{30}t^{20}}$

43. $\left(\frac{5w^4}{-7x^7y^6}\right)^{-2}$ $\frac{49x^{14}y^{12}}{25w^8}$

44. $\left(\frac{9r^3t^2}{13y^5}\right)^{-2}$ $\frac{169y^{10}}{81r^6t^4}$

In Exercises 45 through 52, multiply or divide as indicated.

45. $\frac{w^2x}{wx^3}$ $\frac{w}{x^2}$

46. $\frac{w^3x}{wx^4}$ $\frac{w^2}{x^3}$

47. $(27w^7x)(w^6x^{-3})$ $\frac{27w^{13}}{x^2}$

48. $(36w^5x)(w^{-4}x)$ $36wx^2$

49. $\frac{20wx^2y}{-40w^{-2}y}$ $-\frac{w^3x^2}{2}$

50. $\frac{-48x^3y^{-1}}{26x^2y^{-3}}$ $-\frac{24y^2}{13}$

51. $\left(\frac{12w^2xy^4}{18w^3x^{-2}y^5}\right)^6$ $\frac{64x^{18}}{729w^6y^6}$

52. $\left(\frac{24r^2t^2u}{36r^3t^{-4}u^3}\right)^3$ $\frac{8t^{18}}{27r^3u^6}$

In Exercises 53 through 56, raise each quotient to the indicated power, and then simplify.

53. $\left(\frac{63w^4x^2}{-7w^{-7}x^5y}\right)^3\left(\frac{-6.3w^2x^{-6}y^3}{2.1x^3y^7}\right)^{-2}$ $-81w^{29}x^9y^5$

54. $\left(\frac{-56r^2s^3}{-8r^{-2}s^6t^3}\right)^2\left(\frac{7rs^2t}{2s^3t^4}\right)^{-4}$ $\frac{16r^4t^6}{49s^2}$

55. $\left(\frac{3w^4x^{-2}y^{-3}}{2}\right)^8\left(\frac{3w^5}{2x^5y^{-3}}\right)^{-6}$ $\frac{9w^2x^{14}}{4y^{42}}$

56. $\left(\frac{2w^5x^{-2}y^3}{5w^3y}\right)^4\left(\frac{4w^3x^{-4}y}{5w}\right)^{-3}$ $\frac{w^2x^4y^5}{20}$

In Exercises 57 through 76, estimate the results of the computation. Use the method of Example 10. Write your estimate in scientific notation, rounded to the nearest whole number.

57. $(210,064)(0.003)(0.000001)$ 6.0×10^{-4}

58. $(364,502)(0.02)(0.0002)$ 2.0×10^{0}

59. $(411,000)(0.0008)(0.001)$ 3.0×10^{-1}

60. $(111,110)(0.042)(0.00002)$ 8.0×10^{-2}

61. $\dfrac{21,046,928}{0.0008049}$ 3.0×10^{10}

62. $\dfrac{0.0630482}{3.0843202}$ 2.0×10^{-2}

63. $\dfrac{98.124}{0.02049}$ 5.0×10^{3}

64. $\dfrac{24,163,290}{0.47846200}$ 4.0×10^{7}

65. $\dfrac{(2.604)(0.0002099)}{(0.010064)(0.260894)}$ 2.0×10^{-1}

66. $\dfrac{(10.4659)(0.089926)}{(0.29946)(0.0001)}$ 3.0×10^{4}

67. $\dfrac{(624.064)(1.00996)}{(0.024901)(0.001971)}$ 2.0×10^{7}

68. $\dfrac{(0.0024423)}{(10.42109)(0.010126)}$ 2.0×10^{-2}

69. $(0.004496)(0.000001)(100.964)(0.0002096)$
8.0×10^{-11}

70. $(2.04683)(0.04996)(0.001984)(200.9634)(0.001)$
4.0×10^{-5}

71. $(0.016943)(2.04269)(0.002012)(0.00501046)$
4.0×10^{-7}

72. $(900,421)(0.00199)(0.00206)(0.0000101)$
4.0×10^{-5}

73. $\dfrac{(6.89 \times 10^{-4})(7.38 \times 10^{-5})}{(9.87 \times 10^{-10})}$ 5.0×10^{1}

74. $\dfrac{(5.768 \times 10^{12})(1.897 \times 10^{-6})}{(2.76)(3.999 \times 10^{10})}$ 1.0×10^{-4}

75. $\dfrac{(893.76)(593,000,000)(18.72)}{(1889.3)}$ 5.0×10^{9}

76. $\dfrac{(893,500,000)(0.0000038)(0.00095)}{(18,620,000,000)(13 \times 10^{-12})}$ 2.0×10^{1}

EXCURSIONS

Data Analysis

1. Write a list of data facts that are either given in the following newspaper article or that you can determine by calculation.

> ### Mammoth iceberg may take 10 years to melt completely
>
> SYDNEY, Australia – An iceberg the size of Rhode Island that sheared off the coast of Antarctica could drift for 10 years before it melts, a scientist said Friday.
>
> The ice floe covered more than 1400 square miles when it split from the coast of East Antarctica in May, said Neal Young, an Australian scientist working at the Antarctic Cooperative Research Center.
>
> The berg originally measured about 54 miles by 27 miles. The biggest chunk, which covers about 535 square miles, is grounded off the Eastern coast of Antarctica.
>
> The mammoth iceberg, first observed by a research ship scouting the area, has sheer walls rising 100 feet to 160 feet above the water, and an estimated depth of 1000 feet.
>
> Australia's Antarctic Division is tracking the vast fragments through U.S. weather satellites and European research satellites. The ice chunks were moving with ocean currents at speeds of about 3 miles a day.

Source: Press-Telegram, Long Beach, Calif., 16 November 1996.

Exploring Patterns

2. Evaluate the following expressions on your graphing calculator's Home Screen. Try to find a pattern in the results that will help you write a rule for working with $(-a)^n$ and $-(a^n)$. The work for part (a) is shown at the right.

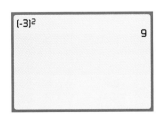

a. $(-3)^2$ **b.** $(-3)^3$ **c.** $(-6)^5$ **d.** $(-6)^6$

e. $(-11)^3$ **f.** $(-11)^8$ **g.** $(-10)^3$ **h.** $(-10)^8$

Now evaluate these expressions. Notice the difference in the position of the parentheses and the negation sign in these exercises and the exercises you just completed.

i. $-(3^3)$ **j.** $-(3^2)$ **k.** $-(6^5)$ **l.** $-(6^6)$ **m.** $-(11^5)$ **n.** $-(11^8)$

✏ Can you now write a rule that a friend can use when he or she is working with $(-a)^n$ and $-(a^n)$?

3. Use your graphing calculator to complete the following lists of values. Give your answers for part (b) in fraction form.

a. $3^1 =$ $3^{11} =$ **b.** $3^{-1} =$ $3^{-11} =$
$3^2 =$ $3^{12} =$ $3^{-2} =$ $3^{-12} =$
$3^3 =$ $3^{13} = 1{,}594{,}323$ $3^{-3} =$ $3^{-13} =$
$3^4 =$ $3^{14} =$ $3^{-4} =$ $3^{-14} =$
$3^5 = 243$ $3^{15} =$ $3^{-5} =$ $3^{-15} =$
$3^6 =$ $3^{16} =$ $3^{-6} =$ $3^{-16} =$
$3^7 =$ $3^{17} =$ $3^{-7} =$ $3^{-17} =$
$3^8 =$ $3^{18} =$ $3^{-8} =$ $3^{-18} =$
$3^9 =$ $3^{19} =$ $3^{-9} =$ $3^{-19} =$
$3^{10} = 59{,}049$ $3^{20} =$ $3^{-10} =$ $3^{-20} =$

c. ✏ Describe the relationship between positive and negative exponents.

Posing Problems

4. Ask and answer four questions using this data.

Year	Super Bowl Champions	Next Season's Results
1967	Green Bay	Repeated, beat Oakland 33–14 in Super Bowl
1968	Green Bay	Placed third in Central Division with a 6–7–1 record
1969	N.Y. Jets	Lost to Kansas City 13–6 in AFL Divisional Playoff
1970	Kansas City	Placed second in Western Division with a 7–5–2 record
1971	Baltimore	Lost to Miami 21–0 in AFC Championship
1972	Dallas	Lost to Washington 26–3 in NFC Championship
1973	Miami	Repeated, beat Minnesota 24–7 in Super Bowl
1974	Miami	Lost to Oakland 28–26 in AFC Divisional Playoff
1975	Pittsburgh	Repeated, beat Dallas 21–17 in Super Bowl
1976	Pittsburgh	Lost to Oakland 24–7 in AFC Championship
1977	Oakland	Lost to Denver 20–17 in AFC Championship
1978	Dallas	Lost to Pittsburgh 35–31 in Super Bowl
1979	Pittsburgh	Repeated, beat Los Angeles Rams 31–19 in Super Bowl
1980	Pittsburgh	Placed third in Central Division with a 9–7 record
1981	Oakland	Placed fourth in Western Division with a 7–9 record
1982	San Francisco	Placed 11th in conference with a 3–6 record
1983	Washington	Lost to Los Angeles Raiders 38–9 in Super Bowl
1984	L.A. Raiders	Lost to Seattle 13–7 in AFC Wild-Card Game
1985	San Francisco	Lost to New York Giants 17–3 in NFC Wild-Card Game

1986	Chicago	Lost to Washington 27–13 in NFC Divisional Playoff
1987	N.Y. Giants	Placed last in NFC Eastern Division with a 6–9 record
1988	Washington	Placed third in NFC Eastern Division with a 7–9 record
1989	San Francisco	Repeated, beat Denver 55–10 in NFC Championship
1990	San Francisco	Lost to New York Giants 15–13 in NFC Championship
1991	N.Y. Giants	Placed fourth in NFC Eastern Division with 8–8 record
1992	Washington	Lost to San Francisco 20–13 in NFC Divisional Playoff
1993	Dallas	Repeated, beat Buffalo 30–13 in Super Bowl
1994	Dallas	Lost to San Francisco 38–28 in NFC Championship
1995	San Francisco	Lost to Green Bay 27–17 in NFC Divisional Playoff
1996	Dallas	Lost to Carolina 26–17 in NFC Divisional Playoff

Source: Associated Press.

CONNECTIONS TO *STATISTICS*

Reading Bar Graphs

Bar Graphs

Information is often presented in the form of a **graph,** a diagram that shows numerical data in visual form. Graphs enable us to "see" relationships that are difficult to describe with numbers alone.

The following graph is called **bar graph.** This one shows the amount of electricity used daily, on the average, by a certain customer in each of 13 successive months.

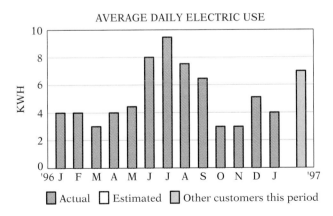

Information that helps you read the graph is given along the bottom and left side of the graph. Along the bottom are the months covered by the graph, abbreviated, from January 1996 to January 1997. There is one bar for each month. The height of the bar gives the average amount of electricity used per day in kilowatt hours (kwh) for its month. The height is read on the scale at the left of the graph.

▪ ▪ ▪

EXAMPLE

a. How many kilowatt hours did this customer use daily during March, to the nearest whole number?

b. During four of the months shown, the customer used the same number of kilowatt hours per day. Which months were these?

SOLUTION

a. The third bar from the left, the bar for March, ends halfway between 2 and 4. Thus, during March this customer used 3 kilowatt hours per day.

b. We must find four bars that are the same height. These occur in January of both years, February, and April. They have a height of 4 kwh.

◀

Double Bar Graphs

Comparative information is often shown on a **double bar graph** such as the following figure. This type of graph usually presents two measurements for each given date or time. The reader then compares the measurements by comparing bars.

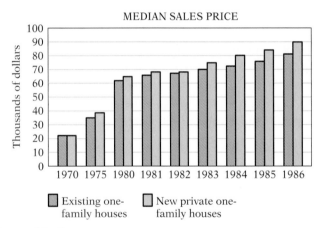

MEDIAN SALES PRICE

■ Existing one-family houses □ New private one-family houses

Source: U.S. Bureau of the Census.

The scale on the left side of this graph shows the median sales prices of one-family houses—existing or new—in thousands of dollars. The scale along the bottom shows the years from 1970 through 1986.

▪ ▪ ▪

EXAMPLE

In what three years were the differences in median sales prices about $10,000?

SOLUTION

The price scale on the left shows that $10,000 is the difference between any two adjacent horizontal lines. The years for which the tops of the two bars are approximately that far apart are 1984, 1985, and 1986.

◀

PRACTICE

Use the Average Daily Electric Use graph to answer Exercises 1 and 2.

1. How many more kilowatt hours were used per day in June than in November?　5 kilowatt hours

2. How many kilowatt hours were used altogether in the months of October, November, and December 1996 and January 1997?　462 kilowatt hours

Use the following figure to answer Exercises 3 and 4.

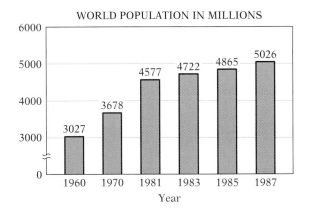

WORLD POPULATION IN MILLIONS

3. Was the increase from 1960 to 1970 more or less than the increase from 1970 to 1981?　less

4. Which two-year time period had the greatest increase?　1985 to 1987

Use the following figure to answer Exercises 5 and 6.

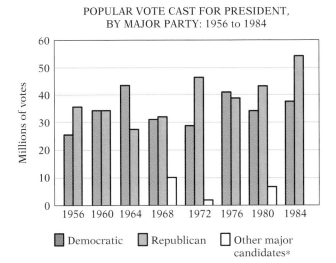

POPULAR VOTE CAST FOR PRESIDENT, BY MAJOR PARTY: 1956 to 1984

*1968 and 1972-American Independent; 1980-John Anderson

Source: U.S. Bureau of the Census.

5. In what year was the difference in votes cast for the Democratic and Republican parties about 2.0 million?   1976

6. What year had the largest total vote?   1984

Use the following figure to answer Exercises 7 and 8.

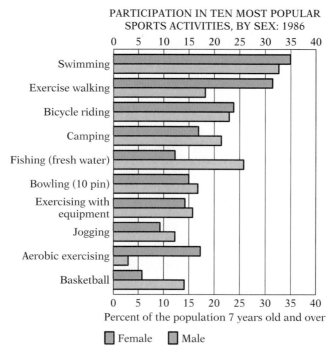

PARTICIPATION IN TEN MOST POPULAR
SPORTS ACTIVITIES, BY SEX: 1986

Percent of the population 7 years old and over

■ Female   ■ Male

Source: U.S. Bureau of the Census.

7. What activity shows the biggest difference in participation between men and women?   aerobic exercising

8. Which three activities are the least popular overall?   aerobic exercising, basketball, and jogging

5.2 Introduction to Polynomials

SECTION LEAD-IN*

The Punnett square has been used to calculate and display genetic information.

There are three genes for blood types A, B, and O. The ways that these three genes can combine can be found by using the Punnett square:

	A	B	O
A	AA	AB	AO
B	BA	BB	BO
O	OA	OB	OO

Possible genotypes

Genes A and B are dominant. When they are paired with O, the dominant genes determine the blood type.

Blood types are A, B, O, or AB in phenotype. (The phenotype is the physical appearance, or, in this case, the chemical appearance.)

What are the blood types in each of the above cells?

*Source: Gregory Fiore, "An Out of Math Experience: Quadratic Equations and Polynomial Multiplication as used in Genetics." *The AMATYC Review,* Volume 17, Number 1 (Fall 1995): pp. 20–27.

SECTION GOALS

- To find the degree of a polynomial
- To add and subtract polynomials
- To find the sum or difference of polynomial functions
- To find the product of a monomial and a polynomial
- To find the quotient of a polynomial and a monomial

In the rest of this chapter, we shall be concerned with sums and differences of monomials, such as $3x^3y - 5x^2 + 8y^2$. These expressions are called *polynomials*.

Polynomial

A **polynomial** is an algebraic expression that is a monomial or the sum or difference of two or more monomials.

A polynomial of one term is a monomial. A polynomial that consists of two terms is called a **binomial**. A polynomial that has three terms is called a **trinomial**. Examples include

$$3y^2 - 8y \quad \text{Binomial}$$
$$5x^3 - 2x^2 + 7x \quad \text{Trinomial}$$
$$16x^2 + 2x - 3xy + 5 \quad \text{Polynomial}$$

A **polynomial in one variable** is a polynomial that includes only one variable. That variable may appear in several terms of the polynomial, and it may have different exponents in different terms.

> **Degree of a Polynomial**
>
> The **degree of a polynomial in one variable** is the greatest exponent on the variable in any term of the polynomial. The degree of a constant term is zero.

So the degree of $5x^3 + 4x - 6$ is 3.

The terms of a polynomial in one variable are usually written with the exponents in descending order from left to right, as in $3x^2 + 9x - 5$.

A **polynomial in two variables** is a polynomial whose terms include two variables. Both variables *need not* appear in all terms. The **degree** of such a polynomial is the greatest sum of the exponents in any term. So the degree of $4x^2y^3 - 8x^2y^5$ is 7.

▪▪▪
EXAMPLE 1

Give the degree of each of these polynomials.

a. $5x^3 + 4x - 6$ **b.** $2x^{15}y^8 - 9x^3y^{12}$

SOLUTION

a. The greatest exponent on the only variable x is 3, so the degree of the polynomial is 3.

b. The greatest sum of the exponents on variables in any term is 23, so the binomial is of degree 23.

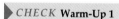 CHECK **Warm-Up 1**

Adding Polynomials

We add polynomials in the same way as we combined like terms in Section 2.1.

> **To add polynomials**
>
> 1. Use the commutative and associative properties to group like terms, moving each term with its sign.
> 2. Use the distributive property to combine like terms.

▪▪▪
EXAMPLE 2

Add: $(7m^2 + 12m - 4) + (-11m^2 - 2m - 5)$

SOLUTION

First we regroup the terms.

$$(7m^2 + 12m - 4) + (-11m^2 - 2m - 5)$$
$$= 7m^2 - 11m^2 + 12m - 2m - 4 - 5$$

Then we combine like terms.

$$7m^2 + (-11)m^2 + 12m - 2m - 4 - 5$$
$$= [7 + (-11)]\,m^2 + (12 - 2)m + [-4 + (-5)]$$
$$= -4m^2 \qquad\quad + 10m \qquad + -9$$

Thus

$$(7m^2 + 12m - 4) + (-11m^2 - 2m - 5) = -4m^2 + 10m - 9$$

▶ CHECK **Warm-Up 2**

The next sum is written vertically. Because the like terms are already lined up, we can simply add the columns.

•••

EXAMPLE 3

Add: $6y^2 + 2y + 7$
       $+\ 3y^2 - 8y - 4$

SOLUTION

Adding the columns gives

$$\begin{array}{r} 6y^2 + 2y + 7 \\ +\ 3y^2 - 8y - 4 \\ \hline 9y^2 - 6y + 3 \end{array}$$

▶ CHECK **Warm-Up 3**

The sum in Example 3 is written in a simplified form. A polynomial is said to be **simplified** when all like terms have been combined.

> **STUDY HINT**
>
> *It's a good idea to count the original terms and the regrouped terms to make sure you haven't lost one!*

Subtracting Polynomials

Recall that we subtract a real number b from a real number a by adding the opposite, or additive inverse, of b to a. That is,

$$a - b = a + (-b)$$

We do the same for polynomials.

To find the opposite of a polynomial, change the sign of every term of the polynomial. Thus, for example, the opposite of $2x^2 - 5x - 3$ is $-2x^2 + 5x + 3$.

> **To subtract polynomial Q from polynomial P**
>
> **1.** Rewrite the problem as the addition of P and the opposite of Q. That is, write $P - Q = P + (-Q)$.
> **2.** Add as usual.

> **! ! !**
> ***ERROR ALERT***
>
> Identify the error and give a correct answer.
>
> Simplify:
>
> $3xy^2 - 6xy +$
> $3 - (xy^2 + 5)$
>
> *Incorrect Solution:*
>
> $3xy^2 - 6xy +$
> $3 - xy^2 + 5$
> $= 3xy^2 - xy^2 - 6xy$
> $\qquad\quad + 3 + 5$
> $= 2xy^2 - 6xy + 8$

▪ ▪ ▪

EXAMPLE 4

Subtract: $(6t^2 + 2t - 1) - (-4t^2 + 7t - 8)$

SOLUTION

INSTRUCTOR NOTE

Students might find both addition and subtraction easier if they rewrite the problems in a vertical format.

We rewrite this as an addition, making sure to change the signs of *all the terms* in the second polynomial.

$$(6t^2 + 2t - 1) - (-4t^2 + 7t - 8)$$
$$= (6t^2 + 2t - 1) + (4t^2 - 7t + 8)$$

We then remove the parentheses and add, obtaining

$$6t^2 + 2t - 1 + 4t^2 - 7t + 8 \qquad \text{Removing parentheses}$$
$$= 6t^2 + 4t^2 + 2t - 7t - 1 + 8 \quad \text{Regrouping}$$
$$= 10t^2 - 5t + 7 \qquad \text{Combining like terms}$$

▶ *CHECK* **Warm-Up 4**

▪ ▪ ▪

EXAMPLE 5

Subtract: $\quad 9x^2 - 5$
$\qquad \underline{-(2x^2 + 3x - 7)}$

SOLUTION

First we align the columns so that each column contains only like terms. Note how we handle the missing x term.

$$9x^2 + 0x - 5$$
$$\underline{- (2x^2 + 3x - 7)}$$

Then we rewrite the problem as an addition, changing the sign of every term in the second polynomial.

$$9x^2 + 0x - 5$$
$$\underline{+ (-2x^2 - 3x + 7)}$$

Finally, we perform the addition, column by column.

$$9x^2 + 0x - 5$$
$$\underline{- 2x^2 - 3x + 7}$$
$$7x^2 - 3x + 2$$

▶ *CHECK* **Warm-Up 5**

▪ ▪ ▪

EXAMPLE 6

Simplify: $(6t^2 + 5t - 4) + (3t^2 - 4t - 1) - (7t^2 - 4)$

Solution

The first two polynomials are to be added, but the third is to be subtracted. Instead, we add its opposite.

$$(6t^2 + 5t - 4) + (3t^2 - 4t - 1) + (-7t^2 + 4)$$

We remove the parentheses, regroup, and combine like terms.

$$6t^2 + 5t - 4 + 3t^2 - 4t - 1 - 7t^2 + 4 \qquad \text{Removing parentheses}$$
$$= 6t^2 + 3t^2 - 7t^2 + 5t - 4t - 4 - 1 + 4 \quad \text{Regrouping}$$
$$= 2t^2 + t - 1 \qquad \text{Combining like terms}$$

▶ *CHECK* **Warm-Up 6**

Introduction to Polynomial Functions

Every polynomial in one variable may be used to define a function. That is, if the function rule is given as a polynomial, as in

$$f(x) = \text{a polynomial in } x$$

then f satisfies the requirement for a function: Its ordered pairs assign exactly one range element to each domain element.

A function of this type is called a **polynomial function**. Its domain is assumed to be the set of real numbers, unless some other domain is stated or implied.

The Algebra of Functions

Polynomial functions can be added, subtracted, multiplied, and divided to form new functions. These new functions are

$$(f + g)(x) = f(x) + g(x)$$
$$(f - g)(x) = f(x) - g(x)$$
$$(fg)(x) = f(x) \cdot g(x)$$
$$\left(\frac{f}{g}\right)(x) = f(x) \div g(x) \quad \text{for } g(x) \text{ not equal to } 0$$

To use these definitions, you also need to know what is meant by equality of functions.

INSTRUCTOR NOTE

Because students have just finished adding, subtracting, multiplying, and dividing polynomials, the transition to work with functions should be smooth. Students continuing in any further courses in mathematics will need exposure to functional notation.

> **Equal Functions**
>
> Two functions f and g are **equal functions** if they specify exactly the same set of ordered pairs.

The functions $f(x) = 3x + 2$ with domain all real numbers and $g(x) = 3x + 2$ with domain all integers are *not* equal. Because of the way their domains are defined, f includes many ordered pairs that are not in g. One such pair is $(1.1, 5.3)$. You should be able to write many more.

Functions can be added or subtracted.

■ ■ ■

EXAMPLE 7

Let functions f and g be given by $f(x) = x^2 + 7$ and $g(x) = x^2 - 2x + 1$. Find $d(0) = f(0) - g(0)$.

SOLUTION

We are told that $d(0)$ is equal to the difference of $f(0)$ and $g(0)$, so we substitute and subtract the polynomials

$$f(0) = 0^2 + 7 = 7$$
$$g(0) = 0^2 - 2(0) + 1 = 1$$

$$d(0) = 7 - 1$$
$$= 6$$

▶ CHECK **Warm-Up 7**

We often use more formal notation when we are dealing with functions.

■ ■ ■

EXAMPLE 8

Let $f(x) = x + 1$ and $g(x) = x$, both with domain all real numbers. Find $(f + g)(x)$ and its domain.

SOLUTION

We substitute in the definition of the "sum" function, obtaining

$$(f + g)(x) = f(x) + g(x)$$
$$= (x + 1) + x$$
$$= 2x + 1$$

The domain of $f + g$ is all real numbers, the same as the domains of f and g.

▶ CHECK **Warm-Up 8**

You already know how to multiply a monomial by a monomial. If you use that procedure and the distributive property, you can find the product of any two polynomials.

Multiplying a Polynomial by a Monomial

Our major task here is to multiply every term of the polynomial by the monomial. In this example, the monomial has a negative coefficient. The minus sign must go everywhere the coefficient goes.

■ ■ ■

EXAMPLE 9

Multiply: $-3r^2(2r^2 + 4r - 8)$

SOLUTION

We first distribute multiplication by $-3r^2$ to each term of the polynomial, taking care not to lose a minus sign.

$$-3r^2(2r^2 + 4r - 8) = (-3r^2)(2r^2) + (-3r^2)(4r) + (-3r^2)(-8)$$

We then perform the monomial multiplications.

$$(-3r^2)(2r^2) + (-3r^2)(4r) + (-3r^2)(-8)$$
$$= -6r^{2+2} + (-12r^{2+1}) + 24r^2$$
$$= -6r^4 - 12r^3 + 24r^2$$

> **STUDY HINT**
>
> *When you apply the distributive property, be very careful to distribute the sign with each term.*

▶ *CHECK* **Warm-Up 9**

Dividing a Polynomial by a Monomial

To divide a monomial by a monomial, we write the division as a fraction and apply the rules of exponents.

> **To divide a polynomial by a monomial**
>
> Divide each term of the polynomial (dividend) by the monomial (divisor) and add the results.

In this example, the divisor is a monomial with a negative coefficient. We deal with the signs as soon as possible. Remember that

$$\frac{-a}{b} = \frac{a}{-b} = \frac{-a}{b}$$

▪ ▪ ▪

EXAMPLE 10

Divide: $\dfrac{16x^2 - 12x + 4}{-4x}$

SOLUTION

The last term on the right in the numerator, 4, has no variable part. To help keep track of the variable, it is a good idea to rewrite this as $4x^0$ (because $4x^0 = 4 \cdot 1 = 4$). Then we have

$$\frac{16x^2 - 12x + 4}{-4x} = \frac{16x^2}{-4x} + \frac{-12x}{-4x} + \frac{4x^0}{-4x}$$

$$= \frac{-16x^2}{4x} + \frac{12x}{4x} + \frac{-4x^0}{4x} \qquad \text{Changing signs to get positive divisors}$$

$$= -4x^{2-1} + 3x^{1-1} + (-1x^{0-1}) \qquad \text{Dividing}$$

$$= -4x + 3x^0 - 1x^{-1} \qquad \text{Simplifying}$$

$$= -4x + 3 - \frac{1}{x}$$

! ! !
ERROR ALERT

Identify the error and give a correct answer.

Divide: $\dfrac{x^3 + 5x + 3}{5x}$

Incorrect Solution:

$$\frac{x^3 + 5x + 3}{5x} = \frac{x^3 + \cancel{5x} + 3}{\cancel{5x}}$$

$$= x^3 + 3$$

Note that the quotient here is not a polynomial because it includes a term with a negative exponent.

▶ CHECK Warm-Up 10

▪▪▪

EXAMPLE 11

Divide: $\dfrac{18r^5t^4u^5 - 27t^5u^7 + 45r^3t^8u^2}{27t^5u^7}$

SOLUTION

Rewriting to show the individual divisions, we get

$$\dfrac{18r^5t^4u^5 - 27t^5u^7 + 45r^3t^8u^2}{27t^5u^7}$$

$$= \dfrac{\overset{2}{18}r^5t^4u^5}{\underset{3}{27}t^5u^7} - \dfrac{\overset{1}{27}t^5u^7}{\underset{1}{27}t^5u^7} + \dfrac{\overset{5}{45}r^3t^8u^2}{\underset{3}{27}t^5u^7}$$

$$= \dfrac{2}{3}r^5t^{4-5}u^{5-7} - 1t^{5-5}u^{7-7} + \dfrac{5}{3}r^3t^{8-5}u^{2-7} \qquad \text{Dividing}$$

$$= \dfrac{2}{3}r^5t^{-1}u^{-2} - 1 + \dfrac{5}{3}r^3t^3u^{-5} \qquad \text{Simplifying}$$

$$= \dfrac{2r^5}{3tu^2} - 1 + \dfrac{5r^3t^3}{3u^5} \qquad \text{Simplifying}$$

▶ CHECK Warm-Up 11

▪▪▪

EXAMPLE 12

Let $f(x) = x + 1$ and $g(x) = x$, both with domain all real numbers. Find $\left(\dfrac{f}{g}\right)(x)$ and its domain.

SOLUTION

$$\left(\dfrac{f}{g}\right)(x) = f(x) \div g(x) = \dfrac{x+1}{x}$$

The domain of $\dfrac{f}{g}$ cannot include any x for which $\left(\dfrac{f}{g}\right)(x)$ is not defined, so the domain cannot include 0. The domain is

$$\{x \,|\, x \text{ is real and } x \text{ is not equal to } 0\}$$

which is *not* the same as the domains of f and g. Note that $\dfrac{f}{g}$ is *not* a polynomial function.

▶ CHECK Warm-Up 12

Practice what you learned.

SECTION FOLLOW-UP

	A	B	O
A	A	AB	A
B	AB	B	B
O	A	B	O

Possible phenotypes ◄——— (What you see is not necessarily what you got!)

a. A mother and father are both type O. What blood type is their son?   Type O

b. A mother is type A; the father is type B. Their daughter is type O. What are the genotypes of the parents?   AO and BO

c. Ask and answer two other questions about these tables.

d. Use the table to multiply $(a^2 + bc - 3)^2$.
    $a^4 + 2a^2bc + b^2c^2 - 6a^2 - 6bc + 9$

5.2 WARM-UPS

Work these problems before you attempt the exercises.

1. What is the degree of this polynomial?

    $2x^2y^2 - 9x^2y + 3xy^2$ degree 4

2. Add: $(7m^2 + 5m + 1) + (5m^2 - 3m + 4)$

    $12m^2 + 2m + 5$

3. Add: $11y - 3y^2$
            $5y - 8$
         $+ 6y^2 - 12$
    _____
         $3y^2 + 16y - 20$

4. Subtract: $(7n^2 + 9n + 4) - (-n^2 + 9n + 1)$
    $8n^2 + 3$

5. Subtract: $8x^2y - 7xy + 5$
              $- (7x^2y + 9y)$
    _____
              $x^2y - 7xy - 9y + 5$

6. Simplify: $(5t^2 - 6t + 2) + (-4t + 5t^2 - 2) - (7t + 3t^2)$
    $7t^2 - 17t$

7. Let $h(y) = g(y) + f(y)$ when $g(y) = 5y^2 - 1$ and $f(y) = 2y + 5$. Find $h(0)$.
    4

8. Let $f(x) = x + 1$ and $g(x) = x^2 - 1$. Find $(f - g)(x)$ and its domain. $(f - g)(x) = -x^2 + x + 2$; domain: {all real numbers}

9. Multiply: $2x^2(x^2 - x + 3)$

$2x^4 - 2x^3 + 6x^2$

10. Divide: $\dfrac{-28x^4 + 12x^3 - 4}{-6x^2}$

$\dfrac{14x^2}{3} - 2x + \dfrac{2}{3x^2}$

11. Divide: $\dfrac{-15n^3x^3 + 3n^2x}{5n^2x}$

$-3nx^2 + \dfrac{3}{5}$

12. Let $f(x) = 3x - 2$ and $g(x) = x^2$. Compare $(f \div g)(-3)$ and $f(-3) \div g(-3)$.

They are equal: $\left(\dfrac{-11}{9}\right)$.

5.2 EXERCISES

Note: Use your graphing calculator to check your results whenever possible.

In Exercises 1 through 4, find the degree of each polynomial.

1. $3t^2 + 5t - 4t^3$
degree 3

2. $18y^4 + 2y^2 - 9xy^2$
degree 4

3. $x - 4x^2y + 9x^3 - y$
degree 3

4. $x^3y^2 + xy^3 + x^2 + 6$
degree 5

In Exercises 5 through 14, add the polynomials.

5. $\begin{aligned} 6x^2y - 4y + 6 \\ + \ 7x^2y - 3x - 6 \end{aligned}$ $\quad 13x^2y - 3x - 4y$

6. $\begin{aligned} 7x + y + 15 \\ + \ 14x^2 - 6y \end{aligned}$ $\quad 14x^2 + 7x - 5y + 15$

7. $(6t - 7) + (5t + 34)$
$11t + 27$

8. $(-2y - 8) + (13y^2 - 6y + 17)$
$13y^2 - 8y + 9$

9. $(3r^2 + 7r) + (2 - 4r^2 - 9r + 6)$
$-r^2 - 2r + 8$

10. $(-8x^2y + 14xy - 9) + (5x^2y - 10x + xy - 9x)$
$-3x^2y + 15xy - 19x - 9$

11. $(5y^3 + 6y) + (1 - 17y^3 - 18y^2 - 19y)$
$-12y^3 - 18y^2 - 13y + 1$

12. $(5x^2 - 4x + 5) + (8x^2 - 2x - 7) + (4x - 12x^2)$
$x^2 - 2x - 2$

13. $(-w^3 - 21w^2 + 13w + 2) + (4w^3 + 3w^2 + 5w - 52)$
$3w^3 - 18w^2 + 18w - 50$

14. $(4y^2 + 3y + 8) + (9y^2 - 7y + 9) + (6y^2 + y - 5)$
$19y^2 - 3y + 12$

In Exercises 15 through 18, find the additive inverse.

15. $-2t^2 - 7t + 4$
$2t^2 + 7t - 4$

16. $2x + 6x^2 - 9xy$
$-2x - 6x^2 + 9xy$

17. $6t + 8t^3 - 5 - 9t^2r$
$-6t - 8t^3 + 5 + 9t^2r$

18. $8r - 3r^2 + 7r^2t$
$-8r + 3r^2 - 7r^2t$

In Exercises 19 through 30, subtract the polynomials.

19. $(7y^2 + 6) - (-5y^2 - 6)$
$12y^2 + 12$

20. $(3y^2 - 11) - (-6y^2 + 5)$
$9y^2 - 16$

21. $(-4w^2 + w) - (-7w^2 + w)$
$3w^2$

22. $(5y^2 - y) - (-5y^2 + 9)$
$10y^2 - y - 9$

23. $(-4y^2 - 2y - 32) - (5y^2 + 13y - 19)$
$-9y^2 - 15y - 13$

24. $(-5y^2 + 36y - 12) - (4y^2 - 4y - 18)$
$-9y^2 + 40y + 6$

25. $(6w^2 - 17w + 30) - (5w^2 + 6w - 4)$
$w^2 - 23w + 34$

26. $(73y^2 + 18y + 25) - (32y^2 - y + 5)$
$41y^2 + 19y + 20$

27. $(26y^2 + 44y) - (83y - 8)$
$26y^2 - 39y + 8$

28. $(-71y^2 - 43y) - (22y - 6)$
$-71y^2 - 65y + 6$

29. $(-35w^2 + 26) - (7w^2 + 12w)$
$-42w^2 - 12w + 26$

30. $(35y^2 - 5) - (19y^2 - 18y)$
$16y^2 + 18y - 5$

In Exercises 31 through 34, simplify these expressions.

31. $(3x^2 + 4x - 8) + (7x^2 - 8x + 5) - (14x^2 + 16x - 5)$ $-4x^2 - 20x + 2$

32. $(-7t^2 + 2t - 6) - (23t^2 + 6t - 19) + (18t^2 + 2t - 11)$ $-12t^2 - 2t + 2$

33. $(13w^3 - 8w^2 - 16w - 11) - (12w^2 + 4w + 7) - (13w^2 - 9w - 17)$ $13w^3 - 33w^2 - 11w - 1$

34. $(12y^2 + 22y - 21) - (15y^2 - 2y - 6) - (20y^2 - 5y - 8)$ $-23y^2 + 29y - 7$

In Exercises 35 through 40, find $h(x) = f(x) + g(x)$. Then evaluate $h(0), f(0)$, and $g(0)$, and verify that $h(0) = f(0) + g(0)$ for the given functions f and g.   Students should verify $h(0)=f(0)+g(0)$.

35. $f(x) = 3x^2 + 1$
$g(x) = 2x^2 - 1$
$h(x) = 5x^2; h(0) = 0; f(0) = 1; g(0) = -1$

36. $f(x) = 3x^2 + 4x$
$g(x) = 4x^2 - 10x$
$h(x) = 7x^2 - 6x; h(0) = 0; f(0) = 0; g(0) = 0$

37. $f(x) = 5x^2 + 2x$
$g(x) = x + 3$
$h(x) = 5x^2 + 3x + 3; h(0) = 3; f(0) = 0;$
$g(0) = 3$

38. $f(x) = 3x - 2x^2$
$g(x) = 5x^2 - 1$
$= 3x^2 + 3x - 1; h(0) = -1; f(0) = 0; g(0) = -1$

39. $f(x) = 2x^2 - 2x$
$g(x) = 1 - 3x^2$
$h(x) = -x^2 - 2x + 1; h(0) = 1; f(0) = 0; g(0) = 1$

40. $f(x) = 4x^2 - 1$
$g(x) = x^3$
$h(x) = x^3 + 4x^2 - 1; h(0) = -1;$
$f(0) = -1; g(0) = 0$

In Exercises 41 through 46, find $h(x) = f(x) - g(x)$. Then evaluate $h(-1), f(-1)$, and $g(-1)$, and verify that $h(-1) = f(-1) - g(-1)$ for the given functions f and g.   Students should verify $h(-1) = f(-1) - g(-1)$.

41. $f(x) = x + 2$
$g(x) = x - 2$
$h(x) = 4; h(-1) = 4; f(-1) = 1;$
$g(-1) = -3$

42. $f(x) = 3x + 1$
$g(x) = 2x$
$h(x) = x + 1; h(-1) = 0;$
$f(-1) = -2; g(-1) = -2$

43. $f(x) = 5x^2 - 6$
$g(x) = 6 - 5x^2$
$h(x) = 10x^2 - 12; h(-1) = -2;$
$f(-1) = -1; g(-1) = 1$

44. $f(x) = 2x^2$
$g(x) = 3x^2 - 5$
$h(x) = -x^2 + 5; h(-1) = 4;$
$f(-1) = 2; g(-1) = -2$

45. $f(x) = 1$
$g(x) = 3x^2 + 2$
$h(x) = -3x^2 - 1; h(-1) = -4;$
$f(-1) = 1; g(-1) = 5$

46. $f(x) = 5 + x$
$g(x) = 2x - 7x^2$
$h(x) = 7x^2 - x + 5; h(-1) = 13;$
$f(-1) = 4; g(-1) = -9$

In Exercises 47 through 52, find the requested function and give its domain.

47. $f(x) = x + 2$ $g(x) = x^2 + 2$
Find $(f + g)(x)$. $(f + g)(x) = x^2 + x + 4$
domain: all real numbers

48. $f(x) = x - 2$ $g(x) = x^3 - 4$
Find $(f + g)(x)$. $(f + g)(x) = x^3 + x - 6$
domain: all real numbers

49. $f(x) = 2x + 3$ $g(x) = -3x - 2$
Find $(f - g)(x)$. $(f - g)(x) = 5x + 5$
domain: all real numbers

50. Let $f(x) = 3x^2 - 3x + 2$ and $g(x) = 3x^2 + 3x - 5$.
Find $(f + g)(x)$. $(f + g)(x) = 6x^2 - 3$
domain: all real numbers

51. $f(x) = 2x^2 - 3$ $g(x) = -2x^2 + 2$
Find $(f + g)(x)$. $(f + g)(x) = -1$
domain: all real numbers

52. $f(x) = 3x^3 - 2x^2$ $g(x) = 2x^2 - 4x^3$
Find $(f + g)(x)$. $(f + g)(x) = -x^3$
domain: all real numbers

In Exercises 53 through 86, multiply or divide by using the distributive property.

53. $7x(x + 4)$ $7x^2 + 28x$

54. $5y(y - 11)$ $5y^2 - 55y$

55. $(x + 9)8x$ $8x^2 + 72x$

56. $(y - 8)6y$ $6y^2 - 48y$

57. $\dfrac{32k^{11} + 18k^{13}}{18k^{13}}$ $\dfrac{16}{9k^2} + 1$

58. $\dfrac{15t^3n - 18t^2}{3t^2n}$ $5t - \dfrac{6}{n}$

59. $-7x(-3x^2 - 5x + 8)$
$21x^3 + 35x^2 - 56x$

60. $-11y(-2y^2 - y + 5)$
$22y^3 + 11y^2 - 55y$

61. $\dfrac{8t^2n^2 + 12tn^2}{8t^2n^2}$ $1 + \dfrac{3}{2t}$

62. $\dfrac{7x^3y^3 + 14xy^3}{7x^3y^3}$ $1 + \dfrac{2}{x^2}$

63. $-9x(-5x^2 - 10x)$ $45x^3 + 90x^2$

64. $-12y(-7y^2 - 9y)$
$84y^3 + 108y^2$

65. $\dfrac{36x^3 + 8x^8}{8x^3}$ $\dfrac{9}{2} + x^5$

66. $\dfrac{32t^5 + 12t^7}{12t^5}$ $\dfrac{8}{3} + t^2$

67. $\dfrac{120n^4 - 45 + 20n^2}{120n^2}$ $n^2 - \dfrac{3}{8n^2} + \dfrac{1}{6}$

68. $\dfrac{36x^6 - 24x^3 - 6x^2}{6x^3}$ $6x^3 - 4 - \dfrac{1}{x}$

69. $\dfrac{30x^4y^2 + 5xy - 45x^3}{45x^3y}$ $\dfrac{2xy}{3} + \dfrac{1}{9x^2} - \dfrac{1}{y}$

70. $\dfrac{80n^4t^4 - 80nt + 10n^2t}{40n^3t^2}$ $2nt^2 - \dfrac{2}{n^2t} + \dfrac{1}{4nt}$

71. $\dfrac{42tx^2 + 12t^2x - 21t^2x^2}{6t^2x^2}$ $\dfrac{7}{t} + \dfrac{2}{x} - \dfrac{7}{2}$

72. $\dfrac{32x^3y^2 - 12x^2y + 84xy^2}{6x^2y^2}$ $\dfrac{16x}{3} - \dfrac{2}{y} + \dfrac{14}{x}$

73. $\dfrac{48ru^2 + 8r^3u - 28r^2u^2}{12r^3u^5}$ $\dfrac{4}{r^2u^3} + \dfrac{2}{3u^4} - \dfrac{7}{3ru^3}$

74. $\dfrac{28x^2y^{12} + 63x^3y - 280xy^{12}}{28x^3y^{12}}$ $\dfrac{1}{x} + \dfrac{9}{4y^{11}} - \dfrac{10}{x^2}$

75. $-11y(-6y^2 - 8y)$ $66y^3 + 88y^2$ **76.** $-10t(-8t^2 - 14t)$ $80t^3 + 140t^2$ **77.** $\dfrac{18x^{12}y^{12} + 2xy^7}{18x^{12}y^{12}}$ $1 + \dfrac{1}{9x^{11}y^5}$

78. $\dfrac{12r^5u^2 + 2ru^2}{12r^5u^2}$ $1 + \dfrac{1}{6r^4}$

79. $-6y^2 + 3y + 2$
$$\underline{\times \hspace{4em} 8y}$$
$$-48y^3 + 24y^2 + 16y$$

80. $-6y^2 + 5y - 8$
$$\underline{\times \hspace{4em} 5y}$$
$$-30y^3 + 25y^2 - 40y$$

81. $\dfrac{120n^4 - 15 + 20n}{20n^2}$ $6n^2 - \dfrac{3}{4n^2} + \dfrac{1}{n}$

82. $\dfrac{30x^6 - 4x - 6x^2}{6x^3}$ $5x^3 - \dfrac{2}{3x^2} - \dfrac{1}{x}$

83. $\dfrac{30x^4y^2 + 5xy - 15x^3}{15x^3y}$ $2xy + \dfrac{1}{3x^2} - \dfrac{1}{y}$

84. $\dfrac{80n^4t^4 - 30nt + 10n^2t}{10n^3t^2}$

$8nt^2 - \dfrac{3}{n^2t} + \dfrac{1}{nt}$

85. $3x^2 + 4x - 3$
$$\underline{\times \hspace{4em} -2x}$$
$$-6x^3 - 8x^2 + 6x$$

86. $-7t^2 + 9t - 12$
$$\underline{\times \hspace{4em} -7t}$$
$$49t^3 - 63t^2 + 84t$$

In Exercises 87 through 92, evaluate the functions at the given point and give the domain.

87. $f(x) = 6x^2$ $g(x) = 3x^3$
    Find $\left(\dfrac{f}{g}\right)(x)$. $\left(\dfrac{f}{g}\right)(x) = \dfrac{2}{x}$
    domain: x is a real number and $x \neq 0$

88. $f(x) = 12x$ $g(x) = 2x^2$
    Find $\left(\dfrac{f}{g}\right)(x)$. $\left(\dfrac{f}{g}\right)(x) = \dfrac{6}{x}$
    domain: x is a real number and $x \neq 0$

89. $f(x) = 12x^2 - 6$ $g(x) = 2x$
    Find $(fg)(x)$. $(fg)(x) = 24x^3 - 12x$
    domain: x is a real number

90. $f(x) = 6x^2 - x$ $g(x) = 6x$
    Find $\left(\dfrac{f}{g}\right)(x)$. $\left(\dfrac{f}{g}\right)(x) = x - \dfrac{1}{6}$
    domain: x is a real number, $x \neq 0$

91. $f(x) = 3x - 6x^2$ $g(x) = 3x$
    Find $(fg)(x)$. $(fg)(x) = 9x^2 - 18x^3$
    domain: x is a real number

92. $f(x) = 4x^2 - 6x^3$ $g(x) = 2x^2$
    Find $\left(\dfrac{f}{g}\right)(x)$. $\left(\dfrac{f}{g}\right)(x) = 2 - 3x$
    domain: x is a real number, $x \neq 0$

MIXED PRACTICE

By doing these exercises, you will practice the topics up to this point in the chapter.

93. Multiply using scientific notation: $(210{,}000)(3000)(0.0000007)$ 441

94. Subtract: $(-7r^2 + 3r - 12) - (8r^2 + 10r + 3)$ $-15r^2 - 7r - 15$

95. Divide: $\dfrac{-8x^5y^5}{24x^3y}$ $-\dfrac{x^2y^4}{3}$

96. Write this number using scientific notation.

$$9^8 = 43{,}046{,}721$$

4.3046721×10^7

97. Multiply: $(5x^2y^3)(9xy^{-3})$ $45x^3$

98. Simplify: $(x^2 + 3x - 11) + (7x^2 - 13x + 9)$ $8x^2 - 10x - 2$

99. Divide using scientific notation: $(0.0000000144) \div (0.0000012)$ 0.012

100. What is the opposite of $x^3 - 3x^2 - 5$? $-x^3 + 3x^2 + 5$

EXCURSIONS

Class Act

1. a. The following graph shows the percent of businesses reporting that information technology has affected each area (customer service, productivity, etc.). The data was analyzed in 1994. How do you believe this graph might have changed in 1996?

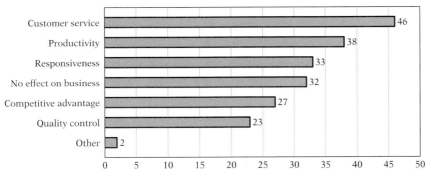

IMPACT OF INFORMATION TECHNOLOGY
ON SMALL AND MID-SIZED BUSINESSES (%)

Source: National Small Business United and Arthur Andersen Enterprises, June 1994, as cited in the *1996 Information Please® Business Almanac* (©1995 Houghton Mifflin Co.), p. 593. All rights reserved. Used with permission by Information Please LLC.

b. Ask and answer four questions using information from this graph.

c. Why is a bar graph a better representation of this data than a line graph?

d. Write a paragraph to a friend giving information from the graph.

Data Analysis

2. a. Make a bar graph of the U.S. Presidents' birth dates by day of the month from the table on the following page. Ask and answer four questions about this data.

b. *Research* How many presidents were born on Monday? Tuesday? etc.?

Presidents

Name and (party)	Term	State of birth	Born
1. Washington (F)	1789–1797	Va.	2/22/1732
2. J. Adams (F)	1797–1801	Mass.	10/30/1735
3. Jefferson (DR)	1801–1809	Va.	4/13/1743
4. Madison (DR)	1809–1817	Va.	3/16/1751
5. Monroe (DR)	1817–1825	Va.	4/28/1758
6. J. Q. Adams (DR)	1825–1829	Mass.	7/11/1767
7. Jackson (D)	1829–1837	S.C.	3/15/1767
8. Van Buren (D)	1837–1841	N.Y.	12/5/1782
9. W. H. Harrison (W)	1841	Va.	2/9/1773
10. Tyler (W)	1841–1845	Va.	3/29/1790
11. Polk (D)	1845–1849	N.C.	11/2/1795
12. Taylor (W)	1849–1850	Va.	11/24/1784
13. Fillmore (W)	1850–1853	N.Y.	1/7/1800
14. Pierce (D)	1853–1857	N.H.	11/23/1804
15. Buchanan (D)	1857–1861	Pa.	4/23/1791
16. Lincoln (R)	1861–1865	Ky.	2/12/1809
17. A. Johnson (U)	1865–1869	N.C.	12/29/1808
18. Grant (R)	1869–1877	Ohio	4/27/1822
19. Hayes (R)	1877–1881	Ohio	10/4/1822
20. Garfield (R)	1881	Ohio	11/19/1831
21. Arthur (R)	1881–1885	Vt.	10/5/1830
22. Cleveland (D)	1885–1889	N.J.	3/18/1837
23. B. Harrison (R)	1889–1893	Ohio	8/20/1833
24. Cleveland (D)	1893–1897	—	—
25. McKinley (R)	1897–1901	Ohio	1/29/1843
26. T. Roosevelt (R)	1901–1909	N.Y.	10/27/1858
27. Taft (R)	1909–1913	Ohio	9/15/1857
28. Wilson (D)	1913–1921	Va.	12/28/1856
29. Harding (R)	1921–1923	Ohio	11/2/1865
30. Coolidge (R)	1923–1929	Vt.	7/4/1872
31. Hoover (R)	1929–1933	Iowa	8/10/1874
32. F. D. Roosevelt (D)	1933–1945	N.Y.	1/30/1882
33. Truman (D)	1945–1953	Mo.	5/8/1884
34. Eisenhower (R)	1953–1961	Tex.	10/14/1890
35. Kennedy (D)	1961–1963	Mass.	5/29/1917
36. L. B. Johnson (D)	1963–1969	Tex.	8/27/1908
37. Nixon (R)	1969–1974	Calif.	1/9/1913
38. Ford (R)	1974–1977	Neb.	7/14/1913
39. Carter (D)	1977–1981	Ga.	10/1/1924
40. Reagan (R)	1981–1989	Ill.	2/6/1911
41. Bush (R)	1989–1993	Mass.	6/12/1924
42. Clinton (D)	1993–	Ark.	8/19/1946

1. The following party abbreviations are used: F = Federalist, DR = Democratic-Republican, D = Democratic, R = Republican, W = Whig, and U = Union. *Source: 1996 Information Please® Almanac* (©1995 Houghton Mifflin Co.), p. 633. All rights reserved. Used with permission by Information Please LLC.

3. Study the following table then answer the questions.

   a. Use a graph to help you analyze this data.

   b. How does your state compare with others in relation to this data?

State Welfare Statistics (1996)

State	Population below poverty line[1]	Maximum monthly welfare grant per family of 3	Maximum allowable monthly income		% reduction in max. welfare benefits from 1970 to 1996 (adjusted for inflation)
			In dollars	As % of poverty line[1]	
Alabama	16.4%	$164	$366	34%	−36%
Alaska	10.2	923	1,662	123	−29
Arizona	15.9	347	641	59	−36
Arkansas	15.3	204	426	39	−42
California	17.9	607	1,215	112	−18
Colorado	9.0	421	752	69	−45
Connecticut	10.8	636	1,428	132	−43
Delaware	8.3	338	627	58	−47
Florida	14.9	303	575	53	−33
Georgia	14.0	280	756	70	−34
Hawaii	8.7	712	1,188	95	−20
Idaho	12.0	317	596	55	−62
Illinois	12.4	377	686	63	−59
Indiana	13.7	288	552	51	−39
Iowa	10.7	426	759	70	−46
Kansas	14.9	429	764	71	−51
Kentucky	18.5	262	909	84	−55
Louisiana	25.7	190	405	37	−45
Maine	9.4	418	950	88	−22
Maryland	10.7	373	680	63	−42
Massachusetts	9.7	565	968	89	−47
Michigan	14.1	459[3]	809[3]	75[3]	−47[3]
Minnesota	11.7	532	918	85	−48
Mississippi	19.9	120	672	62	−46
Missouri	15.6	292	558	52	−29
Montana	11.5	425	932	86	−47
Nebraska	8.8	364	666	62	−46
Nevada	11.1	348	642	59	−27
New Hampshire	7.7	550	945	87	−47
New Jersey	9.2	424	785	73	−65
New Mexico	21.1	389	704	65	−34
New York	17.0	577[2]	986[2]	91[2]	−48[2]
North Carolina	14.2	272	936	87	−53
North Dakota	10.4	431	767	71	−49
Ohio	14.1	341	632	58	−47
Oklahoma	16.7	307	581	54	−49
Oregon	11.8	460	810	75	−37
Pennsylvania	12.5	421	752	69	−60
Rhode Island	10.3	554	951	88	−39
South Carolina	13.8	200	420	39	−41
South Dakota	14.5	430	881	81	−59
Tennessee	14.6	185	995	92	−58
Texas	19.1	188	402	37	−68
Utah	8.0	426	972	90	−39
Vermont	7.6	650	1,095	101	−38
Virginia	10.7	240	480	44	−60
Washington	11.7	546	939	87	−47
Washington, D.C.	21.2	420	750	69	−46
West Virginia	18.6	253	500	46	−44
Wisconsin	9.0	517	896	83	−29
Wyoming	9.3	360	1,005	93	−29

1. The Federal poverty line is $1082 per month for a family of three, except in Alaska and Hawaii. 2. Figures are for New York City only. 3. Figures are for Wayne County, which includes Detroit. *Source:* House Ways and Means Committee, Commerce Dept.

Posing Problems

4. Use this additional information with the State Welfare Statistics table in question 3. Ask and answer four additional questions.

Population by State

State	1990	Percent change, 1980–90	Population per sq. mi., 1990	Population rank, 1990	1980	1950	1900	1790
Alabama	4,040,587	+3.8	79.6	22	3,893,888	3,061,743	1,828,697	—
Alaska	550,403	+36.9	1.0	49	401,851	128,643	63,592	—
Arizona	3,665,228	+34.8	32.3	24	2,718,215	749,587	122,931	—
Arkansas	2,350,725	+2.8	45.1	33	2,286,435	1,909,511	1,311,564	—
California	29,760,021	+25.7	190.4	1	23,667,902	10,586,223	1,485,053	—
Colorado	3,294,394	+14.0	31.8	26	2,889,964	1,325,089	539,700	—
Connecticut	3,287,116	+5.8	674.7	27	3,107,576	2,007,280	908,420	237,946
Delaware	666,168	+12.1	344.8	46	594,338	318,085	184,735	59,096
Washington, D.C.	606,900	−4.9	—	—	638,333	802,178	278,718	—
Florida	12,937,926	+32.7	238.9	4	9,746,324	2,771,305	528,542	—
Georgia	6,478,216	+18.6	109.9	11	5,463,105	3,444,578	2,216,331	82,548
Hawaii	1,108,229	+14.9	172.5	41	964,691	499,794	154,001	—
Idaho	1,006,749	+6.7	12.2	42	943,935	588,637	161,772	—
Illinois	11,430,602	0.0	205.4	6	11,426,518	8,712,176	4,821,550	—
Indiana	5,544,159	+1.0	154.2	14	5,490,224	3,934,224	2,516,462	—
Iowa	2,776,755	−4.7	49.6	30	2,913,808	2,621,073	2,231,853	—
Kansas	2,477,574	+4.8	30.3	32	2,363,679	1,905,299	1,470,495	—
Kentucky	3,685,296	+0.7	92.9	23	3,660,777	2,944,806	2,147,174	73,677
Louisiana	4,219,973	+0.3	94.8	21	4,205,900	2,683,516	1,381,625	—
Maine	1,227,928	+9.2	39.6	38	1,124,660	913,774	694,466	96,540
Maryland	4,781,468	+13.4	486.0	19	4,216,975	2,343,001	1,188,044	319,728
Massachusetts	6,016,425	+4.9	768.9	13	5,737,037	4,690,514	2,805,346	378,787
Michigan	9,295,297	+0.4	163.2	8	9,262,078	6,371,766	2,420,982	—
Minnesota	4,375,099	+7.3	55.0	20	4,075,970	2,982,483	1,751,394	—
Mississippi	2,573,216	+2.1	54.5	31	2,520,638	2,178,914	1,551,270	—
Missouri	5,117,073	+4.1	74.2	15	4,916,686	3,954,653	3,106,665	—
Montana	799,065	+1.6	5.5	44	786,690	591,024	243,329	—
Nebraska	1,578,385	+0.5	20.6	36	1,569,825	1,325,510	1,066,300	—
Nevada	1,201,833	+50.1	10.9	39	800,493	160,083	42,335	—
New Hampshire	1,109,252	+20.5	123.3	40	920,610	533,242	411,588	141,885
New Jersey	7,730,188	+5.0	1,035.1	9	7,364,823	4,835,329	1,883,669	184,139
New Mexico	1,515,069	+16.3	12.5	37	1,302,894	681,187	195,310	—
New York	17,990,455	+2.5	379.7	2	17,558,072	14,830,192	7,268,894	340,120
North Carolina	6,628,637	+12.7	135.7	10	5,881,766	4,061,929	1,893,810	393,751
North Dakota	638,800	−2.1	9.0	47	652,717	619,636	319,146	—
Ohio	10,847,115	+0.5	264.5	7	10,797,630	7,946,627	4,157,545	—
Oklahoma	3,145,585	+4.0	45.8	28	3,025,290	2,233,351	790,391[1]	—
Oregon	2,842,321	+7.9	29.5	29	2,633,105	1,521,341	413,536	—
Pennsylvania	11,881,643	+0.1	264.7	5	11,863,895	10,498,012	6,302,115	434,373
Rhode Island	1,003,464	+5.9	951.1	43	947,154	791,896	428,556	68,825
South Carolina	3,486,703	+11.7	115.4	25	3,121,820	2,117,027	1,340,316	249,073
South Dakota	696,004	+0.8	9.1	45	690,768	652,740	401,570	—
Tennessee	4,877,185	+6.2	118.5	17	4,591,120	3,291,718	2,020,616	35,691
Texas	16,986,510	+19.4	64.8	3	14,229,191	7,711,194	3,048,710	—
Utah	1,722,850	+17.9	20.9	35	1,461,037	688,862	276,749	—
Vermont	562,758	+10.0	60.7	48	511,456	377,747	343,641	85,425
Virginia	6,187,358	+15.7	155.8	12	5,346,818	3,318,680	1,854,184	747,610[2]
Washington	4,866,692	+17.8	73.1	18	4,132,156	2,378,963	518,103	—
West Virginia	1,793,477	−8.0	73.8	34	1,949,644	2,005,552	958,800	—
Wisconsin	4,891,769	+4.0	89.9	16	4,705,767	3,434,575	2,069,042	—
Wyoming	453,588	−3.4	4.7	50	469,557	290,529	92,531	—
Total U.S.	248,709,873	+9.8	—	—	226,545,805	151,325,798	76,212,168	3,929,214

5.3 Multiplying and Dividing Polynomials

SECTION LEAD-IN

We can use the Punnett square to find $(p + q)^6$. We simply have to use it several times, adapting it to accomplish the last step of $(p + q)^6$.

$(p + q)^2$:

	p	q
p	p^2	pq
q	pq	q^2

$= p^2 + 2pq + q^2$

$(p + q)^4$:

	p^2	$2pq$	q^2
p^2	p^4	$2p^3q$	p^2q^2
$2pq$	$2p^3q$	$4p^2q^2$	$2pq^2$
q^2	p^2q^2	$2pq^3$	q^4

$= p^4 + 4p^3q + 6p^2q^2 + 4pq^3 + q^4$

$(p + q)^6$:

	p^4	$4p^3q$	$6p^2q^2$	$4pq^3$	q^4
p^2	p^6	$4p^5q$	$6p^4q^2$	$4p^3q^3$	p^2q^4
$2pq$	$2p^5q$	$8p^4q^2$	$12p^3q^3$	$8p^2q^4$	$2pq^5$
q^2	p^4q^2	$4p^3q^3$	$6p^2q^4$	$4pq^5$	q^6

$= p^6 + 6p^5q + 15p^4q^2 + 20p^3q^3 + 15p^2q^4 + 6pq^5 + q^6$

- Complete these Punnett squares. In a family of exactly 6 children, what is the probability of having 3 boys and 3 girls?

Multiplying Polynomials

The distributive property is very important here. When we multiply two polynomials, every term in the first polynomial must multiply every term in the second. To do that, we apply the distributive property carefully.

•••
EXAMPLE 1

Multiply: $(x + 4)(x^2 + 5x - 9)$

SOLUTION

We use the distributive property first to multiply the second polynomial by each term of the first polynomial.

$$(x + 4)(x^2 + 5x - 9) = x(x^2 + 5x - 9) + 4(x^2 + 5x - 9)$$

Then we use the distributive property again to perform the indicated monomial-polynomial multiplications.

$$x(x^2 + 5x - 9) + 4(x^2 + 5x - 9)$$
$$= x(x^2) + x(5x) + x(-9) + 4(x^2) + 4(5x) + 4(-9)$$

Now, doing the multiplications, we get

$$x^3 + 5x^2 + (-9x) + 4x^2 + 20x + (-36) \qquad \text{Multiplying}$$
$$= x^3 + 5x^2 + 4x^2 + (-9x) + 20x + (-36) \qquad \text{Regrouping}$$
$$= x^3 + 9x^2 + 11x - 36 \qquad \text{Simplifying}$$

Multiplication is commutative, so in this example we could have begun by multiplying the first polynomial by each term of the second.

▶ *CHECK* **Warm-Up 1**

Whichever you decide to do, do it consistently so that it becomes habit. Then there will be less chance of your forgetting a term.

Calculator Corner

You can check your multiplication of polynomials by graphing *each side* of the equation. If your work is correct, then you should get only one graph. If your answer is incorrect, you would have two graphs. For instance, multiply and check $(-7x + 3)(5x - 4)$.

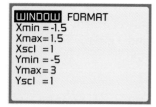
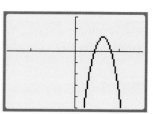

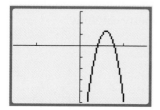

If, however, your multiplication is incorrect, then you will have two graphs, meaning that the two sides of the equation *are not* equal, and, therefore, your equation is not true.

In the next example, we multiply polynomials vertically. This is essentially just like multiplying numbers vertically.

▪▪▪
EXAMPLE 2

Multiply: $11n^2 - 3n + 9$
$\underline{\times \qquad\quad 7n - 2}$

SOLUTION

Just as when we multiply numbers, we multiply the entire polynomial on the top line by the rightmost term below it (-2), then by the next term to the left ($7n$), and so on if there are more terms. We align the resulting partial products by degree, and then add column by column.

$$
\begin{array}{r}
11n^2 - 3n + 9 \\
\underline{\times \qquad\quad 7n - 2} \\
\end{array}
$$

$\phantom{77n^3 -{}}-22n^2 + 6n - 18$	Multiplying by -2
$\underline{77n^3 - 21n^2 + 63n}$	Multiplying by $7n$
$77n^3 - 43n^2 + 69n - 18$	Product

▶ *CHECK* **Warm-Up 2**

Multiplying Binomials: FOIL

In its most general form, the multiplication of two binomials may be written as

$$(a + b)(c + d)$$

where a, b, c, and d are terms of the binomials and may include variables.

If we label the first and last terms of the binomials, and the inner and outer terms of the multiplication, we get the pattern shown in the following equation.

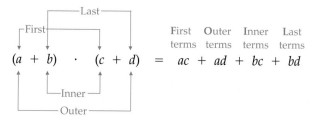

The completed multiplication then looks like this:

$$(a + b)(c + d) = a(c + d) + b(c + d)$$
$$= ac + ad + bc + bd$$

First Outer Inner Last
terms terms terms terms

The product of the two binomials is the sum of the products of the **First**, **Outer**, **Inner**, and **Last** terms. The initials FOIL give this method of multiplying binomials its name.

▪▪▪
EXAMPLE 3

Multiply: $(3x + 4y)(2x - 5y)$

SOLUTION

Using the FOIL method, we can write

$$(3x + 4y)(2x - 5y) = (3x)(2x) + (3x)(-5y) + (4y)(2x) + (4y)(-5y)$$

First   Outer   Inner   Last

Multiplying and then combining like terms give us

$$6x^2 + (-15xy) + 8xy + (-20y^2) = 6x^2 - 7xy - 20y^2$$

▶ CHECK **Warm-Up 3**

The Product of a Sum and Difference

The product of the sum and difference of two terms is a **special product**—one that is easy to find. With practice, you can write it without actually doing the multiplication.

▪▪▪
EXAMPLE 4

Multiply: $(n + 8)(n - 8)$

SOLUTION

Using the FOIL method, we find that

$$(n + 8)(n - 8) \overset{\text{F} \quad \text{O} \quad \text{I} \quad \text{L}}{=} n^2 - 8n + 8n - 64 \qquad \text{Multiplying}$$
$$= n^2 - 64 \qquad \text{Simplifying}$$

▶ *CHECK* **Warm-Up 4**

> The product of the sum and difference of two terms is the difference of the squares of the terms.
> $$(a + b)(a - b) = a^2 - b^2$$

The Square of a Binomial

The square of a binomial is another special product.

▪▪▪

EXAMPLE 5

Simplify: $(y + 3)^2$

SOLUTION

We can rewrite this as the product of identical binomials and then apply FOIL. We get

$$(y + 3)^2 = (y + 3)(y + 3)$$
$$= y^2 + 3y + 3y + 9$$
$$= y^2 + 6y + 9$$

▶ *CHECK* **Warm-Up 5**

> The square of a binomial is a trinomial in which the first and last terms are the squares of the terms in the binomial, and the middle term is twice the product of these terms.
> $$(a + b)^2 = a^2 + 2ab + b^2$$

INSTRUCTOR NOTE

Students should memorize the results of these special multiplications.

Thus, for example

First term squared
Twice product of first and second terms
Second term squared

$$(3x - 7)^2 = (3x)^2 + (2)(3x)(-7) + (-7)^2$$
$$= 9x^2 - 42x + 49$$

▪ ▪ ▪
EXAMPLE 6

Multiply: $(r + 6)^3$

SOLUTION

We know that $(r + 6)^3$ can be rewritten as $(r + 6)^2(r + 6)$.

Using the special-products rule for the square of a binomial, we can simplify this to

$$(r^2 + 12r^2 + 36)(r + 6)$$

Using vertical multiplication, we find that

$$
\begin{array}{r}
r^2 + 12r + 36 \\
\times \qquad r + 6 \\
\hline
6r^2 + 72r + 216 \\
r^3 + 12r^2 + 36r \qquad\quad \\
\hline
r^3 + 18r^2 + 108r + 216
\end{array}
$$

So $(r + 6)^3 = r^3 + 18r^2 + 108r + 216$.

▶ *CHECK* **Warm-Up 6**

▪ ▪ ▪
EXAMPLE 7

Multiply: $(x - 2)^2(x + 2)^2$

SOLUTION

We know an easy way to multiply $(x - 2)(x + 2)$, so we will rearrange the factors and multiply.

$$
\begin{aligned}
(x - 2)^2(x + 2)^2 &= [(x - 2)(x + 2)]^2 && \text{Power rule} \\
&= [x^2 - 4]^2 && \text{Multiplying}
\end{aligned}
$$

Then we square

$$
\begin{aligned}
[x^2 - 4]^2 &= (x^2)^2 + 2(-4)(x^2) + (-4)^2 && \text{Squaring} \\
&= x^4 - 8x^2 + 16 && \text{Simplifying}
\end{aligned}
$$

So $(x - 2)^2(x + 2)^2 = x^4 - 8x^2 + 16$

▶ *CHECK* **Warm-Up 7**

Dividing Polynomials

To divide a polynomial by a polynomial, we use a procedure similar to long division in arithmetic.

> **To divide a polynomial by another polynomial**
>
> 1. Rewrite the division as a long division, with powers of the variable in descending order.
> 2. If there is an exponent missing in the order, list the variable with that exponent and with the coefficient zero.
> 3. Perform the division as with whole numbers.

▪▪▪

EXAMPLE 8

Divide: $(-5x - 14 + x^2) \div (x + 2)$

SOLUTION

We set the problem up as a long division, with the exponents in descending order in both the dividend and the divisor.

$$x + 2 \overline{)x^2 - 5x - 14}$$

To determine the first partial quotient, we divide the first term in the polynomial, x^2, by the first term in the divisor, x. The partial quotient is x, because $\frac{x^2}{x} = x$. So we write x above x^2.

$$
\begin{array}{r}
x \\
x + 2 \overline{)x^2 - 5x - 14}
\end{array}
$$
    First partial quotient

Next we find the product of x and the divisor $x + 2$ and write each term of the product below the like terms of the dividend. Then we subtract by changing signs and adding. (Remember: To subtract means to add the additive inverse.)

$$
\begin{array}{r}
x \\
x + 2 \overline{)x^2 - 5x - 14} \\
x^2 + 2x
\end{array}
$$
    $x(x + 2)$

We subtract and "bring down" the next term of the dividend, -14.

$$
\begin{array}{r}
x \\
x + 2 \overline{)x^2 - 5x - 14} \\
-(x^2 + 2x) \\
\hline
-7x - 14
\end{array}
$$
    Bring down -14

We now divide $-7x$ (the first term of $-7x - 14$) by the first term of the divisor, x. The result is -7, so -7 is the next term of the quotient. We then multiply -7 by the divisor, write the result below the dividend, and subtract.

$$
\begin{array}{r}
x - 7 \\
x + 2 \overline{)x^2 - 5x - 14} \\
x^2 + 2x \\
\hline
-7x - 14 \\
-7x - 14 \\
\hline
0
\end{array}
$$
    $(-7)(x + 2)$
    Remainder

Because there are no other terms to bring down, and the remainder is zero, we are finished.

So $(-5x - 14 + x^2) \div (x + 2)$ is $x - 7$. This means that $-5x - 14 + x^2$ can be factored as $(x + 2)(x - 7)$. To check the answer, we multiply.

$$\text{Divisor} \times \text{quotient} = \text{dividend}$$
$$(x + 2)(x - 7) = x^2 - 7x + 2x - 14$$
$$= x^2 - 5x - 14$$

▶ CHECK **Warm-Up 8**

Calculator Corner

You can check your division work with your graphing calculator. Graph each side of the equation. If your result is correct, you should obtain *only one graph* on your graphing calculator screen. If your answer is incorrect, there will be *two graphs* on the screen.

Divide and check: $(-x - 20 + x^2) \div (x + 4)$.

After you performed the division, you should have gotten the answer $(-x - 20 + x^2) \div (x + 4) = x - 5$.

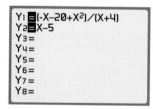
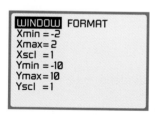
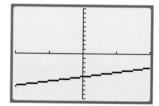

The binomial $(x - 5)$ is the correct answer because you got *only one graph*, meaning the two sides of the equation are equal. If your answer had been incorrect, then you would have obtained two graphs, meaning the two sides of the equation *are not equal!*

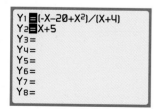
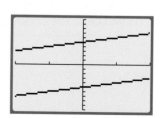

In the next example, we have to insert a term when we arrange the terms in order.

▪ ▪ ▪
EXAMPLE 9

Divide $16t^4 + 8t^3 - 6t + 10$ by $4t^2 + 4t$.

SOLUTION

We arrange the powers in descending order, inserting $0t^2$ in order to include all powers of t in the dividend.

$$4t^2 + 4t \overline{)16t^4 + 8t^3 + 0t^2 - 6t + 10}$$

We then divide step by step until the remainder (if any) has a degree that is less than the degree of the divisor. We get

$$
\begin{array}{r}
4t^2 - 2t + 2 \\
4t^2 + 4t \overline{)16t^4 + 8t^3 + 0t^2 - 6t + 10}
\end{array}
$$

$\underline{16t^4 + 16t^3}$	$4t^2(4t^2 + 4t)$
$-8t^3 + 0t^2$	
$\underline{-8t^3 - 8t^2}$	$-2t(4t^2 + 4t)$
$+8t^2 - 6t$	
$\underline{+8t^2 + 8t}$	$2(4t^2 + 4t)$
$-14t + 10$	Remainder

INSTRUCTOR NOTE

You may find it useful to encourage students to write ⊕ or ⊖ to indicate sign changes.

The quotient is

$$4t^2 - 2t + 2 + \frac{(-14t + 10)}{4t^2 + 4t}$$

▶ CHECK **Warm-Up 9**

Synthetic Division

Synthetic division is a quick, "compact" method for dividing a polynomial by a binomial of the form $x + a$. The coefficient of the x-term in the divisor must be 1, and a can be any real number.

> **To divide using synthetic division**
>
> 1. Arrange the terms of the dividend in decreasing order, inserting a term with a zero coefficient for each missing term.
> 2. Delete all variable parts of the dividend, leaving only the coefficients.
> 3. The divisor $x + a$ is a factor. Set $x + a = 0$ and solve for x. So, $x = -a$. Use $-a$ as your divisor.
> 4. Bring down the first dividend coefficient, multiply by the "divisor" $-a$, insert the result under the next coefficient, and add.
> 5. Repeat for each sum until all coefficients have been used. Write the quotient by combining the sums with variables, beginning with a variable that has a degree that is one less than the degree of the original equation.

▪▪▪
EXAMPLE 10

Divide $9t^3 + 6t + 12$ by $t + 1$.

SOLUTION

We need to add a t^2 term to the dividend, obtaining

$$9t^3 + 0t^2 + 6t + 12.$$

Rewriting the dividend without the variable parts, and the divisor as -1, we have the synthetic division

$$-1 \overline{)9 \quad 0 \quad 6 \quad 12}$$

We first bring down the 9, multiplying it by -1. We insert the result, and add, which yields

$$
\begin{array}{r}
-1 \overline{)9 \quad 0 \quad 6 \quad 12} \\
\underline{-9 } \\
9 \quad -9
\end{array}
$$

Continuing, we obtain

$$
\begin{array}{r}
-1 \overline{)9 \quad 0 \quad 6 \quad 12} \\
\underline{-9 \quad +9 \quad -15} \\
9 \quad -9 \quad 15 \quad -3
\end{array}
$$

Because the degree of the original polynomial was 3, the variable parts of the quotient must begin with x^2. The quotient is $9x^2 - 9x + 15$ remainder -3, or

$$9x^2 - 9x + 15 + \frac{-3}{t+1}$$

▶ CHECK **Warm-Up 10**

▪▪▪
EXAMPLE 11

Divide $x + x^3 + 20$ by $x + 2$.

SOLUTION

First we rewrite the dividend in decreasing order, adding an x^2 term.

$$x^3 + 0x^2 + x + 20$$

We rewrite the dividend without the variable parts and the divisor as -2. We now have the synthetic division

$$-2 \overline{)1 \quad 0 \quad 1 \quad 20}$$

Dividing, we obtain

$$
\begin{array}{r}
-2 \overline{)1 \quad 0 \quad 1 \quad 20} \\
\underline{-2 \quad 4 \quad -10} \\
1 \quad -2 \quad 5 \quad 10
\end{array}
$$

so the quotient is $x^2 - 2x + 5$ remainder 10, or $x^2 - 2x + 5 + \dfrac{10}{x+2}$.

▶ *CHECK* **Warm-Up 11**

▪▪▪

EXAMPLE 12

One factor of $x^3 - x^2 - x + 1$ is $x - 1$. Find the other factor.

SOLUTION

The statement of this example implies that

$$x^3 - x^2 - x + 1 = (x - 1) \text{ (factor)}$$

If we multiply both sides of this equation by $\dfrac{1}{x-1}$, we get:

$$\frac{x^3 - x^2 - x + 1}{x - 1} = \text{factor}$$

so we need to *divide* to find the other factor.

We will use synthetic division to divide.

$$
\begin{array}{r|rrrr}
1) & 1 & -1 & -1 & 1 \\
   &   & 1 & 0 & -1 \\
\hline
   & 1 & 0 & -1 & 0
\end{array}
$$

The other factor is $x^2 - 1$.

▶ *CHECK* **Warm-Up 12**

Practice what you learned.

◣ SECTION FOLLOW-UP

The probability of having exactly 3 boys and 3 girls in a family of 6 children can be found using

$$20p^3q^3$$

where p is the probability of having a girl and q is the probability of having a boy. For any single birth, each of these probabilities is $\frac{1}{2}$. So, the probability of having 3 boys and 3 girls in a family of 6 children is

$$20\left(\frac{1}{2}\right)^3\left(\frac{1}{2}\right)^3 = 20\left(\frac{1}{8}\right)\left(\frac{1}{8}\right) = \frac{20}{64} = \frac{5}{16}$$

a. What is the probability of having 3 boys and 2 girls in a family of 5 children? (*Hint:* Find the results of $(p + q)^5$ and use the p^3q^2 term, with $p = \frac{1}{2}$ and $q = \frac{1}{2}$.) $\quad \frac{5}{16}$

b. Ask and answer two more questions using this technique.

5.3 WARM-UPS

Work these problems before you attempt the exercises.

1. Multiply:
   $(3y - 7)(2y^2 + y - 12)$
   $6y^3 - 11y^2 - 43y + 84$

2. Multiply: $12t^2 + 8t - 7$
   $\underline{\times \qquad 6t - 12}$
   $72t^3 - 96t^2 - 138t + 84$

3. Multiply using FOIL:
   $(4n + 2)(7n - 8)$
   $28n^2 - 18n - 16$

4. Multiply: $(t - 4)(t + 4)$
   $t^2 - 16$

5. Simplify: $(5x - 1)^2$
   $25x^2 - 10x + 1$

6. Simplify: $(6x - 2)^3$
   $216x^3 - 216x^2 + 72x - 8$

7. Simplify: $(3x - 2y)^2(3x + 2y)^2$ $81x^4 - 72x^2y^2 + 16y^4$

8. Divide: $(-9x + x^2 + 20) \div (x - 5)$ $x - 4$

9. Divide: $(3x^2 - 5x + 3) \div (3x - 2)$ $x - 1 + \frac{1}{3x - 2}$

10. Use synthetic division to divide: $(t^4 + 5t^3 - 3t - 15) \div (t + 5)$
    $t^3 - 3$

11. Use synthetic division to divide: $(x^2 + 5x + 4) \div (x + 1)$ $x + 4$

12. One of the factors of $n^4 - 2n^3 + 2n - 4$ is $n - 2$. What is another one? Use synthetic division to find it. $n^3 + 2$

5.3 EXERCISES

Note: Use your graphing calculator to check your results whenever possible.

In Exercises 1 through 6, multiply or divide as indicated.

1. $(-6y + 5)(-9y^2 - 4y + 7)$
   $54y^3 - 21y^2 - 62y + 35$

2. $(-6 - 6t)(-9t^2 - 3t + 6)$
   $54t^3 + 72t^2 - 18t - 36$

3. $\qquad -8x^2 + 6x - 4$
   $\underline{\times \qquad\quad 15x + 3}$
   $-120x^3 + 66x^2 - 42x - 12$

4. $\qquad -3y^2 + 7y - 2$
   $\underline{\times \qquad\qquad 7y + 3}$
   $-21y^3 + 40y^2 + 7y - 6$

5. $\qquad -7y^2 + 9y - 5$
   $\underline{\times \qquad\qquad 12y - 5}$
   $-84y^3 + 143y^2 - 105y + 25$

6. $\qquad -4t^2 + 8t - 6$
   $\underline{\times \qquad\qquad 8 - 9t}$
   $36t^3 - 104t^2 + 118t - 48$

In Exercises 7 through 18, multiply by using FOIL.

7. $(x + 3)(x + 2)$
   $x^2 + 5x + 6$

8. $(x + 5)(x + 7)$
   $x^2 + 12x + 35$

9. $(x - 6)(x - 4)$
   $x^2 - 10x + 24$

10. $(x - 8)(x + 9)$
    $x^2 + x - 72$

11. $(x + 5)(4x - 7)$
    $4x^2 + 13x - 35$

12. $(y - 3)(9y + 10)$
    $9y^2 - 17y - 30$

13. $(5x + 4)(x - 9)$
    $5x^2 - 41x - 36$

14. $(6x - 5)(x + 4)$
    $6x^2 + 19x - 20$

15. $(2x + 1)(3x + 2)$
    $6x^2 + 7x + 2$

16. $(5x + 4)(2x + 3)$
$10x^2 + 23x + 12$

17. $(7x + 9)(3x - 4)$
$21x^2 - x - 36$

18. $(8x + 5)(6x - 1)$
$48x^2 + 22x - 5$

In Exercises 19 through 30, multiply as special products. Do not use FOIL.

19. $(x + 1)(x - 1)$
$x^2 - 1$

20. $(y - 3)(y + 3)$
$y^2 - 9$

21. $(2x + 1)(2x - 1)$
$4x^2 - 1$

22. $(3y - 1)(3y + 1)$
$9y^2 - 1$

23. $(t + 2)^2$
$t^2 + 4t + 4$

24. $(x + 3)^2$
$x^2 + 6x + 9$

25. $(2y - 5)^2$
$4y^2 - 20y + 25$

26. $(3t - 7)^2$
$9t^2 - 42t + 49$

27. $(7x - 4t)(7x + 4t)$
$49x^2 - 16t^2$

28. $(0.8x - 0.7)(0.8x + 0.7)$
$0.64x^2 - 0.49$

29. $(5x + 0.2y)^2$
$25x^2 + 2xy + 0.04y^2$

30. $(0.4x - 0.9y)^2$
$0.16x^2 - 0.72xy + 0.81y^2$

In Exercises 31 through 38, use the special-products rule $(a + b)(a - b) = a^2 - b^2$ to multiply the given numbers.

31. $19 \cdot 21$
399

32. $39 \cdot 41$
1599

33. $78 \cdot 82$
6396

34. $98 \cdot 102$
9996

35. $910 \cdot 890$
$809{,}900$

36. $760 \cdot 640$
$486{,}400$

37. $830 \cdot 970$
$805{,}100$

38. $550 \cdot 650$
$357{,}500$

In Exercises 39 through 46, use the special-products rule $(a + b)^2 = a^2 + 2ab + b^2$ or the special-products rule $(a - b)^2 = a^2 - 2ab + b^2$ to find the given square.

39. 51^2
2601

40. 99^2
9801

41. 78^2
6084

42. 67^2
4489

43. 55^2
3025

44. 73^2
5329

45. 110^2
$12{,}100$

46. 220^2
$48{,}400$

In Exercises 47 through 54, multiply by using any method.

47. $(x + 2)^3$
$x^3 + 6x^2 + 12x + 8$

48. $(x - 2)^3$
$x^3 - 6x^2 + 12x - 8$

49. $(x + 1)^4$
$x^4 + 4x^3 + 6x^2 + 4x + 1$

50. $(x - 1)^4$
$x^4 - 4x^3 + 6x^2 - 4x + 1$

51. $[(x - 1)(x + 1)]^3$
$x^6 - 3x^4 + 3x^2 - 1$

52. $[(x - 3t)(x + 3t)]^3$
$x^6 - 27x^4t^2 + 243x^2t^4 - 729t^6$

53. $(2x - 1)^2(2x + 1)^2$
$16x^4 - 8x^2 + 1$

54. $(5x^2 - 8)^2(5x^2 + 8)^2$
$625x^8 - 3200x^4 + 4096$

In Exercises 55 through 68, divide by using long division.

55. $\dfrac{x^2 + x - 12}{x + 4}$ $x - 3$

56. $\dfrac{x^2 + 7x + 12}{x + 3}$ $x + 4$

57. $\dfrac{x^2 - 4x - 12}{x - 6}$ $x + 2$

58. $\dfrac{x^2 - 2x - 48}{x - 8}$ $x + 6$

59. $\dfrac{x^3 + 2x^2 - 11x - 12}{x + 1}$ $x^2 + x - 12$

60. $\dfrac{x^3 + 6x^2 + 5x - 12}{x - 1}$
$x^2 + 7x + 12$

61. $\dfrac{2x^3 + 11x^2 - 38x + 16}{x - 2}$
$2x^2 + 15x - 8$

62. $\dfrac{3x^3 + 8x^2 - 17x - 42}{x + 2}$
$3x^2 + 2x - 21$

63. $\dfrac{18x^3 - 15x^2 - 16x - 3}{3x + 1}$
$6x^2 - 7x - 3$

64. $\dfrac{96x^3 - 64x^2 - 10x + 3}{4x + 1}$
$24x^2 - 22x + 3$

65. $\dfrac{x^3 - 26x + 5}{x - 5}$ $x^2 + 5x - 1$

66. $\dfrac{y^3 - 4y^2 - 21y}{y + 3}$ $y^2 - 7y$

67. $\dfrac{n^2 - 8n + 5}{n + 2}$ $n - 10 + \dfrac{25}{n + 2}$

68. $\dfrac{3x^2 - 2x + 9}{x - 5}$ $3x + 13 + \dfrac{74}{x - 5}$

In Exercises 69 through 90, use synthetic division to divide.

69. $(x^2 - 3x - 4) \div (x + 1)$
$x - 4$

70. $(x^2 - 6x + 6) \div (x - 2)$
$x - 4 - \dfrac{2}{x - 2}$

71. $(2x^2 + 7x + 3) \div (x + 3)$
$2x + 1$

72. $(3x^2 + 17x - 6) \div (x + 6)$
$3x - 1$

73. $(x^3 - 3x^2 - 2x + 4) \div (x - 1)$
$x^2 - 2x - 4$

74. $(x^3 + x^2 - 7x + 2) \div (x - 2)$
$x^2 + 3x - 1$

75. $(2x^3 - 2x^2 - 9x + 10) \div (x - 2)$
$2x^2 + 2x - 5$

76. $(3x^3 - 4x^2 - 3x + 4) \div (x - 1)$
$3x^2 - x - 4$

77. $y - 1 \overline{)\, y^3 - 1}$
$y^2 + y + 1$

78. $y + 2\overline{)y^3 + 8}$ $y^2 - 2y + 4$

79. $y + 3\overline{)y^3 + 27}$ $y^2 - 3y + 9$

80. $y - 5\overline{)y^3 - 125}$ $y^2 + 5y + 25$

81. $(3x^4 - 15x^3 + x^2 - 4x - 5) \div (x - 5)$ $3x^3 + x + 1$

82. $(2x^4 + 5x^3 + 2x^2 - 4x - 8) \div (x + 2)$ $2x^3 + x^2 - 4$

83. $(x^4 - 1) \div (x - 1)$ $x^3 + x^2 + x + 1$

84. $(x^4 + 1) \div (x + 1)$ $x^3 - x^2 + x - 1 + \dfrac{2}{x + 1}$

85. One factor of $x^3 - 27$ is $x - 3$. Find the other factor. $x^2 + 3x + 9$

86. One factor of $x^6 - 1$ is $x - 1$. Find the other factor. $x^5 + x^4 + x^3 + x^2 + x + 1$

87. One factor of $x^8 - 1$ is $x + 1$. Find the other factor. $x^7 - x^6 + x^5 - x^4 + x^3 - x^2 + x - 1$

88. One factor of $x^{16} - 1$ is $x - 1$. Find the other factor. $x^{15} + x^{14} + x^{13} + x^{12} + x^{11} + x^{10} + x^9 + x^8 + x^7 + x^6 + x^5 + x^4 + x^3 + x^2 + x + 1$

89. One factor of $x^4 + 4x^3 - 2x - 8$ is $x + 4$. Find the other factor. $x^3 - 2$

90. One factor of $y^3 + 11y^2 - 5y - 55$ is $y + 11$. Find the other factor. $y^2 - 5$

MIXED PRACTICE

By doing these exercises, you will practice the topics up to this point in the chapter.

91. Subtract $3x^2 + 8x$ from $-3x + 2x^2 - 8$. $-x^2 - 11x - 8$

92. Subtract $3x^2 - 2x + 1 - (3x^2 - 3x - 2)$ $x + 3$

93. Multiply: $2xy(3x^2y - 4xy + 2)$ $6x^3y^2 - 8x^2y^2 + 4xy$

94. Simplify: $\dfrac{3x^2y - 12xy^2}{6x^2y^2}$ $\dfrac{1}{2y} - \dfrac{2}{x}$

95. Let $f(x) = 2x + 3$ and $g(x) = x^2$. Find $(f - g)(0)$.   3

96. Divide $3x^2 - 300$ by $x - 10$. $3x + 30$

EXCURSIONS

Data Analysis

1. Study the following table, then answer the questions.

    a. Analyze this data. Use graphs to present your results.

    b. State four things you learned from your analysis.

Tallying Grades, State by State

	Standards/ assessments	Teaching quality	School climate	Resources			Achievement	
				Adequacy	Allocation	Equity	4th grade reading[1]	8th grade math[2]
Alabama	A	C	C−	C+	C−	B+	23	10
Alaska	C+	C	C	C+	F	D+		
Arizona	C	D	D	D+	C	C	24	15
Arkansas	B	C+	C−	F	C	B	24	10
California	I	B−	D−	D−	C−	D	18	16
Colorado	I	B−	C	D	C−	B-	28	22
Connecticut	A−	C+	C	B	C	C−	38	26
Delaware	A	C	D	B	D−	A−	23	15
Florida	B	C	D−	B−	C	C+	23	15
Georgia	A	B−	C	B−	B	B	26	13
Hawaii	B	D	D+	D−	C	A	19	14
Idaho	C−	D+	C−	C+	D−	C+		22
Illinois	B	C+	C−	C+	D−	C−		
Indiana	B+	C+	C	B	B	B	33	20
Iowa	F	C	C	C+	C+	B	35	31
Kansas	A	C	C+	B−	C	C		
Kentucky	A	B	C−	B	C+	B+	26	14
Louisiana	B	C	D	D	D+	C	15	7
Maine	A−	C	B	B+	C−	B	41	26
Maryland	A	C−	D+	C+	D−	A−	26	20
Massachusetts	I	B−	C	B−	C−	D+	36	23
Michigan	A−	C	D	B+	C−	B		19
Minnesota	I	B	D+	B+	C−	B−	33	31
Mississippi	B+	C−	D−	C+	D+	B+	18	6
Missouri	A−	C	C−	C	D+	C−	31	20
Montana	D	C−	B−	C	C−	B−	35	
Nebraska	D−	B−	C+	C	D−	C+	34	26
Nevada	C	C−	D	C	D+	B		
New Hampshire	A−	C+	C	B	D−	B−	36	25
New Jersey	A	C	C+	A	C+	D+	33	24
New Mexico	A	C−	D+	B	B−	C−	21	11
New York	A	C−	C−	A	C−	C	27	20
North Carolina	A	C	D	B−	C+	A−	30	12
North Dakota	C	C	C+	D	C	B−	38	30
Ohio	A−	C	D+	B	C	C+		18
Oklahoma	B+	C+	B−	D	D+	B−		17
Oregon	A−	C	C−	C+	D−	C+		
Pennsylvania	B+	C+	D	B+	C	C+	30	22
Rhode Island	C	C	D	B+	D+	D	32	16
South Carolina	B	C	D+	B−	D+	B+	20	15
South Dakota	C+	C	C+	C−	C	B−		
Tennessee	A−	C+	C−	C	B	B	27	12
Texas	A	C−	C+	C+	B−	D	26	18
Utah	A	C	D−	C	C+	B+	30	22
Vermont	A	B−	B+	A−	D+	B−		
Virginia	B	C+	C−	B−	B	B+	26	19
Washington	B	C+	D+	B−	C+	C+	27	
West Virginia	A	C	C	A	C−	A	26	10
Wisconsin	B−	C	C−	A−	C−	B	35	27
Wyoming	F	C	C−	C	C	B−	32	21

1. 1994 figures from National Association of Educational Progress. 2. 1992 figures.
Source: Reprinted with permission by *Education Week,* Pew Charitable Trusts, from *USA Today,* 17 January 1997.

2. Study the following table about speed limits, then answer the questions.

STATES THAT RAISED SPEED LIMITS DURING 1996

State	Rural interstates	Urban interstates
Alabama	70 mph	70 mph
Arizona	75 mph	55 mph
California	70 mph[1]	65 mph
Colorado	75 mph	65 mph
Delaware	65 mph	55 mph
Florida	70 mph	65 mph
Georgia	70 mph	65 mph
Idaho	75 mph	65 mph
Kansas	70 mph	70 mph
Michigan	70 mph[1]	65 mph
Mississippi	70 mph	70 mph
Missouri	70 mph	60 mph
Montana	Unlimited[2]	Unlimited
Nebraska	75 mph	65 mph
Nevada	75 mph	65 mph
New Mexico	75 mph	55 mph
New York	65 mph	65 mph
North Carolina	70 mph	65 mph
North Dakota	70 mph	55 mph
Oklahoma	75 mph	70 mph
Rhode Island	65 mph	55 mph
South Dakota	75 mph	65 mph
Texas	70 mph[3]	70 mph
Utah	75 mph	65 mph
Washington	70 mph[3]	60 mph
Wisconsin	65 mph	65 mph
Wyoming	75 mph	60 mph

[1]Trucks 55 mph   [2]Trucks 65 mph   [3]Trucks 60 mph

a. Graph and compare the state's speed limits on rural interstates and on urban interstates.

b. Ask and answer four questions about this data.

Exploring Numbers

3. a. Assign the digits 1, 2, 3 to the letters a, b, and c so that the following statement is true. *Note:* Each letter stands for a single digit; that is, abc is a 3-digit number.

$$abc = ab + ac + ba + bc + ca + cb$$

b. In words, tell what this statement means.

4. The numbers 242, 243, 244, and 245 each have 6 divisors. Find them.

Perimeter and Area

Perimeter of a Rectangle or Parallelogram

A rectangle has two pairs of equal sides—its length and its width. The same is true for a parallelogram.

> **To find the perimeter of a rectangle or parallelogram**
>
> If ℓ is the length and w is the width, then
>
> $$P = 2\ell + 2w \quad \text{or} \quad P = 2(\ell + w)$$

Note that in order to use either formula, we need to be familiar with the order of operations. Both formulas give the same result when applied properly.

▪▪▪
EXAMPLE

Find the perimeter of a rectangle with length 6 feet and width 7 feet.

SOLUTION

We use the second formula, substituting the actual length and width.

$$P = 2(\ell + w) = 2(6 + 7)$$

Then, using the standard order of operations, we have

$$P = 2(13) = 26 \text{ feet}$$

As a check, we use the first formula, obtaining

$$P = 2\ell + 2w$$
$$= 2(6) + 2(7)$$
$$= 12 + 14 = 26$$

Therefore, the perimeter is 26 feet.

Perimeter of a Square

The formula for the perimeter of a rectangle also will give you the perimeter of a square, but because the length and width of a square are both s, we can derive another formula.

$$2(\ell + w) = 2(s + s)$$
$$= 2(2s)$$
$$= 4s$$

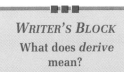

WRITER'S BLOCK
What does *derive* mean?

> **To find the perimeter of a square**
> If s is the length of a side, then
> $$\text{Perimeter} = 4s$$

Area of a Rectangle

> **To find the area of a rectangle**
> If ℓ is the length and w is the width, then
> $$\text{Area} = \ell w$$

▪▪▪

EXAMPLE

Find the area of a rectangle that has width 1.6 inches and length 0.9 inches.

SOLUTION

Using the formula for finding the area of a rectangle, we can substitute the values for ℓ and w:

$$A = \ell w = 0.9(1.6) = 1.44$$

So, the area is 1.44 square inches.

Please note that the length can be shorter, longer, or equal to the width.   ◢

Area of a Parallelogram

We use multiplication to find the area of a parallelogram.

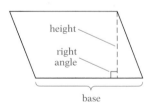

> **To find the area of a parallelogram**
> If b is the length of the base and h is the height, then
> $$\text{Area} = bh$$

The base and height are shown in the accompanying figure.

Area of a Square

Because a square is also a rectangle, we can find its area with the rectangular area formula: Area = $\ell \times w$. But ℓ and w are the same for a square, and they both are equal to the length of a side s. Accordingly,

To find the area of a square

If s is the length of a side of a square, then

$$\text{Area} = s^2$$

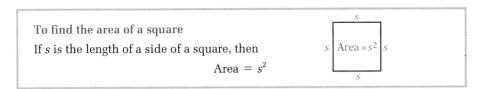

This formula is what gives the name "squared" to the exponent 2.

■ ■ ■

EXAMPLE

Find the area of a parallelogram with height 1.9 feet and base 12 feet. Give your answer in square meters. Use the exact relationship—2.54 centimeters = 1 inch—to convert.

SOLUTION

First, we will change feet to meters.

$$1.9 \text{ ft} \cdot \frac{12 \text{ in.}}{1 \text{ ft}} \cdot \frac{2.54 \text{ cm}}{1 \text{ in.}} \cdot \frac{1 \text{ m}}{100 \text{ cm}}$$

We cancel the units:

$$1.9 \text{ ft} \cdot \frac{12 \text{ in.}}{1 \text{ ft}} \cdot \frac{2.54 \text{ cm}}{1 \text{ in.}} \cdot \frac{1 \text{ m}}{100 \text{ cm}}$$

We multiply and divide:

$$\frac{(1.9)(12)(2.54)(1)\text{m}}{(1)(1)(100)} = 0.57912 \text{ m}$$

We change 12 feet to meters similarly:

$$12 \text{ ft} \cdot \frac{12 \text{ in.}}{1 \text{ ft}} \cdot \frac{2.54 \text{ cm}}{1 \text{ in.}} \cdot \frac{1 \text{ m}}{100 \text{ cm}} = 3.6576 \text{ m}$$

Substituting in the formula, we get:

$$A = b \times h = (3.6576)(0.57912) \approx 2.12$$

So the area of the parallelogram is approximately 2.12 square meters.

> ■ ■ ■
> **WRITER'S BLOCK**
> **Explain why**
> $$\frac{2.54 \text{ cm}}{1 \text{ in.}}$$
> **and**
> $$\frac{1 \text{ in.}}{2.54 \text{ cm}}$$
> **are equivalent.**

PRACTICE

The playing areas described in Exercises 1 through 6 are rectangular. Find their perimeters and areas. Express your answers in meters using the exact equivalency—2.54 centimeters = 1 inch—in your conversions. (*Hint:* Convert lengths before calculating area.) Write your answers to the nearest whole number.

1. U.S. football field: 120 yards long, 53 yards 1 foot wide   P: 317 m; A: 5351 m²

2. Fencing piste: length 46 feet, width 6 feet 6 inches   P: 32 m; A: 28 m²

3. Basketball court: 28 yards by 15 yards 9 inches   P: 79 m; A: 357 m²

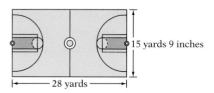

15 yards 9 inches

28 yards

4. Netball court: 33 yards 1 foot by 16 yards 2 feet   P: 91 m; A: 465 m²

5. Ice hockey rink: length 66 yards 2 feet and width 33 yards 1 foot
   P: 183 m; A: 1858 m²

6. Judo court: 52 feet 6 inches on a side   P: 64 m; A: 256 m²

7. Find the perimeter and area of the double-parallelogram shown below. Give the results in meters or square meters as necessary. Treat the figure as one large surface.   P: 10 m; A: 1 m²

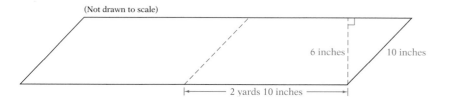

(Not drawn to scale)

6 inches   10 inches

2 yards 10 inches

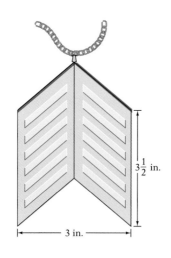

3½ in.

3 in.

8. An arrow-shaped pendant was designed by an artist. See the accompanying figure. The two parallelograms that form the arrow are each $3\frac{1}{2}$ inches long. The distance across the pendant is 3 inches and the outside long edges are parallel to each other. The point of the arrow (emphasized in color in the figure) has a total "length" of 4 inches. Find the perimeter and area in centimeters and square centimeters. (Disregard its "depth.") Round your answers to the nearest whole number.   P: 38 cm; A: 68 cm²

9. A rectangle with a height of 2 feet and an area of 8.2 square feet was sat on until its height was only 18.5 inches. What is its new perimeter (in meters) and area (in square meters)?   P: 4 m; A: 1 m²

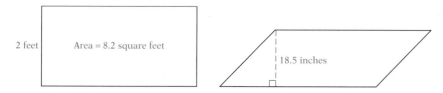

2 feet   Area = 8.2 square feet

18.5 inches

10. Four parallelograms form a trivet (see the accompanying figure). All four sides are equal. The perimeter of the trivet is 12 inches. The area is 6 square inches. Find the perimeter (in centimeters) and the area (in square centimeters) of each parallelogram in the trivet. Round your answers to the nearest whole number. (*Hint:* There are three different sizes of parallelograms. How many of each size are there?)
   1 large: P: 30 cm, A: 39 cm²; 4 medium: P: 23 cm, A: 19 cm²; 4 small: P: 15 cm, A: 10 cm²

5.4 Factoring Polynomials

SECTION LEAD-IN

Many so-called geniuses can wow their friends by applying the rules they will learn in this section.

In your head, multiply

$$(21)(19) = ?$$
$$39 \cdot 39 = ?$$

Having trouble? Check in later!

Factors and Multiples

Let a, b, and c be integers, and let $a \cdot b = c$. Then a and b are **factors,** or **divisors,** of c, and c is a **multiple** of a and of b.

These definitions also apply in a general way to monomials—numbers that are written using variables.

Greatest Common Factor

Greatest Common Factor

The **greatest common factor** (GCF) of two integers is the greatest positive integer that is a factor of both.

This definition also applies in a general way to variable terms: The greatest common factor of two monomials is the greatest term that is a factor of both monomials. It consists of the GCF of the coefficients, multiplied by each common variable raised to its least power. Thus, for example,

The GCF of $18x^3$ and $6x^5y$ is $6x^3$.
The GCF of x^2y^2 and x^5y is x^2y.

▪▪▪
EXAMPLE 1

Find the GCF of $3x^4z^8$ and $-9x^2y^2z^3$.

SOLUTION

The GCF of the coefficients is 3. The common variables are x and z; raised to their least powers, they are x^2 and z^3.

So the GCF is $3x^2z^3$.

SECTION GOALS

- To factor polynomials by removing the greatest common factor
- To factor polynomials by grouping
- To factor the difference of two squares
- To factor perfect-square trinomials
- To factor the sum or difference of two cubes

■■■
WRITER'S BLOCK

Why are 0 and 1 neither prime nor composite?

■■■
WRITER'S BLOCK

Explain to a friend how the GCF and the greatest common divisor are alike.

! ! !
ERROR ALERT

Identify the error and give a correct answer.

Find the factors of 124.

Incorrect Solution:

The factors of 124 are 2 and 31.

▶ *CHECK* **Warm-Up 1**

Factoring Out a Common Factor

The distributive property states that

$$a(b + c) = ab + ac$$

We factor when we reverse this procedure. That is, when we write $ab + ac$ as $a(b + c)$, we have "factored out a." One way to factor an algebraic expression is *by removing the greatest common factor.*

To factor an algebraic expression by removing the greatest common factor

1. Find the GCF of the terms of the expression.
2. Rewrite each term as the product of the GCF and another factor.
3. Simplify the result of step 2.

▪▪▪
EXAMPLE 2

Factor: $6rt^2 + 12t^3$

! ! !

ERROR ALERT

Identify the error and give a correct answer.

Factor: $3x^4 - 12x^2$

Incorrect Solution:

$3x^4 - 12x^2$
$= 3x(x^3) - 3x(4x)$
$= 3x(x^3 - 4x)$

SOLUTION

The GCF of the coefficients, 6 and 12, is 6. The variable parts, rt^2 and t^3, have only t^2 in common. So the GCF of the two terms is $6t^2$.

Now we want to write each term as the product of $6t^2$ and something else. Dividing each term by $6t^2$, we have

$$6rt^2 \div 6t^2 = r \quad \text{so} \quad 6rt^2 = 6t^2(r)$$
$$12t^3 \div 6t^2 = 2t \quad \text{so} \quad 12t^3 = 6t^2(2t)$$

Then the original polynomial may be written as

$$6rt^2 + 12t^3 = 6t^2(r) + 6t^2(2t)$$
$$= 6t^2(r + 2t) \qquad \text{Distributive property}$$

You can multiply out the right member (the factored form) to check the result.

▶ *CHECK* **Warm-Up 2**

You should soon be able to divide out the common factor mentally and write out the factored form immediately. Try it in the next example.

! ! !

ERROR ALERT

Identify the error and give a correct answer.

Factor: $10x^2y - 5xy$

Incorrect Solution:

$5xy(2x)$

▪▪▪
EXAMPLE 3

Factor: $-7r^2t - 21rt$

SOLUTION

The GCF of these terms is $7rt$, so

$$-7r^2t - 21rt = 7rt(-r - 3)$$

▶ *CHECK* **Warm-Up 3**

Factoring an Expression by Grouping

Sometimes, when there are several terms in an expression, it is not possible to find a common factor for all terms—but only for some of the terms. We factor those terms that can be factored and combine factors where possible. This is called **factoring by grouping**, or **partitioning**.

> **To factor an expression by grouping**
>
> 1. Remove the GCF of the expression if there is one.
> 2. By applying the commutative and associative properties, group terms of the remaining expression that have common factors.
> 3. Remove the GCF of each group and rewrite the group as a product.
> 4. Simplify the result of step 3.

▪▪▪

EXAMPLE 4

Factor: $(x - 1)(y + 3) + (x - 1)(y + 5)$

SOLUTION

In this example there are two terms, $(x - 1)(y + 3)$ and $(x - 1)(y + 5)$, and these terms have the GCF $(x - 1)$. Thus the entire expression can be factored to

$$(x - 1)[(y + 3) + (y + 5)]$$

Now the terms in the brackets can be simplified to $y + 3 + y + 5 = 2y + 8$. So the entire expression is equivalent to $(x - 1)(2y + 8)$. We factor out the 2 from $2y + 8$ and rewrite this as $2(x - 1)(y + 4)$.

▶ CHECK **Warm-Up 4**

▪▪▪

EXAMPLE 5

Factor: $3r^2 - 6rt + 2r - 4t$

SOLUTION

There is no GCF. But the first two terms have a common factor, and so do the last two. Therefore, we group them as follows: $(3r^2 - 6rt) + (2r - 4t)$.

Each of these groups has a GCF, so we remove the GCF from each group and rewrite the group as a product: $(3r^2 - 6rt) + (2r - 4t) = 3r(r - 2t) + 2(r - 2t)$.

These two new terms have the common factor $(r - 2t)$, so we simplify the result to $(r - 2t)(3r + 2)$. You can multiply these binomials by the FOIL method to check this result.

Note: We could have grouped $(3r^2 + 2r) + (-6rt - 4t)$ and factored: $r(3r + 2) - 2t(3r + 2) = (3r + 2)(r - 2t)$.

▶ CHECK **Warm-Up 5**

! ! !
ERROR ALERT

Identify the error and give a correct answer.

Factor:
$36x^2y - 6xy - 12xy^2$

Incorrect Solution:

$36x^2y - 6xy - 12xy^2$
$= 6xy(6x - 2y)$

Factoring Special Products

Next, we factor three special kinds of polynomials. Their special forms enable us to use factoring shortcuts.

Difference of Two Squares

The expression $y^2 - 81$ is a **difference of two squares:** The left term is the square of y, and the right term is the square of 9. We could factor it by writing it as

$$y^2 + 0y - 81$$

and then finding two numbers whose product is -81 and whose sum is 0. Those numbers are $+9$ and -9, so

$$y^2 - 81 = y^2 - 9^2 = (y + 9)(y - 9)$$

You may be able to see the shortcut method from this equation.

> **To factor the difference of two squares $(x^2 - a^2)$**
>
> Write the factors as the sum and difference of the terms that were squared: $(x + a)(x - a)$.

We cannot factor the *sum* of two squares into real numbers.

▪▪▪
EXAMPLE 6

Factor: $t^2 - 225$

SOLUTION

Both t^2 and 225 are squares (because 225 is 15^2). This is then a difference of squares, and we factor it as follows:

$$(t + 15)(t - 15)$$

We check using FOIL.

$$(t + 15)(t - 15) = t^2 - 15t + 15t - 225$$
$$= t^2 - 225$$

▶ CHECK **Warm-Up 6**

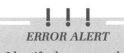

ERROR ALERT

Identify the error and give a correct answer.

Factor: $x^2 + 4$

Incorrect Solution:

$x^2 + 4 = (x + 2)(x + 2)$

In the next example, the variable term has a coefficient.

▪▪▪
EXAMPLE 7

Factor: $49n^2 - 169$

SOLUTION

Although the n^2 term has a coefficient, the two terms have no common factors. They are both squares, however, so we have

$$49n^2 - 169 = (7n)^2 - 13^2 = (7n + 13)(7n - 13)$$

You should check the result.

▶ CHECK **Warm-Up 7**

Perfect-Square Trinomial

Suppose we square the binomial $x + a$ using FOIL.

$$(x + a)(x + a) = x^2 + ax + ax + a^2$$
$$= x^2 + 2ax + a^2$$

The square of a binomial is called a **perfect-square trinomial**. Its first and last terms are the squares of the terms of the binomial, and its middle term is twice the product of the terms of the binomial.

$$x^2 + 2ax + a^2$$

First term   Second term
squared   squared

Twice the first term times
the second term

▪▪▪
EXAMPLE 8

Determine whether each of the following expressions is a perfect-square trinomial.

a. $x^2 + 6x + 9$ **b.** $4x^2 - 4x + 1$ **c.** $4x^2 + 16x + 4$

SOLUTION

a. $x^2 + 6x + 9$ *is* a perfect-square trinomial. Both x^2 and 9 are squares (of x and 3, respectively), and $6x$ is twice x times 3.

$$x^2 + 6x + 9$$

x squared   3 squared
Twice x times 3

b. $4x^2 - 4x + 1$ is also a perfect-square trinomial.

$$4x^2 \qquad -4x \qquad + 1$$
$(2x)^2$ Twice $2x$ times 1 $(1)^2$

Note: the negative sign in front of the middle term indicates the factors are in the form of $(a - b)^2$.

c. $4x^2 + 16x + 4$ is not a perfect-square trinomial. Both $4x^2$ and 4 are squares (of $2x$ and 2, respectively). But twice $2x$ times 2 is $(2)(2x)(2) = 8x$, not $16x$.

▶ CHECK **Warm-Up 8**

! ! !
ERROR ALERT

Identify the error and give a correct answer.

Factor: $x^2 + 4x + 3$

Incorrect Solution:

$x^2 + 4x + 3 =$
$(x + 3)(x + 3)$

The following Study Hint gives another way to tell when a trinomial is *not* a perfect-square trinomial. To factor a perfect-square trinomial, we simply write it as the square of the original binomial.

> **To factor a perfect-square trinomial**
>
> Write the factors as the square of the original binomial:
> $$x^2 + 2ax + a^2 = (x + a)^2$$
> $$x^2 - 2ax + a^2 = (x - a)^2$$

▪▪▪
EXAMPLE 9

Factor: $m^2 - 46m + 529$

SOLUTION

The first and last terms here are squares, m^2 and 23^2. Twice 23 times m is $46m$. Because the middle term is $-46m$, this is a perfect-square trinomial with factors $m - 23$ and $m - 23$. In other words,

$$m^2 - 46m + 529 = (m - 23)^2$$

Check using FOIL.

▶ CHECK **Warm-Up 9**

In this next example, we must use a combination of operations.

▪▪▪
EXAMPLE 10

Factor: $4xy^2 + 104x^2y + 676x^3$

SOLUTION

First, there is a common factor, namely $4x$. We factor it out to get

$$4xy^2 + 104x^2y + 676x^3 = 4x(y^2 + 26xy + 169x^2)$$

Then we check whether the expression in parentheses is a perfect-square trinomial. The first and last terms are perfect squares, y^2 and $(13x)^2$. Twice $y \cdot 13x$ is equal to $26xy$, the middle term. So this is a perfect-square trinomial with factors $(y + 13x)(y + 13x)$. Thus

$$4xy^2 + 104x^2y + 676x^3 = 4x(y + 13x)(y + 13x)$$

When you check by multiplying, be sure to include the factor $4x$.

▶ CHECK **Warm-Up 10**

Sum or Difference of Cubes

A **cube** is the value that results when a term is used as a factor three times. The sum of two cubes $(x^3 + y^3)$ and the difference of two cubes $(x^3 - y^3)$ can be factored into a binomial and a trinomial.

> To factor a polynomial that is a sum or difference of two cubes
>
> Write the factors as a binomial and a trinomial, as follows:
> $$x^3 + y^3 = (x + y)(x^2 - xy + y^2)$$
> $$x^3 - y^3 = (x - y)(x^2 + xy + y^2)$$

The formulas are the same, except that

1. The sum formula has a minus sign in the trinomial.
2. The difference formula has a minus sign in the binomial.

▪ ▪ ▪

EXAMPLE 11

Factor: $8y^3 - 125$

SOLUTION

Again, both terms are cubes, so we can substitute in the difference formula.

$$\begin{aligned} 8y^3 - 125 &= (2y)^3 - 5^3 \\ &= (2y - 5)[(2y)^2 + (2y)(5) + 5^2] \\ &= (2y - 5)(4y^2 + 10y + 25) \end{aligned}$$

Check the factorization by multiplying.

▶ *CHECK* **Warm-Up 11**

Factoring More Complex Polynomials

We can use the procedures of this section to factor polynomials that are in the form of the sum or difference of squares or cubes.

▪ ▪ ▪

EXAMPLE 12

Factor: **a.** $(r + n)^3 + 8$ **b.** $27xy^3 - 125x^4$

SOLUTION

a.   This is a sum of two cubes. It can be rewritten as

$$(r + n)^3 + 2^3$$

Substituting into the formula yields

$$\begin{aligned} (r + n)^3 + 2^3 &= [(r + n) + 2][(r + n)^2 - (2)(r + n) + 2^2] \\ &= (r + n + 2)(r^2 + 2rn + n^2 - 2r - 2n + 4) \end{aligned}$$

b.   First we remove the common factor.

$$27xy^3 - 125x^4 = x(27y^3 - 125x^3)$$

Now we have a difference of two cubes, $(3y)^3$ and $(5x)^3$, on the right. Substituting in the formula gives us

$$27xy^3 - 125x^4 = x[(3y)^3 - (5x)^3]$$
$$= x(3y - 5x)[(3y)^2 + (3y)(5x) + (5x)^2]$$
$$= x(3y - 5x)(9y^2 + 15xy + 25x^2)$$

▶ *CHECK* **Warm-Up 12**

Practice what you learned.

SECTION FOLLOW-UP

We can use what we learned in this chapter to do multiplication in our heads.

$$(21)(19)$$

can be rewritten (mentally) as

$$(20 + 1)(20 - 1)$$

Using what we just learned, we can obtain

$$400 - 1 = 399$$

To multiply $39 \cdot 39$, we rewrite the problem (mentally) as

$$(40 - 1)^2$$

We know this is

$$40^2 - 2(40)(1) + 1^2$$
$$= 1600 - 80 + 1$$
$$= 1521$$

▪ What other numbers can you multiply this way?

5.4 WARM-UPS

Work these problems before you attempt the exercises.

1. Find the GCF of $16x^5y^6$ and $28xy^4$ $4xy^4$

2. Factor: $48x^2y^5 + 36x^4y^3$ $12x^2y^3(4y^2 + 3x^2)$

3. Factor: $24r^4y^6 - 48r^5y^3$ $24r^4y^3(y^3 - 2r)$

4. Factor:
$(y - 3)(y + 4) + (y - 3)(y - 6)$ $2(y - 3)(y - 1)$

5. Factor: $4t^2 + 6t + 10t + 15$
$(2t + 3)(2t + 5)$

6. Factor: $289 - n^2$
$(17 + n)(17 - n)$

7. Factor: $121t^2 - 4$
$(11t + 2)(11t - 2)$

8. Is $t^2 - 18t + 81$ a perfect-square trinomial? yes

9. Factor: $r^2 - 20r + 100$
$(r - 10)^2$

10. Factor: $3t^2 - 66ty + 363y^2$
$3(t - 11y)^2$

11. Factor: $729y^3 - 27x^3$
$27(3y - x)(9y^2 + 3xy + x^2)$

12. **a.** Factor: $(y - 3)^3 + 27$
$y(y^2 - 9y + 27)$
b. Factor: $(x + 8)^2 - (x - 9)^2$
$17(2x - 1)$

5.4 EXERCISES

Note: Use your graphing calculator to check your results whenever possible.

In Exercises 1 through 12, find the greatest common factor of the given monomials.

1. $25x^2, 45x$
$5x$

2. $68x^3, 16x$
$4x$

3. $54t^3, 63t^2$
$9t^2$

4. $96t^2, 48t^3$
$48t^2$

5. $90xy^2, 18x^2y$
$18xy$

6. $24x^2y^2, 400xy$
$8xy$

7. $260t^2u, 200t^2u^2$
$20t^2u$

8. $56tv, 42t^2v^2$
$14tv$

9. x^6y, xy^9, x^5y^5
xy

10. x^4y^2, xy^6, x^2y^2
xy^2

11. $69m^6, 18m, 23m$
m

12. $21m^3t, 63mt, 42m^2t^2$
$21mt$

In Exercises 13 through 36, factor by removing the greatest common factor.

13. $8x^4 - 4x^6$
$4x^4(2 - x^2)$

14. $9t^4 - 18t^2$
$9t^2(t^2 - 2)$

15. $42x + 7x^6$
$7x(6 + x^5)$

16. $21xt + 70t^6$
$7t(3x + 10t^5)$

17. $25xy^6 + 45xy^7$
$5xy^6(5 + 9y)$

18. $28xy^6 + 63x^6y$
$7xy(4y^5 + 9x^5)$

19. $64t^6 - 24x^3t^4$
$8t^4(8t^2 - 3x^3)$

20. $9x^4yz + 15x^6y$
$3x^4y(3z + 5x^2)$

21. $8x^2y^6 - 4x^4y^6 + 2xy$
$2xy(4xy^5 - 2x^3y^5 + 1)$

22. $20x^6y^6 - 90x^4y + 10xy$
$10xy(2x^5y^5 - 9x^3 + 1)$

23. $48x^9y^6 - 24x^4y^4 - 8xy$
$8xy(6x^8y^5 - 3x^3y^3 - 1)$

24. $18x^9y^6 - 12x^6y - 6xy$
$6xy(3x^8y^5 - 2x^5 - 1)$

25. $25t^6x^6 + 30tx^4 - 5t^9x^6$
$5tx^4(5t^5x^2 + 6 - t^8x^2)$

26. $40x^4t^6 + 24xt^4 + 32t^4$
$8t^4(5x^4t^2 + 3x + 4)$

27. $24t^6 - 8tx^4 + 18xt^6$
$2t(12t^5 - 4x^4 + 9xt^5)$

28. $6x^6y^6 - 4x^4y - 2xy^4$
$2xy(3x^5y^5 - 2x^5 - y^3)$

29. $65x^6y^4 - 25x^6y + 5x^9y^6$
$5x^6y(13y^3 - 5 + x^3y^5)$

30. $60x^4y^6 - 20x^4y + 5xy^6$
$5xy(12x^3y^5 - 4x^3 + y^5)$

31. $9tx^2 - 51t^4x + 81tx^3$
$3tx(3x - 17t^3 + 27x^2)$

32. $63txy^4 - 21tx^6y^4 + 6x^4ty^6$
$3txy^4(21 - 7x^5 + 2x^3y^2)$

33. $(2x - 1)(3x + 2) + (2x - 1)(2x - 5)$
$(2x - 1)(5x - 3)$

34. $(3x - 1)(2x + 5) + (3x - 1)(3x - 8)$
$(3x - 1)(5x - 3)$

35. $(2x - 9)(3x - 7) - (3x - 8)(2x - 9)$
$2x - 9$

36. $(3x + 4)(2x - 5) - (2x - 5)(2x - 10)$
$(2x - 5)(x + 14)$

In Exercises 37 through 48, factor by grouping.

37. $x^2 + 4x + 8 + 2x$
$(x + 4)(x + 2)$

38. $y^2 + 1y + 5 + 5y$
$(y + 1)(y + 5)$

39. $r^2 + 3r + 6 + 2r$
$(r + 3)(r + 2)$

40. $8 + x^2 + x + 8x$
$(x + 1)(x + 8)$

41. $6t^2 - 10t + 20 - 12t$
$2(3t - 5)(t - 2)$

42. $7x^2 - x + 3 - 21x$
$(7x - 1)(x - 3)$

43. $r(r - 2) + (2 - r)$
$(r - 2)(r - 1)$

44. $5x(2 - x) + 4(2 - x)$
$(5x + 4)(2 - x)$

45. $4x(7x - 8) + 3(8 - 7x)$
$(7x - 8)(4x - 3)$

46. $7t(6 + t) - 4(t + 6)$
$(t + 6)(7t - 4)$

47. $16n(n - 1) + (1 - n)$
$(n - 1)(16n - 1)$

48. $3t(t + 5) + 2(5 + t)$
$(t + 5)(3t + 2)$

In Exercises 49 through 60, factor each difference of two squares.

49. $t^2 - 25$
$(t + 5)(t - 5)$

50. $x^2 - 441$
$(x - 21)(x + 21)$

51. $169 - r^2$
$(13 + r)(13 - r)$

52. $484 - v^2$
$(22 + v)(22 - v)$

53. $289u^2 - 1$
$(17u + 1)(17u - 1)$

54. $9n^2 - 25$
$(3n + 5)(3n - 5)$

55. $49t^8 - 9$
$(7t^4 + 3)(7t^4 - 3)$

56. $r^{36} - 16x^4$
$(r^{18} + 4x^2)(r^9 + 2x)(r^9 - 2x)$

57. $x^2 - 4y^{36}$
$(x + 2y^{18})(x - 2y^{18})$

58. $64x^8 - 81$
$(8x^4 + 9)(8x^4 - 9)$

59. $25x^{16} - 121$
$(5x^8 + 11)(5x^8 - 11)$

60. $9y^4 - 169$
$(3y^2 + 13)(3y^2 - 13)$

In Exercises 61 through 78, factor each perfect-square trinomial.

61. $y^2 - 10y + 25$
$(y - 5)^2$

62. $y^2 + 6y + 9$
$(y + 3)^2$

63. $y^2 + 4y + 4$
$(y + 2)^2$

64. $9x^2 - 6x + 1$
$(3x - 1)^2$

65. $36x^2 - 12x + 1$
$(6x - 1)^2$

66. $72x^2 + 24x + 2$
$2(6x + 1)^2$

67. $36x^2 + 24x + 4$
$4(3x + 1)^2$

68. $4y^2 - 16y + 16$
$4(y - 2)^2$

69. $3y^2 + 30y + 75$
$3(y + 5)^2$

70. $5y^2 + 10y + 5$
$5(y + 1)^2$

71. $6y^2 - 36y + 54$
$6(y - 3)^2$

72. $9x^2 + 42x + 49$
$(3x + 7)^2$

73. $25x^2 + 20x + 4$
$(5x + 2)^2$

74. $4x^2 - 20x + 25$
$(2x - 5)^2$

75. $49x^2 - 42x + 9$
$(7x - 3)^2$

76. $49y^2 + 112y + 64$
$(7y + 8)^2$

77. $36y^2 - 84y + 49$
$(6y - 7)^2$

78. $16y^2 - 40y + 25$
$(4y - 5)^2$

In Exercises 79 through 84, factor each sum or difference of two cubes.

79. $x^3 - 125$
$(x - 5)(x^2 + 5x + 25)$

80. $r^3 - 8$
$(r - 2)(r^2 + 2r + 4)$

81. $125r^3 - 64$
$(5r - 4)(25r^2 + 20r + 16)$

82. $8t^3 - 27$
$(2t - 3)(4t^2 + 6t + 9)$

83. $12y^3 - 12$
$12(y - 1)(y^2 + y + 1)$

84. $2x^3 + 250$
$2(x + 5)(x^2 - 5x + 25)$

In Exercises 85 through 90, factor completely.

85. $x^4 - 16$
$(x^2 + 4)(x + 2)(x - 2)$

86. $x^4 - 81$
$(x^2 + 9)(x + 3)(x - 3)$

87. $x^4 - 1$
$(x^2 + 1)(x + 1)(x - 1)$

88. $16x^4 - 1$
$(4x^2 + 1)(2x + 1)(2x - 1)$

89. $81x^4 - 1$
$(9x^2 + 1)(3x + 1)(3x - 1)$

90. $256x^4 - 1$
$(16x^2 + 1)(4x + 1)(4x - 1)$

MIXED PRACTICE

By doing these exercises, you will practice the topics up to this point in the chapter.

91. Multiply using scientific notation:
$(810,000,000)(0.009)(50,000)(0.0000002)$
$72,900$ or 7.29×10^4

92. Multiply using FOIL: $(-2y - 5t)(6y + t)$
$-12y^2 - 32ty - 5t^2$

93. Simplify: $(-27x^3y^2 + 18xy) \div (-9xy)$
$3x^2y - 2$

94. Divide: $(x^2 + 6x + 18) \div (x + 2)$
$x + 4 + \dfrac{10}{x + 2}$

95. Factor: $15x^2y^3z^4 - 20x^3y^2z + 25x^5y^6z$
$5x^2y^2z(3yz^3 - 4x + 5x^3y^4)$

96. Multiply: $(2x - 3)^2$
$4x^2 - 12x + 9$

97. What is the additive inverse of $3x^2 - 2x + 1$?
$-3x^2 + 2x - 1$

98. Multiply: $(9x^8y^3)(4xy^{-8})$
$\dfrac{36x^9}{y^5}$

99. Simplify: $(-3x^2 - 8x + 4) - (-8x^2 - 9x - 11)$
$5x^2 + x + 15$

100. Remove the greatest common factor:
$20r^2t^2 - 10r^2t + 5rt^2$
$5rt(4rt - 2r + t)$

EXCURSIONS

Data Analysis

1. An equation often given for your optimum target heart rate in exercise is

$$70\% \, (220 - \text{your age}) = \text{target heart rate}$$

Other people have argued that any level from 70% to 85% is acceptable.

a. Draw a graph showing the target heart rate for exercising adults for maximum health benefits.

b. Use what you know about linear equations to write an equation that describes the upper level of the target zone (at the 85% level).

c. Estimate the equation for an "80% level."

d. Using your graph, discuss the information presented. That is, analyze what data is given and describe it to a friend.

e. Ask and answer four questions about this data.

Posing Problems

2. Ask and answer four questions about this data.

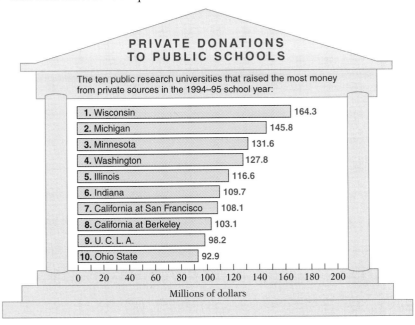

PRIVATE DONATIONS
TO PUBLIC SCHOOLS

The ten public research universities that raised the most money from private sources in the 1994–95 school year:

1. Wisconsin — 164.3
2. Michigan — 145.8
3. Minnesota — 131.6
4. Washington — 127.8
5. Illinois — 116.6
6. Indiana — 109.7
7. California at San Francisco — 108.1
8. California at Berkeley — 103.1
9. U. C. L. A. — 98.2
10. Ohio State — 92.9

Millions of dollars

Class Act

3. Use this data and see if you can predict the winner of the Super Bowl for 1997. Research to find which team really did win.

Team Comparisons

League 1 Playoff

Offense	A	B
GAMES (Won–Lost)	1–0	1–0
FIRST DOWNS	18	15
Rushing	8	10
Passing	7	5
Penalty	3	0
YDS GAINED (tot)	227	210
Avg per Game	227.0	210.0
RUSHING (net)	127	139
Avg. per Game	127.0	139.0
Rushes	37	39
Yards per Rush	3.4	3.6
PASSING (net)	100	71
Avg per Game	100.0	71.0
Passes Att.	22	15
Completed	12	11
Pct Completed	54.5	73.3
Yards Gained	100	79
Sacked	0	1
Yards Lost	0	8
Had Intercepted	1	0
Yards Opp Ret	0	0
Opp TDs on Int	0	0
PUNTS	1	6
Avg Yards	39.0	43.2
PUNT RETURNS	5	2
Avg Return	8.0	39.0
Returned for TD	0	1
KICKOFF RETURNS	5	2
Avg Return	31.0	22.0
Returned for TD	0	0
PENALTIES	5	1
Yards Penalized	38	5
FUMBLES BY	2	5
Fumbles Lost	1	1
Opp Fumbles	1	3
Opp Fum Lost	0	2
POSS. TIME (avg)	29:23	33:58
TOUCHDOWNS	2	5
Rushing	0	2
Passing	2	1
Returns	0	2
EXTRA POINTS (tot)	2	5
Kicks Made/2Pt	2/0	5/0
FIELD GOALS/FGA	4/4	0/0
POINTS SCORED	26	35

Defense	A	B
POINTS ALLOWED	17	14
OPP FIRST DOWNS	21	12
Rushing	5	4
Passing	12	8
Penalty	4	0
OPP YARDS GAINED	244	196
Avg per Game	244.0	196.0
OPP RUSHING (net)	96	68
Avg per Game	96.0	68.0
Rushes	24	18
Yards per Rush	4.0	3.8
OPP PASSING (net)	148	128
Avg per Game	148.0	128.0
Passes Att.	36	41
Completed	18	21
Pct Completed	50.0	51.2
Sacked	2	1
Yards Lost	17	5
INTERCEPTED BY	3	3
Yards Returned	122	5
Returned for TD	0	0
OPP PUNT RETURNS	0	3
Avg return	0.0	7.7
OPP KICKOFF RET	6	4
Avg return	16.8	18.8
OPP TOUCHDOWNS	1	2
Rushing	0	1
Passing	1	1
Returns	0	0

League 2 Playoff

Offense	C	D
GAMES (Won–Lost)	2–0	1–0
FIRST DOWNS	40	17
Rushing	14	9
Passing	24	6
Penalty	2	2
YDS GAINED (tot)	852	346
Avg per Game	426.0	346.0
RUSHING (net)	387	194
Avg. per Game	193.5	194.0
Rushes	71	32
Yards per Rush	5.5	6.1
PASSING (net)	465	152
Avg per Game	232.5	152.0
Passes Att.	62	26
Completed	36	15
Pct Completed	58.1	57.7
Yards Gained	484	167
Sacked	4	2
Yards Lost	19	15
Had Intercepted	2	2
Yards Opp Ret	38	0
Opp TDs on Int	1	0
PUNTS	8	7
Avg Yards	46.5	43.6
PUNT RETURNS	3	7
Avg Return	4.7	10.3
Returned for TD	0	0
KICKOFF RETURNS	9	2
Avg Return	19.3	27.0
Returned for TD	0	0
PENALTIES	9	2
Yards Penalized	60	21
FUMBLES BY	0	0
Fumbles Lost	0	0
Opp Fumbles	6	0
Opp Fum Lost	2	0
POSS. TIME (avg)	32:45	31:21
TOUCHDOWNS	6	4
Rushing	2	3
Passing	3	1
Returns	1	0
EXTRA POINTS (tot)	6	4
Kicks Made/2Pt	6/0	4/0
FIELD GOALS/FGA	6/7	0/0
POINTS SCORED	60	28

Defense	C	D
POINTS ALLOWED	54	3
OPP FIRST DOWNS	40	12
Rushing	12	7
Passing	27	5
Penalty	1	0
OPP YARDS GAINED	659	213
Avg per Game	329.5	213.0
OPP RUSHING (net)	218	123
Avg per Game	109.0	123.0
Rushes	50	27
Yards per Rush	4.4	4.6
OPP PASSING (net)	441	90
Avg per Game	220.5	90.0
Passes Att.	74	39
Completed	47	16
Pct Completed	63.5	41.0
Sacked	4	2
Yards Lost	31	20
INTERCEPTED BY	1	2
Yards Returned	20	14
Returned for TD	1	0
OPP PUNT RETURNS	7	5
Avg return	6.3	6.8
OPP KICKOFF RET	13	5
Avg return	22.7	19.8
OPP TOUCHDOWNS	7	0
Rushing	3	0
Passing	3	0
Returns	1	0

The teams are: A-Carolina; B-Green Bay; C-Jacksonville; D-New England.

Exploring Patterns

4. The following lists were found. On each ticket, the blanks can be filled by one digit and only one digit. Fill in the blanks.

a.　$5\_\_^2 = 313\_\_$

　$52\_\_^2 = 27\_\_,\_\_7\_\_$

　$5\_\_^3 = 175\_\_1\_\_$

　$5\_\_^4 = 9{,}834{,}49\_\_$

b.　$\_\_8^{\_\_} = 5{,}308{,}\_\_16$

　$\_\_6^{\_\_} = \_\_,\_\_77,\_\_56$

　$\_\_\_\_^{\_\_} = 3{,}7\_\_8{,}096$

　$22^{\_\_} = 23\_\_,256$

c.　$\_\_\_\_^5 = \_\_6\_\_,05\_\_$

　$3\_\_^7 = 27{,}5\_\_2{,}6\_\_4{,}_____$

　$4\_\_^5 = \_\_\_\_5{,}856{,}20\_\_$

　$_____^3 = \_\_,367{,}63\_\_$

CONNECTIONS TO *GEOMETRY*

Finding Area

Area of a Triangle

To find the area of a triangle, we must first identify the *base* and the *height* of the triangle. Then:

To find the area of a triangle

If b is the length of the base and h is the height, then

$$\text{Area} = \tfrac{1}{2} \times b \times h, \quad \text{or } A = \tfrac{1}{2}bh$$

Any side can be chosen as the base of a triangle. The height is then the shortest distance from the angle opposite that base. If the triangle is a right triangle, then the height can be the length of any side except the hypotenuse. Usually, though, the height is not the same as the length of a side. But the height always makes a right (90°) angle with the base.

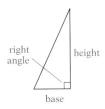

Right triangle

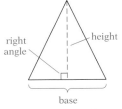

Acute triangle

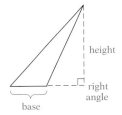

Obtuse triangle

Area of a Trapezoid

In this figure, trapezoid $ABCD$ has been divided into two triangles.

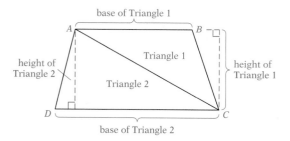

To find the area of the trapezoid, we can add the areas of the triangles. That is,

$$\text{Area of trapezoid} = \text{area}\,①\,+\,\text{area}\,②$$
$$= \tfrac{1}{2}\,\text{base}\,①\times\,\text{height}\,①\,+\,\tfrac{1}{2}\,\text{base}\,②\times\text{height}\,②$$

The heights of the triangles are equal, so we can write this as

$$\text{Area of trapezoid} = \tfrac{1}{2}\,\text{height}\,(\text{base}\,①\,+\,\text{base}\,②).$$

In symbolic form:

> **To find the area of a trapezoid**
>
> If h is the height, and b_1 and b_2 are the lengths of the bases, then
> $$A = \tfrac{1}{2}h(b_1 + b_2)$$

▪▪▪

EXAMPLE

Find the area of a trapezoid with bases of 12 and 36 units and height 10 units.

SOLUTION

We substitute the values for b_1, b_2, and h into the formula

$$A = \tfrac{1}{2}h(b_1 + b_2)$$
$$= \tfrac{1}{2}(10)(12 + 36)$$
$$= \tfrac{1}{2}(10)(48)$$
$$= 5(48)$$
$$= 240$$

So the area is 240 square units.

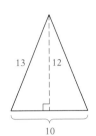

■■■
EXAMPLE

Find the area of the triangle shown in the accompanying figure. The unit of length is feet.

SOLUTION

We choose the 10-foot side as the base; then the height is equal to 12 feet, not 13 feet. Using the formula, we substitute the known values for b and h.

$$A = \frac{1}{2}bh$$
$$= \frac{1}{2} \times 10 \times 12$$
$$= \frac{120}{2}$$
$$= 60$$

So the area of the triangle is 60 square feet. (Remember, area is expressed in *square* units.)

■■■
> **WRITER'S BLOCK**
>
> Explain why 13 feet cannot be the base.

■■■
EXAMPLE

The perimeter of a triangle is 19 centimeters. The length of the second side of the triangle is twice that of the first, and the length of the third side is three centimeters greater than that of the first. Find the lengths of all three sides of the triangle.

SOLUTION

We let a be the length of the first side. Then the second side has length $2a$, and the third side has length $3 + a$. We add these three sides to get the perimeter, 19 cm.

$$P = a + b + c \qquad \text{Formula}$$
$$19 = a + (2a) + (3 + a) \qquad \text{Substituting values}$$
$$19 = a + 2a + 3 + a \qquad \text{Removing parentheses}$$
$$19 = 4a + 3 \qquad \text{Combining like terms}$$
$$19 + (-3) = 4a + 3 + (-3) \qquad \text{Addition principle}$$
$$16 = 4a \qquad \text{Simplifying}$$
$$\left(\tfrac{1}{4}\right)16 = \left(\tfrac{1}{4}\right)(4a) \qquad \text{Multiplication principle}$$
$$4 = a \qquad \text{Simplifying}$$

Going back to the original problem and substituting $a = 4$ gives us the three sides:

$$a = 4$$
$$b = 2a = 8$$
$$c = 3 + a = 7$$

So the sides of the triangle are 4, 7, and 8 centimeters. The check is left for you to do.

PRACTICE

1. ✏ Two triangles have equal bases and equal heights, but one is a right triangle and one is a scalene triangle. Which has the larger area? Justify your answer.   They have equal bases and equal heights, so they have the same area.

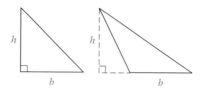

2. Which has the greater area, a triangular sail whose base is $3\frac{1}{5}$ yards and whose height is $7\frac{3}{5}$ yards, or a $5\frac{1}{2}$-yard square sail?   the $5\frac{1}{2}$-yard square sail

3. ✏ Are the terms "$5\frac{1}{2}$-yard square" and "$5\frac{1}{2}$ square yard" equivalent terms? Justify your answer.

4. Find the area of the accompanying triangle. $26\frac{1}{4}$ square yards

5. The length of a rectangular brass door knocker is one-half of the width. If the perimeter is 14.4 inches, find the dimensions of the door knocker.
$w = 4.8$ inches; $\ell = 2.4$ inches

6. The width of a rectangular nameplate is 5 centimeters more than 4 times the length. If the perimeter is 40 centimeters, find the dimensions of the nameplate. $w = 17$ centimeters; $\ell = 3$ centimeters

7. The perimeter of a small triangular brace is $13\frac{4}{5}$ inches. The second side is 1 inch longer than the first, and the third side is 1 inch longer than the second. Find the lengths of the sides.   3.6 inches, 4.6 inches, and 5.6 inches

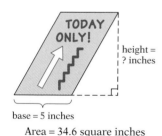

8. The area of a sign shaped like a parallelogram is 34.6 square inches. If the length of the base is 5 inches, find the height.   6.92 inches

9. A base for a sculpture is a trapezoid and has a perimeter of 100 inches. One base is 34 inches long. The other base and the two sides are equal in length. What is this length?   22 inches

10. Find the perimeter of a basketball court that has a length of 28 yards and a width of 15 yards 9 inches.   86.5 yards

5.5   Trinomial Functions and Equations

SECTION LEAD-IN

Mathematics can be used to explore many natural phenomena. Entomologists, for example, use mathematics to study the rhythms of cicadas.

Cicadas are insects. Many cicadas have life cycles that are prime. There is a 13-year cicada and a 17-year cicada as well as others. (You could research this.) Neither of these insects has natural predators for a good reason. A natural predator needs a regular food supply. The 13-year and 17-year cicadas only appear every 13 or 17 years. During the rest of the time, what are they doing?

SECTION GOALS

- To factor trinomials of the form $ax^2 + bx = c$
- To find $f \circ g$ for functions and identify the domain of the composition
- To solve equations by using the principle of zero products

Trinomials with $a = 1$

If we multiply two binomials $x + m$ and $x + n$, we get a trinomial.

$$\overset{\text{F} \quad\text{O}\quad\text{I}\quad\text{L}}{(x + n)(x + m) = x^2 + mx + nx + mn}$$
$$= x^2 + (m + n)x + mn$$

In the product, the coefficient of x is the sum of m and n, and the constant is the product of m and n. Here m and n can take on any value, including zero.

To factor a trinomial $x^2 + bx + c$, we need to write it as the product of two binomial factors.

$$x^2 + bx + c = (x + m)(x + n)$$
$$\overset{\uparrow}{(m + n)} \quad \overset{\uparrow}{(mn)}$$

We can do this by finding m and n such that

$$m + n = b \quad \text{and} \quad m \cdot n = c$$

> **To factor a trinomial in the form $x^2 + bx + c$**
>
> 1. List all pairs of integers whose product is c.
> 2. Of these, find the pair m and n whose sum is b.
> 3. Write the factorization of the trinomial as $(x + m)(x + n)$.

▪▪▪

EXAMPLE 1

Factor: $x^2 + 9x + 20$

SOLUTION

The coefficient of x^2 is 1, so we can use the procedure just outlined. We first want m and n such that $m \cdot n = 20$. Six pairs of integers have the product 20:

$$20 \text{ and } 1 \qquad -20 \text{ and } -1$$
$$10 \text{ and } 2 \qquad -10 \text{ and } -2$$
$$5 \text{ and } 4 \qquad -5 \text{ and } -4$$

We also know that $m + n = 9$. To determine which of the pairs of numbers has the sum of 9, we add

$$20 + 1 = 21 \quad \text{Not good}$$
$$10 + 2 = 12 \quad \text{Not good}$$
$$5 + 4 = 9 \quad \text{Good}$$

Thus m and n are 5 and 4. (We can stop checking as soon as we find a pair that works because only one pair will work). The factors are $(x + 5)(x + 4)$.

We check by multiplying the factors:

$$(x + 5)(x + 4) = x^2 + 4x + 5x + 20$$
$$= x^2 + 9x + 20$$

▶ CHECK **Warm-Up 1**

In factoring a trinomial, you do not always have to try each possible pair of integers m and n. You can eliminate some pairs by considering the signs of the constants b and c in $x^2 + bx + c$.

> Let $x^2 + bx + c = (x + m)(x + n)$. When b and c are *positive*, then m and n are both positive.

Thus, when every constant in the trinomial is positive, as in Example 1, both m and n are positive.

> Let $x^2 + bx + c = (x + m)(x + n)$. When c is positive and b is negative, m and n are both negative.

We use this property in the next example.

▪▪▪

EXAMPLE 2

Factor: $y^2 - 5y + 6$

SOLUTION

The coefficient of y is 1, so we can use the factoring procedure. We want m and n such that $m \cdot n = 6$ and $m + n = -5$. Because $c = m \cdot n$ is positive and $b = m + n$ is negative, we consider only negative values of m and n. We have

Pairs, m and n where $m \cdot n = 6$	Sums, $m + n$
-6 and -1	$-6 + (-1) = -7$
-2 and -3	$-2 + (-3) = -5$ Good

So m and n are -2 and -3; the factors are $(x - 2)(x - 3)$. Check them with a FOIL multiplication.

▶ *CHECK* **Warm-Up 2**

Calculator Corner

You can use graphing to check your answers when factoring trinomials. For instance, check $x^2 + 3x + 2 = (x + 2)(x + 1)$ using your graphing calculator.

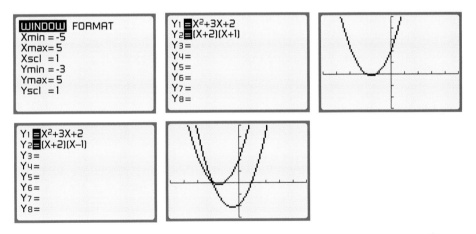

Two different graphs mean that the two sides of the equation *are not equal.*

Let $x^2 + bx + c = (x + m)(x + n)$. When c is *negative*, then either m or n is negative, but not both. Of m and n, the one with the greater absolute value has the same sign as b.

This property is illustrated in the next example.

Removing a common factor: Most trinomials do not have 1 as the coefficient of the x^2 term (called the **leading coefficient**). For some trinomials, however, we can remove a common factor to obtain x^2 as the first term.

▪ ▪ ▪

EXAMPLE 3

Factor: $3x^2 + 6x - 105$

SOLUTION

The leading coefficient is not 1 here, but the factor 3 is common to all three terms. So we first factor out the 3, obtaining

$$3x^2 + 6x - 105 = 3(x^2 + 2x - 35)$$

To factor the trinomial in parentheses, we need to find m and n such that $m \cdot n = -35$ and $m + n = 2$.

The pairs of integers that give a product of -35, along with their sums, are

Pairs, m and n	Sums, $m + n$
35 and -1	$35 + (-1) = 34$
-35 and 1	$-35 + 1 = -34$
7 and -5	$7 + (-5) = 2$ Good
-7 and 5	Not needed

Note that we checked only pairs where the integer with the greater absolute value was positive, like $b = m + n$.

Thus m and n are 7 and -5, and the factors of $x^2 + 2x - 35$ are $(x + 7)$ and $(x - 5)$. But we began by factoring out a 3, and we must keep it, so

$$3x^2 + 6x - 105 = 3(x + 7)(x - 5)$$

You can check by multiplying.

▶ *CHECK* **Warm-Up 3**

Trinomials in two variables: We can use this same procedure to factor certain trinomials in *two* variables. They must have the form $x^2 + bxy + cy^2$.

▪ ▪ ▪

EXAMPLE 4

Factor: $5x^2y^2 - 55xy^3 + 150y^4$

SOLUTION

We first divide each term by the common factor $5y^2$, getting

$$5x^2y^2 - 55xy^3 + 150y^4 = 5y^2(x^2 - 11xy + 30y^2)$$

We can now factor the trinomial in parentheses. We want m and n such that $m \cdot n = 30$ and $m + n = -11$. We know that m and n are both negative, so the choices and their sums are

Pairs, m and n	Sums $m + n$
-30 and -1	$-30 + (-1) = -31$
-6 and -5	$-6 + (-5) = -11$ Good

Thus the correct pair is $-6y$ and $-5y$, and the trinomial can be factored as $(x - 6y)(x - 5y)$. Returning the common factor $5y^2$ gives us

$$5y^2(x - 6y)(x - 5y)$$

Don't forget to check.

▶ CHECK **Warm-Up 4**

Prime trinomials: Certain trinomials in the form $ax^2 + bx + c$ are called **prime trinomials.** These trinomials cannot be factored in the form $(x + m)(x + n)$ such that m and n are integers. One such trinomial is $x^2 + x + 6$. Many more exist.

Trinomials $ax^2 + bx + c$ with $a \neq 1$

Even after a common factor is removed from a trinomial, the leading coefficient may not be 1. Then, to factor the trinomial, we must use a method that is partly trial and error and partly the previous method.

> **To factor a trinomial of the form $ax^2 + bx + c$ when a, b, and c have no common factors other than 1 or -1.**
>
> 1. If the term ax^2 is negative, remove the common factor -1 to make it positive.
> 2. List the possible pairs of positive factors of a.
> 3. List the possible pairs of factors of c (both positive and negative).
> 4. Combine pairs of factors from steps 2 and 3, in turn, to form trial binomial factors. Determine the middle terms that these trial factors would produce when multiplied together.
> 5. The trial factors that produce the proper middle term bx are the factors of the original trinomial.

This procedure is not so involved as it looks. Consider the next example.

▪ ▪ ▪
EXAMPLE 5

Factor: $2x^2 - 9x + 7$

SOLUTION

The terms of the trinomial do not have a common factor, and the x^2 term is not negative.

The coefficient of x^2 is 2, and the only pair of positive factors of 2 is 1 and 2. The constant term is 7, and its possible pairs of factors are 7 and 1, or -7 and -1. We now have

Factors of 2	Factors of 7
1, 2	7, 1
	$-7, -1$

We note, however, that the last term of the trinomial is positive and the middle term is negative. Consequently, we can eliminate the positive factors of 7. We end up with

Factors of 2	Factors of 7
1, 2	~~7, 1~~
	$-7, -1$

Now we must combine these factors into trial binomial factors. The possible combinations are

$$(1x - 7)(2x - 1) \quad \text{or} \quad (1x - 1)(2x - 7)$$

Note that the second pair of trial factors is the same as the first, except that the factors of 7 are interchanged.

By using FOIL, we find that the trial factors $(1x - 7)(2x - 1)$ produce the middle term

$$\overset{\text{O}}{(-1)(1x)} + \overset{\text{I}}{(-7)(2x)} = -15x$$

The trial factors $(1x - 1)(2x - 7)$ give the middle term

$$\overset{\text{O}}{(-7)(1x)} + \overset{\text{I}}{(-1)(2x)} = -9x$$

This is the middle term of the original trinomial, so

$$2x^2 - 9x + 7 = (x - 1)(2x - 7)$$

You can check it by multiplying.

▶ CHECK Warm-Up 5

Calculator Corner

You can check your work with a graphing calculator when you factor a trinomial into two binomials. As an example, suppose you want to factor the trinomial $12x^2 + 8x - 15$. If you did it correctly by paper-and-pencil methods, you should have gotten the result: $(2x + 3)(6x - 5)$. If your work is correct, how should the graphs of the trinomial and the factorization compare? Examine the following graphs.

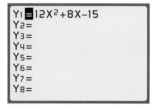
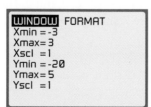
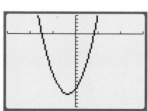

 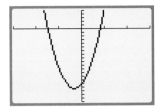

Question: How do the graphs compare? Is this what you would expect? Explain your reasoning.

Now examine the following factorization.

$$6x^2 - 53x - 9 = (6x + 3)(x - 3)$$

Why are these two graphs different?

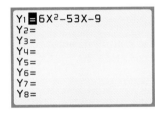 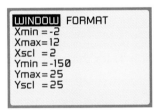 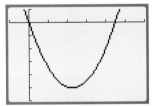

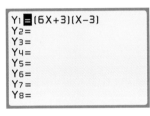 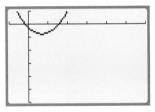

Now use paper-and-pencil methods to factor the following trinomials. Then graph the original trinomial and your answer. If your work is correct, what should you expect to see?

$$x^2 - 8x - 33 \qquad x^2 - 3x + 2 \qquad 8x^2 + 14x + 3$$
$$12x^2 - 11x - 15 \qquad 8x^2 + 46x - 12$$

Remove the common factors from the trinomial first. Then eliminate every trial binomial factor that has a common factor.

In the next example, we use the Study Hints to eliminate some choices.

∎∎∎

EXAMPLE 6

Factor: $3x^2 - 188x - 63$

SOLUTION

There is no common factor. The coefficient of x^2 is 3, with possible positive factors 3 and 1. The constant term is negative, so it has one positive and one negative

STUDY HINT

If the terms of a trinomial do not have a common factor, then the terms of its trial factors cannot have a common factor.

factor. Possibilities are -63 and 1; 63 and -1; -21 and 3; 21 and -3; -9 and 7; or 9 and -7. So we have:

STUDY HINT

If the absolute value of a trial middle term is not the same as the absolute value of the middle term of the trinomial being factored, you can eliminate both those factors and the "opposite" factors.

Factors of 3	Factors of 63
3, 1	$-63, 1$
	$63, -1$
	$-21, 3$
	$21, -3$
	$-9, 7$
	$9, -7$

Now we combine each pair on the left with each pair on the right, in turn, to form trial binomial factors. We begin at the top of the lists and eliminate any trials that produce a common factor (see the Study Hint on page 383). For example, the factors at the top of the two lists would give us the trial factors.

$$(3x - 63)(1x + 1)$$

We can eliminate this trial because 3 and -63 have the common factor 3. We can eliminate $(3x + 63)(x - 1)$ for the same reason.

We are left with

Possible Factors

$$(3x - 7)(x + 9)$$
$$(3x + 7)(x - 9)$$
$$(3x - 1)(x + 63)$$
$$(3x + 1)(x - 63)$$

The goal is a middle term of $-188x$. To reach it, we find the trial middle terms. For the top factors, we get

Factors	Middle term
$(3x - 7)(x + 9)$	$27x - 7x = 20x$

This pair gives $20x$ as a middle term; then, according to the Study Hint at the top of the page, the next pair $(3x + 7)(x - 9)$ will give $-20x$ as a middle term and can be eliminated. We continue:

Factors	Middle term	
$(3x - 1)(x + 63)$	$189x - x = 188x$	
$(3x + 1)(x - 63)$	$-189x + x = -188x$	Good

So $3x^2 - 188x - 63 = (3x + 1)(x - 63)$.

▶ CHECK **Warm-Up 6**

Composite Functions

One way to produce new functions is to form composite functions.

Composite Function

Let f be a function from set X to set Y, and let g be a function from set Y to set Z. Then the **composite function** $g \circ f$ is a function from X to Z defined by $(g \circ f)(x) = g(f(x))$

The following figure shows how this composite function works. At the top of the figure are the two original functions f and g. Function f assigns, to each element x of its domain X, one element $f(x)$ of its range Y. Then function g (whose domain must be Y) assigns one element $g(f(x))$ of its range Z to each element of Y.

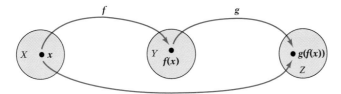

The function $(g \circ f)$ does the same thing in only one step, as you can see at the bottom of the figure: It assigns an element $g(f(x))$ of Z directly to each element x of X. To find $(g \circ f)$, we calculate $g(f(x))$ for f and g given in functional notation.

■ ■ ■

EXAMPLE 7

Let $f(x) = x^2 + 1$ and $g(x) = 2x + 3$.

a. Find $(g \circ f)(x)$ and $(g \circ f)(2)$. **b.** Show that $(g \circ f)(2)$ is equal to $g(f(2))$.

SOLUTION

a. From the definition of composite functions,

$$(g \circ f)(x) = g(f(x))$$

This tells us to evaluate the function g at the value $f(x) = x^2 + 1$, as if it were a value for g. Therefore, we substitute $x^2 + 1$ for x in the expression $g(x) = 2x + 3$. We get

$(g \circ f)(x) = g(x^2 + 1) = 2(x^2 + 1) + 3$ Substituting $x^2 + 1$ in $g(x) = 2x + 3$
$\qquad = 2x^2 + 5$ Simplifying

We use this result to compute $(g \circ f)(2) = 2(2^2) + 5 = 13$.

b. From the original function $f(x) = x^2 + 1$, we compute $f(2) = (2)^2 + 1 = 5$.

Then, from the original function $g(x) = 2x + 3$, we compute

$$g(f(2)) = g(5) = 2(5) + 3 = 13$$

Thus $(g \circ f)(2)$ is equal to $g(f(2))$. This equality holds for all real numbers.

▶ CHECK **Warm-Up 7**

The process of finding a composite function is called a **composition of functions.** Sometimes $(g \circ f)$ is called "the composition g of f." It always means "find $f(x)$ first, and then find $g(f(x))$."

In Example 7, $(f \circ g)$ is not equal to $(g \circ f)$. This is true in general: The process of composition is not commutative.

When we find composite functions, we begin at the rightmost function and work to the left. In Example 8, we find the composition of three polynomial functions. The procedure remains the same.

∎ ∎ ∎

Example 8

Let $f(x) = x^2 - 1$
$g(x) = x + 4$
$h(x) = 2x - 1$

Find $(g \circ h \circ f)(x)$.

Solution

Working from right to left, we begin by finding $(h \circ f)(x)$. To do that, we substitute the expression for $f(x)$, which is $x^2 - 1$, in $h(x)$ as if it were a domain value for h. We get

$$
\begin{aligned}
(h \circ f)(x) &= h(x^2 - 1) \\
&= 2(x^2 - 1) - 1 \quad \text{Substituting } x^2 - 1 \text{ for } x \text{ in } h(x) \\
&= 2x^2 - 2 - 1 \quad \text{Simplifying} \\
&= 2x^2 - 3 \quad \text{Simplifying}
\end{aligned}
$$

Now we use this as the domain value for g, obtaining

$$
\begin{aligned}
(g \circ h \circ f)(x) &= g(2x^2 - 3) \\
&= (2x^2 - 3) + 4 \quad \text{Substituting } 2x^2 - 3 \text{ for } x \text{ in } g(x) \\
&= 2x^2 + 1 \quad \text{Simplifying}
\end{aligned}
$$

▶ *CHECK* **Warm-Up 8**

Using Factoring to Solve Equations

We can solve certain equations by first factoring and then applying the following basic principle:

> **Writer's Block**
>
> What do we mean by zero products?

Principle of Zero Products

If A and B represent real numbers and $A \cdot B = 0$, then

$$A = 0 \quad \text{or} \quad B = 0 \quad \text{or both}$$

To use this principle, we write the equation in the form

$$\text{Polynomial} = 0$$

We then factor the polynomial, set each factor equal to zero, and solve.

∎ ∎ ∎

Example 9

Solve: $36x^2 - 51x = 120$

Solution

We cannot factor the left side of this equation until we have zero on the right. So we add -120 to each side.

$$36x^2 - 51x - 120 = 0$$

Removing the common factor, 3, and then factoring the trinomial, we get

$$3(12x^2 - 17x - 40) = 0$$
$$3(4x + 5)(3x - 8) = 0$$

We then multiply both sides of the equation by $\frac{1}{3}$ to eliminate the constant, 3.

$$(4x + 5)(3x - 8) = 0$$

Now we set each factor equal to zero and solve.

$$
\begin{array}{c|c}
4x + 5 = 0 & 3x - 8 = 0 \\
4x = -5 & 3x = 8 \\
x = \dfrac{-5}{4} & x = \dfrac{8}{3}
\end{array}
$$

We leave the check for you to do.

▶ *CHECK* **Warm-Up 9**

INSTRUCTOR NOTE

Encourage students to check answers always. Although these solutions, if correctly computed, will always check, that situation will change when we work with equations containing radicals or absolute value expressions.

In Example 9, we multiplied by the inverse of a constant factor to remove it. We cannot do that with a variable factor but instead must retain it in the equation; a variable might equal zero, and $\frac{1}{0}$ is *not* a real number.

■ ■ ■
EXAMPLE 10

Solve: $4n^3 + 48n^2 + 4n = -140n$

SOLUTION

We add $140n$ to both sides of this equation to obtain zero on the right side. The resulting equation has a common factor, which we remove.

$$4n^3 + 48n^2 + 4n + 140n = 0 \quad \text{Adding } 140n \text{ to both sides}$$
$$4n^3 + 48n^2 + 144n = 0 \quad \text{Combining like terms}$$
$$4n(n^2 + 12n + 36) = 0 \quad \text{Factoring out } 4n$$

Now we see that the expression in parentheses is a perfect-square trinomial. We factor it and get

$$4n(n + 6)(n + 6) = 0$$

There are three factors here, but the principle of zero products still applies. We set each factor equal to zero and solve.

$$
\begin{array}{c|c|c}
4n = 0 & n + 6 = 0 & n + 6 = 0 \\
n = 0 & n = -6 & n = -6
\end{array}
$$

The solution set is $\{(0, -6, -6)\}$. If we had removed the factor n, we would have lost one of these solutions. We also normally list each different solution once. The solution $\{-6\}$ appears twice, but we give $\{0, -6\}$ as the solution set.

▶ *CHECK* **Warm-Up 10**

▪▪▪
EXAMPLE 11

Solve: $y^2 - 8y + 15 = 0$

SOLUTION

The terms of this trinomial have no common factors. We want integers m and n such that

$$m \cdot n = a \cdot c = (1)(15) = 15$$

and

$$m + n = b = -8$$

So m and n are -3 and -5, in either order.

We now rewrite the trinomial, substituting $-3y - 5y$ for the monomial $-8y$ — the middle term.

$$y^2 - 8y + 15 = y^2 - 3y - 5y + 15$$

Finally, we group and factor.

$$\begin{aligned} y^2 - 3y - 5y + 15 &= (y^2 - 3y) + (-5y + 15) &&\text{Grouping terms} \\ &= y(y - 3) + (-5)(y - 3) &&\text{Factoring each group} \\ &= (y - 5)(y - 3) &&\text{Simplifying} \end{aligned}$$

Setting these factors equal to zero and solving for y, we have

$$\begin{array}{c|c} y - 5 = 0 & y - 3 = 0 \\ y = 5 & y = 3 \end{array}$$

So the solution set is $\{3, 5\}$.

Note that we could have factored without grouping. The result would have been the same.

▶ CHECK **Warm-Up 11**

Often we use factoring to solve word problems.

▪▪▪
EXAMPLE 12

To fit a rectangular picture inside a frame, an artist trims 0.75 inches off each side of the picture. The original length was 4 inches longer than the width. The new area is $152\frac{1}{4}$ square inches. What are the new dimensions?

SOLUTION

We represent the original dimensions using algebra.

Original width $= x$

Original length $= x + 4$

Cut off 0.75

The new dimensions are:

New width $= x - 2(0.75) = x - 1.5$

New length $= (x + 4) - 2(0.75) = (x + 4) - 1.5$

The area of a rectangle is $A = \ell w$, so we have, for the new dimensions,

$$152.25 = [(x + 4) - 1.5][x - 1.5] \quad \text{From } A = \ell w$$
$$152.25 = (x + 2.5)(x - 1.5) \qquad \text{Simplifying}$$
$$152.25 = x^2 + x - 3.75 \qquad \text{Multiplying}$$
$$0 = x^2 + x - 156 \qquad \text{Adding } -152.25 \text{ to both sides}$$

We factor this equation as follows:

$$(x + 13)(x - 12) = 0$$

and solve.

$x + 13 = 0$	$x - 12 = 0$
$x = -13$	$x = 12$

The solution $x = -13$ will give us negative lengths, so we discard it. From $x = 12$ we find that

New width $= 12 - 1.5 = 10.5$ inches

New length $= (12 + 4) - 1.5 = 14.5$ inches

▶ *CHECK* **Warm-Up 12**

Practice what you learned.

SECTION FOLLOW-UP

You can research the answer to the question posed in the Section Lead-In. Here are two more questions.

1. Two (mythical) cicadas, a 13-year and a 17-year cicada, decide to meet again to compare notes. If they first met in 1998, when will they meet again?   in the year 2219

2. A third insect, an 11-year cicada, joins up with the first two. When will they all meet if they first met in 1943?   in the year 4374

5.5 WARM-UPS

Work these problems before you attempt the exercises.

1. Factor: $n^2 + 6n + 5$
   $(n + 5)(n + 1)$

2. Factor: $n^2 - 12n + 20$
   $(n - 10)(n - 2)$

3. Factor: $4x^2 - 12x - 72$
   $4(x - 6)(x + 3)$

4. Factor: $2x^2 - 30xy + 72y^2$
   $2(x - 3y)(x - 12y)$

5. Factor: $3x^2 + 5x + 2$
   $(3x + 2)(x + 1)$

6. Factor: $-244x + 35x^2 - 7$
   $(35x + 1)(x - 7)$

7. Let $f(x) = 7x^2$ and $g(x) = 2x - 1$. Find $(g \circ f)(x)$ and $(g \circ f)(-2)$.
   $(g \circ f)(x) = 14x^2 - 1; (g \circ f)(-2) = 55$

8. Using the polynomial functions in Example 8, find $(f \circ h \circ g)(x)$ and $(g \circ f \circ h)(x)$. $4x^2 + 28x + 48; 4x^2 - 4x + 4$

9. Solve: $12x^2 + 24 = 36x$ $\{1, 2\}$

10. Solve: $-12y^3 + 44y^2 = -16y$ $\left\{-\frac{1}{3}, 0, 4\right\}$

11. Solve: $35x^2 + 48x - 27 = 0$ $\left\{\frac{3}{7}, -\frac{9}{5}\right\}$

12. The product of two positive numbers is 837. One number is four more than the other. What are the numbers?   27 and 31

5.5 EXERCISES

Note: Use your graphing calculator to check your results whenever possible.

In Exercises 1 through 8, find a such that each set of factors yields the given trinomial.

1. $y^2 + 11y + 18 = (y + 2)(y + a)$
   $(y + 2)(y + 9)$

2. $x^2 + 13x + 40 = (x + 5)(x + a)$
   $(x + 5)(x + 8)$

3. $x^2 + x - 12 = (x + 4)(x + a)$
   $(x + 4)[x + (-3)]$

4. $n^2 + 2n - 8 = (n + a)(n + 4)$
   $[n + (-2)](n + 4)$

5. $y^2 + 3y - 40 = (y + a)(y + 8)$
   $[y + (-5)](y + 8)$

6. $x^2 + 5x - 24 = (x + a)(x + 8)$
   $[x + (-3)](x + 8)$

7. $t^2 - 3t + 2 = (t + a)(t - 2)$
   $[t + (-1)](t - 2)$

8. $t^2 - 10t + 9 = (t + a)(t - 1)$
   $[t + (-9)](t - 1)$

In Exercises 9 through 16, supply the correct signs so that the given binomials are factors of the trinomial.

9. $y^2 - 8y - 33$: $(y \underline{\ -\ } 11)(y \underline{\ +\ } 3)$

10. $x^2 - 6x - 72$: $(x \underline{\ +\ } 6)(x \underline{\ -\ } 12)$

11. $-42 + t^2 + t$: $(t \underline{\ +\ } 7)(t \underline{\ -\ } 6)$

12. $2 + x^2 - 3x$: $(x \underline{\ -\ } 2)(x \underline{\ -\ } 1)$

13. $2t^2 + xt - 15x^2$: $(2t \underline{\ -\ } 5x)(t \underline{\ +\ } 3x)$

14. $8x^2 + 14xy + 3y^2$: $(4x \underline{\ +\ } y)(2x \underline{\ +\ } 3y)$

15. $8y^2 + 46y - 12$: $(8y \underline{\ -\ } 2)(y \underline{\ +\ } 6)$

16. $12v^2 - 11v - 15$: $(3v \underline{\ -\ } 5)(4v \underline{\ +\ } 3)$

In Exercises 17 through 46, factor the trinomials. Be sure to remove any common factors first.

17. $x^2 + 7x + 6$
$(x + 6)(x + 1)$

18. $t^2 - 11t + 28$
$(t - 4)(t - 7)$

19. $n^2 - 2n - 35$
$(n + 5)(n - 7)$

20. $n^2 + 5n - 14$
$(n - 2)(n + 7)$

21. $y^2 + 18y + 81$
$(y + 9)(y + 9)$

22. $x^2 - 3x - 40$
$(x + 5)(x - 8)$

23. $45x^2 - 9x - 2$
$(15x + 2)(3x - 1)$

24. $45x^2 + 27x - 2$
$(15x - 1)(3x + 2)$

25. $3x^2 + 33x + 90$
$3(x + 5)(x + 6)$

26. $7x^2 - 7xt - 14t^2$
$7(x + t)(x - 2t)$

27. $8x^2 - 48xt - 128t^2$
$8(x - 8t)(x + 2t)$

28. $4x^2 + 12x - 40$
$4(x + 5)(x - 2)$

29. $5x^2 - 65x + 210$
$5(x - 6)(x - 7)$

30. $9x^2 + 54x - 63$
$9(x + 7)(x - 1)$

31. $6x^2 + 6xy - 432y^2$
$6(x - 8y)(x + 9y)$

32. $30x + 54 - 16x^2$
$-2(8x + 9)(x - 3)$

33. $12v^2 + 10vt - 8t^2$
$2(2v - t)(3v + 4t)$

34. $60v^2 + 147v + 90$
$3(5v + 6)(4v + 5)$

35. $9tx - 9t^2 + 54x^2$
$9(-t + 3x)(t + 2x)$

36. $12t - 32 + 2t^2$
$2(t + 8)(t - 2)$

37. $5t + 5t^2 - 150$
$5(t + 6)(t - 5)$

38. $-(144 + 392v - 392v^2)$
$8(7v - 9)(7v + 2)$

39. $-(287v + 56 - 294v^2)$
$7(6v + 1)(7v - 8)$

40. $4t^2 + 40tx + 36x^2$
$4(t + 9x)(t + x)$

41. $738v - 144v^2 - 405$
$-9(8v - 5)(2v - 9)$

42. $-19x + 15x^2 - 10$
$(3x - 5)(5x + 2)$

43. $162 + 3t^2 - 45t$
$3(t - 6)(t - 9)$

44. $46x + 21x^2 - 7$
$(3x + 7)(7x - 1)$

45. $-(184v + 180 - 252v^2)$
$4(9v + 5)(7v - 9)$

46. $49y^2 - 7t^2 - 42ty$
$-7(t + 7y)(t - y)$

In Exercises 47 through 52, find $h(x) = f(x) \cdot g(x)$.

47. $f(x) = 3x^3 - 5$
$g(x) = x + 4$
$3x^4 + 12x^3 - 5x - 20$

48. $f(x) = x - 5$
$g(x) = x^2 - 8$
$x^3 - 5x^2 - 8x + 40$

49. $f(x) = 7x^2 - 3$
$g(x) = x - 7$
$7x^3 - 49x^2 - 3x + 21$

50. $f(x) = 4x - 3$
$g(x) = 5x^2 - 2x + 6$
$20x^3 - 23x^2 + 30x - 18$

51. $f(x) = -9x^3 - 6$
$g(x) = x - 7$
$-9x^4 + 63x^3 - 6x + 42$

52. $f(x) = -2x + 9$
$g(x) = 2x + 9$
$81 - 4x^2$

In Exercises 53 through 56, find the requested function.

53. $f(x) = 3x + 6$ $g(x) = x + 2$
Find $\left(\frac{f}{g}\right)(x)$. $\left(\frac{f}{g}\right)(x) = 3$

54. $f(x) = 2x^2 - x$ $g(x) = 2x - 1$
Find $\left(\frac{f}{g}\right)(x)$. $\left(\frac{f}{g}\right)(x) = x$

55. $f(x) = 2x - 10x^2$ $g(x) = 1 - 5x$
Find $(fg)(x)$. $(fg)(x) = 2x - 20x^2 + 50x^3$

56. $f(x) = 3x^3 - 2x^2$ $g(x) = 3x^2 - 2x$
Find $\left(\frac{f}{g}\right)(x)$. $\left(\frac{f}{g}\right)(x) = x; x \neq 0$

In Exercises 57 through 64, find the composite functions and then evaluate them for $x = 0$. Let $f(x) = 2x - 5$; $g(x) = x^2 - 1$; and $h(x) = -3x$.

57. $(f \circ g)(x)$
$2x^2 - 7; -7$

58. $(g \circ f)(x)$
$4x^2 - 20x + 24; 24$

59. $(f \circ h)(x)$
$-6x - 5; -5$

60. $(h \circ f)(x)$
$-6x + 15; 15$

61. $(f \circ g \circ h)(x)$
$18x^2 - 7; -7$

62. $(g \circ f \circ h)(x)$
$36x^2 + 60x + 24; 24$

63. $(h \circ g \circ f)(x)$
$-12x^2 + 60x - 72; -72$

64. $(f \circ h \circ g)(x)$
$-6x^2 + 1; 1$

In Exercises 65 through 70, determine whether the first expression is equal to the second expression, using

$$f(x) = 2x \qquad g(x) = x + 1 \qquad h(x) = x^2$$

65. $[f(x) + g(x)] + h(x)$ and $f(x) + [g(x) + h(x)]$
equal

66. $[f(x) \cdot g(x)] \cdot h(x)$ and $f(x) \cdot [g(x) \cdot h(x)]$
equal

67. $[f(x) - g(x)] - h(x)$ and $f(x) - [g(x) - h(x)]$
     not equal

68. $[f(x) - g(x)] + h(x)$ and $h(x) + [f(x) - g(x)]$
     equal

69. $(f \circ g)(x)$ and $(g \circ f)(x)$
     not equal

70. $[(f \circ g) \circ h](x)$ and $[(g \circ f) \circ h](x)$
     not equal

In Exercises 71 through 82, find the solution set by applying the principle of zero products.

71. $(x - 2)(x - 5) = 0$
     $\{2, 5\}$

72. $(x - 6)(x + 4) = 0$
     $\{-4, 6\}$

73. $(2x - 1)(x + 2)(x - 3) = 0$
     $\left\{-2, \frac{1}{2}, 3\right\}$

74. $(x + 7)(2x - 1)(x + 5) = 0$
     $\left\{-7, -5, \frac{1}{2}\right\}$

75. $x^2 + 14x + 49 = 0$
     $\{-7\}$

76. $x^2 - 16x + 64 = 0$
     $\{8\}$

77. $25x^2 + 10x + 1 = 0$
     $\left\{-\frac{1}{5}\right\}$

78. $49x^2 - 56x + 16 = 0$
     $\left\{\frac{4}{7}\right\}$

79. $3x^2 + x - 14 = 0$
     $\left\{-2\frac{1}{3}, 2\right\}$

80. $3x^2 + 2x - 5 = 0$
     $\left\{-1\frac{2}{3}, 1\right\}$

81. $8x^3 + 2x^2 = 3x$
     $\left\{0, -\frac{3}{4}, \frac{1}{2}\right\}$

82. $9x^3 + 12x^2 = 5x$
     $\left\{0, -1\frac{2}{3}, \frac{1}{3}\right\}$

In Exercises 83 through 90, use an equation to solve each problem.

83. A square with a side of 10 units has its sides increased to form a square with an area of 256 square units. How much was added to each side?
     6 units

84. A square has each side decreased by 19. The new square has an area of 121 square inches. What is the area of the original square?   900 square inches

85. A triangle with a base of 8 inches and a height of 7 inches has its base and height increased by the same amount. The result is a triangle with an area of 105 square inches. What are the dimensions of the new triangle?
     base: 15 inches; height: 14 inches

86. A triangle has a base and a height of the same length. The height is increased by 21, giving a new area of 98 square inches. What was the original area? $24\frac{1}{2}$ square inches

87. To find the sum S of the first n positive integers,

$$1 + 2 + 3 + 4 + \cdots + n = S$$

we an use the formula $S = \frac{1}{2}n(n + 1)$. If the sum of the first n positive integers is 528, what is n?   32

88. Find the value of n when S is closest to 1000 (see Exercise 87).   44

89. The number of diagonals in a polygon with n sides can be found by using the formula

$$d = \frac{1}{2}n(n - 3)$$

Find the number of sides of a polygon with 119 diagonals.   17 sides

90. Find the number of sides of a polygon with 230 diagonals (see Exercise 89).
     23 sides

MIXED PRACTICE

By doing these exercises, you will practice the topics up to this point in the chapter.

91. Factor: $36x^2 - 49$
$(6x + 7)(6x - 7)$

92. Factor: $6x^2 - x - 70$
$(2x - 7)(3x + 10)$

93. Add: $13x^2 - 12$ and $-12x^2 + 13$
$x^2 + 1$

94. Divide: $(x^3 - 8) \div (x - 2)$
$x^2 + 2x + 4$

95. Multiply: $(16xy^2 - 1)(3x + 2)$
$48x^2y^2 + 32xy^2 - 3x - 2$

96. Factor: $25t^2 - 70t + 49$
$(5t - 7)^2$

97. Divide: $\dfrac{\frac{3xy - 5x^2 + 15}{15xy}}{\frac{1}{5} - \frac{x}{3y} + \frac{1}{xy}}$

98. Solve: $6x^2 = 216$
$\{6, -6\}$

99. Factor: $8x^2 + 11x - 10$
$(x + 2)(8x - 5)$

100. Solve: $3x^2 + x - 2 = 0$
$\left\{-1, \frac{2}{3}\right\}$

EXCURSIONS

Posing Problems

1. Make up a mathematics problem to give each number in color as an answer. We have given one problem as an example:

What is the result of $a - a$ for any real number?
Answer: 0

Super Stats

    0: The number of punts returned for a touchdown in Super Bowl history.
    4: The number of games played on artificial turf by the New England Patriots in the 1996 season.
    5: The number of games played on artificial turf by the Green Bay Packers in the 1996 season.
  6.7: The weight, in pounds, of the sterling silver Vince Lombardi Trophy awarded to the Super Bowl champions.
    7: The number of rushing yards gained by the New England Patriots in Super Bowl XXI.
    9: The number of Super Bowls played in New Orleans.
  9.7: The area, in acres, of the Superdome roof.
  11: The age of the New England Patriots' quarterback Drew Bledsoe, when the AFC last won a Super Bowl.
  13: The height, in stories, of the New Orleans Superdome.
  29: The number of television cameras Fox had for Super Bowl XXXI.
  37: Super Bowl XXXI was played on the 37th anniversary of Pete Rozelle's appointment as the NFL commissioner.
  257: The number of rushing yards gained by the New England Patriots in Super Bowl XXXI.
103,985: The largest crowd in Super Bowl history (Super Bowl XIV at the Rose Bowl).
1,200,000: The cost, in dollars, of a 30-second commercial on Fox during Super Bowl XXXI.

Source: National Football League.

CHAPTER LOOK-BACK

In the Hardy-Weinberg equation, we have seen that the terms mean something specific.

p^2 = the proportion of the population with 2 Gene Is

q^2 = the proportion of the population with 2 Gene IIs

$2pq$ = the proportion of the population with both Gene I and Gene II

▪ **Research** Find other ways that mathematics can be used in biology or in nature. Write some questions that can be answered by mathematics.

CHAPTER 5
REVIEW PROBLEMS

The following exercises will give you a good review of the material presented in this chapter.

SECTION 5.1

In Exercises 1 through 4, perform the indicated operations and write all variables with positive exponents.

1. $\dfrac{6m^2n^3x^{-5}}{2m^{-3}n^3x^{-3}}$ $\dfrac{3m^5}{x^2}$

2. $(18x^3y^2)\left(\dfrac{1}{2}x^4y^3\right)$ $9x^7y^5$

3. $\left(\dfrac{36x^2yz^3}{42x^{-5}yz^6}\right)^{-2}$ $\dfrac{49z^6}{36x^{14}}$

4. What is the degree of $3xy^2 - 4y^2 + 5x^2y^2$?   4

In Exercises 5 through 8, rewrite each number using scientific notation. Multiply or divide as indicated. Write your answer as a decimal number.

5. $(325,000,000)(0.0002)$ 65,000

6. $380,000,000,000 \div 190,000,000$ 2000

7. $\dfrac{(5,300,000,000)(2.0 \times 10^{-6})(5 \times 10^{16})}{10^{23}}$ 0.0053

8. $\dfrac{(6,300,000)(2.5 \times 10^{-3})}{(9,000,000)(0.007)}$ 0.25

SECTION 5.2

In Exercises 9 through 16, simplify each expression.

9. $5x^2 - 2x + 3 + (-4x - 2x^2 + 10)$ $3x^2 - 6x + 13$

10. $6xy^2 + 2x^2 - 3y^2 - (6x^2 + 2y^2 - 5xy^2)$ $11xy^2 - 4x^2 - 5y^2$

11. $2xy - (3xy^2 + 2xy) + 4xy^2 - (3xy - 2xy^2)$ $-3xy + 3xy^2$

12. $(2y^2 + 3xy - 6y - 3x + 4) + (6y^2 - 2xy + 7y + 3x - 9)$
$xy + 8y^2 + y - 5$

13. $3xy(2x^2y - 3x + 5)$ $6x^3y^2 - 9x^2y + 15xy$

14. $(6x^2 - 2x) \div 12x$ $\frac{x}{2} - \frac{1}{6}$

15. $\frac{15x^3 - 12x^2 + 6}{6x}$ $\frac{5x^2}{2} - 2x + \frac{1}{x}$

16. $(-9x^2y)(-8xy^{-3} + 4)$ $\frac{72x^3}{y^2} - 36x^2y$

Let $f(x) = 2x^2 - 3x$ and $g(x) = 2x^2 - 2x$. Evaluate each of the following.

17. $(f - g)(x)$ $-x$

18. $(f + g)(-1)$ 9

SECTION 5.3

In Exercises 19 through 22, multiply the given polynomials.

19. $(3x + 2)(5x - 2y + 3)$ $15x^2 - 6xy + 19x - 4y + 6$

20. $(3x - 2)(-2x + 6)$ $-6x^2 + 22x - 12$

21. $(x - 1)^3$ $x^3 - 3x^2 + 3x - 1$

22. $(6x - 2y)(3x + y)$ $18x^2 - 2y^2$

In Exercises 23 and 24, use long division to divide.

23. $(x^3 - 1) \div (x - 1)$ $x^2 + x + 1$

24. $(16x^3 - 13x^2 + 10x - 8) \div (x - 1)$
$16x^2 + 3x + 13 + \frac{5}{x - 1}$

SECTION 5.4

In Exercises 25 through 28, factor each trinomial.

25. $-63 + 2x + x^2$ $(x - 7)(x + 9)$

26. $21x + x^2 + 110$ $(x + 10)(x + 11)$

27. $14x + 49 + x^2$ $(x + 7)(x + 7)$

28. $4x^2 - 40x - 800$ $4(x - 20)(x + 10)$

In Exercises 29 through 32, factor each trinomial.

29. $15x^2 - x - 6$ $(3x - 2)(5x + 3)$

30. $-30x^2 + 115x - 35$ $-5(2x - 7)(3x - 1)$

31. $3x^2 - 75$ $3(x + 5)(x - 5)$

32. $25x^2 - 70xy + 49y^2$ $(5x - 7y)^2$

SECTION 5.5

In Exercises 33 through 36, solve by factoring.

33. $2x^2 - 4x - 30 = 0$ $x = 5; x = -3$

34. $18y^2 - 93y + 120 = 0$ $y = 2\frac{1}{2}; y = 2\frac{2}{3}$

35. $5x^2 = 405$ $\{-9, 9\}$

36. $(3x - 5)(x + 2)(2x + 7) = 0$ $\left\{-3\frac{1}{2}, -2, 1\frac{2}{3}\right\}$

Let $f(x) = 2x^2 - 3x$ and $g(x) = 2x^2 - 2x$. Evaluate each of the following.

37. $(f \circ g)(x)$ $8x^4 - 16x^3 + 2x^2 + 6x$

38. $(g \circ f)(x)$ $8x^4 - 24x^3 + 14x^2 + 6x$

MIXED REVIEW

39. Factor: $10x^2 - 9x - 7$ $(5x - 7)(2x + 1)$

40. Factor: $105 - 24x^2 - 18x$ $-3(2x + 5)(4x - 7)$

41. Factor: $2x^2 + 20x + 50$ $2(x + 5)^2$

42. Factor: $100x^2 - 81y^2$ $(10x + 9y)(10x - 9y)$

43. Simplify: $(-8x^3y^7t^3)^{-2}$ $\frac{1}{64x^6 y^{14} t^6}$

44. Factor: $x^3 - 2x^2 - x + 2$ $(x - 2)(x - 1)(x + 1)$

45. Solve: $r^2 - 4r - 12 = 0$ $r = 6; r = -2$ or $\{6, -2\}$

46. Factor: $3t^2 - 39t + 126$ $3(t - 6)(t - 7)$

47. Factor: $3m^2n^3 - 9m^3n^2 + 11mn^2$ $mn^2(3mn - 9m^2 + 11)$

48. Solve: $20r^2 + 28r + 8 = 0$ $r = -\frac{2}{5}; r = -1$

49. Find the degree: $3xy^2 + 2x + 5y - 8$ 3

50. Let $g(y) = 42y^2 + 17y - 4$ and $h(y) = -6y^2 + 24y + 24$. Find $g(y) - h(y)$.
$48y^2 - 7y - 28$

51. Factor: $49y^2 - 144x^2$ $(7y + 12x)(7y - 12x)$

52. Factor: $56x^2 - 13x - 3$ $(8x - 3)(7x + 1)$

CHAPTER 5 TEST

This exam tests your knowledge of the material in Chapter 5.

1. Simplify:
   a. $(-8x^6y^4)(6x^7y^{-3})$
   $-48x^{13}y$
   b. $(5x^2y^3z^{-2})^{-3}$
   $\frac{z^6}{125x^6y^9}$
   c. $\frac{28r^3s^2}{30r^2s^4t} \quad \frac{14r}{15s^2t}$

2. Simplify:
   a. $(-7x^2 + 4x - 8) + 3x^2 + 9$
   $-4x^2 + 4x + 1$
   b. $(x^2 - 7x + 9) - (-8x^2 - 6x - 12)$
   $9x^2 - x + 21$
   c. $-(3x^2 + 5x - 7)$
   $-3x^2 - 5x + 7$

3. Multiply or divide as indicated:
   a. $(9x^8y^6z^2 - 3x^2y^4z - 30x^2y^4z^6) \div 3x^2y^4z$
   $3x^6y^2z - 1 - 10z^5$
   b. $\frac{16r^2v + 22rv - 8rv^2}{-4r^2v}$
   $-4 - \frac{11}{2r} + \frac{2v}{r}$
   c. $8x^2y(-4x^3y^2 - 2x^2y^2 + 8xy)$
   $-32x^5y^3 - 16x^4y^3 + 64x^3y^2$

4. Multiply or divide as indicated:
   a. $(t - 1)(5t^2 - 8t + 3)$
   $5t^3 - 13t^2 + 11t - 3$
   b. $(3 - m)(8 + 3m)$
   $24 + m - 3m^2$
   c. $x - 3\overline{)x^3 - 4x^2 + 7x - 12}$
   $x^2 - x + 4$

5. Factor:
   a. $7x^2 - 7x - 14$
   $7(x - 2)(x + 1)$
   b. $8x^2 - 14x - 15$
   $(4x + 3)(2x - 5)$
   c. $x^3 - 8$
   $(x - 2)(x^2 + 2x + 4)$

6. Factor:
   a. $r^2 - 256$
   $(r + 16)(r - 16)$
   b. $x^2 - 9y^2$
   $(x + 3y)(x - 3y)$
   c. $t^2 - 16t + 64$
   $(t - 8)^2$

7. Find the solution set:
   a. $(x - 3)(2x + 5)(x - 7) = 0$
   $\{-2.5, 3, 7\}$
   b. $6x^2 + 7x - 20 = 0$
   $\left\{-2\frac{1}{2}, 1\frac{1}{3}\right\}$
   c. $19x^2 - 26x = 0$
   $\left\{0, 1\frac{7}{19}\right\}$

8. Let $f(x) = 3x^2 - 2x + 1$ and $g(x) = x - 1$.
   a. Find $(f \circ g)(x)$.
   $3x^2 - 8x + 6$
   b. Find $(f + g)(0)$.
   0
   c. Find $(fg)(-1)$.
   -12

CUMULATIVE REVIEW

CHAPTERS 1–5

The following exercises will help you maintain the skills you have learned in this and previous chapters.

1. Solve: $|r - 4| = 9$ $\quad r = 13; r = -5$

2. Simplify: $(-3x^{-2}y)^{-4}$ $\quad \frac{x^8}{81y^4}$

3. Write $4x - 3(2 - x)$ in words. the product of 4 and x minus the product of 3 and the difference between 2 and x

4. Divide: $(x^2 - 2x - 3) \div (x + 1)$ $\quad x - 3$

5. Simplify:
   $-16.4xy^2 - 24xy - 2.4xy^2 + 11.5xy^2 + 11xy^3 + 3.2$
   $11xy^3 - 7.3xy^2 - 24xy + 3.2$

6. Simplify: $(2.4 - 0.24)^2 - 4(1.6 - 2.5)$ $\quad 8.2656$

7. Solve: $-9t + 6 = -5(t - 2) - t$ $\quad t = \frac{-4}{3}$

8. The difference between the length of a side of a rectangle and its width is 23 inches. The perimeter is 94 inches. What are the dimensions? length: 35 inches; width: 12 inches

9. Solve and graph: $5 < 2 + 3x < 8$ $\{x \mid 1 < x < 2\}$

10. Add: $\left(-7\frac{4}{5}\right) + \left(-5\frac{2}{3}\right)$ $-13\frac{7}{15}$

11. Divide: $\dfrac{-24xy^2 + 36x^2y}{-3xy}$ $8y - 12x$

12. Write the equation of the line that passes through the points $(-3, 2)$ and $(6, -4)$ $y = \dfrac{-2}{3}x$

13. If $f(x) = 2x - 8$, find $f(0)$ and $f(-3)$. $-8; -14$

14. Are the graphs of the equations $y = 3x - 2$ and $3y + x = 4$ parallel, perpendicular, or neither?   perpendicular

15. Solve for m: $y = mx + b$ $\dfrac{y - b}{x} = m$

16. Simplify: $(2)^{-2} + 6^0 - 10^{-1} + 5$ 6.15

17. Let $f(x) = 3x + 2$ and $g(x) = 2x^2$. Find $(f \circ g)(x)$. $6x^2 + 2$

18. Evaluate $xy + 2xy - 3xy^2 + 2x^2y$ when $x = -3$ and $y = -6$. 270

19. Solve: $4(2.4 - 8t) \le 3 + t$ $\{t \mid t \ge 0.2\}$

20. Solve by addition: $14x - 3y = -92$
$50x - 5y = -260$
$x = -4; y = 12$

6

RATIONAL EXPRESSIONS

*H*ave you ever ordered pizza, not knowing exactly how many people would be sharing the meal? Imagine that you have asked some friends to your home but do not know exactly how many will come. Two of those who are coming do not eat pizza. Knowing the servings per pizza—say eight slices—and using polynomials to write a rational expression, you describe the problem as:

$$\frac{8n}{x-2} = c$$

where n = the number of pizzas, x = the number of guests (as yet unknown), less the two who do not eat pizza, and c = the number of slices you determine that each will want.

■ *How would you change the above rational expression if the pizza had six slices?*

SKILLS CHECK

Take this short quiz to see how well prepared you are for Chapter 6. Answers follow the quiz.

1. Divide: $\frac{12}{25} \div \frac{27}{35}$

2. Factor: $y^2 - 7y + 12$

3. Simplify: $-9 + 3 \div 4 - 1$

4. Solve: $2x - 6 = 9$

5. Factor: $y^2 - 64x^2$

6. Factor: $16x^2 - 24x + 9$

7. Simplify: $\frac{-1}{2}(2x - 8)$

8. Multiply: $(x + 3y)(x - 2y)$

9. Subtract: $\frac{2}{3} - \frac{6}{7}$

ANSWERS: **1.** $\frac{28}{45}$ [Section 1.2]   **2.** $(y - 4)(y - 3)$ [Section 5.4]   **3.** $-9\frac{1}{4}$ [Section 1.3]
4. $7\frac{1}{2}$ [Section 2.2]   **5.** $(y + 8x)(y - 8x)$ [Section 5.4]   **6.** $(4x - 3)^2$ [Section 5.4]
7. $-x + 4$ [Section 1.3]   **8.** $x^2 + xy - 6y^2$ [Section 5.3]   **9.** $-\frac{4}{21}$ [Section 1.2]

CHAPTER LEAD-IN

Low fat means less than three calories of fat to every ten calories. Sometimes labels can be confusing. To determine the fat calories in a particular food, you can use the following procedure. Take the number of grams of fat per serving listed on the package. Multiply by 9 (each gram of fat is 9 calories) and divide by the total calories. This is the ratio of fat to total calories. Some fitness experts say that no more than 10% of our diet should be from fat.

Keep a record of what you eat during a typical day. Using the food labels on foods you prepare or purchase and eat, estimate your fat intake for the day. What percent of your calories come from fat? What is your fat ratio?

6.1 Introduction to Rational Expressions

SECTION LEAD-IN

Rational expressions occur in many formulas that can be calculated by graphing calculators. You never see the formula—you simply push a button and get an answer. One such formula follows.

This is a financial formula for computing the present value of money in a certain transaction.

$$PV = \left(\frac{PMT \times G}{i} - FV\right) \times \frac{1}{(1 + i)^N} - \frac{PMT \times G}{i}$$

when $i \neq 0$.

- What does PV equal when $i = 1$?

Another formula is given for this same calculation when $i = 0$.

$$PV = -(FV + PMT \times N)$$

- Why do we need another formula for this special case?

A *rational expression* is a fraction whose numerator and denominator are polynomials rather than integers.

> **Rational Expressions**
> A fraction of the form $\frac{p}{q}$, where p and q are polynomials and $q \neq 0$, is called a **rational expression.**

All the properties that hold for rational numbers also hold for rational expressions.

Meaningful Expressions

The fraction bar in a rational expression indicates division, and division by zero is undefined. Thus when q, the denominator of a rational expression, is equal to zero, the rational expression is said to be **undefined**, or **not meaningful.** Values of the variable that make the denominator zero must be excluded from consideration.

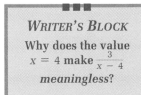

WRITER'S BLOCK
Why does the value $x = 4$ make $\frac{3}{x - 4}$ *meaningless?*

•••

EXAMPLE 1

For which values of the variable are the following expressions defined?

a. $\frac{y - 3}{5y}$ **b.** $\frac{3x - 5}{x - 4}$ **c.** $\frac{t - 5}{t^2 - 36}$

WRITER'S BLOCK

Describe what is meant by a rational expression that has meaning.

SOLUTION

A rational expression is not meaningful for all real numbers that make the denominator 0. We find those real numbers by setting the denominator equal to zero and solving for the variable.

a. The denominator here is $5y$.

$$\text{Let} \quad 5y = 0$$
$$\text{Then} \quad y = 0$$

Thus the expression $\frac{y-3}{5y}$ is not meaningful when $y = 0$. It is **defined** (or **meaningful**) for all real-number values of y except $y = 0$.

b. The denominator is $x - 4$, so we let

$$x - 4 = 0$$
$$x = 4 \quad \text{Adding 4 to both sides}$$

So this expression is not meaningful when x is 4. It is defined for

$$\{x \mid x \text{ is real and } x \neq 4\}$$

c. The denominator is $t^2 - 36$. We let

$$t^2 - 36 = 0$$

and solve by factoring.

$$(t + 6)(t - 6) = 0 \qquad \text{Factoring}$$
$$t + 6 = 0 \quad \bigg| \quad t - 6 = 0 \quad \text{Setting factors equal to 0}$$
$$t = -6 \quad \bigg| \quad t = 6 \quad \text{Solving}$$

So when t is either -6 or 6, the expression is not meaningful. It is meaningful for

$$\{t \mid t \text{ is real and } t \neq 6, -6\}$$

▶ *CHECK* **Warm-Up 1**

ERROR ALERT

Identify the error and give a correct answer.

Incorrect Answer:
$$\frac{3x + 7}{2x}$$
The expression is meaningful because there is no zero in the denominator.

INSTRUCTOR NOTE

Remind students that a factor is something that is multiplied.

$\frac{6a(b + c)}{a}$ The a's can be cancelled

$\frac{3 + a}{a}$ The a's cannot be cancelled.

Simplifying Rational Expressions

To simplify a fraction, we cancel any factor that appears in both numerator and denominator.

Fundamental Property of Fractions

For all integers p, q, and r, with q and $r \neq 0$.

$$\frac{pr}{qr} = \frac{p}{q}$$

WRITER'S BLOCK

Restate the *fundamental property of fractions* in your own words.

This property applies to rational expressions as well as to fractions. If we divide the numerator and denominator of a rational expression by the same non-zero factor, then the value of the rational expression is not changed. This procedure is called **reducing a rational expression to lowest (or simplest) terms**. It works because

$$\frac{pr}{qr} = \frac{p \cdot r}{q \cdot r} = \frac{p}{q} \cdot 1 = \frac{p}{q}$$

Dividing numerator and denominator by the same factor has the effect of dividing the expression by 1.

INSTRUCTOR NOTE

We have used the traditional p, q, and r to represent rational numbers. However, students with learning disabilities might have difficulty discriminating between p and q.

To reduce a rational expression to lowest terms

1. Factor the numerator and denominator.
2. Divide numerator and denominator by the same factors (cancel common factors).

■ ■ ■
EXAMPLE 2

Reduce the following rational expressions to lowest terms

a. $\dfrac{r^3 x^2}{rt}$ **b.** $\dfrac{m^2 - x}{mx}$

SOLUTION

a. Here r is a factor in both numerator and denominator, so we may write

$$\frac{r^3 x^2}{rt} = \frac{r^2 x^2 \cdot r}{t \cdot r} = \frac{r^2 x^2}{t}$$

However, we see from the original expression that we cannot have $t = 0$ or $r = 0$, for then the expression would have no meaning. So we must write

$$\frac{r^3 x^2}{rt} = \frac{r^2 x^2}{t} \text{ for } r \neq 0, t \neq 0$$

b. Here m cannot be cancelled because it is not a factor in the numerator. There is no common factor in all three terms. So $\dfrac{m^2 - x}{mx}$ cannot be simplified in this rational form. This fraction is already in lowest terms.

▶ CHECK **Warm-Up 2**

■ ■ ■
WRITER'S BLOCK

How can you tell when a rational expression is reduced to lowest terms?

In the next example, we have to factor before we can simplify an expression.

■ ■ ■
EXAMPLE 3

Reduce to lowest terms: $\dfrac{24x^2 y - 15xy}{-8x + 5}$

SOLUTION

The numerator and denominator can be factored to yield

$$\frac{24x^2 y - 15xy}{-8x + 5} = \frac{3xy(8x - 5)}{-1(8x - 5)} \quad \begin{array}{l}\text{Factoring out } 3xy \\ \text{Factoring out } -1\end{array}$$

We then cancel the common factors

$$\frac{3xy(\cancel{8x - 5})}{-1(\cancel{8x - 5})} = \frac{3xy}{-1} = -3xy$$

! ! !
ERROR ALERT

Identify the error and give a correct answer.

Simplify: $\dfrac{y - 5}{2y^2 - 10y}$

Incorrect Solution:

$$\frac{y - 5}{2y^2 - 10y} = \frac{1}{2y + 2y} = \frac{1}{4y}$$

The original expression shows that this result has no meaning when

$$-8x + 5 = 0, \text{ or } x = \frac{5}{8}$$

Thus $\dfrac{24x^2y - 15xy}{-8x + 5} = -3xy$ for $x \neq \dfrac{5}{8}$

▶ *CHECK* **Warm-Up 3**

Calculator Corner

You can use your graphing calculator to check your work in simplifying rational expressions. First, however, it might prove helpful to review the graphs of some rational expressions. Some students even say that they look somewhat "weird"—very different from straight lines, absolute values, and parabolas. (*Hint:* On many graphing calculators it is best to use a "friendly" window when viewing rational equations. Why do you think this is true?) The following screens are from the TI-82 graphing calculator. Consult your graphing calculator manual for your model's "friendly" screen.

Simplify the following rational expression: $\dfrac{3x^2 - 6x}{3x^2}$

Doing this by paper-and-pencil methods should give a result of $\dfrac{x - 2}{x}$. Now check your work with your graphing calculator.

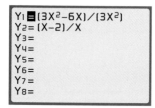 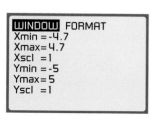 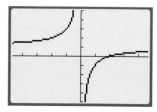

How do the graphs compare? How did you expect them to compare if your paper-and-pencil work had been correct?

Now try the following on your own and check your work using your graphing calculator.

$$\frac{3m^2n - 6m^3n^2}{3m^2n^2} \qquad \frac{8t^2x - 16tx + 24t^3x^3}{12t^2x^2}$$

Multiplying Rational Expressions

We multiply rational expressions just as we multiply numerical factors.

> **Multiplying Rational Expressions**
>
> For polynomials, p, q, r, and s, with q and $s \neq 0$,
> $$\frac{p}{q} \cdot \frac{r}{s} = \frac{pr}{qs}$$

Thus to multiply rational expressions, we multiply the numerators, multiply the denominators, and then simplify.

▪▪▪

EXAMPLE 4

Multiply: $\dfrac{3}{5} \cdot \dfrac{t - 6}{6t + 3}$

SOLUTION

Multiplying the numerators and the denominators gives us

$$\frac{3}{5} \cdot \frac{t - 6}{6t + 3} = \frac{3(t - 6)}{5(6t + 3)} = \frac{3t - 18}{30t + 15}$$

Finally, we can factor out a common factor 3.

$$\frac{3t - 18}{30t + 15} = \frac{3(t - 6)}{3(10t + 5)} = \frac{t - 6}{10t + 5}$$

▶ CHECK **Warm-Up 4**

It is a good idea to factor each numerator and each denominator, as far as it is possible, before multiplying them. Doing so can reduce the amount of work involved.

! ! !
ERROR ALERT

Identify the error and give a correct answer.

Multiply: $\dfrac{4}{3} \cdot \dfrac{3x - 2}{4x + 5}$

Incorrect Solution:

$$= \frac{\cancel{4}}{\cancel{3}} \cdot \frac{\cancel{3}x - 2}{\cancel{4}x + 5}$$

$$= \frac{x - 2}{x + 5}$$

▪▪▪

EXAMPLE 5

Multiply: $\dfrac{n^2 - 3n + 2}{n - 2} \cdot \dfrac{n^2 + 4n + 4}{n + 2}$

SOLUTION

The numerator in each fraction can be factored to give

$$\frac{(n - 2)(n - 1)}{n - 2} \cdot \frac{(n + 2)(n + 2)}{n + 2}$$

Now we can simplify each fraction by cancelling:

$$\frac{\overset{1}{\cancel{(n - 2)}}(n - 1)}{\underset{1}{\cancel{n - 2}}} \cdot \frac{\overset{1}{\cancel{(n + 2)}}(n + 2)}{\underset{1}{\cancel{n + 2}}}$$

This leaves us with the multiplication.

$$\frac{(n - 1)}{1} \cdot \frac{(n + 2)}{1} = n^2 + 2n - n - 2 = n^2 + n - 2$$

▶ CHECK **Warm-Up 5**

Dividing Rational Expressions

Division of rational expressions is like division of fractions.

> **Dividing Rational Expressions**
>
> If p, q, r, and s are polynomials with q, r, and $s \neq 0$, then
> $$\frac{p}{q} \div \frac{r}{s} = \frac{p}{q} \cdot \frac{s}{r} = \frac{ps}{qr}$$

Thus to divide rational expressions, we multiply the dividend by the reciprocal of the divisor.

▪▪▪

EXAMPLE 6

Divide: $\dfrac{2y - 2}{y + 3} \div \dfrac{y - 1}{3y + 9}$

SOLUTION

First we invert the divisor and factor numerator and denominator where possible.

$$\frac{2y - 2}{y + 3} \div \frac{y - 1}{3y + 9} = \frac{2y - 2}{y + 3} \cdot \frac{3y + 9}{y - 1}$$
$$= \frac{2(y - 1)}{y + 3} \cdot \frac{3(y + 3)}{y - 1}$$

Then we multiply and cancel common factors.

$$\frac{2(y - 1)}{y + 3} \cdot \frac{3(y + 3)}{y - 1} = \frac{(2)(\overset{1}{\cancel{y - 1}})(3)(\overset{1}{\cancel{y + 3}})}{(\underset{1}{\cancel{y + 3}})(\underset{1}{\cancel{y - 1}})} = 2 \cdot 3 = 6$$

▶ CHECK **Warm-Up 6**

‼ ‼

ERROR ALERT

Identify the error and give a correct answer.

Divide: $\dfrac{x + 3x}{x^2 + 4x} \div \dfrac{x + 4}{x + 3x}$

Incorrect Solution:

$$\frac{\cancel{x + 3x}}{x^2 + 4x} \div \frac{x + 4}{\cancel{x + 3x}} =$$
$$\frac{x + 4}{x^2 + 4x} = \frac{\cancel{x + 4}}{x(\cancel{x + 4})} = \frac{1}{x}$$

▪▪▪

EXAMPLE 7

Simplify: $\dfrac{n^2 - n - 12}{2n^2 - 3n - 5} \cdot \dfrac{3n^2 - 10n - 8}{-n^2 + 6n - 8} \div \dfrac{n^2 - n - 12}{4n^2 - 4n - 8}$

SOLUTION

We invert the last polynomial and write the division as multiplication and then factor all expressions.

$$\frac{n^2 - n - 12}{2n^2 - 3n - 5} \cdot \frac{3n^2 - 10n - 8}{-n^2 + 6n - 8} \cdot \frac{4n^2 - 4n - 8}{n^2 - n - 12}$$
$$= \frac{(n - 4)(n + 3)}{(n + 1)(2n - 5)} \cdot \frac{(3n + 2)(n - 4)}{-1(n - 4)(n - 2)} \cdot \frac{4(n + 1)(n - 2)}{(n + 3)(n - 4)}$$

Then we multiply and simplify.

$$\frac{(\overset{1}{\cancel{n - 4}})(\overset{1}{\cancel{n + 3}})(3n + 2)(\cancel{n - 4})(4)(\overset{1}{\cancel{n + 1}})(\overset{1}{\cancel{n - 2}})}{(\cancel{n + 1})(2n - 5)(-1)(\underset{1}{\cancel{n - 4}})(\underset{1}{\cancel{n - 2}})(\underset{1}{\cancel{n + 3}})(\underset{1}{\cancel{n - 4}})}$$
$$= \frac{4(3n + 2)}{-1(2n - 5)} = \frac{12n + 8}{5 - 2n}$$

▶ CHECK **Warm-Up 7**

Rational Functions

A function that is given in the form of a rational expression, such as

$$f(x) = \frac{x + 2}{x + 3}$$

is called a **rational function**. The domain of the function f cannot include any real numbers for which the rational expression is not meaningful. If the function f is given by

$$f(t) = \frac{t - 5}{t^2 - 36}$$

we find that the rational expression is not meaningful for $t = 6$ and $t = -6$. Thus the domain of f is

$$\{t \mid t \text{ is real and } t \neq 6, -6\}$$

■ ■ ■

EXAMPLE 8

Let a function g be given by

$$g(x) = \frac{2x^2 + 3x + 1}{x^2 + 2x + 1}$$

with domain $\{x \mid x \text{ is real and } x \neq -1\}$.

a. Rewrite $h(x)$ by reducing the rational expression $g(x)$ to lowest terms.

b. Find $g(-2), g,(y)$, and $g(x^3)$.

SOLUTION

a. The numerator and denominator of the defining rational expression can be factored to give

$$h(x) = \frac{(2x + 1)(x + 1)}{(x + 1)(x + 1)} = \frac{2x + 1}{x + 1} \text{ for } x \neq -1$$

This is the same function as the given function g, because it yields the same exact ordered pairs as g. So here $h = g$.

b. To find $g(-2)$, we can substitute -2 for x in either defining expression. We use the simpler $h(x)$.

$$h(-2) = \frac{2(-2) + 1}{-2 + 1} = \frac{-3}{-1} = 3$$

You might want to calculate $g(-2)$ to see that it is indeed equal to $h(-2)$. To find $g(y)$ and $g(x^3)$, we substitute first y and then x^3 in the simpler $h(x)$.

$$g(y) = h(y) = \frac{2y + 1}{y + 1} \text{ for } y \neq -1$$

$$g(x^3) = h(x^3) = \frac{2x^3 + 1}{x^3 + 1} \text{ for } x \neq -1$$

▶ CHECK **Warm-Up 8**

From this point on, we will assume that expressions are written only for values for which they are meaningful.

> A rational function is equal to zero when the numerator is equal to zero.
> A rational function is undefined when the denominator is equal to zero.

■■■

EXAMPLE 9

Let $f(x) = \dfrac{9x^2 - 25}{x + 8}$.

Find the values of x that makes $f(x) = 0$ and the values of x that make $f(x)$ undefined.

SOLUTION

To find the values that make the function 0, we set the numerator equal to zero and solve for x.

$$9x^2 - 25 = 0$$

$$(3x + 5)(3x - 5) = 0 \quad \text{Factoring}$$

$3x + 5 = 0$	$3x - 5 = 0$	Setting factors equal to zero
$3x = -5$	$3x = 5$	
$x = \dfrac{-5}{3}$	$x = \dfrac{5}{3}$	

So the set $\left\{\dfrac{-5}{3}, \dfrac{5}{3}\right\}$ makes $f(x) = 0$.

To find the values that make the function meaningless, or undefined, we set the denominator equal to zero.

$$x + 8 = 0$$

$$x = -8$$

So -8 makes $f(x)$ undefined.

▶ CHECK **Warm-Up 9**

■■■

EXAMPLE 10

Functions f and g are given by

$$f(y) = \frac{y^2 - 5y + 6}{2y^2 - 2y - 12} \quad \text{and} \quad g(y) = \frac{8y^2 - 24y + 16}{6y^2 - 6y - 12}$$

Find $\left(\dfrac{f}{g}\right)y$. What real numbers must be excluded from the domains of $f(y)$, $g(y)$, and $\left(\dfrac{f}{g}\right)(y)$?

SOLUTION

First we factor each expression.

$$f(y) = \frac{(y - 3)(y - 2)}{2(y - 3)(y + 2)} \qquad g(y) = \frac{8(y - 2)(y - 1)}{6(y - 2)(y + 1)}$$

The denominator factors tell us that the real numbers 3 and -2 must be excluded from the domain of $f(y)$ and that 2 and -1 must be excluded from the domain of $g(y)$.

Then

$$\left(\frac{f}{g}\right)(y) = \frac{(y-3)(y-2)}{2(y-3)(y+2)} \div \frac{8(y-2)(y-1)}{6(y-2)(y+1)}$$

$$= \frac{\overset{1}{(y-3)}(y-2)}{2\underset{1}{(y-3)}(y+2)} \cdot \frac{3(y-2)(y+1)}{4(y-2)(y-1)} \quad \text{Inverting and cancelling}$$

$$= \frac{(y-2)3(y+1)}{2(y+2)4(y-1)} \quad \text{Multiplying}$$

$$= \frac{3(y-2)(y+1)}{8(y+2)(y-1)} \quad \text{Simplifying}$$

The domain of $\left(\frac{f}{g}\right)(y)$ must exclude -2 and 1. We must also exclude any value that makes the denominator of $f(x)$ or $g(x)$ equal to zero. We set each denominator equal to zero and solve. We will use the factored forms:

$$2(y-3)(y+2) = 0$$

so

$$y - 3 = 0 \quad \text{or} \quad y + 2 = 0$$
$$y = 3 \qquad\qquad y = -2$$

and

$$6(y-2)(y+1) = 0$$

so

$$y - 2 = 0 \quad \text{or} \quad y + 1 = 0$$
$$y = 2 \qquad\qquad y = -1$$

So the domain of $\left(\frac{f}{g}\right)(y)$ must exclude $1, -1, 2, -2,$ and 3.

▶ *CHECK* **Warm-Up 10**

Practice what you learned.

SECTION FOLLOW-UP

We start with

$$PV = \left(\frac{PMT \times G}{i} - FV\right) \times \frac{1}{(1+i)^N} - \frac{PMT \times G}{i}$$

When $i = 1$, this formula simplifies to

$$PV = \frac{1}{2^N}(PMT \times G - FV) - PMT \times G$$

Verify this. Now can you answer the question, "Why do we need another formula for the special case $i = 0$?"

We can't have zero in the denominator.

6.1 WARM-UPS

Work these problems before you attempt the exercises.

1. Determine the values of y for which the expression $\dfrac{3}{y-2}$ has meaning. $\{y \mid y \text{ is real and } y \neq 2\}$

2. Reduce to lowest terms: $\dfrac{y^2 n^3}{yn^5}$ $\dfrac{y}{n^2}$

3. Reduce to lowest terms: $\dfrac{m^2 - 25}{m^2 - 3m - 10}$ $\dfrac{m+5}{m+2}$

4. Multiply: $\dfrac{t-7}{v-15} \cdot \dfrac{7}{t-7}$ $\dfrac{7}{v-15}$

5. Multiply: $\dfrac{x^2 - 8x + 12}{x-6} \cdot \dfrac{x-7}{x-2}$ $x-7$

6. Divide: $\dfrac{3x+9}{x-7} \div \dfrac{x+3}{3x-21}$ 9

7. Simplify: $\dfrac{n^2 + 2n - 15}{3n^2 - 9n} \cdot \dfrac{n^2 + n - 30}{n^2 + 12n + 36} \div \dfrac{n^2 - 25}{6n^2}$ $\dfrac{2n}{n+6}$

8. Rewrite $h(x) = \dfrac{x^2 + x - 12}{x^2 + 8x + 16}$ in lowest terms and then find $h(-2)$ and $h(a-2)$. $h(x) = \dfrac{x-3}{x+4}; h(-2) = \dfrac{-5}{2}; h(a-2) = \dfrac{a-5}{a+2}$

9. $f(x) = \dfrac{x^2 - 256}{8x + 16}$
   Find the values for which $f(x) = 0$ and the values for which $f(x)$ is undefined $f(x) = 0$ when $x = -16$ or $x = 16$. $f(x)$ is undefined when $x = -2$.

10. Let $f(y) = \dfrac{y^2 + 2y - 8}{y^2 - 3y - 28}$ and $g(y) = \dfrac{y^2 + 2y - 8}{y^2 - 13y + 42}$. Find $\left(\dfrac{f}{g}\right)(y)$.
    What is the domain of $f(y)$, of $g(y)$, and $\left(\dfrac{f}{g}\right)(y)$?

10. $\left(\dfrac{f}{g}\right)(y) = \dfrac{y-6}{y+4}$
domain $f(y)$: $\{y \mid y$ is real and $y \neq 7, -4\}$
domain $g(y)$: $\{y \mid y$ is real and $y \neq 7, 6\}$
domain $\left(\dfrac{f}{g}\right)(y)$: $\{y \mid y$ is real and $y \neq -4, 2, 4, 6, 7\}$

6.1 EXERCISES

Note: Use your graphing calculator to check your results whenever possible.

In Exercises 1 through 8, determine for which values of x the expressions are defined. Write your answers in set notation.

1. $\dfrac{x-4}{4x}$
   $\{x \mid x \text{ is real and } x \neq 0\}$

2. $\dfrac{x+8}{5x}$
   $\{x \mid x \text{ is real and } x \neq 0\}$

3. $\dfrac{5}{t-2}$
   $\{t \mid t \text{ is real and } t \neq 2\}$

4. $\dfrac{n+5}{n-5}$
   $\{n \mid n \text{ is real and } n \neq 5\}$

5. $\dfrac{8-y}{y^2 - 3y + 2}$
   $\{y \mid y \text{ is real and } y \neq 1, 2\}$

6. $\dfrac{2y}{y^2 + 4y + 4}$
   $\{y \mid y \text{ is real and } y \neq -2\}$

7. $\dfrac{n-4}{n^2 - 9}$
   $\{n \mid n \text{ is real and } n \neq -3, 3\}$

8. $\dfrac{4t}{t^2 + 2t}$
   $\{t \mid t \text{ is real and } t \neq 0, -2\}$

In Exercises 9 through 16, reduce the rational expressions to lowest terms.

9. $\dfrac{x^5}{x^3}$ x^2

10. $\dfrac{y^6}{y^4}$ y^2

11. $\dfrac{n^5 x^5}{n^3 x^7}$ $\dfrac{n^2}{x^2}$

12. $\dfrac{t^5 n^2}{t^2 n^4}$ $\dfrac{t^3}{n^2}$

13. $\dfrac{18n^5m^6}{32n^3m^8}$ $\dfrac{9n^2}{16m^2}$

14. $\dfrac{26t^4s^6}{39t^5s^3}$ $\dfrac{2s^3}{3t}$

15. $\dfrac{105x^2y^3}{100x^3y^4}$ $\dfrac{21}{20xy}$

16. $\dfrac{75m^6n^5}{\text{?}m^7n^4}$ $\dfrac{5n}{6m}$

In Exercises 17 through 26, rewrite the given fraction as the sum or difference of two or three fractions, and reduce to lowest terms.

17. $\dfrac{n^6 + 5}{n^5}$ $n + \dfrac{5}{n^5}$

18. $\dfrac{y^3 + y}{y^2}$ $y + \dfrac{1}{y}$

19. $\dfrac{x^2 + 8x}{x^4}$ $\dfrac{1}{x^2} + \dfrac{8}{x^3}$

20. $\dfrac{i^3 - 3t^2}{t^6}$ $\dfrac{1}{t^3} - \dfrac{3}{t^4}$

21. $\dfrac{x^2y - 6xy^3}{2x^2y^2}$ $\dfrac{1}{2y} - \dfrac{3y}{x}$

22. $\dfrac{3m^2n - 6m^3n^2}{3m^2n^2}$ $\dfrac{1}{n} - 2m$

23. $\dfrac{8t^2x - 16tx + 24t^3x^3}{12t^2x^2}$ $\dfrac{2}{3x} - \dfrac{4}{3tx} + 2tx$

24. $\dfrac{9x^3y - 18x^2y^2 + 27xy}{15x^3y^3}$
$\dfrac{3}{5y^2} - \dfrac{6}{5xy} + \dfrac{9}{5x^2y^2}$

25. $\dfrac{12w^2x^3 - 15w^3x^2 + 20wx}{18w^3x^3}$
$\dfrac{2}{3w} - \dfrac{5}{6x} + \dfrac{10}{9w^2x^2}$

26. $\dfrac{21x^2t - 28x^3t^3 + 35x^4t}{7x^3t}$
$\dfrac{3}{x} - 4t^2 + 5x$

In Exercises 27 through 46, factor the numerator and the denominator separately and reduce to lowest terms.

27. $\dfrac{5x - 15}{3x - 9}$ $\dfrac{5}{3}$

28. $\dfrac{6x + 18}{9x + 27}$ $\dfrac{2}{3}$

29. $\dfrac{8yx - 14x}{20yx - 35x}$ $\dfrac{2}{5}$

30. $\dfrac{t^2y + 5t^2}{10ty + 50t}$ $\dfrac{t}{10}$

31. $\dfrac{15x^2y - 60y^3}{5x + 10y}$ $3y(x - 2y)$

32. $\dfrac{20s^2t^2 - 45t^2}{12s - 18}$ $\dfrac{5t^2(2s + 3)}{6}$

33. $\dfrac{y + 1}{y^2 + 5y + 4}$ $\dfrac{1}{y + 4}$

34. $\dfrac{y + 7}{y^2 + 2y - 35}$ $\dfrac{1}{y - 5}$

35. $\dfrac{2y - 4}{y^2 - 4}$ $\dfrac{2}{y + 2}$

36. $\dfrac{n^2 - 13n + 42}{n - 6}$ $n - 7$

37. $\dfrac{n^2 + 15n + 26}{n + 13}$ $n + 2$

38. $\dfrac{2n^2 - 4n}{n^2 - 17n + 30}$ $\dfrac{2n}{n - 15}$

39. $\dfrac{6y^2 + 24y}{y^2 + 3y - 4}$ $\dfrac{6y}{y - 1}$

40. $\dfrac{y^2 - 9}{y^2 - y - 6}$ $\dfrac{y + 3}{y + 2}$

41. $\dfrac{n^2 + 4n + 4}{n^2 - 4}$ $\dfrac{n + 2}{n - 2}$

42. $\dfrac{y^2 - 6y - 7}{y^2 + 3y + 2}$ $\dfrac{y - 7}{y + 2}$

43. $\dfrac{y^2 + 5y - 6}{y^2 + 4y - 5}$ $\dfrac{y + 6}{y + 5}$

44. $\dfrac{15t^2 - 15}{t^2 + 14t + 13}$ $\dfrac{15(t - 1)}{t + 13}$

45. $\dfrac{27 - 3x^2}{27 - x^3}$ $\dfrac{3(3 + x)}{9 + 3x + x^2}$

46. $\dfrac{8x^2 - 32}{x^3 - 8}$ $\dfrac{8(x + 2)}{x^2 + 2x + 4}$

In Exercises 47 through 92, multiply or divide as indicted. Leave answers in fractional form.

47. $\dfrac{x^5}{n^6} \cdot \dfrac{8n^8}{4x^2}$ $2n^3x^3$

48. $\dfrac{12n^9}{3t^8} \cdot \dfrac{6t^3}{8n^4}$ $\dfrac{3n^5}{t^5}$

49. $\dfrac{7r^{11}}{x^3} \cdot \dfrac{x^6}{14r^3}$ $\dfrac{r^8x^3}{2}$

50. $\dfrac{3t^8}{6r^4} \cdot \dfrac{8t^6}{12r^7}$ $\dfrac{t^{14}}{3r^{11}}$

51. $\dfrac{n}{n - 4} \cdot \dfrac{n - 4}{4n}$ $\dfrac{1}{4}$

52. $\dfrac{y}{2y + 6} \cdot \dfrac{y + 3}{3}$ $\dfrac{y}{6}$

53. $\dfrac{3x - 12}{x - 4} \cdot \dfrac{x - 4}{2}$ $\dfrac{3(x - 4)}{2}$

54. $\dfrac{4x - 28}{4} \cdot \dfrac{7}{x - 7}$ 7

55. $\dfrac{2x^2 - 24x + 40}{2(x - 4)} \cdot \dfrac{5(x - 50)}{5x^2 - 30x - 200}$ $\dfrac{(x - 2)(x - 50)}{(x - 4)(x + 4)}$

56. $\dfrac{25x^2 + 30x + 9}{2x + 22} \cdot \dfrac{x^2 + 22x + 121}{15x + 9}$ $\dfrac{(5x + 3)(x + 11)}{6}$

57. $\dfrac{x^2 + x - 12}{x^2 + 12x + 35} \cdot \dfrac{x^2 + 13x + 40}{2x^2 + 2x - 24}$ $\dfrac{x + 8}{2(x + 7)}$

58. $\dfrac{r^2 - 18r - 40}{r^2 + 15r - 16} \cdot \dfrac{r^2 + 14r - 32}{r^2 - 4}$ $\dfrac{r - 20}{r - 1}$

59. $\dfrac{5x^3}{2t^4} \div \dfrac{15x^4}{6t^5}$ $\dfrac{t}{x}$

60. $\dfrac{16n^4}{3v^7} \div \dfrac{4n^5}{9v^8}$ $\dfrac{12v}{n}$

61. $\dfrac{10x^7}{4t^8} \div \dfrac{5x^8}{16t^7}$ $\dfrac{8}{tx}$

62. $\dfrac{20n^6}{3w^5} \div \dfrac{25n^8}{9w^6}$ $\dfrac{12w}{5n^2}$

63. $\dfrac{x + 3}{3x} \div \dfrac{2x + 6}{x + 6}$ $\dfrac{x + 6}{6x}$

64. $\dfrac{t + 5}{5t} \div \dfrac{t + 5}{t + 10}$ $\dfrac{t + 10}{5t}$

65. $\dfrac{t - 3}{x} \div \dfrac{t - 3}{x + 5}$ $\dfrac{x + 5}{x}$

66. $\dfrac{y - 7}{t - 7} \div \dfrac{y - 7}{t - 7}$ 1

67. $\dfrac{2x^2 - 8x + 6}{2x^2 + 22x - 84} \div \dfrac{2x - 2}{3x - 9}$ $\dfrac{3(x - 3)}{2(x + 14)}$

68. $\dfrac{2y^2 + 16y - 18}{y + 9} \div \dfrac{y^2 - 81}{y - 9}$ $\dfrac{2(y - 1)}{y + 9}$

69. $\dfrac{x^2 + 3x - 28}{x^2 - 18x + 77} \div \dfrac{x^2 + 17x + 70}{x^2 - 15x + 56}$ $\dfrac{(x - 4)(x - 8)}{(x - 11)(x + 10)}$

70. $\dfrac{x^2 - 30x + 189}{x^2 + 26x + 25} \div \dfrac{x^2 - 10x + 9}{x^2 - 13x - 14}$ $\dfrac{(x - 21)(x - 14)}{(x + 25)(x - 1)}$

71. $\dfrac{2x^2 + 3x - 5}{x^2 + 3x} \div \dfrac{2x + 5}{x}$ $\dfrac{x - 1}{x + 3}$

72. $\dfrac{xy - 3x}{x + 2} \cdot \dfrac{3x^2 - x - 14}{y^2 - 9}$ $\dfrac{x(3x - 7)}{y + 3}$

73. $\dfrac{a}{a - b} \cdot \dfrac{b - a}{b^2}$ $\dfrac{-a}{b^2}$

74. $\dfrac{b}{a - b} \div \dfrac{a^2}{b - a}$ $\dfrac{-b}{a^2}$

75. $\dfrac{y^2 + 11y + 18}{y^2 - 5y - 6} \cdot \dfrac{y^2 + 6y - 72}{y^2 + 16y + 63}$ $\dfrac{(y + 2)(y + 12)}{(y + 1)(y + 7)}$

76. $\dfrac{y^2 + 15y + 36}{y^2 - 12y + 27} \div \dfrac{y^2 + 20y + 96}{y^2 - 17y + 72}$ $\dfrac{(y + 3)(y - 8)}{(y - 3)(y + 8)}$

77. $\dfrac{r^2 + 20r + 100}{r - 15} \div \dfrac{r + 10}{r^2 - 225}$ $(r + 10)(r + 15)$

78. $\dfrac{y^2 + y - 42}{y^2 - 10y + 21} \cdot \dfrac{y^2 - 3y - 28}{y^2 - y - 30}$ $\dfrac{(y + 7)(y + 4)}{(y - 3)(y + 5)}$

79. $\dfrac{x^2 + xy - 2y^2}{xy} \cdot \dfrac{1}{5x - 5} \cdot \dfrac{5x^2y^2}{x + 2y}$ $\dfrac{xy(x - y)}{x - 1}$

80. $\dfrac{x^2 - 9}{(x + 3)^2} \cdot \dfrac{x + 3}{(x - 3)^2} \cdot \dfrac{x^3 - 27}{x^3 - 9x}$ $\dfrac{x^2 + 3x + 9}{x(x + 3)(x - 3)}$

81. $\left(\dfrac{(r - m)^2}{(r + m)^2}\right)\left(\dfrac{r^2 - m^2}{r^3 - m^3}\right)$ $\dfrac{(r - m)^2}{(r^2 + rm + m^2)(r + m)}$

82. $\left(\dfrac{x^3 - m^3}{x + m}\right)\left(\dfrac{(x + m)^2}{x^2 - m^2}\right)$ $x^2 + mx + m^2$

83. $\dfrac{x^2 - y^2}{x^2 + 2xy + y^2} \cdot \dfrac{2x^2 + 4xy + 2y^2}{x + y}$ $2(x - y)$

84. $\dfrac{x^3 - y^3}{2x^2 - 2y^2} \cdot \dfrac{(x + y)^2}{y^2 + xy + x^2}$ $\dfrac{x + y}{2}$

85. $\dfrac{x}{x - 3} \div \dfrac{x^3 - x^2 - 12x}{x^2 - 9}$ $\dfrac{1}{x - 4}$

86. $\dfrac{27 - n^3}{n^2} \div \dfrac{3n^2 - 11n + 6}{4 - 6n}$ $\dfrac{2(9 + 3n + n^2)}{n^2}$

87. $\dfrac{4x^2 - 25}{4x + 6} \div \dfrac{4x^2 - 4x - 15}{4x + 10}$ $\dfrac{(2x + 5)^2}{(2x + 3)^2}$

88. $\dfrac{6x + 4}{9x^2 - 64} \div \dfrac{9x^2 - 18x - 16}{6x + 16}$ $\dfrac{4}{(3x - 8)^2}$

89. $\dfrac{xy - xt + wy - wt}{xy + xt + wy + wt} \cdot \dfrac{zy + zt - vy - vt}{zy - zt + vy - vt}$ $\dfrac{z - v}{z + v}$

90. $\dfrac{km + kn - 4pm - 4pn}{km - kn - 4pm + 4pn} \cdot \dfrac{km - kn + pm - pn}{km + kn - pm - pn}$ $\dfrac{k + p}{k - p}$

91. $\dfrac{x^4 + x^2 + x^3 + x}{2x^3 + x + 2x^2 + 1} \div \dfrac{x^3 + x - 4x^2 - 4}{2x^3 - 8x^2 + x - 4}$ x

92. $\dfrac{2x^3 + 2x^2 + 2xy^2 + 2y^2}{x^3 + x + x^2y + y} \cdot \dfrac{x^3 + x + x^2y + y}{2x^2 + 2x - xy^2 - y^2}$ $\dfrac{2(x^2 + y^2)}{2x - y^2}$

In Exercises 93 through 100, evalute each function at $f(0)$. If a function is not defined for $f(0)$, state this.

93. $f(x) = \dfrac{3x^2 - 9}{2x}$ not defined for $x = 0$

94. $f(x) = \dfrac{2x^3 - 4}{3x}$ not defined for $x = 0$

95. $f(x) = \dfrac{x^3 - 27}{x^2 - 9}$ 3

96. $f(x) = \dfrac{3x^3 - 12}{x^2 + 6}$ -2

97. $f(x) = \dfrac{3x^2 - 2x}{4x - 1}$ 0

98. $f(x) = \dfrac{6x^2 + x}{3x + 2}$ 0

99. $f(x) = \dfrac{5x^3 + 2x^2 - 15}{3x^5 + 5}$ -3

100. $f(x) = \dfrac{16x^4 + 3x^3 + 1}{18x^6 - 10}$ $-\dfrac{1}{10}$

EXCURSIONS

Class Act

1. a. Replace A, B, C, and D with different digits chosen from 0 to 9 to give a correct result. You will know when you have a correct answer.

$$\begin{array}{r} A\,B\,C\,D \\ \times\ 9 \\ \hline D\,C\,B\,A \end{array}$$

   b. Replace A, B, C, D, and E with different digits chosen from 0 to 9 to give a correct result. You will know when you have a correct answer.

$$A\ B\ C\ D\ E$$
$$\underline{\times\ 4}$$
$$E\ D\ C\ B\ A$$

c. ✏ Write a set of directions for a friend who is trying to solve the two previous problems.

Data Analysis

2. How have Olympic Marathon winning times changed from 1896 to 1996? Ask four questions and answer them using this data.

Olympic Marathon

Year	Winner	Time	Year	Winner	Time
1896	Spiridon Loues, Greece	2h58m50s	1952	Emil Zatopek, Czechoslovakia	2h23m3.2s
1900	Michel Teato, France	2h59m45s	1956	Alain Mimoun, France	2h25m
1904	Thomas Hicks, United States	3h28m53s	1960	Abebe Bikila, Ethiopia	2h15m16.2s
1906	William J. Sherring, Canada	2h51m23.65s	1964	Abebe Bikila, Ethiopia	2h12m11.2s
1908	John J. Hayes, United States	2h55m18.4s	1968	Mamo Wold, Ethiopia	2h20m26.4s
1912	Kenneth McArthur, South Africa	2h36m54.8s	1972	Frank Shorter, United States	2h12m19.8s
1920	Hannes Kolehmainen, Finland	2h32m35.8s	1976	Walter Cierpinski, East Germany	2h9m55s
1924	Albin Stenroos, Finland	2h41m22.6s	1980	Walter Cierpinski, East Germany	2h11m3s
1928	A.B. El Quafi, France	2h32m57s	1984	Carlos Lopes, Portugal	2h9m55s
1932	Juan Zabala, Argentina	2h31m36s	1988	Gelindo Bordin, Italy	2h10m47s
1936	Kitei Son, Japan	2h29m19.2s	1992	Hwang Young-Cho, South Korea	2h13m23s
1948	Delfo Cabrera, Argentina	2h34m51.6s	1996	Josia Thugwane, South Africa	2h12m36s

Source: 1997 Information Please® Almanac (©1995 Houghton Mifflin Co.), pp. 893–894. All rights reserved. Used with permission by Information Please Almanac LLC.

CONNECTIONS TO *STATISTICS*

Reading a Histogram

Some kinds of graphs use the widths and lengths of vertical bars to represent information. The graph shown here is such a **histogram.** The width of each bar indicates a height *interval* of about 12 inches. Specifically, the intervals are 0–11.9, 12.0–23.9, and so on. The length of each bar indicates how many bushes have heights that fall in the interval. The scale at the left indicates the number of bushes in the height interval.

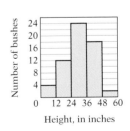

•••
EXAMPLE

In the histogram shown, what height interval contains the most bushes? How many bushes are in that height interval?

SOLUTION

To find the interval with the most bushes, we look for the tallest bar. That bar is the one representing the 24-to-35.9 inch interval. Next we look across the top of this tallest bar to the left scale, and we find that its height represents 24 bushes on that scale. Therefore, 24 bushes are 24.0 to 35.9 inches tall.

◀

PRACTICE

In the following histogram, the number of tenants who live in a small apartment building are graphed by age intervals, 0–14, 15–29, and so on. (A person who is exactly 15, 30, and so on is included in the older group.) Use this information to answer Exercises 1 through 5.

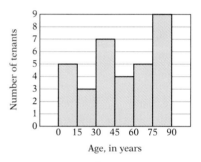

Age, in years

1. What is the width of each age interval in this graph? How many tenants are there altogether? 15 years
   33 tenants

2. What percent of the tenants in this building are less than 60 years of age? Round to the nearest whole percent. 58%

3. What percent of the tenants are at least 30 but less than 60 years of age? Round to the nearest whole percent. 33%

4. What percent of the tenants in this building are less than 90 years of age but at least 60 years old. Round to the nearest whole percent. 42%

5. What is the ratio of the number of tenants under 30 to the number of tenants under 75 years of age? $\frac{1}{3}$

Use the histogram in the Example to answer Exercises 6 and 7.

6. How many bushes are shorter than 24 inches or taller than or equal to 48 inches? 18 bushes

7. What percent of the bushes are 24 to 47.9 inches high? Round to the nearest ten percent. 70%

8. ✉ What are the differences between bar graphs and histograms?
   Answers may vary.

6.2 Adding and Subtracting Rational Expressions

SECTION LEAD-IN

For a 7-inch reel of magnetic recording tape, the remaining playing time t (in minutes) can be determined by measuring the "depth" d (in inches) of tape remaining on the reel. A formula for finding t, when you are given d is

$$\frac{1}{d^2} = \frac{9.38 + \frac{2.25}{d}}{t}$$

a. When $d = 1$, what is the value of t?

b. As t gets larger, what happens to the "depth" of the tape?

c. As d gets smaller, what happens to the amount of time left?

d. What does the answer to part (a) tell you about the tape? Write a sentence that explains what we found out.

SECTION GOALS

▪ To find the least common denominator of rational expressions

▪ To add rational expressions

▪ To subtract rational expressions

Introduction

Rational expressions can be added or subtracted only when they have the same denominator. Then we add or subtract the numerators and place the result over the common denominator.

Adding or Subtracting Rational Expressions

For polynomials p, q, and r, with $q \neq 0$,

$$\frac{p}{q} + \frac{r}{q} = \frac{p + r}{q} \quad \text{and} \quad \frac{p}{q} - \frac{r}{q} = \frac{p - r}{q}$$

In this first problem, we use functional notation to illustrate the addition and subtraction of rational expressions with the same denominators.

■ ■ ■

EXAMPLE 1

Let $f(x) = \frac{5x - 4}{3x^2 - 2}$ and $g(x) = \frac{2x + 5}{3x^2 - 2}$.

Find $f(x) + g(x)$ and $g(x) - f(x)$.

SOLUTION

To find $f(x) + g(x)$, we add the expressions. Because the denominators are the same, we can add the numerators directly. We get

$$f(x) + g(x) = \frac{5x - 4}{3x^2 - 2} + \frac{2x + 5}{3x^2 - 2} = \frac{5x - 4 + 2x + 5}{3x^2 - 2}$$

$$= \frac{7x + 1}{3x^2 - 2}$$

INSTRUCTOR NOTE

Remind students that the additive inverse of a rational expression is simply the opposite of the numerator of that expression.

To find $g(x) - f(x)$, we subtract the expressions. The job is made easy by the common denominator.

$$g(x) - f(x) = \frac{2x + 5}{3x^2 - 2} - \frac{5x - 4}{3x^2 - 2} = \frac{2x + 5 - (5x - 4)}{3x^2 - 2}$$

$$= \frac{2x + 5 - 5x + 4}{3x^2 - 2}$$

$$= \frac{-3x + 9}{3x^2 - 2}$$

▶*CHECK* **Warm-Up 1**

Rational expressions cannot be added or subtracted if they have different denominators. However, they can be rewritten as equivalent expressions with the same denominator—their least common denominator—and *then* added.

The Least Common Denominator of Rational Expressions

To find the least common multiple (LCM) of two or more positive integers, we (1) factor each integer and write the factorization in exponential form, (2) multiply all the bases, and (3) attach to each the greatest exponent it has in any of the factored integers. The product of these bases is the LCM of the integers. We do essentially the same thing to find the LCM of two polynomials.

> **To find the least common multiple of two or more polynomials**
>
> 1. Factor the polynomials completely, and write each in exponential form.
> 2. Multiply together all the bases found in any of the exponential forms.
> 3. Give each of these bases the greatest exponent it has in any of the exponential forms.

The least common denominator (LCD) of rational expressions is the least common multiple (LCM) of the denominators.

▪ ▪ ▪
EXAMPLE 2

Find the LCD of $\dfrac{2x}{x^2 - 3x + 2}$ and $\dfrac{x - 6}{4x - 4}$.

SOLUTION

To find the LCD, we need to find the LCM of the denominators. First we factor them.

$$x^2 - 3x + 2 = (x - 2)(x - 1)$$
$$4x - 4 = 4(x - 1)$$

The factors are $(x - 2)$, $(x - 1)$, and 4. Because they all appear to the first power, the LCD is

$$4(x - 1)(x - 2)$$

▶*CHECK* **Warm-Up 2**

Equivalent Rational Expressions

Two rational expressions are equivalent if they have the same value. According to the fundamental property of fractions (Section 6.1), we can multiply (or divide) the numerator and denominator of a fraction by the same factor without changing its value.

That is,

$$\frac{p}{q} = \frac{pr}{qr}$$

where p, q, and r are polynomials not equal to zero.

If we have a rational expression $\frac{p}{q}$ and we want to rewrite it with an LCD of qr, we just multiply both numerator and denominator by r.

▪ ▪ ▪

EXAMPLE 3

Rewrite $\frac{x}{5(x^2 - 2xy + y^2)}$ and $\frac{y}{x^2 - y^2}$ as equivalent expressions with their LCD. Leave your answer in factored form.

SOLUTION

We first need to find their LCD, so we factor the denominators.

$$5(x^2 - 2xy + y^2) = 5(x - y)(x - y) \quad \text{Perfect-square trinomial}$$
$$= 5(x - y)^2$$
$$(x^2 - y^2) = (x + y)(x - y) \quad \text{Difference of two squares}$$

Writing all the factors with their highest exponents gives us

$$\text{LCD} = 5(x - y)^2(x + y)$$

Now we must rewrite each of the rational expressions with this LCD as the denominator. We work with factored forms so that we can see what factor in the LCD is missing from each original denominator.

For $\frac{x}{5(x - y)^2}$, the missing denominator factor is $(x + y)$, so we multiply numerator and denominator by $(x + y)$.

$$\frac{x}{5(x - y)^2} = \frac{x(x + y)}{5(x - y)^2(x + y)} = \frac{x^2 + xy}{5(x - y)^2(x + y)}$$

For $\frac{y}{(x - y)(x + y)}$, the missing denominator factor is $5(x - y)$, so

$$\frac{y}{(x - y)(x + y)} = \frac{y(5)(x - y)}{(x - y)(x + y)(5)(x - y)}$$
$$= \frac{5xy - 5y^2}{5(x - y)^2(x + y)}$$

▶ CHECK **Warm-Up 3**

> **STUDY HINT**
>
> *If a polynomial factors into identical factors, such as $(x - 3)(x - 3)$, you must use the factor twice in the denominator.*

> ▪ ▪ ▪
> **WRITER'S BLOCK**
> How are the LCD and LCM alike?

Adding or Subtracting Expressions with Different Denominators

To add or subtract two or more rational expressions with different denominators, (1) find their LCD, (2) rewrite them as equivalent expressions with their LCD as denominator, and (3) add or subtract the equivalent expressions.

▪▪▪

EXAMPLE 4

Add:

$$\frac{n+5}{2n-4} + \frac{n}{4-2n}$$

SOLUTION

Factoring each of the denominators yields

$$2n - 4 = 2(n - 2)$$
$$4 - 2n = 2(2 - n), \text{ or } 2(-1)(n - 2)$$

Thus

$$\text{LCD} = 2(-1)(n - 2)$$

> **STUDY HINT**
>
> *Using the least common denominator insures that the resulting computation will be the least complicated.*

Writing equivalent expressions with the LCD as the denominator, and then adding the numerators, we obtain

$$\frac{n+5}{2(n-2)} + \frac{n}{2(-1)(n-2)}$$ Original problem in factored form

$$= \frac{(n+5)(-1)}{2(n-2)(-1)} + \frac{n}{2(-1)(n-2)}$$ Writing equivalent expressions

$$= \frac{(n+5)(-1)+n}{2(n-2)(-1)}$$ Adding numerators

$$= \frac{-n-5+n}{-2(n-2)}$$ Simplifying numerator

$$= \frac{-5}{-2(n-2)} = \frac{5}{2(n-2)}$$ Simplifying

This expression cannot be factored further.

▶ *CHECK* **Warm-Up 4**

▪▪▪

EXAMPLE 5

Add:

$$\frac{y+4}{y^2-5y+4} + \frac{y+1}{y-1}$$

SOLUTION

The first denominator can be factored as

$$y^2 - 5y + 4 = (y - 1)(y - 4).$$

The second cannot be factored. So the LCD is

$$(y - 1)(y - 4)$$

Making equivalent fractions, we obtain

$$\frac{y+4}{(y-1)(y-4)} + \frac{(y+1)(y-4)}{(y-1)(y-4)}$$

> **! ! !**
> **ERROR ALERT**
>
> Identify the error and give a correct answer.
>
> Solve: $\frac{1}{x} + \frac{3x}{2} = -3.5$
>
> *Incorrect Solution:*
>
> $2\left(\frac{1}{x} + \frac{3x}{2}\right) = -3.5(2)$
>
> $2 + 3x = -7$
>
> $3x = -9$
>
>
>
> $x = -3$

Adding the numerators yields

$$\frac{y + 4 + (y + 1)(y - 4)}{(y - 1)(y - 4)} = \frac{y + 4 + y^2 - 3y - 4}{(y - 1)(y - 4)}$$

$$= \frac{y^2 - 2y}{(y - 1)(y - 4)}$$

This expression cannot be simplified further. We write the result as

$$\frac{y^2 - 2y}{y^2 - 5y + 4}$$

▶ *CHECK* **Warm-Up 5**

▪▪▪

EXAMPLE 6

Simplify:

$$\frac{3x - 4}{3x - 6} + \frac{x + 5}{2x - 2} - \frac{2x}{6}$$

SOLUTION

First we factor the denominator in each expression to find the LCD. We get

$$\frac{3x - 4}{3(x - 2)} + \frac{x + 5}{2(x - 1)} - \frac{2x}{2(3)}$$

Thus, the LCD is $6(x - 2)(x - 1)$.

Rewriting each expression as an equivalent fraction with the LCD and simplifying, we have

$$\frac{(3x - 4)(2)(x - 1)}{3(x - 2)(2)(x - 1)} + \frac{(x + 5)(3)(x - 2)}{2(x - 1)(3)(x - 2)} - \frac{2x(x - 2)(x - 1)}{2(3)(x - 2)(x - 1)}$$

$$= \frac{(6x - 8)(x - 1)}{6(x - 2)(x - 1)} + \frac{(3x + 15)(x - 2)}{6(x - 2)(x - 1)} - \frac{(2x^2 - 4x)(x - 1)}{6(x - 2)(x - 1)}$$

$$= \frac{6x^2 - 14x + 8}{6(x - 2)(x - 1)} + \frac{3x^2 + 9x - 30}{6(x - 2)(x - 1)} - \frac{2x^3 - 6x^2 + 4x}{6(x - 2)(x - 1)}$$

$$= \frac{6x^2 - 14x + 8 + 3x^2 + 9x - 30 - 2x^3 + 6x^2 - 4x}{6(x - 2)(x - 1)}$$ Combining

$$= \frac{-2x^3 + 15x^2 - 9x - 22}{6(x - 2)(x - 1)}$$

▶ *CHECK* **Warm-Up 6**

! ! !
ERROR ALERT
Identify the error and
give a correct answer.
Solve: $\dfrac{x + 3}{2} - 4 = \dfrac{-3}{x}$

Incorrect Solution:

$$2x\left(\frac{x + 3}{2}\right) - 4 = \frac{-3}{x}(2x)$$

$$x(x + 3) - 4 = -6$$

$$x^2 + 3x - 4 = -6$$

$$x^2 + 3x + 2 = 0$$

$$(x + 1)(x + 2) = 0$$

$$x = -1 \,|\, x = -2$$

▪▪▪

EXAMPLE 7

Let $f(y)$ be $\dfrac{y}{y - 2}$ and let $g(y)$ be $\dfrac{y - 6}{3}$. Find $f(y) - g(y)$.

SOLUTION

$$f(y) - g(y) = \frac{y}{y - 2} - \frac{y - 6}{3}$$

Because neither of the denominators can be factored, each must be used as part of the new denominator. Accordingly,

$$LCD = 3(y - 2)$$

Writing equivalent expressions with this LCD, we obtain

$$\frac{y(3)}{(y - 2)(3)} - \frac{(y - 6)(y - 2)}{3(y - 2)} = \frac{3y}{3(y - 2)} - \frac{y^2 - 8y + 12}{3(y - 2)}$$

The operation is subtraction, so we must add the additive inverse of the second polynomial.

$$\frac{3y}{3(y - 2)} - \frac{y^2 - 8y + 12}{3(y - 2)}$$

$$= \frac{3y + (-y^2 + 8y - 12)}{3(y - 2)} \qquad \text{Adding the additive inverse}$$

$$= \frac{3y - y^2 + 8y - 12}{3(y - 2)} \qquad \text{Simplifying}$$

$$= \frac{-y^2 + 11y - 12}{3(y - 2)} \qquad \text{Simplifying numerator}$$

We cannot reduce the rational expression further. We leave the result in this form. So

$$f(y) - g(y) = \frac{-y^2 + 11y - 12}{3(y - 2)}$$

▶ CHECK **Warm-Up 7**

Practice what you learned.

SECTION FOLLOW-UP

The questions asked in the Section Lead-In explore the meaning of the formula

$$\frac{1}{d^2} = \frac{9.38 + \frac{2.25}{d}}{t}$$

a. When $d = 1$, $t = 11.63$ minutes.

b. As t gets larger, d gets larger.

c. As d gets smaller, t gets smaller.

d. When the "depth" of the tape is 1 inch, there are 11.63 minutes of playing time left.

You should choose some very large and very small positive values to check the results in parts (b) and (c). (Why are we only interested in positive values here?)

6.2 WARM-UPS

Work these problems before you attempt the exercises.

1. Find $f(y) + g(y)$ when $f(y) = \frac{6y}{8}$ and $g(y) = \frac{y}{8}$. $\frac{7y}{8}$

2. Find the LCD of $\frac{3x}{x^2 - 2x + 1}$ and $\frac{x - 4}{24x - 24}$. $24(x - 1)^2$, or $24x^2 - 48x + 24$

3. Rewrite $\frac{3x - 4}{x^2 - 4}$ and $\frac{2x + 5}{x - 2}$ as equivalent fractions with their LCD.

4. Add: $\frac{t}{t - 5} + \frac{t - 3}{15 - 3t}$ $\frac{2t + 3}{3(t - 5)}$

5. Add: $\frac{x + 3}{3x - 9} + \frac{x + 8}{x^2 + 5x - 24}$ $\frac{x + 6}{3(x - 3)}$

6. Simplify: $\frac{x}{x + 5} + \frac{4x}{4x - 20} - \frac{2x^2}{4x^2 - 100}$ $\frac{3x^2}{2(x + 5)(x - 5)}$

7. Let $f(y) = \frac{y}{y - 2}$ and $g(y) = \frac{y - 6}{3}$. Find $g(y) - f(y)$. $\frac{y^2 - 11y + 12}{3(y - 2)}$

6.2 EXERCISES

Note: Use your graphing calculator to check your results whenever possible.

In Exercises 1 through 8, find $f(x) + g(x)$ and $f(x) - g(x)$.

1. $f(x) = \frac{5x}{4}$

   $g(x) = \frac{3x}{4}$ $2x; \frac{x}{2}$

2. $f(x) = \frac{7x}{8}$

   $g(x) = \frac{17x}{8}$ $3x; \frac{-5x}{4}$

3. $f(x) = \frac{6(x + 2)}{5}$

   $g(x) = \frac{4(x + 2)}{5}$ $2(x + 2); \frac{2(x + 2)}{5}$

4. $f(x) = \frac{5(x^2 - 1)}{9}$

   $g(x) = \frac{4(x^2 - 1)}{9}$ $x^2 - 1; \frac{x^2 - 1}{9}$

5. $f(x) = \frac{8(x - 3)}{5x}$

   $g(x) = \frac{2x + 14}{5x}$ $\frac{2(x - 1)}{x}; \frac{6x - 38}{5x}$

6. $f(x) = \frac{9(x - 1)}{7x}$

   $g(x) = \frac{4 - 2x}{7x}$ $\frac{7x - 5}{7x}; \frac{11x - 13}{7x}$

7. $f(x) = \frac{6x^2 - 2x}{5x^2}$

   $g(x) = \frac{12x + 4x^2}{5x^2}$ $\frac{2(x + 1)}{x}; \frac{2(x - 7)}{5x}$

8. $f(x) = \frac{18x + 5x^2}{9x^2}$

   $g(x) = \frac{13x^2}{9x^2}$ $\frac{2(1 + x)}{x}; \frac{2(9 - 4x)}{9x}$

In Exercises 9 through 16, find the least common multiple.

9. $6xy, 24y, 8y^2$ $24xy^2$

10. $9xy, 27x, 18x^2$ $54x^2y$

11. $15x^2y, 18xy, 24y$ $360x^2y$

12. $12xy^2, 21y^2, 9xy$ $252xy^2$

13. $25xyz, 15yz^2, 10xz$ $150xyz^2$

14. $36xyz, 18xz^2, 24xyz^2$ $72xyz^2$

15. $14xy^2, 21xy^2z^2, 28xy^2z$ $84xy^2z^2$

16. $32x^2y, 36xyz, 28xy^2z$ $2016x^2y^2z$

In Exercises 17 through 24, rewrite these fractions with their least common denominator.

17. $\frac{8y}{5x}$ and $\frac{16x}{6y}$ $\frac{48y^2}{30xy}; \frac{80x^2}{30xy}$

18. $\frac{6x}{5z}$ and $\frac{2z}{7x}$ $\frac{42x^2}{35xz}; \frac{10z^2}{35xz}$

19. $\frac{6}{36x^2y}$ and $\frac{y + 1}{32xy}$ $\frac{48}{288x^2y}; \frac{9xy + 9x}{288x^2y}$

20. $\dfrac{2}{42xyz}$ and $\dfrac{3 + x}{45xz^2}$

$\dfrac{30z}{630xyz^2}, \dfrac{14y(3 + x)}{630xyz^2}$

21. $\dfrac{2x}{x - 1}$ and $\dfrac{x}{x + 1}$

$\dfrac{2x^2 + 2x}{(x + 1)(x - 1)}, \dfrac{x^2 - x}{(x + 1)(x - 1)}$

22. $\dfrac{2y}{y - 3}$ and $\dfrac{y + 1}{y - 5}$

$\dfrac{2y^2 - 10y}{(y - 5)(y - 3)}, \dfrac{y^2 - 2y - 3}{(y - 5)(y - 3)}$

23. $\dfrac{5}{x - 6}$ and $\dfrac{x - 5}{x^2 - 8x + 12}$

$\dfrac{5x - 10}{(x - 6)(x - 2)}, \dfrac{x - 5}{(x - 6)(x - 2)}$

24. $\dfrac{7x}{x + 3}$ and $\dfrac{x + 2}{x^2 - 16}$

$\dfrac{7x^3 - 112x}{(x + 3)(x + 4)(x - 4)}, \dfrac{x^2 + 5x + 6}{(x + 3)(x + 4)(x - 4)}$

In Exercises 25 through 66, add or subtract as indicated.

25. $\dfrac{r}{11x} + \dfrac{r}{9}$

$\dfrac{9r + 11rx}{99x}$

26. $\dfrac{y}{5} + \dfrac{y}{9x}$

$\dfrac{9xy + 5y}{45x}$

27. $\dfrac{11t}{6t^2} + \dfrac{7}{9t}$

$\dfrac{47}{18t}$

28. $\dfrac{3x}{9xy} + \dfrac{2}{3x}$

$\dfrac{x + 2y}{3xy}$

29. $\dfrac{x}{11} - \dfrac{x}{8y}$

$\dfrac{x(8y - 11)}{88y}$

30. $\dfrac{t}{5x} - \dfrac{t}{15}$

$\dfrac{t(3 - x)}{15x}$

31. $\dfrac{5}{3y} - \dfrac{2y}{4y}$

$\dfrac{10 - 3y}{6y}$

32. $\dfrac{8}{4r^2} - \dfrac{6r}{8r}$

$\dfrac{8 - 3r^2}{4r^2}$

33. $\dfrac{2 - 5t}{4t} + \dfrac{2t - 5}{3}$

$\dfrac{8t^2 - 35t + 6}{12t}$

34. $\dfrac{1 + 6x^2}{5xy} + \dfrac{5 - 4x}{3xy}$

$\dfrac{18x^2 - 20x + 28}{15xy}$

35. $\dfrac{9 - 2r}{7r} - \dfrac{5}{2r^3}$

$\dfrac{-4r^3 + 18r^2 - 35}{14r^3}$

36. $\dfrac{3y - 4}{9y^2} - \dfrac{6}{4y}$

$\dfrac{-8 - 21y}{18y^2}$

37. $\dfrac{r + 3}{r + 5} + \dfrac{r - 5}{r - 3}$

$\dfrac{2r^2 - 34}{(r - 3)(r + 5)}$

38. $\dfrac{y - 5}{y - 4} + \dfrac{y + 4}{y + 5}$

$\dfrac{2y^2 - 41}{(y - 4)(y + 5)}$

39. $\dfrac{x - 6}{x - 5} - \dfrac{x + 5}{x + 6}$

$\dfrac{-11}{(x - 5)(x + 6)}$

40. $\dfrac{x + 9}{x - 8} - \dfrac{x + 8}{x - 9}$

$\dfrac{-17}{(x - 9)(x - 8)}$

41. $\dfrac{3}{r - 8} + \dfrac{5}{r + 3}$

$\dfrac{8r - 31}{(r + 3)(r - 8)}$

42. $\dfrac{6}{y - 14} + \dfrac{2}{y + 5}$

$\dfrac{8y + 2}{(y - 14)(y + 5)}$

43. $\dfrac{4}{t - 5} - \dfrac{3}{t + 5}$

$\dfrac{t + 35}{(t - 5)(t + 5)}$

44. $\dfrac{9}{x - 3} - \dfrac{8}{x + 3}$

$\dfrac{x + 51}{(x - 3)(x + 3)}$

45. $\dfrac{8r + 16}{r - 2} + \dfrac{3 + r}{r + 2}$

$\dfrac{9r^2 + 33r + 26}{(r + 2)(r - 2)}$

46. $\dfrac{3y + 6}{y - 3} + \dfrac{2y + 1}{y + 3}$

$\dfrac{5y^2 + 10y + 15}{(y + 3)(y - 3)}$

47. $\dfrac{4t + 1}{t - 3} - \dfrac{3t - 2}{t + 2}$

$\dfrac{t^2 + 20t - 4}{(t + 2)(t - 3)}$

48. $\dfrac{2x - 5}{x + 5} - \dfrac{3x + 2}{x - 3}$

$\dfrac{-x^2 - 28x + 5}{(x + 5)(x - 3)}$

49. $\dfrac{12r}{2r - 3} - \dfrac{r - 5}{4r - 6}$

$\dfrac{23r + 5}{2(2r - 3)}$

50. $\dfrac{11}{2y - 10} - \dfrac{4y - 1}{y - 5}$

$\dfrac{13 - 8y}{2(y - 5)}$

51. $\dfrac{3t + 1}{6t + 12} + \dfrac{2t - 5}{t + 2}$

$\dfrac{15t - 29}{6(t + 2)}$

52. $\dfrac{4x + 3}{12x + 9} + \dfrac{3x - 4}{8x + 6}$

$\dfrac{17x - 6}{6(4x + 3)}$

53. $\dfrac{2y - 2}{y^2 - 4} - \dfrac{3y + 4}{y + 2}$

$\dfrac{-3y^2 + 4y + 6}{(y + 2)(y - 2)}$

54. $\dfrac{5 + r}{r - 3} - \dfrac{2r - 6}{r^2 - 9}$

$\dfrac{r^2 + 6r + 21}{(r + 3)(r - 3)}$

55. $\dfrac{t + 5}{t - 5} - \dfrac{10t^2 - 3}{t^2 - 25}$

$\dfrac{-9t^2 + 10t + 28}{(t + 5)(t - 5)}$

56. $\dfrac{x + 8}{x - 8} - \dfrac{16x + 64}{x^2 - 64}$

$\dfrac{x^2}{(x + 8)(x - 8)}$

57. $\dfrac{5}{t^2 - 8t + 16} - \dfrac{t}{t^2 - 16}$

$\dfrac{-(t^2 - 9t - 20)}{(t - 4)^2(t + 4)}$

58. $\dfrac{x}{x^2 - 22x + 121} - \dfrac{x + 11}{x^2 - 121}$

$\dfrac{11}{(x - 11)^2}$

59. $\dfrac{r - 12}{r^2 + 24r + 144} - \dfrac{1}{r^2 - 144}$

$\dfrac{r^2 - 25r + 132}{(r + 12)^2(r - 12)}$

60. $\dfrac{y - 7}{y^2 - 14y + 49} + \dfrac{y + 7}{y^2 - 49}$

$\dfrac{2}{y - 7}$

61. $\dfrac{3}{t^2 - 4t - 5} - \dfrac{2}{t^2 - 2t - 3}$

$\dfrac{1}{(t - 3)(t - 5)}$

62. $\dfrac{1}{t^2 - 5t + 6} + \dfrac{1}{2t^2 - 3t - 2}$

$\dfrac{3t - 2}{(t - 2)(t - 3)(2t + 1)}$

63. $\dfrac{t - 1}{3t^2 - 5t - 2} - \dfrac{t}{3t^2 + 7t + 2} + \dfrac{3}{(3t + 1)(t - 2)(t + 2)}$

$\dfrac{1}{(t - 2)(t + 2)}$

64. $\dfrac{x - 1}{x^2 - x - 6} - \dfrac{1}{x^2 - 2x - 3} - \dfrac{3}{(x - 3)(x + 2)(x + 1)}$

$\dfrac{1}{x + 1}$

65. $\dfrac{y + 1}{y^2 - 2y - 3} - \dfrac{7}{y^2 + 6y + 9} + \dfrac{8y}{(y + 3)^2}$

$\dfrac{9y^2 - 25y + 30}{(y + 3)^2(y - 3)}$

66. $\dfrac{2}{(t - 1)^2} + \dfrac{5}{t + 1} - \dfrac{6t}{t^2 - 1}$

$\dfrac{-t^2 - 2t + 7}{(t^2 - 1)(t - 1)}$

In Exercises 67 through 80, find the sum or difference as requested.

67. Let $f(x) = \dfrac{x - 7}{x^2 - x - 2}$

$g(x) = \dfrac{x + 1}{x - 2}$

Find $f(x) - g(x)$ $\quad \dfrac{-(x^2 + x + 8)}{(x - 2)(x + 1)}$

68. Let $f(x) = \dfrac{x}{x + 2}$

$g(x) = \dfrac{x - 6}{x - 4}$

Find $f(x) - g(x)$ $\quad \dfrac{12}{(x + 2)(x - 4)}$

69. Let $f(x) = \dfrac{x}{x - 2}$

$g(x) = \dfrac{x - 6}{3}$

Find $g(x) - f(x)$ $\quad \dfrac{x^2 - 11x + 12}{3(x - 2)}$

$f(x) + g(x)$ $\quad \dfrac{x^2 - 5x + 12}{3(x - 2)}$

70. Let $f(x) = \dfrac{x + 1}{x^2 - 8x + 15}$

$g(x) = \dfrac{x - 3}{x - 5}$

Find $f(x) - g(x)$ $\quad \dfrac{-(x^2 - 7x + 8)}{(x - 3)(x - 5)}$

$f(x) + g(x)$ $\quad \dfrac{x^2 - 5x + 10}{(x - 3)(x - 5)}$

71. Let $f(x) = \dfrac{6x - 5}{-3x^2}$

$g(x) = \dfrac{6}{9x}$

Find $f(x) + g(x)$ $\quad \dfrac{5 - 4x}{3x^2}$

$f(x) - g(x)$ $\quad \dfrac{5 - 8x}{3x^2}$

72. Let $f(x) = \dfrac{x}{x - 5}$

$g(x) = \dfrac{x - 3}{15 - 3x}$

Find $f(x) + g(x)$ $\quad \dfrac{2x + 3}{3(x - 5)}$

$f(x) - g(x)$ $\quad \dfrac{4x - 3}{3(x - 5)}$

73. Let $f(x) = \dfrac{x - 2}{x^2 - 7x + 10}$

$g(x) = \dfrac{x - 4}{2x - 10}$

Find $f(x) + g(x)$ $\quad \dfrac{x - 2}{2(x - 5)}$

$g(x) - f(x)$ $\quad \dfrac{x - 6}{2(x - 5)}$

74. Let $f(x) = \dfrac{x + 3}{3x - 9}$

$g(x) = \dfrac{x + 8}{x^2 + 5x - 24}$

Find $g(x) + f(x)$ $\quad \dfrac{x + 6}{3(x - 3)}$

$f(x) - g(x)$ $\quad \dfrac{x}{3(x - 3)}$

75. Let $f(x) = \dfrac{x - 3}{x^2 - 4}$

$g(x) = \dfrac{3x - 2}{x^2 - 4x + 4}$

Find $f(x) + g(x)$ $\quad \dfrac{4x^2 - x + 2}{(x + 2)(x - 2)^2}$

$g(x) - f(x)$ $\quad \dfrac{2x^2 + 9x - 10}{(x + 2)(x - 2)^2}$

76. Let $f(y) = \dfrac{2y + 8}{2y^2 - 32}$

$g(y) = \dfrac{4y - 8}{y^2 + 2y - 8}$

Find $f(y) + g(y)$ $\quad \dfrac{5y - 12}{(y + 4)(y - 4)}$

$f(y) - g(y)$ $\quad \dfrac{-3y + 20}{(y + 4)(y - 4)}$

77. Let $f(x) = \dfrac{4x}{4x - 20}$

$g(x) = \dfrac{2x^2}{4x^2 - 100}$

$h(x) = \dfrac{x}{x + 5}$

Find $f(x) + g(x) - h(x)$ $\dfrac{x(x + 20)}{2(x + 5)(x - 5)}$

$f(x) - g(x) - h(x)$ $\dfrac{-x(x - 20)}{2(x + 5)(x - 5)}$

$g(x) - f(x) + h(x)$ $\dfrac{x(x - 20)}{2(x + 5)(x - 5)}$

78. Let $f(x) = \dfrac{2x}{4x - 12}$

$g(x) = \dfrac{8}{2x - 4}$

$h(x) = \dfrac{1}{4x^2}$

Find $f(x) - g(x) + h(x)$ $\dfrac{2x^4 - 20x^3 + 49x^2 - 5x + 6}{4x^2(x - 2)(x - 3)}$

$f(x) + g(x) - h(x)$ $\dfrac{2x^4 + 12x^3 - 49x^2 + 5x - 6}{4x^2(x - 2)(x - 3)}$

$g(x) - h(x) - f(x)$ $\dfrac{-2x^4 + 20x^3 - 49x^2 + 5x - 6}{4x^2(x - 2)(x - 3)}$

79. Let $f(y) = \dfrac{8}{(y + 3)^2}$

$g(y) = \dfrac{y - 1}{y^2 + 2y - 3}$

$h(y) = \dfrac{7}{y^2 + 6y + 9}$

Find $f(y) - g(y) + h(y)$ $\dfrac{12 - y}{(y + 3)^2}$

$f(y) + g(y) - h(y)$ $\dfrac{y + 4}{(y + 3)^2}$

$f(y) + g(y) + h(y)$ $\dfrac{y + 18}{(y + 3)^2}$

80. Let $f(t) = \dfrac{2}{(t - 1)^2}$

$g(t) = \dfrac{5}{t + 1}$

$h(t) = \dfrac{6t}{t^2 - 1}$

Find $f(t) - g(t) + h(t)$ $\dfrac{t^2 + 6t - 3}{(t - 1)^2(t + 1)}$

$f(t) + g(t) - h(t)$ $\dfrac{-(t^2 + 2t - 7)}{(t - 1)^2(t + 1)}$

$g(t) - h(t) - f(t)$ $\dfrac{-(t^2 + 6t - 3)}{(t - 1)^2(t + 1)}$

MIXED PRACTICE

By doing these exercises, you will practice the topics up to this point in the chapter.

81. Reduce $\dfrac{x - 4}{x^2 - 8x + 16}$ to lowest terms. $\dfrac{1}{x - 4}$

82. Divide: $\dfrac{x^2 + 17x + 66}{x + 6} \div \dfrac{x^2 + 13x + 22}{x^2 - 4}$ $x - 2$

83. Subtract: $\dfrac{4}{x + 4} - \dfrac{3 - x}{x^2 - 16}$ $\dfrac{5x - 19}{x^2 - 16}$

84. Simplify: $\dfrac{y^2 + 3y}{y}$ $y + 3$

85. Divide: $\dfrac{x - 5}{x^2 + 10x + 25} \div \dfrac{x^2 - 25}{x + 5}$ $\dfrac{1}{(x + 5)^2}$

86. Subtract: $\dfrac{x + 4}{x - 2} - \dfrac{x - 5}{3x - 6}$ $\dfrac{2x + 17}{3(x - 2)}$

87. Let $f(x) = \dfrac{x^2 - 5x + 6}{x^2 - 6x + 8}$. Evaluate $f(x)$ when $x = 0$. $f(0) = \dfrac{3}{4}$

88. Multiply: $\dfrac{x^2 - 4x}{x^2 + 3x + 2} \cdot \dfrac{x + 2}{x - 4}$ $\dfrac{x}{x + 1}$

89. Add: $\dfrac{4}{x - 5} + \dfrac{x + 5}{x^2 - 25}$ $\dfrac{5}{x - 5}$

90. Subtract: $\dfrac{9 - x^2}{x - 5} - \dfrac{x^2 - 25}{15 - 3x}$ $\dfrac{-2(x^2 - 1)}{3(x - 5)}$

EXCURSIONS

Posing Problems

1. Use the following graph to ask and answer four questions.

NUMBER OF CARS LEASED FOR BUSINESS USE (MILLIONS)

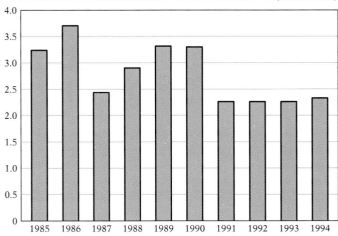

2. Study the following postage tables and then answer the questions.

Airmail Rates
All Countries except Canada & Mexico

Weight not over (oz)	Cost	Weight not over (oz)	Cost
0.5	$ 0.60	9.0	$ 7.40
1.0	1.00	9.5	7.80
1.5	1.40	10.0	8.20
2.0	1.80	10.5	8.60
2.5	2.20	11.0	9.00
3.0	2.60	11.5	9.40
3.5	3.00	12.0	9.80
4.0	3.40	12.5	10.20
4.5	3.80	13.0	10.60
5.0	4.20	13.5	11.00
5.5	4.60	14.0	11.40
6.0	5.00	14.5	11.80
6.5	5.40	15.0	12.20
7.0	5.80	15.5	12.60
7.5	6.20	16.0	13.00
8.0	6.60	16.5	13.40
8.5	7.00		

Airmail Rates
Canada and Mexico

Weight not over (lbs)	(oz)	Cost Canada	Cost Mexico
0	0.5	$.48	$.40
0	1	.52	.48
0	1.5	.64	.66
0	2	.72	.86
0	3	.95	1.26
0	4	1.14	1.66
0	5	1.33	2.06
0	6	1.52	2.46
0	7	1.71	2.86
0	8	1.90	3.26
0	9	2.09	3.66
0	10	2.28	4.06
0	11	2.47	4.46
0	12	2.66	4.86
1	0	3.42	6.46
1	8	4.30	9.66

a. On the same axes, graph the airmail rates for Canada, Mexico, and all other countries.

b. Ask and answer four questions about your graphs. Use line graphs.

c. *Research* Are these prices current? If not, find the new ones and add them to your graph in another color.

d. What happens when you send a package weighing slightly more or slightly less than a weight shown in the table?

3. Using these data, ask and answer four questions that you think are interesting.

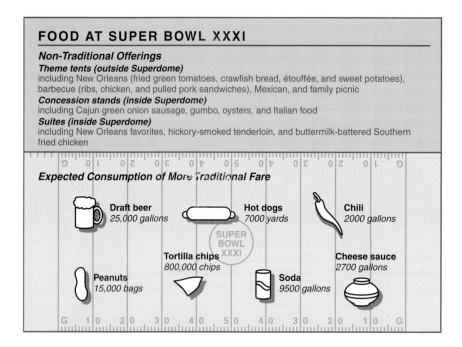

FOOD AT SUPER BOWL XXXI

Non-Traditional Offerings
Theme tents (outside Superdome)
including New Orleans (fried green tomatoes, crawfish bread, étouffée, and sweet potatoes), barbecue (ribs, chicken, and pulled pork sandwiches), Mexican, and family picnic
Concession stands (inside Superdome)
including Cajun green onion sausage, gumbo, oysters, and Italian food
Suites (inside Superdome)
including New Orleans favorites, hickory-smoked tenderloin, and buttermilk-battered Southern fried chicken

Expected Consumption of More Traditional Fare

Draft beer 25,000 gallons

Hot dogs 7000 yards

Chili 2000 gallons

Tortilla chips 800,000 chips

Cheese sauce 2700 gallons

Peanuts 15,000 bags

Soda 9500 gallons

CONNECTIONS TO *GEOMETRY*

Circles

> **Circle**
>
> A **circle** is the set of points that are equidistant from a fixed point called the **center.**

This means that any point on a circle is the same distance from the center as any other point. That distance is the measure of the radius.

Three radii are shown in the accompanying figure.

> **Radius**
>
> A **radius** (plural: radii) is a line segment drawn from the center of a circle to any point on the circle.

> **Diameter**
>
> A **diameter** is a line segment drawn from one point on a circle through the center to another point on the circle.

In the next figure, we can see that a diameter has measure equivalent to that of two radii of the same circle. The radius of a circle is equal in measure to one-half that of the diameter. In symbols,

> If a circle has radius of measure r and diameter of measure d, then
>
> $$r = \frac{d}{2} \quad \text{and} \quad d = 2r$$

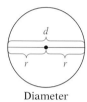

Diameter

Circumferences of Circles

> **Circumference**
>
> The **circumference** of a circle is the distance around the circle—its perimeter.

To calculate the circumference C, we use the fact that $C/d = \pi$ (the Greek letter pi) for any circle. This fact gives us the following formulas.

> **To find the circumference C of a circle**
>
> If d is the diameter or r is the radius of the circle, then
>
> $$C = \pi d \quad \text{or} \quad C = 2\pi r$$

■■■
WRITER'S BLOCK

The circumference of any circle divided by its diameter always gives the same constant value. Why is this an important idea?

The symbol π represents a constant—a fixed number like 7 or 342—but its value cannot be determined exactly. To ten decimal places, it is 3.1415926536. This is usually approximated as 3.14 or $\frac{22}{7}$.

Circumferences have units of length—that is, feet, inches, miles, and such.

Area A

Areas of Circles

The formula for the area of a circle also includes the constant π.

> **To find the area A of a circle**
>
> If r is the radius of the circle, then
> $$A = \pi r^2$$

Areas are measured in square units.

▪ ▪ ▪
EXAMPLE

A unit circle is inscribed within a square. Find the area of the square that lies outside of the circle.

SOLUTION

A unit circle is a circle with a radius of one. By *inscribed* we mean that the circle is entirely within the square and touches it on four sides. We make a sketch and note that the diameter of the circle is the same as the length of one side of the square.

The diameter is twice the length of the radius. (All the formulas we need for solving this problem are on the inside covers of this text.)

$$d = 2r$$
$$d = 2(1)$$
$$d = 2 \text{ units}$$

Next we calculate the area of the square. (Area of the square $= s^2$.)

$$A = 2^2$$
$$A = 4 \text{ square units}$$

Now, we calculate the area of the circle.

$$A = \pi r^2$$
$$A = (3.14)(1^2)$$
$$A = 3.14 \text{ square units}$$

To complete this problem, we find the difference in the area of the square and the area of the circle:

$$4 - 3.14 = 0.86 \text{ square inches}$$

PRACTICE

Use the formulas you learned in this Connection to work the following problems. If you need an estimate for pi, use either 3.14 or $\frac{22}{7}$. Write all answers as decimals and round to the nearest hundredth.

1. When baseball was first played, a regulation baseball had a diameter of $2\frac{3}{84}$ inches. Find the radius of a circle with the same diameter.   1.02 inches

2. Find the radius of a quarter if its diameter is $\frac{11}{12}$ inches.   0.46 inch

3. The diameter of the moon is approximately 2160 miles. If a space vehicle were to move in a straight path all the way around the moon, how far would it travel?   6782.40 miles

4. The widest point on a baseball bat can be no larger than 2.75 inches in diameter. What is the largest that the circumference can be?   8.64 inches

2.75 inches
maximum

5. The largest sunspot recorded was 124,274 miles across. What was the radius of this sunspot?   62,137 miles

6. Find the area of a Diamond Jubilee New York City subway token: diameter 0.90 inch   0.64 square inch

7. One of the first bicycles made in 1876 had a front wheel that measured 5.19 feet in diameter. What was the radius of the wheel?   2.60 feet

8. Earth's equator is a circle with a radius of 3963.49 miles. What is its diameter?   7926.98 miles

d = 5.19 ft

9. The actual wrestling area in a wrestling ring has a diameter of 29.5 feet. What does this area measure?   683.15 square feet

10. A regulation basketball hoop measures 18 inches in diameter. What is the measure of the distance around the rim of the hoop? Use $\frac{22}{7}$ as an estimate for pi in your calculations.   56.57 inches

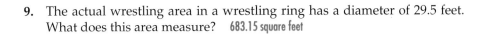
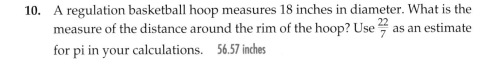

CONNECTIONS TO *STATISTICS*

Reading a Pie Chart

A common type of data display in business reports, newspapers, and magazines is a **circle chart,** or **pie chart.** In a circle chart, the total is represented as a circle, and the parts that make up the total are shown as "slices." The numbers are usually given in the form of percents; the total is 100%.

■ ■ ■

EXAMPLE

The following chart shows how a family spends its yearly income of $31,000. How much money does this family spend on transportation?

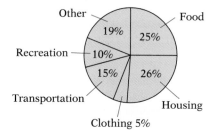

SOLUTION

The chart indicates that 15% of the income is spent on transportation. We must answer the question: 15% of $31,000 is what?

Writing as an equation and solving, we get

$$n = 0.15 \times 31{,}000 = 4650$$

So the family spends $4650 on transportation yearly.

◢

PRACTICE

Pie chart A represents the way a third-world government intends to spend money received on the sale of oil to the United States. (The total amount of money received is $2 billion.) Use this information to answer Exercises 1 and 2.

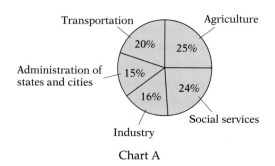

Chart A

1. How much will the country spend on social services and agriculture together? $980,000,000

2. How much more will be spent on industry than on the administration of states and cities? $20,000,000

Circle chart B shows how individuals spent their money for recreation in 1986. Use this information to answer Exercises 3 and 4 for a total annual recreation budget of $2000.

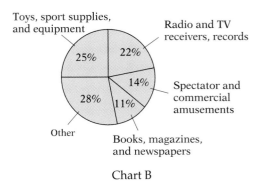

Chart B

3. How much money is left each month after the money for spectator and commercial amusements is spent? Assume the same amount is spent each month. $143.33

INSTRUCTOR NOTE

We always round up when we talk about money.

4. How much is spent per week on radio and TV receivers and records if the same amount is spent each week? $8.47

Pie chart C shows how a farmer in a certain third-world country spends his time. Assume that the entire circle represents a 16-hour day (he has to sleep), and use this information to answer Exercises 5 and 6.

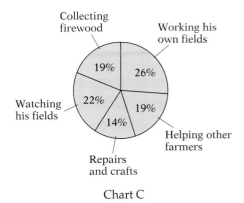

Chart C

5. How many hours does he spend working his own fields and helping other farmers? 7.2 hours

6. How many minutes does he spend collecting firewood? 182.4 minutes

7. See pie chart A. What is the ratio of the amount of money spent on social services to that spent on transportation? $\frac{6}{5}$

8. See the circle chart given in the Connections example. What is the ratio of the amount of money spent on transportation to that spent on food? $\frac{3}{5}$

For Exercises 9 and 10, see pie chart C.

9. What is the ratio of the amount of time spent on repairs and crafts to the time spent working the farmer's own fields and helping other farmers? $\frac{14}{45}$

10. What is the ratio of the amount of time a farmer spends sleeping to the amount of time shown in the chart? $\frac{1}{2}$

SECTION GOALS

- *To simplify complex rational expressions*
- *To solve equations that contain rational expressions*
- *To solve literal equations and formulas*

6.3 More Operations with Rational Expressions

SECTION LEAD-IN

We do not need to understand the meaning or use of a formula to be able to investigate how that formula reacts when the value of a variable changes. The following formula is used by a certain calculator to find the degrees of freedom (*df*) for a two-sample *t*-test in statistics when the variances are "pooled." What does the formula simplify to if

$$Sx_1^2 = Sx_2^2 \quad \text{The variances are equal.}$$

and

$$n_1 = n_2 \qquad \text{The sample sizes are equal.}$$

$$df = \frac{\left(\frac{Sx_1^2}{n_1} + \frac{Sx_2^2}{n_2}\right)^2}{\frac{1}{n_1 - 1}\left(\frac{Sx_1^2}{n_1}\right)^2 + \frac{1}{n_2 - 1}\left(\frac{Sx_2^2}{n_2}\right)^2}$$

Simplifying Complex Fractions

A rational expression that has a fraction or rational expression in the numerator, in the denominator, or in both, is called a **complex fraction.**

Complex fractions must be simplified before they can be added, subtracted, multiplied, or divided. That is, the fractions or rational expressions must be removed from the numerator and the denominator. Two different methods can be used to do this.

Rewriting Complex Fractions as Division

In the first example, both the numerator and the denominator contain a single fraction. So, we *rewrite the complex fraction as a division problem* and then do the division.

▪ ▪ ▪
EXAMPLE 1

Simplify: $\dfrac{\dfrac{y + 6}{x^2 y}}{\dfrac{2y + 12}{x^3 y^2}}$

> **WRITER'S BLOCK**
>
> What makes a fraction *simple*?

SOLUTION

We first rewrite this fraction as a division problem.

$$\frac{\dfrac{y + 6}{x^2 y}}{\dfrac{2y + 12}{x^3 y^2}} = \frac{y + 6}{x^2 y} \div \frac{2y + 12}{x^3 y^2}$$

We then divide.

$$\frac{y + 6}{x^2 y} \div \frac{2y + 12}{x^3 y^2} = \frac{y + 6}{x^2 y} \cdot \frac{x^3 y^2}{2y + 12} = \frac{(y + 6)(x^3 y^2)}{(x^2 y)(2)(y + 6)} = \frac{xy}{2}$$

▶ CHECK **Warm-Up 1**

In the next example, the numerator and denominator of the complex fraction have two distinct terms each. Before we can divide, we must combine those terms.

▪ ▪ ▪
EXAMPLE 2

Simplify: $\dfrac{\dfrac{2x}{4} + 8}{\dfrac{x}{3} - x}$

SOLUTION

We combine the terms in the numerator, and then we separately combine the terms in the denominator. To do so, we find a common denominator for each.

For the numerator:

$$\frac{2x}{4} + 8 = \frac{2x}{4} + \frac{8(4)}{1(4)} = \frac{2x}{4} + \frac{32}{4} \quad \text{The LCD of } \frac{2x}{4} \text{ and } \frac{8}{1} \text{ is 4.}$$
$$= \frac{2x + 32}{4}$$
$$= \frac{2(x + 16)}{4} = \frac{x + 16}{2}$$

For the denominator:

$$\frac{x}{3} - x = \frac{x}{3} - \frac{x(3)}{1(3)} = \frac{x}{3} - \frac{3x}{3} = \frac{-2x}{3} \quad \text{The LCD of } \frac{x}{3} \text{ and } \frac{x}{1} \text{ is 3.}$$

This leaves us with a single fraction in the numerator and in the denominator. We can now simplify, as in Example 1, by rewriting as division.

$$\frac{\dfrac{2x}{4} + 8}{\dfrac{x}{3} - x} = \frac{\dfrac{x + 16}{2}}{\dfrac{-2x}{3}}$$

$$= \frac{x + 16}{2} \div \frac{-2x}{3} = \frac{x + 16}{2} \cdot \frac{3}{-2x} = \frac{-3(x + 16)}{4x}$$

▶ *CHECK* **Warm-Up 2**

Multiplying by the LCD

In the second method of simplifying a complex fraction, we multiply each term in the numerator and denominator by the LCD of all the terms in both numerator and denominator. This changes the complex fraction to a rational expression. The process is equivalent to multiplying by $\frac{\text{LCD}}{\text{LCD}}$ or 1.

▪ ▪ ▪
EXAMPLE 3

Simplify: $\dfrac{\dfrac{3}{n + 4} + \dfrac{5}{2n}}{n + \dfrac{6n}{5}}$

SOLUTION

The denominators are $n + 4$ and $2n$ in the numerator and 5 in the denominator. Their LCD is $10n(n + 4)$. Multiplying each term of the numerator and denominator by the LCD gives us

$$\frac{(10n)(n + 4)\dfrac{3}{n + 4} + \dfrac{5}{2n}(10n)(n + 4)}{(10n)(n + 4)\dfrac{n}{1} + \dfrac{6n}{5}(10n)(n + 4)}$$

Multiplying out and simplifying, we get

$$\frac{(10n)(3) + 5(5)(n + 4)}{(10n)(n + 4)(n) + 6n(2n)(n + 4)}$$

$$= \frac{30n + 25n + 100}{10n^3 + 40n^2 + 12n^3 + 48n^2} = \frac{55n + 100}{22n^3 + 88n^2}$$

The expression is in lowest terms.

▶ *CHECK* **Warm-Up 3**

▪ ▪ ▪
EXAMPLE 4

Simplify: $\dfrac{\dfrac{4}{y + 7} + \dfrac{9y}{y^2 - 49}}{\dfrac{2}{y - 7} - 5}$

SOLUTION

The LCD of the four terms is $(y + 7)(y - 7)$. Multiplying each term by the LCD gives us

$$\frac{(y + 7)(y - 7)\dfrac{4}{y + 7} + \dfrac{9y}{y^2 - 49}(y + 7)(y - 7)}{(y + 7)(y - 7)\dfrac{2}{y - 7} - 5(y + 7)(y - 7)}$$

Then simplifying and multiplying, we get

$$\frac{4(y - 7) + 9y}{2(y + 7) - 5(y^2 - 49)} = \frac{4y - 28 + 9y}{2y + 14 - 5y^2 + 245} = \frac{13y - 28}{-5y^2 + 2y + 259}$$

▶ *CHECK* **Warm-Up 4**

Solving Equations that Contain Rational Expressions

Suppose you need to solve an equation containing rational expressions, say

$$\frac{3}{5} = \frac{x + 10}{27}$$

One way to solve such an equation is with a graphing calculator.

Calculator Corner

Equations containing rational expressions can be solved on your graphing calculator by graphing each side of the equation. The solution to the equation will be the point of intersection of the two graphs. The **WINDOW** chosen here is a "friendly window" for the *TI-82* graphing calculator. Check your calculator manual for the dimensions of a "friendly screen" for your particular graphing calculator. (A "friendly window" is one that will give *nice numbers* when you **TRACE**.)

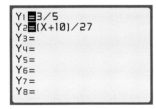
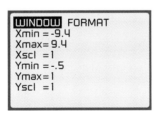
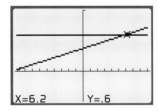

Notice that when you **TRACE**, the two lines intersect at the point $x = 6.2$ and $y = 0.6$. Some graphing calculators have the ability to find the point of intersection for you. (Note: See the Calculator Corner on pages 89 and 90 to review how to find the points of intersection of two graphs.)

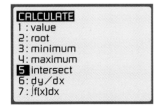
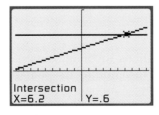

So the solution set is $\{(6.2, 0.6)\}$.

In this section, we will explore some ways to solve equations containing rational expressions algebraically.

To solve an equation that includes rational expressions

1. Multiply each term of the equation by the least common multiple of all the denominators (this is the LCD).
2. Simplify the equation and solve for the variable as usual.
3. Check to be sure that the solution is meaningful.

▪▪▪

EXAMPLE 5

Solve: $\dfrac{6}{y-4} + 6 = \dfrac{1}{3}$

SOLUTION

The denominators, from left to right, are $y-4$, 1, and 3. Their LCM is $3(y-4)$.

Multiplying each term by this LCM will clear the equation of fractions.

$$3(y-4)\frac{6}{y-4} + 3(y-4)6 = 3(y-4)\frac{1}{3} \qquad \text{Multiplying and cancelling}$$

$$3(6) + 3(y-4)6 = y-4 \qquad \text{Rewriting}$$

Simplifying this equation and solving yield

$$18 + 18y - 72 = y - 4 \qquad \text{Multiplying out}$$

$$-54 + 18y = y - 4 \qquad \text{Combining like terms}$$

$$-54 + 17y = -4 \qquad \text{Adding } -y \text{ to both sides}$$

$$17y = 50 \qquad \text{Adding 54 to both sides}$$

$$y = \frac{50}{17} \qquad \text{Multiplying by } \frac{1}{17}$$

Substituting $\frac{50}{17}$ in the denominator $y-4$ does not make it zero, so $y = \frac{50}{17}$ is a meaningful solution. (You should also check to be sure that $\frac{50}{17}$ is a *true* solution.)

▶ CHECK **Warm-Up 5**

Calculator Corner

Your graphing calculator can be used to solve equations involving rational expressions by graphing each side of the equation and then finding the point of intersection. Examine the graphing calculator work for the following problem. (Note: See the Calculator Corner on pages 89 and 90 to review how to find the intersection points of two graphs.)

$$\frac{6}{x-4} + 6 = \frac{1}{3}$$

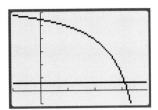

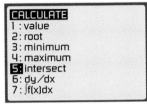

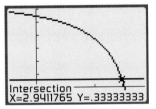

The answer obtained by paper-and-pencil methods is $\frac{50}{17}$, which is approximately 2.941176471.

Now try the following on your own using both paper-and-pencil methods and your graphing calculator.

$$\frac{2x}{x+3} = \frac{x}{x-2} - \frac{12}{x^2+x-6} \qquad \frac{5x^2+1}{2x^2+x-1} - \frac{x}{2x-1} = \frac{2x}{x+1} \qquad \frac{1}{x^2-4} + \frac{1}{x+2} = \frac{2}{x-2}$$

▪ ▪ ▪

EXAMPLE 6

Solve: $6 + \frac{t}{t-12} = \frac{2t}{2t-24}$

SOLUTION

The denominator $2t - 24$ can be factored into $2(t - 12)$. Then, because the denominators are 1, $t - 12$, and $2(t - 12)$, the LCD is $2(t - 12)$.

Multiplying by the LCD clears the equation of fractions.

$$2(t-12)6 + 2(t-12)\frac{t}{t-12} = 2(t-12)\frac{2t}{2(t-12)}$$
$$12(t-12) + 2t = 2t$$

Solving for t gives us

$$
\begin{array}{ll}
12t - 144 + 2t = 2t & \text{Distributive property} \\
14t - 144 = 2t & \text{Combining like terms} \\
12t = 144 & \text{Adding 144 and } -2t \text{ to both sides} \\
t = 12 & \text{Multiplying both sides by } \frac{1}{12}
\end{array}
$$

Going back to the original equation, we see that the only restriction on t is that t cannot be 12. Because our result is $t = 12$, there is no solution. So the solution set is the empty set.

▶ *CHECK* **Warm-Up 6**

Calculator Corner

What happens when you attempt to graph

$$Y_1 = 6 + X/(X - 12)$$
$$Y_2 = 2X/(2X - 24)$$

and calculate the point of intersection?

(*Hint:* Use the window $[-2, 30]$ for x and $[-20, 20]$ for y. You may also want to try **DOT** Mode.) Explain.

▪▪▪

EXAMPLE 7

Solve: $\dfrac{n}{n - 3} - \dfrac{3n}{n + 2} = \dfrac{-6}{n^2 - n - 6}$

SOLUTION

In order to find the LCM of the denominators, we factor $n^2 - n - 6$. the denominators are then $(n - 3)$, $(n + 2)$, and $(n - 3)(n + 2)$. Thus, the LCM of the denominators is $(n - 3)(n + 2)$. Multiplying each term by it, we get

$$(n - 3)(n + 2)\frac{n}{n - 3} - (n - 3)(n + 2)\frac{3n}{n + 2} = (n - 3)(n + 2)\frac{-6}{(n - 3)(n + 2)}$$
$$n(n + 2) - 3n(n - 3) = -6$$

Simplifying yields

$$n^2 + 2n - 3n^2 + 9n = -6$$
$$-2n^2 + 11n = -6$$

We can solve this equation by factoring. But first we must add 6 to both sides to get zero on the right.

$$-2n^2 + 11n + 6 = 0$$

Factoring out the -1, and continuing, we have

$$(-1)(2n^2 - 11n - 6) = 0$$
$$2n^2 - 11n - 6 = 0 \quad \text{Multiplying both sides by } -1$$
$$(2n + 1)(n - 6) = 0 \quad \text{Factoring the trinomial}$$

$2n + 1 = 0$	$n - 6 = 0$	Setting factors equal to 0
$2n = -1$	$n = 6$	Solving for n
$n = \dfrac{-1}{2}$		

Checking the original equation, we see that $-\frac{1}{2}$ and 6 are meaningful solutions.

▶ CHECK **Warm-Up 7**

Calculator Corner

Enter each side of the equation in Example 7 in the $Y =$ screen.

```
Y₁ ▪ X/(X−3)−3X/(X+2)
Y₂ ▪ −6/(X²−X−6)
Y₃ =
Y₄ =
Y₅ =
Y₆ =
Y₇ =
Y₈ =
```

You can check your results by using your graphing calculator's Home Screen to evaluate each equation at the value $x = 6$. Y1 and Y2 should be equal for each value of x.

 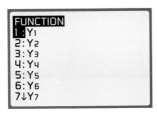 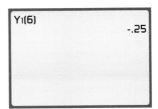

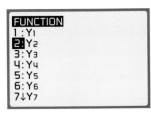 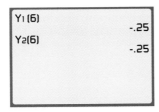

Because both equations give a result of $-.25$ when $x = 6$, one point of intersection is $(6, -0.25)$. There are two solutions. Check the solution $x = \dfrac{-1}{2}$ in the same way.

Formulas that Contain Rational Expressions

Formulas, or literal equations, that include rational expressions are solved just as we solve other equations with rational expressions. We clear the equation of fractions and solve for the required variable.

...

EXAMPLE 8

If two electrical resistances a and b (expressed in ohms) are connected in a parallel circuit, their combined, or effective, resistance R is given by the equation

$$\frac{1}{R} = \frac{1}{a} + \frac{1}{b}$$

Find the values of these resistances if the larger is 5 ohms greater than the smaller, and the combined resistance is $\frac{4}{5}$ of the smaller resistance.

SOLUTION

We let

$$\text{Smaller resistance} = x$$

then

$$\text{Larger resistance} = x + 5$$

and

$$\text{Combined resistance} = \frac{4}{5}x$$

Substituting in the formula gives us

$$\frac{1}{\frac{4}{5}x} = \frac{1}{x} + \frac{1}{x + 5}$$

To solve this equation, we multiply each term by the LCM of the denominators, which is $\frac{4}{5}x(x + 5)$.

We get

$$\left(\frac{4}{5}x\right)(x + 5)\frac{1}{\frac{4}{5}x} = \left(\frac{4}{5}x\right)(x + 5)\frac{1}{x} + \left(\frac{4}{5}x\right)(x + 5)\frac{1}{x + 5}$$

$$(x + 5) = \frac{4}{5}(x + 5) + \frac{4}{5}x$$

We simplify and solve the equation, obtaining

$$x + 5 = \frac{4}{5}x + \frac{4}{5}(5) + \frac{4}{5}x \qquad \text{Distributive property}$$

$$x + 5 = \frac{8}{5}x + 4 \qquad \text{Combining like terms}$$

$$x = \frac{8}{5}x - 1 \qquad \text{Adding } -5 \text{ to both sides}$$

$$-\frac{3}{5}x = -1 \qquad \text{Adding } -\frac{8}{5}x \text{ to both sides}$$

$$x = \frac{5}{3} \qquad \text{Multiplying by } -\frac{5}{3}$$

Going back to the original equation, we see that this solution has meaning. However, we are not finished. We must find the values of the parallel resistances.

Because x represents the smaller resistance, its value is $1\frac{2}{3}$ ohms.

The larger resistance, $x + 5$, is then $6\frac{2}{3}$ ohms.

The combined resistance is

$$\frac{4}{\cancel{5}} \cdot \frac{\cancel{5}}{3} = \frac{4}{3}$$

This is actually smaller than either parallel resistance, as is always the case.

Check by substituting these values in the parallel resistance equation. You will find they are both reasonable and correct.

▶ *CHECK* **Warm-Up 8**

Practice what you learned.

SECTION FOLLOW-UP

The formula

$$df = \frac{\left(\dfrac{Sx_1^2}{n_1} + \dfrac{Sx_2^2}{n_2}\right)^2}{\dfrac{1}{n_1 - 1}\left(\dfrac{Sx_1^2}{n_1}\right)^2 + \dfrac{1}{n_2 - 1}\left(\dfrac{Sx_2^2}{n_2}\right)^2}$$

simplifies to

$$df = 2n - 2$$

when $Sx_1^2 = Sx_2^2$ and $n_1 = n_2$. Verify this.

6.3 WARM-UPS

Work these problems before you attempt the exercises.

1. Simplify: $\dfrac{\dfrac{5x - 25}{x}}{\dfrac{x - 5}{4x}}$ 20

2. Simplify: $\dfrac{\dfrac{y}{7} + 3}{\dfrac{y}{8} - 2}$ $\dfrac{8(y + 21)}{7(y - 16)}$

3. Simplify: $\dfrac{\dfrac{5}{r} - \dfrac{2}{10r}}{r + \dfrac{r}{5}}$ $\dfrac{4}{r^2}$

4. Simplify: $\dfrac{\dfrac{x - 3}{x + 1} + \dfrac{1}{x - 2}}{\dfrac{2}{3} - \dfrac{x}{x + 1} - \dfrac{7}{x - 2}}$ $\dfrac{3x^2 - 12x + 21}{-x^2 - 17x - 25}$

5. Solve: $\dfrac{x + 3}{2x + 2} = \dfrac{x}{x + 1}$ $x = 3$

6. Solve: $\dfrac{3}{n - 5} + \dfrac{2}{n - 2} = \dfrac{7}{n^2 - 7n + 10}$ $n = 4\frac{3}{5}$

7. A formula to determine the power P generated by a windmill is

$$\frac{P}{K} = v^3$$

where P is in watts and v is in miles per hour. Solve this equation for K, and then find the value of K when $P = 9.41$ and $v = 8.6$. $\dfrac{P}{v^3} = K; K \approx 0.0148$

6.3 EXERCISES

Note: Use your graphing calculator to check your results whenever possible.

In Exercises 1 through 32, simplify each complex rational expression by using whichever method is more convenient.

1. $\dfrac{\frac{nr}{r}}{\frac{nr^2}{n}}$ $\dfrac{n}{r^2}$

2. $\dfrac{\frac{k^3 m}{k^3}}{\frac{km^3}{m^2}}$ $\dfrac{1}{k}$

3. $\dfrac{\frac{k-6}{3k}}{\frac{3k-18}{9}}$ $\dfrac{1}{k}$

4. $\dfrac{\frac{r-8}{7}}{\frac{2r-16}{21r}}$ $\dfrac{3r}{2}$

5. $\dfrac{\frac{r+4}{4}}{\frac{5r+20}{8r}}$ $\dfrac{2r}{5}$

6. $\dfrac{\frac{r-5}{6r}}{\frac{4r-20}{12}}$ $\dfrac{1}{2r}$

7. $\dfrac{\frac{n}{6}+3}{-4+\frac{n}{9}}$ $\dfrac{3n+54}{-72+2n}$

8. $\dfrac{\frac{r}{4}-5}{6+\frac{r}{8}}$ $\dfrac{2r-40}{48+r}$

9. $\dfrac{9+\frac{k}{3}}{\frac{k}{7}+8}$ $\dfrac{189+7k}{3k+168}$

10. $\dfrac{\frac{m}{7}-7}{5+\frac{m}{6}}$ $\dfrac{6m-294}{210+7m}$

11. $\dfrac{\frac{r}{3}+r}{\frac{3r}{2}-5}$ $\dfrac{8r}{9r-30}$

12. $\dfrac{\frac{k}{6}-k}{\frac{2k}{8}+3}$ $\dfrac{-10k}{3k+36}$

13. $\dfrac{\frac{12-r}{2r}}{\frac{4r}{r-8}}$ $\dfrac{-r^2+20r-96}{8r^2}$

14. $\dfrac{\frac{r-10}{r}+4}{r-2}$ $\dfrac{5}{r}$

15. $\dfrac{\frac{k}{9}+k}{\frac{k}{18}+\frac{k}{2}}$ 2

16. $\dfrac{\frac{y}{7}+y}{\frac{y}{63}-\frac{y}{7}}$ -9

17. $\dfrac{\frac{7k}{k+3}}{\frac{7k}{4k+12}}$ $\dfrac{7k}{4(k+3)^2}$

18. $\dfrac{\frac{4n}{3-n}}{\frac{6n}{n^2-9}}$ $\dfrac{-2(n+3)}{3}$

19. $\dfrac{\frac{4+y}{6}}{\frac{16-y^2}{3}}$ $\dfrac{1}{2(4-y)}$

20. $\dfrac{\frac{rn}{r^2 n^2}}{\frac{r}{r-n}}$ $\dfrac{r-n}{r^2 n}$

21. $\dfrac{\frac{4}{y}-\frac{3}{2y}}{\frac{5}{3y}+\frac{9}{4y}}$ $\dfrac{30}{47}$

22. $\dfrac{\frac{6}{5y}+\frac{4}{3y}}{\frac{7}{7y}-\frac{6}{2y}}$ $\dfrac{-19}{15}$

23. $\dfrac{x+\frac{5}{x}}{x-\frac{2}{x^2}}$ $\dfrac{x^3+5x}{x^3-2}$

24. $\dfrac{x+\frac{2}{x}}{x-\frac{3}{x^2}}$ $\dfrac{x^3+2x}{x^3-3}$

25. $\dfrac{1-\frac{4}{t}-\frac{5}{t^2}}{1+\frac{4}{t}+\frac{3}{t^2}}$ $\dfrac{t-5}{t+3}$

26. $\dfrac{1-\frac{6}{t^2}+\frac{1}{t}}{1+\frac{2}{t}-\frac{3}{t^2}}$ $\dfrac{t-2}{t-1}$

27. $\dfrac{\frac{2}{t}-\frac{1}{t+1}}{1+\frac{2}{t}}$ $\dfrac{1}{t+1}$

28. $\dfrac{\frac{3}{m}-\frac{2}{m-1}}{\frac{1}{m(m-4)}+\frac{1}{m}}$ $\dfrac{m-4}{m-1}$

29. $\dfrac{1-\frac{1}{k+1}}{1+\frac{1}{k-1}}$ $\dfrac{k-1}{k+1}$

30. $\dfrac{1+\frac{1}{m+1}}{1-\frac{1}{m-1}}$ $\dfrac{(m+2)(m-1)}{(m-2)(m+1)}$

31. $\dfrac{\frac{x}{1+x}-\frac{1-x}{x}}{\frac{x}{1+x}+\frac{1-x}{x}}$ $2x^2-1$

32. $\dfrac{\frac{1}{x}+\frac{x}{x+2}}{\frac{2}{x+2}-\frac{x+1}{x}}$ -1

In Exercises 33 through 80, solve each equation for the indicated variable. If there is no solution, indicate that.

33. $\dfrac{y}{3}=\dfrac{y+6}{5}$ $y=9$

34. $\dfrac{r-3}{5}=\dfrac{r}{7}$ $r=10.5$

35. $\dfrac{x-2}{5}=\dfrac{x+4}{20}$ $x=4$

36. $\dfrac{y+4}{6}=\dfrac{y-6}{8}$ $y=-34$

37. $\dfrac{2y-3}{4}=y+1$ $y=-3.5$

38. $\dfrac{6y-2}{2}=y+5$ $y=3$

39. $\dfrac{y+3}{4}+\dfrac{y+4}{6}=\dfrac{7}{2}$ $y=5$

40. $\dfrac{1}{y}+\dfrac{3}{4}=\dfrac{10}{4y}$ $y=2$

41. $\dfrac{23}{3y}-\dfrac{3}{y}=\dfrac{2}{3}$ $y=7$

42. $\dfrac{3}{4x}+\dfrac{3}{6x}=\dfrac{1}{2}$ $x=\dfrac{5}{2}$

43. $\dfrac{4}{3x}=\dfrac{5}{6}+\dfrac{1}{2x}$ $x=1$

44. $\dfrac{1}{5}+\dfrac{3}{2x}=\dfrac{17}{10x}$ $x=1$

45. $\dfrac{6}{x-2}=\dfrac{x^2+x}{x-2}$ $x=-3$

46. $\dfrac{3}{x-3}=\dfrac{x}{x-3}-2$ no solution

47. $\dfrac{x}{x-4}=2+\dfrac{4}{x-4}$ no solution

48. $\dfrac{w}{w-3} = \dfrac{3}{w-3} - 2$ no solution

49. $\dfrac{1}{5x} - \dfrac{3}{2x} + \dfrac{1}{x} = \dfrac{1}{5}$ $x = -1.5$

50. $\dfrac{4}{3x} - \dfrac{1}{2} = \dfrac{-4}{3x} - \dfrac{1}{x}$ $x = 7\tfrac{1}{3}$

51. $\dfrac{21}{6y} - 9 = \dfrac{6}{3y}$ $y = \tfrac{1}{6}$

52. $8 + \dfrac{6}{16x} = \dfrac{-1}{8x} + 7$ $x = \tfrac{-1}{2}$

53. $\dfrac{2y}{y+2} = \dfrac{1+2y}{y}$ $y = \tfrac{-2}{5}$

54. $\dfrac{x+4}{x-2} = \dfrac{x+2}{x}$ $x = -1$

55. $\dfrac{x-1}{x+2} = \dfrac{x-3}{x+1}$ $x = -5$

56. $3 = \dfrac{2x}{x+3} + \dfrac{x}{x+9}$ $x = -5.4$

57. $\dfrac{x}{x+2} + \dfrac{1}{x+3} = 1$ $x = -4$

58. $\dfrac{3}{y+7} + 3 = \dfrac{4}{y+5}$ $y = -4;$ $y = -7\tfrac{2}{3}$

59. $\dfrac{3}{t-1} = 1 - \dfrac{4}{t+1}$ $t = 0;$ $t = 7$

60. $2 - \dfrac{5}{y-5} = \dfrac{2}{y-8}$ $y = 10;\, y = 6.5$

61. $2 - \dfrac{5}{y-3} = \dfrac{9}{y+1}$ $y = 8;\, y = 1$

62. $\dfrac{3}{t-5} - 1 = \dfrac{4}{5-t}$ $t = 12$

63. $\dfrac{t+7}{2t+4} + 3 = \dfrac{3t-5}{t-2}$ $t = -6;\, t = 3$

64. $6 - \dfrac{1}{x-5} = \dfrac{6}{x-4} + 2$ $x = 6;\, x = 4.75$

65. $4 - \dfrac{5}{x-3} = 6 + \dfrac{9}{x-11}$ $x = 8;\, x = -1$

66. $\dfrac{5}{t+2} + \dfrac{10}{t^2+2t} = -2$ $t = \tfrac{-5}{2}$

67. $\dfrac{1}{t-2} = 1 + \dfrac{2}{t^2-2t}$ $t = 1$

68. $\dfrac{10}{2x+6} + \dfrac{2}{x+3} = \dfrac{1}{2}$ $x = 11$

69. $\dfrac{y}{y-1} + \dfrac{2}{y} = \dfrac{3y-2}{y^2-y}$ no solution

70. $\dfrac{3t+2}{t-1} + \dfrac{2t}{t+1} = \dfrac{7t+3}{t^2-1}$ $t = \tfrac{-1}{5}$

71. $\dfrac{3}{2(x-2)} - \dfrac{2}{x+2} = \dfrac{35}{x^2-4}$ $x = -56$

72. $\dfrac{2n}{n+3} = \dfrac{n}{n-2} - \dfrac{12}{n^2+n-6}$ $n = 4;\, n = 3$

73. $\dfrac{3t}{t+2} - \dfrac{t}{t-1} = \dfrac{18}{t^2+t-2}$ $t = 4.5$

74. $\dfrac{35}{x^2+2x+1} - \dfrac{3}{x+1} = \dfrac{4}{x+1}$ $x = 4$

75. $\dfrac{6x}{x+1} = \dfrac{6x-1}{x+3} - \dfrac{12}{x^2+4x+3}$ no solution

76. $\dfrac{5w}{w+2} = \dfrac{30}{w^2+7w+10} + \dfrac{5w}{w+5}$ $w = 2$

77. $\dfrac{2t}{t-3} - \dfrac{t}{t+4} = \dfrac{24+t^2}{t^2+t-12}$ $t = \tfrac{24}{11}$

78. $\dfrac{5x^2+1}{2x^2+x-1} - \dfrac{x}{2x-1} = \dfrac{2x}{x+1}$ no solution

79. $\dfrac{t+7}{t-1} + \dfrac{t-1}{t+1} = \dfrac{4}{t^2-1}$ $t = -2$

80. $\dfrac{1}{y^2-4} + \dfrac{1}{y+2} = \dfrac{2}{y-2}$ $y = -5$

81. *Mechanics*   The efficiency E of a jack is determined from the pitch p of the jack's thread with the formula

$$E = \dfrac{\tfrac{p}{2}}{p + \tfrac{1}{2}}$$

What is the efficiency of a jack with $p = 0.85$ millimeters? $E = 0.315$

82. Solve the equation in Exercise 81 for p. $p = \dfrac{E}{1-2E}$

83. *Music Production*   For a 7-inch reel of magnetic recording tape, the playing time remaining t can be determined by measuring the "depth" d of tape remaining on the reel. The formula is

$$\dfrac{1}{d^2} = \dfrac{9.38 + \tfrac{2.25}{d}}{t}$$

Solve this formula for t. $t = 9.38\,d^2 + 2.25\,d$

84. Evaluate the formula in Exercise 83 when $d = 0.5$ inches. What does the answer mean in "real life"?   There are 3.47 minutes of playing time left on this reel of tape.

85. $\frac{1}{f} = k\left(\frac{1}{r_1} - \frac{1}{r_2}\right)$ is known as "the lens-maker's equation," where f is the focal length of a thin lens, k is a constant, and r_1 and r_2 are the radii of the lens' spherical surfaces. Solve the equation for r_1. Write as a simple fraction. $\quad r_1 = \frac{kfr_2}{r_2 + kf}$

86. Evaluate the expression in Exercise 85 in terms of k and f when $r_2 = 2r_1$. Substitute for r_2 and solve for f. $\quad f = \frac{2r_1}{k}$

MIXED PRACTICE

By doing these exercises, you will practice the topics up to this point in the chapter.

87. Divide: $\frac{x^2 + 4x + 3}{3x - 9} \div \frac{x + 3}{x^2 - 9}$ $\quad \frac{x^2 + 4x + 3}{3}$

88. Determine the domain of $f(x) = \frac{x - 2}{x + 4}$.
$\{x \mid x \text{ real and } x \neq -4\}$

89. Reduce to lowest terms $\frac{x^2 - 16}{x^2 - 4x}$. $\quad \frac{x + 4}{x}$

90. Add: $\frac{x - 3}{2x} + \frac{x + 2}{5x}$ $\quad \frac{7x - 11}{10x}$

91. Evaluate $\frac{r^2t - x}{tx^2}$ when $r = 1$, $t = -2$, and $x = 4$. $\frac{3}{16}$

92. Solve: $\frac{1}{n} + \frac{4}{n + 6} = \frac{2}{n}$ $\quad n = 2$

93. Simplify: $\dfrac{\dfrac{n}{n + 1}}{\dfrac{1}{n^2 - 1}}$ $\quad n^2 - n$

94. Subtract: $\frac{3}{x + 6} - \frac{4 - x}{x^2 + 6x}$ $\quad \frac{4(x - 1)}{x(x + 6)}$

95. Solve: $\frac{13}{x^2 - x} - \frac{2}{x} = \frac{1}{x - 1}$ $\quad x = 5$

96. Multiply: $\frac{x + 9}{x^2 + x - 56} \cdot \frac{x^2 + 3x - 40}{x^2 - 81}$ $\quad \frac{x - 5}{(x - 7)(x - 9)}$

97. Subtract: $\frac{3}{v + 2} - \frac{4}{v - 2}$ $\quad \frac{-v - 14}{(v + 2)(v - 2)}$

98. Reduce $\frac{x^2yt^3}{y^2x^3t^2}$ to lowest terms. $\quad \frac{t}{xy}$

99. Multiply: $\frac{t^2 + 5t + 6}{t + 2} \cdot \frac{t^3 - 3t^2}{t^2 - 9}$ $\quad t^2$

100. Simplify: $\dfrac{2x + \dfrac{x}{3}}{\dfrac{7x^2}{9x}}$ $\quad 3$

EXCURSIONS

Data Analysis

1. Use this picture and caption to estimate the answers to parts (a) and (b).

Nearly two million people attended the Paris Exposition of 1889 to see and perhaps ride the Ferris Wheel. Invented by George Ferris, the steel circle towered 265 feet in the air. Thirty-six cars revolved slowly, allowing passengers to see newly built skyscrapers in the northern part of the city.

a. What is the circumference of the wheel?

b. If one complete trip (without a stop) took 10 minutes, at what speed would you travel?

c. *Research* How many people could be carried at one time?

2. Study the table and then answer the following questions.

History of the Record for the Mile Run

Time	Athlete	Country	Year	Location
4:36.5	Richard Webster	England	1865	England
4:29.0	William Chinnery	England	1868	England
4:28.8	Walter Gibbs	England	1868	England
4:26.0	Walter Slade	England	1874	England
4:24.5	Walter Slade	England	1875	London
4:23.2	Walter George	England	1880	London
4:21.4	Walter George	England	1882	London
4:18.4	Walter George	England	1884	Birmingham, England
4:18.2	Fred Bacon	Scotland	1894	Edinburgh, Scotland
4:17.0	Fred Bacon	Scotland	1895	London
4:15.6	Thomas Conneff	United States	1895	Travers Island, N.Y.
4:15.4	John Paul Jones	United States	1911	Cambridge, Mass.
4:14.4	John Paul Jones	United States	1913	Cambridge, Mass.
4:12.6	Norman Taber	United States	1915	Cambridge, Mass.
4:10.4	Paavo Nurmi	Finland	1923	Stockholm
4:09.2	Jules Ladoumegue	France	1931	Paris
4:07.6	Jack Lovelock	New Zealand	1933	Princeton, N.J.
4:06.8	Glenn Cunningham	United States	1934	Princeton, N.J.
4:06.4	Sydney Wooderson	England	1937	London
4:06.2	Gunder Hägg	Sweden	1942	Goteborg, Sweden
4:06.2	Arne Andersson	Sweden	1942	Stockholm
4:04.6	Gunder Hägg	Sweden	1942	Stockholm
4:02.6	Arne Andersson	Sweden	1943	Goteborg, Sweden
4:01.6	Arne Andersson	Sweden	1944	Malmo, Sweden
4:01.4	Gunder Hägg	Sweden	1945	Malmo, Sweden
3:59.4	Roger Bannister	England	1954	Oxford, England
3:58.0	John Landy	Australia	1954	Turku, Finland
3:57.2	Derek Ibbotson	England	1957	London
3:54.5	Herb Elliott	Australia	1958	Dublin
3:54.4	Peter Snell	New Zealand	1962	Wanganui, N.Z.
3:54.1	Peter Snell	New Zealand	1964	Auckland, N.Z.
3:53.6	Michel Jazy	France	1965	Rennes, France
3:51.3	Jim Ryun	United States	1966	Berkeley, Calif.
3:51.1	Jim Ryun	United States	1967	Bakersfield, Calif.
3:51.0	Filbert Bayi	Tanzania	1975	Kingston, Jamaica
3:49.4	John Walker	New Zealand	1975	Goteborg, Sweden
3:49.0	Sebastian Coe	England	1979	Oslo
3:48.8	Steve Ovett	England	1980	Oslo
3:48.53	Sebastian Coe	England	1981	Zurich, Switzerland
3:48.40	Steve Ovett	England	1981	Koblenz, W. Ger.
3:47.33	Sebastian Coe	England	1981	Brussels
3:46.31	Steve Cram	England	1985	Oslo
3:44.39	Noureddine Morceli	Algeria	1993	Rieti, Italy

Source: USA Track & Field, as cited in the *1996 Information Please*® *Almanac* (©1995 Houghton Miffin Co.), p. 957. All rights reserved. Used with permission by Information Please LLC.

a. Graph these data and comment on the shape of the graph. Ask and answer four questions about these data.

b. *Research* In 3:44.39 the fastest time ever for the mile run? Add any additional records to your graph.

3. ✏ Using these two pie charts, discuss how the Japanese and Americans are alike and different in their savings and investments strategies.

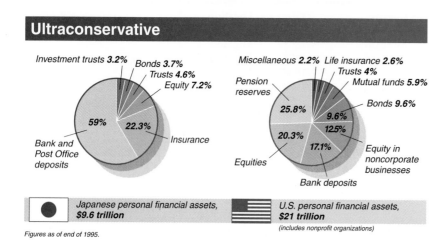

Source: Reprinted by permission of FORBES Magazine © Forbes Inc., 1996.

Exploring Geometry

4. Use the data in the three pie charts that follow and construct tables. We started the first for you. Interpret the results you find in the tables and in the pie charts.

Number of Programs

Category	Number of People Reporting out of 7000	Equation
1–2	420	6% of 7000 = 0.06 × 7000 = 420
3–4		
5–6		
7–8		
9 or more		
Total	**7000**	

> **WRITER'S BLOCK**
>
> The Number of Programs pie chart does not sum to 100%. How is this possible?

For frequent flyer tickets and ticket type, choose your own base number of respondents.

Number of Programs

"In how many frequent flyer programs are you enrolled?"

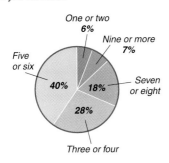

Frequent Flyer Tickets

"How many free tickets did you actually claim in the last twelve months?"

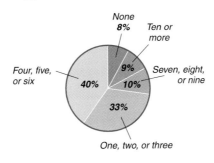

Ticket Type

"How do you fly most frequently when traveling internationally?"

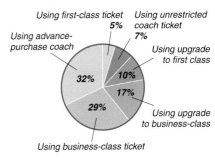

Source: Frequent Flyer, December 1994, as cited in the *1996 Business Information Please® Almanac* (©1995 Houghton Mifflin Co.), pp. 539–540. Reprinted by permission of *Frequent Flyer.*

6.4 Applying Rational Expressions: Word Problems

SECTION LEAD-IN

The efficiency E of a jack is determined from the pitch of the jack's thread with the formula

$$E = \frac{\frac{p}{2}}{p + \frac{1}{2}}$$

The pitch is the distance between the threads.

1. When $p = 1$, what is E?

2. When the efficiency is 1, what is the (approximate) value of p?

3. Keeping p positive, as p gets smaller, what happens to the value of E?

SECTION GOALS

- *To solve word problems that lead to equations with rational expressions or proportions*

- *To solve word problems that involve variation*

Introduction

In this section, we solve word problems that lead to equations involving rational expressions.

▪▪▪

EXAMPLE 1

The numerator of a fraction is twice the denominator of the fraction. If the numerator is increased by five and the denominator is tripled, the resulting fraction is $\frac{13}{12}$. Find the original fraction.

SOLUTION

We begin by writing down what we know. If the original denominator is d, then the numerator is twice d, so we have

$$\text{Denominator} = d$$
$$\text{Numerator} = 2d$$
$$\text{Fraction} = \frac{2d}{d}$$

Now, with the numerator increased by 5 and the denominator tripled, we get

$$\text{New fraction} = \frac{2d + 5}{3d}$$

The new fraction is equal to $\frac{13}{12}$, so we have an equation to solve:

$$\frac{2d + 5}{3d} = \frac{13}{12}$$

The LCM of the denominators is $12d$. Multiplying and then solving the equation, we have

$$(\overset{4}{\cancel{12}d})\frac{2d + 5}{\cancel{3d}} = (\cancel{12}d)\frac{13}{\cancel{12}}$$
$$4(2d + 5) = 13d$$
$$8d + 20 = 13d$$
$$20 = 5d$$
$$4 = d$$

Substituting 4 for d gives us the original fraction.

$$\frac{2d}{d} = \frac{8}{4}$$

We check by returning to the original problem statement to see that this gives the new fraction. And indeed, $\frac{8 + 5}{3 \cdot 4} = \frac{13}{12}$.

▶ CHECK **Warm-Up 1**

Problems that Involve Proportions

A proportion is a simple rational equation. It is an equation showing that two ratios (fractions) are equal, as in

$$\frac{a}{b} = \frac{c}{d}$$

Cross-Product Property

If $\frac{a}{b} = \frac{c}{d}$, then the **cross products** ad and bc are equal.

We can use this fact to solve simple rational equations instead of using the LCM. We use this method to solve the next example.

■ ■ ■

EXAMPLE 2

Rolando had 60 photographs. Thirty-six of these were black-and-white photos, and the rest were color photos. After he mounted some of the black-and-white photographs, the ratio of the remaining black-and-white photos to the total remaining photographs was $\frac{1}{4}$. How many black-and-white photographs did he mount?

SOLUTION

We can write the ratio of the number of black-and-white photographs (36) to the total of photographs as

$$\frac{\text{Black-and-white photos}}{\text{Total photos}} = \frac{36}{60}$$

Rolando mounted a number of black-and-white photos; call that number x. In doing so, he decreased both the number of unmounted black-and-whites and the number of unmounted photos by x. So he changed the ratio to

$$\frac{36 - x}{60 - x}$$

We are told that this ratio is equal to $\frac{1}{4}$, so we have

$$\frac{36 - x}{60 - x} = \frac{1}{4}$$

Using the cross-products property, we know that

$$
\begin{aligned}
4(36 - x) &= 1(60 - x) & \text{Cross products} \\
144 - 4x &= 60 - x & \text{Distributive property} \\
-3x &= -84 & \text{Simplifying} \\
x &= 28
\end{aligned}
$$

He mounted 28 black-and-white photographs. You should check this result with the problem statement.

ERROR ALERT
Identify the error and give a correct answer.
Solve: $\frac{y + 5}{5} = y$

Incorrect Solution:
$\frac{y + 5}{5} = y$ is already solved for y.

▶ *CHECK* **Warm-Up 2**

Rate Problems

Many rate problems, using the formula

$$\text{Rate} \times \text{time} = \text{distance}$$

lead to rational equations.

▪▪▪
EXAMPLE 3

A plane travels 675 miles per hour in still air. Against the wind, this plane can travel 4410 miles in the same time as it can travel 5040 miles with the wind. What is the speed of the wind?

SOLUTION

Let w represent the speed of the wind; then $675 + w$ is the speed of the plane with the wind, and $675 - w$ is the speed against the wind. We know this much so far:

	Rate	$\times$	Time	= Distance
With wind	$675 + w$		t	5040
Against wind	$675 - w$		t	4410

The times, although unknown, can be written as the quotients of the distances traveled and the speeds $\left(t = \dfrac{D}{r}\right)$. They are also equal.

Thus, with the wind, $t = \dfrac{5040}{(675 + w)}$.

And, against the wind, $t = \dfrac{4410}{(675 - w)}$.

We are told that these times are the same, so we can equate them to obtain the equation

$$\frac{5040}{(675 + w)} = \frac{4410}{(675 - w)}$$

$$(675 - w)(5040) = (675 + w)(4410) \qquad \text{Cross products}$$

$$3{,}402{,}000 - 5040w = 2{,}976{,}750 + 4410w \qquad \text{Distributive property}$$

$$425{,}250 = 9450w \qquad \text{Simplifying}$$

$$45 = w$$

So the wind speed is 45 miles per hour. The check is left for you to do.

▶ *CHECK* **Warm-Up 3**

Work Problems

The next problem is called a **work problem**, even though it may not concern work specifically. We solve such problems by applying the **work principle**.

Work Principle

If it takes t minutes to complete 1 task, then it takes 1 minute to complete $\dfrac{1}{t}$ task.

■ ■ ■

EXAMPLE 4

To fill a washing machine with only hot water takes 4.5 minutes. To fill it with only cold water takes 3 minutes. How long is needed to fill it with warm water (hot and cold) simultaneously?

SOLUTION

We apply the work principle to each faucet.

If hot water fills the machine in 4.5 minutes ($t_h = 4.5$), then in 1 minute it fills $\dfrac{1}{4.5}$ of the machine $\left(\dfrac{1}{t_h}\right)$.

If cold water fills the machine in 3 minutes ($t_c = 3$), then in 1 minute it fills $\dfrac{1}{3}$ of the machine $\left(\dfrac{1}{t_c}\right)$.

Suppose that they can fill the machine in x minutes; then in 1 minute they can fill $\dfrac{1}{x}$ of the machine. Thus

$$\begin{array}{ccc} \text{Amount both fill} & \text{Amount hot fills} & \text{Amount cold fills} \\ \text{in 1 minute} & = \text{in 1 minute} + & \text{in 1 minute} \\ \dfrac{1}{x} & = \dfrac{1}{4.5} & + \quad \dfrac{1}{3} \end{array}$$

This is the equation that we have to solve.

The LCD of the fractions is $x(3)(4.5) = 13.5x$. So we have

$$(13.5)(x)\frac{1}{x} = (13.5)(x)\frac{1}{4.5} + (13.5)(x)\frac{1}{3}$$

Simplifying and solving for x gives us

$$13.5 = 3x + 4.5x$$

$$13.5 = 7.5x$$

$$1.8 = x$$

So the washing machine will fill with both hot and cold water in 1.8 minutes.

▶ *CHECK* **Warm-Up 4**

▪▪▪

EXAMPLE 5

It takes Pat 90 minutes to rake the leaves in front of the house. It takes his wife 120 minutes to rake the same leaves, and their neighbor Gladys takes 60 minutes to rake the same amount of leaves. How long would it take if they all worked together?

SOLUTION

We apply the work principle to each person.

1. If Pat takes 90 minutes to rake, then in 1 minute, he rakes $\frac{1}{90}$ of the leaves.
2. If his wife takes 120 minutes to rake, then in 1 minute, she rakes $\frac{1}{120}$ of the leaves.
3. If Gladys takes 60 minutes to rake, then in 1 minute, she rakes $\frac{1}{60}$ of the leaves.

Together they can rake the leaves in x minutes, so in 1 minute they can take $\frac{1}{x}$ of the leaves. Therefore,

$$\frac{1}{90} + \frac{1}{120} + \frac{1}{60} = \frac{1}{x}$$

We then solve the equation. The LCD of the fractions is $360x$,

$$(360x)\frac{1}{90} + (360x)\frac{1}{120} + (360x)\frac{1}{60} = (360x)\frac{1}{x}$$

Cancelling and then solving for x, we find that

$$4x(1) + 3x(1) + 6x(1) = 360(1)$$

$$4x + 3x + 6x = 360$$

$$13x = 360$$

$$x = 27\frac{9}{13}$$

Thus, it would take the three people approximately 28 minutes to rake the leaves together.

▶ *CHECK* **Warm-Up 5**

You can also work Examples 4 and 5 using proportions.

Problems that Involve Variation

Variation means "change." In mathematics, variation refers specifically to the way the dependent variable changes when the independent variable changes.

Direct Variation

An equation of the form

$$y = kx$$

is said to exhibit **direct variation**. The dependent variable y increases when the independent variable x increases, and y decreases when x decreases. We also say y **is directly proportional to** x The constant k is called the **proportionality constant**, or constant of variation.

▪ ▪ ▪

EXAMPLE 6

In an electric circuit, the voltage V varies directly as the current I. If the voltage is 60 volts when the current is 15 amps, find the voltage at 10 amps.

SOLUTION

Here V is the dependent variable because it "varies directly as I." The general equation is

$$V = kI$$

To find k, we substitute 60 for V and 15 for I and solve.

$$60 = k \cdot 15$$
$$4 = k$$

Thus the equation of variation is $V = 4I$.

Substituting $I = 10$ and solving for V:

$$V = 4I = 4 \cdot 10 = 40$$

So at 10 amps, the voltage is 40 volts.

▶ CHECK Warm-Up 6

Inverse Variation

An equation of the form

$$y = \frac{k}{x}$$

shows **inverse variation**. The dependent variable y increases when the independent variable x decreases, and y decreases when x increases. We also say y **is inversely proportional to** x.

▪ ▪ ▪
EXAMPLE 7

The time t that it takes for a "standard" ice cube to melt in a liquid is inversely proportional to the temperature T of the liquid, provided that T is above $50°$. It takes 1.6 minutes to melt an ice cube when the water is at $75°$ F. How long will it take to melt a cube if the liquid is at $90°$ F?

SOLUTION

Since t is inversely proportional to T, we have

$$t = \frac{k}{T}$$

To find k, we substitute the given values 1.6 for t and 75 for T.

$$1.6 = \frac{k}{75}$$
$$120 = k$$

So the equation of variation is

$$t = \frac{120}{T}$$

Substituting $T = 90$ and solving for t, we have

$$t = \frac{120}{90} = 1\frac{1}{3}$$

It will take $1\frac{1}{3}$ minutes for a standard ice cube to melt in water that is $90°$ F.

▶ *CHECK* **Warm-Up 7**

Joint Variation

An equation of the form $y = kxz$, where $x, y,$ and z are all variables, is said to exhibit **joint variation:** y varies directly as the product of x and z varies.

▪ ▪ ▪
EXAMPLE 8

The area A of a triangle varies jointly as its base b and height h. If the area is 36 inches when the height is 8 inches and the base is 9 inches, find the area when the base is 7 inches and the height is 22 inches.

SOLUTION

The dependent variable is A, so the general equation is

$$A = kbh$$

Substituting $A = 36, h = 8,$ and $b = 9$ and simplifying, we get

$$36 = k(9)(8)$$
$$\frac{1}{2} = k$$

So the equation of joint variation is

$$A = \left(\frac{1}{2}\right)bh$$

We now substitute $b = 7$ and $h = 22$ and simplify.

$$A = \left(\tfrac{1}{2}\right)(7)(22) = \tfrac{154}{2} = 77$$

▶ *CHECK* **Warm-Up 8**

Combined Variation

Real situations often involve combinations of direct and indirect variation. This combined variation is described by $y = \frac{kx}{z}$.

▪▪▪

EXAMPLE 9

The frequency of vibration f of a musical string varies directly as the square root of the tension T and inversely as the length L of the string. How does f change when T is quadrupled?

SOLUTION

The general equation for this variation is

$$f = \frac{k\sqrt{T}}{L}$$

We substitute $4T$ for T. The general equation becomes

$$f = \frac{k\sqrt{4T}}{L} = \frac{2k\sqrt{T}}{L}$$

Comparing this last equation with the general equation, we see that f is doubled when T is quadrupled.

▶ *CHECK* **Warm-Up 9**

Practice what you learned.

SECTION FOLLOW-UP

The questions asked in the Section Lead-In investigate the formula

$$E = \frac{\frac{p}{2}}{p + \frac{1}{2}}$$

1. When $p = 1$, $E = \frac{1}{3}$.

2. When the efficiency is 1, $p \approx -1$.

3. For $p > 0$, as p gets smaller, E gets smaller.

6.4   WARM-UPS

Work these problems before you attempt the exercises.

1. One number is 8 times another number. The sum of their reciprocals is three-sixteenths. Find the numbers.   6 and 48

2. Out of 50 tourists, 10 have been to the Empire State Building. A tandem bus full of tourists just returning from the Empire State Building joins the original group, and now 25% have not been to that site. How many tourists were on the bus?   110 tourists

3. An ice skater normally skates at a rate of 8 kilometers per hour. In a long-distance skate-a-thon on a windy day, she found that she could skate 20 kilometers with the wind at her back in $\frac{2}{3}$ the time it took to skate the same distance into the wind. How fast was the wind? What was her total time for the 40-kilometer skate-a-thon?   The wind speed was 1.6, or $1\frac{3}{5}$, kilometers per hour. It took the skater $5\frac{5}{24}$ hours to complete the skate-a-thon.

4. A tub can be filled by the hot water faucet in 8 minutes and by the cold water faucet in 6 minutes. It can be emptied by the drain in 4 minutes. Both faucets are turned on but the drain is left open. How long will it take to fill the tub?   24 minutes

5. Two hoses are used to fill a backyard children's pool. One hose fills the pool three times as fast as another. If together they can fill the pool in 24 minutes, how long would each hose take alone?   Hose 1 fills the pool in 32 minutes. Hose 2 fills the pool in 96 minutes.

6. The Social Security tax t deducted from your biweekly paycheck varies directly as your biweekly wage w. Write an equation that describes this variation. $t = kw$

7. The angular speed f of a gear varies inversely as the number of teeth t. Write an equation that describes this variation. $f = \frac{k}{t}$

8. The total force F exerted on a surface by a liquid of constant density varies jointly as the area of the surface s and the height of the liquid h. Write the equation that describes this variation. $F = ksh$

9. A measurement used by anthropologists to study early humans is the cephalic index C. This measurement varies directly as the head width w and inversely as the length n of the head. The general equation is $C = \frac{kw}{n}$. If the cephalic index is 70 for a width of 7 and a length of 10, find the index for a head with a width of 6 and a length of 8.   75

6.4 EXERCISES

Note: Use your graphing calculator to check your results whenever possible.

In Exercises 1 through 28, work the word problems by first writing an equation.

1. **Optics** A formula used in optics is

$$\frac{1}{p} + \frac{1}{q} = \frac{1}{f}$$

In this formula, p is the distance of an object from the lens, q is the distance of the image from the lens, and f is the focal length of 12 centimeters. How far from the lens will the image appear when the object is 36 centimeters from the lens?   **18 centimeters**

2. **Electricity** When three certain resistors are wired in parallel, the total resistance R_T of the circuit is 280 ohms. The formula is

$$\frac{1}{R_T} = \frac{1}{R_1} + \frac{1}{R_2} + \frac{1}{R_3}$$

   a. Find the value of each resistor when $R_1 = R_2 = R_3$.   **R = 840 ohms**

   b. Find R_T in terms of R_3 when R_1 is twice as large as R_2 and R_2 is twice $R_T = \frac{4R_3}{7}$
   as large as R_3.

3. **Psychology** A baby armadillo was taught to run through mazes. The armadillo's maze time in minutes t is

$$t = 6 + \frac{20}{n + 2}$$

where n is the number of previous times the armadillo has run that maze. Find n when t is 10 minutes.   **n = 3**

4. **Speed** The average speed of a round trip is

$$s = \frac{2d}{\dfrac{d}{r_1} + \dfrac{d}{r_2}}$$

where d is the one-way distance, r_1 is the rate of speed going, and r_2 is the rate returning. Use this formula to determine whether it is possible to have $s = 60$ miles per hour when d is 50 miles and the rate one way is 40 miles per hour.   **There is no way (legally) that one could average 60 mph if one traveled the first 50 miles at 40 mph. You would have to drive 120 mph.**

5. **Speed** A boat traveled 21 miles up the river (against the current) and then back down the river (with the current). The entire trip took 10 hours. If the current is 2 miles per hour, what is the speed of the boat in still water?   **5 mph**

6. **Wind Speed** An airplane that flies 650 miles per hour in calm air can cover 2160 miles with the wind in the same time that it can cover 2000 miles against the wind. Find the speed of the wind.   **25 mph**

7. **Number Problem** In a fraction, the numerator is 4 less than the denominator. If the numerator is doubled, and the denominator is increased by 20, the $\frac{15}{19}$
resulting fraction is equivalent to $\frac{10}{13}$. Find the original fraction.

8. *Travel to Greenland* The ratio of usable airports to total airports in Greenland gives a fraction with a numerator 3 less than the denominator. If you triple the numerator and decrease the denominator by thirteen, the result is equivalent to -12. How many airports are there in Greenland How many of those are usable? There are 11 airports in Greenland; 8 are usable.

9. *Geometry* The width of a rectangle is 5 inches more than $\frac{2}{3}$ of its length. The ratio of the length to the width is $\frac{3}{4}$. Find the perimeter. 35 inches

10. *Geometry* The longest side of a triangle is $\frac{3}{4}$ unit plus the length of the shortest side, and the third side is equivalent to $\frac{1}{2}$ unit plus the length of the shortest side. The ratio of the third side to the longest side is $\frac{4}{5}$. Find the perimeter of the triangle. 2.75 units

11. *Work* Splash Down is a game played by two teams of 2 players. The first team that increases or decreases the original water level in a tank by 20 gallons wins. Players A and B are on one team. A can add 20 gallons in 8 minutes; and B can add 20 gallons in 16 minutes. Players C and D are on another team. C can empty 20 gallons in 9 minutes and D can empty 20 gallons in 12 minutes. Which team will win, and how long will it take? Players C and D will win in 2 hours and 24 minutes.

12. *Work* To fill a washing machine with only hot water takes 4.5 minutes. To fill it with only cold water takes 3 minutes. A drain that can empty the water in 5 minutes is accidently opened 1 minute after the machine starts filling. With this drain open, how much longer (than originally expected) will it take the machine to fill with hot and cold water? about 27 seconds longer

13. *Number Problem* One number is five times another number. The sum of their reciprocals is three-fifths. Find the numbers 2 and 10

14. *Number Problem* The denominator of a number is $\frac{1}{5}$ its numerator. The denominator is increased by six and the numerator is decreased by eight. This results in a fraction whose reciprocal is equivalent to $\frac{5}{6}$. Find the original fraction. $\frac{20}{4}$

15. *Water* The cost C (in thousands of dollars) of eliminating x percent of the impurities in a water supply is given by the formula
$$C = \frac{4x}{100 - x}$$
   a. How much will it cost to remove 90% of the impurities? to remove 95% of the impurities The cost to remove 90% of the impurities is $36,000. The cost to remove 95% of the impurities is $76,000.
   b. What happens to the value of C as x gets close to 100? As x gets larger (and closer to 100), C also gets larger (at an increasing rate).
   c. What is the domain of x? domain: $\{x \mid 0 \le x < 100\}$

16. *Vaccinations* The cost C (in millions of dollars) to vaccinate x percent of the population against an airborne strain of influenza is given by
$$C = \frac{125x}{100 - x}$$

a. How much will it cost to vaccinate 50% of the population? $125,000,000

b. How much more will it cost to vaccinate 85% of the population?
   $583,333,333.33 more

c. What do you think would happen to C as x gets closer and closer to 100? What does this mean in real life?   The cost gets larger and larger. It is prohibitively expensive to vaccinate everyone.

17. **Printing**   A printing company that has to print a certain number of copies of a book has two printing plants. Together they can print the books in 8 days. Working alone, plant B would take 12 days longer than plant A. Find the time that each plant alone requires to complete this same job.   Plant A requires 12 days. Plant B requires 24 days.

18. **Bake Sale**   Two men are baking cookies for a bake sale on pre-school conference day. Juan can bake several dozen in $1\frac{1}{2}$ hours. Benoit has a larger oven and pans and can bake the same amount in 27 minutes. If they work together, can they bake 4 times this amount in less time than $1\frac{1}{2}$ hours.   yes

19. **Food Services**   Hillary baked some cookies and iced $\frac{2}{3}$ of the cookies she baked. Then she baked 2 dozen more cookies and iced half. Now $\frac{3}{5}$ of all the cookies she baked have been iced. How many cookies did Hillary bake altogether?   60 cookies

20. **Etching**   Jan had 50 etchings. Twenty of these were of her cat, Tuna. The rest were of her husband, Justin. She sold some of the etchings of Tuna. Six-sevenths of the remaining etchings were of Justin. How many etchings did Jan sell?   15 etchings

21. **Prime Numbers**   Pierre de Fermat (1601–1665) proved that the following relationship is true of every odd prime number p:

$$p + \left(\frac{p-1}{2}\right)^2 = \left(\frac{p+1}{2}\right)^2$$

Simplify to show that the two sides are equal. $\frac{p^2 + 2p + 1}{4} = \frac{p^2 + 2p + 1}{4}$

22. **Efficiency**   A car jack's efficiency E can be measured by the formula

$$E = \frac{\frac{h}{2}}{h + \frac{1}{2}}$$

The value of h is determined by the pitch of the jack's thread. Solve for h. $h = \frac{E}{1 - 2E}$

23. **Work Problem**   One pipe can fill a pool in 4 hours. A second pipe can empty the pool in 2.5 hours. The first pipe is turned on. After 2 hours, the other pipe is opened. With both pipes open, how many hours will it take to empty the pool? $3\frac{1}{3}$ hours

24. **Mowing Grass**   A woman can mow a small lawn in 14 minutes. Her son can mow it in 7 minutes. How long will it take them to do this work together? $4\frac{2}{3}$ minutes

25. **Number Problem**   A fraction has a numerator that is 3 more than the denominator. If 3 is added to the numerator of the number, and five is subtracted from the denominator of the number, the result is equivalent to 3.5. Find the original fraction. $\frac{12.4}{9.4}$

26. **Number Problem**   The reciprocal of 8 less than a number is two times the reciprocal of the number. Find the number.   16

27. **Business, Payroll**   It takes Suzanne 5 hours to do the payroll, Veronica $3\frac{1}{2}$ hours, and Norma $2\frac{1}{2}$ hours. If they could all work on the payroll at one time, how long would it take. (Give your answer to the nearest minute.)   1 hour and 8 minutes

28. **Sorting**   Kevin takes 10 minutes to complete a sorting job by himself; Marijke takes $2\frac{1}{2}$ minutes by herself; and Mei-Ling takes 5 minutes by herself. How long would the sorting job take if all three worked together?   1.43 minutes

In Exercises 29 through 40, using the given variables, write the general equation representing each statement. Use k as the constant.

29. **Travel**   The scaled distances s on a map vary directly as the actual distances d. $s = kd$

30. **Biology**   In biology, a rule states that the number of days d after a given time that insects begin to appear in a certain geographic location varies directly as the change in altitude h. $d = kh$

31. **Physics**   The gravitational force of attraction F between Earth and an object varies directly as the mass m of the object. $F = km$

32. **Physics**   The rate of work in horsepower H varies jointly as the number of pounds moved p and the distance d through which they are moved in a fixed time. $H = kpd$

33. **Biology**   Reaumur, a biologist, suggested in 1735 that the length of time t it takes fruit to ripen during the growing season varies inversely as the sum T of the average daily temperatures during the growing season. $t = \frac{k}{T}$

34. **Physics**   The electrical resistance R of a cable of given length is inversely proportional to the square of its diameter d. $R = \frac{k}{d^2}$

35. **Sociology**   The sociologist Joseph Cavanaugh found that the number of long-distance phone calls n between two cities in a given time period varied (approximately) jointly as the population P_1 and P_2 of the two cities, and inversely as the distance d between the two cities. $n = \frac{kP_1P_2}{d}$

36. **Psychology**   In their study of intelligence, psychologists often use an index called IQ. This index varies directly as mental age MA and inversely as chronological age CA up to the age of 15. $IQ = \frac{kMA}{CA}$

In Exercises 37 through 40, find the change described.

37. *Physics*   The illuminance I of a light source varies inversely as the square of the distance d from the light source. Write the equation and find how I changes when d doubles $I = \frac{k}{d^2}$. When d doubles, I is $\frac{1}{4}$ its original value.

38. *Physics*   In an automobile accident, the destructive force F of a car varies jointly as the weight w of the car and the square of the speed v of the car. If the speed of the car is halved, how does F change? F is cut to 25% of the original force.

39. *Chemistry*   The volume V of a given mass varies directly as the temperature T and inversely as the pressure P. How does V change if the pressure is increased 1.5 times and the temperature is tripled? V is doubled.

40. *Physics*   When a ball is being twirled on the end of a string, the tension T in the string is directly proportional to the square of the speed v of the ball and inversely proportional to the length r of the string. If the string length is doubled, how does T change? T is cut in half.

In Exercises 41 through 50, solve using an equation of variation.

Sports   Use the following information to answer Exercises 41 and 42. The weight of a ball is directly proportional to it circumference. Assume that the materials used to manufacture the ball account for the differences in weight.

	Circumference	Weight
Basketball	78 cm	650 g
Volleyball	67 cm	280 g

41. If a basketball were shrunk to the size of the volleyball, how would their weights compare? Explain.

42. If the material used to make a volleyball was used for a ball with a circumference of 100 centimeters, what would it weigh?   about 418 g

41. If a basketball were shrunk to the size of a volleyball, it would be much heavier. For a basketball, $k = 8\frac{1}{3}$ g/cm, while for the volleyball, $k \approx 4.2$ g/cm. The difference in weight is probably due to the weight of materials in the balls.

43. *Travel*   On a map, the length of a line varies directly as the number of miles represented by the line. A map of Hamilton, Ontario, uses $\frac{1}{4}$ inch to represent 5 miles. A particular road is $3\frac{1}{2}$ inches long on the map. What is the length of the actual road?   70 miles

44. *Physics*   If we neglect air resistance, the distance D that a body falls from rest varies directly as the square of time t it falls. It takes 2 seconds for a ball to fall 64 feet. How many seconds will it take to fall 324 feet?   4.5 seconds

45. *Physics*   The distance that a person can see to the horizon from a point above the surface of Earth varies directly as the square root of altitude. A person standing 900 feet above sea level can see $25\frac{1}{2}$ miles to the horizon. How high would you have to be to see 50 miles? Round to the nearest 10 feet.   3460 feet

46. *Physics* The range R of a projectile varies directly as the square of its velocity v. A skateboarder can make a jump of 20 feet coming off a ramp at 15 miles per hour. How far can he jump if he comes off the ramp at 18 miles per hour? 28.8 feet

47. *Physics* The weight of an object above the surface of Earth varies inversely as the square of its distance from the center of Earth. A person weighs 120 pounds on the surface of Earth. How much would she weigh in an airplane 6.0 miles above Earth? (Use 4000 miles for the radius of Earth. Round your answer to the nearest tenth). 119.6 pounds

48. *Pendulum* The period of a pendulum f (the length of time it takes to make one complete cycle) varies directly as the square root of its length L. A pendulum 12 feet long has a period of 3.6 seconds. What is the period of a pendulum 3 feet long? 1.8 seconds

49. *Projected Area* The area of a projected image P from an overhead projector varies directly as the square of the distance d from the projector to the screen. A projector 8 feet from the screen projects a 16-square-foot image. What area is projected when the overhead projector is 6 feet from the screen? 9 square feet

50. ✏ *Pendulum* The period p of a pendulum, or the time it takes for a pendulum to make one complete swing, varies directly as the square root of the length L of the pendulum. Write an equation to represent this variation. Solve this equation for L, explaining each step in your own words. Answers may vary.

6.4 MIXED PRACTICE

By doing these exercises, you will practice the topics up to this point in the chapter.

51. Add: $\dfrac{9}{4x} + \dfrac{8}{12x}$ $\dfrac{35}{12x}$

52. Simplify: $\dfrac{\dfrac{1}{x-2}}{\dfrac{2}{2-x}}$ $-\dfrac{1}{2}$

53. Solve: $9 - \dfrac{15}{t} = \dfrac{9t-7}{t+2}$ $t = 3$

54. Divide: $\dfrac{m^6 n^7}{x^2 y^3} \div \dfrac{m^5 n^3}{x^4 y^5}$ $mn^4 x^2 y^2$

55. Solve: $\dfrac{2}{3x} + \dfrac{1}{3} = \dfrac{11}{6x} - \dfrac{1}{4}$ $x = 2$

56. Add: $\dfrac{2}{m^2 - 16} + \dfrac{3}{m - 4} + \dfrac{m + 1}{m^2 + 8m + 16}$

$\dfrac{4m^2 + 23m + 52}{(m-4)(m+4)(m+4)}$

57. *Geometry* The perimeter of a square varies directly as the side of the square. The perimeter is 500 inches when a side is 125 inches. Find the length of a side when the perimeter is 750. 187.5 inches

58. ✏ *Travel* A boat can go 30 kilometers upstream in the same time as it took to go downstream 50 kilometers. The speed of the boat in still water is 10 kilometers per hour. Find the speed of the current. In your own words, explain how the total travel time will change if the speed of the current increases. 2.5 kilometers per hour. Travel time increases. Explanations may vary.

59. **Mowing Grass** Buzzie can mow a lawn in 4 hours, Bobby can mow it in 5 hours, and Lenora can mow it in 3 hours. How long will it take them to mow the lawn if they are working together? *about 1.28 hours*

60. **Retail Sales** At a boutique, 70% of the 50 dresses in stock were made of cotton. At a special sale, only cotton dresses were sold, and afterwards, only 40% of the remaining dresses were cotton. How many cotton dresses were sold? *25 dresses*

61. **Pendulum** The period p of a pendulum, or the time it takes for a pendulum to make one complete swing, varies directly as the square root of the length L of the pendulum. Write an equation to represent this variation. What happens to p when L quadruples? *$p = k\sqrt{L}$; p doubles*

62. Reduce $\dfrac{18x^2 + 6x}{4x^2 + 8x}$ to lowest terms. *$\dfrac{3(3x + 1)}{2(x + 2)}$*

63. Evaluate $\dfrac{r^2 t v^3}{(rt)^2}$ when $r = -2$, $t = 8$, and $v = 0$. *0*

64. Solve: $\dfrac{3x}{2x} + \dfrac{4 + x}{6x} = \dfrac{8x}{12x}$ *$x = -\dfrac{2}{3}$*

65. Determine when $\dfrac{x - 4}{x^2 + 16x + 64}$ is meaningless. *It is meaningless when $x + 8 = 0$ and $x = -8$.*

66. Divide: $\dfrac{x^2 + 10x + 9}{x^2 + 7x - 18} \div \dfrac{x^2 - 6x - 7}{x^2 + 5x - 14}$ *$\dfrac{x + 7}{x - 7}$*

67. Simplify: $\dfrac{3 + \frac{4}{x}}{2 + \frac{x}{7}}$ *$\dfrac{21x + 28}{14 + x^2}$*

68. Solve: $\dfrac{1}{x + 3} - \dfrac{2}{x - 3} = \dfrac{x}{x^2 - 9}$ *$x = -4\dfrac{1}{2}$*

69. Simplify: $\dfrac{\frac{x + y}{x} + 6}{\frac{x + y}{6x}}$ *$\dfrac{42x + 6y}{x + y}$*

70. Add: $\dfrac{8}{y} + \dfrac{y - 3}{y(y - 2)}$ *$\dfrac{9y - 19}{y(y - 2)}$*

71. Solve for a: $\dfrac{c - 5}{b} = \dfrac{a + 8}{c}$ *$\dfrac{c^2 - 5c}{b} - 8 = a$*

72. Multiply: $\dfrac{x + 8}{x - 3} \cdot \dfrac{x^2 - 9}{x^2 + 11x + 24}$ *1*

EXCURSIONS

Class Act

1. Use the information from these two graphs to answer the questions.

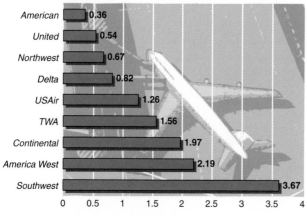

Denied Boarding (Involuntary, per 10,000 Passengers)

American	0.36
United	0.54
Northwest	0.67
Delta	0.82
USAir	1.26
TWA	1.56
Continental	1.97
America West	2.19
Southwest	3.67

Mishandled Baggage (per 1000 Passengers)

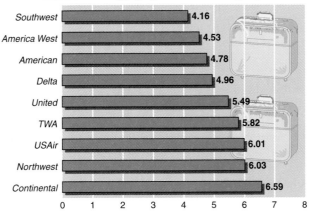

Airline	per 1000
Southwest	4.16
America West	4.53
American	4.78
Delta	4.96
United	5.49
TWA	5.82
USAir	6.01
Northwest	6.03
Continental	6.59

a. Complete the table.

	Rank Denied Boarding	Rank Mishandled Baggage
American	1	3
America West		
Continental		
Delta		
Northwest		
TWA		
Southwest		
United		
USAir		

b. Plot the points from part (a) on the graph.

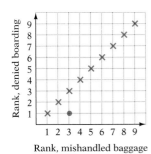

c. ✏ If the points line up perfectly with the ×s, a perfect correlation would have occurred. Comment on how well correlated these two rankings are. If these two rankings had been perfectly correlated, what would that have meant in terms of how "denied" and "mishandled" were related?

d. Use the information from the Airlines Ranked by Quality table and graph those rankings against each of the other two rankings.

e. Which of the other two rankings is most closely correlated to the Quality ranking? Justify your decision.

2. Here is information about two research sites.

a. Graph both sets of this data and compare and contrast the information about the sites. They are both located 450 m above sea level. Their latitudes are 43° S and 45° N; their longitudes are 103° W and 147° E.

b. See if you can determine what countries these sites are in.

Airlines Ranked by Quality

Rank	Airline
1	Southwest
2	American
3	United
4	Delta
5	USAir
6	Northwest
7	TWA
8	America West
9	Continental

Source: National Institute for Aviation Research, as cited in the *1996 Information Please® Business Almanac* (©1995 Houghton Mifflin Co.), p. 540. All rights reserved. Used with permission by Information Please LLC.

Mean Monthly Temperature (°C)

Site 1		Site 2	
January	−17.9	January	12.3
February	−13.8	February	12.8
March	−7.8	March	11.0
April	3.1	April	8.6
May	10.6	May	5.9
June	15.7	June	4.3
July	18.6	July	3.5
August	17.5	August	4.0
September	11.5	September	5.4
October	5.3	October	7.4
November	−5.1	November	8.9
December	−12.9	December	10.5

Mean Monthly Precipitation Including Snow (mm)

Site 1		Site 2	
January	21	January	88
February	18	February	82
March	22	March	97
April	28	April	132
May	50	May	142
June	74	June	124
July	53	July	142
August	56	August	139
September	42	September	144
October	25	October	138
November	17	November	132
December	22	December	117

Exploring Geometry

3. a. How does the circumference of a circle change when the radius is doubled?

b. How does the square of the radius of a circle (that is, r^2) change when the area is doubled? (Use the fact that $r^2 = \frac{A}{\pi}$. Consider circles with areas of 31.4 and 62.8 square units.)

c. How does the circumference of a circle change when the diameter is doubled?

d. How does the area of a circle change when the diameter is doubled?

CHAPTER LOOK-BACK

	Product A		Product B	
	Nutrition Facts	Percent Daily Value	Nutrition Facts	Percent Daily Value
Serving size	1/4 cup (28g)		1/2 cup (56g)	
Servings per container	about 4		about 6	
Amount per serving				
Calories	80		95	
Calories from fat	45		54	
Total fat	5g	8%	6g	9.6%
Saturated fat	3.5g	18%	3.5g	18%
Cholesterol	15mg	5%	24mg	8%

1. Which food item (A or B) has the greatest fat ratio per package? per cup?   *B (36 grams of fat per package); A (20 grams of fat per cup)*

2. Give an example of a fat ratio less than 10%.

CHAPTER **6**

REVIEW PROBLEMS

The following exercises will give you a good review of the material presented in this chapter.

SECTION 6.1

1. Reduce: $\dfrac{32x^3y^4z^6}{36x^4y^2z^3}$ *$\dfrac{8y^2z^3}{9x}$*

2. Reduce: $\dfrac{n^3tx^4}{n^2tx^2}$ *nx^2*

3. Simplify: $\dfrac{x^2-4x+3}{x-3}$ *$x-1$*

4. Reduce: $\dfrac{x^2-36}{x^2+12x+36}$ *$\dfrac{x-6}{x+6}$*

5. Let $h(x)=\dfrac{x^2+2x-3}{x+3}$. Simplify the right side and then find $h(-5)$ and $h(x+1)$.
   $x-1$; $h(-5)=-6$; $h(x+1)=x$

6. Let $h(x)=\dfrac{3x^2+19x-14}{2-3x}$. Simplify the right side and then find $h(-3)$ and $h(x-7)$.
   $-(x+7)$; $h(-3)=-4$; $h(x-7)=-x$

7. Divide: $\dfrac{1}{x^2+8x+15}\div\dfrac{3}{x+5}$ *$\dfrac{1}{3(x+3)}$*

8. Divide: $\dfrac{(x+2y)^2}{12}\div\dfrac{4x^2-16y^2}{9}$ *$\dfrac{3(x+2y)}{16(x-2y)}$*

9. Simplify: $(64x^3y^5\div16x^5y^7)32xy^2$ *$\dfrac{128}{x}$*

10. Multiply: $(3x^2y^3)(9x^5y^5)(63x^5y^9)$ *$1701x^{12}y^{17}$*

11. Multiply: $\dfrac{63x^3y^2}{49x^5y^3}\cdot\dfrac{8x^5y^3}{56x^3y^5}$ *$\dfrac{9}{49y^3}$*

12. Divide: $\dfrac{100x^5y^3}{500x^2y^4}\div\dfrac{200x^7y^5}{1000x^6y^5}$ *$\dfrac{x^2}{y}$*

SECTION 6.2

13. Let $f(x)=\dfrac{x+3}{2}$ and $g(x)=\dfrac{2}{x+3}$. Find $g(x)-f(x)$.   *$\dfrac{-x^2-6x-5}{2(x+3)}$*

14. Combine: $\dfrac{4}{3t-2}-\dfrac{1}{t-4}+5$ *$\dfrac{15t^2-69t+26}{(3t-2)(t-4)}$*

15. Combine: $\dfrac{2x}{x-3} + \dfrac{36}{x^2-9} - \dfrac{2x}{x+3}$ $\dfrac{12}{x-3}$

16. Combine: $\dfrac{8}{15x} + \dfrac{3}{20x}$ $\dfrac{41}{60x}$

17. Combine: $\dfrac{2}{x-3} + \dfrac{4}{x-1}$ $\dfrac{6x-14}{(x-3)(x-1)}$

18. Let $f(x) = \dfrac{x+3}{2}$ and $g(x) = \dfrac{2}{x+3}$. $\dfrac{x^2+6x+13}{2(x+3)}$
Find $f(x) + g(x)$.

SECTION 6.3

19. Simplify: $\dfrac{\dfrac{x}{x+1}-1}{\dfrac{2x+1}{x+1}}$ $-\dfrac{1}{2x+1}$

20. Simplify: $\dfrac{1+\dfrac{m}{m+1}}{\dfrac{2m+1}{m-1}}$ $\dfrac{m-1}{m+1}$

21. Simplify: $\dfrac{\dfrac{2x+1}{x-1}}{\dfrac{x}{x+1}+1}$ $\dfrac{x+1}{x-1}$

22. Simplify: $\dfrac{\dfrac{9}{n}+\dfrac{3}{n^2}}{3+\dfrac{1}{n}}$ $\dfrac{3}{n}$

23. Simplify: $\dfrac{\dfrac{x^2}{y}-y}{\dfrac{y^2}{x}-x}$ $-\dfrac{x}{y}$

24. Simplify: $\dfrac{\dfrac{3}{5}+\dfrac{2}{7}}{\dfrac{1}{7}+\dfrac{6}{15}}$ $\dfrac{31}{19}$

25. Solve: $\dfrac{4}{m} + \dfrac{3}{2m} = \dfrac{1}{2} + \dfrac{5}{12}$ $m=6$

26. Solve: $\dfrac{20}{t+4} + 2 = \dfrac{2t-4}{t-4}$ $t=6$

27. Solve: $\dfrac{3x-2}{x+1} - 4 = -\dfrac{x+2}{x-1}$ $x=4$

28. Solve: $\dfrac{16}{t-4} = \dfrac{t^2}{t-4}$ $t=-4$

29. Solve: $\dfrac{1}{4} = \dfrac{m+1}{8}$ $m=1$

30. Solve: $\dfrac{4m+5}{6} = \dfrac{7}{2}$ $m=4$

SECTION 6.4

31. *Biking*   Ute rode her bike for 60 miles. The next day she rode 45 miles in $1\frac{1}{2}$ hours less time. How fast was she riding if her speed was the same both day? What was her total biking time for the two days?   10 mph; 10.5 hours

32. *Copy Machines*   Two old copy machines and one new one are being put to work on a job. Each of the old machines would take twice as much time as the new one to do the job. Together, the three copies do the job in 2 days. How long would each take to do the job alone?
The new copier would take 4 days. The old would copiers would take 8 days each.

33. *Running*   Out of 16 runners, 13 are women. Some of the women leave for a track meet. Out of the remaining runners, only half are women. How many women went to the track meet?   10 women

34. *Pizza*   At Alix and A.J.'s pizza parlor, there were 60 slices of pizza, and 70% of them had pepperoni on them. After a group of students came and ordered only slices with pepperoni, 40% of the remaining slices had pepperoni. How many slices did the students eat?   30 slices

35. *Motorboat*   A motorboat travels 400 kilometers. If the boat went 18 kilometers per hour faster, it could have traveled 600 kilometers in the same amount of time. What was its original speed? How long did the original trip take?   36 kilometers per hour; $11\frac{1}{9}$ hours

36. *Time on task*   Georgette can do a certain job in 3.15 hours. If she works with Sebastian, together they can do the job in 2.05 hours. How long would it take for Sebastian to do the job alone? (Round your answer to the nearest hundredth of a hour.) What proportion of this job could Georgette complete in one minute?   5.87 hours; $\frac{1}{189}$ job per minute

37. *Sound*   The time T that elapses before a sound is heard varies directly as the distance D from the source of the sound. If you hear the sound of an explosion 8.8 seconds after you see the flash, the explosion is 1.8 miles away from you. How long will it take you to hear an explosion 2 miles away? $9\frac{7}{9}$ seconds

38. *Sports Equipment*   The weight of a ball is directly proportional to its diameter.

	Diameter	Weight
Golf ball	1.68 inches	1.62 ounces
Bowling Ball	8.5 inches	16 pounds

If a golf ball were made the size of a bowling ball, how would their weights compare? Explain.   The golf ball would weigh 8.2 ounces, which is much less than the the bowling ball weighs. The difference is due to the difference in the weights of the materials they are made from.

39. *Light*   The intensity of light L varies inversely as the square of the distance d between the object illuminated and the source of light. How does L change when the distance is halved? L is 4 times greater.

40. *Electricity*   The resistance R of a wire varies directly as the length L of the wire and inversely as the square of the diameter d. How does R change as L doubles? R doubles

41. *Weight*   According to Alvin Shemesh, the threshold weight T for men between the ages of 40 and 49, above which the death rate rises astronomically, is directly proportional to the cube of the man's height h in inches. Write an equation representing this relationship. $T = kh^3$

42. *Construction*   The maximum safe load L on a horizontal beam supported at both ends varies directly as the beam's width w and the square of its height h and inversely as the distance d between the supports. Write an equation for this relation. $L = \frac{kwh^2}{d}$.

MIXED REVIEW

43. Solve: $\frac{6m + 7}{10} = \frac{2m + 9}{6}$ $m = 3$

44. Let $f(x) = \frac{x + 2}{x - 4}$ and $g(x) = \frac{x - 4}{x + 2}$.
Find $f(x) - g(x)$. $\frac{12(x - 1)}{(x - 4)(x + 2)}$

45. Simplify: $\frac{x^2 + 11x + 30}{x^2 + 12x + 36} \cdot \frac{6x^2}{x^2 - 25} \div \frac{3x^2 - 9x}{x^2 + 2x - 15}$ $\frac{2x(x + 5)}{(x + 6)(x - 5)}$

46. Reduce: $\frac{m - 2}{m^2 - 6m + 8}$ $\frac{1}{m - 4}$

47. Multiply: $\frac{16x^3y^2}{24x^5y^6} \cdot \frac{15x^7y^2}{32x^6y}$ $\frac{5}{16xy^3}$

48. Reduce: $\frac{n - 2}{n^2 - 4}$ $\frac{1}{n + 2}$

49. Combine: $\frac{3}{x - 5} + \frac{6}{x + 5}$ $\frac{9x - 15}{x^2 - 25}$

50. Solve: $\frac{15x - 2}{28} = \frac{5x - 3}{7}$ $x = 2$

51. *Roofing*   It takes Max 9 hours longer to roof a building than it takes Jokie. If they work together, they can complete the roofing in 20 hours. How long would it take each working alone to roof the building?
Jokie would take 36 hours. Max would take 45 hours.

52. **Retail Sales** Rae Ann had a total of 80 belts for sale, 50% of them being burlap and the rest leather. After a closeout sale on burlap belts, only 20% of the remaining belts were burlap. How many burlap belts were sold? 30 burlap belts

53. **Weaving** A weaver can complete 10 articles in a certain amount of time. In 20 hours more, he can complete a total of twice that many. How many articles can he complete per hour? $\frac{1}{2}$ article per hour

54. **Houseboat** A houseboat can travel 50 miles with the current in the same time that it can travel 15 miles against the current. The speed of the houseboat is 10 miles per hour.

   a. Find the speed of the current. $5\frac{5}{13}$ miles per hour

   b. How would the time of the houseboat change if its total speed were doubled for the same distance ? The same distance would be covered in half the time.

55. **Geometry** The area of a rhombus A varies jointly as the length of its diagonals d_1 and d_2. Write the general equation. $A = kd_1d_2$

56. **Assigning Grades** The grade on an exam varies directly as the number of questions correct. Elizabeth got 90 on an exam with 20 questions correct. Her friend Nuve got 18 questions right. What was Nuve's grade? 81

57. **Work** Hot water will fill a basin in 3 minutes; cold water will fill it in 6 minutes. The drain, when open, can empty the basin. The hot water and cold water are turned on, and the drain is opened. It takes 5 minutes to fill the basin. How quickly can the basin be emptied? $3\frac{1}{3}$ minutes

58. **Inclined plane** The distance s a ball can roll down an inclined plane is directly proportional to the square of the time t it rolls. Write an equation for this relationship. $s = kt^2$

59. **Prescription Service** Twenty out of thirty doctors subscribe to a service that delivers prescriptions directly to their offices. Some of the doctors dropped this service until only half of the doctors used it. How many of the original 30 doctors dropped the service? 10 doctors

60. Let $f(x) = \frac{x-3}{3}$ and $g(x) = \frac{x+3}{2}$. Find $f(x) + g(x)$. $\frac{5x+3}{6}$

61. Simplify: $\dfrac{\frac{2}{m} - \frac{1}{2m}}{m + \frac{m}{2}}$ $\frac{1}{m^2}$

62. Solve: $\frac{t-3}{5} = \frac{8}{t}$ $t = -5; t = 8$

63. Simplify: $\dfrac{\frac{1+x}{y}}{x - \frac{x}{y}}$ $\frac{1+x}{xy - x}$

64. Simplify: $\frac{x^2 n^2}{x^3 n}$ $\frac{n}{x}$

65. Reduce: $\frac{3n-18}{8n-48}$ $\frac{3}{8}$

66. Reduce: $\frac{y^2 + 5y + 6}{y^2 + 7y + 10}$ $\frac{y+3}{y+5}$

CHAPTER 6 TEST

This exam tests your knowledge of the material in Chapter 6.

1. a. Rewrite $h(x) = \dfrac{x^2 - 6x + 5}{x^2 - 10x + 25}$ in simplest form and then find $h(-2)$ and $h(a-1)$. $\quad \dfrac{x-1}{x-5}; \dfrac{3}{7}; \dfrac{a-2}{a-6}$

   b. Reduce: $\dfrac{8n^4 x^3 y^6}{2n^{-3} x^5 y^7}$ $\quad \dfrac{4n^7}{x^2 y}$

   c. Reduce: $\dfrac{5m-5}{5-5m}$ $\quad -1$

2. Simplify: a. $\dfrac{60x^5 y^6}{120x^3 y^5} \div \dfrac{30x^5 y^2}{40x^3 y^5}$ $\quad \dfrac{2y^4}{3}$

   b. $\dfrac{m^6 n^5}{x^3 y^4} \cdot \dfrac{x^6 y^5}{m^5 n^4}$ $\quad mnx^3 y$

   c. $\dfrac{x^2 - y^2}{x+y} \div \dfrac{2x^2 - xy - y^2}{y + 2x}$ $\quad 1$

3. Combine: a. $\dfrac{3}{x-2} + \dfrac{2}{x+2}$ $\quad \dfrac{5x+2}{x^2-4}$

   b. $\dfrac{5}{15x} - \dfrac{2}{10x}$ $\quad \dfrac{2}{15x}$

   c. $\dfrac{3}{m-1} + \dfrac{4}{m+1} + \dfrac{m+2}{m^2-1}$ $\quad \dfrac{8m+1}{(m+1)(m-1)}$

4. Simplify: a. $\dfrac{2 + \frac{1}{5}}{1 + \frac{3}{5}}$ $\quad 1\frac{3}{8}$

   b. $\dfrac{x - \frac{x}{y}}{\frac{1+x}{y}}$ $\quad \dfrac{x(y-1)}{1+x}$

   c. $\dfrac{6 - \frac{1}{x}}{\frac{4}{x} + \frac{2}{x^2}}$ $\quad \dfrac{x(6x-1)}{4x+2}$

5. Solve: a. $\dfrac{m-3}{m+1} = \dfrac{m-6}{m+5}$ $\quad m = \dfrac{9}{7}$

   b. $\dfrac{3m}{m+1} + 2 = \dfrac{5m}{m-1}$ $\quad m = \dfrac{-1}{4}$

   c. $2 - \dfrac{3y-6}{y-3} = \dfrac{4}{3-y}$ $\quad y = 4$

6. Solve:

   a. **Rowing** A rower can travel against the current for 24 miles in the same time it would take him to travel 36 miles with the current. If his speed in still water is 12 miles per hour, what is the speed of the current? 2.4 mph

   b. **Food Services** Nephrites baked 20 cakes. Seventy percent of them were chocolate. She sold some chocolate cakes until only 40% of the remaining cakes were chocolate. How many cakes did she sell? 10 chocolate cakes

   c. **Pigeons** A racing pigeon can fly 22 miles with an 8 mile-per-hour tailwind in the same time in which it can fly 15 miles into an 8 mile-per-hour wind. What is the pigeon's speed in still air? $42\frac{2}{7}$ mph

7. Solve:

   a. **Centrifugal Force** Write an equation for this relationship: An automobile rounding a curve is pulled toward the outside of the curve by a centrifugal force f that varies directly as the square of the speed v and inversely as the radius of the curve r. $f = \dfrac{kv^2}{r}$

   b. **Books** The discounted price of books in the Student Book Center varies directly with the list price of the books. You pay $50 for books that list for $65. How much will you pay for books that list for $156? $120

   c. **Magnetism** The repulsive force f between the north poles of two magnets is inversely proportional to the square of the distance c between them. The magnets are relocated at a distance that is $\frac{1}{4}$ of the original distance. How does f change? f becomes 16 times greater

CUMULATIVE REVIEW

CHAPTERS 1–6

The following exercises will help you maintain the skills you have learned in this and previous chapters.

1. Write -0.003^2 without an exponent. $\quad -0.000009$

2. Simplify: $1\frac{1}{2} \div 3 - 15 \div 5 + 2 \div 3 \quad -1\frac{5}{6}$

3. Write the equation of the line that is parallel to the x-axis and passes through the point $(3, -6)$. $\quad y = -6$

4. Rewrite using only positive exponents: $\left(\dfrac{2x^{-3}y^2z^3}{xy^3z}\right)^{-1} \quad \dfrac{x^4y}{2z^2}$

5. Factor: $2x^2 + 20x + 50 \quad 2(x + 5)^2$

6. Divide: $0.7315 \div 3.5 \quad 0.209$

7. Let $f(x) = 2x^3 - 2x^2 + 5x - 4$, and find $f(-1)$. $\quad -13$

8. Solve: $|t| = 6 \quad t = 6; t = -6$

9. Define two sets A and B such that $A \cap B = \{5\}$. Answers may vary. The two sets should be disjoint except for the element 5, which must be in both sets.

10. Write the equation of the line that passes through the points $(1, 2)$ and $(-2, 5)$. $\quad y = -x + 3$

11. Write $\left(\dfrac{2}{3}\right)^{-2}$ without an exponent. $\quad \dfrac{9}{4}$

12. Solve: $\begin{aligned} 2x + y &= 3 \\ 5x - y &= 2 \end{aligned} \quad \left\{\frac{5}{7}, 1\frac{4}{7}\right\}$

13. Add: $-3.28 + 5.72 \quad 2.44$

14. Divide: $\dfrac{6x^2y - 3xy^2 + 2x^2}{2x^2} \quad 3y - \dfrac{3y^2}{2x} + 1$

15. Evaluate the determinant: $\begin{vmatrix} 2 & 3 \\ 5 & 6 \end{vmatrix} \quad -3$

16. Factor: $6(x^2 - y^2) + 2(x^2 - y^2) \quad 8(x + y)(x - y)$

17. Write $2\frac{1}{8}$ as a decimal. $\quad 2.125$

18. Solve: $\dfrac{3x}{2} = 198 \quad x = 132$

19. *Pricing* Solange bought two dresses, one at $\frac{4}{5}$ of the price of the other. Altogether she paid \$100.80. How much was each dress? $\quad$ \$56 and \$44.80

20. *Movies* During the early fifties in Fitzgerald, Georgia, tickets to the Saturday movie cost 18 cents for children and 42 cents for adults. One Saturday, a total of \$69.12 was collected for 328 tickets. How many adults attended? $\quad$ 42 adults

RADICALS AND EXPONENTS

*T*ightrope walkers must anticipate and compensate for the movements of their partners. When one aerialist jumps onto the cable, a pulse, or wave, travels through the cable. If unprepared, the cable's motion will tumble the second acrobat. The speed of a wave on a cable depends on many variables: the cable's mass, its length and tension, and the mass of the two acrobats. The formula for determining the speed *v* of a wave is

$$v = \sqrt{\frac{T}{\frac{M}{L}}}$$

where *T* is the tension, *M* is the mass, and *L* is the length of the cable.

■ *Where else might you encounter motion similar to that of the tightrope cable?*

SKILLS CHECK

Take this short quiz to see how well prepared you are for Chapter 7. The answers follow the quiz.

1. Simplify: $(2^2)(3^2)(5^2)$

2. Simplify: $(2^2 3xyz^2)^2$

3. Write 1240 as a product of prime factors.

4. Find: $(2x + 6)(2x - 6)$

5. Find: $(0.1x - 8y)^2$

6. Reduce to lowest terms:

   $$\frac{6rx^2y^4}{18r^2x^3y}$$

7. Find the square of 100.

8. Find the principal square root of 81.

ANSWERS: **1.** 900 [Section 1.3]   **2.** $144x^2y^2z^4$ [Section 5.1]   **3.** $2^3 \cdot 5 \cdot 31$ [Section 5.4]
4. $4x^2 - 36$ [Section 5.3]   **5.** $0.01x^2 - 1.6xy + 64y^2$ [Section 5.3]   **6.** $\frac{y^3}{3rx}$ [Section 5.1]
7. 10,000 [Section 1.3]   **8.** 9 [Section 1.3]

CHAPTER LEAD-IN

In the past decade, more and more legislation has led to equal access for the disabled. Ramps permit equal access for those who use wheelchairs or who have difficulty climbing stairs. The ramps must fit the available space, yet must be usable—a ramp too long or too steep could be impossible or dangerous to use. Planning a safe ramp may require the use of the Pythagorean theorem—the square of the ramp's length is equal to the sum of the squares of the height and the base of the ramp—and radicals. The radical, or square root, of the squared length gives the ramp's actual length.

Architecture makes use of many familiar geometric shapes. The triangle is a "rigid" shape that is often used as a brace for other shapes. What characteristics of a triangle make it useful for this purpose? What shapes other than squares or rectangles can you find in building structures?

7.1 Roots and Numbers

SECTION LEAD-IN

A 30-foot water tower sits in back of a farmhouse. It is in bad repair. What length of metal is needed for the four legs of this tower? In this section, we will learn how to calculate such measures.

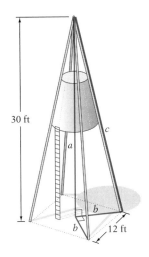

To raise a number to a given power, we use it as a factor the stated number of times. For example,

$$x^2 = \underbrace{x \cdot x}_{2 \text{ times}} \qquad y^3 = \underbrace{y \cdot y \cdot y}_{3 \text{ times}} \qquad t^4 = \underbrace{t \cdot t \cdot t \cdot t}_{4 \text{ times}}$$

To find a root of a given number, we find a number that can be multiplied by itself to produce the given number. In our examples, x is the **square** (second) **root** of x^2, y is the **cube** (third) **root** of y^3, and t is the **fourth root** of t^4. In general,

$$\text{if } \underbrace{a \cdot a \ldots a}_{n \text{ times}} = b$$

then a is the ***n*th root** of b.

Square Roots

Every positive real number has a positive real square root and a negative real square root. These square roots are additive inverses.

> If a and b are positive real numbers and $a \cdot a = b$, then a and $-a$ are the square roots of b.

Negative real numbers do not have real square roots. The square root of zero is zero.

INSTRUCTOR NOTE

When dealing with square roots of decimals, students sometimes make errors in calculation. This would be a good time to review multiplication of decimals.

Principal Square Roots

The positive real square root of a number b is called the **principal square root** of b and is written $\sqrt{b}$. The negative real square root of b is then $-\sqrt{b}$. Thus

$$\sqrt{16} = 4 \quad \text{and} \quad -\sqrt{16} = -4$$

INSTRUCTOR NOTE

We cover imaginary numbers in Section 7.4.

(But be careful of $\sqrt{-16}$, which is not a real number—it is imaginary.)

The $\sqrt{}$ sign is called the **radical sign**, or **square root sign.** A term or expression under the square root, or radical sign, is called a **radicand.** Together, they are called a **radical.**

$$\text{Radical sign} \longrightarrow \sqrt{402} \longleftarrow \text{Radicand}$$

An expression that includes a radical, such as $\sqrt{x - 7} + 1$, is called a **radical expression.**

The square roots of an unknown real number x^2 are x and $-x$, because

$$x \cdot x = x^2 \quad \text{and} \quad (-x)(-x) = x^2$$

But which is the principal square root? If we don't know whether x is positive or negative, we must write the principal square root as an absolute value,

$$\sqrt{x^2} = |x|$$

to be sure that the principle square root is positive.

If we *know* that $x \geq 0$, then we may write

$$\sqrt{x^2} = x$$

▪▪▪

EXAMPLE 1

Find the principal square root of:

a. $\frac{9}{16}a^2$ **b.** $y + 2$ **c.** $0.25x^2y^2$ **d.** $(-2)^2$

SOLUTION

▪▪▪
WRITER'S BLOCK
How can the variable
a
represent a negative
number? It looks
positive. Explain.

a. The principal square root is either $\frac{3}{4}a$ or $\frac{-3}{4}a$, depending on whether a represents a positive or a negative number. We must use absolute value signs to ensure that $\sqrt{\frac{9}{16}a^2}$ is positive.

$$\sqrt{\frac{9}{16}a^2} = \left|\frac{3}{4}a\right| = \frac{3}{4}|a|$$

b. We do not know the value of y, so we cannot evaluate this square root. We simply indicate the principal, or positive, square root as $\sqrt{y + 2}$.

c. The principal square root may be $0.5xy$ or $-0.5xy$. Here again, we must use absolute value signs.

$$\sqrt{0.25x^2y^2} = 0.5|xy|$$

d. Because $\sqrt{(-2)^2} = \sqrt{(-2)(-2)} = \sqrt{4}$, and because the principal square root of 4 is 2, the principal square root of $(-2)^2$ is also 2. So $\sqrt{(-2)^2} = 2$.

▶ CHECK **Warm-Up 1**

Irrational Square Roots

Positive integers that have integers for square roots are called **perfect squares;** those that are not perfect squares have irrational numbers as square roots. One example is $\sqrt{5} = 2.23606797\ldots$. Recall that an irrational number is a decimal number whose decimal part does not terminate or repeat. An irrational number cannot be written exactly. But the irrational square root of a number can be approximated on a calculator. On a scientific calculator, key in the number and then press $\boxed{\sqrt{}}$. (On some graphing calculators, this keying sequence may be reversed.) Round the displayed approximation to the required number of places.

Higher Roots

We use the radical sign, written with a positive integer called the **index,** to show principal roots higher than the second. For example, $\sqrt[3]{a}$ represents the cube root of the real number a. Thus

$$\sqrt[3]{a} = b \quad \text{means that} \quad b \cdot b \cdot b = a$$

Every real number (positive, negative, or zero) has just one cube root. For that reason, we don't usually speak of a "principal" cube root. As examples,

$$\sqrt[3]{8} = 2 \quad \text{because } 2 \cdot 2 \cdot 2 = 8$$
$$\sqrt[3]{-8} = -2 \quad \text{because } -2 \cdot -2 \cdot -2 = -8$$

Similarly, $\sqrt[4]{a}$ represents the principal fourth root of the real number a. Every positive real number has two real fourth roots, one positive and one negative. The principal fourth root is the positive fourth root. So, for example,

$$\sqrt[4]{81} = 3 \quad \text{because} \quad 3 \cdot 3 \cdot 3 \cdot 3 = 81$$

We use $-\sqrt[4]{81}$ to indicate the negative fourth root, -3. Negative real numbers do not have real fourth roots, so $\sqrt[4]{-81}$ does not exist in the real numbers. The fourth root of zero is zero.

We can generalize these properties of higher roots by considering odd roots and even roots separately: If n is an odd integer, then a real number a has one real nth root. This root is positive if a is positive and is negative if a is negative. Absolute value signs are *not* used to represent odd roots.

If n is an even integer, then a positive real number a has two real nth roots, one positive and one negative. The principal nth root is positive, and absolute value signs are needed to specify it.
In symbols,

> ▪▪▪
> **WRITER'S BLOCK**
> Use $\sqrt{a^b}$ to give examples of a <u>radical sign</u>, <u>radical</u>, <u>radicand</u>, <u>base</u>, and <u>exponent</u>.

For an odd index n and real b,
$$\sqrt[n]{b^n} = b$$

For an even index n and real b,
$$\sqrt[n]{b^n} = |b|$$
when n is even; $-|b|$ is also an nth root of b^n.

INSTRUCTOR NOTE

Students with learning disabilities may need to circle each index to distinguish it from the coefficient of the radical.

▪▪▪
EXAMPLE 2

Find the principal root.

a. $\sqrt[7]{-1}$ **b.** $\sqrt[6]{x^6}$ **c.** $\sqrt[9]{(y^2-7)^9}$ **d.** $\sqrt[6]{(y^2-7)^6}$

SOLUTION

a. $\sqrt[7]{-1} = -1$. All odd roots of -1 are -1.

b. $\sqrt[6]{x^6} = |x|$. We need the absolute value sign for an even root of a variable.

c. $\sqrt[9]{(y^2-7)^9} = y^2 - 7$

d. $\sqrt[6]{(y^2-7)^6} = |y^2 - 7|$. Again, we need the absolute value sign to ensure that the even principal root is positive.

▶ CHECK **Warm-Up 2**

Simplifying Radicals

A radical or a radical expression is simplified when

1. The radicand does not contain a power that is larger than the index of the radical.
2. No denominator contains a radical.
3. All fractions are reduced to lowest terms.

From here to the end of this chapter, we shall assume that all variables represent positive real numbers, so that absolute value signs are not needed to show principal square roots.

We use two properties of radicals to simplify radicals and radical expressions— the product property and the quotient property.

> **The Product Property of Radicals**
>
> If a and b are real numbers with real roots, and n is a positive integer, then
> $$\sqrt[n]{a} \cdot \sqrt[n]{b} = \sqrt[n]{ab}$$

In words, the product of two roots with the same index is equal to the root of the product. Also, because these radicals are real numbers, we can use properties of real numbers (such as commutativity and associativity) in solving problems.

When it is read from right to left, the product property equation tells us how to *simplify radicals by factoring*.

▪▪▪
EXAMPLE 3

INSTRUCTOR NOTE

Remind students that we will no longer use absolute value marks to ensure that variables are positive. We will assume they represent positive real numbers.

Simplify by finding the principal root:

a. $\sqrt{300}$ **b.** $\sqrt{x^5}$ **c.** $\sqrt{3x^4y^6}$ **d.** $\sqrt[3]{8y^5}$

SOLUTION

We apply the product property, rewriting each radical as a product of radicals that can be simplified.

a. $\sqrt{300} = \sqrt{100} \cdot \sqrt{3} = \sqrt{10^2} \cdot \sqrt{3} = 10\sqrt{3}$

b. $\sqrt{x^5} = \sqrt{x^4} \cdot \sqrt{x^1} = \sqrt{(x^2)^2} \cdot \sqrt{x^1} = x^2\sqrt{x}$

c. $\sqrt{3x^4y^6} = \sqrt{3} \cdot \sqrt{x^4} \cdot \sqrt{y^6} = \sqrt{3} \cdot \sqrt{(x^2)^2} \cdot \sqrt{(y^3)^2} = x^2y^3\sqrt{3}$

d. $\sqrt[3]{8y^5} = \sqrt[3]{8} \cdot \sqrt[3]{y^5} = \sqrt[3]{2^3} \cdot \sqrt[3]{y^3 \cdot y^2} = \sqrt[3]{2^3} \cdot \sqrt[3]{y^3} \cdot \sqrt[3]{y^2} = 2y\sqrt[3]{y^2}$

▶ *CHECK* **Warm-Up 3**

When it is read from left to right, the product property equation tells us how to *multiply radicals with the same index.*

▪▪▪

EXAMPLE 4

Multiply: **a.** $\sqrt{24} \cdot \sqrt{60}$ **b.** $\sqrt[3]{4} \cdot \sqrt[3]{128}$ **c.** $\sqrt{18y} \cdot \sqrt{16x} \cdot \sqrt{12xy^2}$

SOLUTION

We apply the product property and then factor and simplify.

a. $\sqrt{24} \cdot \sqrt{60} = \sqrt{24 \cdot 60}$ Product property

$\qquad\qquad\quad = \sqrt{2^5 \cdot 3^2 \cdot 5}$ Factoring

$\qquad\qquad\quad = \sqrt{2^4 \cdot 2 \cdot 3^2 \cdot 5}$ Rearranging

$\qquad\qquad\quad = 2^2 \cdot 3\sqrt{2 \cdot 5}$ Taking square roots

$\qquad\qquad\quad = 12\sqrt{10}$ Multiplying

b. $\sqrt[3]{4} \cdot \sqrt[3]{128} = \sqrt[3]{4 \cdot 128}$ Product property

$\qquad\qquad\quad = \sqrt[3]{2^2 \cdot 2^7}$ Factoring

$\qquad\qquad\quad = \sqrt[3]{2^9}$ Multiplying

$\qquad\qquad\quad = 2^3$ Taking cube root

$\qquad\qquad\quad = 8$ Writing in standard notation

c. $\sqrt{18y} \cdot \sqrt{16x} \cdot \sqrt{12xy^2} = \sqrt{18y \cdot 16x \cdot 12xy^2}$ Product property

$\qquad\qquad\qquad\qquad = \sqrt{2 \cdot 3^2y \cdot 2^4x \cdot 2^2 \cdot 3xy^2}$ Factoring

$\qquad\qquad\qquad\qquad = \sqrt{2^7 \cdot 3^3x^2y^3}$ Multiplying

$\qquad\qquad\qquad\qquad = \sqrt{2^6 \cdot 2 \cdot 3^2 \cdot 3 \cdot x^2 \cdot y^2 \cdot y}$ Rearranging

$\qquad\qquad\qquad\qquad = 2^3 \cdot 3 \cdot x \cdot y\sqrt{2 \cdot 3 \cdot y}$ Taking square roots

$\qquad\qquad\qquad\qquad = 24xy\sqrt{6y}$ Simplifying

▶ *CHECK* **Warm-Up 4**

> **The Quotient Property of Radicals**
>
> If a and b are real numbers with real roots, b is not equal to zero, and n is a positive integer, then
>
> $$\frac{\sqrt[n]{a}}{\sqrt[n]{b}} = \sqrt[n]{\frac{a}{b}}$$

When it is read from left to right, the quotient property equation tells us that if the indexes are the same, *a quotient of radicals can be written as the root of the quotient.*

▪ ▪ ▪
EXAMPLE 5

Simplify:

a. $\dfrac{\sqrt[3]{64}}{\sqrt[3]{8}}$ b. $\dfrac{\sqrt{1620xy^6}}{\sqrt{180y^2}}$ c. $\sqrt{\dfrac{100}{49}}$ d. $\sqrt[3]{\dfrac{8x^3}{27}}$

SOLUTION

a. $\dfrac{\sqrt[3]{64}}{\sqrt[3]{8}} = \sqrt[3]{\dfrac{64}{8}}$ Quotient property

       $= \sqrt[3]{8} = 2$ Simplifying

b. $\dfrac{\sqrt{1620xy^6}}{\sqrt{180y^2}} = \sqrt{\dfrac{1620xy^6}{180y^2}}$ Quotient property

       $= \sqrt{\dfrac{2^2 \cdot 3^4 \cdot 5xy^6}{2^2 \cdot 3^2 \cdot 5y^2}}$ Factoring

       $= \sqrt{3^2xy^4} = 3y^2\sqrt{x}$ Simplifying

c. $\sqrt{\dfrac{100}{49}} = \sqrt{\dfrac{10^2}{7^2}}$ Factoring

       $= \dfrac{\sqrt{10^2}}{\sqrt{7^2}}$ Quotient property

       $= \dfrac{10}{7}$ Simplifying

d. $\sqrt[3]{\dfrac{8x^3}{27}} = \sqrt[3]{\dfrac{2^3x^3}{3^3}}$ Factoring

       $= \dfrac{\sqrt[3]{2^3x^3}}{\sqrt[3]{3^3}}$ Quotient property

       $= \dfrac{2x}{3}$ Simplifying

▶ CHECK **Warm-Up 5**

Rational Exponents

Now we extend the ideas of exponents to include not only integer exponents but also exponents that are rational numbers. Recall that a rational number is one that can be written as a ratio $\frac{m}{n}$, where m and n are integers and n is not equal to zero.

Defining Rational Exponents

To ensure that all the properties of integer exponents apply to rational exponents, we first define the number $a^{\frac{1}{n}}$ as follows:

> If a and its nth root are real numbers, and n is a positive integer, then
> $$a^{\frac{1}{n}} = \sqrt[n]{a}$$

This definition also provides a link between rational exponents and roots. We can use it to "translate" numbers from exponent form to radical form, and vice versa.

▪▪▪

EXAMPLE 6

Write in radical form and simplify where possible:

a. $3^{\frac{1}{4}}$ **b.** $(-27)^{\frac{1}{3}}$ **c.** $(x^2 y^4)^{\frac{1}{2}}$

SOLUTION

We rewrite each expression according to the definition and simplify where possible.

a. $3^{\frac{1}{4}} = \sqrt[4]{3}$

b. $(-27)^{\frac{1}{3}} = \sqrt[3]{-27} = \sqrt[3]{(-3)^3} = -3$

c. $(x^2 y^4)^{\frac{1}{2}} = \sqrt{x^2 y^4} = xy^2$

▶ *CHECK* **Warm-Up 6**

▪▪▪

EXAMPLE 7

Write in exponent form: **a.** $\sqrt{37x}$ **b.** $\sqrt[9]{\frac{a}{2}}$ **c.** $\sqrt[4]{x+y}$

SOLUTION

We rewrite according to the definition.

a. $\sqrt{37x} = (37x)^{\frac{1}{2}}$

b. $\sqrt[9]{\frac{a}{2}} = \left(\frac{a}{2}\right)^{\frac{1}{9}}$

c. $\sqrt[4]{x+y} = (x+y)^{\frac{1}{4}}$

▶ *CHECK* **Warm-Up 7**

So far, we have defined rational exponents only for the rational number $\frac{1}{n}$. To include all possible rational exponents $\frac{m}{n}$ in the definition, we apply the power rule for exponents to $a^{\frac{1}{n}}$ in two ways:

$$a^{\frac{m}{n}} = \left(a^{\frac{1}{n}}\right)^m = \left(\sqrt[n]{a}\right)^m$$

and

$$a^{\frac{m}{n}} = \left(a^m\right)^{\frac{1}{n}} = \sqrt[n]{a^m}$$

This tells us that

> If a and its nth root are real numbers, and m and n are positive integers, then
> $$a^{\frac{m}{n}} = \left(\sqrt[n]{a}\right)^m = \sqrt[n]{a^m}$$

■ ■ ■

EXAMPLE 8

Translate to exponential form:

a. $\left(\sqrt[3]{27}\right)^4$ **b.** $\sqrt[4]{6^3}$ **c.** $\sqrt[5]{8^4}$

SOLUTION

We use the definition.

a. $\left(\sqrt[3]{27}\right)^4 = 27^{\frac{4}{3}}$ **b.** $\sqrt[4]{6^3} = 6^{\frac{3}{4}}$ **c.** $\sqrt[5]{8^4} = 8^{\frac{4}{5}}$

▶ CHECK **Warm-Up 8**

■ ■ ■

EXAMPLE 9

Write as an integer using both methods shown in the definition.

a. $8^{\frac{2}{3}}$ **b.** $16^{\frac{3}{4}}$ **c.** $(-1)^{\frac{3}{5}}$

SOLUTION

We write each number $a^{\frac{m}{n}}$ as $\left(a^m\right)^{\frac{1}{n}}$ and as $\left(a^{\frac{1}{n}}\right)^m$.

a. $8^{\frac{2}{3}} = \left(8^2\right)^{\frac{1}{3}} = \sqrt[3]{8^2} = \sqrt[3]{64} = 4$ and $8^{\frac{2}{3}} = \left(8^{\frac{1}{3}}\right)^2 = \left(\sqrt[3]{8}\right)^2 = 2^2 = 4$

b. $16^{\frac{3}{4}} = [(16)^3]^{\frac{1}{4}} = \sqrt[4]{16^3} = \sqrt[4]{4096} = 8$ and
$16^{\frac{3}{4}} = [(16)^{\frac{1}{4}}]^3 = \left(\sqrt[4]{16}\right)^3 = 2^3 = 8$

c. $(-1)^{\frac{3}{5}} = [(-1)^3]^{\frac{1}{5}} = \sqrt[5]{(-1)^3} = \sqrt[5]{(-1)} = -1$ and
$(-1)^{\frac{3}{5}} = [(-1)^{\frac{1}{5}}]^3 = \left(\sqrt[5]{-1}\right)^3 = (-1)^3 = -1$

▶ CHECK **Warm-Up 9**

Practice what you learned.

SECTION FOLLOW-UP

The calculations needed for the water tower require simplifying radicals. We will use the Pythagorean theorem:

In a right triangle with legs of lengths a and b and hypotenuse of length c, $a^2 + b^2 = c^2$.

(You can learn more about the Pythagorean theorem on page 485.) We want to find c in the figure. To find that, we must first find b. We use $c^2 = a^2 + b^2$. We know that $a = 30$ feet. So $c^2 = 30^2 + b^2$. But also, $12^2 = b^2 + b^2$. (Explain.) So

$$2b^2 = 12^2$$
$$2b^2 = 144$$
$$b^2 = 72$$

Substituting in $c^2 = 30^2 + b^2$, we get $c^2 = 30^2 + 72$. So

$$c^2 = 972$$
$$c \approx 31.2$$

We need about 125 feet of this metal. (Explain.)

7.1 WARM-UPS

Work these problems before you attempt the exercises.

1. Find: $\sqrt{(-9)^2}$ and $\sqrt{\frac{25}{36}x^4}$ $9; \frac{5}{6}x^2$

2. Find the indicated roots: $\sqrt[8]{(y^2-3)^8}$ and $\sqrt[6]{-1}$ $|y^2-3|$; no real root

3. Simplify by finding the principal roots: $\sqrt{400}$, $\sqrt{260x^4y}$, and $\sqrt[3]{27y^7}$ $20; 2x^2\sqrt{65y}; 3y^2\sqrt[3]{y}$

4. Find the product: $\sqrt{50} \cdot \sqrt{40}$ and $\sqrt[3]{125} \cdot \sqrt[3]{27}$ $20\sqrt{5}; 15$

5. Divide: $\frac{\sqrt[3]{64x^3}}{\sqrt[3]{8x^6}}$ and $\frac{\sqrt{150xy^4}}{\sqrt{180x^3}}$ $\frac{2}{x};\frac{y^2}{x}\sqrt{\frac{5}{6}}$

6. Write in radical form and simplify: $\left(x^3y^4\right)^{\frac{1}{3}}$ and $\left(3x^2\right)^{\frac{1}{4}}$
   $\sqrt[3]{x^3y^4} = xy\sqrt[3]{y}; \sqrt[4]{3x^2}$

7. Write in exponent form: $\sqrt{74x}$ and $\sqrt[3]{\frac{x}{3}}$ $74^{\frac{1}{2}}x^{\frac{1}{2}}; 3^{-\frac{1}{3}}x^{\frac{1}{3}}$

8. Rewrite with rational exponents: $\sqrt[5]{x^2y^3}$ $x^{\frac{2}{5}}y^{\frac{3}{5}}$

9. Simplify: $64^{\frac{2}{4}}$ 8

7.1 EXERCISES

Note: Use your graphing calculator to check your results whenever possible.

In Exercises 1 through 16, find the principal square root of the given term. Use absolute value signs where necessary.

1. $16x^4$ $4x^2$

2. $100t^8$ $10t^4$

3. $49y^8$ $7y^4$

4. $64w^4$ $8w^2$

5. $256x^2y^2$ $16\,|\,xy\,|$

6. $324w^2x^2$ $18\,|\,wx\,|$

7. $400t^4v^2$ $20t^2\,|\,v\,|$

8. $625u^6v^2$ $25\,|\,u^3v\,|$

9. $289s^2t^4$ $17\,|\,s\,|\,t^2$

10. $169w^4x^6$ $13w^2\,|\,x^3\,|$

11. $1600v^4x^4$ $40v^2x^2$

12. $4900y^4z^8$ $70y^2z^4$

13. $121x^2y^4z^4$ $11\,|\,x\,|\,y^2z^2$

14. $196x^6y^6z^6$ $14\,|\,x^3y^3z^3\,|$

15. $144w^4y^4z^2$ $12w^2y^2\,|\,z\,|$

16. $8100r^2s^2t^4$ $90\,|\,rs\,|\,t^2$

In Exercises 17 through 32, find the principal square root. Use absolute value signs where needed.

17. $\frac{4}{9}$ $\frac{2}{3}$

18. $\frac{25}{64}$ $\frac{5}{8}$

19. $\frac{121}{169}$ $\frac{11}{13}$

20. $\frac{144}{289}$ $\frac{12}{17}$

21. $\frac{16}{625}x^4y^8$ $\frac{4}{25}x^2y^4$

22. $\frac{81}{100}t^8y^4$ $\frac{9}{10}t^4y^2$

23. $0.0001t^4z^4$ $0.01t^2z^2$

24. $0.00000001x^2y^2$ $0.0001\,|\,xy\,|$

25. $0.81x^2y^4z^4$ $0.9\,|\,x\,|\,y^2z^2$

26. $0.49x^6y^6z^6$ $0.7\,|\,x^3y^3z^3\,|$

27. $0.04w^4y^4z^2$ $0.2w^2y^2\,|\,z\,|$

28. $0.0036r^2s^4t^4$ $0.06\,|\,r\,|\,s^2t^2$

29. $\frac{25}{361}(xyz)^6$ $\frac{5}{19}\,|\,x^3y^3z^3\,|$

30. $\frac{256x^2}{49y^2}$ $\frac{16}{7}\left|\frac{x}{y}\right|$

31. $0.0289x^2y^4z^6$ $0.17\,|\,xz^3\,|\,y^2$

32. $0.0361(x^2yz^2)^2$ $0.19x^2z^2\,|\,y\,|$

Note: In the remaining exercises in this chapter, we assume that all variables are positive real numbers.

In Exercises 33 through 52, find the principal indicated root, if it exists.

33. $\sqrt[3]{125x^6}$ $5x^2$

34. $\sqrt[3]{216x^{18}}$ $6x^6$

35. $\sqrt[4]{1296x^8}$ $6x^2$

36. $\sqrt[4]{2401x^{12}}$ $7x^3$

37. $\sqrt[3]{-343x^{12}y^{15}}$ $-7x^4y^5$

38. $\sqrt[3]{729x^{18}y^{21}}$ $9x^6y^7$

39. $\sqrt[4]{6561x^{16}y^{20}}$ $9x^4y^5$

40. $\sqrt[4]{4096x^{28}y^{32}}$ $8x^7y^8$

41. $\sqrt[3]{2744(xyz)^6}$ $14x^2y^2z^2$

42. $\sqrt[3]{-1728(xy^2z^3)^3}$ $-12xy^2z^3$

43. $\sqrt[3]{-1331x^3y^3z^6}$ $-11xyz^2$

44. $\sqrt[3]{2197x^6y^6z^3}$ $13x^2y^2z$

45. $\sqrt[3]{343,000}$ 70

46. $\sqrt[3]{-1,000,000}$ -100

47. $\sqrt[3]{729,000}$ 90

48. $\sqrt[3]{512,000}$ 80

49. $\sqrt[4]{160,000}$ 20

50. $\sqrt[4]{40,960,000}$ 80

51. $\sqrt[4]{12,960,000}$ 60

52. $\sqrt[4]{146,410,000}$ 110

In Exercises 53 through 68, simplify the radical by finding the principal root.

53. $\sqrt{240x^2}$ $4x\sqrt{15}$

54. $\sqrt{350y^2}$ $5y\sqrt{14}$

55. $\sqrt{1000x^2y^2}$ $10xy\sqrt{10}$

56. $\sqrt{1260xy^4}$ $6y^2\sqrt{35x}$

57. $\sqrt{356xy^3}$ $2y\sqrt{89xy}$

58. $\sqrt{4000x^2y^2}$ $20xy\sqrt{10}$

59. $\sqrt{1296xy}$ $36\sqrt{xy}$

60. $\sqrt{3608x^2y^2}$ $2xy\sqrt{902}$

61. $\sqrt{456x^2y^2}$ $2xy\sqrt{114}$

62. $\sqrt{550x^2y}$ $5x\sqrt{22y}$

63. $\sqrt{1440x^2y^2}$ $12xy\sqrt{10}$

64. $\sqrt{1656x^2y^4}$ $6xy^2\sqrt{46}$

65. $\sqrt{981x^2}$ $3x\sqrt{109}$

66. $\sqrt{396xy^4}$ $6y^2\sqrt{11x}$

67. $\sqrt{5280x^2y^2}$ $4xy\sqrt{330}$

68. $\sqrt{2430x^2y^4}$ $9xy^2\sqrt{30}$

In Exercises 69 through 84, find the product.

69. $\sqrt{65x}\cdot\sqrt{104x}$ $26x\sqrt{10}$

70. $\sqrt{56x^2}\cdot\sqrt{64x^3}$ $16x^2\sqrt{14x}$

71. $\sqrt{49x^3}\cdot\sqrt{84x}$ $14x^2\sqrt{21}$

72. $\sqrt{120x^4}\cdot\sqrt{90x}$ $60x^2\sqrt{3x}$

73. $\sqrt{63x^2} \cdot \sqrt{63x^2}$
$63x^2$

74. $\sqrt{94xy} \cdot \sqrt{94xy}$
$94xy$

75. $\sqrt{120xy} \cdot \sqrt{120xy}$
$120xy$

76. $\sqrt{653xy} \cdot \sqrt{653xy}$
$653xy$

77. $\sqrt{105xy} \cdot \sqrt{35xy}$
$35xy\sqrt{3}$

78. $\sqrt{216xy} \cdot \sqrt{48xy}$
$72xy\sqrt{2}$

79. $\sqrt{57xy} \cdot \sqrt{19x^2y}$
$19xy\sqrt{3x}$

80. $\sqrt{150xy} \cdot \sqrt{180x}$
$30x\sqrt{30y}$

81. $\sqrt{90xy} \cdot \sqrt{60x^2y}$
$30xy\sqrt{6x}$

82. $\sqrt{150xy} \cdot \sqrt{50xy}$
$50xy\sqrt{3}$

83. $\sqrt{250xy} \cdot \sqrt{125x}$
$125x\sqrt{2y}$

84. $\sqrt{300xy} \cdot \sqrt{30xy}$
$30xy\sqrt{10}$

In Exercises 85 through 100, divide and simplify.

85. $\dfrac{\sqrt{1600}}{\sqrt{324}}$ $\dfrac{20}{9}$

86. $\dfrac{\sqrt{4096xy}}{\sqrt{784xy}}$ $\dfrac{16}{7}$

87. $\dfrac{\sqrt{2744}}{\sqrt{196}}$ $\sqrt{14}$

88. $\sqrt{\dfrac{2025}{225}}$ 3

89. $\sqrt{\dfrac{625x^2y^2}{144}}$ $\dfrac{25xy}{12}$

90. $\sqrt{\dfrac{289y^2}{36x^2}}$ $\dfrac{17y}{6x}$

91. $\sqrt{\dfrac{400x}{20xy^2}}$ $\dfrac{2\sqrt{5}}{y}$

92. $\dfrac{\sqrt{2744x^2}}{\sqrt{512x^2y^2}}$ $\dfrac{7\sqrt{7}}{8y}$

93. $\sqrt{\dfrac{256x^2}{18y^2}}$ $\dfrac{8x\sqrt{2}}{3y}$

94. $\sqrt{\dfrac{1056x^2}{600y^2}}$ $\dfrac{2x\sqrt{11}}{5y}$

95. $\sqrt{\dfrac{640x^2}{648y^2}}$ $\dfrac{4x\sqrt{5}}{9y}$

96. $\sqrt{\dfrac{1000x^2y^2}{900x}}$ $\dfrac{y\sqrt{10x}}{3}$

97. $\dfrac{\sqrt{1080}}{\sqrt{1440x^2}}$ $\dfrac{\sqrt{3}}{2x}$

98. $\dfrac{\sqrt{1734x^2y}}{\sqrt{392y}}$ $\dfrac{17x\sqrt{3}}{14}$

99. $\dfrac{\sqrt{2268x^2}}{\sqrt{1029y^2}}$ $\dfrac{6x\sqrt{3}}{7y}$

100. $\dfrac{\sqrt{1536xy^2}}{\sqrt{768x}}$ $y\sqrt{2}$

EXCURSIONS

Posing Problems

1. **a.** Pick a country from the figure below.

   b. Choose an item that you just purchased here. How much did you pay?

   c. If you could have purchased the item in (b) in your chosen country, what would it have cost in 1997? in 1996?

   d. Ask and answer four other questions using this data.

$1 EQUALS . . .*

	January 6, 1997	January 6, 1996
AFRICA		
Kenya (shilling)	42.08	42.22
Morocco (dirham)	8.01	7.31
Senegal (C.F.A. franc)	495.29	495.29
South Africa (rand)	4.30	3.31
THE AMERICAS		
Argentina (peso)	.89	.91
Brazil (real)	.88	.82
Canada (dollar)	1.27	1.26
Mexico (peso)	7.02	6.70
ASIA-PACIFIC		
Australia (dollar)	1.16	1.22
Hong Kong (dollar)	7.11	7.11
India (rupee)	30.83	32.24
Japan (yen)	112.27	101.15
New Zealand (dollar)	1.29	1.39
ASIA-MIDDLE EAST		
Egypt (pound)	2.85	2.72
Israel (shekel)	2.84	2.59
Turkey (lira)	83,333.00	43,478.00

* For example, $1 equals 4.30 rands in South Africa in 1997 and 3.31 rands in 1996.

2. There must be a number of questions you have as you look at the following information. Ask and answer four questions using this data. Share your questions with a classmate.

Home Entertainment

CD-ROM

Reference	$156 million based on end-user spending
Educational Software	$522 million based on end-user spending
Video Games	$2.93 billion based on end-user spending

On-Line Services

$1.43 billion in subscription fees of consumer-oriented services (Dialog, Lexis/Nexis, Dow Jones News Retrieval, and CompuServe are not included)

Video

Rentals	$9.39 billion based on 1994 consumer spending
Sales	$4.64 billion based on 1994 consumer spending

Sources: Association of American Publishers; Magazine Publishers of America; Opera America; Radio Business Report; Recording Industry Association of America; Theatre Communications Group; Variety, and Veronis Suhler and Associates, Inc., as cited in the *1996 Information Please® Business Almanac* (©1995 Houghton Mifflin Co.), p. 7. All rights reserved. Used with permission by Information Please LLC.

Top Entertainment Companies

Rank*	Company	1994 Sales (in millions)
1.	Time Warner Inc.	15,905
2.	Walt Disney Co.	10,055
3.	Sony Corp.	8,726
4.	News Corp.	8,640
5.	Viacom Inc.	7,363
6.	Capital Cities/ABC Inc.	6,379
7.	Pioneer Electronic Co.	5,128
8.	PolyGram NV	4,943
9.	Tele-Communications Inc.	4,936
10.	MCA Inc.	4,818
11.	CBS Inc.	3,712
12.	Rank Organization PLC	3,585
13.	Turner Broadcasting System Inc.	2,809
14.	Carlton Communications PLC	2,215
15.	Tribune Co.	2,155
16.	Comcast Corp.	1,375
17.	Grupo Televisa SA	1,288
18.	Home Shopping Network Inc.	1,126
19.	British Sky Broadcasting	860
20.	Cablevision Systems Corp.	837

*Rank as of July 6, 1995

Source: The Hollywood Reporter, as cited in the *1996 Information Please® Business Almanac* (©1995 Houghton Mifflin Co.), p. 7. All rights reserved. Used with permission by Information Please LLC.

3. Ask and answer four questions using this data.

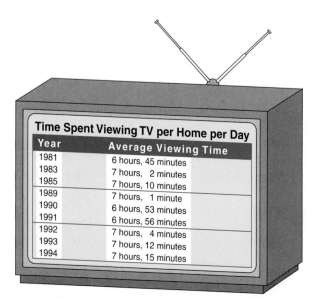

Time Spent Viewing TV per Home per Day

Year	Average Viewing Time
1981	6 hours, 45 minutes
1983	7 hours, 2 minutes
1985	7 hours, 10 minutes
1989	7 hours, 1 minute
1990	6 hours, 53 minutes
1991	6 hours, 56 minutes
1992	7 hours, 4 minutes
1993	7 hours, 12 minutes
1994	7 hours, 15 minutes

Source: TVB & Nielsen Media Research, as cited in the *1996 Information Please® Business Almanac* (©1995 Houghton Mifflin Co.), p. 483. All rights reserved. Used with permission by Information Please LLC.

CONNECTIONS TO *GEOMETRY*

The Pythagorean Theorem

The ancient Egyptians knew that if the sides of a triangle are in the ratio of 3 to 4 to 5, then that triangle contains a right angle (of measure 90°). They also knew that the right angle is always opposite the longest side, the **hypotenuse.** They used this idea in a variety of ways in everyday life—for example, to ensure that vertical walls were actually vertical.

Later, in the sixth century B.C., the Greek mathematician Pythagoras showed that the 3:4:5 triangle contains a right angle because

$$3^2 + 4^2 = 5^2$$

More important, he proved that a relationship holds among the sides of every right triangle:

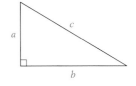

> **The Pythagorean Theorem**
>
> In a right triangle with legs of lengths a and b and hypotenuse of length c, $a^2 + b^2 = c^2$.
>
> The sum of the squares of the lengths of the two legs of a right triangle is equal to the square of the length of the hypotenuse.

We can find the third side of any right triangle if we know the other two sides by using this theorem or its variations, $c^2 - a^2 = b^2$ and $c^2 - b^2 = a^2$.

▪▪▪

EXAMPLE

Find the length of the third side of this triangle.

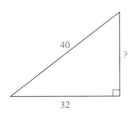

SOLUTION

$$
\begin{aligned}
c^2 - a^2 &= b^2 &&\text{Pythagorean theorem} \\
40^2 - 32^2 &= b^2 &&\text{Substituting known values} \\
576 &= b^2 &&\text{Simplifying} \\
\sqrt{576} &= b \\
\sqrt{576} = 24 &= b &&\text{Finding the square root}
\end{aligned}
$$

So the length of the third side is 24 units.

Calculator Corner

As shown at the right, you can use your calculator to do the computation for the previous example.

```
40²-32²
              576
√576
               24
```

■ ■ ■
EXAMPLE

A wooden brace 1.5 meters in length is used to secure the top of a vertical post that is 0.9 meter tall. How far from the base of the post is the base of the brace?

SOLUTION

We will first make a sketch of the situation and label the brace and the post.

From the diagram, we can see that the brace, the post, and the ground form a right triangle. We are looking for the length of one leg. We substitute $a = 0.9$ and $c = 1.5$ into the formula and simplify.

$$a^2 + b^2 = c^2$$
$$(0.9)^2 + b^2 = 1.5^2$$
$$0.81 + b^2 = 2.25$$
$$b^2 = 1.44$$
$$b = 1.2$$

So the base of the brace is 1.2 meters from the base of the post. (The other theoretical answer, -1.2 meters, makes no sense.)

◢

■ ■ ■
EXAMPLE

Two children walk diagonally across a square lawn. If the perimeter of the lawn is 140 meters, about how far did they walk?

SOLUTION

First we draw and label a diagram.

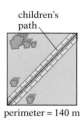

children's path

perimeter = 140 m

Note that the diagonal forms two right triangles. We want to find the length of the hypotenuse of either of these triangles. Because the perimeter of the square is 140 meters, the length of each side is

$$\frac{140}{4} = 35 \text{ meters}$$

We substitute $a = b = 35$ in the Pythagorean theorem to find the length of the diagonal.

$$a^2 + b^2 = c^2$$
$$35^2 + 35^2 = c^2$$
$$1225 + 1225 = c^2$$
$$2450 = c^2$$
$$49.50 \approx c$$

So the children walked between 49 and 50 meters.

◢

The Distance Formula

The distance between two points in the coordinate plane is the length of the line segment connecting them. Thus the distance between points A and B in the figure is the length of the line segment AB. We write $\overline{AB}$ to indicate this line segment.

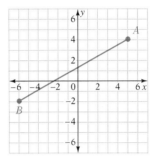

> **Distance Formula**
>
> The distance d between two points with coordinates (x_1, y_1) and (x_2, y_2) is
> $$d = \sqrt{(x_1 - x_2)^2 + (y_1 - y_2)^2}$$

It does not matter which of the coordinates we call (x_1, y_1) and which we call (x_2, y_2). We must, however, be careful about the signs of the coordinates.

▪▪▪
EXAMPLE

Find the distance between the points $(3, -2)$ and $(-1, 1)$.

SOLUTION

First we must identify the coordinates. It's a good idea to write down the identification:

$$\begin{array}{cc} (x_1, y_1) & (x_2, y_2) \\ \downarrow\downarrow & \downarrow\downarrow \\ (3, -2) & (-1, 1) \end{array}$$

Next we substitute these values in the distance formula.

$$d = \sqrt{(x_1 - x_2)^2 + (y_1 - y_2)^2}$$
$$= \sqrt{[3 - (-1)]^2 + (-2 - 1)^2}$$

Then we solve the equation by simplifying the radical.

$$d = \sqrt{(3 + 1)^2 + (-2 - 1)^2}$$
$$= \sqrt{(4)^2 + (-3)^2} = \sqrt{16 + 9} = \sqrt{25} = 5$$

So the distance from $(3, -2)$ to $(-1, 1)$ is 5 units.

◢

INSTRUCTOR NOTE

Caution students that they must simplify the entire expression before taking the square root.

Calculator Corner

You can use your graphing calculator's Home Screen to evaluate the distance formula. In the previous example, use (a, b) and (c, d) on your Home Screen to stand for (x_1, y_1) and (x_2, y_2) as follows. Then rewrite the distance formula using these new variables. (Most graphing calculators will not let you use subscripts. Check the manual of your graphing calculator.)

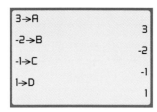 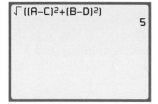

You can now **STO**re new values in a, b, c, and d and use the **RECALL** (2nd ENTER) feature of your graphing calculator to recall the distance formula and find the distance between a new pair of points.

Try finding the distance between the points $(-4, 6)$ and $(8, -12)$.

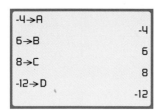 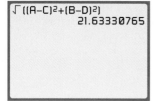

What is the length of line $\overline{AB}$ in the graph on page 487? Give the answer in terms of a radical. Use your calculator (or estimate the square root) to give an answer without the radical.

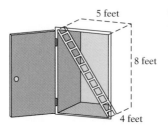

5 feet

8 feet

4 feet

PRACTICE

Solve the following word problems. Sketch a diagram where none is given.

1. I want to store a ladder in a closet that is 8 feet tall, 4 feet deep, and 5 feet wide. What is the length of the longest ladder (in the whole numbers of feet) that can be stored in this closet?   9 feet

2. You need a piece of ribbon long enough to go diagonally across a package that is 11 inches by 15 inches by 3 inches. How long must the ribbon be to go all the way around this package?   43.2 inches

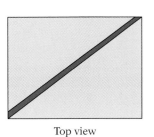

Top view

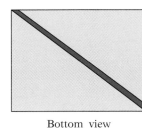

Bottom view

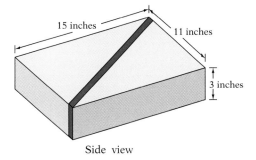

Side view

3. *Points on a Line*   Do the points A (3, −1), B (9, −3), and C (6, −2) lie on the same line? (*Hint*: They lie on the same line if the distance between the two outermost points is equal to the sum of the distances from the middle point to each of the two outer points.)   yes

4. *Height of a Tower*   A man sits on top of a tower that is 60 feet tall. He tosses a 100-foot rope to a man who walks away from the tower holding on to the rope until it is taut. How far away from the center of the base of the tower is the second man then standing?   80 feet

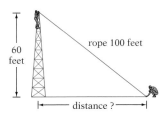

5. *Triangle Area*   What is the area of a triangle that has sides of 3 inches, 4 inches, and 5 inches? (*Hint:* Check to see if this is a right triangle. What then are the base and height?)   Yes, it is a right triangle. The area is 6 square inches.

6. *Telephone Pole*   A telephone pole is 35 feet high. If I stand 40 feet from the pole and I am 5 feet tall, how far away is the top of my head from the top of the pole?   50 feet

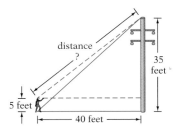

7. *Tree House* The distance from where I am sitting to the tree house my son built is 75 yards. If the tree house is 15 feet directly above the base of the tree, how far away from the tree, to the nearest yard, am I sitting?
75 yards

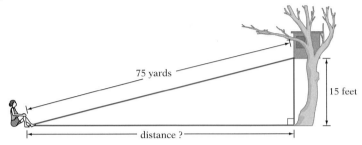

8. *Roof Height* The edge of my roof is 16 feet above the ground. The closest to the house that I can put the foot of a ladder is 12 feet. How long a ladder will I need to touch the roof?   20 feet

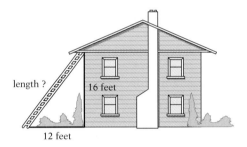

9. *Train Travel* Two trains leave a certain location at the same time. One travels east 5 miles and stops, and the other travels south. How far south must the second train travel in order to be 13 miles away from the first?
12 miles

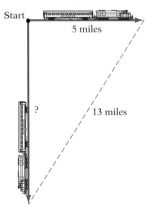

10. *Traffic Control* A "Don't Block the Box" sign at an intersection in a certain city consists of an X painted from corner to corner within a square. Each of the lines that make up the X is 60 feet long. How wide are the streets? (*Hint:* Use the distance formula, the Pythagorean theorem, or a combination of the two.)   about 42.4 feet

7.2 Operations with Radical Expressions

*SECTION LEAD-IN**

One factor that determines the color of the glaze on a piece of pottery is the temperature that the piece is fired to. The temperature may vary depending on how many pieces are on each shelf of the kiln, as well as the heat applied. To determine the temperature, potters often use cones of clay designed to melt at certain temperatures. A potter says that she fires to "cone 8" or "cone 10" when she fires her pots instead of indicating an actual temperature.

Potters use radicals when they glaze their pottery in an oven. The following formula is used to calculate the gas flow needed for heating a pottery kiln:

$$D = V \times A \times \overline{K} \times 1655\sqrt{\frac{H}{G}}$$

where

D = BTU per hour is the flow output
A = area of orifice opening in square inches
$\overline{K}$ = orifice coefficient
H = pressure in units WC (inches of water in a tube)
G = specific gravity
V = gas flow input

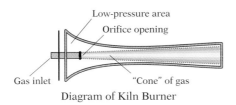

Low-pressure area
Orifice opening
Gas inlet
"Cone" of gas
Diagram of Kiln Burner

▪ How does the value for D change as A or V or H get larger?

Adding and Subtracting Radical Expressions

Only *like* radicals can be combined by adding or subtracting. Like radicals are expressions that have the same radicand and index. Here are some examples:

Like Radicals   *Unlike Radicals*
$\sqrt{5}$ and $3\sqrt{5}$ $\sqrt{5}$ and $\sqrt{3}$ Radicands differ
$\sqrt[4]{7}$ and $\sqrt[4]{7}$ $\sqrt[3]{7}$ and $\sqrt[4]{7}$ Indices differ
                    $\sqrt[5]{3}$ and $\sqrt[5]{5}$ Radicands differ

Source: Marc Ward, "The Mysterious Hole," *Ceramics Monthly* 43:5 (May 1995), pp. 50–53.

We use the distributive property to combine like radicals. For instance, we can write

$$3\sqrt[5]{7} + 6\sqrt[5]{7} = (3 + 6)\sqrt[5]{7} = 9\sqrt[5]{7}$$

but we cannot simplify

$$\sqrt{7} - 5\sqrt[3]{7} \quad \text{or} \quad \sqrt{7} + \sqrt{6}$$

▪ ▪ ▪

EXAMPLE 1

Add or subtract as indicated:

a. $\sqrt{13} + 6\sqrt{13} + 9\sqrt{13} - 7\sqrt{13}$

b. $\sqrt[4]{3xy} + 6\sqrt[4]{3xy} - 8\sqrt[4]{3xy}$

c. $\sqrt[4]{23} + 7\sqrt[3]{18} - 9\sqrt[5]{18} + \sqrt[4]{23} + 2\sqrt[5]{18}$

SOLUTION

a. Because these are all like radicals, we can apply the distributive property and combine.

$$\sqrt{13} + 6\sqrt{13} + 9\sqrt{13} - 7\sqrt{13}$$
$$= (1 + 6 + 9 - 7)\left(\sqrt{13}\right) = 9\sqrt{13}$$

b. Applying the distributive property yields

$$\sqrt[4]{3xy} + 6\sqrt[4]{3xy} - 8\sqrt[4]{3xy}$$
$$= (1 + 6 - 8)\sqrt[4]{3xy} = -1\sqrt[4]{3xy} \quad \text{or} \quad -\sqrt[4]{3xy}$$

c. We use the commutative and associative properties to rearrange the radicals so that we can combine. Then we simplify.

$$\sqrt[4]{23} + 7\sqrt[3]{18} - 9\sqrt[5]{18} + \sqrt[4]{23} + 2\sqrt[5]{18}$$

$$= \sqrt[4]{23} + \sqrt[4]{23} + 7\sqrt[3]{18} - 9\sqrt[5]{18} + 2\sqrt[5]{18} \quad \text{Rearranging}$$

$$= (1 + 1)\sqrt[4]{23} + 7\sqrt[3]{18} + (-9 + 2)\sqrt[5]{18} \quad \text{Combining}$$

$$= 2\sqrt[4]{23} + 7\sqrt[3]{18} - 7\sqrt[5]{18} \quad \text{Simplifying}$$

▶ CHECK **Warm-Up 1**

In the next example, we must simplify the radicals first so that they can be combined.

▪ ▪ ▪

EXAMPLE 2

Simplify:

a. $\sqrt{45} + \sqrt{125} - \sqrt{75}$

b. $\sqrt{64x^2} + \sqrt{36x^4} - \sqrt{72x^5}$

c. $\sqrt[3]{125x^4y^5} + \sqrt[3]{1000x^7y^5} + \sqrt[3]{8x^4y^5}$

SOLUTION

We rewrite the radicals in simplest form to see if there are any like radicals that can be combined.

a. $\sqrt{45} + \sqrt{125} - \sqrt{75}$

$\qquad = \sqrt{9}\sqrt{5} + \sqrt{25}\sqrt{5} - \sqrt{25}\sqrt{3}$ Product property

$\qquad = 3\sqrt{5} + 5\sqrt{5} - 5\sqrt{3}$ Simplifying

We can then combine the like radicals.

$$3\sqrt{5} + 5\sqrt{5} - 5\sqrt{3} = (3 + 5)\sqrt{5} - 5\sqrt{3}$$
$$= 8\sqrt{5} - 5\sqrt{3}$$

b. $\sqrt{64x^2} + \sqrt{36x^4} - \sqrt{72x^5}$

$\qquad = \sqrt{64x^2} + \sqrt{36x^4} - \sqrt{36 \cdot 2 \cdot x^4 \cdot x}$ Product property

$\qquad = 8x + 6x^2 - 6x^2\sqrt{2x}$ Simplifying

c.   We first simplify each radical.

$\sqrt[3]{125x^4y^5} + \sqrt[3]{1000x^7y^5} + \sqrt[3]{8x^4y^5}$

$\qquad = 5xy\sqrt[3]{xy^2} + 10x^2y\sqrt[3]{xy^2} + 2xy\sqrt[3]{xy^2}$ Simplifying

$\qquad = (5xy + 10x^2y + 2xy)\sqrt[3]{xy^2}$ Distributive property

$\qquad = (7xy + 10x^2y)\sqrt[3]{xy^2}$ Simplifying

▶ *CHECK* **Warm-Up 2**

▪ ▪ ▪

EXAMPLE 3

Simplify:

a. $\sqrt{28} - \left(2\sqrt{63} + \sqrt{112} + \sqrt{56}\right)$

b. $\left(2\sqrt{2x} - 7\sqrt{12y}\right) - \left(\sqrt{18x} + 4\sqrt{3y}\right)$

SOLUTION

a.   We first eliminate the parentheses.

$$\sqrt{28} - \left(2\sqrt{63} + \sqrt{112} + \sqrt{56}\right)$$
$$= \sqrt{28} - 2\sqrt{63} - \sqrt{112} - \sqrt{56}$$

Now we simplify the radicals where we can.

$$\sqrt{28} - 2\sqrt{63} - \sqrt{112} - \sqrt{56}$$
$$= 2\sqrt{7} - 6\sqrt{7} - 4\sqrt{7} - 2\sqrt{14}\quad\text{Simplifying}$$

And finally, we combine like radicals.

$$2\sqrt{7} - 6\sqrt{7} - 4\sqrt{7} - 2\sqrt{14} = -8\sqrt{7} - 2\sqrt{14}$$

b.   We simplify each radical.

$$\left(2\sqrt{2x} - 7\sqrt{12y}\right) - \left(\sqrt{18x} + 4\sqrt{3y}\right)$$
$$= \left(2\sqrt{2x} - 14\sqrt{3y}\right) - \left(3\sqrt{2x} + 4\sqrt{3y}\right)\quad\text{Simplifying}$$

Removing parentheses and applying the commutative property we have

$$2\sqrt{2x} - 3\sqrt{2x} - 14\sqrt{3y} - 4\sqrt{3y}$$
$$= -\sqrt{2x} - 18\sqrt{3y} \qquad \text{Simplifying}$$

▶ *CHECK* **Warm-Up 3**

Multiplying Radical Expressions

Radical expressions are multiplied in the same way as other expressions made up of real numbers—by using FOIL and the special-products formulas where appropriate. The results are then simplified.

▪▪▪

EXAMPLE 4

Simplify: **a.** $\left(\sqrt{2x} + \sqrt{y}\right)\left(\sqrt{3x} - \sqrt{4y}\right)$ **b.** $\left(6\sqrt{10} - 10\sqrt{2}\right)^2$

SOLUTION

a. We simplify by using FOIL.

$$\left(\sqrt{2x} + \sqrt{y}\right)\left(\sqrt{3x} - \sqrt{4y}\right)$$

$$\overset{\text{F}}{} \qquad \overset{\text{O}}{} \qquad \overset{\text{I}}{} \qquad \overset{\text{L}}{}$$

$$= \left(\sqrt{2x}\right)\left(\sqrt{3x}\right) + \left(\sqrt{2x}\right)\left(-\sqrt{4y}\right) + \left(\sqrt{y}\right)\left(\sqrt{3x}\right) + \left(\sqrt{y}\right)\left(-\sqrt{4y}\right)$$

$$= \sqrt{6x^2} - \sqrt{8xy} + \sqrt{3xy} - \sqrt{4y^2} \quad \text{Product property}$$

$$= x\sqrt{6} - 2\sqrt{2xy} + \sqrt{3xy} - 2y \qquad \text{Simplifying}$$

This cannot be simplified further.

b. We square this binomial by using the special-products formula

$$(a - b)^2 = a^2 - 2ab + b^2$$

We get

$$\left(6\sqrt{10} - 10\sqrt{2}\right)^2 = \left(6\sqrt{10}\right)^2 - 2\left(6\sqrt{10}\right)\left(10\sqrt{2}\right) + \left(10\sqrt{2}\right)^2$$

$$= 36\left(\sqrt{10}\right)^2 - 2\left(6\sqrt{10}\right)\left(10\sqrt{2}\right) + 100\left(\sqrt{2}\right)^2$$

$$= (36 \cdot 10) - \left(2 \cdot 6 \cdot 10\sqrt{10}\sqrt{2}\right) + (100 \cdot 2)$$

$$= 360 - 120\sqrt{20} + 200$$

$$= 560 - 120\left(2\sqrt{5}\right)$$

$$= 560 - 240\sqrt{5}$$

▶ *CHECK* **Warm-Up 4**

In the next example, we combine several operations.

▪▪▪

EXAMPLE 5

Simplify: $\left(\sqrt{6x} + \sqrt{3y}\right)^2 - 2\sqrt{50xy}$

SOLUTION

We simplify the squared expression by applying the special-products formula.

$$\left(\sqrt{6x} + \sqrt{3y}\right)^2 = \left(\sqrt{6x}\right)^2 + 2\left(\sqrt{6x}\right)\left(\sqrt{3y}\right) + \left(\sqrt{3y}\right)^2$$

$$= 6x + 2\sqrt{18xy} + 3y$$

$$= 6x + 2\left(3\sqrt{2xy}\right) + 3y$$

$$= 6x + 6\sqrt{2xy} + 3y$$

Next we substitute this expression in the original problem.

$$\left(\sqrt{6x} + \sqrt{3y}\right)^2 - 2\sqrt{50xy}$$

$$= 6x + 6\sqrt{2xy} + 3y - 2\left(5\sqrt{2xy}\right)$$

$$= 6x + 6\sqrt{2xy} + 3y - 10\sqrt{2xy}$$

$$= 6x - 4\sqrt{2xy} + 3y \qquad \text{Combining like radicals}$$

▶ *CHECK* **Warm-Up 5**

Dividing Rational Expressions

To divide one radical expression by another, we write the division as a fraction and then simplify.

▪▪▪
EXAMPLE 6

Divide: $\left(\sqrt{56} + \sqrt{84}\right) \div \sqrt{7}$

SOLUTION

We first write the division as

$$\frac{\sqrt{56} + \sqrt{84}}{\sqrt{7}}$$

We now rewrite each term of the dividend over the divisor and then simplify. We get

$$\frac{\sqrt{56} + \sqrt{84}}{\sqrt{7}} = \frac{\sqrt{56}}{\sqrt{7}} + \frac{\sqrt{84}}{\sqrt{7}}$$

$$= \sqrt{\frac{56}{7}} + \sqrt{\frac{84}{7}} \qquad \text{Quotient property}$$

$$= \sqrt{8} + \sqrt{12} \qquad \text{Reducing}$$

$$= 2\sqrt{2} + 2\sqrt{3} \qquad \text{Simplifying}$$

▶ *CHECK* **Warm-Up 6**

▪▪▪
EXAMPLE 7

Simplify: $\dfrac{\sqrt[3]{21x^2} + \sqrt[3]{27y^5}}{\sqrt[3]{3x^2}}$

SOLUTION

We rewrite the division with each term of the dividend over the divisor and simplify.

$$\frac{\sqrt[3]{21x^2} + \sqrt[3]{27y^5}}{\sqrt[3]{3x^2}} = \frac{\sqrt[3]{21x^2}}{\sqrt[3]{3x^2}} + \frac{\sqrt[3]{27y^5}}{\sqrt[3]{3x^2}} = \sqrt[3]{\frac{21x^2}{3x^2}} + \sqrt[3]{\frac{27y^5}{3x^2}}$$

$$= \sqrt[3]{7} + y\sqrt[3]{\frac{9y^2}{x^2}}$$

▶ CHECK **Warm-Up 7**

The quotient in Example 7 cannot be simplified further with the properties we have discussed so far, but it is not in simplest terms because one denominator contains a radical. Next, we solve this dilemma with a "new" method.

Rationalizing the Denominator

An expression that has a radical in the denominator is not in simplest terms. Here we discuss a method for finding an equivalent expression without the denominator radical. This method, which is called **rationalizing the denominator,** works for any fraction. Briefly, it involves multiplying numerator and denominator by the same expression.

Denominator with a Monomial Square Root

> **To rationalize a denominator that contains a monomial square root**
>
> For a fraction written in the form $\dfrac{a}{\sqrt{b}}$, we multiply numerator and denominator by $\sqrt{b}$ to obtain the equivalent fraction $\dfrac{a \cdot \sqrt{b}}{\sqrt{b} \cdot \sqrt{b}}$. Then we simplify the result.

▪ ▪ ▪

EXAMPLE 8

Simplify: $\dfrac{2\sqrt{12}}{\sqrt{5}}$

SOLUTION

First we simplify the fraction whenever possible. Here, we have

$$\frac{2\sqrt{12}}{\sqrt{5}} = \frac{2(2\sqrt{3})}{\sqrt{5}} = \frac{4\sqrt{3}}{\sqrt{5}}$$

Now, to rationalize the denominator, we multiply numerator and denominator by $\sqrt{5}$.

$$\frac{4\sqrt{3} \cdot \sqrt{5}}{\sqrt{5} \cdot \sqrt{5}} = \frac{4\sqrt{15}}{(\sqrt{5})^2} = \frac{4\sqrt{15}}{5}$$

▶ CHECK **Warm-Up 8**

▪▪▪
WRITER'S BLOCK

What do we do when we <u>rationalize</u> the denominator of a fraction?

Denominators with Higher Roots

To rationalize a denominator of the form $a\sqrt[n]{b}$, multiply the numerator and the denominator by a radical that will change the denominator to the form $\sqrt[n]{b^n}$. Then simplify the resulting fraction.

■■■

EXAMPLE 9

Simplify: $\dfrac{3}{2\sqrt[3]{5}}$

SOLUTION

The denominator radical is a cube root, so here $n = 3$. To rationalize the denominator, we multiply numerator and denominator by $\sqrt[3]{5^2}$ to obtain

$$\frac{3}{2\sqrt[3]{5}} = \frac{3 \cdot \sqrt[3]{5^2}}{2\sqrt[3]{5} \cdot \sqrt[3]{5^2}} = \frac{3\sqrt[3]{25}}{2\sqrt[3]{5^3}} = \frac{3\sqrt[3]{25}}{2 \cdot 5} = \frac{3\sqrt[3]{25}}{10}$$

▶ CHECK **Warm-Up 9**

The division problem we couldn't complete in Example 7 can be completed now.

$$\sqrt[3]{7} + y\sqrt[3]{\frac{9y^2}{x^2}} = \sqrt[3]{7} + y\,\frac{\sqrt[3]{9y^2}}{\sqrt[3]{x^2}} \cdot \frac{\sqrt[3]{x}}{\sqrt[3]{x}}$$

$$= \sqrt[3]{7} + \frac{y\sqrt[3]{9xy^2}}{x}$$

Now this result is in simplest terms.

Binomial Denominators with Square Roots

The binomial expressions $a + b$ and $a - b$ are called **conjugates** of each other; they differ only in the sign connecting the two terms. If we multiply conjugates together, we get the special product

$$(a + b)(a - b) = a^2 - b^2$$

Thus, if a or b contains a square root, it is eliminated in the product.

$$\left(\sqrt{x} + \sqrt{y}\right)\left(\sqrt{x} - \sqrt{y}\right) = \left(\sqrt{x}\right)^2 - \left(\sqrt{y}\right)^2 = x - y$$

We use this property to rationalize a binomial denominator that contains square roots.

To rationalize a binomial denominator that contains square roots

For a fraction $\dfrac{a}{b + c}$, where b or c or both contain square roots, multiply numerator and denominator by the conjugate $b - c$. Then simplify the resulting fraction.

▪▪▪

EXAMPLE 10

Simplify: $\dfrac{\sqrt{3}}{\sqrt{3} + \sqrt{10}}$

SOLUTION

To simplify this fraction, we need to rationalize the denominator. We multiply numerator and denominator by the conjugate $\sqrt{3} - \sqrt{10}$ and simplify.

$$\frac{\sqrt{3}}{\sqrt{3} + \sqrt{10}} = \frac{\sqrt{3}\left(\sqrt{3} - \sqrt{10}\right)}{\left(\sqrt{3} + \sqrt{10}\right)\left(\sqrt{3} - \sqrt{10}\right)} \quad \text{Multiplying by conjugate}$$

$$= \frac{\sqrt{3} \cdot \sqrt{3} - \sqrt{3} \cdot \sqrt{10}}{\left(\sqrt{3}\right)^2 - \left(\sqrt{10}\right)^2} \quad \text{Multiplying}$$

$$= \frac{3 - \sqrt{30}}{3 - 10} \quad \text{Simplifying}$$

$$= \frac{3 - \sqrt{30}}{-7} = \frac{\sqrt{30} - 3}{7} \quad \text{Simplifying}$$

▶ CHECK **Warm-Up 10**

Operations with Rational Exponents

Our definitions of rational exponents enable us to use the rules of exponents. Here are the major rules, with examples that include rational exponents.

Product rule: $a^m \cdot a^n = a^{m+n}$

$$3^{\frac{3}{5}} \cdot 3^{\frac{1}{5}} = 3^{\left(\frac{3}{5} + \frac{1}{5}\right)} = 3^{\frac{4}{5}} = \sqrt[5]{3^4}$$

Quotient rule: $a^m \div a^n = a^{m-n}$

$$3^{\frac{3}{5}} \div 3^{\frac{2}{5}} = 3^{\left(\frac{3}{5} - \frac{2}{5}\right)} = 3^{\frac{1}{5}} = \sqrt[5]{3}$$

Power rule: $(a^m)^n = a^{mn}$

$$\left(3^{\frac{4}{5}}\right)^{\frac{1}{2}} = 3^{\left(\frac{4}{5}\right)\left(\frac{1}{2}\right)} = 3^{\frac{2}{5}} = \sqrt[5]{3^2}$$

Product power rule: $(a \cdot b)^m = a^m \cdot b^m$

$$(3^2 y)^{\frac{1}{2}} = (3^2)^{\frac{1}{2}} \cdot y^{\frac{1}{2}} = 3^1 \cdot y^{\frac{1}{2}} = 3\sqrt{y}$$

Quotient power rule: $\left(\dfrac{a}{b}\right)^m = \dfrac{a^m}{b^m}$

$$\left(\frac{3x}{8}\right)^{\frac{1}{3}} = \frac{(3x)^{\frac{1}{3}}}{8^{\frac{1}{3}}} = \frac{\sqrt[3]{3x}}{\sqrt[3]{8}} = \frac{1}{2}\sqrt[3]{3x}$$

Negative exponent rule: $a^{-m} = \dfrac{1}{a^m}$

$$(x + 2)^{-\frac{3}{2}} = \frac{1}{[(x + 2)^{\frac{3}{2}}]}$$

As with all statements of equality, the *reverse* of each statement is also true.

In order to multiply or divide radicals, they *must* have the same index; that is why we must get a common denominator (i.e., a common index) for our rational exponents. Otherwise, they can't be combined!

Be careful not to confuse $a^{\frac{1}{n}}$ and $\frac{1}{a^n}$. They are *not* equivalent, and they are *not* reciprocals.

▪ ▪ ▪

EXAMPLE 11

Use the rules of rational exponents to simplify as much as possible.

$$\sqrt[4]{8^2 9^3}$$

SOLUTION

$$\sqrt[4]{8^2 9^3} = \sqrt[4]{(2^3)^2 (3^2)^3}$$

$$= \sqrt[4]{2^6 3^6}$$

$$= 2^{\frac{6}{4}} 3^{\frac{6}{4}}$$

$$= 2^{\frac{3}{2}} 3^{\frac{3}{2}}$$

$$= 2^{1\frac{1}{2}} 3^{1\frac{1}{2}} \qquad \text{Rewriting exponents}$$

$$= 2^1 \cdot 2^{\frac{1}{2}} \cdot 3^1 \cdot 3^{\frac{1}{2}} \qquad \text{Product rule}$$

$$= 2 \cdot 3 \cdot 2^{\frac{1}{2}} \cdot 3^{\frac{1}{2}}$$

$$= 6 \cdot (2 \cdot 3)^{\frac{1}{2}}$$

$$= 6\sqrt{2 \cdot 3} \qquad \text{Rewriting as radical}$$

$$= 6\sqrt{6}$$

▶ *CHECK* **Warm-Up 11**

▪ ▪ ▪

EXAMPLE 12

Simplify: $\dfrac{\sqrt[3]{6^2} \cdot \sqrt[4]{6^3}}{\sqrt[12]{6^5}}$

SOLUTION

We first rewrite each with rational exponents. Then, we use the product and quotient rules.

$$6^{\frac{2}{3}} \cdot 6^{\frac{3}{4}} \div 6^{\frac{5}{12}} \qquad \text{Rewriting radical expressions}$$

$$= 6^{\left(\frac{2}{3} + \frac{3}{4}\right)} \div 6^{\frac{5}{12}} \qquad \text{Multiplying}$$

$$= 6^{\left(\frac{8}{12} + \frac{9}{12}\right)} \div 6^{\frac{5}{12}}$$

$$= 6^{\frac{17}{12}} \div 6^{\frac{5}{12}} \qquad \text{Simplifying}$$

$$= 6^{\frac{17}{12} - \frac{5}{12}} \qquad \text{Dividing}$$

$$= 6^{\frac{12}{12}} = 6$$

Check this computation using your calculator.

▶ *CHECK* **Warm-Up 12**

Practice what you learned.

SECTION FOLLOW-UP

The value for D gets larger as each of the variables A, V, or H gets larger. The flow of gas in BTUs must be carefully controlled, however, because the temperature of the kiln and how fast the kiln reaches the maximum desired temperature affects the color of the glazes.

7.2 WARM-UPS

Work these problems before you attempt the exercises.

1. Combine: $\sqrt{5} + 3\sqrt{5} - 2\sqrt{5} - 8\sqrt{5} + 14\sqrt{5}$ $8\sqrt{5}$

2. Add: $\sqrt[3]{16x^3y^4} + 2xy\sqrt[3]{2y} + \frac{1}{5}x\sqrt{200y^3}$ $4xy\sqrt[3]{2y} + 2xy\sqrt{2y}$

3. Simplify: $3\sqrt{48x} - 2\sqrt{24y} - \left(\sqrt{108x} - \sqrt{75y} + 2\sqrt{27x}\right)$
   $-4\sqrt{6y} + 5\sqrt{3y}$

4. Simplify: $\left(3\sqrt{10} - 2\sqrt{3}\right)^2$ $102 - 12\sqrt{30}$

5. Simplify: $\left(3\sqrt{2} + 4\right)^2 - \left(4\sqrt{2} - 1\right)^2$ $32\sqrt{2} + 1$

6. Divide: $\dfrac{\sqrt{48} - \sqrt{24}}{\sqrt{8}}$ $\sqrt{6} - \sqrt{3}$

7. Why are we not able to simplify $\dfrac{\sqrt[3]{9x} - \sqrt[3]{24x}}{\sqrt{14x}}$?   We cannot simplify through division because the radicals do not have the same index.

8. Simplify: $\dfrac{2\sqrt{3}}{3\sqrt{18}}$ $\dfrac{\sqrt{6}}{9}$ 9. Simplify: $\dfrac{6x}{\sqrt[5]{(6x)^3}}$ $\sqrt[5]{(6x)^2}$

10. Simplify: $\dfrac{3\sqrt{5} - 5\sqrt{15}}{\sqrt{5} + \sqrt{6}}$ 11. Simplify: $\sqrt[4]{2^6 \cdot 3^8}$ $18\sqrt{2}$

12. Find $4^{\frac{3}{5}} \cdot 16^{\frac{2}{3}} \cdot 2^{\frac{2}{15}}$ 16 10. $-15 + 3\sqrt{30} + 25\sqrt{3} - 15\sqrt{10}$

7.2 EXERCISES

Note: Use your graphing calculator to check your results whenever possible.

In Exercises 1 through 12, add or subtract as indicated.

1. $\sqrt{2} - 3\sqrt{3} + 2\sqrt{2} + 6\sqrt{3}$ $3\sqrt{2} + 3\sqrt{3}$

2. $5\sqrt{6} - 2\sqrt{5} + 7\sqrt{5} - 8\sqrt{6}$ $-3\sqrt{6} + 5\sqrt{5}$

3. $10\sqrt{3} - 2\sqrt{7} + 8\sqrt{3} + 5\sqrt{3} - 6\sqrt{7}$ $23\sqrt{3} - 8\sqrt{7}$

4. $7\sqrt{5} - 3\sqrt{7} - \left(6\sqrt{7} + 2\sqrt{5}\right)$ $5\sqrt{5} - 9\sqrt{7}$

5. $6\sqrt{13} - 6\sqrt{11} - \left(8\sqrt{13} + 2\sqrt{11} - 3\sqrt{13}\right)$
   $\sqrt{13} - 8\sqrt{11}$

6. $8\sqrt{5} + 6\sqrt{11} - \left(2\sqrt{11} + 2\sqrt{5} + 10\sqrt{5}\right)$
   $-4\sqrt{5} + 4\sqrt{11}$

7. $\sqrt{2} - \sqrt[3]{2} + 2\sqrt{2} - \left(\sqrt[3]{2} - 5\sqrt{2}\right)$
$8\sqrt{2} - 2\sqrt[3]{2}$

8. $3\sqrt[3]{3} + 2\sqrt{5} - \left(6\sqrt[3]{3} - 12\sqrt{5} + 3\sqrt{3}\right)$
$-3\sqrt[3]{3} + 14\sqrt{5} - 3\sqrt{3}$

9. $6\sqrt[4]{8} - 2\sqrt{2} + 3\sqrt[3]{4} - \left(\sqrt{2} - \sqrt[3]{4} + \sqrt[4]{8}\right)$
$5\sqrt[4]{8} - 3\sqrt{2} + 4\sqrt[3]{4}$

10. $10\sqrt[3]{3} - 2\sqrt{3} - \left(5\sqrt{3} + 6\sqrt[3]{3} + 6\sqrt{3}\right)$
$4\sqrt[3]{3} - 13\sqrt{3}$

11. $8\sqrt{7} + 8\sqrt[3]{7} - \left(6\sqrt[3]{7} + 21\sqrt[3]{7} - 6\sqrt{7}\right)$
$14\sqrt{7} - 19\sqrt[3]{7}$

12. $8\sqrt{5} - 16\sqrt[3]{5} - \left(18\sqrt{5} - 18\sqrt[3]{5} + 2\sqrt{5}\right)$
$-12\sqrt{5} + 2\sqrt[3]{5}$

In Exercises 13 through 24, simplify each radical and then add or subtract as indicated.

13. $2\sqrt{45} + \sqrt{20} - \sqrt{80}$ $4\sqrt{5}$

14. $4\sqrt{24} + \sqrt{54} - 10\sqrt{6}$ $\sqrt{6}$

15. $\sqrt{75} + \sqrt{108} - \sqrt{3}$ $10\sqrt{3}$

16. $-6\sqrt{75} + 4\sqrt{125} - 2\sqrt{5}$ $-30\sqrt{3} + 18\sqrt{5}$

17. $5\sqrt{8} - 3\sqrt{72} + 2\sqrt{50}$ $2\sqrt{2}$

18. $2\sqrt{5} - 3\sqrt{20} - 4\sqrt{45}$ $-16\sqrt{5}$

19. $4\sqrt{32} - \sqrt{18} + 2\sqrt{128}$ $29\sqrt{2}$

20. $\sqrt{8} - \sqrt{12} + 5\sqrt{2}$ $7\sqrt{2} - 2\sqrt{3}$

21. $3\sqrt{27x^2} - 2x\sqrt{108} - \sqrt{48x^2}$ $-7x\sqrt{3}$

22. $3\sqrt{50x^2} - 8x\sqrt{18} - 3\sqrt{72x^2}$ $-27x\sqrt{2}$

23. $2\sqrt{5x} + 4\sqrt{45x} - 3\sqrt{20x}$ $8\sqrt{5x}$

24. $3\sqrt{45x^3} + x\sqrt{5x} - x\sqrt{125x}$ $5x\sqrt{5x}$

In Exercises 25 through 32, multiply as indicated and then simplify.

25. $5\sqrt{3}\left(5\sqrt{6} + 2\sqrt{15}\right)$ $75\sqrt{2} + 30\sqrt{5}$

26. $6\sqrt{5}\left(6\sqrt{20} + 5\sqrt{30}\right)$ $360 + 150\sqrt{6}$

27. $\sqrt{2x}\left(3y\sqrt{12xy} - 4\sqrt{48xy^3}\right)$ $-10xy\sqrt{6y}$

28. $4\sqrt{3}\left(2\sqrt{6} - 3\sqrt{12}\right)$ $24\sqrt{2} - 72$

29. $2\sqrt{3}\left(7\sqrt{15} - 3\sqrt{6}\right)$ $42\sqrt{5} - 18\sqrt{2}$

30. $5\sqrt{6}\left(7\sqrt{8} - 3\sqrt{12}\right)$ $140\sqrt{3} - 90\sqrt{2}$

31. $3\sqrt{2}\left(8\sqrt{10} - 2\sqrt{20}\right)$ $48\sqrt{5} - 12\sqrt{10}$

32. $\sqrt{3x}\left(2y\sqrt{12xy} + 2\sqrt{12xy^3}\right)$ $24xy\sqrt{y}$

In Exercises 33 through 40, multiply and then simplify.

33. $\left(\sqrt{3} + 5\right)\left(\sqrt{3} - 4\right)$ $-17 + \sqrt{3}$

34. $\left(\sqrt{3} - 1\right)\left(\sqrt{3} + 6\right)$ $-3 + 5\sqrt{3}$

35. $\left(\sqrt{3} + \sqrt{2}\right)\left(\sqrt{3} - \sqrt{2}\right)$ 1

36. $\left(\sqrt{6} - \sqrt{5}\right)\left(\sqrt{6} + \sqrt{5}\right)$ 1

37. $\left(2\sqrt{3} - \sqrt{2}\right)\left(2\sqrt{3} + \sqrt{2}\right)$ 10

38. $\left(4\sqrt{5} + \sqrt{3}\right)\left(4\sqrt{5} - \sqrt{3}\right)$ 77

39. $\left(6\sqrt{3} - 2\right)^2$ $112 - 24\sqrt{3}$

40. $\left(2\sqrt{3} - 3\sqrt{2}\right)^2$ $30 - 12\sqrt{6}$

In Exercises 41 through 48, divide and then simplify.

41. $\dfrac{6\sqrt{7} - 2\sqrt{14}}{\sqrt{7}}$ $6 - 2\sqrt{2}$

42. $\dfrac{2\sqrt{18} - 3\sqrt{12}}{\sqrt{6}}$ $2\sqrt{3} - 3\sqrt{2}$

43. $\dfrac{-4\sqrt{90} + 3\sqrt{40} + 2\sqrt{10}}{\sqrt{10}}$ -4

44. $\dfrac{8\sqrt{45} + 7\sqrt{20} - 3\sqrt{5}}{\sqrt{5}}$ 35

45. $\dfrac{4\sqrt{32} - \sqrt{18} + 2\sqrt{128}}{\sqrt{2}}$ 29

46. $\dfrac{6\sqrt{32} + 2\sqrt{8} - 7\sqrt{128}}{\sqrt{8}}$ -14

47. $\dfrac{20\sqrt{x^3y^2} - 10\sqrt{x^5y^6} + 15\sqrt{x^9y^6}}{5x^2y^3\sqrt{x}}$ $\dfrac{4}{xy^2} - 2 + 3x^2$

48. $\dfrac{6\sqrt{x^3y^3z} - 16xy\sqrt{xyz^2} + 18x\sqrt{xyz}}{6\sqrt{xyz}}$ $xy - \dfrac{8xy\sqrt{z}}{3} + 3x$

In Exercises 49 through 56, simplify each fraction.

49. $\dfrac{1}{\sqrt{2}}$ $\quad \dfrac{\sqrt{2}}{2}$

50. $\dfrac{1}{\sqrt{3}}$ $\quad \dfrac{\sqrt{3}}{3}$

51. $\dfrac{3}{\sqrt{5}}$ $\quad \dfrac{3\sqrt{5}}{5}$

52. $\dfrac{2}{\sqrt{7}}$ $\quad \dfrac{2\sqrt{7}}{7}$

53. $\dfrac{1}{2\sqrt{3}}$ $\quad \dfrac{\sqrt{3}}{6}$

54. $\dfrac{1}{3\sqrt{5}}$ $\quad \dfrac{\sqrt{5}}{15}$

55. $\dfrac{\sqrt{14t}}{t\sqrt{3t}}$ $\quad \dfrac{\sqrt{42}}{3t}$

56. $\dfrac{3\sqrt{5x}}{x\sqrt{2x}}$ $\quad \dfrac{3\sqrt{10}}{2x}$

In Exercises 57 through 72, simplify the fraction.

57. $\dfrac{\sqrt{26}+\sqrt{12}}{\sqrt{6}}$ $\quad \dfrac{\sqrt{39}}{3}+\sqrt{2}$

58. $\dfrac{\sqrt{24}+\sqrt{10}}{\sqrt{14}}$ $\quad \dfrac{2\sqrt{21}}{7}+\dfrac{\sqrt{35}}{7}$

59. $\dfrac{\sqrt{42x}-\sqrt{33x}}{\sqrt{6x}}$ $\quad \sqrt{7}-\dfrac{\sqrt{22}}{2}$

60. $\dfrac{\sqrt{50x}-\sqrt{40x}}{\sqrt{15x}}$ $\quad \dfrac{\sqrt{30}}{3}-\dfrac{2\sqrt{6}}{3}$

61. $\dfrac{3\sqrt{15}-2\sqrt{18}}{\sqrt{45}}$ $\quad \sqrt{3}-\dfrac{2\sqrt{10}}{5}$

62. $\dfrac{8\sqrt{14}-6\sqrt{28}}{\sqrt{35}}$ $\quad \dfrac{8\sqrt{10}}{5}-\dfrac{12\sqrt{5}}{5}$

63. $\dfrac{6\sqrt{20x}+3\sqrt{35x}}{\sqrt{28x}}$ $\quad \dfrac{6\sqrt{35}}{7}+\dfrac{3\sqrt{5}}{2}$

64. $\dfrac{7\sqrt{18x}+7\sqrt{27x}}{\sqrt{54x}}$ $\quad \dfrac{7\sqrt{3}}{3}+\dfrac{7\sqrt{2}}{2}$

65. $\dfrac{\sqrt{3}}{\sqrt{6}-3}$ $\quad -\sqrt{2}-\sqrt{3}$

66. $\dfrac{\sqrt{5}}{4-\sqrt{15}}$ $\quad 4\sqrt{5}+5\sqrt{3}$

67. $\dfrac{\sqrt{5}}{\sqrt{10}+\sqrt{30}}$ $\quad \dfrac{\sqrt{6}}{4}-\dfrac{\sqrt{2}}{4}$

68. $\dfrac{\sqrt{3}}{\sqrt{15}-\sqrt{21}}$ $\quad -\dfrac{\sqrt{5}}{2}-\dfrac{\sqrt{7}}{2}$

69. $\dfrac{2\sqrt{15}}{\sqrt{20}+\sqrt{30}}$ $\quad 3\sqrt{2}-2\sqrt{3}$

70. $\dfrac{6\sqrt{14}}{\sqrt{21}+\sqrt{28}}$ $\quad 12\sqrt{2}-6\sqrt{6}$

71. $\dfrac{3\sqrt{10}}{2\sqrt{30}-5\sqrt{2}}$ $\quad \dfrac{6\sqrt{3}}{7}+\dfrac{3\sqrt{5}}{7}$

72. $\dfrac{6\sqrt{35}}{2\sqrt{14}-3\sqrt{21}}$ $\quad -\dfrac{12\sqrt{10}}{19}-\dfrac{18\sqrt{15}}{19}$

In Exercises 73 through 76, rewrite each result without a radical.

73. $\dfrac{4\sqrt{169}}{3\sqrt{81}-2\sqrt{225}}$ $\quad -17\tfrac{1}{3}$

74. $\dfrac{7\sqrt{625}}{3\sqrt{121}+5\sqrt{100}}$ $\quad 2\tfrac{9}{83}$

75. $\dfrac{8\sqrt{289}}{9\sqrt{144}-12\sqrt{8100}}$ $\quad -\dfrac{34}{243}$

76. $\dfrac{19\sqrt{100}}{23\sqrt{36}+12\sqrt{441}}$ $\quad \dfrac{19}{39}$

In Exercises 77 through 82, simplify each fraction.

77. $\dfrac{\sqrt{7}-\sqrt{21}}{\sqrt{7}+\sqrt{14}}$ $\quad -1+\sqrt{2}+\sqrt{3}-\sqrt{6}$

78. $\dfrac{\sqrt{10}-\sqrt{5}}{2\sqrt{5}-\sqrt{10}}$ $\quad \tfrac{1}{2}\sqrt{2}$

79. $\dfrac{\sqrt{12}-\sqrt{24}}{2\sqrt{6}-\sqrt{12}}$ $\quad -1$

80. $\dfrac{\sqrt{5y}+\sqrt{15y}}{\sqrt{5y}-\sqrt{15y}}$ $\quad -2-\sqrt{3}$

81. $\dfrac{\sqrt{10}+\sqrt{5}}{\sqrt{10}-\sqrt{5}}$ $\quad 3+2\sqrt{2}$

82. $\dfrac{\sqrt{8}-\sqrt{18}}{3\sqrt{2}-\sqrt{8}}$ $\quad -1$

In Exercises 83 through 90, rationalize each denominator.

83. $\dfrac{\sqrt[3]{4}}{\sqrt[3]{2}}$ $\quad \sqrt[3]{2}$

84. $\dfrac{\sqrt[3]{6}}{\sqrt[3]{3}}$ $\quad \sqrt[3]{2}$

85. $\dfrac{\sqrt[3]{5}}{\sqrt[3]{25}}$ $\quad \dfrac{\sqrt[3]{25}}{5}$

86. $\dfrac{\sqrt[3]{7}}{\sqrt[3]{49}}$ $\quad \dfrac{\sqrt[3]{49}}{7}$

87. $\dfrac{\sqrt[4]{9}}{\sqrt[4]{6}}$ $\quad \dfrac{\sqrt[4]{24}}{2}$

88. $\dfrac{\sqrt[4]{2}}{\sqrt[4]{8}}$ $\quad \dfrac{\sqrt[4]{4}}{2}$

89. $\dfrac{2\sqrt[3]{2x}}{\sqrt[3]{x}}$ $\quad 2\sqrt[3]{2}$

90. $\dfrac{4\sqrt[3]{2x}}{\sqrt[3]{x}}$ $\quad 4\sqrt[3]{2}$

MIXED PRACTICE

By doing these exercises, you will practice the topics up to this point in the chapter. Rationalize all denominators.

91. Simplify: $\sqrt{250x^2y^3}$ $\quad 5xy\sqrt{10y}$

92. Find the product: $\left(y+\sqrt{2x}\right)\left(\sqrt{2x}-2y\right)$
$\quad -y\sqrt{2x}-2y^2+2x$

93. Simplify: $\dfrac{2}{\sqrt{2}}$ $\quad \sqrt{2}$

94. Simplify: $\sqrt{121x^2y^2}$ $\quad 11xy$

95. Combine and simplify: $2\sqrt[3]{x^7y^7}-3x\sqrt[3]{x^4y^7}$ $\quad -x^2y^2\sqrt[3]{xy}$

96. Simplify: $\dfrac{2}{\sqrt{2}+\sqrt{3}}$ $\quad 2\sqrt{3}-2\sqrt{2}$

97. Simplify: $\sqrt[3]{-8x^3y^3}$ $\quad -2xy$

98. Simplify: $\sqrt{\dfrac{1020x^3y^2}{200xy}}$ $\quad \dfrac{x}{10}\sqrt{510y}$

99. Simplify: $\left(\sqrt{3}-2\right)^2$ $\quad 7-4\sqrt{3}$

100. Simplify: $\dfrac{\sqrt{7}+1}{\sqrt{7}-1}$ $\quad \dfrac{4+\sqrt{7}}{3}$

■ ■

EXCURSIONS

Posing Problems

1. Using any of the following data, ask and answer four questions. Compare with those of other students.

Basketball (National Collegiate A.A. Men's Rules)

Playing court: College: 94 feet long by 50 feet wide (ideal dimensions). High School: 84 feet long by 50 feet wide (ideal inside dimensions).

Baskets: Rings 18 inches in inside diameter, with white cord 12-mesh nets, 15 to 18 inches in length. Each ring is made of metal, is not more than 5/8 of an inch in diameter, and is bright orange in color.

Height of basket: 10 feet (upper edge).

Weight of ball: Not less than 20 ounces nor more than 22.

Circumference of ball: Not greater than 30 inches and not less than 29-1/2.

Free-throw line: 15 feet from the face of the backboard, 2 inches wide.

Three-point field goal line: 19 feet, 9 inches from the center of the basket. In the National Basketball Association, the distance is 22 feet.

Baseball

Home plate to pitcher's box: 60 feet 6 inches.

Plate to second base: 127 feet 3-3/8 inches.

Distance from base to base (home plate included): 90 feet.

Size of bases: 15 inches by 15 inches.

Pitcher's plate: 24 inches by 6 inches.

Batter's box: 4 feet by 6 feet.

Home plate: Five-sided, 17 inches by 8-1/2 inches by 8-1/2 inches by 12 inches by 12 inches, cut to a point at rear.

Home plate to backstop: Not less than 60 feet (recommended).

Weight of ball: Not less than 5 ounces nor more than 5-1/4 ounces.

Circumference of ball: Not less than 9 inches nor more than 9-1/4 inches.

Bat: Must be one piece of solid wood, round, not over 2-3/4 inches in diameter at thickest part, nor more than 42 inches in length.

Tennis

Size of court: 120 feet long by 60 feet wide, with rectangle marked off at 78 feet long by 27 feet wide (singles) and 78 feet long by 36 feet wide (doubles).

Height of net: 3 feet in center, gradually rising to reach 3-foot 6-inch posts at a point 3 feet outside each side of court.

Ball: Shall be more than 2-1/2 inches and less than 2-5/8 inches in diameter and weigh more than 2 ounces and less than 2-1/16 ounces.

Service line: 21 feet from net.

Bowling

Lane dimensions: Overall length 62 feet 10-3/16 inches, measuring from foul line to pit (not including tail plank), with + or - 1/2 inch tolerance permitted. Foul line to center of No. 1 pinspot 60 feet, with + or - 1/2 tolerance permitted. Lane width, 41-1/2 inches with a tolerance of + or - 1/2 inch permitted. Approach, not less than 15 feet. Gutters, 9-5/16 inches wide with plus 3/16 inch or minus 5/16 inch tolerances permitted.

Ball: Circumference, not more than 27.002 inches. Weight, 16 pounds maximum.

Source: 1996 Information Please® Almanac (©1995 Houghton Mifflin Co.), p. 1004. All rights reserved. Used with permission of Information Please LLC.

Data Analysis

2. Find out what you can about these locations by comparing information about their temperatures—highs, lows, and range. Ask and answer four questions using these data. Are the temperatures given in Fahrenheit or Celsius measurement? Justify your answer.

Temperatures of North America

Location	December—March (high/low)	June—August (high/low)	Location	December—March (high/low)	June—August (high/low)
Acapulco	87/70	89/75	Mexico City	72/43	75/53
Albuquerque	72/40	91/62	Miami	76/59	88/75
Austin	63/42	93/72	Montréal	24/10	72/54
Bermuda	68/58	84/73	Nassau	77/67	88/76
Boston	40/22	80/58	New Orleans	65/48	90/76
Cancún	87/70	89/75	New York	41/27	80/65
Chicago	34/18	82/64	Palm Beach	79/43	95/73
Dallas	58/37	92/72	Philadelphia	42/29	83/64
Denver	43/17	85/57	Port-au-Prince	86/68	70/73
Dominican Rep.	85/69	88/72	St. Thomas	85/72	89/76
Honolulu	76/68	84/72	San Juan	82/72	87/76
Jackson Hole, WY	36/11	80/52	Tucson	65/39	97/71
Lake Tahoe	50/16	89/40	Vancouver	44/36	67/53
Las Vegas	65/34	103/71	Washington, D.C.	45/29	85/64
Los Angeles	66/47	76/58			

Source: National Weather Service, as cited in the *1996 Information Please® Business Almanac* (©1995 Houghton Mifflin Co.), p. 414. All rights reserved. Used with permission by Information Please LLC.

CONNECTIONS TO *STATISTICS*

Finding the Variance and Standard Deviation

In an earlier statistics Connection, we discussed one measure of variation, or spread—the range. Recall that the range is the difference between the upper and lower limits of the data. While this is important, it does have one major disadvantage. It does not describe the variation among the variables. For instance, both of these sets of data have the same range, yet their values are definitely different.

90, 90, 90, 98, 90   Range = 8
1, 6, 8, 1, 9, 5   Range = 8

To better describe the variation, we will introduce two other measures of variation—*variance* and *standard deviation* (the variance is the square of the standard deviation). These measures tell us how much the actual values differ from the mean. The larger the standard deviation, the more spread out the values. The smaller the standard deviation, the less spread out the values. This measure is particularly helpful to teachers as they try to find whether their students' scores on a certain test are closely related to the class average.

To find the standard deviation of a set of values:

a. Find the mean of the data.
b. Find the difference (deviation) between each of the scores and the mean.
c. Square each deviation.
d. Sum the squares.
e. Dividing by one less than the number of values, find the "mean" of this sum (the **variance***).
f. Find the square root of the variance (the **standard deviation**).

*This formula is for the **sample variance**.

Follow along in the first example.

▪ ▪ ▪

EXAMPLE

Find the variance and standard deviation of the following scores on an exam:

$$92, 95, 85, 80, 75, 50$$

SOLUTION

First we find the mean of the data:

$$\text{Mean} = \frac{92 + 95 + 85 + 80 + 75 + 50}{6} = \frac{477}{6} = 79.5$$

Then we find the difference between each score and the mean (deviation).

Score	Score − Mean	Difference from mean
92	92 − 79.5	+12.5
95	95 − 79.5	+15.5
85	85 − 79.5	+5.5
80	80 − 79.5	+0.5
75	75 − 79.5	−4.5
50	50 − 79.5	−29.5

Next we square each of these differences and then sum them.

Difference	Difference squared
+12.5	156.25
+15.5	240.25
+5.5	30.25
+0.5	0.25
−4.5	20.25
−29.5	870.25
	1317.50 ◄——— sum of the squares

The sum of the squares is 1317.50.

Next we find the "mean" of this sum (the variance).

$$\frac{1317.50}{5} = 263.5$$

Finally, we find the square root of this variance.

$$\sqrt{263.5} \approx 16.2$$

So, the standard deviation of the scores is 16.2; the variance is 263.5.

◢

■■■

EXAMPLE

Find the standard deviation of the average temperatures recorded over a five-day period last winter:

$$18, 22, 19, 25, 12$$

SOLUTION

This time we will use a table for our calculations.

Temp.		Temp. − mean = deviation	Deviation squared
18		18 − 19.2 = −1.2	1.44
22		22 − 19.2 = 2.8	7.84
19		19 − 19.2 = −0.2	0.04
25	mean	25 − 19.2 = 5.8	33.64
12	↓	12 − 19.2 = −7.2	51.84
96 ÷ 5 = 19.2			94.80 ←—sum of squares

To find the variance, we divide by $5 - 1 = 4$.

$$\frac{94.8}{4} = 23.7$$

Finally we find the square root of this variance.

$$\sqrt{23.7} \approx 4.9$$

So the standard deviation for the temperatures recorded is 4.9; the variance is 23.7.

◢

Note that the values in the second example were much closer to the mean than those in the first example. This resulted in a smaller standard deviation.

We can write the formula for the standard deviation as

$$s = \sqrt{\frac{\Sigma(x_i - \bar{x})^2}{n - 1}}$$

where

Σ   means "the sum of."

x_i represents each value x in the data.

$\bar{x}$ is the mean of the x_i values.

n is the total number of x_i values.

PRACTICE

1. The five states with the most covered bridges are

    Oregon: 106
    Vermont: 121
    Indiana: 152
    Ohio: 234
    Pennsylvania: 347

    Find the variance and standard deviation for the number of covered bridges in these states.   9956.5; 99.8

2. The five tallest skyscrapers in the United States are

    Sears Tower: 1454 feet
    World Trade Center: 1353 feet
    Empire State Building: 1250 feet
    Standard Oil Building: 1136 feet
    Hancock Insurance Building: 1127 feet

    Find the variance and standard deviation of the heights of these five skyscrapers.   19842.5; 140.9

3. Four fast-food restaurants serve a fish sandwich. The caloric count for each is as follows: 486, 468, 402, 445. Find the variance and standard deviation of these caloric counts.   1316.25; 36.3

4. Scores on the most recent reading test are 7.7, 7.4, 7.3, and 7.9. Find the variance and standard deviation of these scores.   0.758; 0.275

5. The highest temperatures recorded in eight specific states are 112, 100, 127, 120, 134, 118, 105, and 110. Find the variance and standard deviation of these scores.   127.6; 11.3

6. TV viewing time during a certain week in millions was 95.2, 94.8, 91.7, 97.7, and 92.4. Find the variance and standard deviation of the numbers of viewers during this week.   5.74; 2.4

7. Using data from Practice Exercises 1 through 6, ask and answer four more questions.   Answers may vary

8. Which data from Practice Exercises 1 through 6 had the greatest variation?   the skyscraper heights

7.3 Solving Equations that Contain Radicals

SECTION LEAD-IN*

The size of the orifice (round opening) used for supplying gas to a kiln must be converted from a drill size or pipe size to an area given in square inches. The openings are small, usually less than $\frac{1}{2}$ inch. How do the areas vary? Write a formula to find the diameter of an orifice, given its area in square inches.

Equations that Contain Square Roots

Equations that contain radicals can be solved by applying the following property of real numbers:

Equality Property of Powers

For real numbers x and a, and positive integer n,

$$\text{if } \sqrt[n]{x} = a, \text{ then } x = a^n$$

In other words, if two real numbers are equal, then their nth powers are equal. The same property holds, of course, for variables or expressions that represent real numbers.

To apply the property, we must first isolate the radical on one side of the given equation. Then we raise both sides to the proper power and solve as usual.

▪▪▪

EXAMPLE 1

Find the solution: $\sqrt{x} - 5 = 7$

SOLUTION

INSTRUCTOR NOTE

Remind students that they must square the entire side of the equation, not just the radical. Again, encourage the use of parentheses.

Adding 5 to both sides of this equation will isolate the radical, yielding $\sqrt{x} = 12$. Squaring both sides gives

$$\left(\sqrt{x}\right)^2 = (12)^2 \quad \text{Equality property of powers}$$
$$x = 144 \quad \text{Multiplying}$$

We check the solution in the *original* equation.

$$\sqrt{x} - 5 = 7$$
$$\sqrt{144} - 5 = 7$$
$$12 - 5 = 7 \quad \text{True}$$

Source: Marc Ward, "The Mysterious Hole," *Ceramics Monthly* 43:5 (May 1995), pp. 50–53.

The principal square root of 144, is, indeed, 12.

▶*CHECK* **Warm-Up 1**

▪ ▪ ▪

EXAMPLE 2

Find the solution: $\sqrt{2x + 2} - 6 = 0$

SOLUTION

We isolate the radical and then apply the powers property.

$$\sqrt{2x + 2} - 6 = 0 \qquad \text{Original equation}$$
$$\sqrt{2x + 2} = 6 \qquad \text{Adding 6 to both sides}$$
$$\left(\sqrt{2x + 2}\right)^2 = 6^2 \qquad \text{Squaring both sides}$$
$$2x + 2 = 36 \qquad \text{Multiplying}$$

We continue solving this equation by adding -2 to both sides, obtaining

$$2x = 34$$
$$x = 17 \qquad \text{Multiplying both sides by } \tfrac{1}{2}$$

Check:
$$\sqrt{2x + 2} - 6 = 0$$
$$\sqrt{2(17) + 2} - 6 = 0$$
$$\sqrt{36} - 6 = 0$$
$$6 - 6 = 0 \qquad \text{True}$$

So the solution set is {17}.

> **STUDY HINT**
>
> *Be careful when you write radicals to include only those numbers that belong within the radical. It may be safer to write $\sqrt{x + 3} - 8$ as $-8 + \sqrt{x + 3}$.*

▶*CHECK* **Warm-Up 2**

Some equations have no solutions. For example, the equation $\sqrt{x} = -2$ has no real solution because no real number has a principal square root that is negative. Furthermore, the equality property of powers can produce false solutions. So *every* solution must be checked.

▪ ▪ ▪

EXAMPLE 3

Solve: $x + 1 = \sqrt{x + 3}$

SOLUTION

The radical is already isolated, so we apply the powers property and solve.

$$x + 1 = \sqrt{x + 3} \qquad \text{Original equation}$$
$$(x + 1)^2 = \left(\sqrt{x + 3}\right)^2 \qquad \text{Squaring both sides}$$
$$x^2 + 2x + 1 = x + 3 \qquad \text{Multiplying}$$
$$x^2 + x - 2 = 0 \qquad \text{Adding } -x - 3 \text{ to both sides}$$

We can factor this expression into $(x - 1)(x + 2) = 0$. Using the property of zero products, we set each factor equal to zero and solve for x.

$$(x - 1)(x + 2) = 0$$

$$\begin{array}{c|c} x - 1 = 0 & x + 2 = 0 \\ x = 1 & x = -2 \end{array}$$

It seems as though 1 and -2 are solutions here, but both must be checked in the original equation.

$$\begin{array}{c|c} x + 1 = \sqrt{x + 3} & x + 1 = \sqrt{x + 3} \\ 1 + 1 = \sqrt{1 + 3} & (-2) + 1 = \sqrt{(-2) + 3} \\ 2 = \sqrt{4} \quad \text{True} & -1 = \sqrt{1} \quad \text{False} \end{array}$$

So the solution set is $\{1\}$. The number -2 is not a solution because it gives a false statement.

▶ CHECK **Warm-Up 3**

Calculator Corner

Let's examine and solve the equation $\sqrt{x + 3} - 8 = 0$ in three ways. This is often referred to as the **rule of three.**

a. symbolically   **b.** numerically   **c.** graphically

i. We solve this equation *symbolically.*

$$\sqrt{x + 3} - 8 = 0$$
$$\sqrt{x + 3} - 8 + 8 = 0 + 8 \qquad \text{Adding 8}$$
$$\sqrt{x + 3} = 8$$
$$\left(\sqrt{x + 3}\right)^2 = 8^2 \qquad \text{Squaring}$$
$$x + 3 = 64$$
$$x + 3 - 3 = 64 - 3 \qquad \text{Adding } -3$$
$$x = 61 \qquad \text{Simplifying}$$

ii. You can use your graphing calculator to examine and solve the equation for x *numerically* by using the **TABLE** feature of the caculator. (Note: See the Calculator Corner on page 162 to review how to set up tables.)

```
Y₁ =√(X+3)-8
Y₂=
Y₃=
Y₄=
Y₅=
Y₆=
Y₇=
Y₈=
```

```
TABLE SETUP
 TblMin=58
 ∆Tbl=1
Indpnt:  Auto  Ask
Depend:  Auto  Ask
```

X	Y₁
58	-.1898
59	-.126
60	-.0627
61	0
62	.06226
63	.12404
64	.18535

X=61

You can see from the **TABLE** that when $x = 61$, $y = 0$.

It is also possible to use the **TABLE** to examine another form of the equation, namely, $\sqrt{x + 3} = 8$. Now use the **TABLE** to find what x-value results in a y-value of 8.

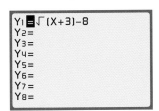

This gives an x-value of 61 when $y = 8$.

iii. There are two methods for solving an equation *graphically*. Let's solve the equation by *both* methods and see how they compare. We will first graph the equation exactly as it is written: $\sqrt{x + 3} - 8 = 0$. After obtaining the solution using the original equation, let's then solve it by rewriting the equation so that the radical is alone on one side: $\sqrt{x + 3} = 8$.

Question: Will the graphs for the two methods *look the same*? If you work the problem correctly, how many solutions should you obtain?

Some people refer to this method as the **x-intercept method** because you are looking for the x-value when $y = 0$. In other words, where does the graph cross the x-axis? Use the **TRACE** feature of your graphing calculator to estimate where the graph crosses the x-axis.

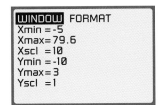
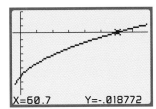
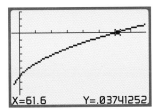

It appears that the x-value is somewhere between $x = 60.7$ and $x = 61.6$.

If you have a **CALCULATE** and **ROOT/ZERO** utility on your graphing calculator, use it to find the exact solution.

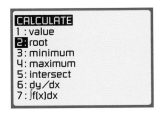

Move the cursor to the left of the root. Press **ENTER**.

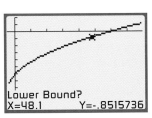

Move the cursor to
the right of the root.
Press **ENTER**.

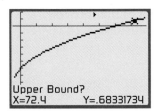 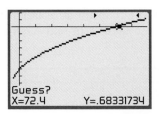 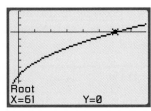

The solution is $x = 61$.

Now find the solution using the method sometimes referred to as the **multi-graph method.** Where does $\sqrt{x + 3} = 8$?

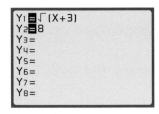

TRACE to estimate
what the x-value at the
point of intersection
would be.

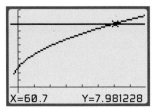

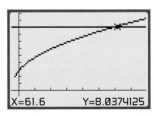

It appears that the correct answer for x is somewhere between 60.7 and 61.6.
Now use the **CALCULATE** and **INTERSECT** feature of your graphing calculator
to find the exact solution. (Note: See the Calculator Corner on pages 89 and 90
to review how to find the points of intersection of two graphs.)

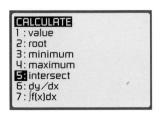 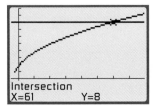

Once again you get the result $x = 61$.

Your graphing calculator can also be used to check your algebraic solution in
several ways. One of these would be to use *function notation.*

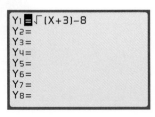 

Go to your
Home Screen
and evaluate
Y1 at $x = 61$.

Now you have confirmed that when $x = 61$, Y1 is indeed equal to 0.

Sometimes we must apply the powers property more than once to eliminate all the radicals.

▪ ▪ ▪
EXAMPLE 4

Solve: $\sqrt{x} + 1 = \sqrt{x + 9}$

SOLUTION

One of the two radicals is already isolated (on the right), so we can apply the powers property by squaring both sides. We get

$$
\begin{array}{ll}
\sqrt{x} + 1 = \sqrt{x + 9} & \text{Original equation} \\
\left(\sqrt{x} + 1\right)^2 = \left(\sqrt{x + 9}\right)^2 & \text{Squaring both sides} \\
x + 2\sqrt{x} + 1 = x + 9 & \text{Multiplying} \\
2\sqrt{x} = 8 & \text{Simplifying} \\
\sqrt{x} = 4 & \text{Multiplying by } \frac{1}{2}
\end{array}
$$

Now we apply the powers property again.

$$
\begin{array}{ll}
\left(\sqrt{x}\right)^2 = (4)^2 & \text{Squaring both sides} \\
x = 16 & \text{Multiplying}
\end{array}
$$

Check your answer in the original equation.

▶ *CHECK* **Warm-Up 4**

Equations that Contain Higher Roots

When an equation contains an nth root, we apply the same principle by raising both sides of the equation to the nth power. We then should be able to simplify.

▪ ▪ ▪
EXAMPLE 5

Solve: $\sqrt[4]{x^2 + 7} - 2 = 0$

SOLUTION

In order to isolate the radical, we add 2 to each side.

$$
\begin{array}{ll}
\sqrt[4]{x^2 + 7} - 2 = 0 & \text{Original equation} \\
\sqrt[4]{x^2 + 7} = 2 & \text{Adding 2 to each side}
\end{array}
$$

Then we must remove the fourth root by raising both sides of the equation to the fourth power. This gives us

$$
\begin{array}{ll}
\sqrt[4]{x^2 + 7} = 2 & \\
\left(\sqrt[4]{x^2 + 7}\right)^4 = 2^4 & \text{Raising to fourth power} \\
x^2 + 7 = 16 & \text{Simplifying} \\
x^2 = 9 & \text{Adding } -7 \text{ to both sides} \\
x = 3 \quad \text{or} \quad x = -3 & \text{Taking square roots}
\end{array}
$$

Note that the equation $x^2 = 9$ has two solutions because we are not looking only for the *principal* square root of 9. Both solutions must be checked in the original equation.

▶ *CHECK* **Warm-Up 5**

▪▪▪
EXAMPLE 6

Solve: $\sqrt[3]{3y + 4} - \sqrt[3]{2y - 5} = 0$

SOLUTION

We add the inverse of $-\sqrt[3]{2y - 5}$ to both sides so that we can apply the equality property of powers.

$$\sqrt[3]{3y + 4} - \sqrt[3]{2y - 5} = 0 \qquad \text{Original equation}$$
$$\sqrt[3]{3y + 4} = \sqrt[3]{2y - 5} \qquad \text{Adding } \sqrt[3]{2y - 5} \text{ to both sides}$$
$$\left(\sqrt[3]{3y + 4}\right)^3 = \left(\sqrt[3]{2y - 5}\right)^3 \qquad \text{Cubing each side}$$
$$3y + 4 = 2y - 5 \qquad \text{Simplifying}$$
$$y = -9 \qquad \text{Simplifying}$$

Check your results.

▶ *CHECK* **Warm-Up 6**

Equations that Contain Rational Exponents

Solving equations that contain rational exponents is no more difficult than solving equations that contain radicals. We use the equality property of powers to remove the rational exponent on the variable and then solve.

▪▪▪
EXAMPLE 7

Solve: $x^{\frac{3}{4}} = 8$

SOLUTION

The right side of the given equation can be written as a power. We do so because that will make the solution simpler. We get

$$x^{\frac{3}{4}} = 2^3$$

We then raise each side to the $\frac{4}{3}$ power (this is the key step) and solve.

$$\left(x^{\frac{3}{4}}\right)^{\frac{4}{3}} = \left(2^3\right)^{\frac{4}{3}} \qquad \text{Raising to the } \frac{4}{3} \text{ power}$$
$$x^{\frac{3}{4} \cdot \frac{4}{3}} = 2^{3 \cdot \frac{4}{3}} \qquad \text{Power rule}$$
$$x = 2^4 \qquad \text{Simplifying}$$
$$= 16 \qquad \text{Simplifying}$$

▶ *CHECK* **Warm-Up 7**

Applications

Very often, real-life situations lead to equations with radicals. The next two examples show that it is not difficult to solve them.

▪ ▪ ▪
EXAMPLE 8

When two forces, F_1 and F_2 act at a right angle to each other, the resultant force R (the effective force) can be found with the equation

$$R = \sqrt{(F_1)^2 + (F_2)^2}$$

See the accompanying figure.

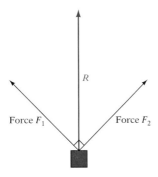

Two groups of students are attempting to pull a trailer out of a muddy playing field by pulling on two ropes that are perpendicular to each other. If one group is exerting a force of 1010 pounds, and the resultant of the two forces is 1453 pounds, what is the force exerted by the second group?

SOLUTION

First we substitute $R = 1453$ and $F_1 = 1010$ into the "resultant formula."

$$1453 = \sqrt{(1010)^2 + (F_2)^2}$$

To find F_2, we square both sides of this equation and then solve.

$$(1453)^2 = \left(\sqrt{(1010)^2 + (F_2)^2} \right)^2 \quad \text{Squaring both sides}$$
$$2{,}111{,}209 = (1010)^2 + (F_2)^2 \quad \text{Simplifying}$$
$$2{,}111{,}209 = 1{,}020{,}100 + (F_2)^2 \quad \text{Squaring 1010}$$
$$2{,}111{,}209 - 1{,}020{,}100 = (F_2)^2 \quad \text{Adding } -1{,}020{,}100 \text{ to both sides}$$
$$1{,}091{,}109 = (F_2)^2 \quad \text{Simplifying}$$

Then we take the square roots of both sides, obtaining

$$F_2 = 1044.56 \quad \text{and} \quad F_2 = -1044.56$$

We retain only the positive force in this situation. So the second force is about 1045 pounds.

▶ *CHECK* **Warm-Up 8**

▪ ▪ ▪
EXAMPLE 9

The velocity V required for a spacecraft to escape the gravitational pull g of a planet can be found by using the formula $V = \sqrt{2gR}$. For Earth, the escape velocity is about 11,145 meters per second. If Earth's radius R is 6,370,000 meters, what is the value of g?

SOLUTION

We will use a calculator to solve this. First we substitute, obtaining

$$V = \sqrt{2gR}$$
$$11{,}145 = \sqrt{2g(6{,}370{,}000)} \qquad \text{Substituting}$$
$$(11{,}145)^2 = \left(\sqrt{2g(6{,}370{,}000)}\right)^2 \qquad \text{Squaring both sides}$$
$$124{,}211{,}025 = 2g(6{,}370{,}000)$$
$$\frac{124{,}211{,}025}{(2)(6{,}370{,}000)} = g$$
$$9.75 = g$$

The units in which we express g are meters per second per second.

▶ CHECK **Warm-Up 9**

Practice what you learned.

SECTION FOLLOW-UP

To find a formula for the area in terms of the diameter of an orifice, we first solve for r. Because $d = 2r$ implies $r = \frac{d}{2}$ (diameter = d) and $A = \pi r^2$, we have

$$A = \pi\left(\frac{d}{2}\right)^2 = \frac{\pi d^2}{4}$$

for the area written in terms of the diameter.

7.3 WARM-UPS

Work these problems before you attempt the exercises.

1. Find the solution:
   $6 = 11 - \sqrt{r}$ $r = 25$

2. Find the solution:
   $\sqrt{5r + 3} - 7 = 0$ $r = 9\frac{1}{5}$

3. Solve: $\sqrt{2r + 6} = r + 3$
   $r = -1$ or $r = -3$

4. Solve: $5 = \sqrt{t + 3} + \sqrt{t}$
   $t = \frac{121}{25}$

5. Solve: $\sqrt[3]{x^2 - 5} = 4$
   $x = \sqrt{69}$ or $x = -\sqrt{69}$

6. Let $f(x) = 2x + 3$ and $g(x) = 4x - 5$. Find x so that $\sqrt{f(x) + 3} = \sqrt{g(x) - 9}$. $x = 10$

7. Solve: $2x^{\frac{1}{3}} = 8$ $x = 64$

8. The length of the diagonal of a rectangular solid is given by the formula $d = \sqrt{a^2 + b^2 + c^2}$, where the dimensions of the rectangle are a = length, b = width, and c = depth. Find the length to two decimal places when the width is 12 inches, the depth is 8 inches, and the diagonal is 23.1 inches. $a = 18.04$ inches

9. A formula similar to the one in Example 9, $V = \sqrt{2gh}$, gives the velocity V attained by an object dropped from a height of h feet. Here g represents the acceleration due to gravity, 32 feet per second per second. An object dropped from the top of Peachtree Center Plaza in Atlanta, which is 723 feet high, would attain what velocity? $V = 215.11$ feet per second

7.3 EXERCISES

Note: Use your graphing calculator to check your results whenever possible.

In Exercises 1 through 48, solve each equation and check the solutions. If there are no real solutions, state that.

1. $\sqrt{x} = 6$ $x = 36$

2. $\sqrt{x} = 7$ $x = 49$

3. $\sqrt{x} - 3 = 0$ $x = 9$

4. $\sqrt{x} - 5 = 0$ $x = 25$

5. $\sqrt[3]{x} = 2$ $x = 8$

6. $\sqrt[3]{x} = 3$ $x = 27$

7. $\sqrt[3]{x} + 5 = 0$ $x = -125$

8. $\sqrt[3]{x} - 10 = 0$ $x = 1000$

9. $\sqrt[4]{x} = 1$ $x = 1$

10. $\sqrt[5]{x} = 1$ $x = 1$

11. $\sqrt[5]{x} = -1$ $x = -1$

12. $\sqrt[4]{x} = -1$ no real solution

13. $\sqrt{x} + 3 = 8$ $x = 25$

14. $\sqrt[3]{x} + 2 = 10$ $x = 512$

15. $\sqrt[3]{x} + 8 = 9$ $x = 1$

16. $\sqrt[3]{x} + 8 = 10$ $x = 8$

17. $\sqrt{7n - 10} = \sqrt{n + 2}$ $n = 2$

18. $\sqrt{4y - 3} = \sqrt{5y - 5}$ $y = 2$

19. $\sqrt{2t + 5} = \sqrt{3t + 2}$ $t = 3$

20. $\sqrt{2x + 8} = \sqrt{5x - 3}$ $x = \frac{11}{3}$

21. $\sqrt{x - 7} = 7 - \sqrt{x}$ $x = 16$

22. $\sqrt{x - 8} = \sqrt{x} - 2$ $x = 9$

23. $\sqrt{x + 3} = -3 + \sqrt{x}$ no real solution

24. $\sqrt{x + 1} = 1 + \sqrt{x}$ $x = 0$

25. $\sqrt{x^2 + 3x + 9} = x$ no real solution

26. $\sqrt{x^2 + 9x + 3} = -x$ $x = \frac{-1}{3}$

27. $\sqrt{m - 3} = \sqrt{m + 2} - 1$ $m = 7$

28. $\sqrt{2t - 3} + 1 = \sqrt{2t}$ $t = 2$

29. $\sqrt{x + 16} - 2 = \sqrt{x}$ $x = 9$

30. $\sqrt{x + 3} + 3 = \sqrt{x}$ no real solution

31. $\sqrt{x - 3} - 4 = \sqrt{x - 3}$ no real solution

32. $\sqrt{2x + 8} = 2 + \sqrt{2x - 4}$ $x = 4$

33. $\sqrt{y + 1} + \sqrt{y} = 2$ $y = \frac{9}{16}$

34. $\sqrt{x + 7} + \sqrt{x - 8} = 5$ $x = 9$

35. $\sqrt{m - 3} - \sqrt{m + 5} = 4$ no real solution

36. $\sqrt{y + 1} + 1 = \sqrt{y + 2}$ $y = -1$

37. $\sqrt{x + 2} - 3 = -\sqrt{x - 1}$ $x = 2$

38. $\sqrt{x + 2} + \sqrt{x - 3} = 1$ no real solution

39. $\sqrt{x - 4} - 4 = -\sqrt{x + 4}$ $x = 5$

40. $\sqrt{x + 20} - 2 = \sqrt{x + 4}$ $x = 5$ **41.** $\sqrt[4]{3x + 1} = 2$ $x = 5$ **42.** $\sqrt[3]{3x - 1} = -4$ $x = -21$

43. $\sqrt[3]{4x - 3} = 3$ $x = \frac{15}{2}$ **44.** $\sqrt[3]{2x + 11} = 3$ $x = 8$ **45.** $\sqrt[3]{6x - 3} = 3$ $x = 5$

46. $\sqrt[3]{x - 12} = \sqrt[3]{5x + 16}$ $x = -7$ **47.** $\sqrt[3]{6x + 1} = \sqrt[3]{2x + 5}$ $x = 1$ **48.** $\sqrt[4]{x + 8} = \sqrt[4]{2x}$ $x = 8$

In Exercises 49 through 60, let $f(x) = x + 1$ and $g(x) = x - 1$. Find values of x that satisfy each of the following.

49. $f(x) - 4\sqrt{f(x)} + 4 = 0$ $x = 3$

50. $g(x) - 4\sqrt{g(x)} + 4 = 0$ $x = 5$

51. $3f(x) + 6\sqrt{f(x)} - 9 = 0$ $x = 0$

52. $g(x) - 6\sqrt{g(x)} + 9 = 0$ $x = 10$

53. $\sqrt{f(x) + 1} = \sqrt{2 - f(x)}$ $x = -\frac{1}{2}$

54. $\sqrt{g(x) + 1} = \sqrt{3 + g(x)}$ no real solution

55. $3\sqrt{g(x) - 2} = 2\sqrt{g(x) + 3}$ $x = 7$

56. $3\sqrt{f(x) - 1} = \sqrt{f(x) + 6}$ $x = \frac{7}{8}$

57. $\sqrt{f(x) - 4} + 2 = \sqrt{f(x)}$ $x = 3$

58. $g(x) - 2 = \sqrt{g(x) - 2}$ $x = 3$ and $x = 4$

59. $\sqrt{f(x) + 1} = \sqrt{g(x) + 1}$ no real solution

60. $\sqrt{f(x) - 3} = \sqrt{g(x) - 3}$ no real solution

In Exercises 61 through 72, solve each equation.

61. $x^{\frac{1}{2}} = 8$ $x = 64$

62. $x^{\frac{1}{3}} = 2$ $x = 8$

63. $x^{\frac{1}{3}} = -1$ $x = -1$

64. $x^{\frac{1}{2}} = 4$ $x = 16$

65. $3x^{\frac{1}{2}} = 18$ $x = 36$

66. $2x^{\frac{1}{3}} = 8$ $x = 64$

67. $\frac{1}{2}x^{\frac{1}{2}} = 6$ $x = 144$

68. $\frac{1}{3}x^{\frac{1}{5}} = -1$ $x = -243$

69. $\frac{2}{3}x^{-\frac{1}{2}} = 2$ $x = \frac{1}{9}$

70. $\frac{3}{4}x^{-\frac{1}{3}} = -2$ $x = -\frac{27}{512}$ **71.** $x^{\frac{1}{2}} + 3 = 5$ $x = 4$

72. $x^{\frac{1}{3}} - 2 = 4$ $x = 216$

73. *Trout Farming*   Fifty thousand baby trout are released each year into a fishing lake and are protected until they grow to the legal length at which they can be kept if caught. The number of trout N that survive for t months after release is given by

$$N = 5000\sqrt{100 - t}$$

For how many months will 40,000 of the trout survive?   36 months

74. *Recycling*   The formula $P = \dfrac{7500}{2.5\sqrt{x + 1}}$ gives the number of tons P of paper that was recycled in the xth year after a recycling program was started. In which year was 1000 tons of paper recycled?   the 8th year

75. *Memory Training*   The formula $R = 200\sqrt{t} - 10$ gives the number of restaurants and stores R that a taxi driver is expected to be able to locate t weeks after beginning the job. How many weeks will it take for a driver to know the location of 1190 stores and restaurants?   36 weeks

76. *IQ Testing*   A formula used by psychologists to predict the number of nonsense syllables N a person with an IQ of Q can repeat in a row is;

$$N = 2\sqrt{Q} - 9$$

A person who can repeat 13 nonsense syllables has what IQ?   about 121

MIXED PRACTICE

By doing these exercises, you will practice the topics up to this point in the chapter. Rationalize all denominators.

77. Simplify: $\dfrac{\sqrt{350x^3y}}{\sqrt{140x}}$ $\dfrac{x\sqrt{10y}}{2}$

78. Simplify: $\left(\sqrt{3} - \sqrt{2}\right)\left(\sqrt{3} + \sqrt{2}\right)$ 1

79. Simplify: $\sqrt{\dfrac{4x}{5y}}$ $\dfrac{2\sqrt{5xy}}{5y}$

80. Solve: $\sqrt{5n - 1} - \sqrt{3n - 2} = 1$
$n = 1$ and $n = 2$

81. Simplify: $\sqrt{49x^2y^2}$ $7xy$

82. Simplify: $\dfrac{3}{\sqrt{6}}$ $\dfrac{\sqrt{6}}{2}$

83. Solve: $\sqrt{x - 5} + \sqrt{x - 8} = 3$ $x = 9$

84. Simplify: $\sqrt{80x^2y}$ $4x\sqrt{5y}$

85. Simplify: $8\sqrt{45} + 7\sqrt{20} + 2\sqrt{5}$ $40\sqrt{5}$

86. Simplify: $\dfrac{3}{\sqrt{6} - 5}$ $\dfrac{3\sqrt{6} + 15}{-19}$

■ ▬▬▬▬▬▬▬▬▬▬▬▬▬▬▬▬▬▬▬▬▬▬▬▬▬ ■

EXCURSIONS

Posing Problems

1. Ask and answer four questions using this data.

English Language Daily and Sunday U.S. Newspapers
(number of newspapers as of Feb. 1, 1995; circulation as reported for Sept. 30, 1994)

State	Morning papers and circulation		Evening papers and circulation		Sunday papers and circulation	
Alabama	17	345,279	9	404,772	19	770,875
Alaska	5	101,310	2	10,122	4	129,843
Arizona	10	548,530	13	236,349	17	925,451
Arkansas	12	349,343	19	114,699	16	510,843
California	66	5,189,250	37	997,510	65	6,530,267
Colorado	14	909,023	14	141,545	11	1,239,124
Connecticut	13	666,404	9	155,389	12	844,838
Delaware	3	87,713	1	62,871	2	182,720
District of Columbia	2	905,201	0	0	2	1,206,622
Florida	33	2,920,911	8	198,738	35	3,997,910
Georgia	16	735,100	18	347,935	19	1,324,716
Hawaii	3	134,387	3	106,143	5	261,311
Idaho	7	141,446	5	78,719	8	237,058
Illinois	18	1,866,072	51	658,723	28	2,696,205
Indiana	16	715,785	57	762,497	22	1,355,052
Iowa	13	451,552	25	241,835	10	712,964
Kansas	7	273,375	40	228,580	16	458,731
Kentucky	8	460,510	15	196,643	12	686,119
Louisiana	12	503,398	14	247,256	20	861,846
Maine	5	235,810	2	25,388	2	191,795

Maryland	10	459,085	5	173,909	7	676,805
Massachusetts	10	1,261,636	29	609,910	14	1,760,883
Michigan	10	724,786	41	1,359,293	27	2,376,106
Minnesota	14	803,606	11	139,107	14	1,189,155
Mississippi	6	218,465	16	187,913	15	388,425
Missouri	11	811,922	34	226,572	21	1,329,583
Montana	6	154,563	5	41,014	7	201,863
Nebraska	5	212,905	14	258,251	7	444,063
Nevada	4	221,943	4	53,616	4	309,359
New Hampshire	8	141,204	5	102,445	6	200,407
New Jersey	14	1,286,467	7	280,875	17	1,887,067
New Mexico	7	203,402	11	103,065	13	301,301
New York	27	5,528,255	46	1,230,685	43	5,583,199
North Carolina	18	1,002,872	31	400,596	37	1,520,487
North Dakota	7	170,685	3	13,709	7	187,654
Ohio	17	1,630,400	67	1,034,189	36	2,891,971
Oklahoma	10	472,090	35	217,665	40	862,099
Oregon	5	319,155	15	365,431	10	701,119
Pennsylvania	41	2,175,658	48	905,051	38	3,335,787
Rhode Island	3	143,483	4	131,626	3	305,341
South Carolina	11	587,696	5	76,494	14	765,832
South Dakota	5	126,482	6	43,454	4	139,972
Tennessee	11	637,999	17	282,320	16	1,113,766
Texas	37	2,556,151	56	622,833	85	4,240,610
Utah	1	125,037	5	189,289	6	359,147
Vermont	4	96,195	4	33,336	3	105,660
Virginia	17	2,287,883	13	257,000	15	1,007,917
Washington	10	692,563	14	507,752	16	1,290,044
West Virginia	10	251,797	13	160,588	11	402,531
Wisconsin	10	465,867	26	680,944	21	1,224,020
Wyoming	6	70,927	3	19,219	4	68,336
Totals	**635**	**43,381,578**	**935**	**15,923,865**	**886**	**62,294,799**
Total U.S., Sept. 30, 1993	623	43,093,866	954	16,717,737	884	62,565,574
Total U.S., Sept. 30, 1990	559	41,311,167	1,084	21,016,795	863	62,634,512
Total U.S., Sept. 30, 1985	482	36,361,561	1,220	26,404,671	798	58,825,978

2. Use the newspaper data from Excursion 1 and the population data from Excursion 4 on page 340. Ask and answer four questions that can be answered using this combined data.

3. Study the following information and ask and answer four questions. Try to ask questions that other students would want to answer.

The First Ten Minor Planets (Asteroids)

Name	Year of discovery	Mean distance from sun (millions of miles)	Orbital period (years)	Diameter (miles)	Magnitude
1. Ceres	1801	257.0	4.60	485	7.4
2. Pallas	1802	257.4	4.61	304	8.0
3. Juno	1804	247.8	4.36	118	8.7
4. Vesta	1807	219.3	3.63	243	6.5
5. Astræa	1845	239.3	4.14	50	9.9
6. Hebe	1847	225.2	3.78	121	8.5
7. Iris	1847	221.4	3.68	121	8.4
8. Flora	1847	204.4	3.27	56	8.9
9. Metis	1848	221.7	3.69	78	8.9
10. Hygeia	1849	222.6	5.59	40 (?)	9.5

The Asteroids

Between the orbits of Mars and Jupiter are an estimated 30,000 pieces of rocky debris, known collectively as the asteroids, or planetoids. The first and, incidentally, the largest (Ceres) was discovered during the New Year's night of 1801 by the Italian astronomer Father Piazzi (1746–1826), and its orbit was calculated by the German mathematician Karl Friedrich Gauss (1777–1855). Gauss invented a new method of calculating orbits on that occasion. A German amateur astronomer, the physician Olbers (1748–1840), discovered the second asteroid, Pallas. The number now known, catalogued, and named is over 6,000 and could reach 10,000 by the end of the 20th century. A few asteroids do not move in orbits beyond the orbit of Mars, but in orbits which cross the orbit of Mars. The first of them was named Eros because of this peculiar orbit. It had become the rule to bestow female names on the asteroids, but when it was found that Eros crossed the orbit of a major planet, it received a male name. Since then around two dozen orbit-crossers have been discovered, and they are often referred to as the "male asteroids." A few of them—Albert, Adonis, Apollo, Amor, and Icarus—cross the orbit of the Earth, and two of them may have come closer than our Moon; but the crossing is like a bridge crossing a highway, not two highways intersecting. Hence there is very little danger of collision from these bodies. They are all small, three to five miles in diameter, and therefore very difficult objects to identify, even when quite close. Some scientists believe the asteroids represent the remains of an exploded planet. Asteroid 1992 AD, discovered January 1992, is the outermost asteroid known. It takes 93 years to orbit the Sun. This minor planet's orbit crosses the paths of Saturn, Uranus, and Neptune.

On Oct. 29, 1991, the Galileo spacecraft took a historic photograph of asteroid 951 Gaspra from a distance of 10,000 miles (16,200 kilometers) away. It was the first close-up photo ever taken of an asteroid in space.

Gaspra was discovered to be an irregular, potato-shaped object about 12.5 mi (20 km) by 7.5 mi (12 km) by 7 mi (11 km) in size. Its surface is covered with a layer of loose rubble and its terrain is covered with several dozen small craters.

Scientists believe that Gaspra is a fragment of a larger body which was shattered from collisions with other asteroids. The asteroid was named after a Black Sea retreat favored by Russian astronomer Grigoriy N. Neujmin who discovered it in 1916.

Closeup photos of Asteroid 243 Ida taken by the Galileo spacecraft on Aug. 28, 1993, revealed that Ida had a tiny egg-shaped moon measuring 0.9 miles by 0.7 miles (1.6 by 1.2 kilometers). The moon has been named Dactyl.

Source: 1996 Information Please® Almanac (©1995 Houghton Mifflin Co.), pp. 343–344. All rights reserved. Used with permission by Information Please LLC.

4. Study the following table and ask and answer four questions. Try to ask questions that other students would want to answer.

Active Pilot Certificates Held

Year	Total	Airline transport	Commercial	Private
1970	720,028	31,442	176,585	299,491
1980	814,667	63,652	182,097	343,276
1985	722,376	79,192	155,929	320,086
1990	702,659	107,732	149,666	299,111
1991	692,095	112,167	148,365	293,306
1992	682,959	115,855	146,385	288,078
1993	665,069	117,070	143,014	283,700
1994	654,088	117,434	138,728	284,236

Source: 1996 Information Please® Almanac (©1995 Houghton Mifflin Co.), p. 359. All rights reserved. Used with permission by Information Please LLC.

CONNECTIONS TO *GEOMETRY*

Calculating Volume

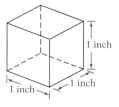

1 inch
1 inch
1 inch

The volume of a three-dimensional object is equal to the amount of space it encloses. The units of volume are cubic units: cubic inches, cubic feet, and so on. A cubic inch is a cube with its length, width, and height all equal to 1 inch; a cubic foot is a cube with its length, width, and height all equal to 1 foot.

> **Cube**
>
> The geometric figure called a **cube** is a solid whose length, width, and height are equal in measure. Each side of a cube is called a face. Two faces meet in an **edge.**

To find the volume of any cube, we multiply its length, width, and height together. Because they all have the same measure:

> **To find the volume of a cube**
>
> If s is the length of an edge, then
> $$V = s^3$$

▪▪▪

EXAMPLE

Find the volume of a cube with an edge of $4\frac{1}{4}$ feet.

SOLUTION

We use the formula $V = s^3$.

$$V = s^3 \qquad \text{The formula}$$
$$= \left(4\frac{1}{4}\right)^3 \qquad \text{Substituting } s = 4\frac{1}{4}$$
$$= \frac{17}{4} \times \frac{17}{4} \times \frac{17}{4} \qquad \text{Multiplying out}$$
$$= \frac{4913}{64} \quad \text{or} \quad 76\frac{49}{64} \qquad \text{Simplifying}$$

So the volume is $76\frac{49}{64}$ cubic feet.

◢

A cube has six square surfaces that are called its **faces.** A solid whose faces are rectangles is called a **rectangular solid** (sometimes called a **rectangular prism**). A rectangular solid looks somewhat like a cube, but its dimensions—length, width, and height—are not necessarily all equal.

To find the volume of a rectangular solid, we multiply its length, width, and height. In symbols,

---

To find the volume of a rectangular solid

If ℓ is the length, w is the width, and h is the height, then

$$V = \ell \times w \times h$$

∎∎∎

EXAMPLE

Find the volume of a rectangular solid with length 14 inches, width 13 inches, and height 8 inches.

SOLUTION

We substitute the values for ℓ, w, and h in the formula and solve.

$$\begin{aligned} V &= \ell \times w \times h && \text{The formula} \\ &= 14 \times 13 \times 8 && \text{Substituting} \\ &= 1456 && \text{Multiplying} \end{aligned}$$

So the volume is 1456 cubic inches.

Calculating Surface Area

Next we will find the surface area of a cube and a rectangular solid.

A cube has six faces, each in the shape of a square.

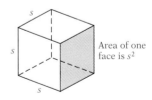

To find the surface area of a cube, we simply find the area of one of the square faces and multiply it by 6.

To find the surface area of a cube

If s is the length of an edge of a cube, then

$$SA = 6s^2$$

∎∎∎

EXAMPLE

Find the surface area of a cube whose edge is $3\frac{1}{2}$ inches long.

SOLUTION

We substitute into the formula and solve.

$$SA = 6s^2 \qquad \text{The formula}$$

$$= 6\left(3\tfrac{1}{2}\right)^2 \quad \text{Substituting}$$

$$= 6\left(\tfrac{7}{2}\right)^2 \quad \text{Changing to improper fraction}$$

$$= 6\left(\tfrac{49}{4}\right) \quad \text{Removing the exponent}$$

$$= 73\tfrac{1}{2} \qquad \text{Multiplying and simplifying}$$

The surface area of the cube with an edge of $3\frac{1}{2}$ inches is $73\frac{1}{2}$ square inches.

◀

Notice that surface areas are given in square units.

To find the surface area of a rectangular solid, we find the areas of the faces and add these areas together. The solid shown in the following figure has two faces of area ℓh, two faces of area wh, and two faces of area $w\ell$.

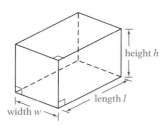

To find the surface area of a rectangular solid

If ℓ is the length, w is the width, and h is the height, then

$$SA = 2(\ell h + wh + w\ell)$$

▪▪▪

EXAMPLE

Find the surface area of a rectangular solid with a length of $2\frac{1}{2}$ inches, a width of 1 inch, and a height of $1\frac{1}{4}$ inches.

SOLUTION

We substitute in the formula and solve.

$$SA = 2(\ell h + wh + w\ell) \qquad \text{The formula}$$

$$SA = 2\left[\left(2\tfrac{1}{2}\right)\left(1\tfrac{1}{4}\right) + (1)\left(1\tfrac{1}{4}\right) + (1)\left(2\tfrac{1}{2}\right)\right] \quad \text{Substituting}$$

$$= 2\left[\left(\tfrac{5}{2}\right)\left(\tfrac{5}{4}\right) + \tfrac{5}{4} + \tfrac{5}{2}\right] \qquad \text{Changing to improper fractions}$$

$$= 2\left[\left(\tfrac{25}{8}\right) + \tfrac{5}{4} + \tfrac{5}{2}\right] \qquad \text{Multiplying}$$

$$= 2\left[\frac{25}{8} + \frac{10}{8} + \frac{20}{8}\right] \qquad \text{Rewriting with LCD}$$

$$= 2\left(\frac{55}{8}\right) \qquad \text{Adding}$$

$$= \frac{110}{8} = 13\frac{3}{4} \qquad \text{Multiplying and simplifying}$$

Thus the surface area of the rectangular solid is $13\frac{3}{4}$ square inches.

PRACTICE

1. A toy box is advertised as holding 40 cubic feet of toys. If the length of the toy box is 5 feet and its height is 3 feet, how wide must the toy box be? Use the formula $w = V/(\ell h)$. If you paint the toy box on the outside, what surface area must the paint cover? Include the top. $2\frac{2}{3}$ feet wide; surface area, $72\frac{2}{3}$ square feet

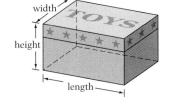

2. Find the number of cubic feet in a freezer that has a height of $3\frac{1}{2}$ feet, a length of 8 feet, and a width of $2\frac{1}{4}$ feet.   63 cubic feet

3. To find the required capacity of an air conditioner (in BTUs), we multiply the volume of the room it is to cool by 3. How much capacity is needed for a room that measures $17\frac{1}{3}$ feet long, $11\frac{1}{2}$ feet wide, and 9 feet high?   5382 BTUs

4. A mini-storage facility advertises bins that are 15 feet wide, $10\frac{1}{2}$ feet long, and $8\frac{1}{4}$ feet wide. What is the volume of each bin? $1299\frac{3}{8}$ cubic feet

5. Find the volume of a cube that has an edge of $1\frac{1}{2}$ inches. What is its surface area?   volume, $3\frac{3}{8}$ cubic inches; surface area, $13\frac{1}{2}$ square inches

6. A refrigerator is said to have a volume of 32 cubic feet. If the length of the interior of the refrigerator is 3 feet and the interior width is 2 feet, how high must the interior of the refrigerator be? Use the formula $h = V/w\ell$. $5\frac{1}{3}$ feet

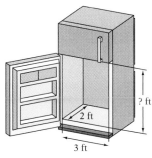

7. Find three things in your environment and calculate their surface area and volume.   Answers may vary.

8. ✏ Do surface area and volume have a relationship to each other that is constant? Justify your answer with data. You may use your results from Exercises 1 through 6 or make up other cubes and rectangular solids to test your ideas.   Answers may vary.

9. Use your graphing calculator to investigate the maximum volume of an open box that you can make from a piece of cardboard that measures 15 inches by 12 inches. It might be helpful to first sketch the piece of cardboard and imagine how you would construct the open box. Then write the equation for the volume of the box. Graph the equation and find the maximum volume. Then give the measurements for the width and length of the bottom of the box and the height of the box. Finally, sketch the resulting box and indicate the appropriate dimensions.

7.4 Complex Numbers

SECTION LEAD-IN

Find the number of BTUs that are produced per hour with a $\frac{1}{8}$-inch (diameter) orifice in a burner using natural gas:

$$D = V \times A \times \overline{K} \times 1655\sqrt{\frac{H}{G}}$$

where

A = area of orifice opening in square inches
$\overline{K} = 0.80$
$H = 7$ WC (inches of water in a tube)
$G = 0.65$
$V = 1000$ cubic feet per hour (for natural gas)
D = flow output in BTUs per hour

This formula will not give you complex numbers, but it is somewhat complex.

The Number i

In order to solve equations such as $x^2 = -1$, mathematicians defined a new number that is a solution to that equation. The new number, which is not a real number, has the symbol i. And because it is a solution to $x^2 = -1$, we know that

$$i = \sqrt{-1} \quad \text{and} \quad i^2 = -1$$

That is, i is the principal (positive) square root of -1. Then $-i$ is the negative square root of -1.

$$-i = -\sqrt{-1} \quad \text{and} \quad (-i)^2 = -1$$

The number i has interesting powers. Using $i^1 = i$ and $i^2 = -1$, we can evaluate all other powers of i:

INSTRUCTOR NOTE

This section may be particularly difficult for your students with learning disabilities. The symbol i and 1 are easily confused. Students may need to box the i's or in some other way distinguish them.

$$i^3 = i^2 \cdot i = -1 \cdot i = -i$$
$$i^4 = i^2 \cdot i^2 = (-1)(-1) = 1$$
$$i^5 = i^4 \cdot i = 1 \cdot i = i$$
$$i^6 = i^4 \cdot i^2 = (1)(-1) = -1$$
$$i^7 = i^4 \cdot i^3 = (1)(-i) = -i$$
$$i^8 = i^4 \cdot i^4 = 1 \cdot 1 = 1$$

and so on

The integer powers of i continually repeat the four values i, -1, $-i$, and 1 in that order.

■■■

EXAMPLE 1

Simplify: **a.** i^{16} **b.** i^{67} **c.** i^{-5}

SOLUTION

The best way to simplify high powers of i is to begin by removing powers of $i^4 = 1$.

a. $i^{16} = (i^4)^4 = (1)^4 = 1$

b. $i^{67} = i^{64} \cdot i^3 = (i^4)^{16} \cdot i^3 = 1 \cdot i^3 = -i$

c. $i^{-5} = \dfrac{1}{i^5} = \dfrac{1}{i^4 i} = \dfrac{1}{i}$

We rationalize the denominator

$$\frac{1}{i} \cdot \frac{i}{i} = \frac{i}{-1} = -i$$

▶ *CHECK* **Warm-Up 1**

The Imaginary Numbers

Using the new number i, mathematicians define two entire sets of numbers. The simpler of the two is the *imaginary numbers*.

> A number of the form bi, where b is real and not equal to zero, is an **imaginary number.**

For example, $i\sqrt{3}$ is an imaginary number, it is another way of writing the positive square root of -3.

$$\sqrt{-3} = \sqrt{-1} \cdot \sqrt{3} = i\sqrt{3}$$

In general, $\sqrt{-n} = i\sqrt{n}$ if n is real and non-negative.

Note that we write $i\sqrt{n}$ rather than $\sqrt{n}i$, to make sure that i is not mistakenly placed under the radical sign.

▪▪▪
EXAMPLE 2

Rewrite as multiplies of i: **a.** $\sqrt{-27}$ **b.** $\sqrt{-36}$

SOLUTION

We rewrite each radical as the product of $\sqrt{-1}$ and $\sqrt{n}$, for positive n; then we simplify.

a. $\sqrt{-27} = \sqrt{-1} \cdot \sqrt{27} = i \cdot \sqrt{9}\sqrt{3} = 3i\sqrt{3}$

b. $\sqrt{-36} = \sqrt{-1} \cdot \sqrt{36} = 6i$

▶ *CHECK* **Warm-Up 2**

∎∎∎
EXAMPLE 3

Rewrite without i: **a.** $15i$ **b.** $i^2\sqrt{13}$ **c.** $i^8\sqrt{52}$

SOLUTION

a. Because $i = \sqrt{-1}$, we rewrite $15i$ as

$$15 \cdot \sqrt{-1} = 15\sqrt{-1}$$

b. Because $i^2 = -1$, we rewrite $i^2\sqrt{13}$ as

$$-1 \cdot \sqrt{13} = -\sqrt{13}$$

c. Because $i^8 = i^4 \cdot i^4 = 1 \cdot 1$, we have

$$i^8\sqrt{52} = 1\sqrt{52} = 2\sqrt{13}$$

▶ *CHECK* **Warm-Up 3**

The imaginary numbers are defined such that the associative, commutative, and distributive properties hold. For example,

$$ai + bi = (a + b)i$$
$$ai(bi + ci) = (ai)(bi) + (ai)(ci)$$
$$= abi^2 + aci^2$$
$$= -ab - ac$$

∎∎∎
EXAMPLE 4

Simplify: **a.** $3i + 2.5i - i$ **b.** $(7i)(4i)$ **c.** $\frac{6}{3i}$ **d.** $\sqrt{-8} \cdot \sqrt{-2}$

SOLUTION

a. By the distributive property,

$$3i + 2.5i - i = (3 + 2.5 - 1)i = 4.5i$$

b. By the associative and commutative properties,

$$(7i)(4i) = 7 \cdot 4 \cdot i \cdot i = 28i^2 = 28(-1) = -28$$

c. Dividing numerator and denominator by 3, we have

$$\frac{6}{3i} = \frac{2}{i}$$

However, because i represents a radical, $\sqrt{-1}$, this result is not in simplest terms. So we multiply numerator and denominator by i to remove i from the denominator. This gives us

$$\frac{2}{i} = \frac{2 \cdot i}{i \cdot i} = \frac{2i}{-1} = -2i$$

d. We first translate each radical into imaginary notation and then combine.

$$\sqrt{-8} \cdot \sqrt{-2} = i\sqrt{8} \cdot i\sqrt{2} = i^2 \cdot \sqrt{16}$$
$$= -1 \cdot 4 = -4$$

Note: we cannot multiply $(-8)(-2)$ under the radical sign to get $\sqrt{(-8)(-2)} = \sqrt{16}$. The multiplication property for radicals holds only for numbers with *real* roots. So we must first rewrite the imaginary factors such that they have real radicands. Operations with imaginary numbers differ from operations with real numbers in this way.

▶ *CHECK* **Warm-Up 4**

Complex Numbers

The second set of numbers that resulted from the definition of i is called the *complex numbers*.

> A number of the form $a + bi$, where a and b are real numbers, is a **complex number**.

INSTRUCTOR NOTE

Remind students to write complex numbers in this form for consistency in notation and ease in calculation.

Examples of complex numbers are

$$5 + i \qquad 1.75i \qquad 5 \qquad 3i$$

The real number 5 is a complex number because it can be written as $5 + 0i$. The imaginary number $3i$ is a complex number because it can be written $0 + 3i$. Thus the set of complex numbers includes both the real numbers and the imaginary numbers as subsets.

By carefully defining the arithmetic operations on complex numbers, we ensure that the properties of real numbers apply to the larger set of complex numbers. As you will see, the four operations are very straightforward.

Addition and Subtraction

To add two complex numbers, add the real parts and add the imaginary parts.

▪▪▪

EXAMPLE 5

Add: **a.** $3 + 4i$ and $-6 + 3i$ **b.** $36 - 27i$ and $-32 + 14i$

SOLUTION

We can use the commutative and associative properties to make the addition easier.

a. $(3 + 4i) + (-6 + 3i) = [3 + (-6)] + (4i + 3i) = -3 + 7i$

b. Complex numbers can easily be added vertically. Just remember to write the real and complex parts in separate columns.

> **■■■**
> **WRITER'S BLOCK**
> Comment on the two uses of the word "complex" in the Lead-In to this section. Which use has a specific mathematical meaning? How can you tell?

Add each column separately

↓   ↓

$$36 - 27i$$
$$\underline{-32 + 14i}$$
$$4 - 13i$$

▶ *CHECK* **Warm-Up 5**

The additive inverse of a complex number $a + bi$ is the complex number $-(a + bi)$. To subtract a complex number, we add its additive inverse (just as we do with real numbers).

▪▪▪

EXAMPLE 6

a. Find the additive inverse of $-3 + 2i$.

b. Subtract $3 + 4i$ from $2 - 12i$.

SOLUTION

a. The additive inverse of $-3 + 2i$ is
$$-(-3 + 2i) = 3 - 2i$$

b. $(2 - 12i) - (3 + 4i) = 2 - 12i + (-3 - 4i)$
$$= [2 + (-3)] + (-12i) + (-4i)$$
$$= -1 - 16i$$

▶ *CHECK* **Warm-Up 6**

Multiplication

Complex numbers are constructed like binomials. Thus to multiply two complex numbers together, you can simply apply FOIL. If one part of one factor is zero, apply the distributive property instead.

▪▪▪

EXAMPLE 7

Multiply: **a.** $(3 + 2i)(2 + 3i)$ **b.** $3i(2 + 2i)$ **c.** $3(3 - 2i)$

SOLUTION

a. Using FOIL, we have

$$
\begin{array}{cccc}
\text{F} & \text{O} & \text{I} & \text{L}
\end{array}
$$
$$(3 + 2i)(2 + 3i) = 3 \cdot 2 + 3 \cdot 3i + 2 \cdot 2i + 2i \cdot 3i$$
$$= 6 + 9i + 4i + 6i^2$$
$$= 6 + 13i + 6(-1)$$
$$= 13i$$

b.   Applying the distributive property, we obtain

$$3i(2 + 2i) = (3i)(2) + (3i)(2i)$$
$$= 6i + 6i^2$$
$$= 6i + 6(-1)$$
$$= -6 + 6i$$

c. $3(3 - 2i) = 9 - 6i$

▶ *CHECK* **Warm-Up 7**

The complex numbers $a + bi$ and $a - bi$ are *conjugates* of each other. If we multiply them together, we obtain a special product:

$$\overset{\text{F} \qquad \text{O} \qquad \text{I} \qquad \text{L}}{(a + bi)(a - bi) = a^2 - abi + abi - (bi)^2}$$
$$= a^2 - (bi)^2$$
$$= a^2 - b^2 i^2$$
$$= a^2 + b^2$$

So, for example, $(-3 + 5i)(-3 - 5i) = 9 + 25 = 34$. *The product of a complex number and its conjugate is always a real number.*

▪▪▪
EXAMPLE 8

Find, and then multiply by, the conjugate:

a. $\sqrt{3} + 2i$ **b.** $\sqrt{3}$ **c.** $2i$

SOLUTION

a.   The conjugate of $\sqrt{3} + 2i$ is $\sqrt{3} - 2i$. Multiplying, we have

$$\left(\sqrt{3} + 2i\right)\left(\sqrt{3} - 2i\right) = \left(\sqrt{3}\right)^2 - (2i)^2 = 3 - 4i^2 = 3 + 4 = 7$$

b. $\sqrt{3}$ can be rewritten as $\sqrt{3} + 0i$. The conjugate of $\sqrt{3} + 0i$ is $\sqrt{3} - 0i$. The product is $\left(\sqrt{3} + 0i\right)\left(\sqrt{3} - 0i\right)$.

$$\left(\sqrt{3} + 0i\right)\left(\sqrt{3} - 0i\right) = \left(\sqrt{3}\right)^2 + 0^2 = 3$$

c. $2i$ can be written as $0 + 2i$. Its conjugate is $0 - 2i$. The product is $(0 + 2i)(0 - 2i)$ or $(2i)(-2i) = -4i^2 = 4$.

$$(0 + 2i)(0 - 2i) = 0^2 + 2^2 = 4$$

▶ *CHECK* **Warm-Up 8**

Division

To divide a complex number by a complex number, write the division as a fraction. Next multiply numerator and denominator by the conjugate of the denominator (to remove i from the denominator). Then simplify.

▪▪▪
EXAMPLE 9

Divide: $6 \div \left(2 + i\sqrt{2}\right)$

SOLUTION

In fraction form, this problem becomes

$$\frac{6}{2 + i\sqrt{2}}$$

The conjugate of the denominator is $2 - i\sqrt{2}$, so we have

$$\frac{6}{2 + i\sqrt{2}} = \frac{(6)\left(2 - i\sqrt{2}\right)}{\left(2 + i\sqrt{2}\right)\left(2 - i\sqrt{2}\right)}$$

$$= \frac{12 - 6i\sqrt{2}}{2^2 + \left(\sqrt{2}\right)^2}$$

$$= \frac{12 - 6i\sqrt{2}}{6}$$

$$= 2 - i\sqrt{2}$$

▶ CHECK **Warm-Up 9**

Graphs and Magnitudes of Complex Numbers

We graph a complex number $a + bi$ on the coordinate plane by graphing the ordered pair (a, b). When we do so, we usually relabel the x-axis as the **real axis** and the y-axis as the **imaginary axis.** We then refer to the coordinate plane as the **complex plane.** The figure on the left below shows the graph of $-2 - 3i$ on the complex plane.

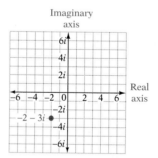
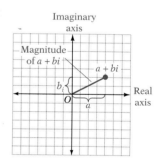

The **magnitude of a complex number** $a + bi$ is the distance from the origin 0 of the complex plane to the graph of $a + bi$ (see the figure on the right above). This is similar to the **absolute value** of a real number, which is the distance from zero to its graph on a number line. In fact, we write the magnitude of a complex number with an absolute value sign, as $|a + bi|$. The magnitude of a complex number can be calculated as

$$|a + bi| = \sqrt{a^2 + b^2}$$

Magnitudes are always positive real numbers. As an example,

$$|-3 + 4i| = \sqrt{(-3)^2 + 4^2} = \sqrt{9 + 16} = 5$$

▪▪▪

EXAMPLE 10

Graph the complex numbers (a) $3 - 5i$, (b) $5i$, and (c) -4. Also calculate their magnitudes and show those on the graph.

SOLUTION

The graphs are shown below.

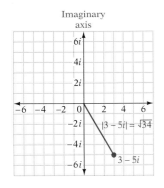
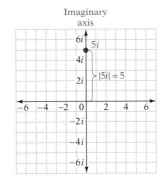
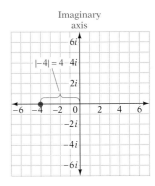

The magnitudes are

a. $|3 - 5i| = \sqrt{3^2 + (-5)^2} = \sqrt{9 + 25} = \sqrt{34}$

b. $|5i| = \sqrt{0^2 + 5^2} = \sqrt{25} = 5$

c. $|-4| = \sqrt{(-4)^2 + 0^2} = \sqrt{16} = 4$

▶ CHECK **Warm-Up 10**

Practice what you learned.

SECTION FOLLOW-UP

When we complete our calculations, using 3.14 for π, we obtain 53,292 BTUs.

If we change to a high-pressure propane line at 3 psi (equivalent to 83.1 WC), we replace $V = 2500$ and $G = 1.52$. We substitute in the formula

$$D = V \times A \times \overline{K} \times 1655\sqrt{\frac{H}{G}}$$

$$D = 2500 \times 0.0122654 \times 0.80 \times 1655\sqrt{\frac{83.1}{1.52}}$$

$$D = 300{,}185 \text{ BTUs}$$

7.4 WARM-UPS

Work these problems before you attempt the exercises.

1. Simplify: i^{26} and i^{-17}
   $-1; -i$

2. Rewrite as multiples of i:
   $\sqrt{-25}$ and $\sqrt{-32}$ $5i; 4i\sqrt{2}$

3. Rewrite without i: $27i$
   and $i^6\sqrt{26}$ $27\sqrt{-1}; -\sqrt{26}$

4. Simplify: $(10i)(2i)(3i)$ and $\frac{28}{12i}$
   $-60i; \frac{-7i}{3}$

5. Add: $16i - 49$ and $23 + 2i$
   $-26 + 18i$

6. Subtract: $5i - \sqrt{-16}$ from
   $\sqrt{-81} + 12i$. $20i$

7. Multiply: $(2i + 12)(2 - 12i)$
   $48 - 140i$

8. Multiply $15 - 3i$ by its
   conjugate and simplify. 234

9. Divide 3 by $5 + 3i$.
   $\frac{15 - 9i}{34}$

10. Graph and find the magnitude
    of $-2 + i$, $-3i$, and $\sqrt{-25}$.
    $|-2 + i| = \sqrt{5}; |0 - 3i| = 3;$
    $|5i| = 5$

7.4 EXERCISES

Note: Use your graphing calculator to check your results whenever possible.

In Exercises 1 through 20, simplify.

1. i^5 i
2. i^6 -1
3. i^8 1
4. i^7 $-i$
5. i^{18} -1

6. i^{14} -1
7. i^{22} -1
8. i^{42} -1
9. i^{150} -1
10. i^{198} -1

11. i^{273} i
12. i^{316} 1
13. i^{-3} i
14. i^{-1} $-i$
15. i^{-2} -1

16. i^{-4} 1
17. i^{-8} 1
18. i^{-12} 1
19. i^{-11} i
20. i^{-10} -1

In Exercises 21 through 32, rewrite each as a multiple of i.

21. $\sqrt{-49}$ $7i$
22. $\sqrt{-81}$ $9i$
23. $\sqrt{-64}$ $8i$
24. $\sqrt{-36}$ $6i$

25. $-\sqrt{-100}$ $-10i$
26. $\sqrt{-144}$ $12i$
27. $-\sqrt{-225}$ $-15i$
28. $-\sqrt{-400}$ $-20i$

29. $-\sqrt{-40}$ $-2i\sqrt{10}$
30. $-\sqrt{-32}$ $-4i\sqrt{2}$
31. $\sqrt{-12}$ $2i\sqrt{3}$
32. $-\sqrt{-50}$ $-5i\sqrt{2}$

In Exercises 33 through 44, write each without i.

33. $16i^2$ -16
34. $23i^3$ $-23\sqrt{-1}$
35. $15i^5$ $15\sqrt{-1}$
36. $18i^4$ 18

37. $30i^7$ $-30\sqrt{-1}$
38. $8i$ $8\sqrt{-1}$
39. $19i^6$ -19
40. $14i^8$ 14

41. $-16i^3$ $16\sqrt{-1}$
42. $-52i^5$ $-52\sqrt{-1}$
43. $-44i^2$ 44
44. $-49i^4$ -49

In Exercises 45 through 76, perform the indicated operations.

45. $-2i + 6i - 3$ $-3 + 4i$

46. $3i + 2 - 7i$ $2 - 4i$

47. $16i - 21i + 7i - 12$ $-12 + 2i$

48. $15i - 3i + 2 - 17i$ $2 - 5i$

49. $(2i)(50i)$ -100

50. $(16i)(3i^2)$ $-48i$

51. $(4i)(6i^2)$ $-24i$

52. $(18i)(2i)$ -36

53. Add $-12 + 3i$ and $13 + 16i$. $1 + 19i$

54. Add $-7 + 16i$ and $13 - 4i$. $6 + 12i$

55. Subtract $12 + 10i$ from $-12 + 13i$. $-24 + 3i$

56. Subtract $23 - 14i$ from $25 + 19i$. $2 + 33i$

57. Multiply: $6i(12i^2 - 3i + 2)$ $18 - 60i$

58. Multiply: $5i(10i - 9i^2 - 6)$ $-50 + 15i$

59. Multiply: $(3i - 5)(6i + 2)$ $-28 - 24i$

60. Multiply: $(15i + 1)(3i - 6)$ $-51 - 87i$

61. Simplify: $(3 - 2i)^2$ $5 - 12i$

62. Simplify: $(5 + 6i)^2$ $-11 + 60i$

63. Multiply: $(3 - i)(3 + i)$ 10

64. Multiply: $(4 - 2i)(4 + 2i)$ 20

65. Simplify: $\frac{2 + 3i}{2i}$ $\frac{3}{2} - i$

66. Simplify: $\frac{5 - 2i}{2i}$ $-1 - \frac{5}{2}i$

67. Divide 1 by $2 + 3i$. $\frac{2}{13} - \frac{3}{13}i$

68. Divide $4i$ by $6 - 3i$. $-\frac{4}{15} + \frac{8}{15}i$

69. Divide 8 by $3 + i\sqrt{7}$. $\frac{3}{2} - \frac{i\sqrt{7}}{2}$

70. Divide $6 - 2i$ by $6 - 3i$. $\frac{14}{15} + \frac{2i}{15}$

71. Divide $4 - 2i$ by $3 + 5i$. $\frac{1}{17} - \frac{13i}{17}$

72. Divide $11 + 3i$ by $6 - 5i$. $\frac{51}{61} + \frac{73i}{61}$

73. Calculate: $|-5 + 5i|$ $5\sqrt{2}$

74. Calculate: $|6 - 4i|$ $2\sqrt{13}$

75. Calculate: $|10 + 4i|$ $2\sqrt{29}$

76. Calculate: $|16 - 2i|$ $2\sqrt{65}$

77. Of $2 - 3i$, $6 + 5i$, and $7 + i$, which has the greatest magnitude? $6 + 5i$

78. Of $5 + 2i$, $7 - 3i$, and $5 - i$, which has the least magnitude? $5 - i$

79. Of $2 - 3i$, $5 + 6i$, and $2 + 2i$, which has the least magnitude? $2 + 2i$

80. Of $4i + 3$, $-6 + 6i$, and $3 - 3i$, which has the greatest magnitude? $-6 + 6i$

In Exercises 81 through 88, multiply each complex number by its conjugate.

81. $12 + 3i$ 153

82. $6 - 10i$ 136

83. $16 - 15i$ 481

84. $5 + 13i$ 194

85. $120 + 100i$ $24{,}400$

86. $30 - 25i$ 1525

87. $-\frac{1}{5} + \frac{1}{8}i$ $\frac{89}{1600}$

88. $\frac{1}{3} + \frac{1}{2}i$ $\frac{13}{36}$

In Exercises 89 through 92, use the fact that the product of a number and its multiplicative inverse is 1 to answer the questions.

89. Show that $5 + 6i$ and $\frac{5}{61} - \frac{6i}{61}$ are multiplicative inverses.
Because their product is 1, these factors are multiplicative inverses.

90. Show that $3 - 2i$ and $\frac{3}{13} + \frac{2}{13}i$ are multiplicative inverses.
Because their product is 1, the factors are multiplicative inverses.

91. What is the multiplicative inverse of $-5 + 2i$? $\quad -\frac{5}{29} - \frac{2}{29}i$

92. What is the multiplicative inverse of $6 + 3i$? $\quad \frac{2}{15} - \frac{i}{15}$

MIXED PRACTICE

By doing these exercises, you will practice the topics up to this point in the chapter.

93. Simplify: $\left(\sqrt{10} + 2\right)^2 \quad 14 + 4\sqrt{10}$

94. Find the solution set:
$\sqrt{n + 8} = \sqrt{3n + 12} - 2 \quad \{8\}$

95. Simplify: $(8 - 2i)(8 + 2i) \quad 68$

96. Simplify using rational exponents:
$\sqrt{2xy^2} \cdot \sqrt{14x^2y} \quad 2xy(7xy)^{\frac{1}{2}}$

97. Simplify: $\frac{\sqrt{2} - 1}{\sqrt{2} + 1} \quad 3 - 2\sqrt{2}$

98. Simplify: $i^{35} \quad -i$

99. Simplify using rational exponents:
$\sqrt[3]{4x^2y^6}\left(\sqrt[3]{9x^8y^5} - \sqrt[3]{7x^4y}\right) \quad x^3y^3(36xy^2)^{\frac{1}{3}} - x^2y^2(28y)^{\frac{1}{3}}$

100. Simplify: $\sqrt{60x^3y} \quad 2x\sqrt{15xy}$

EXCURSIONS

Exploring Numbers

1. **a.** Between any two fractions on a number line, there is another fraction that is halfway between them. We call this property **betweenness.** Use the idea of the mean to find the fraction that lies halfway between each of the following pairs of fractions.

    i. $\frac{1}{5}$ and $\frac{1}{125}$ **ii.** $\frac{1}{7}$ and $\frac{1}{8}$ **iii.** $\frac{1}{100}$ and $\frac{1}{10}$ **iv.** $\frac{11}{13}$ and $\frac{12}{13}$

    b. This same property (betweenness) holds for decimal fractions. By finding the mean of the pair, find the decimal fraction that lies halfway between each of the following pairs of decimal fractions.

    i. 0.21 and 0.22   **ii.** 0.004 and 0.005   **iii.** 0.231 and 0.232   **iv.** 0.10097 and 0.10098

2. Discuss the implications of these required public accommodations on the construction (or modification) of a theatre, parking garage, hotel, or school. What mathematical questions can you ask and answer using the following data and other information you can easily gather? Find four questions you think are interesting. Answer them too.

Accessibility for the Handicapped

Many areas and items must be addressed in an accessibility audit of public accommodations. The following is a sample accessibility checklist:

- Accessible routes (paths or walks) at least 3 feet wide and with at least 80 inches of headroom;
- Ramps that are at least 3 feet wide and with a maximum slope equal to 1 inch in 12 and maximum rise equal to 30 inches;
- Stairs with treads at least 11 inches wide and having a tactile warning at the top of the stairs;
- Parking facilities with spaces at least 8 feet wide and having special reserved spaces for the handicapped;
- Passenger loading zone that is at least 4 feet wide and 20 feet long;
- Drinking fountain: spout 3 feet high or less;
- Public telephones that are controlled by push button;
- Seating and tables that are 27 to 34 inches wide and 19 inches deep;
- Corridors with carpet pile one half inch or less;
- Door openings at least 32 inches wide.

Source: 1996 Information Please® Business Almanac (©1995 Houghton Mifflin Co.), p. 324. All rights reserved. Used with permission by Information Please LLC.

Exploring Geometry

3. In these diagrams, the distance between any two dots horizontally or vertically is one unit. Use the distance formula and the Pythagorean theorem to label all the sides with their lengths.

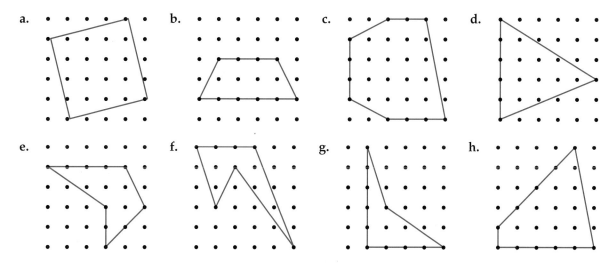

Data Analysis

4. Study the table on the following page. Pick the temperature you think best for room temperature. What combinations of room temperature and relative humidity will give the apparent temperature you picked?

Apparent Temperature for Values of Room Temperature and Relative Humidity

	Relative Humidity (%)										
Room Temperature (F)	0	10	20	30	40	50	60	70	80	90	100
75	68	69	71	72	74	75	76	76	77	78	79
74	66	68	69	71	72	73	74	75	76	77	78
73	65	67	68	70	71	72	73	74	75	76	77
72	64	65	67	68	70	71	72	73	74	75	76
71	63	64	66	67	68	70	71	72	73	74	75
70	63	64	65	66	67	68	69	70	71	72	73
69	62	63	64	65	66	67	68	69	70	71	72
68	61	62	63	64	65	66	67	68	69	70	71
67	60	61	62	63	64	65	66	67	68	68	69
66	59	60	61	62	63	64	65	66	67	67	68
65	59	60	61	61	62	63	64	65	65	66	67
64	58	59	60	60	61	62	63	64	64	65	66
63	57	58	59	59	60	61	62	62	63	64	64
62	56	57	58	58	59	60	61	61	62	63	63
61	56	57	57	58	59	59	60	60	61	61	62
60	55	56	56	57	58	58	59	59	60	60	61

Source: National Oceanic and Atmospheric Administration, Environmental Data and Information Service and National Climatic Center, as cited in the *1996 Information Please® Almanac* (©1995 Houghton Mifflin Co.), p. 400. All rights reserved. Used with permission by Information Please LLC.

CHAPTER LOOK-BACK

Triangles are used as braces because, given three sides of a triangle, one and only one such figure exists. A rhombus can be formed whose sides are identical to a given square; a parallelogram can have sides identical in length with those of a rectangle.

Other shapes you will find in building structures are hexagons (honeycombs—a building of sorts), pentagons (geodesic domes), or trapezoids (found on certain roofs).

CHAPTER 7
REVIEW PROBLEMS

The following exercises will give you a good review of the material presented in this chapter.

SECTION 7.1

In Exercises 1 through 6, simplify.

1. $\sqrt{25x^2y^2}$ $5xy$

2. $\sqrt{441x^2y^4}$ $21xy^2$

3. $\sqrt[3]{-1000x^3y^9}$ $-10xy^3$

4. $\sqrt[4]{256x^8y^{12}}$ $4x^2y^3$

5. $\sqrt{(x+4)^4}$ $(x+4)^2$

6. $\sqrt{(x-3)^8}$ $(x-3)^4$

In Exercises 7 through 12, multiply, divide, or simplify as indicated.

7. $\sqrt{125x^5} \cdot \sqrt{250x^2y} \cdot \sqrt{150x^2y}$ $1250x^4y\sqrt{3x}$

8. $\sqrt{\dfrac{550x^4y}{110x}}$ $x\sqrt{5xy}$

9. $\sqrt{200x^3y^3}$ $10xy\sqrt{2xy}$

10. $\sqrt{26} \cdot \sqrt{39}$ $13\sqrt{6}$

11. $(16x^3y^5z^6)^{\frac{1}{4}}$ $2yz\sqrt[4]{x^3yz^2}$

12. $(24x^4y^6z^8)^{\frac{1}{3}}$ $2xy^2z^2\sqrt[3]{3xz^2}$

SECTION 7.2

In Exercises 13 through 18, perform the indicated operations.

13. $4\sqrt{32} - \sqrt{18} + 2\sqrt{128}$ $29\sqrt{2}$

14. $3\sqrt[3]{25}\left(12\sqrt[3]{15} + 5\sqrt[3]{15} - 8\sqrt[3]{15}\right)$ $135\sqrt[3]{3}$

15. $\left(3\sqrt{5} - \sqrt{7}\right)\left(3\sqrt{5} + \sqrt{7}\right)$ 38

16. $\left(\sqrt{10} - \sqrt{2}\right)\left(\sqrt{10} + \sqrt{2}\right)$ 8

17. $\left(\sqrt{x} + \sqrt{3}\right)\left(\sqrt{x} - 12\right)$ $x - 12\sqrt{x} + \sqrt{3x} - 12\sqrt{3}$

18. $\left(\sqrt{5} + 6\right)^2$ $41 + 12\sqrt{5}$

In Exercises 19 through 24, write in simplest terms.

19. $\sqrt{\dfrac{3x}{4y}}$ $\dfrac{\sqrt{3xy}}{2y}$

20. $\dfrac{2\sqrt{3}}{\sqrt{32}}$ $\dfrac{\sqrt{6}}{4}$

21. $\dfrac{5}{\sqrt{x} - 3}$ $\dfrac{5\sqrt{x} + 15}{x - 9}$

22. $\dfrac{\sqrt{10} + 3}{\sqrt{10} - 2}$ $\dfrac{16 + 5\sqrt{10}}{6}$

23. Find the product by using rational exponents:
$\sqrt{2x^4y^5} \cdot \sqrt{8xy^3} \cdot \sqrt{16y^6}$ $16x^2y^7\sqrt{x}$

24. Find the product by using rational exponents:
$\sqrt{2xy^4} \cdot \sqrt{20x^4y^8} \cdot 2\sqrt{6xy^9}$ $8x^3y^{10}\sqrt{15y}$

SECTION 7.3

In Exercises 25 through 30, find the solution set.

25. $\sqrt{x + 5} - \sqrt{x - 3} = 2$ $\{4\}$

26. $\sqrt{2x - 3} = x - 3$ $\{6\}$

27. $\sqrt{2t + 4} = \sqrt{t + 3} + 1$ $\{6\}$

28. $\sqrt{x + 3} + 1 = \sqrt{2x - 1}$ $\{13\}$

29. $x^{\frac{1}{4}} = 10$ $x = 10,000$

30. $x^{\frac{2}{5}} = 4$ $x = 32$

SECTION 7.4

31. Simplify: i^{25} i

32. Find: $|6 - 3i|$ $3\sqrt{5}$

In Exercises 33 through 36, find the product and simplify.

33. $(6i - 8)(2i + 2)$ $-28 - 4i$

34. $(3i - 2)(4i + 2)$ $-16 - 2i$

35. $(4i - 6)(4i + 6)$ -52

36. $(3i - 12)(3i + 12)$ -153

MIXED REVIEW

37. Find the principal square root: $\sqrt{36x^2y^2}$
$6xy$

38. Find the product: $\sqrt[3]{5xy^2} \cdot \sqrt[3]{25x^4y^{12}}$
$5xy^4 \sqrt[3]{x^2y^2}$

39. Simplify: $\sqrt{120x^3y}$ $2x\sqrt{30xy}$

40. Simplify: $\dfrac{2\sqrt{3}}{\sqrt{5}}$ $\dfrac{2\sqrt{15}}{5}$

41. Solve: $x^{\frac{1}{3}} = 3$ $x = 27$

42. Find: $|10i - 5|$ $5\sqrt{5}$

43. Simplify: i^{20} 1

44. Find the product: $\sqrt{3y} \cdot \sqrt{27y^2} \cdot \sqrt{y}$ $9y^2$

45. Simplify: $\dfrac{8}{\sqrt{2}}$ $4\sqrt{2}$

46. Find the product: $\left(2\sqrt{3x} - 5\right)\left(3\sqrt{3x} + 1\right)$
$18x - 13\sqrt{3x} - 5$

47. Simplify: $\sqrt{30} \cdot \sqrt{60}$ $30\sqrt{2}$

48. Find the solution set: $x^{\frac{3}{2}} = 8$ $x = 4$

49. Simplify: $\dfrac{3}{1 + \sqrt{2}}$ $3\sqrt{2} - 3$

50. Simplify: $(9 - 3i)(9 + 3i)$ 90

51. Simplify and write using rational exponents:
$\sqrt{2x^3y^2} \cdot \sqrt{4xy^5} \cdot \sqrt{12x^{10}y}$ $4x^7y^4(6)^{\frac{1}{2}}$

52. Simplify: $\dfrac{5}{\sqrt{10}}$ $\dfrac{\sqrt{10}}{2}$

53. Simplify: $\sqrt{400x^4}$ $20x^2$

54. Simplify: $\dfrac{\sqrt{1200x}}{\sqrt{80xy}}$ $\dfrac{\sqrt{15y}}{y}$

55. Combine: $x\sqrt[3]{27x^5y^2} - x^2\sqrt[3]{x^2y^2} + 2\sqrt[3]{x^8y^2}$
$4x^2\sqrt[3]{x^2y^2}$

56. Simplify: i^{18} -1

57. Write using rational exponents and simplify:
$\sqrt[3]{9x^7y^{12}} \cdot \sqrt[3]{6x^4y}$ $3x^3y^4(2x^2y)^{\frac{1}{3}}$

58. Simplify: i^{19} $-i$

59. Simplify: $\sqrt{\dfrac{5000xy^2}{250x^3y^6}}$ $\dfrac{2\sqrt{5}}{xy^2}$

60. Write using rational exponents and
simplify: $\sqrt{15xy^4} \cdot \sqrt{6xy^3}$ $3xy^3(10y)^{\frac{1}{2}}$

CHAPTER 7 TEST

This exam tests your knowledge of the material in Chapter 7.

1. Simplify:

   a. $\sqrt{225x^2y^2}$ $15xy$ **b.** $\sqrt[5]{-1}$ -1 **c.** $\sqrt{16x^4}$ $4x^2$

2. Find the product or quotient:

   a. $\sqrt{20x} \cdot \sqrt{60x}$ $20x\sqrt{3}$ **b.** $\sqrt{\dfrac{100xy}{25x}}$ $2\sqrt{y}$ **c.** $\dfrac{\sqrt{2}}{\sqrt{5}} \cdot \dfrac{\sqrt{20}}{\sqrt{18}}$ $\dfrac{2}{3}$

3. Simplify:
   a. $\dfrac{3\sqrt{20} - 2\sqrt{45} + \sqrt{5}}{\sqrt{5}}$
   b. $\sqrt{6}\left(\sqrt{96} - 3\sqrt{24} - 5\sqrt{12}\right)$
   $-12 - 30\sqrt{2}$
   c. $\dfrac{\sqrt{17} - 3\sqrt{34}}{2\sqrt{17}}$ $\dfrac{1 - 3\sqrt{2}}{2}$

4. Rationalize the denominator and simplify.
   a. $\dfrac{3\sqrt{2}}{\sqrt{2} - 1}$ $6 + 3\sqrt{2}$
   b. $\dfrac{4\sqrt{6}}{\sqrt{15}}$ $\dfrac{4\sqrt{10}}{5}$
   c. $\dfrac{2\sqrt{3} - 1}{2\sqrt{3} + 1}$ $\dfrac{13 - 4\sqrt{3}}{11}$

5. Solve:
   a. $\sqrt{3x + 4} = 8$
   $x = 20$
   b. $\sqrt{x - 5} - \sqrt{x - 8} = 3$
   no real solution
   c. $\sqrt{x + 5} + \sqrt{x} = 5$
   $x = 4$

6. Perform the indicated operations.
   a. Divide: $3^{\frac{3}{2}} x^{\frac{2}{3}} \div 3^{\frac{2}{3}} x^{\frac{1}{4}}$
   $\sqrt[12]{3^{10} x^5}$
   b. Simplify: $\dfrac{30x^{\frac{4}{3}}y^{\frac{1}{3}} - 20x^{\frac{1}{3}}y^{\frac{1}{3}}}{10x^{\frac{1}{3}}y^{\frac{1}{3}}}$
   $3x - 2$
   c. Find the solution set: $x^{\frac{1}{2}} = 3$
   $\{9\}$

7. Simplify:
   a. $(3 - 2i)(3 + 2i)$ 13
   b. $(2 + 3i)(6 - 5i)$ $27 + 8i$
   c. $\dfrac{1}{6 - 2i}$ $\dfrac{3}{20} + \dfrac{i}{20}$

CUMULATIVE REVIEW

CHAPTERS 1–7

The following exercises will help you maintain the skills you have learned in this and previous chapters.

1. Simplify: $3 \div 2 + 7 \div 8 - 10 \div 16$ $1\frac{3}{4}$

2. Define two sets A and B such that $A \cup B = \{2, 3\}$.   Answers may vary, but neither set may have elements other than 2 or 3.
   Examples: $A = \{2\}, B = \{3\}; A = \{2, 3\}, B = \{2\}; A = \{2, 3\}, B = \{2, 3\}; A = \{\ \}, B = \{2, 3\}$

3. A rule for determining medicine dosage for children when the adult dosage has been determined is Young's rule. It states
$$c = \dfrac{A}{A + 12}d$$
   where A = age of the child
   d = adult dose
   c = child's dose

   Find A when c is one ounce and d is 2.5 ounces. $A = 8$ years

4. Arrange in order from least to greatest:
   0.0019   0.00099   0.00100
   $0.00099 < 0.00100 < 0.0019$

5. Divide using scientific notation: $\dfrac{(3,800,000)}{(19,000)}$
   $(2)(10^2)$ or 200

6. Rewrite using only positive exponents:
   $\left(\dfrac{3x^5 y^{-3} z^{-2}}{x^4 y^{-2} z^5}\right)^{-1}$ $\dfrac{yz^7}{3x}$

7. Solve the system: $x + y = 10$
   $5x - 3y = 26$
   $x = 7, y = 3$, or $\{(7, 3)\}$

8. Simplify: $\dfrac{x^2 - 3x - 4}{\dfrac{3x - 12}{x + 1}}$ $\dfrac{(x + 1)^2}{3}$

9. Subtract 0.0035 from 2.1.   2.0965

10. Write $16\frac{1}{2}$ as a percent.   1650%

11. Find and graph the solution set: $|x - 2| > 4$
    $\{x \mid x < -2 \text{ and } x > 6\}$

12. Write the equation of the line that passes through the points $(0, 2)$ and $(5, 3)$.
    $y = \frac{1}{5}x + 2$

13. Solve for r: $\frac{I}{r} = pt$ $r = \frac{I}{pt}$

14. What is the slope of $6x - 3y = 2$?
    slope $= 2$

15. Simplify: $16(x - 5) + 18(x + 5)$ $34x + 10$

16. Solve: $|n - 7| = 9$ $\{-2, 16\}$

17. Simplify: $3x^2(5xy - 2x + 5y) - 16x^3y - 5x^3$ $-x^3y - 11x^3 + 15x^2y$

18. Norbert purchases some three-cent stamps and some one-cent stamps for $3.05. There are 19 more three-cent stamps than one-cent stamps. How many of each kind does he buy?   81 three-cent stamps and 62 one-cent stamps

19. Orcen bought h hot dogs for $1.25 each and d sodas for 75¢ each. How much, c in dollars, did he spend? $c = 1.25h + 0.75d$

20. The weight w of an object on or beneath Earth's surface varies directly as its distance d from the center of Earth. If an object weights 128 pounds on the surface of Earth, how far beneath the surface would the object have to be to weight 80 pounds? (Use 4000 miles as Earth's radius.)   1500 miles

8

QUADRATIC EQUATIONS

*T*he demand for a product, a certain type of denim jeans, for example, can be described as a function of their cost.

Manufacturers' surveys would try to determine the number of jeans suppliers would be willing to offer at different prices and how much customers would be willing to spend. Both the supply and the demand relationships could be represented by quadratic equations. This information, along with use of the quadratic formula can help the manufacturer determine the market price of the jeans when supply equals demand.

■ *Explain why knowledge of supply and demand—and quadratic equations—is useful in business.*

SKILLS CHECK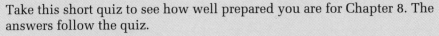

Take this short quiz to see how well prepared you are for Chapter 8. The answers follow the quiz.

1. Find: $(a - 2b)^2$

2. Simplify: $(2x - 3)(2x + 3)$

3. Solve: $-3x < -12$

4. Solve $x^2 - 2x - 3 = 0$ by using the principle of zero products.

5. Factor: $8x^2 - 26x + 21$

6. Combine: $-3x + 2x - 7(3x - 2)$

7. Let $f(x) = 5x^2 - 3x + 2$, and find $f(-1)$.

8. Simplify: $\left(\dfrac{5x^3}{4x}\right)^2$

ANSWERS: **1.** $a^2 - 4ab + 4b^2$ [Section 5.3] **2.** $4x^2 - 9$ [Section 5.3] **3.** $x > 4$ [Section 2.4]
4. $x = 3, -1$ [Section 5.5] **5.** $(4x - 7)(2x - 3)$ [Section 5.5] **6.** $-22x + 14$ [Section 2.1]
7. 10 [Section 5.2] **8.** $\dfrac{25x^4}{16}$ [Section 5.1]

CHAPTER LEAD-IN

The Summer Olympic Games typically include events such as the

1. hammer throw
2. high jump
3. javelin throw

Sketch the "shape" of each of these events. What do the shapes have in common?

8.1 Solving Quadratic Equations by Factoring or Taking Square Roots

SECTION GOALS

- *To solve quadratic equations by factoring*
- *To solve quadratic equations by taking square roots*
- *To solve quadratic equations by completing the square.*

SECTION LEAD-IN

A diver dives off of a platform and reaches the water in 2.0 seconds. Describe the shape of the path of the dive.

A second-degree equation in one variable is called a **quadratic equation.**

> The **standard form of a quadratic equation in one variable** is $ax^2 + bx + c = 0$, where a, b, and c are real numbers, and a is not equal to zero.

We shall discuss several different methods for solving quadratic equations.

■ ■ ■
WRITER'S BLOCK
What is meant by the *standard form* of an equation? What is useful about a standard form?

Solving by Factoring

When a quadratic equation is in standard form, the expression on the left is called a **quadratic polynomial.** As you saw in Section 5.5, we can solve some quadratic equations by factoring the quadratic polynomial and then applying the principle of zero products. That principle holds for complex numbers as well as for real numbers.

> Principle of Zero Products (expanded)
> If A and B represent complex numbers and $A \cdot B = 0$, then $A = 0$, or $B = 0$, or both.

INSTRUCTOR NOTE
Remind students that integers, rational numbers, and real numbers can be classified as complex numbers also.

To use this principle, set each factor of the quadratic polynomial equal to zero and solve. You will obtain one solution for each factor. Here's an example to illustrate the procedure:

▪ ▪ ▪

E**XAMPLE** 1

Solve: $5x^2 = 17x + 12$

S**OLUTION**

We first rewrite the equation in standard form, obtaining

$$5x^2 - 17x - 12 = 0$$

The terms of the quadratic polynomial have no common factors, so now we must factor the polynomial. Thus the original equation becomes

$$(x - 4)(5x + 3) = 0$$

Now we set each factor equal to zero and solve.

$$
\begin{array}{c|c}
x - 4 = 0 & 5x + 3 = 0 \\
x = 4 & 5x = -3 \\
& x = \dfrac{-3}{5}
\end{array}
$$

We check by substituting in the original equation. So the solution set is $\left\{4, \dfrac{-3}{5}\right\}$.

STUDY HINT

Check your results every time.

▶ *CHECK* **Warm-Up 1**

Calculator Corner

Use your graphing calculator to graph the equation $Y1 = 6x^2 + x - 15$ on a "friendly" grid or screen. Check your graphing calculator manual to see how to do this on your particular model. The screens that follow are from the *TI-82/83*.

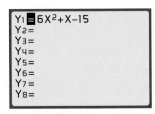

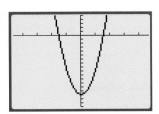

This graph is called a **parabola.**

Now **TRACE** on the graph to find the two points where the parabola crosses the *x*-axis.

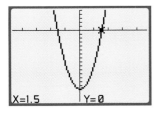 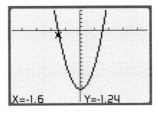 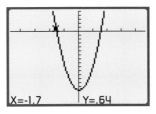

The first screen above shows that the parabola crosses the *x*-axis at the point $(1.5, 0)$. The second and third screens above show that it also crosses the *x*-axis somewhere between the point $(-1.7, 0.64)$ and the point $(-1.6, -1.24)$. If you find the points where the parabola crosses the *x*-axis by the paper-and-pencil method, you will get the point $\left(-\frac{5}{3}, 0\right)$, which can also be written as $(-1.667, 0)$.

Some graphing calculators have a **CALC**ulation utility that can calculate the *x*-intercept directly on the graph screen. The following screens are for the *TI-82/83* graphing calculators. (Note: See the Calculator Corner on pages 510–512 to review how to find the roots of an equation.)

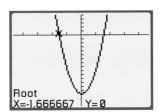

Because $x = -\frac{5}{3}$ equals -1.666667 when it is rounded off, you have obtained the same point as you did when you used the paper-and-pencil method.

Our next equation has complex solutions.

▪▪▪
EXAMPLE 2

The factored form of $x^2 + 4x + 8 = 0$ is $(x + 2 + 2i)(x + 2 - 2i) = 0$. Solve the equation.

SOLUTION

To solve, we simply set each factor equal to zero.

$$
\begin{array}{c|c}
x + 2 + 2i = 0 & x + 2 - 2i = 0 \\
x = -2 - 2i & x = -2 + 2i
\end{array}
$$

INSTRUCTOR NOTE

You may want to comment that we do not have procedures for factoring equations like $x^2 + 4x + 8 = 0$. We obtain these factors by first using the quadratic formula and then using the roots to construct the factors.

Check:

$$(x + 2 + 2i)(x + 2 - 2i) = 0$$

$$(-2 - 2i + 2 + 2i)(-2 - 2i + 2 - 2i) = 0$$

$$(0)(-4i) = 0 \quad \text{True}$$

$$(x + 2 + 2i)(x + 2 - 2i) = 0$$

$$(-2 + 2i + 2 + 2i)(-2 + 2i + 2 - 2i) = 0$$

$$(4i)(0) = 0 \quad \text{True}$$

So the solution set is $\{-2 - 2i, -2 + 2i\}$.

▶ CHECK **Warm-Up 2**

The solutions of an equation are also called its **roots.** We can use the principle of zero products to write an equation when we know its roots.

▪▪▪
EXAMPLE 3

Find a quadratic equation that has the roots $2 + i$ and $2 - i$.

SOLUTION

The roots of the equation are $2 + i$ and $2 - i$, so we can write

$$x = 2 + i \quad \text{and} \quad x = 2 - i$$

These equations can also be written as

$$x - 2 - i = 0 \quad \text{and} \quad x - 2 + i = 0$$

Because zero times zero equals zero, we can now write

$$(x - 2 - i)(x - 2 + i) = 0$$

Multiplying out on the left side yields

$$x^2 - 2x + xi - 2x + 4 - 2i - xi + 2i - i^2 = 0$$

$$x^2 - 4x + 4 - i^2 = 0$$

$$x^2 - 4x + 5 = 0$$

▶ CHECK **Warm-Up 3**

The solution to Example 3 is not unique. Any multiple of the equation $x^2 - 4x + 5 = 0$ has the roots $2 + i$ and $2 - i$. So an equation such as

$$3x^2 - 12x + 15 = 0$$

also has those roots.

Solving by Taking Square Roots

When the coefficient of the x-term is zero, a quadratic equation can be put in the form $x^2 = $ a number. Then the following property can be used to solve it.

> **Equality Property of Square Roots**
> For any real number c, if $a^2 = c$, then $a = \sqrt{c}$ or $a = -\sqrt{c}$. We say $a = \pm\sqrt{c}$.

■■■
WRITER'S BLOCK
Contrast the *equality property of squaring* and the *equality property of square roots*. When is each used?

In an equation, a can be any expression that represents a real or complex number. Applying this property is like taking the square roots of both sides of an equation. We get two solutions, except when $c = 0$; in the case of $c = 0$, we get only one solution, $a = 0$.

▪▪▪
EXAMPLE 4

Solve the equation: $y^2 + 81 = 0$

SOLUTION

Adding -81 to both sides of the given equation yields

$$y^2 = -81$$

This equation is in proper form to apply the equality property of square roots, which tells us that

$y = \sqrt{-81}$ and	$y = -\sqrt{-81}$	Equality property of square roots
$y = i\sqrt{81}$	$y = -i\sqrt{81}$	Definition of $i = \sqrt{-1}$
$y = 9i$	$y = -9i$	Simplifying

So the solution set is $\{9i, -9i\}$ or $\{\pm 9i\}$. Check your work.

▶ *CHECK* **Warm-Up 4**

❗❗❗
ERROR ALERT
Identify the error and give a correct answer.
Solve: $(x + 3)^2 = 16$
Incorrect Solution:
$(x + 3)^2 = 16$
$x + 3 = 4$
$x = 1$

▪▪▪
EXAMPLE 5

Solve: $5y^2 = 32$

SOLUTION

This equation is not in the form $x^2 = c$, but we can put it in that form by multiplying both sides by $\frac{1}{5}$.

$$y^2 = \frac{32}{5}$$

Now we apply the equality property of square roots.

$y = \sqrt{\frac{32}{5}}$ and	$y = -\sqrt{\frac{32}{5}}$	
$y = \sqrt{16}\sqrt{\frac{2}{5}}$	$y = -\sqrt{16}\sqrt{\frac{2}{5}}$	Rewriting
$y = 4\sqrt{\frac{2}{5}}$	$y = -4\sqrt{\frac{2}{5}}$	Simplifying

We can rationalize the denominators by writing

$$\sqrt{\frac{2}{5}} = \frac{\sqrt{2}}{\sqrt{5}} = \frac{\sqrt{2} \cdot \sqrt{5}}{\sqrt{5} \cdot \sqrt{5}} = \frac{\sqrt{10}}{5}$$

STUDY HINT
All quadratic equations have 2 roots. However, sometimes the roots may be identical. We say these are double roots.

Thus the solution set is $\left\{\frac{4}{5}\sqrt{10}, -\frac{4}{5}\sqrt{10}\right\}$. You should check both solutions by substituting in the original equation.

▶ *CHECK* **Warm-Up 5**

Calculator Corner

It is possible to solve $(2x + 3)^2 = 16$ by graphing each side of the equation. This method is called the *multi-graph method* because you are graphing more than one equation in order to find the point(s) where the two functions are *equal (intersect)*.

Use the following steps to find the point(s) of intersection on your graphing calculator.

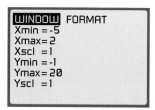
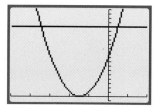

Now you can use **TRACE** to estimate the two points of intersection, *or*, if your graphing calculator has the capability, you could **CALC**ulate the two points of intersection. (Note: See the Calculator Corner on pages 89 and 90 to review how to find the points of intersection of two graphs.)

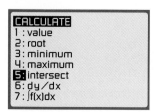
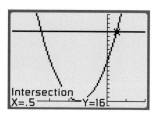
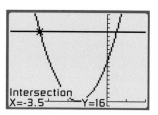

The equation in the next example is in a form different from the others we've looked at so far; but it is in the correct form to solve by taking square roots.

▪▪▪
EXAMPLE 6

Solve: $(3y + 3)^2 = -16$

SOLUTION

The left side of the equation is written as a square, so we may write

$$3y + 3 = \sqrt{-16} \quad \text{or} \quad 3y + 3 = -\sqrt{-16}$$

$$3y + 3 = 4i \qquad\qquad 3y + 3 = -4i$$

$$3y = -3 + 4i \qquad\qquad 3y = -3 - 4i$$

$$y = -1 + \tfrac{4}{3}i \qquad\qquad y = -1 - \tfrac{4}{3}i$$

Check these results in the original equation.

▶ CHECK **Warm-Up 6**

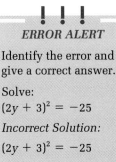

! ! !
ERROR ALERT

Identify the error and give a correct answer.

Solve:
$(2y + 3)^2 = -25$

Incorrect Solution:

$(2y + 3)^2 = -25$

$2y + 3 = -5$

$2y = -8$

$y = -4$

Completing the Square

Most quadratic equations are not easy to factor. If we could transform such equations so that one side was the square of a binomial, as in the equation $(2y + 3)^2 = 16$, then we could solve it by taking square roots. Actually, we can do exactly that, and the method we use is called **completing the square.**

To complete the square for a quadratic equation

1. Write the equation in the form $x^2 + bx = c$, where b and c are real numbers.
2. Find $\frac{b}{2}$, square it, and add the result to both sides of the equation.
3. Write the left side of the equation as $\left(x + \frac{b}{2}\right)^2$, and simplify the right side of the equation.
4. Use the equality property of square roots to complete the solution.

▪▪▪

EXAMPLE 7

Solve by completing the square:

$$x^2 - 6x + 8 = 0$$

SOLUTION

STEP 1: This equation is in standard form. But we want the terms that contain the variable to be on the left and the constant on the right. So we add -8 to both sides, obtaining

$$x^2 - 6x = -8$$

The equation is now in the proper form for completing the square.

STEP 2: Because b (the coefficient of x) is -6

$$\frac{b}{2} \text{ is } \frac{-6}{2} \quad \text{and} \quad \left(\frac{b}{2}\right)^2 \text{ is } (-3)^2$$

INSTRUCTOR NOTE

Encourage your students to become proficient with many different ways to solve quadratic equations. One way is not enough.

INSTRUCTOR NOTE

Students should always use parentheses in substituting so the entire $\left(\frac{b}{2}\right)$ is squared, not simply b.

We add this value, $(-3)^2$, to both sides of the equation.

$$x^2 - 6x + (-3)^2 = -8 + (-3)^2$$

WRITER'S BLOCK

Why do we want to "complete the square" to solve a quadratic equation?

STEP 3: We have transformed the left side of the equation into the square of the binomial $[x + (-3)]^2$. To see that this is so, multiply it out. You get

$$[x + (-3)]^2 = x^2 + 2(x)(-3) + (-3)^2$$
$$= x^2 + (-6)x + (-3)^2$$
$$= x^2 - 6x + (-3)^2$$

We have "completed the square" of the left side.
We write the left side as $[x + (-3)]^2$ and simplify the right side, obtaining

$$[x + (-3)]^2 = -8 + (-3)^2$$
$$= -8 + 9 \qquad \text{Squaring}$$
$$= 1 \qquad \text{Combining}$$

STEP 4: To solve

$$[x + (-3)]^2 = 1$$

INSTRUCTOR NOTE

Reinforce to students that we keep the left side of the equation in the form $\left[x + \left(\dfrac{-b}{2}\right)\right]^2$. Then we can easily take its square root.

we take the square root of both sides of the equation.

$$x + (-3) = 1 \quad \text{or} \quad x + (-3) = -1$$

Thus x has two values, which we find as follows:

$$
\begin{array}{l|ll}
x - 3 = 1 & x - 3 = -1 & \\
\quad = 1 + 3 & \quad = -1 + 3 & \text{Adding 3 to both sides} \\
\quad = 4 & \quad = 2 & \text{Combining like terms}
\end{array}
$$

So $x = 4$ and $x = 2$ are solutions to the equation $x^2 - 6x + 8 = 0$. You should check both solutions to make sure they satisfy the original equation.

▶ CHECK **Warm-Up 7**

▪ ▪ ▪

EXAMPLE 8

Solve: $n^2 - 6n - 10 = 0$

WRITER'S BLOCK

Write a checklist that a student can follow when he or she is "completing a square."

SOLUTION

We must first rewrite the equation in such a way that the variable terms appear on one side of the equation and the constant term on the other side. To do so, we add 10 to both sides.

$$n^2 - 6n = 10$$

To complete the square, we need to add $\left(\dfrac{b}{2}\right)^2$ to both sides. Because b in this equation is -6, $\dfrac{b}{2}$ is $\dfrac{-6}{2} = -3$. Then we add $(-3)^2$ to both sides.

$$n^2 - 6n + (-3)^2 = 10 + (-3)^2$$

We have "completed the square," so the left side of this equation is equivalent to $[n + (-3)]^2$. We must also simplify the right side.

$$[n + (-3)]^2 = 10 + (-3)^2$$
$$= 10 + 9$$
$$= 19$$

Now we apply the equality property of square roots. Taking the square roots of both sides of $[n + (-3)]^2 = 19$, we obtain

$$n - 3 = \sqrt{19} \quad \text{or} \quad n - 3 = -\sqrt{19}$$

Adding 3 to both sides of each equation gives

$$n = 3 + \sqrt{19} \quad \text{or} \quad n = 3 - \sqrt{19}$$

We will check $n = 3 - \sqrt{19}$ in the original equation; you should check $n = 3 + \sqrt{19}$ yourself.

$$n^2 - 6n - 10 = 0 \quad \text{Original equation}$$
$$\left(3 - \sqrt{19}\right)^2 - 6\left(3 - \sqrt{19}\right) - 10 = 0 \quad \text{Substituting } n = 3 - \sqrt{19}$$
$$9 - (2)\left(3\sqrt{19}\right) + 19 - 18 + 6\sqrt{19} - 10 = 0 \quad \text{Multiplying out}$$
$$9 - 6\sqrt{19} + 19 - 18 + 6\sqrt{19} - 10 = 0 \quad \text{Simplifying}$$
$$28 - 28 - 6\sqrt{19} + 6\sqrt{19} = 0 \quad \text{Combining like terms}$$
$$0 = 0 \quad \text{True}$$

▶ *CHECK* **Warm-Up 8**

▪ ▪ ▪

EXAMPLE 9

Solve $4x^2 - 12x + 13 = 0$ by completing the square.

SOLUTION

We rewrite the equation and divide both sides by 4, the coefficient of the x^2 term.

$$4x^2 - 12x = -13$$
$$x^2 - 3x = \frac{-13}{4}$$

We then add $\left(\frac{b}{2}\right)^2$ to both sides of the equation and simplify.

$$x^2 - 3x + \left(\frac{-3}{2}\right)^2 = \frac{-13}{4} + \left(\frac{-3}{2}\right)^2$$
$$x^2 - 3x + \left(\frac{-3}{2}\right)^2 = \frac{-13}{4} + \frac{9}{4}$$
$$x^2 - 3x + \left(\frac{-3}{2}\right)^2 = -1$$

We write the left side of the equation as a square.

$$\left(x - \frac{3}{2}\right)^2 = -1$$

STUDY HINT

The squared term must have a coefficient of 1 before you can complete the square. If it does not, divide each term by the coefficient of the squared term before you complete the square.

▶CHECK **Warm-Up 9**

Applying the equality property of square roots, we obtain

$$x - \frac{3}{2} = i \quad \text{or} \quad x - \frac{3}{2} = -i$$
$$x = \frac{3}{2} + i \qquad\qquad x = \frac{3}{2} - i$$

Remember to check the solutions.

Practice what you learned.

SECTION FOLLOW-UP

The path of her dive is in the shape of a **parabola.** The equation of this path is, roughly, quadratic.

8.1 WARM-UPS

Work these problems before you attempt the exercises.

1. Solve by factoring: $2x^2 - 9x = 5$ $\left\{\frac{-1}{2}, 5\right\}$

2. The factored form of $x^2 + 6x + 18 = 0$ is $(x + 3 + 3i)(x + 3 - 3i) = 0$. Solve the equation.
    $x = -3 - 3i$ and $x = -3 + 3i$

3. Find three quadratic equations that have the roots $3 - 2i$ and $3 + 2i$. $x^2 - 6x + 13 = 0;\ 2x^2 - 12x + 26 = 0;\ -5x^2 + 30x - 65 = 0.$
    Answers may vary.

! ! !

ERROR ALERT

Identify the error and give a correct answer.

Complete the square:
$8x^2 + 4x = 9$

Incorrect Solution:

$8x^2 + 4x = 9$

$8x^2 + 4x + 2^2$

$\quad = 9 + 2^2$

$(8x + 2)^2 = 13$

$8x + 2 = \sqrt{13}$

$8x = -2 + \sqrt{13}$

$x = \frac{-2\sqrt{13}}{8}$

! ! !

ERROR ALERT

Identify the error and give a correct answer.

Complete the square of the equation
$2x^2 - 8x - 64 = 0.$

Incorrect Solution:

$2x^2 - 8x - 64 = 0$

$\frac{b}{2} = \frac{-8}{2} = -4$

$\left(\frac{b}{2}\right)^2 = (-4)^2 = 16$

$2x^2 - 8x + 16$

$\quad = 64 + 16$

$(x - 4)^2 = 80$

$x - 4 = \pm\sqrt{80}$

$x = 4 \pm 4\sqrt{5}$

4. Solve: $x^2 + 64 = 0$ $x = \pm 8i$

5. Solve: $6x^2 = 48$ $x = \pm 2\sqrt{2}$

6. Solve: $(-3x + 2)^2 + 49 = 0$ $x = \frac{2 \pm 7i}{3}$

7. Solve by completing the square: $49x^2 + 7x - 20 = 0$ $\left\{\frac{4}{7}, \frac{-5}{7}\right\}$

8. Solve by completing the square: $x^2 + 6x + 12 = 0$ $x = -3 \pm i\sqrt{3}$

9. Solve $16x^2 - 8x + 145 = 0$ by completing the square.
   $x = \frac{1}{4} + 3i$ and $x = \frac{1}{4} - 3i$

8.1 EXERCISES

Note: Use your graphing calculator to check your results whenever possible.

In Exercises 1 through 18, find the solution set for each of the quadratic equations by factoring.

1. $x^2 + 7x + 6 = 0$ $\{-1, -6\}$
2. $x^2 - 3x - 28 = 0$ $\{-4, 7\}$
3. $x^2 + 5x = 36$ $\{-9, 4\}$

4. $x^2 + 6x = 72$ $\{-12, 6\}$
5. $6x^2 - x - 15 = 0$ $\left\{\frac{5}{3}, \frac{-3}{2}\right\}$
6. $10x^2 + 61x + 72 = 0$ $\left\{\frac{-9}{2}, \frac{-8}{5}\right\}$

7. $24x^2 + 29x + 7 = 0$ $\left\{\frac{-7}{8}, \frac{-1}{3}\right\}$
8. $10x^2 - 3x - 4 = 0$ $\left\{\frac{4}{5}, \frac{-1}{2}\right\}$
9. $4x^2 - 17x - 15 = 0$ $\left\{\frac{-3}{4}, 5\right\}$

10. $30x^2 + 28x - 2 = 0$ $\left\{-1, \frac{1}{15}\right\}$
11. $6x^2 + 11x - 7 = 0$ $\left\{\frac{-7}{3}, \frac{1}{2}\right\}$
12. $35x^2 + 11x - 6 = 0$ $\left\{\frac{-3}{5}, \frac{2}{7}\right\}$

13. $27x^2 + 3x - 2 = 0$ $\left\{\frac{2}{9}, \frac{-1}{3}\right\}$
14. $35x^2 + 46x = 16$ $\left\{\frac{-8}{5}, \frac{2}{7}\right\}$
15. $-20x - 32 = -63x^2$ $\left\{\frac{-4}{7}, \frac{8}{9}\right\}$

16. $-23x + 6 = -7x^2$ $\left\{\frac{2}{7}, 3\right\}$
17. $-32 - 12x = -27x^2$ $\left\{\frac{-8}{9}, \frac{4}{3}\right\}$
18. $10 - 13x = -4x^2$ $\left\{\frac{5}{4}, 2\right\}$

In Exercises 19 through 34, find an equation that has the given roots.

19. 6 and 3
    $x^2 - 9x + 18 = 0$
20. 2 and -5
    $x^2 + 3x - 10 = 0$
21. -6 and -4
    $x^2 + 10x + 24 = 0$

22. -5 and 6
    $x^2 - x - 30 = 0$
23. -3 and $\frac{2}{3}$
    $3x^2 + 7x - 6 = 0$
24. 5 and $\frac{-1}{2}$
    $2x^2 - 9x - 5 = 0$

25. -0.3 and -2.5
    $x^2 + 2.8x + 0.75 = 0$
26. -6.8 and 0.2
    $x^2 + 6.6x - 1.36 = 0$
27. $3 + \sqrt{2}$ and $3 - \sqrt{2}$
    $x^2 - 6x + 7 = 0$

28. $5 + \sqrt{3}$ and $5 - \sqrt{3}$
    $x^2 - 10x + 22 = 0$
29. $1 - 2\sqrt{3}$ and $1 + 2\sqrt{3}$
    $x^2 - 2x - 11 = 0$
30. $2 + 4\sqrt{3}$ and $2 - 4\sqrt{3}$
    $x^2 - 4x - 44 = 0$

31. $5 + i$ and $5 - i$
    $x^2 - 10x + 26 = 0$
32. $6 - i$ and $6 + i$
    $x^2 - 12x + 37 = 0$
33. $3 - 2i$ and $3 + 2i$
    $x^2 - 6x + 13 = 0$
34. $2 + 5i$ and $2 - 5i$
    $x^2 - 4x + 29 = 0$

In Exercises 35 through 55, solve the equations by taking square roots.

35. $x^2 - 289 = 0$
    $x = 17$ and -17
36. $x^2 - 441 = 0$
    $x = 21$ and -21
37. $25x^2 - 36 = 0$
    $x = \frac{6}{5}$ and $\frac{-6}{5}$
38. $9x^2 - 100 = 0$
    $x = \frac{10}{3}$ and $\frac{-10}{3}$

39. $64x^2 - 49 = 0$
$x = \frac{7}{8}$ and $\frac{-7}{8}$

40. $81x^2 - 121 = 0$
$x = \frac{11}{9}$ and $\frac{-11}{9}$

41. $x^2 = 35$
$x = -\sqrt{35}$ and $\sqrt{35}$

42. $r^2 = 56$
$r = 2\sqrt{14}$ and $-2\sqrt{14}$

43. $x^2 - 6 = 0$
$x = \pm\sqrt{6}$

44. $x^2 - 23 = 0$
$x = \pm\sqrt{23}$

45. $x^2 + 25 = 0$
$x = \pm 5i$

46. $x^2 + 36 = 0$
$x = \pm 6i$

47. $(x + 1)^2 - 6 = 0$
$x = -1 + \sqrt{6}$ and $-1 - \sqrt{6}$

48. $(x - 2)^2 - 8 = 0$
$x = 2 \pm 2\sqrt{2}$

49. $(2x + 3)^2 - 5 = 0$
$x = \frac{-3 \pm \sqrt{5}}{2}$

50. $(3x + 2)^2 - 8 = 0$
$x = \frac{-2 \pm 2\sqrt{2}}{3}$

51. $(x - 6)^2 + 25 = 0$
$x = 6 \pm 5i$

52. $(x + 5)^2 + 36 = 0$
$x = -5 \pm 6i$

In Exercises 53 through 56, write each equation in the form $x^2 + bx = c$ and find $\left(\frac{b}{2}\right)^2$.

53. $-x^2 + 16x - 180 = 0$ $x^2 - 16x = -180; 64$

54. $2y^2 = 12y + 84$ $y^2 - 6y = 42; 9$

55. $6 - x^2 + \frac{1}{3}x = 0$ $x^2 - \frac{1}{3}x = 6; \frac{1}{36}$

56. $2x^2 + 3x + 8 = 0$ $x^2 + \frac{3}{2}x = -4; \frac{9}{16}$

In Exercises 57 through 60, write each equation in the form $(x + a)^2 = 0$ or $(x - a)^2 = 0$.

57. $x^2 + 3x + \left(\frac{3}{2}\right)^2 = 0$ $\left(x + \frac{3}{2}\right)^2 = 0$

58. $x^2 - 5x + \left(\frac{-5}{2}\right)^2 = 0$ $\left(x - \frac{5}{2}\right)^2 = 0$

59. $x^2 - 20x + \left(\frac{-20}{2}\right)^2 = 0$ $(x - 10)^2 = 0$

60. $x^2 - 16x + \left(\frac{-16}{2}\right)^2 = 0$ $(x - 8)^2 = 0$

In Exercises 61 through 84, solve the equations by completing the square.

61. $x^2 - 20x = 261$ $\{29, -9\}$

62. $x^2 - 10x - 75 = 0$ $\{15, -5\}$

63. $x^2 + 20x - 96 = 0$ $\{4, -24\}$

64. $x^2 + 36x + 224 = 0$ $\{-8, -28\}$

65. $x^2 + 26x - 87 = 0$ $\{3, -29\}$

66. $x^2 - 40x - 129 = 0$ $\{43, -3\}$

67. $10x^2 - x - 3 = 0$ $\left\{\frac{3}{5}, \frac{-1}{2}\right\}$

68. $4x^2 + 12x - 27 = 0$ $\left\{\frac{3}{2}, \frac{-9}{2}\right\}$

69. $9t^2 - 18t + 8 = 0$ $\left\{1\frac{1}{3}, \frac{2}{3}\right\}$

70. $-4x^2 - 26x - 36 = 0$ $\left\{-2, \frac{-9}{2}\right\}$

71. $6x^2 - x - 12 = 0$ $\left\{1\frac{1}{2}, -1\frac{1}{3}\right\}$

72. $9x^2 - 3x - 2 = 0$ $\left\{\frac{2}{3}, \frac{-1}{3}\right\}$

73. $100x^2 - 100x - 39 = 0$
$\{1.3, -0.3\}$

74. $50x^2 + 100x + 48 = 0$
$\{-0.8, -1.2\}$

75. $10x^2 - 204x + 80 = 0$ $\{0.4, 20\}$

76. $x^2 + 8x - 4 = 0$
$\left\{-4 + 2\sqrt{5}, -4 - 2\sqrt{5}\right\}$

77. $x^2 + 4x - 8 = 0$
$\left\{-2 + 2\sqrt{3}, -2 - 2\sqrt{3}\right\}$

78. $x^2 + 3x + \frac{1}{4} = 0$
$\left\{\frac{-3 + 2\sqrt{2}}{2}, \frac{-3 - 2\sqrt{2}}{2}\right\}$

79. $x^2 + 10x + 34 = 0$
$\{-5 + 3i, -5 - 3i\}$

80. $x^2 - 16x + 100 = 0$
$\{8 + 6i, 8 - 6i\}$

81. $x^2 - 6x + 25 = 0$
$\{3 + 4i, 3 - 4i\}$

82. $10x^2 + 40x + 50 = 0$
$\{-2 + i, -2 - i\}$

83. $4t^2 + 12t + 13 = 0$
$\left\{\frac{-3}{2} + i, \frac{-3}{2} - i\right\}$

84. $x^2 - 8x + 17 = 0$
$\{4 + i, 4 - i\}$

EXCURSIONS

Data Analysis

1. The concept of a "moving average" has been used in business to determine if a stock is increasing or decreasing in price when viewed several months (or years) at a time. To calculate a three-year moving average of snow accumulation, we average the three-year periods 1987–1990,

1988–1991, 1989–1992, and so on. When you have completed the calculations, analyze the results and write a paragraph reporting them.

SNOW ACCUMULATION IN CENTRAL PARK		
Year	Through January *(inches)*	Total for season *(inches)*
1995-96	40.5	75.6
1994-95	0.2	11.8
1993-94	18.9	53.4
1992-93	1.9	24.5
1991-92	2.2	12.6
1990-91	15.6	24.9
1989-90	7.9	13.4
1988-89	5.3	8.1
1987-88	17.6	19.1

Source: Weather Services Corporation, Lexington, Mass.

2. Information can be given in a number of ways. The figure shows an address locator for Manhattan. Use it to find where each of the following is located—that is, between which two cross streets.

Guggenheim Museum   1071 5th Avenue
Hunter College   695 Park Avenue
Morgans Hotel   237 Madison Avenue
Sylvia's Restaurant   328 Lenox Avenue
Sign of the Dove   1110 Third Avenue

A R I T H M E T I C   O F   T H E   A V E N U E S

Hunting for an address on an avenue in Manhattan without knowing the cross street does not have to be difficult. Just drop the last figure, divide by 2 and add or subtract as indicated below. The answer is the nearest numbered cross street. This key does not apply to Broadway below Eighth Street because of the many streets with names rather than numbers.

Ave. A, B, C, D   add 3

1st and 2nd Ave.   add 3

3rd Ave.   add 10

4th Ave.   add 8

5th Ave.
Up to 200 add 13
Up to 400 add 16
Up to 600 add 18
Up to 775 add 20
775 to 1286 drop last figure and subtract 18
Up to 1500 add 45
Up to 2000 add 24

Broadway
754 to 858 subtract 29
858 to 958 subtract 25
Above 1000 subtract 30

Amsterdam Ave.   add 60

Audubon Ave.   add 165

W. 55TH ST.
W. 54TH ST.
W. 53RD ST.
W. 52ND ST.
BROADWAY
SEVENTH AVENUE
AVE. OF THE AMERICAS

Ave. of the Americas   subtract 12

7th Ave.   add 12
   Above 110th St. add 20

8th Ave.   add 10

9th Ave.   add 13

10th Ave.   add 14

Central Park West
divide street number by 10 and add 60

Columbus Ave.   add 60

Convent Ave.   add 127

Edgecombe Ave.   add 134

Ft. Washington Ave.   add 158

Lenox Ave.   add 110

Lexington Ave.   add 22

Madison Ave.   add 26

Manhattan Ave.   add 100

Park Ave.   add 35

Pleasant Ave.   add 101

Riverside Drive
divide house number by 10 and add 72 up to 165th St.

St. Nicholas Ave.   add 110

Wadsworth Ave.   add 173

West End Ave.   add 60

Source: New York Times, 19 January 1997. Copyright ©1997 by The New York Times Co. Reprinted by permission.

How do you locate addresses in your town?

3.* a. Choose the graph that most closely represents the situation. Justify your choice by analyzing the graph and writing a paragraph about it.

    i. A boy drags a sled up a hill and then slides back down.

    ii. A car pulls up to the curb and a passenger gets in.

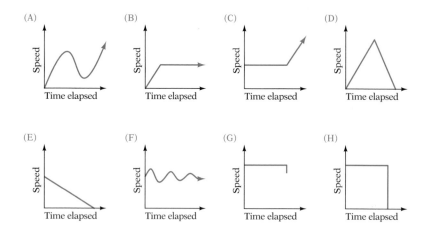

b. Write a verbal "scene" that could represent one of the remaining graphs. Give it to a classmate and see if he or she can identify the graph you described.

*Adapted with permission from "Relating to Graphs in Introductory Algebra," by Frances Van Dyke, Mathematics Teacher, copyright September 1994 by the National Council of Teachers of Mathematics.

Posing Problems

4. Ask and answer four questions using this data.

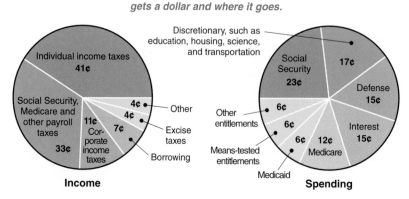

SLICING IT UP

Where the federal government gets a dollar and where it goes.

Source: New York Post, 7 February 1997.

Spheres

Volume of a Sphere

A sphere is shown in the accompanying figure; it is a three-dimensional circle. Every point on the sphere is at a distance r (the radius) from the center.

The volume of a sphere is found with a formula:

Sphere

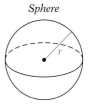

> **To find the volume of a sphere**
>
> If r is the radius of the sphere, then
> $$V = \frac{4}{3}\pi r^3$$

▪ ▪ ▪

EXAMPLE

Find the volume of a sphere with a radius of 6 inches.

SOLUTION

To find the volume, we substitute and simplify.

$$V = \frac{4}{3}\pi r^3$$
$$= \frac{4}{3}(3.14)(6)^3$$
$$= \frac{4}{3}(3.14)(216)$$
$$= 904.32$$

So the volume of the sphere is about 904 cubic inches. Note that volume is always measured in cubic units.

Surface Area of a Sphere

If we think about the amount of surface "covering" a sphere or cylinder, we are considering the "surface area." Area is always expressed in square units. The surface area of a sphere is four times the area of a circle with the same radius as the sphere.

> **To find the surface area of a sphere**
>
> If r is the radius of the sphere, then
> $$SA = 4\pi r^2$$

▪ ▪ ▪

EXAMPLE

Find the surface area of a sphere with a radius of 2.1 units.

SOLUTION

To find the surface area, we substitute in the formula and simplify.

$$SA = 4\pi r^2$$
$$= 4(3.14)(2.1)^2 = 12.56(4.41)$$
$$= 55.3896$$

The surface area of this sphere is about 55 square units.

◢

PRACTICE

Use the following information to answer Exercises 1 and 2. Neither the sun nor the planets are perfect spheres; they are flattened somewhat at their poles. Use their given diameters to determine their approximate surface areas. Give your answers to the nearest 10 million square miles. Use π on your calculator.

1. a. Sun: diameter 865,500 miles
        2,353,330,000,000 mi^2
    b. Pluto: diameter 3700 miles
        40,000,000 mi^2
    c. Venus: diameter 7521 miles
        180,000,000 mi^2
    d. Neptune: diameter 30,800 miles
        2,980,000,000 mi^2
    e. Saturn: diameter 74,600 miles
        17,480,000,000 mi^2

2. a. Mercury: diameter 3032 miles
        30,000,000 mi^2
    b. Mars: diameter 4217 miles
        60,000,000 mi^2
    c. Earth: diameter 7926 miles
        200,000,000 mi^2
    d. Uranus: diameter 32,200 miles
        3,260,000,000 mi^2
    e. Jupiter: diameter 88,700 miles
        24,720,000,000 mi^2

3. a. 3.39×10^{17} mi^3
    b. 2.65×10^{10} mi^3
    c. 2.23×10^{11} mi^3
    d. 1.53×10^{13} mi^3
    e. 2.17×10^{14} mi^3
4. a. 1.46×10^{10} mi^3
    b. 3.93×10^{10} mi^3
    c. 2.61×10^{11} mi^3
    d. 1.75×10^{13} mi^3
    e. 3.65×10^{14} mi^3

3. Find the approximate volume of the sun and of each planet in Exercise 1. Give your answers in scientific notation.

4. Find the approximate volume of each planet in Exercise 2. Give your answers in scientific notation.

In Exercises 5 and 6, find the surface area and the volume of each ball. Round answers to the nearest ten. Use your calculator if you wish.
Answers may vary according to rounding.

5. a. A basketball has a circumference of 78 centimeters.
        $SA = 1940$ cm^2; $V = 8010.70$ cm^3
    b. A cricket ball has a circumference of 9 inches.
        $SA = 26$ in.^{2}; $V = 12$ in.^{3}

6. a. A soccer ball has a circumference of 28 inches.
        $SA = 250$ in.^{2}; $V = 370$ in.^{3}
    b. A volleyball has a circumference of 67 centimeters.
        $SA = 1430$ cm^2; $V = 5080$ cm^3

7. The surface of a sphere with a radius of $\frac{1}{5}$ unit is to be painted. How much paint will be needed if 1 gallon of the paint covers exactly 12.5 square units?   0.04 gallon

8. You double the radius of a sphere. How does its surface area change?
It becomes 4 times as large.

9. A beach ball with a radius of 12 inches is filled with air. Another beach ball with a radius of 15 inches is also filled with air. What is the difference in volume between the two balls?   about 6900 cubic inches

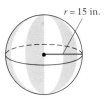

10. Find the surface area of (a) a sphere with a radius of 0.001 unit and (b) a sphere whose radius is 4 times greater. How do these surface areas compare?   (a) 0.00001257 square unit; (b) 0.00020106 square unit; (b) is 16 times (a).

8.2 Solving Quadratic Equations

SECTION LEAD-IN

A volleyball player serves the ball. Describe the path of the volleyball.

Using the Quadratic Formula

The **quadratic formula,** can be used quickly to find the roots of any quadratic equation. You should memorize this formula:

Quadratic Formula

The quadratic equation

$$ax^2 + bx + c = 0$$

where a, b and c are real numbers and $a \neq 0$, has solutions

$$x = \frac{-b \pm \sqrt{b^2 - 4ac}}{2a}$$

To solve a given equation, we substitute the values of a, b, and c, from the equation into the formula and then simplify.

SECTION GOALS

- To solve quadratic equations by using the quadratic formula
- To use the discriminant of a quadratic equation to describe the solutions
- To rewrite rational equations as quadratic equations and solve them
- To solve equations that can be put in quadratic form
- To rewrite radical equations as quadratic equations and solve them

▪▪▪

EXAMPLE 1

Solve $2x^2 - 5x + 2 = 0$ by using the quadratic formula.

SOLUTION

This equation is in standard form, and we identify a, b, and c, as follows: $a = 2$, $b = -5$, and $c = 2$.

Substituting these values into the quadratic formula, we get

$$x = \frac{-b \pm \sqrt{b^2 - 4ac}}{2a} = \frac{-(-5) \pm \sqrt{(-5)^2 - 4(2)(2)}}{2(2)} \quad \text{Substituting known values}$$

$$= \frac{5 \pm \sqrt{25 - 16}}{4} = \frac{5 \pm \sqrt{9}}{4} \quad \text{Simplifying}$$

$$= \frac{5 \pm 3}{4} \quad \text{Taking the square root}$$

Then we break this last fraction into two expressions (for the two roots) and continue.

$$x = \frac{5 + 3}{4} \qquad \bigg| \qquad x = \frac{5 - 3}{4}$$

$$x = \frac{8}{4} = 2 \qquad \bigg| \qquad x = \frac{2}{4} = \frac{1}{2}$$

As always, check your solutions by substituting in the original equation. These solutions check, so the solution set is $\left\{2, \frac{1}{2}\right\}$.

▶ CHECK **Warm-Up 1**

Calculator Corner

You can use your graphing calculator's Home Screen to verify the numerical results you obtain using the quadratic formula. Verify the results for $4x^2 + 5x - 6 = 0$ with your graphing calculator. First store the values of a, b, and c. (Note: See the Calculator Corner on page 79 to review how to store values for variables.)

The quadratic formula gives

$$x = \frac{-5 \pm \sqrt{5^2 - 4(4)(-6)}}{2(4)}$$

$$= \frac{-5 \pm \sqrt{121}}{8} = \frac{3}{4}, -2$$

Now type in the quadratic formula on the Home Screen.

Use the calculator's **2nd ENTRY** feature to recall the previous line and change the $+$ sign after the $-B$ to a $-$ sign (see the second screen at the right).

The results $x = 0.75$ and $x = -2$ verify the results obtained using the paper-and-pencil method of calculation.

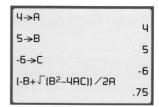

Remember that you can now **STO**re new values for a, b, and c and then use **2nd ENTRY** to recall the two forms of the quadratic formula in order to work more problems using the quadratic formula.

...

EXAMPLE 2

Solve: $-4y^2 - 5y + 3 = 0$

SOLUTION

This equation is in standard form, and $a = -4$, $b = -5$, and $c = 3$. We substitute these values in the quadratic equation and simplify, obtaining

$$y = \frac{-b \pm \sqrt{b^2 - 4ac}}{2a} = \frac{-(-5) \pm \sqrt{(-5)^2 - 4(-4)(3)}}{2(-4)} \qquad \text{Substituting}$$

$$= \frac{5 \pm \sqrt{73}}{-8} \qquad \text{Simplifying}$$

We separate this into the two expressions

$$y = \frac{5 + \sqrt{73}}{-8} \quad \text{and} \quad y = \frac{5 - \sqrt{73}}{-8}$$

We then can do little more than rewrite these roots as

$$y = \frac{-5 - \sqrt{73}}{8} \quad \text{and} \quad y = \frac{-5 + \sqrt{73}}{8}$$

to remove the minus sign from the denominators. Obviously, not all roots can be reduced to neat expressions. These two roots do check; try them.

▶ *CHECK* **Warm-Up 2**

Which of the four methods should you use to solve each type of quadratic equation? Use the method that best suits the equation and the solver! Here's one way to decide:

> **To solve a quadratic equation** $ax^2 + bx + c = 0$
>
> 1. If $b = 0$, write the equation in the form $x^2 = -\frac{c}{a}$, and solve by taking the square roots of both sides.
> 2. If b is not zero, try to factor the left side. (This is very easy if $c = 0$). If you can, solve by setting each factor equal to zero and solving the resulting equation.
> 3. If factoring is difficult, solve by using the quadratic formula. (That usually requires less work than completing the square.)

INSTRUCTOR NOTE

When substituting into the quadratic formula, students should always use parentheses to make sure that the proper sign is substituted.

ERROR ALERT

Identify the error and give a correct answer.

Solve using the quadratic formula:
$x^2 - 3x - 4 = 0$

Incorrect Solution:

$$\frac{-3 \pm \sqrt{(-3)^2 - 4(1)(-4)}}{2}$$

$$\frac{-3 \pm \sqrt{25}}{2}$$

$$\frac{-3 \pm 5}{2}$$

So $x = 1$ and $x = -4$.

The Discriminant

The quantity $b^2 - 4ac$, which appears under the radical sign in the quadratic formula, is called the **discriminant** of the quadratic equation $ax^2 + bx + c$. It provides helpful information about the solutions of an equation.

If $b^2 - 4ac$ is	Then the equation has
Zero	One real (rational) solution
Positive and a perfect square	Two real (rational) solutions
Positive but not a perfect square	Two real (irrational) solutions
Negative	Two complex (non-real) solutions

Computing the discriminant before you try to solve a quadratic equation can help you decide which solution method to use.

▪▪▪

EXAMPLE 3

Describe the solutions of:

a. $x^2 - 2x + 2 = 0$ **b.** $x^2 + 4x + 3 = 0$ **c.** $2x^2 - 5x - 6 = 0$

SOLUTION

a. Here $a = 1, b = -2, c = 2$, and

$$b^2 - 4ac = (-2)^2 - 4(1)(2)$$
$$= 4 - 8 = -4$$

The discriminant is negative, so the equation has two solutions that are not real numbers. The quadratic formula should be used to solve it.

b. Here $a = 1, b = 4, c = 3$, and

$$b^2 - 4ac = (4)^2 - 4(1)(3)$$
$$= 16 - 12 = 4$$

The discriminant is a perfect square, so the equation has two rational solutions. Equations with such solutions can often be solved by factoring.

c. Here $a = 2, b = -5, c = -6$, and

$$b^2 - 4ac = (-5)^2 - 4(2)(-6)$$
$$= 25 + 48 = 73$$

The discriminant is positive but not a perfect square; the two solutions are real numbers. Because $\sqrt{73}$ is irrational, it would be best to solve this equation with the quadratic formula.

▶ *CHECK* **Warm-Up 3**

Equations that Can Be Solved Like Quadratic Equations

Some types of equations that don't look like quadratic equations do turn out to be quadratic when they are simplified.

■ ■ ■

EXAMPLE 4

Solve: $\dfrac{2}{x^2} + \dfrac{7}{x(x + 1)} + \dfrac{4}{x + 1} = 0$

SOLUTION

The LCD of the three denominators is $x^2(x + 1)$. Multiplying each term by the LCD gives us

$$x^2(x + 1)\frac{2}{x^2} + x^2(x+1)\frac{7}{x(x+1)} + x^2(x+1)\frac{4}{x+1} = 0$$

$$\begin{aligned} 2(x + 1) + 7x + 4x^2 &= 0 \quad \text{Simplifying} \\ 2x + 2 + 7x + 4x^2 &= 0 \quad \text{Multiplying} \\ 4x^2 + 9x + 2 &= 0 \quad \text{Combining like terms} \end{aligned}$$

This is a quadratic equation. We solve it with the quadratic formula, with $a = 4$, $b = 9$, and $c = 2$. We get

$$x = \frac{-b \pm \sqrt{b^2 - 4ac}}{2a} = \frac{-9 \pm \sqrt{9^2 - 4(4)(2)}}{2(4)} \quad \text{Substituting}$$

$$= \frac{-9 \pm \sqrt{49}}{8} = \frac{-9 \pm 7}{8} \quad \text{Simplifying}$$

The roots are

$$x = \frac{-2}{8} = \frac{-1}{4} \quad \text{and} \quad x = \frac{-16}{8} = -2 \quad \text{Separating and solving}$$

So the solution set is $\left\{-\dfrac{1}{4}, -2\right\}$. Don't forget to check the roots in the original equation.

> ▶ *CHECK* **Warm-Up 4**

INSTRUCTOR NOTE

Caution students to check that the original equation is still meaningful with these solutions. That is, a solution may not yield a denominator equal to zero.

Equations of Quadratic Form

Some equations that contain higher or lower powers of the variables can be solved like quadratic equations. Their powers must have a certain relationship to each other, however, as we will see shortly.

■ ■ ■

EXAMPLE 5

Solve: $x^4 + x^2 - 6 = 0$

SOLUTION

This is not a quadratic equation. However, if we let $u = x^2$, then

$$u^2 = (x^2)^2 = x^4$$

Now we can rewrite the original equation as

$$u^2 + u - 6 = 0$$

This equation *is* quadratic, and we can solve it by factoring. We get

$$(u - 2)(u + 3) = 0$$

so that

$$u - 2 = 0 \quad \big| \quad u + 3 = 0$$
$$u = 2 \quad \big| \quad u = -3$$

These are the solutions only of the *rewritten* equation. To find the solutions of the *original* equation, we must substitute x^2 back for u in both solutions. We then get

$$x^2 = 2 \qquad \big| \qquad x^2 = -3 \qquad \text{Substituting } x^2 \text{ for } u$$
$$x = \pm\sqrt{2} \quad \big| \quad x = \pm\sqrt{-3} \quad \text{Solving}$$
$$\big| \qquad\quad = \pm i\sqrt{3} \quad \text{Simplifying}$$

The solution set for the original equation is $\left\{\sqrt{2}, -\sqrt{2}, i\sqrt{3}, -i\sqrt{3}\right\}$. You should verify all four solutions in the original equation.

▶ *CHECK* **Warm-Up 5**

You can solve any equation of the form

$$ax^{2n} + bx^n + c = 0$$

where n is a rational number, by using the method of Example 5. We substitute u for x^n and u^2 for x^{2n} and solve the resulting quadratic equation. To solve the original equation, we then substitute x^n for u in the solutions and solve once again.

▪▪▪

Example 6

Solve: $4x^{\frac{1}{2}} - 12x^{\frac{1}{4}} + 9 = 0$

Solution

Here $u = x^{\frac{1}{4}}$ so

$$u^2 = \left(x^{\frac{1}{4}}\right)^2$$
$$= x^{\frac{2}{4}} = x^{\frac{1}{2}}$$

Then the original equation becomes

$$4u^2 - 12u + 9 = 0$$

This equation can be factored as

$$(2u - 3)(2u - 3) = 0$$

Then we obtain

$$2u - 3 = 0 \quad \big| \quad 2u - 3 = 0$$
$$2u = 3 \quad \big| \quad 2u = 3$$
$$u = \frac{3}{2} \quad \big| \quad u = \frac{3}{2}$$

This is a double root.

Substituting $x^{\frac{1}{4}}$ for u in these solutions gives us

$$x^{\frac{1}{4}} = \frac{3}{2}$$

$$\left(x^{\frac{1}{4}}\right)^4 = \left(\frac{3}{2}\right)^4 \qquad \text{Equality property of powers}$$

$$x = \frac{3^4}{2^4} = \frac{81}{16}$$

So the solution set is $\left\{\frac{81}{16}\right\}$.

▶ *CHECK* **Warm-Up 6**

▪ ▪ ▪

EXAMPLE 7

Solve: $2x^{-6} - 5x^{-3} + 2 = 0$

SOLUTION

Here we let $u = x^{-3}$, so

$$u^2 = (x^{-3})^2 = x^{-6}$$

Then the original equation becomes

$$2u^2 - 5u + 2 = 0$$

This equation can be factored as

$$(2u - 1)(u - 2) = 0$$

Then we obtain

$$
\begin{array}{c|c}
2u - 1 = 0 & u - 2 = 0 \\
2u = 1 & u = 2 \\
u = \frac{1}{2} &
\end{array}
$$

Substituting x^{-3} for u in these solutions gives us

$$
\begin{array}{l|ll}
x^{-3} = \frac{1}{2} & x^{-3} = 2 & \text{Substituting } x^{-3} \text{ for } u \\
(x^{-3})^{-\frac{1}{3}} = \left(\frac{1}{2}\right)^{-\frac{1}{3}} & (x^{-3})^{-\frac{1}{3}} = 2^{-\frac{1}{3}} & \text{Taking the } -\frac{1}{3} \text{ powers} \\
x = \left[\left(\frac{1}{2}\right)^{-1}\right]^{\frac{1}{3}} & x = [(2)^{-1}]^{\frac{1}{3}} & \text{Rewriting the powers} \\
= 2^{\frac{1}{3}} & = \left(\frac{1}{2}\right)^{\frac{1}{3}} & \text{Definition of negative exponent} \\
= \sqrt[3]{2} & = \frac{1}{(\sqrt[3]{2})} & \text{Definition of rational power} \\
& = \frac{(\sqrt[3]{4})}{2} & \text{Rationalizing the denominator}
\end{array}
$$

So the solution set is $\left\{\sqrt[3]{2}, \frac{\sqrt[3]{4}}{2}\right\}$.

INSTRUCTOR NOTE

Remind students that second degree equations have at most 2 solutions, third degree have 3 possible solutions, fourth degree have 4 and so on.

We can also use this method to solve any equation of the form

$$a(\,f(x))^{2n} + b(\,f(x))^n + c = 0$$

if *f(x)* is not too complicated.

▶ *CHECK* **Warm-Up 7**

■ ■ ■

EXAMPLE 8

Solve: $(x^2 + 1)^2 + 2(x^2 + 1) + 1 = 0$

SOLUTION

Let $u = (x^2 + 1)$ and $u^2 = (x^2 + 1)^2$. Then the given equation becomes

$$u^2 + 2u + 1 = 0$$

This equation can be factored as

$$(u + 1)(u + 1) = 0$$

so it has the double root $u = -1$.

We substitute $x^2 + 1$ for u and obtain

$$x^2 + 1 = -1$$
$$x^2 = -2 \quad \text{Adding } -1$$

Thus

$$x = \sqrt{-2} \text{ and } -\sqrt{-2}$$
$$= i\sqrt{2} \text{ and } -i\sqrt{2}$$

The solution set for the original equation is $\left\{i\sqrt{2}, -i\sqrt{2}\right\}$. Both solutions happen to be double roots.

▶ *CHECK* **Warm-Up 8**

Radical Equations

Some equations that contain radicals can be solved as quadratic equations. That is true, for example, for a radical equation that is quadratic in form.

■ ■ ■

EXAMPLE 9

Solve the equation $2x - 5\sqrt{x} + 2 = 0$.

SOLUTION

Because $x = \left(\sqrt{x}\right)^2$, we can write this equation in the form of a quadratic equation:

$$2\left(\sqrt{x}\right)^2 - 5\sqrt{x} + 2 = 0$$

Now we can let $u = \sqrt{x}$, so $u^2 = \left(\sqrt{x}\right)^2 = x$.

With these substitutions, the given equation becomes

$$2u^2 - 5u + 2 = 0$$

which we solved in Example 7. Its solutions are $u = \frac{1}{2}$ and $u = 2$. Substituting $\sqrt{x}$ for u in these solutions yields

$$\sqrt{x} = \frac{1}{2} \qquad \qquad \sqrt{x} = 2$$

$$\left(\sqrt{x}\right)^2 = \left(\tfrac{1}{2}\right)^2 \qquad \left(\sqrt{x}\right)^2 = 2^2 \qquad \text{Equality property of square roots}$$

$$x = \frac{1}{4} \qquad \qquad x = 4 \qquad \text{Simplifying}$$

Both solutions check.

▶ *CHECK* **Warm-Up 9**

▪ ▪ ▪

EXAMPLE 10

Solve: $\sqrt{x + 1} + 2 = x - 3$

SOLUTION

This equation does not have a quadratic form, so we cannot substitute for x. The only other possibility is to isolate the radical, square both sides, and then decide how to proceed. We get

$$\sqrt{x + 1} = x - 5 \qquad \text{Adding } -2 \text{ to both sides}$$

$$\left(\sqrt{x + 1}\right)^2 = (x - 5)^2 \qquad \text{Equality property of square roots}$$

$$x + 1 = x^2 - 10x + 25 \qquad \text{Squaring}$$

$$0 = x^2 - 11x + 24 \qquad \text{Adding } -x - 1 \text{ to both sides}$$

This is a quadratic equation that can be solved by factoring or by use of the quadratic formula. With the formula, we get

$$x = \frac{-b \pm \sqrt{b^2 - 4ac}}{2a} = \frac{-(-11) \pm \sqrt{(-11)^2 - 4(1)(24)}}{2(1)}$$

$$= \frac{11 \pm \sqrt{25}}{2} = \frac{11 \pm 5}{2}$$

So $x = \frac{16}{2} = 8$ and $x = \frac{6}{2} = 3$. Checking in the original equation shows that $x = 3$ is a false solution. The solution set is $\{8\}$.

▶ *CHECK* **Warm-Up 10**

INSTRUCTOR NOTE

Caution students that equations in radical form often yield solutions that do not check. They must always verify their answers.

Practice what you learned.

SECTION FOLLOW-UP

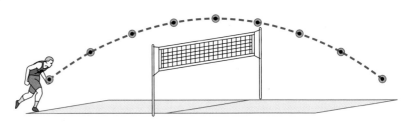

The track of the volleyball is a parabola. The equation of this path is quadratic.

8.2 WARM-UPS

Work these problems before you attempt the exercises.

1. Solve by using the quadratic formula: $12x^2 = 7x + 10$ $\left\{\frac{5}{4}, -\frac{2}{3}\right\}$

2. Solve by using the quadratic formula: $t^2 - 10t - 9 = 0$
$$\left\{5 + \sqrt{34}, 5 - \sqrt{34}\right\}$$

3. Describe the solutions and then solve using the quadratic formula: $x^2 + 2x + 3 = 0$ two complex solutions; $\left\{-1 + i\sqrt{2}, -1 - i\sqrt{2}\right\}$

4. Solve: $\dfrac{16 - 3y}{y + 6} + \dfrac{y + 3}{5} = \dfrac{3y - 2}{15}$ $\{9\}$

5. Solve: $x^4 - 5x^2 + 6 = 0$ 6. Solve: $9r^{\frac{1}{4}} + 6r^{\frac{1}{8}} - 48 = 0$ $\{256\}$
$$\left\{\pm\sqrt{2}, \pm\sqrt{3}\right\}$$

7. Solve: $12x^{-12} + x^{-6} - 1 = 0$ 8. Solve:
$$\left\{\sqrt[6]{4}, \sqrt[6]{-3}\right\}$$ $6(x + 2)^2 + 11(x + 2) = -4$
$$\left\{-2\tfrac{1}{2}, -3\tfrac{1}{3}\right\}$$

9. Solve: $x - 6x^{\frac{1}{2}} + 5 = 0$ $\{25, 1\}$ 10. Solve: $2x - 5\sqrt{x} - 3 = 0$ $\{9\}$

8.2 EXERCISES

Note: Use your graphing calculator to check your results whenever possible.

In Exercises 1 through 24, use the quadratic formula to solve each equation.

1. $x^2 - 8x + 15 = 0$ $\{5, 3\}$

2. $x^2 - 8x + 12 = 0$ $\{6, 2\}$

3. $x^2 - 6x + 9 = 0$
$\{3\}$; double root

4. $x^2 - 2x - 24 = 0$ $\{6, -4\}$

5. $0 = x^2 + 23x + 120$ $\{-8, -15\}$

6. $0 = x^2 - 22x + 96$ $\{16, 6\}$

7. $12x^2 + 17x + 6 = 0$ $\left\{-\frac{2}{3}, -\frac{3}{4}\right\}$ **8.** $40x^2 + 23x + 3 = 0$ $\left\{-\frac{1}{5}, -\frac{3}{8}\right\}$ **9.** $10x^2 + 7x = -1$ $\left\{-\frac{1}{5}, -\frac{1}{2}\right\}$

10. $11x = 15x^2 + 2$ $\left\{\frac{2}{5}, \frac{1}{3}\right\}$ **11.** $12x^2 - 17x + 6 = 0$ $\left\{\frac{3}{4}, \frac{2}{3}\right\}$ **12.** $48t^2 + 3 = 26t$ $\left\{\frac{3}{8}, \frac{1}{6}\right\}$

13. $0 = x^2 - 10$ $\left\{\sqrt{10}, -\sqrt{10}\right\}$ **14.** $0 = x^2 - 7$ $\left\{\sqrt{7}, -\sqrt{7}\right\}$ **15.** $2x^2 = 90$ $\left\{3\sqrt{5}, -3\sqrt{5}\right\}$

16. $x^2 - 6x + 7 = 0$ $\left\{3 + \sqrt{2}, 3 - \sqrt{2}\right\}$ **17.** $x^2 - 4x + 1 = 0$ $\left\{2 + \sqrt{3}, 2 - \sqrt{3}\right\}$ **18.** $x^2 - 10x + 26 = 0$ $\{5 + i, 5 - i\}$

19. $x^2 - 8x + 17 = 0$ $\{4 + i, 4 - i\}$ **20.** $x^2 + 25 = 0$ $\{5i, -5i\}$ **21.** $x^2 + 36 = 0$ $\{6i, -6i\}$

22. $0.3x^2 = 1.2 - 2.4x$ $\left\{-4 + 2\sqrt{5}, -4 - 2\sqrt{5}\right\}$ **23.** $x^2 + 4x = 8$ $\left\{-2 + 2\sqrt{3}, -2 - 2\sqrt{3}\right\}$ **24.** $0.03x^2 + 0.04x = 0.02$ $\left\{\frac{-2 + \sqrt{10}}{3}, \frac{-2 - \sqrt{10}}{3}\right\}$

In Exercises 25 through 36, use the discriminant $b^2 - 4ac$ to describe the roots.

25. $x^2 - 16x + 18 = 0$
two real, irrational roots

26. $x^2 - 19x - 12 = 0$
two real, irrational roots

27. $x^2 + 6x + 9 = 0$
one real, rational root (double root)

28. $x^2 - 16x + 64 = 0$
one real, rational root (double root)

29. $3x^2 - 8x = 9$
two real, irrational roots

30. $6x^2 - 5x = 8$
two real, irrational roots

31. $16x^2 - 32x + 1 = 0$
two real, irrational roots

32. $19x - 3x^2 + 100 = 0$
two real, irrational roots

33. $\frac{1}{3}x^2 - \frac{1}{2}x + 6 = 0$
two complex roots

34. $\frac{1}{4}x^2 - \frac{2}{3}x - 8 = 0$
two real, irrational roots

35. $0.8x^2 + 0.5x - 1.3 = 0$
two real, rational roots

36. $9.2x^2 - 1.8x + 2 = 0$
two complex roots

In Exercises 37 through 42, rewrite each equation as a quadratic equation and solve.

37. $\frac{x + 8}{2} = \frac{2x - 11}{x - 9}$
$x^2 - 5x - 50 = 0; \{-5, 10\}$

38. $\frac{3x + 2}{4x - 3} = \frac{5x}{2x + 15}$
$7x^2 - 32x - 15 = 0; \left\{-\frac{3}{7}, 5\right\}$

39. $\frac{17}{x} + \frac{15}{x^2} = 4$
$4x^2 - 17x - 15 = 0; \left\{5, -\frac{3}{4}\right\}$

40. $x + \frac{1}{2} = \frac{1}{9x}$
$18x^2 + 9x - 2 = 0; \left\{\frac{1}{6}, -\frac{2}{3}\right\}$

41. $\frac{1}{x} - \frac{1}{x + 6} = \frac{2}{105}$
$x^2 + 6x - 315 = 0; \{15, -21\}$

42. $\frac{20}{x + 2} + \frac{20}{x - 2} = \frac{15}{2}$
$3x^2 - 16x - 12 = 0; \left\{6, -\frac{2}{3}\right\}$

In Exercises 43 through 48, rewrite each equation, as a quadratic equation by substituting the given u. Do not solve.

43. $3x^{-4} - 2x^{-2} + 8 = 0$
$u = x^{-2}$ $3u^2 - 2u + 8 = 0$

44. $5x^{-8} + 2x^{-4} - 9 = 0$
$u = x^{-4}$ $5u^2 + 2u - 9 = 0$

45. $2x^{\frac{2}{3}} + 2x^{\frac{1}{3}} - 7 = 0$
$u = x^{\frac{1}{3}}$ $2u^2 + 2u - 7 = 0$

46. $5x^{-\frac{2}{3}} + x^{-\frac{1}{3}} - 5 = 0$
$u = x^{-\frac{1}{3}}$ $5u^2 + u - 5 = 0$

47. $3\left(\sqrt{3x + 2}\right)^2 + 2\sqrt{3x + 2} - 7 = 0$
$u = \sqrt{3x + 2}$ $3u^2 + 2u - 7 = 0$

48. $2(5 - 3x) - 3\sqrt{5 - 3x} + 2 = 0$
$u = \sqrt{5 - 3x}$ $2u^2 - 3u + 2 = 0$

In Exercises 49 through 82, use substitution of a new variable u to rewrite each equation as a quadratic equation. Then solve it.

49. $x^4 - 10x^2 + 9 = 0$ $\{1, -1, 3, -3\}$ **50.** $x^6 - 2x^3 + 1 = 0$ $\{1\}$ **51.** $x^{\frac{1}{2}} - 7x^{\frac{1}{4}} + 12 = 0$ $\{256, 81\}$

52. $x^{-6} - 16x^{-3} + 64 = 0$ $\left\{\frac{1}{2}\right\}$ **53.** $x^{-4} - 8x^{-2} - 9 = 0$ $\left\{\frac{1}{3}, -\frac{1}{3}, i, -i\right\}$ **54.** $x^8 - 17x^4 + 16 = 0$ $\{1, -1, 2, -2\}$

55. $x^4 - 29x^2 + 100 = 0$ $\{-2, 2, -5, 5\}$ **56.** $x^{\frac{2}{3}} - x^{\frac{1}{3}} - 6 = 0$ $\{27, -8\}$ **57.** $x^{\frac{1}{2}} - 10x^{\frac{1}{4}} + 9 = 0$
$\{6561, 1\}$

58. $x^{-2} - 4x^{-1} + 3 = 0$ $\left\{\frac{1}{3}, 1\right\}$ **59.** $x^{-6} + 7x^{-3} - 8 = 0$ $\left\{-\frac{1}{2}, 1\right\}$ **60.** $x + 2\sqrt{x} - 24 = 0$ $\{16\}$

61. $x + 3\sqrt{x} - 4 = 0$ $\{1\}$ **62.** $2 + \sqrt{x + 8} = \sqrt{3x + 12}$ $\{8\}$ **63.** $\sqrt{5x + 9} - x + 1 = 0$ $\{8\}$

64. $x + \sqrt{5x - 1} - 5 = 0$ **65.** $\sqrt{7x + 23} = 2 + \sqrt{3x + 7}$ **66.** $\sqrt{5x - 11} = 1 + \sqrt{3x - 8}$
$\{2\}$ $\{-2, -1\}$ $\{3, 4\}$

67. $x - 1 = \sqrt{x + 11}$ $\{5\}$ **68.** $x + \sqrt{x - 4} - 4 = 0$ $\{4\}$ **69.** $x^4 - 13x^2 + 36 = 0$
$\{3, -3, 2, -2\}$

70. $6 - x^{\frac{1}{3}} - x^{\frac{2}{3}} = 0$ **71.** $x^{\frac{1}{2}} + x^{\frac{1}{4}} - 12 = 0$ **72.** $x^4 - 17x^2 + 16 = 0$
$\{-27, 8\}$ $\{81\}$ $\{1, -1, 4, -4\}$

73. $(x + 2)^2 + 7(x + 2) + 12 = 0$ $\{-5, -6\}$ **74.** $3(2x - 3)^2 - 7(2x - 3) = 6$ $\left\{3, \frac{7}{6}\right\}$

75. $2(4x + 1)^2 + 5(4x + 1) + 2 = 0$ $\left\{-\frac{3}{8}, -\frac{3}{4}\right\}$ **76.** $(x^2 + x)^2 + 12 = 8(x^2 + x)$ $\{-3, 2, -2, 1\}$

77. $(x - 5) - 11\sqrt{x - 5} + 30 = 0$ $\{41, 30\}$ **78.** $x - 3 + \sqrt{x - 3} - 20 = 0$ $\{19\}$

79. $(2x - 1)^{\frac{1}{2}} = -2(2x - 1)^{\frac{1}{4}} + 8$ $\left\{\frac{17}{2}\right\}$ **80.** $2(x - 1)^{\frac{1}{2}} - 5(x - 1)^{\frac{1}{4}} + 2 = 0$ $\left\{1\frac{1}{16}, 17\right\}$

81. $(3x^2 - x)^2 - 14(3x^2 - x) + 40 = 0$ $\left\{\frac{4}{3}, -1, -\frac{5}{3}, 2\right\}$ **82.** $(2x^2 - 5x)^2 - 10(2x^2 - 5x) - 24 = 0$ $\left\{-\frac{3}{2}, 4, \frac{1}{2}, 2\right\}$

MIXED PRACTICE

By doing these exercises, you will practice the topics up to this point in the chapter.

83. Find the solution set: $8x^2 - 4x = -1$ $\left\{\frac{1 + i}{4}, \frac{1 - i}{4}\right\}$ **84.** Find the solution set: $25x^2 - 50x + 24 = 0$
$\{1.2, 0.8\}$

85. Find the solution set: **86.** Solve: $(5 - 2n)^2 = 200$ $\left\{\frac{5 \pm 10\sqrt{2}}{2}\right\}$
$\left(\sqrt{x + 2}\right) - 21\left(\sqrt{x + 2}\right)^{\frac{1}{2}} + 110 = 0$ $\{9998, 14{,}639\}$

87. Find the solution set: $3x^2 - 30x + 123 = 0$ $\{5 \pm 4i\}$ **88.** Solve: $18x^2 - \frac{6}{8}x - \frac{1}{16} = 0$ $\left\{-\frac{1}{24}, \frac{1}{12}\right\}$

89. Solve: $5x^{-2} - 6x^{-1} = -1$ $\{5, 1\}$ **90.** Solve: $(x - 4)^2 - 20 = 0$ $\left\{4 \pm 2\sqrt{5}\right\}$

91. Solve: $-5x = x^2 - 8$ $\left\{\frac{-5 \pm \sqrt{57}}{2}\right\}$ **92.** Describe the roots of $6x^2 + 15 = 0$. two imaginary roots

93. Solve: $x^2 + 11x + 18 = 0$ $\{-2, -9\}$ **94.** Solve: $(16x - 18)^2 = 8$ $\left\{\frac{9 + \sqrt{2}}{8}, \frac{9 - \sqrt{2}}{8}\right\}$

95. Solve: $2x^2 + 3x - 5 = 0$ $\left\{1, \frac{-5}{2}\right\}$ **96.** Solve: $(x + 3)^2 - 16 = 0$ $x = 1$ and $x = -7$

97. Solve: $3x^2 - 2x = 6$ $\left\{\frac{1 \pm \sqrt{19}}{3}\right\}$ **98.** Find the quadratic equation whose solutions are $6 + 2i$ and $6 - 2i$. $x^2 - 12x + 40 = 0$

In Exercises 99 and 100, rewrite each equation as a quadratic equation and solve.

99. $1 - \frac{3}{x} = \frac{10}{x^2}$ $x^2 - 3x - 10 = 0; \{5, -2\}$ **100.** $10x + 1 - \frac{2}{x} = 0$ $10x^2 + x - 2 = 0; \left\{\frac{2}{5}, \frac{-1}{2}\right\}$

EXCURSIONS

Posing Problems

1. Ask and answer four questions using the following data.

World Population, Land Areas, and Elevations

Area	Estimated population, mid-1955	Approximate land area (sq. mi.)	Percent of total land area	Population density per sq. mi.	Elevation (feet)	
					Highest	Lowest
WORLD	5,734,106,000	58,433,000	100.0	98.1[1]	Mt. Everest, Asia, 29,028	Dead Sea, Asia, 1290 below sea level
ASIA, incl. Philippines, Indonesia, and European and Asiatic Turkey; excl. Asiatic former U.S.S.R.	3,403,451,000	10,644,000	18.2	319.7	Mt. Everest, Tibet-Nepal, 29,028	Dead Sea, Israel-Jordan, 1290 below sea level
AFRICA	721,472,000	11,707,000	20.0	61.2	Mt. Kilimanjaro, Tanzania, 19,340	Lake Assal, Djibouti, 571 below sea level
NORTH AMERICA, including Hawaii, Central America, and Caribbean region	454,187,000	9,360,000	16.0	48.5	Mt. McKinley, Alaska, 20,320	Death Valley, Calif., 282 below sea level
SOUTH AMERICA	319,553,000	6,883,000	11.8	46.2	Mt. Aconcagua, Arg.-Chile, 23,034	Valdes Peninsula, 131 below sea level
ANTARCTICA	—	6,000,000	10.3	—	Vinson Massif, Sentinel Range, 16,863	Sea level
EUROPE, incl. Iceland; excl. European former U.S.S.R. and European Turkey	509,254,000	1,905,000	3.3	267.3	Mont Blanc, France, 15,781	Sea level
OCEANIA, incl. Australia, New Zealand, Melanesia, Micronesia, and Polynesia[2]	28,680,000	3,284,000	5.6	8.7	Wilhelm, Papua, New Guinea, 14,793	Lake Eyre, Australia, 38 below sea level
Former U.S.S.R., both European and Asiatic	297,508,000	8,647,000	14.8	34.4	Communism Peak, Pamir, 24,590	Caspian Sea, 96 below sea level

1. In computing density per square mile, the area of Antarctica is omitted. 2. Although Hawaii is geographically part of Oceania, its population is included in the population figure for North America. *Note:* The land area of Asia including the Asiatic portion of the former U.S.S.R. is 17,240,000 sq. miles. *Source:* U.S. Bureau of the Census, International Data Base, as cited in the *1996 Information Please® Almanac* (©1995 Houghton Mifflin Co.), p. 481. All rights reserved. Used with permission by Information Please LLC.

Class Act

2. Organize this data by continent. Which continent contributed the most to New York City immigrant numbers from 1990 to 1994? Ask and answer four other questions using this data. Share your questions with another group.

Immigrants To New York City, 1990–1994

	Country of Birth	Number of Immigrants		Country of Birth	Number of Immigrants
1.	Dominican Republic	110,140	21.	El Salvador	4,099
2.	Former Soviet Union*	66,301	22.	Vietnam	3,917
3.	China (including Taiwan and Hong Kong	59,798	23.	Mexico	3,449
			24.	Romania	3,301
4.	Jamaica	32,918	25.	Japan	3,197
5.	Guyana	30,764	26.	Barbados	3,101
6.	Poland	19,537	27.	Yemen	2,943
7.	Philippines	17,378	28.	Egypt	2,888
8.	Trinidad and Tobago	15,878	29.	Iran	2,711
9.	Haiti	14,957	30.	Guatemala	2,615
10.	India	14,486	31.	Grenada	2,575
11.	Ecuador	13,980	32.	Panama	2,398
12.	Ireland	12,403	33.	Former Yugoslavia	2,388
13.	Colombia	11,309	34.	Canada	2,335
14.	Bangladesh	9,556	35.	Nigeria	2,148
15.	Korea (North and South)	8,626	36.	Afghanistan	2,107
16.	Pakistan	7,465	37.	St. Vincent and Grenadines	2,057
17.	Peru	6,275			
18.	Honduras	6,182	38.	Italy	2,024
19.	Britain	5,935	39.	Brazil	2,014
20.	Israel	4,827	40.	Ghana	1,696

*Includes 4,306 from Belarus; 15,347 from Ukraine; 4,343 from Uzbekistan; and others. *Source*: Data from *The Newest New Yorkers 1990–1994*, *New York Times*, 12 January 1997. New York Department of City Planning.

SECTION GOALS

▪ *To solve verbal problems involving quadratic equations*

▪ *To use a calculator to estimate solutions to quadratic equations*

8.3 Applying Quadratic Equations: Word Problems

SECTION LEAD-IN

The shot put is an Olympic event. A men's shot is a 16-pound ball (women's = 7.6 pounds) made of iron or brass. It is put, not thrown, from shoulder level and can travel almost 80 feet. Its path of flight is a parabola. The equation of that flight is quadratic. Find another "throwing" sport with a quadratic relationship.

Many situations are modeled by quadratic equations. To solve a problem involving such a situation, we translate the situation into an equation, solve the equation, and then translate the solution back to the language of the situation.

▪ ▪ ▪
EXAMPLE 1

The Doyle log model is a formula that is used in the lumber industry to determine the number of board feet N that can be obtained from a tree of length y feet and diameter of x inches.

$$N = \left(\frac{x-4}{4}\right)^2 \cdot y$$

Find the diameter of a tree 32 feet tall that produces 120 board feet of lumber.

SOLUTION

We first substitute the values that we are given.

$$N = \left(\frac{x-4}{4}\right)^2 \cdot y$$
$$120 = \left(\frac{x-4}{4}\right)^2 \cdot 32$$

Then we simplify the resulting equation.

$$120 = \frac{x^2 - 8x + 16}{16} \cdot 32 \quad \text{Squaring as indicated}$$
$$120 = 2x^2 - 16x + 32 \quad \text{Simplifying}$$

Now we have a quadratic equation. We add -120 to both sides and divide by 2 to get

$$0 = x^2 - 8x - 44$$

and then use the quadratic formula to solve.

$$x = \frac{-(-8) \pm \sqrt{(-8)^2 - 4(1)(-44)}}{2} \quad \text{Substituting into quadratic formula}$$
$$= \frac{-(-8) \pm \sqrt{64 + 176}}{2} \quad \text{Simplifying}$$
$$= \frac{8 \pm \sqrt{240}}{2} \quad \text{Simplifying}$$
$$\approx \frac{8 \pm 15.5}{2} \quad \text{Approximating the square root}$$

INSTRUCTOR NOTE

Here, we estimate the value of the radical to find a "real-life" answer. If we want to check our solution, we should use the result

$$x = \frac{8 \pm \sqrt{240}}{2}$$

INSTRUCTOR NOTE

Note for your students that just because an equation yields an answer does not mean that this answer solves the real-life problem. We have to use the problem situation to help us interpret the results.

Then

$$x \approx \frac{8 + 15.5}{2} \quad \text{and} \quad x \approx \frac{8 - 15.5}{2}$$
$$x \approx 11.75 \qquad \bigg| \qquad x \approx -3.75$$

Because the diameter of a tree cannot be negative, we reject the second solution. The diameter of the tree is 11.75 inches.

▶ CHECK **Warm-Up 1**

In the next example, we will need to set up the equation first and then use the information from this solution to solve a second equation.

▪ ▪ ▪
EXAMPLE 2

A group of college students decided to rent a summer home for two weeks at the New Jersey shore. The total cost for the home, regardless of the number of people staying, was $360.60. Before they left for their trip, two more students joined the group. This reduced the cost per person by $48.08. How many people were originally renting the home? What was the original cost per person?

SOLUTION

Let n be the number of college students originally sharing the rent. The cost per student is

$$\frac{360.60}{n}$$

When two additional students join the group, the new cost per student is

$$\frac{360.60}{n + 2}$$

This amount is 48.08 less than the original cost per person. Therefore,

$$\frac{360.60}{n} - 48.08 = \frac{360.60}{n + 2}$$

We solve this equation to find n.

$$(n)(n + 2)\left[\frac{360.60}{n} - 48.08\right] = \left[\frac{360.60}{n + 2}\right](n)(n + 2)$$

$$(\cancel{n})(n + 2)\left(\frac{360.60}{\cancel{n}}\right) - (n)(n + 2)(48.08) = \left(\frac{360.60}{\cancel{n + 2}}\right)(n)(\cancel{n + 2})$$

$$360.60(n + 2) - 48.08(n)(n + 2) = 360.60n$$

We divide both sides by 360.60, obtaining

$$n + 2 - \frac{2}{15}(n^2 + 2n) = n$$

We multiply by 15, obtaining

$$15n + 30 - 2n^2 - 4n = 15n$$
$$-2n^2 - 4n + 30 = 0 \qquad \text{Simplifying}$$
$$n^2 + 2n - 15 = 0 \qquad \text{Dividing by } -2$$

The left side of this equation can be factored as

$$(n + 5)(n - 3) = 0$$

with solutions $n = 3$ and $n = -5$.

Because we cannot have a negative number of people, we reject $x = -5$ as a solution. Thus 3 people originally decided to rent the home. And the original cost per person was $\frac{\$360.60}{3} = \$120.20$.

▶ *CHECK* **Warm-Up 2**

▪ ▪ ▪
EXAMPLE 3

Using your calculator, find the solution of $(3y - 2)^2 = 37$ to three decimal places.

SOLUTION

To "undo" the left side of the equation so that y is alone there, we must

first take the square root;
then add 2;
then divide by 3.

To find the value of y, we must do the same things, in the same order, to the right side of the equation. So using our graphing calculator, we enter

| 2nd | | √ | (37) | + | 2 | ENTER | ÷ | 3 | ENTER |

and obtain $y = 2.694254177$ on the screen. Rounded, this solution is $y = 2.694$.

Note that we must press | ENTER | immediately after adding the constant 2, in order to make sure that the entire right side of the equation is divided by 3 in the next step.

To obtain the second solution, we must operate on the negative square root of 37. We key in

| − | 2nd | √ | (37) | + | 2 | ENTER | ÷ | 3 | ENTER |

The display shows -1.360920843. Rounded, this is -1.361. So the solution set is $\{-1.361, 2.694\}$.

▶ *CHECK* **Warm-Up 3**

Practice what you learned.

SECTION FOLLOW-UP

Which of these sports have "quadratic" relationships?

a. javelin

b. basketball

c. discus

d. baseball

8.3 WARM-UPS

Work these problems before you attempt the exercises.

1. Another version of the Doyle log model is

$$\sqrt{\frac{N}{y}} = \frac{x-4}{4}$$

A tree that produces 150 board feet is 1.5 feet in diameter. Use this formula to determine how tall it is.   12.24 feet tall

2. A rectangular plot of grass in the park has a length that is 15 feet greater than the width. A border of flowers is planted 8 inches in from the outside edge, all the way around the plot. The border encloses an area of $1251\frac{7}{9}$ square feet. What is the length of the entire border? (*Hint*: Think of the border as the perimeter of a smaller plot, 8 inches in from the original.) $144\frac{2}{3}$ feet

3. Using your calculator, find the solutions of $(2y - 2)^2 = 40$ to two decimal places. $y = 4.16$ or $y = -2.16$

8.3 EXERCISES

Note: Use your graphing calculator to check your results whenever possible.

Solve each of the following verbal problems.

1. The product of two consecutive positive odd integers is 899. What are the integers?   29 and 31

2. **Braking Distance**   The formula $D = 0.054x^2 + 0.058x$ describes the distance in feet D that it takes to stop a vehicle traveling x miles per hour on dry pavement.

   a. How fast can you drive if you wish to be able to stop your car within 65 feet?   34 mph

   b. On black ice, a truck's stopping distance is 3 times its stopping distance on dry pavement. A truck traveling 5 miles per hour applies the brakes, on black ice, at a distance of 65 feet in front of a rubber traffic cone. Will the truck hit the cone?   no

3. **Dropping the Ball**   The World Trade Center buildings in New York City are 1353 feet tall. A ball is thrown upward from the side of the rooftop observation deck, at an initial velocity of 128 feet per second, at the same time that a ball is dropped from the same spot. Which ball will

hit the ground first, and how many seconds later will the second ball hit the ground? The equation giving the height h of the ball that was tossed upward is

$$h = -16t^2 + 128t + 1353$$

the equation for the falling ball is

$$h = 1353 - 16t^2$$

In both equations, t is the time in seconds. Use $h = 0$ for the ground.   The ball that was simply dropped lands first (after 9 seconds), then about 5 seconds later the "thrown" ball lands (14 seconds from time "zero").

4. **Deep Well**   Ahmed has a well on his land. He drops a stone over the opening and 2.86 seconds later hears the stone hit the water. He wants to find the depth of the well. An engineer gives him the formula

$$T = \frac{\sqrt{d}}{4} + \frac{d}{1100}$$

where d is the depth of the well and T is the time between dropping a stone and hearing it hit the water. How deep is Ahmed's well?   121 feet

5. Find two numbers whose sum is 10 and whose product is 40. $5 + i\sqrt{15}$ and $5 - i\sqrt{15}$

6. **Gardening**   Lydia, an avid gardener, wants to use 131 feet of fencing to make two rectangular gardens side by side. The area enclosed is 713 square feet. What are the dimensions of the area?   23 feet by 31 feet

7. **Income for Married Couples**   The equation
$$I = \frac{3t^2 - 13t + 505}{100}$$

describes the median income for married couples in the United States t years after 1955, where I is in thousands of dollars. Find the year in which the median income was $10,000.   about 1970

8. Find two numbers whose sum is 16 and whose product is 73. $8 - 3i$ and $8 + 3i$

9. **Bacteria Growth**   The formula
$$N = \frac{14400 - 120t - 100t^2}{144 + t^2}$$

gives the number of bacteria present in a culture t hours after 100 bacteria are treated with an antibacterial agent. How many hours will it take before the population is 0?   11.4 hours

10. **Nuclear Waste**   The amount of nuclear waste, in gallons, flowing into a stream t days after a leak in the cooling system is $w = 8t^2 + 4t$. After how many days is the waste level 840 gallons?   after 10 days

11. **Roller Coaster Speed**   A roller coaster at an amusement park in California has an initial 70-foot vertical drop. The speed v of the last car t seconds after it starts this descent can be approximated as $v = 8.12t$, where v is in feet per second. The height of the last car from the bottom of this drop, in feet, can be estimated by $h = -0.8t^2 - 10t + 70$. At what time does the last car reach the bottom of the vertical drop? What is the speed of the last car at this time?   The last car reaches the bottom 5 seconds after it starts. Its speed at that time is 40.6 feet per second.

12. ***Box Sizes*** Two cube-shaped boxes have the same surface area. One is closed and one has an open top. Each edge of the open cube is $\frac{2}{3}$ inches longer than each edge of the closed cube. Find the dimensions of each cube. side of closed cube $= 6.98$ inches; side of open cube $= 7.65$ inches

13. ***Electric Current*** The electric current I (in amperes) across a certain electric circuit varies according to the formula $I = t^2 - 7t + 12$, where t is the time (in seconds) since the switch was closed. When is the current equal to 42 amperes? after 10 seconds

14. ***Controlling Bacteria*** The formula $C = 20t^2 - 200t + 600$ describes the concentration C of bacteria per cubic centimeter in a body of water t days after treatment to slow bacterial growth. After how many days will the concentration be 100 bacteria per cubic centimeter? after 5 days

15. ***Manufacturing Cost*** A manufacturing company produces n items per week at a cost C of

$$C = (n - 100)^2 + 400$$

It also sells every item it produces at $27.16 each. From these sales, the company makes a profit P according to the equation

$$P = 27.16n$$

Find, to the nearest whole number, the values of n that cause P and C to be equal. $n = 64$ or $n = 164$

16. ***Pollution*** The number of tons of pollutants P dumped into a flowing stream by a factory in its tth year of operation is represented by the equation $P = 8\sqrt[3]{t} + 12$. What year is the first in which 36 tons are dumped? the 27th year of operation

17. ***Explosion Shock Waves*** The formula $d = t^2 + 2t$ describes the distance d in miles that shock waves from an explosion travel in t seconds. How long will it take for shock waves to travel 15 miles? 3 seconds

18. ***Resort Temperature*** The temperature F in degrees Fahrenheit t hours after 8 A.M. in a mountain resort is given by
$$\frac{-(5t^2 - 48t - 78)}{36} = F$$
At what times is the temperature 5 degrees? at 11:36 A.M. and 2:00 P.M.

19. ***Virus Contagion*** The formula $N = 40x - x^2$ describes the number of cases N of a virus reported x days after exposure. How long does it take for 400 people to get the virus? 20 days

20. ***Geometry*** The area of a trapezoid is 49.5 square inches. The shorter base is 3 inches longer than the height. The longer base is 3.5 inches shorter than 4 times the height. Find the dimensions. height $= 4.5$ inches; shorter side $= 7.5$ inches; longer side $= 14.5$ inches

21. ***Office Renovation*** In the planning stages of the renovation of an 1800-square-foot space, the amount of office space allocated per person is increased by 25 square feet when 6 people are removed to an adjoining building. How many people share this renovated space after the move? How much space does each person have? 18 people; 100 square feet

22. *Course Fees* A mathematics class must split a $100 computer fee. During add/drop, 15 more students signed up for the class, and the share per student dropped $1.50. How many students were in the class before and after add/drop?   40 students after add/drop; 25 students before add/drop

In Exercises 23 through 28, solve the word problems.

23. *Football* Big Dave kicked a football at an angle. To find how long it took to hit the ground, solve

$$12.25t - \frac{1}{2}(9.80)t^2 = 0$$

The answer will be in seconds. $t \approx 2.5$ seconds

24. *Football* To find how far the football (in Exercise 23) traveled, use the formula

$$d = 65t - 4.9t^2$$

Replace t with your solution from Exercise 23 and solve for d, the distance traveled. Your answer is in feet. $131\frac{7}{8}$ feet

25. Suppose a rock is dropped off a cliff into a stream 80 meters below. How far will it have fallen after 1, 2, 3, and 4 seconds? Use

$$y = \frac{1}{2}(9.8)s^2$$

where y represents the distance the rock falls in s seconds. Let $s = 1.00$, $s = 2.00$, and so on. Your answer for y will be in meters.
$s = 1, y = 4.9$ meters; $s = 2, y = 19.6$ meters; $s = 3, y = 44.1$ meters; $s = 4, y = 78.4$ meters

26. Use the information in Exercise 25 and find out how long it will take for the rock to hit the water. $t \approx 4$ seconds

80 meters

27. ✏ "I threw a ball into the air; it fell to Earth I know not where...."

But, I can tell how long it is in the air by using

$$0 = 15.0t + \frac{1}{2}(-9.80)t^2$$

The answer t will be in seconds. This problem has only one "useful" answer. Explain why. $t \approx 3$ seconds; At $t = 0$, the ball is not yet in the air.

28. ✏ We can tell when the ball in Exercise 27 passed a height of 8 meters by solving

$$8.0 = 15.0t + \frac{1}{2}(-9.8)t^2$$

This problem has two answers. Both are correct. Explain why.
0.7 second and 2.4 seconds; The ball passed 8 meters twice; on the way up and on the way down.

In Exercises 29 through 46, solve the equations using your calculator. Write each answer to the nearest tenth.

29. $7x^2 - 289 = 0$ $x = \pm 6.4$

30. $5x^2 - 441 = 0$ $x = \pm 9.4$

31. $25x^2 - 38 = 0$ $x = \pm 1.2$

32. $9x^2 - 700 = 0$ $x = \pm 8.8$

33. $64x^2 - 49 = 0$ $x = \pm 0.9$

34. $81x^2 - 121 = 0$ $x = \pm 1.2$

35. $x^2 = 35$ $x = \pm 5.9$

36. $r^2 = 56$ $r = \pm 7.5$

37. $x^2 - 6 = 0$ $x = \pm 2.4$

38. $x^2 - 23 = 0$ $x = \pm 4.8$

39. $(x + 1)^2 - 6 = 0$ $x = 1.4$ or $x = -3.4$

40. $(x + 2)^2 - 8 = 0$ $x = 0.8$ or $x = -4.8$

41. $(2x + 3)^2 - 5 = 0$ $x = -0.4$ or $x = -2.6$

42. $(3x + 2)^2 - 18 = 0$ $x = 0.7$ or $x = -2.1$

43. $(x - 6)^2 - 26 = 0$ $x = 11.1$ or $x = 0.9$

44. $(x + 5)^2 - 38 = 0$ $x = 1.2$ or $x = -11.2$

45. $(2x + 5)^2 - 491 = 0$ $x = 8.6$ or $x = -13.6$

46. $(3x - 7)^2 - 69 = 0$ $x = 5.1$ or $x = -0.4$

MIXED PRACTICE

By doing these exercises, you will practice the topics up to this point in the chapter.

47. Solve: $10x^2 - 110x = -300$
$\{6, 5\}$

48. Find the solution set: $\sqrt{2x - 5} - \sqrt{x - 2} = 2$
$\{27\}$

49. Find the solution set: $7x^4 - 13x^2 - 2 = 0$
$\left\{ \sqrt{2}, -\sqrt{2}, \frac{i\sqrt{7}}{7}, \frac{-i\sqrt{7}}{7} \right\}$

50. Describe the roots of $16x^2 + 29 = -23x$. Do not solve.   two complex roots

51. Find the roots of $(x + 15)^2 + 49 = 0$.
$\{-15 + 7i$ and $-15 - 7i\}$

52. Find the solution set: $3x^2 + 16x = 18$
$\left\{ \frac{-8 \pm \sqrt{118}}{3} \right\}$

53. The solution set of a quadratic equation is $\left\{ \frac{(3 - 2i)}{3}, \frac{(3 + 2i)}{3} \right\}$. Find the quadratic equation.
$9x^2 - 18x + 13 = 0$

54. Solve: $\left(\sqrt{x + 2} \right)^{\frac{1}{2}} - 9 = 0$ $\{6559\}$

55. Solve: $(2x + 2)^2 = 361$ $x = 8.5$ or $x = -10.5$

56. The profit earned in a certain business is represented by the equation $P = n^2 - 10n - 400$, where $26 \leq n \leq 75$. In this equation, n is the number of lots of a particular item that can be sold per week. How many lots must the business sell to earn a profit of $2500?   59 lots

57. Regis purchased some Cat Stevens compact discs before a sale and got 3 fewer than he could have bought for the same total price during the sale. On sale, the discs cost two dollars less than the original price. The total cost was $180. How many compact discs did he purchase?   15 compact discs

58. A quiltmaker wants to sew a quilt made up of squares 8 inches by 8 inches, and the quilt must have a diagonal measurement of 10.4 feet. The length will have 2 more squares than the width. How many squares will there be in the quilt?   120 squares

EXCURSIONS

Class Act

1. a. A certain mathematics text uses $x = -\dfrac{b}{2a} \pm \sqrt{\dfrac{b^2}{4a^2} - \dfrac{c}{a}}$ as its quadratic formula.

Is this formula equivalent to $x = \dfrac{-b \pm \sqrt{b^2 - 4ac}}{2a}$?

Justify your answer.

b. ✏ Which formula is easiest to use and why?

2. Graphs take many forms. Ask four questions that can be answered by this graph. Answer your own questions or swap and answer someone else's questions.

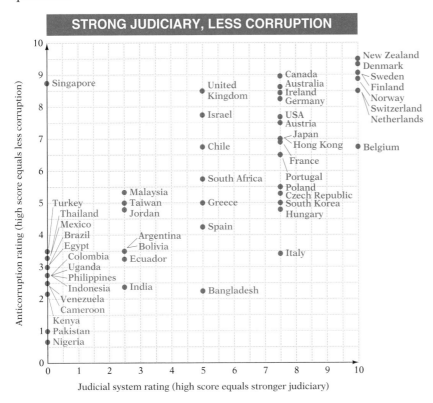

Source: Reprinted by permission of *FORBES Magazine* © Forbes Inc., 1996.

3. A company produces n items per week that cost c dollars to produce. The relationship between the cost and the number produced is given by the equation

$$\frac{1}{n^2} = \frac{1 - \dfrac{200}{n}}{c - 10{,}400}$$

This equation, rational in form, is actually a quadratic equation. Verify this.

8.4 Quadratic Functions and Their Graphs

SECTION LEAD-IN

A long jumper runs between 40 and 45 meters and then jumps. The world record distance is about 30 feet. The path of the jump is a parabola. The equation of that path is quadratic. Find another "jumping" sport with a quadratic relationship.

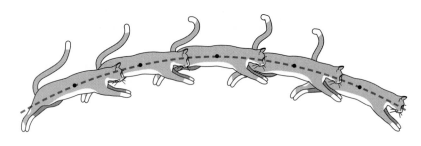

Note that this leaping cat's center of mass moves in a parabolic arc.

Introduction

A function *f* given by

$$f(x) = ax^2 + bx + c$$

where *a*, *b*, and *c* are real numbers, and *a* is not equal to zero, is called a **quadratic function.** (It is also a polynomial function of degree 2.)

Recall that the graph of a function is the graph of all its ordered pairs. We can graph a quadratic function by graphing several ordered pairs and then sketching in the curve that they suggest.

To determine ordered pairs $(x, f(x))$ of a quadratic function, we substitute values for *x* and find corresponding values of $f(x)$.

▪▪▪

EXAMPLE 1

Find six ordered pairs $(x, f(x))$ of the function *f*, given $f(x) = x^2 - 2x - 1$, and sketch its graph.

SOLUTION

Let us find the ordered pairs in which *x* is 0, 1, −1, 2, −2, and 3. We first substitute 0 for *x* in the function equation and find

$$f(0) = 0^2 - 2(0) - 1 = -1$$

Thus $(0, -1)$ is one ordered pair of the function. You should verify that the other first elements we chose produce the ordered pairs

$$(1, -2), (-1, 2), (2, -1), (-2, 7), (3, 2)$$

WRITER'S BLOCK

What is the relationship between graphing a parabola and graphing a quadratic equation?

WRITER'S BLOCK

How does a quadratic equation differ from a linear equation?

We graph the six ordered pairs we just computed as shown in figure at the left below. The points are obviously not in a straight line, so we draw a smooth curve through them. The result is shown in the figure at the right below. Note the arrowheads, which we add to suggest that the graph goes on indefinitely.

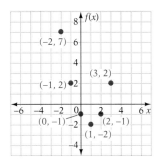
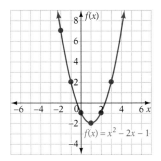

▶ *CHECK* **Warm-Up 1**

The curve sketched in the figure at the right above is a vertical **parabola,** and every quadratic function has a vertical parabola as its graph. The following figure shows some of the features of parabolas.

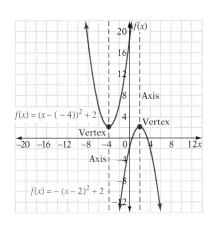

1. A parabola *opens upward* if the coefficient of x^2 in its quadratic polynomial is positive; it *opens downward* if that coefficient is negative.
2. The **vertex** is the lowest point of a parabola that opens upward or the highest point of a parabola that opens downward.
3. The **axis** of a vertical parabola is the vertical line that passes through its vertex.
4. Every parabola is **symmetric** about its axis. That is, the part of the parabola that lies on either side of the axis is the mirror image of the part that lies on the other side of the axis.

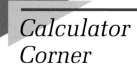

Calculator Corner

Use your graphing calculator to find the vertex of $\left[x + \left(-\frac{1}{2}\right)\right]^2 = \frac{25}{4}$. Set this equation equal to zero in order to obtain:

$$\left[x + \left(-\frac{1}{2}\right)\right]^2 - \frac{25}{4} = 0$$

Now graph this parabola on your graphing calculator using a "friendly" window. **TRACE** to estimate the *vertex* of the parabola.

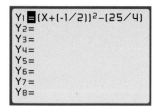
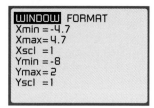
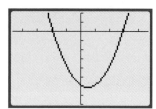

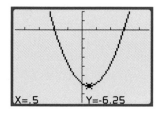

Question: Is $\left(\frac{1}{2}, -\frac{25}{4}\right)$ the same as $(0.5, -6.25)$?

Now graph the following parabolas on your graphing calculator using "friendly" windows.

a. $(x + 4)^2 - 7 = 0$ **b.** $(x + 3)^2 - (7/2) = 0$ **c.** $[x + (-3/2)]^2 + (5/2) = 0$

Can you state a conjecture about this form of the equation of a parabola and the vertex of the parabola? We will learn more about parabolas in Chapter 9.

◼◼◼

WRITER'S BLOCK

Define *intercept* in your own words. Can a graph of a parabola have three x-intercepts? Justify your answer.

The **intercepts** of a parabola are points where it touches, or crosses, the axes. Knowing these points can help us graph the parabola, as we will see in Chapter 9. In addition, the x-intercepts, where the graph touches (or crosses) the x-axis, provide important information about the equation $f(x) = 0$.

Calculator Corner

Use your graphing calculator to solve the equation $(x)(x + 2) = 0$.

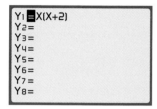 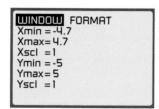 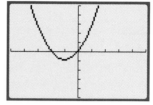

Now use **TRACE** to find out where the equation crosses the x-axis (because that is where $y = 0$). The values of x where $y = 0$ are called the **roots,** or solutions, of the equation.

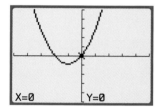 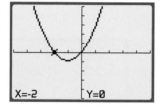

Some graphing calculators also have the ability to find the root, or zero, of an equation. (Note: See the Calculator Corner on pages 510–512 to review how to find the roots of an equation.)

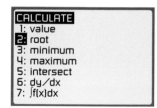

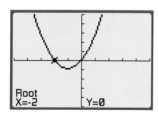

Press **ENTER.** Repeat to find the second root.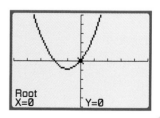

> Let $f(x) = ax^2 + bx + c$ be a quadratic function. If the graph of f has x-intercepts $(s, 0)$ and $(t, 0)$, then s and t are real number solutions, or roots, of the equation $f(x) = 0$. If the graph of $f(x)$ has only one x-intercept, then $f(x) = 0$ has only one real solution. If the graph of f has no x-intercepts, then $f(x) = 0$ has no real solutions.

Calculator Corner

You can use your graphing calculator to solve $20x^2 + 22x = 12$ by graphing each side of the equation and finding where the two graphs intersect. Some graphing calculators can find the point of intersection of two graphs. (Note: See the Calculator Corner on pages 89 and 90 to review how to find the points of intersection of two graphs.)

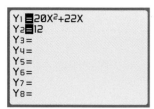
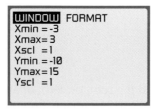
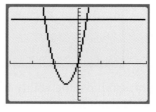

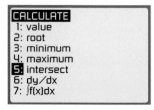
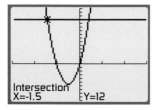
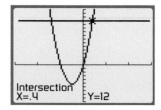

The values of x that we found here, $x = -1.5$ and $x = 0.4$, are zeros, or solutions, of our original equation.

WRITER'S BLOCK

How do the roots of an equation and the x-intercept of a graph relate to each other?

∙∙∙
EXAMPLE 2

Graph $-\frac{1}{2}x^2 - 2x + 3 = y$. Use the graph to estimate the roots, and verify the roots by solving the equation.

SOLUTION

We graph by finding 6 points: $(0, 3)$; $(1, 0.5)$; $(-1, 4.5)$; $(-2, 5)$; $(2, -3)$; $(-4, 3)$. The graph is shown in the following figure.

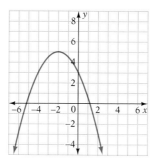

The roots appear to be at $\left(1\frac{1}{3}, 0\right)$ and $\left(-5\frac{1}{3}, 0\right)$. The quadratic formula gives us

$$x = \frac{-(-2) \pm \sqrt{(-2)^2 - 4\left(\frac{-1}{2}\right)(3)}}{2\left(\frac{-1}{2}\right)}$$

$$x = \frac{2 \pm \sqrt{10}}{-1}$$

Using a calculator, we get two solutions: $(-5.2, 0)$ and $(1.2, 0)$. Our estimates were reasonable.

▶ *CHECK* **Warm-Up 2**

INSTRUCTOR NOTE

Remind students that the value of the square root of a number that is not a perfect square is always an approximation. The exact answer must be given in radical form. For real-life situations, the approximate answer is sufficient.

Calculator Corner

Use your graphing calculator to solve $10x^3 - 49x = x^3$. This equation actually factors to give $x(9x^2 - 49) = 0$. We can find where the graphs of the two sides of the equation have the same values for x and y by finding where they intersect. Those will be the values that solve the equation. (Note: See the Calculator Corner on pages 89 and 90 to review how to find the points of intersection of two graphs.)

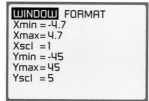
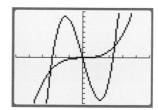
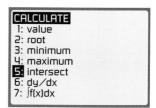

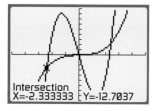

TRACE close to another intersection point and repeat the process twice more.

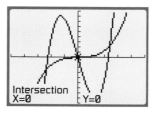
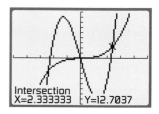

The graphs of the two sides of the equation intersect at the values of x for which y is equal to zero. These are the zeros, or solutions, of the original equation. Here the solutions are $x = 0$, $x = -2\frac{1}{3}$, and $x = 2\frac{1}{3}$. You can see the intersection points on the graph. Check the solutions in the original equation.

Optional Topic: Solving Quadratic and Rational Inequalities

Quadratic Inequalities

An inequality that can be put in the standard form

$$ax^2 + bx + c > 0$$

is a **quadratic inequality.** The expression on the left is a quadratic polynomial. The symbol $<$, $\leq$, or $\geq$ can appear in place of $>$. Examples include

$$x^2 \geq 4 \qquad 3x^2 - 4x - 7 < 0 \qquad \tfrac{1}{2}x^2 > \tfrac{1}{2}x$$

The solution of a quadratic inequality is the set of all real numbers that make the inequality true. As is usual for inequalities with one variable, the graph of the solution set is one or more sections of the real number line. But how do we find them?

Consider the inequality $x^2 \geq 4$. We can place it in standard form by adding -4 to both sides. We then have

$$x^2 - 4 \geq 0$$

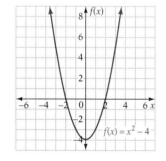

The left side of this inequality is the quadratic function $f(x) = x^2 - 4$; if we graph it, we get the parabola shown in the accompanying figure.

This parabola intersects the x-axis at $x = \sqrt{4}$ and at $x = -\sqrt{4}$; that is, at $x = 2$ and $x = -2$. These are the solutions of the corresponding equation $x^2 - 4 = 0$.

But what about the *inequality?* The graph in this figure shows us that the graph of the quadratic function $f(x) = x^2 - 4$ splits the x-axis, or the real number line, into three sets of points:

$$\text{Set } A = \{x \,|\, x \leq -2\}$$

$$\text{Set } B = \{x \,|\, -2 \leq x \leq 2\}$$

$$\text{Set } C = \{x \,|\, x \geq 2\}$$

We show these portions of the real number line in the following figure.

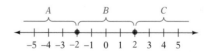

To determine which of these sets represent solutions to $x^2 - 4 \geq 0$, we use the same procedure that we have used previously to graph linear inequalities. We test values of x in the original inequality to determine whether they are in the solution set of the inequality.

Set A: Test $x = -5$ Set B: Test $x = 0$ Set C: Test $x = 5$

$$x^2 \geq 4$$ $$0^2 \geq 4$$ $$5^2 \geq 4$$

$$(-5)^2 \geq 4$$ $$0 \geq 4 \quad \text{False}$$ $$25 \geq 4 \quad \text{True}$$

$$25 \geq 4 \quad \text{True}$$

So $x \leq -2$ and $x \geq 2$ are both parts of the solution to $x^2 \geq 4$. The solution set is $\{x \mid x \leq -2 \text{ or } x \geq 2\}$. It is graphed in the following figure.

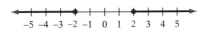

Formally, we follow this procedure:

> **To solve a quadratic inequality** $ax^2 + bx + c > 0$
>
> 1. Solve the corresponding equation $ax^2 + bx + c = 0$.
> 2. Graph the solutions from step 1 on a real number line. These solutions break the real number line into several sections, or sets of points.
> 3. Choose one point from each set, substitute its value in the original inequality, and simplify. If one point in the set satisfies the inequality, all points in the set satisfy it.
> 4. Write the solution set in such a way that it includes all solutions, being careful to use the proper inequality symbols.

▪ ▪ ▪
EXAMPLE 3

Solve the inequality $\frac{1}{2}x^2 > \frac{1}{2}x$ and then graph the solution set.

SOLUTION

The corresponding equation is $\frac{1}{2}x^2 = \frac{1}{2}x$. We solve it by factoring, to get

$$\frac{1}{2}x^2 - \frac{1}{2}x = 0 \quad \text{Adding } {}^{-}\tfrac{1}{2}x \text{ to both sides}$$

$$x^2 - x = 0 \quad \text{Multiplying by 2}$$

$$x(x - 1) = 0 \quad \text{Factoring}$$

$$x = 0 \quad \text{and} \quad x - 1 = 0 \quad \text{Setting the factors equal to 0}$$

$$x = 1$$

The solutions to the *equation* are $x = 0$ and $x = 1$. We graph them as shown in the following figure. They break the number line into three sets of points that we arbitrarily label A, B, and C:

$$A = \{x \mid x < 0\}$$

$$B = \{x \mid 0 < x < 1\}$$

$$C = \{x \mid x > 1\}$$

Note that we use $<$ and $>$ in writing these sets because the original inequality does not contain the equality. The points that solved the equation, then, will not be part of the solution set. Remember that open circles mean the corresponding points are not included.

We test one number from each set, using the original inequality, $\frac{1}{2}x^2 > \frac{1}{2}x$

Set A: Test $x = -1$ Set B: Test $x = \frac{1}{2}$ Set C: Test $x = 2$

$\left(\frac{1}{2}\right)(-1)^2 > \left(\frac{1}{2}\right)(-1)$ $\left(\frac{1}{2}\right)\left(\frac{1}{2}\right)^2 > \left(\frac{1}{2}\right)\left(\frac{1}{2}\right)$ $\left(\frac{1}{2}\right)(2)^2 > \left(\frac{1}{2}\right)(2)$

$\frac{1}{2} > \frac{-1}{2}$ True $\frac{1}{8} > \frac{1}{4}$ False $2 > 1$ True

Our tests of sets A and C give true results, so those sets are part of the solution set. In set A, $x < 0$; in set C, $x > 1$. So the solution set of the original inequality is

$$\{x \mid x < 0 \text{ or } x > 1\}$$

We do not include the equalities in this solution set because the original inequality does not include them. The graph is shown in the following figure.

▶ *CHECK* **Warm-Up 3**

Calculator Corner

Enter $Y_1 = \frac{1}{2}x^2 > \frac{1}{2}x$. Find $>$ on the **TEST** menu. Your resulting graph will look almost like the last figure in Example 3. Use **TRACE** to verify that the points $x = 0$ and $x = 1$ are not part of the solution. (Note: Use **DOT** mode.)

∎∎∎

EXAMPLE 4

Solve: $3x^2 - 4x - 7 < 0$

SOLUTION

The corresponding equation is

$$3x^2 - 4x - 7 = 0$$

Solving by factoring, we have

$$3x^2 - 4x - 7 = 0$$
$$(3x - 7)(x + 1) = 0$$
$$3x = 7 \quad \text{and} \quad x = -1$$
$$x = \frac{7}{3}$$

The solutions to the equation are $x = \frac{7}{3}$ and $x = -1$. We graph these points as shown in the following figure and arbitrarily label the sets of points they define:

$A = \{x \mid x < -1\}$

$B = \left\{x \mid -1 < x < \frac{7}{3}\right\}$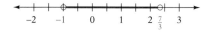

$C = \left\{x \mid x > \frac{7}{3}\right\}$

Again, the points themselves are *not* included in the sets because the original inequality does not include the equality.

Now we can test points from each set.

Set A: Test $x = -2$ Set B: Test $x = 1$

$3(-2)^2 - 4(-2) - 7 < 0$ $3(1)^2 - 4(1) - 7 < 0$

$\qquad\qquad 13 < 0$ False $\qquad\qquad -8 < 0$ True

Set C: Test $x = 3$

$3(3)^2 - 4(3) - 7 < 0$

$\qquad\quad 8 < 0$ False

Only the numbers in set B represent solutions, so the solution set is $\left\{x \mid -1 < x < \frac{7}{3}\right\}$. The graph of the solution set is shown below.

▶ *CHECK* **Warm-Up 4**

Rational Inequalities

A rational inequality is an inequality that contains one or more rational expressions, such as

$$\frac{1}{x} + \frac{1}{x^2} > 1 \quad \text{and} \quad \frac{x+5}{x-1} \geq 3$$

We use the same procedure to solve rational inequalities as we use to solve quadratic inequalities, but we include one additional step: We must also find values of the variable that make denominators zero, because they too break the x-axis, or number line, into segments.

To solve a rational inequality

1. Solve the corresponding equation.
2. Set each denominator equal to zero, and solve the resulting equation(s).
3. Graph the solutions from steps 1 and 2 on a real number line. These break the number line into several sets of points.
4. Substitute the value of one point from each set, in turn, in the original inequality, and simplify. If one point in a set gives a true result, then the entire set represents solutions to the inequality.
5. Write the solution set in such a way that it includes all solutions. Do not include a number that makes a denominator zero.

■ ■ ■

EXAMPLE 5

Solve: $\frac{x + 5}{x - 1} \geq 3$

SOLUTION

We write and solve the equation that corresponds to the given inequality.

$$\frac{x + 5}{x - 1} = 3 \qquad \text{The equation}$$

$$(x - 1)\frac{x + 5}{x - 1} = (x - 1)3 \quad \text{Multiplying by } x - 1$$

$$x + 5 = 3x - 3 \qquad \text{Simplifying}$$

$$8 = 2x \qquad \text{Adding } -x + 3 \text{ to both sides}$$

$$4 = x \qquad \text{Multiplying by } \frac{1}{2}$$

Setting the denominator $x - 1$ equal to zero and solving, we get

$$x - 1 = 0$$

$$x = 1 \quad \text{Adding 1 to both sides}$$

These two values, $x = 4$ and $x = 1$, are graphed in the following figure. The point $x = 1$ is graphed with an empty circle; it cannot be a solution because it makes the denominator zero and thus renders the rational expression meaningless. The two points divide the number line into three sections, which are labeled A, B, and C.

Set $A = \{x \mid x < 1\}$

Set $B = \{x \mid 1 < x \leq 4\}$

Set $C = \{x \mid x \geq 4\}$

Note that we include the point 4 by using $\leq$ and $\geq$ because the original inequality includes the equality.

We test a number from each set in the original inequality: $\frac{x + 5}{x - 1} \geq 3$

Set A: Test $x = 0$ Set B: Test $x = 2$ Set C: Test $x = 5$

$\frac{0 + 5}{0 - 1} \geq 3$ $\frac{2 + 5}{2 - 1} \geq 3$ $\frac{5 + 5}{5 - 1} \geq 3$

$-5 \geq 3$ False $7 \geq 3$ True $2\frac{1}{2} \geq 3$ False

Thus only the points in set B are part of the solution set; the solution set is $\{x \mid 1 < x \leq 4\}$. The graph is shown below.

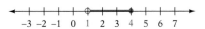

CHECK **Warm-Up 5**

Calculator Corner

Using the **TEST** menu on your *TI-83*, you can enter

$$Y_1 = (x + 5)/(x - 1) \geq 3$$

and get the screen shown at the right.

Using a "friendly" window, you can **TRACE** the graphed line to confirm that

$$1 < x \leq 4$$

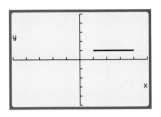

Practice what you learned.

SECTION FOLLOW-UP

Research Which of these sports have quadratic relationships?

   a. pole vault

   b. hurdles

   c. ski jump

8.4 WARM-UPS

Work these problems before you attempt the exercises.

1. Find and plot six ordered pairs $(x, f(x))$ of the function f, given $f(x) = x^2 + 4x + 4$.

2. Graph $-\frac{1}{4}x^2 + 2x - 1 = y$. Use the graph to estimate the roots, and verify your estimate by solving the equation. Estimates may vary but should be close to $x = 7.5$ and $x = 0.5$.

3. Find and graph the solution set of $y^2 - 9y - 190 \geq 0$. $\{y \mid y \leq -10 \text{ or } y \geq 19\}$

4. Find and graph the solution set of $(x - 3)(2x + 6)(4x) > 0$. $\{x \mid -3 < x < 0 \text{ or } x > 3\}$

5. Find and graph the solution set of $\frac{2x - 3}{3x + 1} \geq -3$. $\left\{x \mid x < \frac{-1}{3} \text{ or } x \geq 0\right\}$

8.4 EXERCISES

Note: Use your graphing calculator to check your results whenever possible.

In Exercises 1 through 6, find six ordered pairs $(x, f(x))$ of the given function f.

1. $f(x) = 2x^2 - 3x + 1$ $(0, 1)$,
$(-1, 6), (-2, 15), (1, 0), (2, 3), (3, 10)$

2. $f(x) = 3x^2 + x - 2$ $(0, -2)$,
$(-1, 0), (-2, 8), (1, 2), (2, 12), (3, 28)$

3. $f(x) = -x^2 + 3x + 1$ $(0, 1)$,
$(-1, -3), (-2, -9), (1, 3), (2, 3), (3, 1)$

4. $f(x) = -x^2 - 2x + 3$
$(0, 3), (-1, 4), (-2, 3), (1, 0),$
$(2, -5), (3, -12)$

5. $f(x) = 8x^2 - 5x - 2$ $(0, -2)$,
$(-1, 11), (-2, 40), (1, 1), (2, 20), (3, 55)$

6. $f(x) = 2x^2 + 8x - 5$
$(0, -5), (-1, -11), (-2, -13),$
$(1, 5), (2, 19), (3, 37)$

In Exercises 7 through 18, graph each function by locating and graphing six points $(x, f(x))$ of the function.

7. $y = -x^2$

8. $y = x^2$

9. $f(x) = x^2 - 2$

10. $f(x) = -x^2 - 2$

11. $f(x) = -x^2 + 3$

12. $f(x) = x^2 - 3$

13. $f(x) = x^2 - x + 1$

14. $f(x) = 4x^2 - x + 1$

15. $f(x) = x^2 - 3x + 5$

16. $f(x) = x^2 + 3x + 5$

17. $f(x) = -x^2 + 3x + 2$

18. $f(x) = x^2 + 3x + 2$

For Exercises 19 through 30, graph each function and estimate the roots of $f(x) = 0$. Then check your estimate by setting the quadratic polynomial equal to zero and solving.

19. $f(x) = x^2 - 3$
$\left\{\left(\sqrt{3}, 0\right), \left(-\sqrt{3}, 0\right)\right\} \approx \{(1.7, 0), (-1.7, 0)\}$

20. $f(x) = x^2 - 5$
$\left\{\left(\sqrt{5}, 0\right), \left(-\sqrt{5}, 0\right)\right\} \approx \{(2.2, 0), (-2.2, 0)\}$

21. $y = x^2 - 2x - 2$
$\left\{\left(1 + \sqrt{3}, 0\right), \left(1 - \sqrt{3}, 0\right)\right\} \approx \{(2.7, 0), (-0.7, 0)\}$

22. $f(x) = 8x^2 - 4x$
$\left\{\left(\tfrac{1}{2}, 0\right), \left(0, 0\right)\right\}$

23. $f(x) = x^2 + 4x - 1$

$\left\{\left(-2 + \sqrt{5}, 0\right), \left(-2 - \sqrt{5}, 0\right)\right\} \approx \{(0.2, 0), (-4.2, 0)\}$

24. $f(x) = x^2 - 5x - 1$

$\left\{\left(\dfrac{5 + \sqrt{29}}{2}, 0\right),\right.$

$\left.\left(\dfrac{5 - \sqrt{29}}{2}, 0\right)\right\} \approx \{(5.2, 0), (-0.2, 0)\}$

25. $f(x) = 6x^2 + x - 1$

$\left\{\left(\dfrac{1}{3}, 0\right), \left(-\dfrac{1}{2}, 0\right)\right\}$

26. $f(x) = 4x^2 - 2x$

$\left\{\left(\dfrac{1}{2}, 0\right), (0, 0)\right\}$

27. $f(x) = 2x^2 - x - 1$

$\left\{(1, 0), \left(-\dfrac{1}{2}, 0\right)\right\}$

28. $f(x) = 3x^2 + 3x - 2$

$\left\{\left(\dfrac{-3 + \sqrt{33}}{6}, 0\right), \left(\dfrac{-3 - \sqrt{33}}{6}, 0\right)\right\} \approx$

$\{(0.5, 0), (-1.5, 0)\}$

29. $f(x) = 2x^2 + 3x - 1$

$\left\{\left(\dfrac{-3 + \sqrt{17}}{4}, 0\right), \left(\dfrac{-3 - \sqrt{17}}{4}, 0\right)\right\} \approx$

$\{(0.3, 0), (-1.8, 0)\}$

30. $f(x) = x^2 + 5x - 5$

$\left\{\left(\dfrac{-5 + 3\sqrt{5}}{2}, 0\right), \left(\dfrac{-5 - 3\sqrt{5}}{2}, 0\right)\right\} \approx$

$\{(0.9, 0), (-5.9, 0)\}$

In Exercises 31 and through 42, graph each solution set.

31. $\{x \mid 5 \le x \le 8\}$

32. $\{x \mid 3 \le x \le 9\}$

33. $\{x \mid -8 < x < -3\}$

34. $\{x \mid -2 < x \le 3\}$

35. $\{x \mid -3 \le x < 5\}$

36. $\{x \mid -1 \le x < 4\}$

37. $\{x \mid x \le -2 \text{ or } x \ge 3\}$

38. $\{x \mid x < -2 \text{ or } x > 10\}$

39. $\{x \mid x < -2 \text{ or } x \ge 3\}$

40. $\{x \mid x < 5 \text{ or } x \ge 8\}$

41. $\{x \mid x \le 0 \text{ or } x > 6\}$

42. $\{x \mid x \le -1 \text{ or } x > 0\}$

In Exercises 43 through 54, write inequalities that describe the given graph.

43.

$-3 \le x \le 6$

44.

$x \le -4 \text{ or } x \ge -1$

45.

$x < -28 \text{ or } x > 17$

46.

$-6 \le x < 15$

47.

$18 \le x \le 40$

48.

$-16 < x \le 0$

49.

$x \le -29 \text{ or } x > 15$

50.

$x \le -32 \text{ or } x > 15$

51.

$1 < x \le 6$

52.

$-4 < x \le -3$

53.

$x \le -18 \text{ or } x > 100$

54.

$-72 \le x < 108$

In Exercises 55 through 66, solve and graph each inequality.

55. $x^2 > 49$ $\quad \{x \mid x < -7 \text{ or } x > 7\}$

56. $x^2 \ge 36$ $\quad \{x \mid x \le -6 \text{ or } x \ge 6\}$

57. $(x - 3)^2 < 1$ $\{x \mid 2 < x < 4\}$

58. $(x - 2)^2 > 4$ $\{x \mid x < 0 \text{ or } x > 4\}$

59. $(x + 2)^2 \geq 9$ $\{x \mid x \leq -5 \text{ or } x \geq 1\}$

60. $(x + 2)^2 \leq 4$ $\{x \mid -4 \leq x \leq 0\}$

61. $x^2 - 3x + 2 \geq 0$ $\{x \mid x \leq 1 \text{ or } x \geq 2\}$

62. $x^2 + 13x + 40 > 0$ $\{x \mid x < -8 \text{ or } x > -5\}$

63. $x^2 + 2x + 1 > 0$ $\{x \mid x < -1 \text{ or } x > -1\}$

64. $x^2 + 11x + 18 < 0$ $\{x \mid -9 < x < -2\}$

65. $5x^2 - 90x + 385 > 0$ $\{x \mid x < 7 \text{ or } x > 11\}$

66. $12x^2 - 7x - 10 \leq 0$ $\left\{x \mid -\frac{2}{3} \leq x \leq \frac{5}{4}\right\}$

In Exercises 67 through 78, solve each of the rational inequalities and graph the solution.

67. $\frac{x - 8}{2x + 4} \leq 0$ $\{x \mid -2 < x \leq 8\}$

68. $\frac{x + 5}{3x - 9} \geq 0$ $\{x \mid x \leq -5 \text{ or } x > 3\}$

69. $\frac{2x - 10}{x + 9} \leq 0$

$\{x \mid -9 < x \leq 5\}$

70. $\frac{5x - 2}{3x + 6} \leq 0$ $\left\{x \mid -2 < x \leq \frac{2}{5}\right\}$

71. $\frac{3x + 6}{x - 1} < 0$ $\{x \mid -2 < x < 1\}$

72. $\frac{2x - 3}{x - 2} > 0$ $\left\{x \mid x < \frac{3}{2} \text{ or } x > 2\right\}$

73. $\frac{x + 5}{x - 5} \leq 6$ $\{x \mid x < 5 \text{ or } x \geq 7\}$

74. $\frac{x - 3}{x + 2} \geq 5$ $\left\{x \mid -\frac{13}{4} \leq x < -2\right\}$

75. $\frac{(x + 3)(x - 2)}{x - 5} \geq 0$

$\{x \mid -3 \leq x \leq 2 \text{ or } x > 5\}$

76. $\frac{(x - 7)(x + 2)}{x - 4} \leq 0$

$\{x \mid x \leq -2 \text{ or } 4 < x \leq 7\}$

77. $\frac{(2x - 3)(x + 2)}{5 - x} \leq 0$

$\left\{x \mid -2 \leq x \leq \frac{3}{2} \text{ or } x > 5\right\}$

78. $\frac{(x - 2)(2x - 1)}{x + 4} \leq 0$

$\left\{x \mid x < -4 \text{ or } \frac{1}{2} \leq x \leq 2\right\}$

MIXED PRACTICE

By doing these exercises, you will practice the topics up to this point in the chapter.

79. Find the solution set:
$5(x + 1)^2 + 6(x + 1) + 1 = 0$ $\left\{-\frac{6}{5}, -2\right\}$

80. Find the solution set: $\frac{x + 3}{2x - 1} \leq 1$
$\left\{x \mid x < \frac{1}{2} \text{ or } x \geq 4\right\}$

81. The solutions of a certain quadratic equation are $\frac{(2 + 3\sqrt{2})}{4}$ and $\frac{(2 - 3\sqrt{2})}{4}$. Write the equation.
$8x^2 - 8x - 7 = 0$

82. Find the solution set: $x^2 + 16x + 12 = 0$
$\left\{-8 + 2\sqrt{13}, -8 - 2\sqrt{13}\right\}$

83. Graph the solution set: $(x + 3)^2 \leq 16$.
$\{x \mid -7 \leq x \leq 1\}$

84. Find the solution set: $2x^{\frac{2}{3}} + 7x^{\frac{1}{3}} - 15 = 0$
$\left\{\frac{27}{8}, -125\right\}$

85. Find six points that lie on the graph of
$y = x^2 - 6$. $(0, -6), (-1, -5), (-2, -2), (1, -5),$
$(2, -2), (3, 3)$; Answers may vary.

86. Graph the solution set: $\left\{x \mid x \leq 4 \text{ or } x > 19\frac{1}{2}\right\}$

87. Find the solution set: $x^2 + 3x - \frac{1}{4} = 0$
$\left\{\dfrac{-3 + \sqrt{10}}{2}, \dfrac{-3 - \sqrt{10}}{2}\right\}$

88. Solve: $2x^2 - 2x = 3$
$\left\{\dfrac{1 + \sqrt{7}}{2}, \dfrac{1 - \sqrt{7}}{2}\right\}$

89. Graph $f(x) = -3x^2 - 5x + 4$, estimate the roots, and then check by solving the quadratic equation when $f(x) = 0$. $\{-2.26, 0.59\}$

90. Find six points that lie on the graph of
$y = -x^2 + 3$.   Answers may vary. Some possibilities are:
$(0, 3), (1, 2), (-1, 2), (2, -1), (-2, -1), (3, -6)$

91. Find the solution set: $\dfrac{1}{x + 1} - \dfrac{1}{x} = \dfrac{1}{2}$
$\left\{\dfrac{-1 + i\sqrt{7}}{2}, \dfrac{-1 - i\sqrt{7}}{2}\right\}$

92. Find the solution set: $8x^2 = 64x + 176$
$\{4 + \sqrt{38}, 4 - \sqrt{38}\}$

93. Find the solution set: $(x + 5)^2 + 100 = 0$
$\{-5 \pm 10i\}$

94. Find the solution set: $7x^4 - 30x^2 + 8 = 0$
$\left\{2, -2, \dfrac{\sqrt{14}}{7}, \dfrac{-\sqrt{14}}{7}\right\}$

95. Describe the roots of $5x^2 - 27 = 0$.
two real roots

96. Graph the function $f(x) = 3x^2 + 8x + 5$; estimate the roots, then check by solving the quadratic equation.
$\{(-1, 0), (-1\frac{2}{3}, 0)\}$

97. The difference of two numbers is 2, and their product is 2303. What are the numbers?   47 and 49 or -47 and -49

98. *Running Times* A model that describes the time t in seconds for an average person who is a years old to run the 100-yard dash is
$$t = \frac{3(a - 20)^2}{80} + 10$$
At what age could an average person run the 100-yard dash in 25 seconds?   40 years of age

99. *Matting Pictures* A picture measures 8 inches by 10 inches. An artist wants to put a square mat around the outside of the picture. The picture and the new mat together will have twice the area of the picture alone. What are the dimensions of the mat?   12.65 inches by 12.65 inches

100. *Watering Plants* The formula that describes the amount of water y, in cups, that a certain plant requires is $36y = -5x^2 + 60x + 144$, where x represents the calendar months (January = 1 through December = 12). Determine the amount of water needed in June and December.   9 cups; 4 cups

EXCURSIONS

Class Act

1. ✏ A student claims that if a quadratic equation has two intercepts, the distance between those two intercepts is the same as the distance from the vertex to the *x*-axis.

Is the student correct? Justify your answer. Show any work you did to prove or disprove this idea.

2. In Experiments 1–3, we will graph data from the following table and predict whether the result is linear or quadratic. The data is from a football kicking experiment.

FOOTBALL KICKING DATA

VELOCITY (feet per second)	HANG TIME (seconds)	DISTANCE (feet)	HEIGHT (feet)
65	2.9	132	33.0
66	2.9	136	34.0
67	2.95	140	35.1
68	3.0	145	36.1
69	3.05	149	37.2
70	3.1	153	38.2
71	3.15	158	39.4
72	3.2	162	40.0
73	3.2	167	41.6
74	3.3	171	42.8
75	3.3	176	43.9
76	3.4	181	45.1
77	3.4	185	46.3
78	3.45	190	47.5
79	3.5	195	48.7

In all the equations that follow, use

$$g = 32.15 \text{ feet per second}$$

and

$$V_{x_0} = V_{y_0} = (\text{velocity})\left(\frac{\sqrt{2}}{2}\right)$$

a. *Experiment 1* Verify the relationship

$$\text{Maximum height} = V_{y_0}t - \frac{1}{2}gt^2 \quad (\text{for } t = \frac{V_{y_0}}{g})$$

by following these steps.

i. Choose a velocity in the table, then calculate

$$t = \frac{V_{y_0}}{g}$$

ii. Substitute t in the following equation and solve for y (the height)

$$y = V_{y_0}t - \frac{1}{2}gt^2$$

iii. Check to see if y is equal to the height in the table that corresponds to your chosen velocity.

iv. Is the relation between maximum height and velocity linear or quadratic? Why?

b. *Experiment 2* Verify the relationship

$$0 = V_{y_0}(\text{hang time}) - \frac{1}{2}g (\text{hang time})^2$$

Use t to stand for hang time.

i. Choose a velocity from the table and solve the equation for t.

$$0 = V_{y_0}t - \frac{1}{2}gt^2$$

    ii. Compare the calculated hang time with that in the table.

    iii. Is the relationship linear or quadratic? Justify your answer.

c. *Experiment 3* Verify the relationship

$$\text{Distance} = (V_{x_0})\,(\text{hang time})$$

    i. Choose a velocity from the table with its hang time (t), then calculate

$$d = V_{x_0} t$$

    ii. Compare the calculated d with the distance in the table.

    iii. Is this relationship linear or quadratic?

d. *Discussion*

    i. Did your graphs give you enough information to decide what kind of relationship there was between these variables? Why or why not?

    ii. Just because your data fits certain kinds of equations, can you conclude that the equation explains the relationship? Why or why not?

3. In the National Hockey League, there are two conferences and four divisions. To determine which teams are in the playoffs, teams are given 2 points for each game they win, 1 point for each tie game, and no points for a loss. So a team with a Win-Tie-Lose record of 39-10-3 has

$$39(2) + 10(1) + 3(0) = 78 + 10 + 0$$
$$= 88 \text{ points}$$

Then the teams are ranked according to those points. The regular season division champs (two in each conference) qualify. Then, the next six terms with the highest number of points qualify.

Research What other numerical methods are used to choose playoff positions for sports teams?

CHAPTER LOOK-BACK

The path followed by any projectile (something thrown, shot, or propelled in some way) is a parabola. The equation that represents the path followed by

    1. a thrown hammer

    2. a high jumper

    3. a thrown javelin

is a quadratic equation.

CHAPTER 8
REVIEW PROBLEMS

The following exercises will give you a good review of the material presented in this chapter.

SECTION 8.1

Factor and solve.

1. $18y^2 = 216$ $\left\{2\sqrt{3}, -2\sqrt{3}\right\}$

2. $x^2 + 12x + 35 = 0$ $\{-5, -7\}$

3. $-10x + 21 = -x^2$ $\{7, 3\}$

4. $(x - 12)^2 - 121 = 0$ $x = 23$ and $x = 1$

Complete the square and solve.

5. $5n^2 - 45n - 350 = 0$ $\{14, -5\}$

6. $4x^2 - 4x - 15 = 0$ $\left\{2\frac{1}{2}, -1\frac{1}{2}\right\}$

SECTION 8.2

Use the quadratic formula and solve.

7. $x^2 - 2x + 1 = 0$ $\{1, 1\}$

8. $63t^2 = 8t + 16$ $\left\{\frac{4}{7}, -\frac{4}{9}\right\}$

9. $12x^2 - 7x - 10 = 0$ $\left\{1\frac{1}{4}, -\frac{2}{3}\right\}$

10. Two solutions of quadratic equation are $5 - 2i$ and $5 + 2i$. What is the equation? $x^2 - 10x + 29 = 0$

11. Find the solution set: $\sqrt{2x - 5} - \sqrt{x - 2} = 2$ $\{27\}$

12. Find the solution set: $\left(\sqrt{x + 6}\right)^{\frac{1}{2}} = 2$ $\{10\}$

13. Find the solution set: $(x + 2)^2 - 1.8(x + 2) + 0.45 = 0$ $\{-0.5, -1.7\}$

14. Find the roots: $x^{\frac{1}{2}} - 1.1x^{\frac{1}{4}} + 0.24 = 0$ $x = 0.4096$ and $x = 0.0081$

15. Find the solution set: $y^{-\frac{2}{3}} - y^{-\frac{1}{3}} - 6 = 0$ $\left\{\frac{1}{27}, -\frac{1}{8}\right\}$

16. Find the solution set: $2 = \frac{5}{x} - \frac{6}{x^2}$

SECTION 8.3

$\left\{\frac{5 + i\sqrt{23}}{4}, \frac{5 - i\sqrt{23}}{4}\right\}$

17. Half of the area of a rectangular field is 285 square yards. The length is 8 yards less than twice the width. What are the dimensions of the field? 19 yards by 30 yards

18. An equation that describes the average volume of mail handled by a small regional U.S. Post Office is

$$f(x) = -325x^2 + 2100x$$

where x is the day of the week: Monday $= 1$, Tuesday $= 2$, and so on. On what day is an average of 3200 pieces of mail processed? Thursday

19. A blood cell in an artery flows at a speed v determined by the cell's distance r from the center of the artery. An equation that describes this is

$$v = -0.95 + 18{,}500r^2$$

Find r when v is 0.975 centimeters per second. 0.0102 centimeter

20. Find two numbers whose sum is 20 and whose product is 125.
$10 + 5i$ and $10 - 5i$

SECTION 8.4

21. Find six ordered pairs that satisfy $f(x) = 2x^2 - 4x$.
$(0, 0), (-1, 6), (-2, 16), (1, -2), (2, 0), (3, 6)$

22. Find six points from the graph of $f(x) = 18x^2 + 9x - 2$.
$(0, -2), (-1, 7), (-2, 52), (1, 25), (2, 88), (3, 187)$

23. Graph the function $f(x) = 8x^2 - 5x$. Estimate the roots from the graph. Then check your estimate by setting the quadratic polynomial equal to zero and solving. $\{(0.625, 0), (0, 0)\}$

24. Find the roots of the function $f(x) = 6x^2 + x - 2$. $\left\{\frac{1}{2}, -\frac{2}{3}\right\}$

OPTIONAL TOPIC

25. Graph the solution set: $(x - 3)^2 \geq 81$
$\{x \mid x \leq -6 \text{ or } x \geq 12\}$

26. Graph the solution set: $x^2 \leq 0$
$\{x \mid x = 0\}$

27. Graph the solution set: $6x^2 + 30x - 36 > 0$
$\{x \mid x < -6 \text{ or } x > 1\}$

28. Write the inequality described by the following graph. $\{x \mid \sqrt{2} < x \leq 5\}$

$$\sqrt{2}$$

0 1 2 3 4 5 6 7 8 9 10

29. Graph the solution set: $\{x \mid x < -5 \text{ or } x \geq 10\}$

30. Graph the solution set: $\frac{x}{x - 1} \geq -3$
$\left\{x \mid x \leq \frac{3}{4} \text{ or } x > 1\right\}$

MIXED REVIEW

31. Find the solution set: $(x + 7)^2 + 81 = 0$
$\{-7 + 9i, -7 - 9i\}$

32. Find the solution set: $8x^2 + 64x + 125 = 0$
$\left\{\frac{-16 + \sqrt{6}}{4}, \frac{-16 - \sqrt{6}}{4}\right\}$

33. Find the solution set: $18x^2 - 72x + 40 = 0$ $\left\{3\frac{1}{3}, \frac{2}{3}\right\}$

34. Find the roots:
$$(2x + 3)^{\frac{2}{3}} - 13(2x + 3)^{\frac{1}{3}} + 36 = 0$$
$x = 30.5$ or $x = 363$

35. Graph the function $y = -2x^2 - 2x + 1$. Describe the roots by observation and estimation. Then check your estimates by setting the quadratic polynomial equal to zero and solving.
Estimates may vary. $\{-1.37, 0.37\}$

36. Graph the solution set: $x^2 - 4x - 5 \leq 0$
$\{x \mid -1 \leq x \leq 5\}$

37. Find the solution set: $(3x - 2)^2 = 144$ $\left\{4\frac{2}{3}, -3\frac{1}{3}\right\}$

38. Find the solution set: $40x^2 - 24x + 52 = 0$
$\{0.3 + 1.1i, 0.3 - 1.1i\}$

39. Describe the roots of $24x^2 - 15 + 2x = 0$.
two real, rational roots

40. Graph the solution set: $2x^2 + 46x + 264 < 0$
$\{x \mid -12 < x < -11\}$

41. The roots of a quadratic equation are $3 - \sqrt{5}$ and $3 + \sqrt{5}$. Find the equation. $x^2 - 6x + 4 = 0$

42. Find the solution set: $2\sqrt{x^2 - 7} = \sqrt{x^2 + 8}$
$\left\{2\sqrt{3}, -2\sqrt{3}\right\}$

43. Graph the solution set: $\{x \mid 3 \leq x < 6\}$

44. Find the solution set: $x^2 - 15x - 324 = 0$
$\{27, -12\}$

CHAPTER 8 TEST

This exam tests your knowledge of the material in Chapter 8.

1. Find the solution set by factoring or taking square roots.

    a. $10x^2 - 7x + 1 = 0$ $\left\{\frac{1}{2}, \frac{1}{5}\right\}$ **b.** $15x^2 - 125 = 0$ $\left\{\frac{5\sqrt{3}}{3}, \frac{-5\sqrt{3}}{3}\right\}$ **c.** $(x - 3)^2 - 25 = 0$ $\{8, -2\}$

2. Find the solution set by completing the square.

    a. $x^2 - 4x = 21$ $\{7, -3\}$ **b.** $x^2 + 14x - 120 = 0$ $\{6, -20\}$ **c.** $4x^2 - 52x + 25 = 0$ $\left\{12\frac{1}{2}, \frac{1}{2}\right\}$

3.

    a. Identify a, b, and c for use in the quadratic formula:

    $9x^2 - 15x - 6 = 0$
    $a = 9, b = -15, c = -6$

    b. Solve using the quadratic formula:

    $x^2 = 10x - 21$
    $\{7, 3\}$

    c. Describe the roots of

    $x^2 = 5x - 29$
    two complex roots

4.

    a. Solve: $\left(\sqrt{x + 6}\right)^{\frac{1}{2}} = 2$
    $x = 10$

    b. Let $u = x^{-3}$, and rewrite the equation:

    $x^{-6} - 7x^{-3} - 8 = 0$
    $u^2 - 7u - 8 = 0$

    c. Solve: $y^{\frac{2}{3}} - 5y^{\frac{1}{3}} + 6.25 = 0$
    $\{15.625\}$

5.

    a. A rectangular area has a length three times its width. It is bordered by a sidewalk that is 2 feet wide and goes all the way around it. The area of the larger rectangle is 207 square feet. What is the area of the small rectangle (enclosed by the sidewalk) to the nearest square foot?   99 square feet

    b. When an object is dropped, the distance in feet that it falls in t seconds is given by the equation $d = 16t^2$. If an object were dropped from an airplane at an altitude of 35,000 feet, how long would it take to fall to the ground?   46.8 seconds

    c. The sum of two numbers is 22 and their product is 221. What are the numbers? $11 + 10i$ and $11 - 10i$

6.

    a. Find six points that lie on the graph of:

    $f(x) = -x^2 - x - 1$
    $(0, -1), (-1, -1), (-2, -3), (1, -3),$
    $(2, -7), (3, -13)$

    b. Find six points that lie on the graph of:

    $f(x) = 0.5x^2 - 0.1x - 0.2$
    $(0, -0.2), (-1, 0.4), (-2, 2), (1, 0.2),$
    $(2, 1.6), (3, 4)$

    c. Graph: $f(x) = 2x^2 - 3x - 4$

7. Solve and graph the solution set:

    a. $x^2 \geq 100$
    $\{x \mid x \leq -10 \text{ or } x \geq 10\}$

    b. $x^2 - x - 6 \geq 0$
    $\{x \mid x \leq -2 \text{ or } x \geq 3\}$

    c. $\frac{x - 5}{x + 6} \geq 0$ $\{x \mid x < -6 \text{ or } x \geq 5\}$

CUMULATIVE REVIEW

CHAPTERS 1–8

The following exercises will help you maintain the skills you have learned in this and previous chapters.

1. The height of a triangle is 3.5 feet less than the base. If the area is $32\frac{1}{2}$ square feet, find the dimensions of the triangle.   base = 10 feet; height = 6.5 feet

2. Multiply: $(-4.5x^2y)(-2.3xy)(2xy^3)$ $20.7x^4y^5$

3. Find the equation of the line that passes through the points $(-13, -4.4)$ and $(1.4, -4.4)$ $y = -4.4$

4. Multiply: $\left(\dfrac{2x + 6}{x^2 - 25}\right)\left(\dfrac{x - 5}{x + 3}\right)$ $\dfrac{2}{x + 5}$

5. Add: $0.6\sqrt{0.01a^2} + 5\sqrt{0.04b^2}$ $0.06a + b$

6. Divide: $(x^3 + 2x^2 - 6x - 4) \div (x - 2)$
   $x^2 + 4x + 2$

7. Subtract $15\sqrt{18} - 2\sqrt{8}$ from $13\sqrt{32}$. $11\sqrt{2}$

8. Solve: $3t = \sqrt{9t^2 - 6t + 32}$ $t = 5\frac{47}{93}$

9. Solve: $5n^2 = 46n - 9$ $\left\{9, \frac{1}{5}\right\}$

10. Solve: $\sqrt{t - 12} = 14$ $t = 208$

11. Divide: $\dfrac{-24xy^2 + 36x^2y}{-3xy}$ $8y - 12x$

12. Simplify: $\dfrac{\frac{t(r - 1)}{r^2 - 5r + 4}}{\frac{3rt}{}}$

    Simplify: $\dfrac{t(r - 1)}{r^2 - 5r + 4} \div 3rt$ $\dfrac{3r}{(r - 1)^2(r - 4)}$

13. Write the equation of a line that is perpendicular to the x-axis and passes through the point $(7, -2)$
    $x = 7$

14. Multiply: $2\left(\sqrt[3]{7x}\right)\sqrt[3]{98x^{-8}}$ $\dfrac{14\sqrt[3]{2x^2}}{x^3}$

15. Solve for r: $3rt - 4s = r$ $r = \dfrac{4s}{3t - 1}$

16. Simplify: $2^3 - 6 + 7 \times 2 - 6$ 10

17. Find the equation of the line that passes through the points $(3, -4)$ and has slope -3. $y = -3x + 5$

18. Simplify: $\sqrt{28t} \cdot \sqrt{14t}$ $14t\sqrt{2}$

19. Subtract: $\dfrac{10}{18x} - \dfrac{3}{9x}$ $\dfrac{2}{9x}$

20. Add: $\dfrac{x}{x + 1} + \dfrac{1}{x - 1}$ $\dfrac{x^2 + 1}{x^2 - 1}$

CONIC SECTIONS

9

*W*hat do the shape of a satellite's orbit, the flight of a baseball, and the relationship between the volume and pressure of a gas have in common? All are curves derived from conic sections. The "flattened circle" that some satellites travel is an ellipse. The open curve of a baseball in flight forms a parabola, and the relationship between volume and pressure of a gas is described by a hyperbola. The relative positions of the cone and the plane that intersects it determine the type of curve formed.

■ *Examine several everyday objects. How would you describe the shapes of the sections formed if you could "slice through" at various angles?*

SKILLS CHECK

Take this short quiz to see how well prepared you are for Chapter 9. The answers follow the quiz.

1. Find the solution set for
   $x^2 - 16x + 64 = 0$.

2. Solve for y:
   $-y + 7 = 3x^2 - 10$

3. Find $f(0)$ and $f(3)$ if
   $f(x) = x^2 - 12x + 10$

4. Find the solution set:
   $x = \sqrt{x^2 - 4}$

5. Graph: $f(x) = 2x^2 - 3$

6. Find three points that lie on the graph of $3x + 2y = 5$. Use $x = 0$, $x = 1$, and $y = 0$.

7. Solve this system of equations:
   $8x - 3y = 2$
   $-12y + 23x = -1$

8. Solve by completing the square: $x^2 + 3x + 4 = 0$

ANSWERS: **1.** $\{8\}$ [Section 8.1]   **2.** $y = -3x^2 + 17$ [Section 2.3]   **3.** $f(0) = 10; f(3) = -17$
[Section 3.5]   **4.** No real number solution exists [Section 7.3]   **5.** [Section 8.4]
6. $(0, 2.5), (1, 1), \left(\frac{5}{3}, 0\right)$ [Section 3.1]   **7.** $(1, 2)$ [Section 4.1]
8. $x = \frac{-3 + i\sqrt{7}}{2}$ and $x = \frac{-3 - i\sqrt{7}}{2}$ [Section 8.2]

CHAPTER LEAD-IN

Many designs are pleasing to look at. We use them as decorations—in tiles, flooring, fabric decorations, and quilts, for example.

The following figure can tessellate—form a pattern with no holes. It is a square.

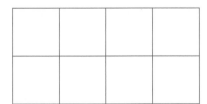

Note that it takes four squares to surround a point. Each right angle measures 90°, therefore it takes 360° to surround each point.

- Can you find examples of designs that use geometric shapes?

9.1 The Parabola

SECTION LEAD-IN

Which of the following shapes tessellate? Justify your answers geometrically.

1. a right triangle

2. a parallelogram

3. a trapezoid

4. a regular hexagon

5. an equilateral triangle

6. a regular pentagon

SECTION GOALS

- *To graph vertical parabolas*

- *To determine the locations of the vertex, intercepts, axis, and other points of a parabola from its equation*

- *To write the equation of a parabola in graphing form*

- *To graph horizontal parabolas*

Introduction

In this chapter, we discuss four types of curves called *conic sections.* They are all graphs of second-degree equations in two variables. And they can all be formed as the intersection of an infinite double cone and a plane.

Axis of cone

Parabola

The Parabola

When the plane is parallel to one edge of the cone, as shown above, their intersection is a **parabola.** In Section 8.4 we discussed parabolas as the graphs of quadratic functions. Here we consider them as graphs of quadratic equations in two variables that have the *general* form

$$y = ax^2 + bx + c \quad \text{where } a \neq 0$$

The graph of a quadratic equation in the form $y = ax^2 + bx + c$ is a parabola and has the following characteristics:

The x-coordinate of the vertex is $-\dfrac{b}{2a}$.

The axis of the parabola is the line $x = -\dfrac{b}{2a}$.

The graph opens upward if $a > 0$.

The graph opens downward if $a < 0$.

We shall look at some special cases of this general equation before we graph it in this form.

INSTRUCTOR NOTE

In this chapter, we carefully develop skills that students can use to sketch graphs quickly and accurately. The skills used in Section 9.1 will be used throughout the chapter.

The Graphs of $y = x^2$ and $y = -x^2$

The simplest quadratic equation in two variables is $y = x^2$, so we begin with its graph.

▪▪▪

EXAMPLE 1

Graph the equation $y = x^2$.

SOLUTION

This equation has the general form with $a = 1$ and $b = c = 0$. Its graph is a parabola; the x-coordinate of its vertex is

$$x = \frac{-b}{2a} = \frac{-0}{2} = 0$$

To find the y-coordinate, we substitute in the equation.

$$y = x^2 = 0^2 = 0$$

So the vertex is located at (0, 0). The axis of the parabola is the line $x = 0$, the y-axis. The parabola is symmetrical about its axis; that is, if we fold our graph along the line $x = 0$, the "sides" of the parabola will coincide. Finally because the coefficient of x is greater than 0 when written in this general form, the graph opens upward. Now we need a few more points to graph $y = x^2$. A table of values and the graph are shown in the following figure.

x	y
1	1
−1	1
2	4
−2	4

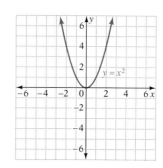

❗❗❗

ERROR ALERT

Identify the error and give a correct answer.

Determine whether the graph of
$y - x^2 + 3 = 0$ opens upward or downward.

Incorrect Solution:

The graph of this equation opens downward because the coefficient of x^2 is −1.

Calculator Corner

Use your graphing calculator to graph $y = 2x^2$ and $y = -2x^2$ on the same axes; note that these graphs are exactly the same except for their orientation. Now do the same for $y = 3x^2$ and $y = -3x^2$; again, only the orientation (opening upward or downward) is different.

Parabolas with Vertical Shift

INSTRUCTOR NOTE

We encourage the use of graphing calculators for this chapter.

> If a quadratic equation has the form
>
> $$y = x^2 + k \quad \text{or} \quad y = -x^2 + k$$
>
> then its graph has exactly the same shape as that in the figures in Examples 1 and 2. However, the entire graph is shifted *vertically* by k units.

Adding k to the right side of the equation $y = x^2$ *increases* each y-value by k, so the graph is shifted upward (toward positive y-values) if k is positive. The graph is shifted downward if k is negative. The vertex is then at $(0, k)$.

▪▪▪

EXAMPLE 2

Sketch the graph of the equation $y = x^2 - 4$ without plotting points.

SOLUTION

Rewriting the equation as $y = x^2 + (-4)$ shows that $k = -4$. Thus the graph is a parabola with the same shape as the graph of $y = x^2$, shifted down by 4 units. The vertex is at $(0, -4)$. The graph is sketched in the following figure.

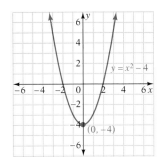

▶ *CHECK* **Warm-Up 2**

We say that the graph in the figure in Example 2 is a sketch because we didn't graph specific points, except for the vertex.

Calculator Corner

Use your graphing calculator to graph $y = x^2 + k$ for $k = -3, -1, 1,$ and 3 and see how the graph shifts as k changes. Then repeat for $y = -x^2 + k$.

Parabolas with Horizontal Shift

> If a quadratic equation has the form
>
> $$y = (x - h)^2 \quad \text{or} \quad y = -(x - h)^2$$
>
> then its graph has the same shape as that in the figures in Examples 1 and 2. However, the entire graph is shifted *horizontally* by h units.

The graph is shifted to the right (toward positive x-values) if h is positive; it is shifted to the left if h is negative. Then the vertex is at $(h, 0)$. Be careful to account for the minus sign in $(x - h)^2$ when you determine the sign of h.

■■■
EXAMPLE 3

Sketch the graph of the equation $y = -(x + 2)^2$ without plotting points.

SOLUTION

Rewriting the equation in the form $y = -(x - h)^2$ gives us

$$y = -[x - (-2)]^2$$

So h is -2 and the vertex is at $(-2, 0)$. The graph opens downward (because the x^2 coefficient is -1). It has the same shape as that in the figures in Examples 1 and 2 and is sketched below.

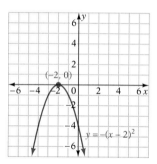

CHECK **Warm-Up 3**

Calculator Corner

Use your graphing calculator to graph $y = (x - h)^2$ for $h = 3, 1, -1$, and -3 and see how the parabola shifts when h changes.

Wider and Narrower Parabolas

If a quadratic equation has the form

$$y = ax^2 \quad \text{for } a \neq 0$$

then its graph has the general ∪ shape of that in the figures in Examples 1 and 2. However, the graph is wider if $|a| < 1$ and is narrower if $|a| > 1$.

As usual, the graph opens upward if a is positive and downward if a is negative. The vertex is at $(0, 0)$.

▪▪▪

EXAMPLE 4

Graph $y = 2x^2$ and $y = \frac{1}{2}x^2$ on the same set of axes.

SOLUTION

Both graphs are parabolas that open upward and have the vertex at $(0, 0)$. We construct a short table of values for each graph to get an idea of its specific shape.

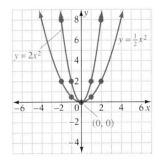

$$y = 2x^2 \qquad\qquad y = \frac{1}{2}x^2$$

x	y		x	y
0	0	vertex	0	0
1	2		1	$\frac{1}{2}$
−1	2		−1	$\frac{1}{2}$
2	8		2	2
−2	8		−2	2

We plot these points, one set at a time, and connect them with smooth curves to get the graphs shown in the accompanying figure. Note that the graph of $y = 2x^2$ is much narrower (closer to its y-axis) than the graph of $y = \frac{1}{2}x^2$.

▶ *CHECK* **Warm-Up 4**

Calculator Corner

Use your graphing calculator to graph $y = ax^2$ for $a = \frac{1}{5}, \frac{1}{3}, 1, 3$, and 5 and see how the parabola changes when a varies.

The Graphing Form of the Equation of a Parabola

All the ideas we have discussed so far can apply to a single quadratic equation.

INSTRUCTOR NOTE

The graphing form of the equation of a parabola is very important. Encourage your students to memorize it and rewrite all their parabolas in this form before they graph them.

> The graphing form of the equation of a parabola is
>
> $$y = a(x - h)^2 + k \quad \text{when } a \neq 0$$
>
> The graph of this equation has the same shape as the graph of $y = ax^2$. The graph opens upward if a is positive and opens downward if a is negative. The vertex of the parabola is at (h, k). The axis of the parabola is the line $x = h$.

▪▪▪

EXAMPLE 5

Describe the graph of the equation $y = -2(x + 2)^2 - 3$.

SOLUTION

In graphing form, this equation is

$$y = -2[x - (-2)]^2 + (-3)$$

Thus its graph is a parabola with its vertex at $(-2, -3)$. The parabola opens downward (because the coefficient a, -2, is negative), and it is symmetric about its axis, the line $x = -2$.

We also know that the parabola has the same shape as the graph of $y = 2x^2$. It is sketched in the accompanying figure.

INSTRUCTOR NOTE

We attempt to show students equations of conic sections in many different forms. However, we encourage them to rewrite the equations in graphing form first, and then sketch the graph.

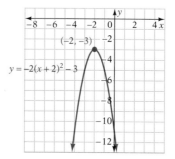

▶ CHECK **Warm-Up 5**

▪▪▪
WRITER'S BLOCK

Write a set of rules for "completing" a square.

The Graph of $y = ax^2 + bx + c$

A quadratic equation can be graphed quickly and easily when it is in graphing form. But quadratic equations usually are in the general form $y = ax^2 + bx + c$. Fortunately, we can transform an equation from the general form to the graphing form by completing the square.

▪ ▪ ▪

EXAMPLE 6

Describe the graph of the equation

$$y = -6x^2 + 12x - 2$$

by writing the equation in graphing form.

SOLUTION

To write this equation in graphing form, we must complete a square on the right side. To simplify the process, we first move any constant term to the left side. Then we factor the terms that contain the variable to get x^2 with the coefficient 1. This gives us

$$y + 2 = -6(x^2 - 2x)$$

Now we can complete the square within the parentheses. To do so, we note that the coefficient of x there is -2. This is the coefficient b, and we want

$$\left(\frac{b}{2}\right)^2 = \left(\frac{-2}{2}\right)^2 = 1$$

So we need to add 1 inside the parentheses to complete the square. Note, however, that this will then be multiplied by the factor -6, so we will in fact be adding $(-6)(1) = -6$. We must do the same on the other side of the equation in order not to change the equation. We get

$$y + 2 = -6(x^2 - 2x) \qquad \text{The equation}$$
$$y + 2 + (-6)(1) = -6(x^2 - 2x + 1) \qquad \text{Completing the square}$$
$$y - 4 = -6(x - 1)^2 \qquad \text{Adding like terms; writing the square}$$
$$y = -6(x - 1)^2 + 4 \qquad \text{Adding 4 to both sides}$$

With the equation in this form, we can tell that its graph is a parabola that opens downward and is relatively narrow (close to its axis), because $a = -6$. Its vertex is at $(1, 4)$, and it is symmetric about the line $x = 1$.

▶ CHECK **Warm-Up 6**

▪ ▪ ▪

EXAMPLE 7

Graph the equation $y = -6x^2 + 12x - 2$ *without* completing the square.

SOLUTION

The x-coordinate of the vertex is

$$x = \frac{-b}{2a} = \frac{-12}{2(-6)} = 1$$

The y-coordinate is then

$$y = -6x^2 + 12x - 2$$
$$y = -6(1)^2 + 12(1) - 2$$
$$= -6 + 12 - 2 = 4$$

INSTRUCTOR NOTE

Remind students of the procedure for completing a square in a quadratic equation. Then show how it is similar to the procedure here. Students will need much practice.

STUDY HINT

If you would rather not complete a square to graph a quadratic equation
$$y = ax^2 + bx + c,$$
there is another way. You can, instead:

1. *Set $x = 0$ and solve for y to obtain the y-intercept.*

2. *Set $y = 0$ and solve for x to obtain the x-intercepts (if there are any).*

3. *Compute the x-coordinate of the vertex as $= \frac{-b}{2a}$; then substitute that in the given equation to find the y-coordinate of the vertex.*

4. *Graph the points found in steps 1 to 3. Graph other points found by symmetry, noting that the axis of the parabola is the line $x = \frac{-b}{2a}$.*

5. *Note whether the graph opens upward or downward, depending on the sign of a, and draw the parabola.*

You can use any or all of these procedures in graphing.

So the vertex is at (1, 4). The graph opens downward because a is negative. It is symmetric about the line $x = 1$. We find two more points and graph the equation as in the following figure.

x	y	
0	-2	y-intercept
2	-2	

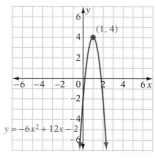

CHECK **Warm-Up 7**

We worked with the same quadratic equation in Examples 6 and 7. Compare the description of the graph in Example 6 and the graph we obtained in Example 7. Do they agree? In graphing equations, the method we use doesn't matter as long as it is applied correctly.

Calculator Corner

A parabola can be defined as the set of all points in a plane that are equidistant from a fixed point (*the focus*) and a fixed line (*the directrix*) that does not contain the focus. To the right is an example of the parabola $x^2 = 8y$. You can easily prove that points on the parabola are equidistant from the focus and directix by using the distance formula. What change can you make to the equation in order to make the parabola open downward instead of upward? Is there only one answer for this question? If there is more than one answer, how many are there?

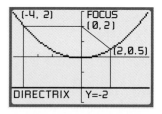

Graphing a parabola opening to the right or left on a graphing calculator is not as straightforward because the calculator works in **Function Mode** and the parabola to the right is *not a function*. To "fake out" the calculator, simply graph $y^2 = 8x$ as Y1 $= \sqrt{8x}$ and Y2 $= -$Y1.

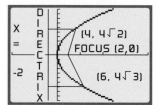

Care should be taken when you use your graphing calculator to see the visual support of the definition of a parabola. If you do not select a **squared viewing rectangle,** how could this affect your visual conclusions about parabolas?

Now explore the following "families" of parabolas and make a conjecture in each case as to what is happening in each exploration.

Vertical Transformation Explorations of the Parabola "Family"

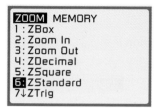

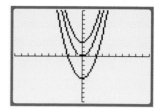

Horizontal Transformation Explorations of the Parabola "Family"

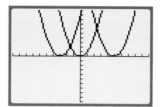

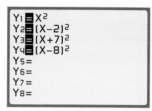

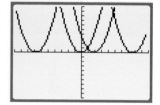

Combining Transformations

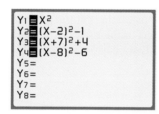

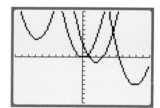

Sketch each of the following parabolas on graph paper. Then support your graph with your graphing calculator.

a. $x^2 = 5y$ **b.** $x^2 = -2y$ **c.** $y = x^2 - 8x + 19$ **d.** $y^2 = -9x$

▪▪▪

EXAMPLE 8

Find the equation of the parabola that has its vertex at $(2, 3)$ and passes through the point $(5, -1)$. In which direction does the graph open?

SOLUTION

We start with the graphing form of the parabola and substitute the coordinates (h and k) of the vertex.

$$y = a(x - h)^2 + k$$
$$= a(x - 2)^2 + 3$$

Then we substitute the coordinates of the given point and solve for a.

$$-1 = a(5 - 2)^2 + 3$$
$$-1 = 9a + 3$$
$$-4 = 9a$$
$$-\frac{4}{9} = a$$

Now, we substitute this value for a, obtaining the equation $y = -\frac{4}{9}(x - 2)^2 + 3$. The graph opens downward.

▶ *CHECK* **Warm-Up 8**

Horizontal Parabolas

INSTRUCTOR NOTE

We have included this topic to highlight the fact that parabolas have many possible orientations. Most of the graphing we do now and later, however, involves only vertical parabolas.

Quadratic equations of the form $y = ax^2 + bx + c$ are vertical parabolas; they are functions. But you could very well encounter an equation of the general form

$$x = ay^2 + by + c$$

The corresponding graphing form is

$$x = a(y - k)^2 + h$$

WRITER'S BLOCK

Can a horizontal parabola ever be a function? Explain.

The graph of $x = a(y - k)^2 + h$ is a *horizontal* parabola like that shown below. The vertex of the parabola is at the point (h, k), and its axis is the line $y = k$. The parabola opens to the right if a is positive and opens to the left if a is negative. It is relatively narrow if $|a| > 1$ and relatively wide if $|a| < 1$.

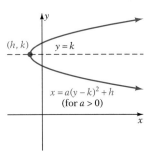

···

EXAMPLE 9

Graph the equation $x = -\frac{1}{3}(y + 2)^2 + 3$.

SOLUTION

We note immediately that

The graph is a parabola that opens to the left (because $-\frac{1}{3}$ is negative).

The parabola is wider than the graph of $x = y^2$ (because $\left|-\frac{1}{3}\right| < 1$).

The vertex is at $(3, -2)$.

The axis is the line $y = -2$.

And solving for x when $y = 0$, we find that the x-intercept is the point $\left(1\frac{2}{3}, 0\right)$. Symmetry then yields the point $\left(1\frac{2}{3}, -4\right)$.

By setting $x = 0$ in the given equation and solving for y, we find that there are y-intercepts at $(0, 1)$ and $(0, -5)$. Setting $y = 2$ (arbitrarily chosen) in the given equation yields the point $\left(-2\frac{1}{3}, 2\right)$, and symmetry yields the point $\left(-2\frac{1}{3}, -6\right)$.

Using this information, we obtain the graph shown in the following figure.

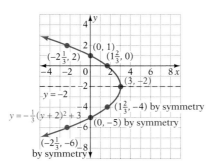

▶ *CHECK* **Warm-Up 9**

> ### ❗❗❗
> *ERROR ALERT*
>
> Identify the error and give a correct answer.
>
> *Incorrect Solution:*
>
> The equation $x = 3(y - 2)^2 - 8$ is a function because it is in graphing form.

Practice what you learned.

SECTION FOLLOW-UP

Here is our geometric justification for a parallelogram that tessellates.

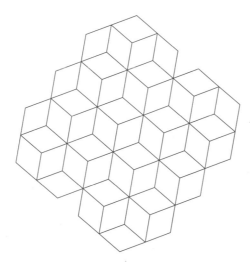

▪ Did you find any figures that do not tessellate? which one(s)?

9.1 WARM-UPS

Work these problems before you attempt the exercises.

1. Graph $y = -x^2$.

   What is the major difference between your graph and the graph in Example 1? In what ways are the graphs similar?
   The graphs are different in that one opens upward and the other opens downward. The graphs have the same vertex and the same shape.

2. **a.** Sketch $y = x^2 + 4$ without plotting the points.

   b. What is the major difference between your graph and the graph in Example 2? In what ways are the graphs similar?
   The graphs have the same shape, and both open upward. They are different in that one has its vertex at $(0, 4)$ and the other at $(0, -4)$. One begins 4 units up the y-axis and the other begins 4 units down the y-axis.

3. Sketch $y = -(x - 2)^2$.   4. Graph $y = 3x^2$.

5. Describe the graph of $y = (x - 2)^2 - 3$.
   The graph opens upward and has its vertex at $(2, -3)$. The shape is the same as that of $y = x^2$.

6. Rewrite $y = 5x^2 + 10x - 3$ in graphing form by completing the square. Describe its graph and find the vertex. $y = 5(x + 1)^2 - 8$.
   The graph opens upward and is very narrow. It is the same shape as $y = 5x^2$. The vertex is at $(-1, -8)$.

7. Graph $y = \frac{1}{2}x^2 + \frac{1}{2}x - 2$ without completing the square.

8. Write the equation of a parabola that has its vertex at $(2, 5)$ and passes through the origin. $y = \frac{-5}{4}(x - 2)^2 + 5$

9. Write the equation of a horizontal parabola with vertex $(2, -3)$ and that passes through point $(4, -2)$.
   $x = 2(y + 3)^2 + 2$

9.1 EXERCISES

Note: Use your graphing calculator to check your results whenever possible.

In Exercises 1 through 4, graph each group of four equations. Then, for each exercise, comment on the similarities and differences among these four graphs.   See the Solutions Manual.

1. **a.** $3x^2 - y = 0$ **b.** $y = -3x^2$ **c.** $y = \frac{1}{3}x^2$ **d.** $-y = \frac{1}{3}x^2$

2. **a.** $y = \frac{1}{2}x^2$ **b.** $y = -2x^2$ **c.** $2x^2 - y = 0$ **d.** $-y = \frac{1}{2}x^2$

3. **a.** $-y = \frac{1}{4}x^2$ **b.** $y = \frac{1}{4}x^2$ **c.** $y = -4x^2$ **d.** $4x^2 - y = 0$

4. **a.** $5x^2 - y = 0$ **b.** $y = \frac{1}{5}x^2$ **c.** $y = -5x^2$ **d.** $-y = \frac{1}{5}x^2$

In Exercises 5 through 16, sketch the graph of each equation by identifying the vertex, the direction in which the graph opens, and the axis. Also graph at least four points on the parabola.

5. $y = x^2 - 2$

6. $y = -x^2 - 2$

7. $y = (x + 2)^2$

8. $y = (x - 2)^2$

9. $y = (x - 1)^2 + 1$

10. $y = (x + 1)^2 - 2$

11. $y = -(x - 1)^2 + 1$

12. $y = -(x + 1)^2 - 2$
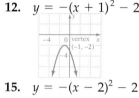

13. $y = (x + 3)^2 - 1$

14. $y = (x - 3)^2 - 1$

15. $y = -(x - 2)^2 - 2$

16. $y = -(x + 2)^2 - 2$
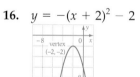

In Exercises 17 through 24, write an equation of the vertical parabola that fits the conditions given.

17. vertex: $(1, 3)$
     passes through
     $(-2, 5)$ $y = \frac{2}{9}(x - 1)^2 + 3$

18. vertex: $(2, -3)$
     passes through $(0, 4)$
     $y = \frac{7}{4}(x - 2)^2 - 3$

19. vertex: $(-1, -2)$
     passes through $(1, 5)$
     $y = \frac{7}{4}(x + 1)^2 - 2$

20. vertex: $(0, -2)$
     passes through
     $(-3, 5)$
     $y = \frac{7}{9}x^2 - 2$

21. vertex: $(1, -3)$
     passes through
     $(-2, -8)$
     $y = \frac{-5}{9}(x - 1)^2 - 3$

22. vertex: $(-1, -4)$
     passes through
     $(-4, 2)$ $y = \frac{2}{3}(x + 1)^2 - 4$

23. vertex: $(0, 0)$
     passes through
     $(1, -3)$ $y = -3x^2$

24. vertex: $(-6, 2)$
     passes through
     $(-5, 1)$
     $y = -(x + 6)^2 + 2$

In Exercises 25 through 32, rewrite the equation in graphing form, completing the square where necessary. Where is the vertex of its graph? In which direction does its graph open?

25. $2x^2 - y = -18$ $y = 2(x - 0)^2 + 18$
     $(0, 18)$; upward

26. $6y - 12 = x^2$ $y = \frac{1}{6}(x - 0)^2 + 2$
     $(0, 2)$; upward

27. $2x^2 + x + 8 = y$ $y = 2\left[x - \left(\frac{-1}{4}\right)\right]^2 + 7\frac{7}{8}$
     $\left(\frac{-1}{4}, 7\frac{7}{8}\right)$; upward

28. $9x^2 + 8x + 8 = y$ $y = 9\left[x - \left(\frac{-4}{9}\right)\right]^2 + 6\frac{2}{9}$
     $\left(\frac{-4}{9}, 6\frac{2}{9}\right)$; upward

29. $\frac{1}{6} - 6y = x^2$ $y = -\frac{1}{6}(x - 0)^2 + \frac{1}{36}$;
     $\left(0, \frac{1}{36}\right)$; downward

30. $-(x - 2)^2 + 3 + y = 0$ $y = (x - 2)^2 - 3$
     $(2, -3)$; upward

31. $4x^2 + 16x - 55 = y$ $y = 4[x - (-2)]^2 - 71$
     $(-2, -71)$; upward

32. $3x^2 - 12x - 15 = y$ $y = 3(x - 2)^2 - 27$
     $(2, -27)$; upward

In Exercises 33 through 40, find the vertex of the graph of the given equation, and determine in which direction the graph opens. Do not rewrite the equation in graphing form.

33. $y = 3x^2 + 6x - 12$ $(-1, -15)$; upward

34. $y = 2x^2 - 24x + 140$ $(6, 68)$; upward

35. $y = 2x^2 + 8x + 5$ $(-2, -3)$; upward

36. $4y = 8x^2 + 16x + 24$ $(-1, 4)$; upward

37. $y = 2x^2 + 2x - 12$ $\left(-\frac{1}{2}, -12\frac{1}{2}\right)$; upward

38. $-5y = 10x^2 - 15x + 20$ $\left(\frac{3}{4}, -\frac{23}{8}\right)$; downward

39. $y = -2x^2 + 6x - 5$ $(1.5, -0.5)$; downward

40. $-3y = 9x^2 + 12x - 15$ $\left(-\frac{2}{3}, \frac{19}{3}\right)$; downward

In Exercises 41 through 44, use any method to find the vertex of the graph of the given equation, and determine in which direction the graph opens.

41. $-2y = 6x^2 - 10x + 8$ $\left(\frac{5}{6}, -\frac{23}{12}\right)$; downward

42. $y = -x^2 + 5x + 10$ $(2.5, 16.25)$; downward

43. $y - 12 = -x^2$ $(0, 12)$; downward

44. $18 + y = -x^2$ $(0, -18)$; downward

In Exercises 45 through 48, write an equation for the horizontal parabola with the given vertex and point.

45. vertex: $(-3, 3)$
point: $(5, -2)$
$x = \frac{8}{25}(y - 3)^2 - 3$

46. vertex: $(-3, 5)$
point: $(0, 0)$
$x = \frac{3}{25}(y - 5)^2 - 3$

47. vertex: $(-2, -1)$
point: $(-5, -4)$
$x = -\frac{1}{3}(y + 1)^2 - 2$

48. vertex: $(3, 2)$
point: $(-1, -3)$
$x = -\frac{4}{25}(y - 2)^2 + 3$

In Exercises 49 through 56, locate the vertex of each parabola, and then graph using a table of values.

49. $y = 2(x + 2)^2 - 1$

50. $y = 0.5(x + 2)^2 + 4$

51. $y = -5(x + 1)^2 - 1$

52. $y = -0.5(x - 1)^2 - 1$

53. $x = y^2 - 4$

54. $x = y^2 - 3$

EXCURSIONS

Posing Problems

1. Ask and answer four questions about the following information. Try to ask questions that other students would want to answer.

23rd Iditarod Trail Sled Dog Race—1995
(Alaska, March 4–16, 1995)

The annual 1159-mile race stretches from Anchorage to Nome, Alaska. Begun in 1973, the course follows an old frozen river route and is named after a deserted mining town along the way. The Iditarod also commemorates a famous midwinter emergency mission to get medical supplies to Nome during a 1925 diphtheria epidemic. Men and women mushers compete together.

Course: 1159 miles, Anchorage to Nome.

1995 Champion—Doug Swingly became the first non-Alaskan to win, in record time of 9 days, 2 hours, 43 minutes, breaking last year's record set by Martin Buser.

Total Purse: $350,000.

Winner's purse: $52,500.

Winning times since 1980: 1980, Joe May, 14 days–7 hours–11 minutes; 1981, Swenson, 12–8–45; 1982, Swenson, 16–4–40; 1983, Rick Mackey, 12–14–10; 1984, Dean Osmar, 12–15–7; 1985, Libby Riddles, 18–00–20; 1986, Butcher, 11–15–6; 1987, Butcher, 11–2–5; 1988, Butcher, 11–11–41; 1989, Joe Runyan, 11–5–24; 1990, Butcher, 11–1–53; 1991, Swenson, 12–16–34; 1992, Martin Buser, 10–19–17; 1993, Jeff King, 10–15–30; 1994, Martin Buser, 10–13–2; 1995, Doug Swingly, 9–2–43.

Source: 1996 Information Please® Almanac (©1995 Houghton Miffin Co.), p. 1003. All rights reserved. Used with permission by Information Please LLC.

2. Have you ever watched Olympic gymnastics? The gymnasts receive decimal scores for their performances. For example, at the 1996 Summer Olympics, six women gymnasts from the United States won the gold medal in the team competition. Their score of 389.225 was calculated by adding scores received by the six gymnasts in various events. The difference between their score and the score of the women who won the silver metal was only 0.821, less than nine tenths of a point. There was only a 0.158 difference between silver and bronze.

   Ask and answer two questions about this information. Name other sports where decimal numbers are used.

3. Ask and answer at least four questions about the following data.

Indianapolis 500

Year	Winner	Car	Time	Average mph	Second place
1985	Danny Sullivan	Miller March–Cosworth	3:16:06.069	152.982	Mario Andretti
1986	Bobby Rahal	Budweiser March–Cosworth	2:55:43.48	170.722	Kevin Cogan
1987	Al Unser, Sr.	Cummins March–Cosworth	3:04:59.147	162.175	Roberto Guerrero
1988	Rick Mears	Pennzoil Penske P.C.17–Chevrolet	3:27:10.204	144.809	Emerson Fittipaldi
1989	Emerson Fittipaldi	Marlboro Penske–Cosworth	2:59:01.04	167.581	Al Unser, Jr.
1990	Arie Luyendyk	Domino's Pizza Lola–Cosworth	2:41:18.248	185.987	Bobby Rahal
1991	Rick Mears	Marlboro Penske–Cosworth	2:50:01.018	176.460	Michael Andretti
1992	Al Unser, Jr.	Valvoline–Chevrolet	3:43:05.148	134.477	Scott Goodyear
1993	Emerson Fittipaldi	Penske–Chevrolet	3:10:49.860	157.207	Arie Luyendyk
1994	Al Unser, Jr.	Penske–Mercedes	3:06:29.006	160.872	Jacques Villeneuve
1995	Jacques Villeneuve	Reynard–Ford	3:15:17.561	156.616	Christian Fittipaldi

Source: 1996 Information Please® Almanac (©1995 Houghton Mifflin Co.), p. 977. All rights reserved. Used with permission by Information Please LLC.

Data Analysis

4. **a.** Discuss the data in this graph. Ask and answer two questions using this data.

   b. *Research* What is a carburetor restrictor plate and its significance in racing speeds?

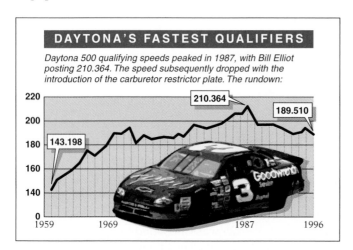

Exploring Geometry

5. If B is the area of the base and h is the height of a pyramid, its volume is

$$V = \frac{1}{3}Bh$$

For example, the Transamerica Tower in San Francisco is a pyramid with a square base. Each side of the base has a length of 52.13 meters. The height of the building is 259.8 meters. Find the volume of the building.

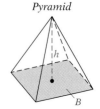

Pyramid

B is the area of the base.

Another formula for the volume of a pyramid is $V = \frac{1}{3}a^2h$ where a is the length of a side and h is the height.

The great pyramid of Egypt stands on a square base each side of which is 764 feet; and its height is 480 feet. Find the number of cubic feet of stone used in its construction. Refer to the first pyramid volume formula. What is the relationship between a and B?

CONNECTIONS TO *GEOMETRY*

Using Diagrams

In solving word problems, we must sometimes combine topics and ideas from geometry, fractions, percents, or other areas of mathematics. It is also an excellent idea to make a sketch for a problem when one is not given.

Let's look at some examples.

▪▪▪

EXAMPLE

Five circles, all of radius 1 inch, are cut out of a rectangular piece of metal that measures 2.3 by 9.8 inches. None of the circles overlap. Find the area of the metal that remains.

SOLUTION

There is no figure, so we draw one. We begin with the rectangle, because circles will be *cut out* of it.

Next we draw in the five circles. We know that they are all circles and that they *don't overlap*. But we can place them wherever we like. Finally, we shade in the area we must find, as a reminder.

We can find the area of the metal that remains by subtracting the area of the five circles from the area of the rectangle.

$$\text{Area(shaded)} = \text{area(rectangle)} - \text{area(five circles)}$$

First we find the area of the rectangle.

$$A = \ell \times w = 2.3 \times 9.8 = 22.54 \text{ square inches}$$

Then we find the area of one circle.

$$A = \pi r^2$$

Because the radius of the circle is 1 inch, the area is

$$A = 3.14 \times 1^2 = 3.14 \text{ square inches}$$

And the area of five circles is 5 times 3.14.

$$5 \times 3.14 = 15.70 \text{ square inches}$$

So the area of the remaining metal is $22.54 - 15.70 = 6.84$ square inches.

◀

In the next example, we are given information about a geometric figure along with a percent, and we are asked to find another piece of information.

▪ ▪ ▪

EXAMPLE

A rectangular area rug is 4 feet wide by 6 feet long. A rectangular piece equivalent to 40% of the area of the original rug is cut off its length. If the width remains 4 feet, what is the length of the cut piece? What are the dimensions of the remaining portion of the original rug?

SOLUTION

This problem will definitely be easier to solve if we first make a sketch. We draw a rectangle, label one side 4 feet and one 6 feet, and draw a dashed line to show where the cut is made. The sketch also shows what we know about the percents. See the accompanying figure.

Let's first find the area of the cut-off piece. To do that, we must find the area of the original rug.

$$\text{Area} = \ell w = 6 \times 4 = 24 \text{ square feet}$$

We know that 40% of the rug has been cut off, so we ask (and then must answer) the question "40% of 24 square feet is what?"

The percent is 40, the base is 24, and the amount is the unknown value.

$$\text{Percent (as a decimal)} \times \text{base} = \text{amount}$$
$$0.40 \times 24 = n$$
$$9.60 = n \qquad \text{Multiplying out}$$

So 9.60 square feet of the rug has been cut off. We can use this area to find the length that was cut off, knowing its width is 4 feet. We have

$$A = \ell \times w$$
$$9.6 = \ell \times 4 \qquad \text{Substituting for } A \text{ and } w$$
$$\frac{9.6}{4} = \frac{\ell \times 4}{4} \qquad \text{Dividing both sides by 4}$$
$$2.4 = \ell \qquad \text{Dividing out}$$

So the length of the cut-off piece is 2.4 feet.

To find the length of the remaining rug, we subtract.

$$6 - 2.4 = 3.6 \text{ feet}$$

Its "width" is still 4 feet, so the dimensions of the remainder of the original rug are 3.6 feet by 4 feet.

◢

PRACTICE

1. **Quilt Panels** A quilt has panels shaped as shown in the figure at the right. The inner panel is a square $9\frac{1}{2}$ square inches in area. The perimeter of the large square is 21 inches. What is the total area of the trapezoids?
   18.1 square inches

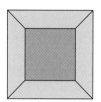

2. **Chicken Wire** Chicken wire is to be placed along a straight fence. There are 21 fence poles that are 27 inches apart. The poles are 3 inches in diameter. How much chicken wire is needed to reach exactly from the first pole to the last?   603 inches

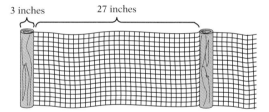

3 inches   27 inches

3. **Geometry** A cube with an edge of 3 inches has a spherical ball placed inside of it. The ball has a diameter of 2 inches. The cube is then filled with sand right up to and level with the top, covering the ball. What volume of sand is in the cube? Use 3.14 for π and round to the nearest hundredth.
   22.81 cubic inches

4. **Geometry** A square card table that is $3\frac{1}{4}$ feet on a side is serving as the workspace for a round jigsaw puzzle with a radius of $18\frac{1}{4}$ inches. What percent of the card table will be exposed when the puzzle is complete? Round to the nearest whole percent.   31%

$3\frac{1}{4}$ feet

$18\frac{1}{4}$ inches

$3\frac{1}{4}$ feet

5. *Area of a Table* The diameter of a round tablecloth is 20.6 centimeters. If $15\frac{1}{2}\%$ of the tablecloth is hanging off the table, what is the actual area of the top of the table in square centimeters? Round to the nearest tenth. 281.5 cm^2

6. *Stone Border* Stones are to be placed around a rectangular swimming pool. There will be 19 stones on each side along the longer sides of the pool, and 14 stones on each side along the shorter ends. How many stones will there be in all? 62 stones

7. A bathroom floor has tiles that are 1 foot square. Aubrey has put 14 tiles in place, covering 14% of the bathroom floor area. How large is the bathroom floor? 100 square feet

8. Books are being placed on a shelf. Each book is $2\frac{1}{2}$ inches thick. Each cover is $\frac{1}{4}$ inch thick. How much shelf space will 2 dozen books take up on the shelf? 72 inches

9. A city law states that classrooms must contain at least 20.5 square feet for every student in the class. If a room is designed to hold 45 students, how big must its area be? 922.5 square feet

10. A rectangular room that has dimensions of $10\frac{1}{2}$ feet by $12\frac{3}{4}$ feet contains 100 tiles that are identical in size. What is the area of each tile? (Round your answer to the nearest tenth of a square foot.) 1.3 ft^2

SECTION GOALS

▪ To write the equation of a circle centered at the origin

▪ To sketch the graph of a circle, an ellipse, and a hyperbola centered at the origin

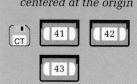

9.2 The Circle, Ellipse, and Hyperbola Centered at the Origin

SECTION LEAD-IN

We can use the triangle and a method of "triangulation" to find the area of an irregular figure.

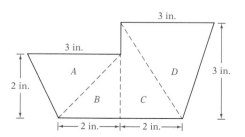

$$\Delta A = \tfrac{1}{2}b \cdot h = \tfrac{1}{2}(3)(2) = 3$$
$$\Delta B = \tfrac{1}{2}b \cdot h = \tfrac{1}{2}(2)(2) = 2$$
$$\Delta C = \tfrac{1}{2}b \cdot h = \tfrac{1}{2}(2)(3) = 3$$
$$\Delta D = \tfrac{1}{2}b \cdot h = \tfrac{1}{2}(3)(3) = 4\tfrac{1}{2}$$
$$\overline{}$$
$12\tfrac{1}{2}$ square inches

Justify the measurements used for ΔC and ΔD.

Draw some irregular figures using a ruler. Use your method to find the area.

The Circle

When an infinite cone is cut by a plane that is perpendicular to the cone's axis, the intersection is a **circle** (see the figure on the left below).

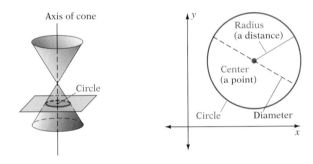

A circle may also be defined as the set of all points in the plane that are a given distance r from a given point. The distance r is called the **radius** of the circle; the point is called its **center** (see the figure on the right above).

A **diameter** of a circle is a line segment that extends from a point on the circle through the center to another point on the circle. The diameter has a length of $2r$.

Circles Centered at the Origin

When circles are centered at the origin, it is easy to graph them.

> The equation of a circle with its center at the origin and radius r has the form
> $$x^2 + y^2 = r^2$$

INSTRUCTOR NOTE

We emphasize sketching circles using basic information we can obtain easily from the equation. Urge your students to rely on this method and not on making a table of values.

▪▪▪

EXAMPLE 1

a. Write an equation for the circle that has a radius of 10 units and is centered at the origin.

b. Sketch the graph of $x^2 = 32 - y^2$.

SOLUTION

a. With $r = 10$, the circle has the equation $x^2 + y^2 = 10^2$, or

$$x^2 + y^2 = 100$$

b. Adding y^2 to both sides of the given equation yields

$$x^2 + y^2 = 36$$

Thus the graph is a circle with its center at the origin and with

$$r = \sqrt{36} = 6 \text{ units}$$

To sketch the circle, first mark off

$$r = 6 \text{ units}$$

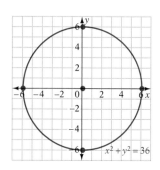

from the origin in all four directions. Then, as best you can, draw a circle that passes through these four points (see the accompanying figure). Or, with a compass set for 6 units, draw a circle centered at the origin.

▶ *CHECK* **Warm-Up 1**

The Ellipse

When a plane cuts completely through a cone and is not perpendicular to the cone's axis, the intersection is an ellipse (see the figure on the left below).

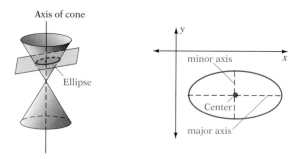

An ellipse is shown on the coordinate plane in the figure on the right above. Note the horizontal and vertical line segments cut off by the ellipse (dashed in the figure); the longer one is called the **major axis** and the shorter one the **minor axis.** They intersect at the center of the ellipse.

The ellipse can be defined as the set of all points for which the sum of the distances to two fixed points is a constant. The fixed points (points F_1 and F_2 in the next figure) are called the foci (singular: focus) of the ellipse.

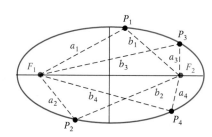

In the previous figure, the sum of the distances to any point P on the ellipse is a constant. So $a_1 + b_1 = a_2 + b_2 = a_3 + b_3 = a_4 + b_4$.

A circle is a special case of an ellipse in which the major and minor axes are equal in length. It also has both foci located at the center.

In the following figure, think of a_1 and b_1 as two lengths of one string tacked at F_1 and F_2 and pulled tight by a pencil. The figure that can be drawn by this pencil, keeping the string taut, is an ellipse.

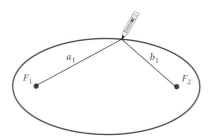

Ellipses Centered at the Origin

> The equation of an ellipse with its center at the origin has the standard form
>
> $$\frac{x^2}{a^2} + \frac{y^2}{b^2} = 1$$
>
> The x-intercepts of the ellipse are at $(a, 0)$ and $(-a, 0)$. The y-intercepts of the ellipse are at $(0, b)$ and $(0, -b)$. When the major axis is horizontal, the foci are at $(c, 0)$ and $(-c, 0)$, where $c^2 = a^2 - b^2$. When the major axis is vertical, the foci are at $(0, c)$ and $(0, -c)$, where $c^2 = b^2 - a^2$. The length of the horizontal axis is $2a$; the length of the vertical axis is $2b$.

EXAMPLE 2

Graph the equation $\frac{x^2}{9} + \frac{y^2}{4} = 1$. Find the lengths of the major and minor axes.

SOLUTION

By inspection, or after rewriting the given equation as

$$\frac{x^2}{3^2} + \frac{y^2}{2^2} = 1$$

we see that $a = 3$ and $b = 2$. So the x-intercepts are at $(-3, 0)$ and $(3, 0)$, and the y-intercepts are at $(0, 2)$ and $(0, -2)$. These four points are usually sufficient for sketching an ellipse as in the accompanying figure.

The major axis is the horizontal axis because $a > b$. It is $2a$, or 6, units long. The minor axis is then the vertical axis. It is $2b$, or 4, units long.

▶ CHECK **Warm-Up 2**

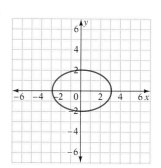

! ! !

ERROR ALERT

Identify the error and give a correct answer.

Find the length of the major and minor axes of the ellipse whose equation is $\frac{x^2}{25} + \frac{y^2}{16} = 1$.

Incorrect Solution:

Because $a^2 = 25, a = 5$
$\qquad\quad b^2 = 16, b = 4$

The length of the major axis is 5, and the length of the minor axis is 4.

Calculator Corner

Consider the following ellipse centered at the origin: $\frac{x^2}{25} + \frac{y^2}{49} = 1$. Solving this equation for y gives $y = \sqrt{49\left(1 - \frac{x^2}{25}\right)}$. In order to graph the ellipse on a graphing calculator, *both the positive and the negative square roots* must be entered into the calculator as is illustrated in the first screen below. The remaining screens show the ellipse in three different viewing rectangles. The window dimensions given here are for the *TI-82/83* graphing calculator. Choose the decimal window appropriate to your calculator model if it is different from the TI-82/83.

Question: Which graph below most *appropriately* represents this ellipse as we are used to seeing it? Explain why each graph is different and why you chose the graphical representation that you did.

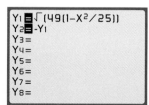 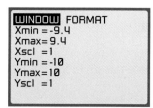 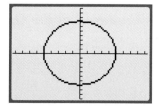

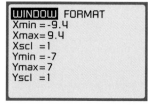 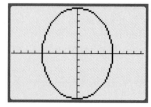 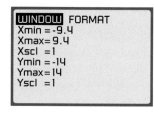 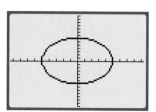

Locate three different points on the ellipse. Use the distance formula to find the sum of the distance from each point to the foci. Record your results in this form.

Point 1 = (_____, _____)   Distance to $(0, c)$ _____   Distance to $(0, -c)$ _____   Sum _____

Point 2 = (_____, _____)   Distance to $(0, c)$ _____   Distance to $(0, -c)$ _____   Sum _____

Point 3 = (_____, _____)   Distance to $(0, c)$ _____   Distance to $(0, -c)$ _____   Sum _____

What do you notice about the sums?

Now try the following on your own with a graphing calculator. First write the equation in standard ellipse form. Sketch your results on graph paper. Find the lengths of the major and minor axes as well as the x- and y-intercepts.

a. $x^2 + 4y^2 = 9$ **b.** $9x^2 + y^2 = 9$

c. $16x^2 + 9y^2 = 144$ **d.** $4x^2 + 25y^2 = 100$

The Hyperbola

When a plane cuts both parts of an infinite double cone and is parallel to the cone's axis, the intersection is a **hyperbola** (see the figure on the left below).

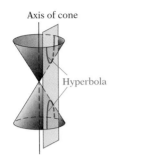

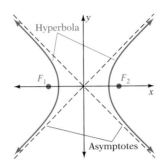

Note in the figure on the right above that a hyperbola has two identical parts and that these parts are not parabolas. The vertex of each part is the point that is closest to the other part. The dashed lines in the figure are the **asymptotes** of the hyperbola—lines to which the hyperbola comes closer and closer but never touches. The asymptotes intersect at the center of the hyperbola.

Hyperbolas Centered at the Origin

A hyperbola with the standard equation

$$\frac{x^2}{a^2} - \frac{y^2}{b^2} = 1$$

has its center at the origin. Its vertices are its x-intercepts; they are located at $(a, 0)$ and $(-a, 0)$. A hyperbola with the standard equation

$$\frac{y^2}{b^2} - \frac{x^2}{a^2} = 1$$

is also centered at the origin. Its vertices are its y-intercepts; they are located at $(0, b)$ and $(0, -b)$.

In these standard equations, if the y^2-term is subtracted, the vertices are on the x-axis; if the x^2-term is subtracted, the vertices are on the y-axis.

A hyperbola centered at the origin can be graphed from its standard equation:

1. Graph the points (a, b), $(a, -b)$, $(-a, b)$, and $(-a, -b)$. Then sketch the *rectangle* that has these points as corner points.
2. Draw and extend the diagonals of this rectangle. These are the *asymptotes* of the graph.
3. Graph the vertices of the hyperbola.
4. Sketch the graph of the hyperbola, with each part passing through one vertex and approaching, *but not touching*, the asymptotes.

...

EXAMPLE 3

Sketch the graph of

$$\frac{y^2}{9} - \frac{x^2}{16} = 1$$

SOLUTION

We begin by writing the equation in standard form.

$$\frac{y^2}{3^2} - \frac{x^2}{4^2} = 1$$

Now we see that $a = 4$ and $b = 3$, so we first graph the points $(4, 3)$, $(4, -3)$, $(-4, 3)$, and $(-4, -3)$ and draw the rectangle they define. Then we draw the diagonals of this rectangle and extend them beyond the rectangle — they will serve as asymptotes. This is done in the figure on the left below.

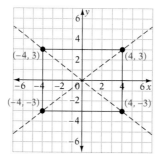 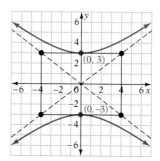

Because the x^2-term of the standard equation is subtracted, the graph has its vertices on the y-axis. They are located at $(0, 3)$ and $(0, -3)$, and we graph them next.

Finally, we sketch the graphs as in the figure on the right above. Each part passes through the y-axis at a vertex and approaches, but does not touch, the asymptotes. The line $x = 0$ that passes through both $(0, 3)$ and $(0, -3)$ is called the **transverse** (or **focal**) **axis.** The line $y = 0$ that passes through both $(4, 0)$ and $(-4, 0)$ is called the **conjugate axis.**

▶ *CHECK* **Warm-Up 3**

Practice what you learned.

SECTION FOLLOW-UP

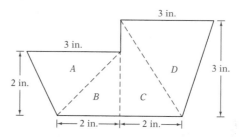

Look at ΔD.

3 in.

D 3 in.

Using the base as the side marked "3 in." and the height as 3 in., the area is clearly

$$\tfrac{1}{2}bh = \tfrac{1}{2}(3)(3) = 4\tfrac{1}{2} \text{ square inches}$$

Now you justify the measurements of ΔC. How would you have determined the one-inch segment if it had not been marked?

9.2 WARM-UPS

Work these problems before you attempt the exercises.

1. **a.** Write an equation for the circle that has a radius of 7 units and whose center is located at the origin. $x^2 + y^2 = 49$

   b. Sketch the graph of $x^2 - 6 + y^2 = 0$.

2. Graph: $\dfrac{x^2}{25} + \dfrac{y^2}{36} = 1$

   Find the lengths of the major and minor axes.   major axis: 12; minor axis: 10

3. Graph: $\dfrac{x^2}{1} - \dfrac{y^2}{9} = 1$

9.2 EXERCISES

Note: Use your graphing calculator to check your results whenever possible.

In Exercises 1 through 8, write an equation for the circle that fits the given conditions.

1. center at origin
   radius: 5
   $x^2 + y^2 = 25$

2. center at origin
   radius: 11
   $x^2 + y^2 = 121$

3. center at origin
   radius: $3\sqrt{2}$
   $x^2 + y^2 = 18$

4. center at origin
   radius: $2\sqrt{5}$
   $x^2 + y^2 = 20$

5. center at origin
   radius: $5\sqrt{2}$
   $x^2 + y^2 = 50$

6. center at origin
   radius: $6\sqrt{3}$
   $x^2 + y^2 = 108$

7. center at origin
   radius: $10\sqrt{5}$
   $x^2 + y^2 = 500$

8. center at origin
   radius: $4\sqrt{10}$
   $x^2 + y^2 = 160$

In Exercises 9 through 12, determine the center and radius of the circle from its equation, and then sketch the circle by graphing four points on it.

9. $x^2 + y^2 = 16$
center: (0, 0)
radius: 4

10. $x^2 + y^2 = 25$
center: (0, 0)
radius: 5

11. $x^2 - 1 = -y^2$
center: (0, 0)
radius: 1

12. $x^2 - 4 = -y^2$
center: (0, 0)
radius: 2

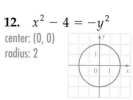

In Exercises 13 through 20, sketch the ellipse. Find the lengths of the major and minor axes, and find the x- and y-intercepts.

13. $\dfrac{x^2}{16} + \dfrac{y^2}{4} = 1$
major axis: 8 (horizontal)
minor axis: 4 (vertical)
x-intercepts: (4, 0), (−4, 0)
y-intercepts: (0, 2), (0, −2)

14. $\dfrac{x^2}{36} + \dfrac{y^2}{9} = 1$
major axis: 12 (horizontal)
minor axis: 6 (vertical)
x-intercepts: (6, 0), (−6, 0)
y-intercepts: (0, 3), (0, −3)

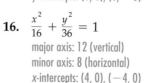

15. $\dfrac{x^2}{25} + \dfrac{y^2}{4} = 1$
major axis: 10 (horizontal)
minor axis: 4 (vertical)
x-intercepts: (5, 0), (−5, 0)
y-intercepts: (0, 2), (0, −2)

16. $\dfrac{x^2}{16} + \dfrac{y^2}{36} = 1$
major axis: 12 (vertical)
minor axis: 8 (horizontal)
x-intercepts: (4, 0), (−4, 0)
y-intercepts: (0, 6), (0, −6)

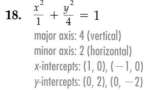

17. $\dfrac{x^2}{9} + \dfrac{y^2}{1} = 1$
major axis: 6 (horizontal)
minor axis: 2 (vertical)
x-intercepts: (3, 0), (−3, 0)
y-intercepts: (0, 1), (0, −1)

18. $\dfrac{x^2}{1} + \dfrac{y^2}{4} = 1$
major axis: 4 (vertical)
minor axis: 2 (horizontal)
x-intercepts: (1, 0), (−1, 0)
y-intercepts: (0, 2), (0, −2)

19. $\dfrac{x^2}{9} + \dfrac{y^2}{25} = 1$
major axis: 10 (vertical)
minor axis: 6 (horizontal)
x-intercepts: (3, 0), (−3, 0)
y-intercepts: (0, 5), (0, −5)

20. $\dfrac{x^2}{25} + \dfrac{y^2}{36} = 1$
major axis: 12 (vertical)
minor axis: 10 (horizontal)
x-intercepts: (5, 0), (−5, 0)
y-intercepts: (0, 6), (0, −6)

In Exercises 21 through 28, graph the hyperbola.

21. $\dfrac{x^2}{4} - \dfrac{y^2}{4} = 1$

22. $\dfrac{y^2}{9} - \dfrac{x^2}{1} = 1$

23. $\dfrac{x^2}{9} - \dfrac{y^2}{1} = 1$

24. $\dfrac{y^2}{16} - \dfrac{x^2}{1} = 1$

25. $\dfrac{y^2}{4} - \dfrac{x^2}{1} = 1$

26. $\dfrac{x^2}{16} - \dfrac{y^2}{25} = 1$

27. $\dfrac{x^2}{25} - \dfrac{y^2}{9} = 1$

28. $\dfrac{y^2}{9} - \dfrac{x^2}{25} = 1$

MIXED PRACTICE

By doing these exercises, you will practice the topics up to this point in the chapter.

29. Graph the equation $y = 1.5(x - 3)^2$.

30. Rewrite $y = -3x^2 + 18x + 2$ in graphing form. Then find the vertex of the parabola and determine the direction in which it opens.
    $y = -3(x - 3)^2 + 29$; vertex: $(3, 29)$; opens downward

31. Graph: $\dfrac{y^2}{9} - \dfrac{x^2}{4} = 1$

32. Graph the equation $\dfrac{x^2}{4} + \dfrac{y^2}{9} = 1$.
    Find the lengths of the major and minor axes, the center, and the x- and y-intercepts.

33. Rewrite $y - 49 = x^2 + 14x$ in graphing form. Then find the vertex of the parabola and determine the direction in which it opens.
    $y = (x + 7)^2$; vertex: $(-7, 0)$; opens upward

34. Find the vertex of the parabola whose equation is $y = 3(x - 3)^2 + 1$. Then graph it.   vertex: $(3, 1)$

35. Graph: $\dfrac{y^2}{4} - \dfrac{x^2}{25} = 1$

36. Find the equation of the parabola that has its vertex at $(1, -2)$ and passes through the point $(0, 0)$. $y = 2(x - 1)^2 - 2$

32.

major axis: 6
minor axis: 4
center at $(0, 0)$
x-intercepts: $(-2, 0)$, $(2, 0)$
y-intercepts: $(0, 3)$, $(0, -3)$

34.

EXCURSIONS

Exploring Problem Solving

1. The moon has a nearly circular orbit about Earth. Use the formulas

$$v = \frac{2\pi r}{T} \quad \text{and} \quad a = \frac{v^2}{r}$$

where v = speed of the moon as it revolves about Earth
  r = radius of the orbit so $2\pi r$ is the path (circumference) of the orbit
  T = time it takes for the moon to circle Earth
  a = acceleration of the moon toward Earth

Find the time T it takes the moon to complete one orbit. Use

$$a = 0.00271 \text{ meters/second}^2$$
$$v = 1020 \text{ meters/second}$$

Your result will be in terms of seconds. Convert your answer to days.

Posing Problems

2. Ask and answer four questions that can be answered using this data.

Basic Planetary Data

	Mercury	Venus	Earth	Mars	Jupiter
Mean distance from sun					
(millions of kilometers)	57.9	108.2	149.6	227.9	778.3
(millions of miles)	36.0	67.24	92.9	141.71	483.88
Period of revolution	88 days	224.7 days	365.2 days	687 days	11.86 years
Rotation period	59 days	243 days retrograde	23 hr 56 min 4 sec	24 hr 37 min	9 hr 55 min 30 sec
Inclination of axis	Near 0°	3°	23°27'	25°12'	3°5'
Inclination of orbit to ecliptic	7°	3.4°	0°	1.9°	1.3°
Eccentricity of orbit	0.206	0.007	0.017	0.093	0.048
Equatorial diameter					
(kilometers)	4,880	12,100	12,756	6,794	142,800
(miles)	3,032.4	7,519	7,926.2	4,194	88,736
Atmosphere (main components)	Virtually none	Carbon dioxide	Nitrogen oxygen	Carbon dioxide	Hydrogen helium
Satellites	0	0	1	2	16
Rings	0	0	0	0	1

	Saturn	Uranus	Neptune	Pluto
Mean distance from sun				
(millions of kilometers)	1,427	2,870	4,497	5,900
(millions of miles)	887.14	1,783.98	2,796.46	3,666
Period of revolution	29.46 yrs	84 yrs	165 yrs	248 yrs
Rotation period	10 hr 40 min 24 sec	16.8 hr(?) retrograde	16 hr 11 min(?)	6 days 9 hr 18 mins retrograde
Inclination of axis	26°44'	97°55'	28°48'	60°(?)
Inclination of orbit to ecliptic	2.5°	0.8°	1.8°	17.2°
Eccentricity of orbit	0.056	0.047	0.009	0.254
Equatorial diameter				
(kilometers)	120,660	51,810	49,528	2,290(?)
(miles)	74,978	32,193	30,775	1,423(?)
Atmosphere (main components)	Hydrogen helium	Helium hydrogen methane	Hydrogen helium methane	None detected
Satellites	18+[1]	15	8	1
Rings	1,000(?)	11	4	?

[1]Two additional moons were found outside the outer *F* ring by the Hubble Space Telescope in 1995. They were temporarily named *S3* and *S4*.

Source: 1996 Information Please Almanac® (©1995 Houghton Mifflin Co.), p. 331. All rights reserved. Used with permission by Information Please LLC.

3. **a.** Which cities have the greatest range of temperatures in January?

  b. Ask two other questions about range using this data.

  c. Ask two other questions about this data. Answer these questions your-self or have your friends answer them.

World Weather in January

City	Average High/Low	Wet Days	City	Average High/Low	Wet Days
Athens	55/44	16	Mexico City	66/42	4
Beijing	34/14	3	Moscow	15/3	18
Bermuda	68/58	14	New York	37/24	12
Buenos Aires	85/63	7	Paris	43/34	17
Cairo	65/47	1	Rio de Janeiro	84/73	13
Delhi	70/44	2	Rome	52/40	8
Dublin	46/34	13	San Francisco	55/45	11
Frankfurt	38/29	17	San Juan	80/70	20
Hong Kong	64/56	4	Stockholm	30/23	16
Jerusalem	55/41	9	Sydney	78/65	14
Johannesburg	78/58	12	Tokyo	47/29	5
London	43/36	15	Toronto	30/16	16
Madrid	47/35	8	Vancouver	41/32	20

Source: The Times Books World Weather Guide.

Data Analysis

4. **a.** Analyze this data. Can this be linear or quadratic in form? exponential?

  b. Give an estimate for the snow "Stored" by a 6-foot fence. Justify your answer. Research (Internet possibly) to check your answer.

Snow Fence	"Stored" Snow
0 feet	0 tons
4 feet	4 tons
8 feet	18 tons

SNOW PROTECTION FOR ROADWAYS

Based on a season's predicted snowfall, engineers design snow fences which effectively shield roadways from additional snow that can be deposited by the wind. This not only helps decrease accumulation, but also serves to increase driver visibility.

  A snow fence four feet high can 'store' four tons of snow per foot of fence length; a fence eight feet high can store 18 tons of snow per foot of fence length.

The wind carries both falling snow plus any fallen snow that it is able to pick up.

The snow fence effectively blocks and slows the wind, allowing the snow to settle into a controlled drift.

Once past the fence, the reduced wind is unable to pick up much snow and therefore little snow reaches the roadway.

The faster the wind, the more snow that can be carried.

Source: USA TODAY Weather Book by Jack Williams. Copyright 1997, USA TODAY, reprinted with permission.

Exploring Geometry

5. How might one estimate the volume of a stream? What geometric figure might model a cross section of a stream? Justify your choice and explain your reasoning.

CONNECTIONS TO *GEOMETRY*

Cylinders

Volume of a Cylinder

Right Circular Cylinder

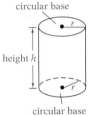

circular base

height *h*

r

r

circular base

A right circular cylinder is shown here. (We shall simply call it a cylinder.) Its bases are circles with the same radius, *r*.

Cylinders are solid figures. (Think of them as being closed on the ends.) The volume of a cylinder is equal to the area of one of its bases times its height. Because the area of a base is πr^2, we have the following formula:

To find the volume of a cylinder

If *r* is the radius of the base and *h* is the height, then
$$V = \pi r^2 h$$

▪ ▪ ▪
EXAMPLE

Find the volume of a can that has a radius of 8 inches and a height of 12 inches.

SOLUTION

First we substitute the values into the formula for the volume of a cylinder.
$$V = \pi r^2 h = (3.14)(8^2)(12)$$

Then we simplify the resulting expression.
$$V = (3.14)(64)(12) = 2411.52$$

The volume of the cylinder is 2411.52 cubic inches.

Surface Area of a Cylinder

The following figure shows that we can think of the cylinder as being made up of three parts: two circular ends and a rectangle rolled up to form the side. The length of the rectangle is the circumference of a base, or $2\pi r$; its width is the height h. The surface area of the cylinder can be found by adding together the areas of the two end circles (πr^2 each) and the area of the rectangular part ($h \times 2\pi r$).

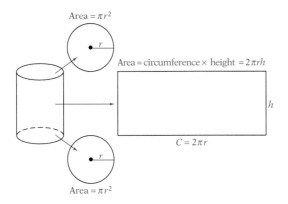

To find the surface area of a cylinder

If r is the radius and h is the height, then

$$SA = \pi r^2 + \pi r^2 + 2\pi rh \quad \text{or} \quad 2\pi r^2 + 2\pi rh \quad \text{or} \quad 2\pi r(r + h)$$

▪ ▪ ▪

EXAMPLE

Find the surface area of a cylinder with a radius of 6 inches and a height of 12 inches.

SOLUTION

We substitute into the formula to find the surface area.

$$SA = 2\pi r(r + h)$$
$$= 2(3.14)(6)(6 + 12)$$
$$= 678.24$$

The surface area of the cylinder is 678.24 square inches.

◢

PRACTICE

Give your answers to the nearest tenth.

1. A can of soup is emptied into a pan. Both the can and the pan are shaped like cylinders. The soup can is 4.5 inches tall with a diameter of 2.5 inches. The pan is 3 inches tall with a diameter of 5 inches. Is there enough space in the pan for a second can of soup?   yes

2. Find the volume of a soda can that has a radius of $1\frac{1}{4}$ inches and a height of $4\frac{7}{8}$ inches. Use $\frac{22}{7}$ as an approximation for π.   23.9 cubic inches

3. You are cutting a label for a can. The can has a diameter of 2.5 inches and is 5.5 inches tall. What is the area of the label that fits exactly on this can? 43.2 square inches

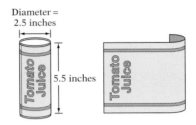

Diameter = 2.5 inches

5.5 inches

4. A container shaped like a cylinder holds glue for woodworking projects. If the container sits 9.5 inches high and has a diameter of 3 inches, how much glue can it hold?   67.1 cubic inches

5. What is the surface area of a cylinder with a radius of 3.5 inches and a height of 2.5 inches?   131.9 square inches

6. Modeling clay is shaped into a long cylinder with a diameter of 4 inches. A piece 6 inches long is cut off. What volume of clay is in this piece? 75.4 cubic inches

7. What is the surface area of a cylinder with a radius of 0.5 unit and a height of 2.5 units?   9.4 square units

8. Find the volume of a pipe that is cylindrical in shape and has a length of 8 feet and a radius of $4\frac{1}{2}$ feet.   508.7 cubic feet

9. A can shaped like a cylinder is filled with water. If it has a radius of 3 inches and a height of 10 inches, how much water can it hold? 282.6 cubic inches

10. Water weighs 62.4 pounds per cubic foot. You have a bucket that is shaped like a cylinder with a diameter of 1 foot and a height of 1 foot, and it is filled with water. How much does the water weigh?   49.0 pounds

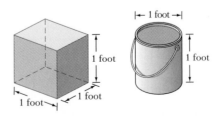

1 foot

1 foot

1 foot

1 foot

1 foot

9.3 The Circle, Ellipse, and Hyperbola Not Centered at the Origin

SECTION LEAD-IN

We know there are 180° in a triangle. You can prove this to yourself.

1. Cut out a triangle.

2. Number the angles.

3. Tear the angles off and arrange them so the vertices touch.

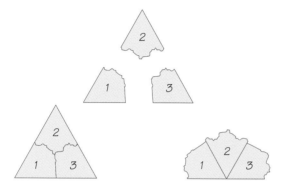

Because they form a straight line (we know a straight angle is 180°), the triangle's angles total 180°.

Use this fact (angles of a triangle total 180°) to find the number of degrees in each figure given below:

    Square
    Rectangle
    Trapezoid
    Pentagon
    Hexagon

SECTION GOALS

- *To write an equation for a circle given its center and radius*

- *To graph the equation of a circle*

- *To recognize and graph the equation of an ellipse*

- *To write the equation of an ellipse from its description*

- *To graph the equation of a hyperbola*

- *To write the equation of a hyperbola from its description*

- *To recognize equations of conic sections without graphing*

Circles Not Centered at the Origin

Suppose a circle of radius r has its center at some arbitrary point (h, k), as in the accompanying figure. Then, according to the distance formula, the distance r between this center and any point (x, y) on the circle is

$$\sqrt{(x - h)^2 + (y - k)^2} = r$$

Squaring both sides gives the graphing form of the equation of a circle.

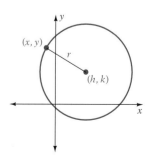

INSTRUCTOR NOTE

Point out to students the similarities of this equation with those for other conic sections. The graph of each can be determined by the equation itself. We do not need to plot points as we did with linear equations.

Graphing Form of the Equation of a Circle

A circle of radius r with its center at the point (h, k) has a graphing form

$$(x - h)^2 + (y - k)^2 = r^2$$

■■■

EXAMPLE 1

a. Write an equation for a circle of radius 2.5 units that is centered at the point $(-2, -1)$.

b. Graph the equation $(x + 3)^2 + (y - 1)^2 = 4$.

SOLUTION

a. Using the graphing form of the equation of the circle, we substitute $r = 2.5$, $h = -2$, and $k = -1$. We get

$$[x - (-2)]^2 + [y - (-1)]^2 = (2.5)^2 \quad \text{Substituting}$$
$$(x + 2)^2 + (y + 1)^2 = 6.25 \quad \text{Simplifying}$$

So the equation of the circle is

$$(x + 2)^2 + (y + 1)^2 = 6.25$$

b. Rewriting the equation in graphing form to identify the center and radius gives us

$$[x - (-3)]^2 + (y - 1)^2 = 2^2$$

From the equation, we can see that the graph is a circle that has its center at $(-3, 1)$ and a radius of 2 (see figure).

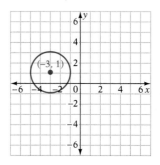

CHECK **Warm-Up 1**

■■■

EXAMPLE 2

A circle centered at $(2, -1)$ passes through the point $(5, 6)$. Write the graphing form of the equation of this circle.

SOLUTION

The radius is the distance between the center $(2, -1)$ and the point $(5, 6)$ on the circle. We use the distance formula to find it.

$$d = r = \sqrt{(x_2 - x_1)^2 + (y_2 - y_1)^2}$$
$$r = \sqrt{(5 - 2)^2 + [6 - (-1)]^2}$$
$$r = \sqrt{9 + 49}$$
$$r = \sqrt{58}$$

Actually, we need $r^2 = 58$. The equation of the circle is then

$$(x - h)^2 + (y - k)^2 = r^2$$

or

$$(x - 2)^2 + (y + 1)^2 = 58 \quad \text{Substituting for } h, k, r^2$$

▶ *CHECK* **Warm-Up 2**

The General Form of the Equation of a Circle

If we multiply out the left side of the equation $(x + 3)^2 + (y - 1)^2 = 4$ in Example 1(b), we get

$$x^2 + 6x + 9 + y^2 - 2y + 1 = 4$$

which can be simplified to

$$x^2 + 6x + 9 + y^2 - 2y + 1 - 4 = 0$$
$$x^2 + y^2 + 6x - 2y + 6 = 0$$

This is the *general form* of the original equation.

The general form of the equation of a circle is $x^2 + y^2 + ax + by + c = 0$, where a, b, and c are real numbers.

We could graph an equation in general form by making a table of values, plotting points, and then connecting the points with a smooth curve. However, it is usually easier to transform a general equation to its graphing form; we do so by completing two squares on the left side.

▪ ▪ ▪

EXAMPLE 3

Find the center and radius of the circle whose equation is

$$x^2 - 2x + y^2 + 6y + 1 = 0$$

SOLUTION

We start by separating the terms containing x and the terms containing y and setting them equal to the constant terms.

■ ■ ■
WRITER'S BLOCK

Tell a friend how to find the center and radius of the graph of a circle by looking at its equation in graphing form.

$$(x^2 - 2x) + (y^2 + 6y) = -1$$

Next we complete a square for the x-terms and one for the y-terms. For x,

$$\left(\frac{b}{2}\right)^2 = \left(\frac{-2}{2}\right)^2 = 1$$

We add 1 to each side of the equation to get

$$(x^2 - 2x + 1) + (y^2 + 6y) = -1 + 1$$

For y,

$$\left(\frac{b}{2}\right)^2 = \left(\frac{6}{2}\right)^2 = 9$$

We add 9 to each side to get

$$(x^2 - 2x + 1) + (y^2 + 6y + 9) = -1 + 1 + 9$$

This gives us the equivalent equation

$$(x^2 - 2x + 1) + (y^2 + 6y + 9) = 9$$

Writing the terms in parentheses as squares gives us the graphing form:

$$(x - 1)^2 + (y + 3)^2 = 9$$

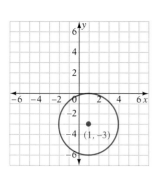

This equation represents a circle with its center at $(1, -3)$ and radius 3. It is graphed in the accompanying figure.

▶ CHECK **Warm-Up 3**

Ellipses Not Centered at the Origin

The graphing equation for an ellipse takes a slightly different form when its center is not at the origin.

Graphing Form of the Equation of an Ellipse

The equation of an ellipse with its center at the point (h, k) has the graphing form

$$\frac{(x - h)^2}{a^2} + \frac{(y - k)^2}{b^2} = 1$$

The constants a and b have the same meaning no matter where the center of an ellipse is located. The extreme upper and lower points of the ellipse are at $(h, k + b)$ and $(h, k - b)$. The extreme right and left points are at $(h + a, k)$ and $(h - a, k)$.

▪▪▪

EXAMPLE 4

Write the graphing form of the equation of an ellipse with a horizontal axis of length 16, a vertical axis of length 12, and its center at $(3, -7)$.

SOLUTION

Because the horizontal axis has length 16, $a = \frac{16}{2} = 8$. Because the vertical axis has length 12, $b = \frac{12}{2} = 6$. Because the center is at $(3, -7)$, $h = 3$ and $k = -7$.

We substitute these four values in the graphing form and obtain

$$\frac{(x - 3)^2}{8^2} + \frac{[y - (-7)]^2}{6^2} = 1 \quad \text{or} \quad \frac{(x - 3)^2}{64} + \frac{(y + 7)^2}{36} = 1$$

▶ *CHECK* **Warm-Up 4**

Calculator Corner

a. Graph the ellipse $\frac{(x - 3)^2}{16} + \frac{(y + 2)^2}{4} = 1$ using your knowledge of the major and minor axes and the center of the ellipse. Label all parts and support with your graphing calculator.

b. Determine whether $16x^2 + 25y^2 - 32x + 50y + 31 = 0$ is an ellipse. Graph your results on a piece of graph paper and label all parts. Support with your graphing calculator.

c. Is $36x^2 + 9y^2 + 48x - 36y + 43 = 0$ an ellipse? Graph your results on a piece of graph paper and label all parts. Support with your graphing calculator.

d. Is $2x^2 + 2xy + y^2 = 5$ an ellipse? Graph your results on a piece of graph paper and label all parts. Support with your graphing calculator.

▪▪▪

EXAMPLE 5

Graph the ellipse that has the equation

$$25(y + 4)^2 + (x - 3)^2 - 100 = 0$$

Find the center of the ellipse and the lengths of the major and minor axes.

SOLUTION

First we rewrite the equation in graphing form.

$25(y + 4)^2 + (x - 3)^2 - 100 = 0$	Original equation
$(x - 3)^2 + 25(y + 4)^2 = 100$	Adding 100 to each side and reordering
$\dfrac{(x - 3)^2}{100} + \dfrac{25(y + 4)^2}{100} = \dfrac{100}{100}$	Multiplying each term by $\frac{1}{100}$
$\dfrac{(x - 3)^2}{100} + \dfrac{(y + 4)^2}{4} = 1$	Simplifying
$\dfrac{(x - 3)^2}{10^2} + \dfrac{[y - (-4)]^2}{2^2} = 1$	Graphing form

The graphing form of the equation now shows that $h = 3, k = -4, a = 10$, and $b = 2$.

The center of the ellipse is at (h, k), or $(3, -4)$

The extreme points are at $(h, k + b)$, or $(3, -2)$; $(h, k - b)$, or $(3, -6)$; $(h + a, k)$ or $(13, -4)$; and $(h - a, k)$ or $(-7, -4)$.

We graph these four points and then sketch the ellipse as in the following figure.

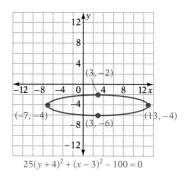

$$25(y + 4)^2 + (x - 3)^2 - 100 = 0$$

The sketch confirms that the ellipse has a major axis of length $2a = 20$ and a minor axis of length $2b = 4$. They intersect at the center $(3, -4)$.

▶ CHECK **Warm-Up 5**

Hyperbolas Not Centered at the Origin

As usual, the standard equation takes a slightly different form when the graph is not centered at the origin.

Graphing Form of the Equation of a Hyperbola

A hyperbola with the equation
$$\frac{(x - h)^2}{a^2} - \frac{(y - k)^2}{b^2} = 1$$
has its center at the point (h, k). Its vertices are located at $(h + a, k)$ and $(h - a, k)$.

A hyperbola with the equation
$$\frac{(y - k)^2}{b^2} - \frac{(x - h)^2}{a^2} = 1$$
also has its center at (h, k), Its vertices are located at $(h, k + b)$ and $(h, k - b)$.

To graph these hyperbolas, we need their asymptotes. The corners of the rectangles that define the asymptotes are at $(h + a, k + b)$, $(h + a, k - b)$, $(h - a, k + b)$, $(h - a, k - b)$.

▪▪▪

EXAMPLE 6

Sketch the graph of the equation

$$\frac{(x-2)^2}{16} - \frac{(y+3)^2}{9} = 1$$

SOLUTION

By inspection of the given equation, we see that $h = 2$, $k = -3$, $a = 4$, and $b = 3$. Thus the asymptotes are defined by the rectangle with corner points

$$(h + a, k + b), \text{ or } (6, 0)$$
$$(h + a, k - b), \text{ or } (6, -6)$$
$$(h - a, k + b), \text{ or } (-2, 0)$$
$$(h - a, k - b), \text{ or } (-2, -6)$$

These points, the rectangle they determine, and its extended diagonals (the asymptotes of the hyperbola) are shown in the accompanying figure.

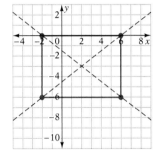

Because the y^2-term is subtracted in the given equation, the vertices are located at

$$(h + a, k) \text{ and } (h - a, k)$$

or

$$(6, -3) \text{ and } (-2, -3)$$

We graph these points; then we sketch each part of the hyperbola as a smooth curve passing through one vertex and approaching the asymptotes closer as we move farther from the vertex. This is done in the figure below.

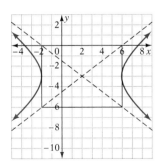

▶ *CHECK* **Warm-Up 6**

▪▪▪

EXAMPLE 7

Write the equation for, and identify the vertices of, the hyperbola centered at $(2, -3)$ with $a = 17$ and $b = 12$. For this hyperbola, assume that the y^2-term is subtracted.

SOLUTION

The general equation for this hyperbola is

$$\frac{(x-h)^2}{a^2} - \frac{(y-k)^2}{b^2} = 1$$

Replacing (h, k) with the center $(2, -3)$, we get

$$\frac{(x-2)^2}{a^2} - \frac{[y-(-3)]^2}{b^2} = 1$$

Now we replace a and b

$$\frac{(x-2)^2}{17^2} - \frac{[(y-(-3)]^2}{12^2} = 1$$

This is the equation.

The vertices for this hyperbola are $(h + a, k)$ and $(h - a, k)$. Substituting $h = 2$, $k = -3$, and $a = 17$, we get

$$(h + a, k) = (2 + 17, -3) = (19, -3)$$

and

$$(h - a, k) = (2 - 17, -3) = (-15, -3)$$

▶CHECK **Warm-Up 7**

Calculator Corner

The graph of a hyperbola can be sketched by following these four steps.

1. Determine the direction of the hyperbola.
2. Determine the values of a and b and draw a box.
3. Draw the asymptotes along the diagonals of the box you just drew.
4. Use the box and asymptotes to sketch in the hyperbola.

As an example, the hyperbola of $9x^2 - 4y^2 = 36$ is shown at the right:

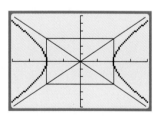

Follow this simple procedure to sketch each of the following hyperbolas. Show all of your work and label all parts of the hyperbola on a piece of graph paper. Confirm your sketching results with your graphing calculator.

a. $4x^2 - y^2 + 16 = 0$ **b.** $36y^2 - 100x^2 = 225$

c. $16x^2 - 9y^2 = -36$ **d.** $3x - 2y^2 - 6x - 12y - 27 = 0$

e. What graph do you get when you use the completing the square method to graph $x^2 - y^2 - 6x + 12y - 27 = 0$?

What happens to the graph if you change the 27 to 26 as follows?

$$x^2 - y^2 - 6x + 12y - 26 = 0$$

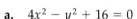

WRITER'S BLOCK

Describe the changes in graphs that you saw in part (e).

What happens to the graph if you change the 27 to 28 as follows?

$$x^2 - y^2 - 6x + 12y - 28 = 0$$

What happens to the graph if you change the 27 to 30 as follows?

$$x^2 - y^2 - 6x + 12y - 30 = 0$$

Recognizing Equations of Conic Sections

Often we have to examine a second-degree equation in general form and determine what conic section it represents. To do this, we write each equation in its graphing form, completing squares where necessary.

▪▪▪

EXAMPLE 8

Equations (a) through (d) are the equations of a circle, an ellipse, a hyperbola, and a parabola—but not necessarily in that order. Identify the conic section described by each equation.

a. $x - y^2 - 4y + 9 = 0$ **b.** $2x^2 - 9y^2 - 8x + 36y - 46 = 0$

c. $x^2 + y^2 - 6x + 4y - 3 = 0$ **d.** $4x^2 + y^2 - 24x - 8y + 48 = 0$

SOLUTION

a. The equation $x - y^2 - 4y + 9 = 0$ has an x-term but no x^2-term. Of the four conic sections, this can only be the equation of a parabola.

$$
\begin{aligned}
x - y^2 - 4y + 9 &= 0 \\
x + 9 &= y^2 + 4y && \text{Rewriting} \\
x + 9 + 2^2 &= y^2 + 4y + 2^2 && \text{Completing the square} \\
x + 13 &= (y + 2)^2 && \text{Simplifying} \\
x &= (y + 2)^2 - 13 \\
x &= [y - (-2)]^2 - 13 && \text{Writing in graphing form}
\end{aligned}
$$

This is a parabola.

b.

$$
\begin{aligned}
2x^2 - 9y^2 - 8x + 36y - 46 &= 0 \\
2x^2 - 8x - 9y^2 + 36y &= 46 && \text{Rewriting} \\
2(x^2 - 4x\quad) - 9(y^2 - 4y\quad) &= 46 && \text{Factoring} \\
2[x^2 - 4x + (-2)^2] - 9[y^2 - 4y + (-2)^2] &= 46 + 2(-2)^2 - 9(-2)^2 && \text{Completing the square} \\
2(x - 2)^2 - 9(y - 2)^2 &= 18 && \text{Simplifying} \\
\frac{(x - 2)^2}{9} - \frac{(y - 2)^2}{2} &= 1 && \text{Multiplying by } \tfrac{1}{18} \\
\frac{(x - 2)^2}{(3)^2} - \frac{(y - 2)^2}{(\sqrt{2})^2} &= 1 && \text{Writing in graphing form}
\end{aligned}
$$

This is a hyperbola.

c.
$$x^2 + y^2 - 6x + 4y - 3 = 0$$
$$x^2 - 6x + y^2 + 4y = 3 \qquad \text{Rewriting}$$
$$(x^2 - 6x \quad) + (y^2 + 4y \quad) = 3$$
$$[x^2 - 6x + (-3)^2] + (y^2 + 4y + 2^2) = 3 + (-3)^2 + (2)^2 \quad \text{Completing the square}$$
$$(x - 3)^2 + (y + 2)^2 = 16 \qquad \text{Simplifying}$$
$$(x - 3)^2 + [y - (-2)]^2 = 4^2 \qquad \text{Writing in graphing form}$$

This is a circle.

d.
$$4x^2 + y^2 - 24x - 8y + 48 = 0$$
$$4x^2 - 24x + y^2 - 8y = -48 \qquad \text{Rewriting}$$
$$4(x^2 - 6x \quad) + (y^2 - 8y \quad) = -48$$
$$4\left[x^2 - 6x + \left(\frac{-6}{2}\right)^2\right] + \left[y^2 - 8y + \left(\frac{-8}{2}\right)^2\right] = -48 + 4\left(\frac{-6}{2}\right)^2 + \left(\frac{-8}{2}\right)^2 \quad \text{Completing the square}$$
$$\underset{(-3)^2}{} \qquad \underset{(-4)^2}{} \qquad \underset{(-3)^2}{} \quad \underset{(-4)^2}{}$$
$$4(x - 3)^2 + (y - 4)^2 = 4 \qquad \text{Simplifying}$$
$$(x - 3)^2 + \frac{(y - 4)^2}{4} = 1 \qquad \text{Writing in graphing form}$$

This is an ellipse.

▶ *CHECK* **Warm-Up 8**

Practice what you learned.

SECTION FOLLOW-UP

The following figure shows that a square has a total of 360° in its angles.

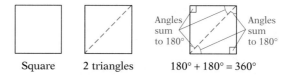

| Square | 2 triangles | $180° + 180° = 360°$ |

What did you find out about the other figures?

What about this one—or any irregular quadrilateral?

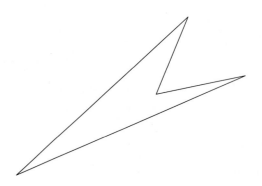

9.3 WARM-UPS

Work these problems before you attempt the exercises.

1. Write an equation for the circle that has a radius of 2 and its center at $(12, -3)$. $(x - 12)^2 + (y + 3)^2 = 4$

2. Write the equation of a circle that passes through $(1, -1)$ and has its center at $(10, 0)$. $(x - 10)^2 + y^2 = 82$

3. Find the center and the radius of the graph of $x^2 - 4x + y^2 + 10y + 1 = 0$.   center: $(2, -5)$; radius: $2\sqrt{7}$

4. Describe the graph of $\dfrac{(x - 6)^2}{81} + \dfrac{(y - 18)^2}{49} = 1$. Give the lengths of the major and minor axes, and determine which is the horizontal axis.   center at $(6, 18)$; major axis: $2 \cdot 9 = 18$ (horizontal); minor axis: $2 \cdot 7 = 14$ (vertical)

5. Graph the ellipse that has the equation $4(x + 2)^2 + y^2 = 4$.

6. Graph: $\dfrac{(y - 1)^2}{4} - \dfrac{(x + 2)^2}{1} = 1$

7. A hyperbola with center $(5, 2)$ and vertices $(5, 0)$ and $(5, 4)$ has $a = 12$ and $b = 2$. Write the equation. $\dfrac{(y - 2)^2}{4} - \dfrac{(x - 5)^2}{144} = 1$

8. The equation $5x^2 + 9y^2 - 20x + 54y + 56 = 0$ represents what conic section?   ellipse

9.3 EXERCISES

Note: Use your graphing calculator to check your results whenever possible.

In Exercises 1 through 8, write an equation for the circle that fits the given conditions.

1. center at $(4, -1)$
   radius: 5
   $(x - 4)^2 + (y + 1)^2 = 25$

2. center at $(2, 5)$
   radius: 3
   $(x - 2)^2 + (y - 5)^2 = 9$

3. center at $(-1, 2)$
   radius: $3\sqrt{3}$
   $(x + 1)^2 + (y - 2)^2 = 27$

4. center at $(-3, -1)$
   radius: $2\sqrt{5}$
   $(x + 3)^2 + (y + 1)^2 = 20$

5. center at $(0, 2)$
   diameter: 0.8
   $x^2 + (y - 2)^2 = 0.16$

6. center at $(3, 0)$
   diameter: 1.2
   $(x - 3)^2 + y^2 = 0.36$

7. center at $(-1, -2)$
   diameter: $2\frac{1}{2}$
   $(x + 1)^2 + (y + 2)^2 = \dfrac{25}{16}$

8. center at $(-3, 4)$
   diameter: $3\frac{1}{4}$
   $(x + 3)^2 + (y - 4)^2 = \dfrac{169}{64}$

10.

In Exercises 9 through 16, determine the center and radius of the circle by completing the square in its general equation. Then sketch the circle by graphing four of its points.

9. $x^2 + y^2 + 2y + 1 = 9$
   center: $(0, -1)$
   radius: 3

10. $x^2 + y^2 - 4x + 4 = 16$
    center: $(2, 0)$
    radius: 4

11. $x^2 + 8x + 16 + y^2 = 4$
    center: $(-4, 0)$
    radius: 2

12. $x^2 - 2x + 1 + y^2 = 9$
center: (1, 0)
radius: 3

13. $x^2 + y^2 + 4x + 4y = 4$
center: (−2, −2)
radius: $2\sqrt{3}$

14. $x^2 + y^2 - 2x - 6y = 6$
center: (1, 3)
radius: 4

15. $x^2 - 2x + y^2 - 2y = 2$
center: (1, 1)
radius: 2

16. $x^2 - 6x + y^2 - 6y + 2 = 0$
center: (3, 3)
radius: 4

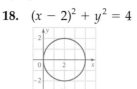

In Exercises 17 through 20, use the equation to sketch the graphs of each circle.

17. $(x - 1)^2 + y^2 = 25$

18. $(x - 2)^2 + y^2 = 4$

19. $(x - 1)^2 + (y - 2)^2 = 16$

20. $(x + 2)^2 + (y + 1)^2 = 4$

In Exercises 21 through 28, use the distance formula to find the radius of the circle that satisfies the given conditions. Then write the equation of the circle.

21. center at (0, 0) $r = 5$
passes through (3, 4) $x^2 + y^2 = 25$

22. center at (0, 0) $r = 25$
passes through (15, 20) $x^2 + y^2 = 625$

23. center at (3, −5) $r = \sqrt{34}$
passes through (0, 0) $(x - 3)^2 + (y + 5)^2 = 34$

24. center at (−8, 9) $r = \sqrt{145}$
passes through (0, 0) $(x + 8)^2 + (y - 9)^2 = 145$

25. center at (2, 3) $r = 13$
passes through (7, 15) $(x - 2)^2 + (y - 3)^2 = 169$

26. center at (−2, −3) $r = 20$
passes through (10, 13) $(x + 2)^2 + (y + 3)^2 = 400$

27. center at (2, −3) $r = 3\sqrt{2}$
passes through (5, −6) $(x - 2)^2 + (y + 3)^2 = 18$

28. center at (−10, 5) $r = \sqrt{149}$
passes through (−3, −5) $(x + 10)^2 + (y - 5)^2 = 149$

In Exercises 29 through 36, write the equation in graphing ellipse form. Then find the lengths of the major and minor axes, and the x- and y-intercepts for the graph of the equation.

29. $9x^2 + y^2 = 9$ $\frac{x^2}{1^2} + \frac{y^2}{3^2} = 1$; 6(v.), 2(h.)
(1, 0), (−1, 0); (0, 3), (0, −3)

30. $x^2 + y^2 = 25$ $\frac{x^2}{5^2} + \frac{y^2}{5^2} = 1$; 10 (both axes)
(5, 0), (−5, 0); (0, −5), (0, 5)

31. $4x^2 + y^2 = 36$ $\frac{x^2}{3^2} + \frac{y^2}{6^2} = 1$; 12 (v.), 6(h.)
(3, 0), (−3, 0); (0, 6), (0, −6)

32. $25x^2 + y^2 = 25$ $\frac{x^2}{1^2} + \frac{y^2}{5^2} = 1$; 10 (v.), 2 (h.)
(1, 0), (−1, 0); (0, 5), (0, −5)

33. $25x^2 + 4y^2 = 100$ $\frac{x^2}{2^2} + \frac{y^2}{5^2} = 1$; 10(v.), 4(h.)
(2, 0), (−2, 0); (0, 5), (0, −5)

34. $4x^2 + 9y^2 = 36$ $\frac{x^2}{3^2} + \frac{y^2}{2^2} = 1$; 6(h.), 4(v.)
(3, 0), (−3, 0); (0, 2), (0, −2)

35. $9x^2 + y^2 = 36$　$\frac{x^2}{2^2} + \frac{y^2}{6^2} = 1$; 12(v.), 4(h.)　　　　**36.** $49x^2 + y^2 = 49$　$\frac{x^2}{1^2} + \frac{y^2}{7^2} = 1$; 14(v.), 2(h.)

$(2, 0), (-2, 0); (0, 6), (0, -6)$　　　　　　　　　$(1, 0), (-1, 0); (0, 7), (0, -7)$

In Exercises 37 through 40, find the center of the graph of the equation and the lengths of the horizontal and vertical axes.

37. $\dfrac{(x-1)^2}{4} + \dfrac{(y-3)^2}{49} = 1$　(1, 3); 4, 14　　**38.** $\dfrac{(x-1)^2}{25} + \dfrac{(y+5)^2}{64} = 1$　(1, −5); 10, 16

39. $\dfrac{(x+3)^2}{16} + \dfrac{(y-6)^2}{9} = 1$　(−3, 6); 8, 6　　**40.** $\dfrac{(x-5)^2}{4} + (y+6)^2 = 1$　(5, −6); 4, 2

In Exercises 41 through 44, find an equation of the ellipse that has the given values of a and b and the given center.

41. $a = 2, b = 1$
center at (3, 4)　$\dfrac{(x-3)^2}{4} + \dfrac{(y-4)^2}{1} = 1$

42. $a = 3, b = 2$
center at (−1, 3)　$\dfrac{(x+1)^2}{9} + \dfrac{(y-3)^2}{4} = 1$

43. $a = 2, b = 3$
center at (−1, 4)　$\dfrac{(x+1)^2}{4} + \dfrac{(y-4)^2}{9} = 1$

44. $a = 5, b = 1$
center at (−5, −2)　$\dfrac{(x+5)^2}{25} + \dfrac{(y+2)^2}{1} = 1$

In Exercises 45 through 48, write an equation of the ellipse that has a horizontal axis of length n, a vertical axis of length p, and its center at the given point.

45. $n = 12, p = 2$
center at (3, 4)　$\dfrac{(x-3)^2}{36} + \dfrac{(y-4)^2}{1} = 1$

46. $n = 30, p = 12$
center at (−1, 3)　$\dfrac{(x+1)^2}{225} + \dfrac{(y-3)^2}{36} = 1$

47. $n = 24, p = 50$
center at (2, 4)　$\dfrac{(x-2)^2}{144} + \dfrac{(y-4)^2}{625} = 1$

48. $n = 36, p = 24$
center at (2, −2)　$\dfrac{(x-2)^2}{324} + \dfrac{(y+2)^2}{144} = 1$

In Exercises 49 though 54, rewrite the ellipse in graphing form, and find the center and the lengths of the major and minor axes for its graph.

49. $16x^2 + y^2 - 144 = 0$　$\dfrac{x^2}{3^2} + \dfrac{y^2}{12^2} = 1$
(0, 0); 24(v.), 6(h.)

50. $9x^2 + 16y^2 = 144$　$\dfrac{x^2}{4^2} + \dfrac{y^2}{3^2} = 1$
(0, 0); 8(h.), 6(v.)

51. $9y^2 + 4(x + 3)^2 - 36 = 0$　$\dfrac{[x-(-3)]^2}{3^2} + \dfrac{y^2}{2^2} = 1$
(−3, 0); 6(h.), 4(v.)

52. $(y - 8)^2 + 4x^2 - 16 = 0$　$\dfrac{x^2}{2^2} + \dfrac{(y-8)^2}{4^2} = 1$
(0, 8); 8(v.), 4(h.)

53. $9(x - 6)^2 + (y + 2)^2 = 36$　$\dfrac{(x-6)^2}{2^2} + \dfrac{[y-(-2)]^2}{6^2} = 1$
(6, −2); 12(v.), 4(h.)

54. $6(x + 5)^2 + 24(y - 1)^2 = 24$　$\dfrac{[x-(-5)]^2}{2^2} + \dfrac{(y-1)^2}{1} = 1$
(−5, 1); 4(h.), 2(v.)

In Exercises 55 through 60, rewrite the ellipse equation in graphing form, and find the values of a and b and the center of its graph.

55. $(x + 1)^2 + \dfrac{(y-2)^2}{9} = 1$　$\dfrac{[x-(-1)]^2}{1^2} + \dfrac{(y-2)^2}{3^2} = 1$
$a = 1, b = 3$; center at (−1, 2)

56. $\dfrac{x^2}{25} + \dfrac{(y-3)^2}{4} = 1$　$\dfrac{x^2}{5^2} + \dfrac{(y-3)^2}{2^2} = 1$
$a = 5, b = 2$; center at (0, 3)

57. $\dfrac{(x-1)^2}{4} + \dfrac{(y+2)^2}{9} = 1$　$\dfrac{(x-1)^2}{2^2} + \dfrac{[y-(-2)]^2}{3^2} = 1$
$a = 2, b = 3$; center at (1, −2)

58. $\dfrac{x^2}{4} + \dfrac{(y-1)^2}{9} = 1$　$\dfrac{x^2}{2^2} + \dfrac{(y-1)^2}{3^2} = 1$
$a = 2, b = 3$; center at (0, 1)

59. $\dfrac{(x-1)^2}{4} + \dfrac{y^2}{9} = 1$　$\dfrac{(x-1)^2}{2^2} + \dfrac{y^2}{3^2} = 1$
$a = 2, b = 3$; center at (1, 0)

60. $\dfrac{(x-1)^2}{9} + \dfrac{(y-2)^2}{4} = 1$　$\dfrac{(x-1)^2}{3^2} + \dfrac{(y-2)^2}{2^2} = 1$
$a = 3, b = 2$; center at (1, 2)

In Exercises 61 through 66, graph each hyperbola.

61. $\dfrac{(y+1)^2}{4} - \dfrac{(x+1)^2}{9} = 1$　　**62.** $\dfrac{y^2}{4} - \dfrac{(x-2)^2}{9} = 1$

63. $\dfrac{(y+2)^2}{25} - \dfrac{x^2}{16} = 1$

64. $\dfrac{y^2}{9} - \dfrac{(x-1)^2}{25} = 1$

65. $\dfrac{x^2}{4} - \dfrac{(y-1)^2}{4} = 1$

66. $\dfrac{(y-3)^2}{4} - \dfrac{(x+4)^2}{1} = 1$

In Exercises 67 through 74, write the equation for, and identify the vertices of, each hyperbola described.

67. $a = 22, b = 53$
center at $(10, -11)$
y^2-term subtracted
$\dfrac{(x-10)^2}{484} - \dfrac{(y+11)^2}{2809} = 1$
$(-12, -11), (32, -11)$

68. $a = 12, b = 1$
center at $(-17, 10)$
y^2-term subtracted
$\dfrac{(x+17)^2}{144} - \dfrac{(y-10)^2}{1} = 1$
$(-5, 10), (-29, 10)$

69. $a = 24, b = 50$
center at $(2, 3)$
x^2-term subtracted
$\dfrac{(y-3)^2}{2500} - \dfrac{(x-2)^2}{576} = 1$
$(2, -47), (2, 53)$

70. $a = 36, b = 24$
center at $(-21, 38)$
y^2-term subtracted
$\dfrac{(x+21)^2}{1296} - \dfrac{(y-38)^2}{576} = 1$
$(15, 38), (-57, 38)$

71. $a = 2, b = 3$
center at $(3, 5)$
x^2-term subtracted
$\dfrac{(y-5)^2}{9} - \dfrac{(x-3)^2}{4} = 1$
$(3, 2), (3, 8)$

72. $a = 2, b = 5$
center at $(-1, 4)$
x^2-term subtracted
$\dfrac{(y-4)^2}{25} - \dfrac{(x+1)^2}{4} = 1$
$(-1, -1), (-1, 9)$

73. $a = 3, b = 4$
center at $(-5, -2)$
y^2-term subtracted
$\dfrac{(x+5)^2}{9} - \dfrac{(y+2)^2}{16} = 1$
$(-8, -2), (-2, -2)$

74. $a = 5, b = 1$
center at $(-1, -3)$
y^2-term subtracted
$\dfrac{(x+1)^2}{25} - (y+3)^2 = 1$
$(-6, -3), (4, -3)$

In Exercises 75 through 86, write the hyperbola equation in graphing form. Then identify the center, the values of a and b, and the vertices of its graph.

75. $9y^2 - x^2 = -36$ $\dfrac{x^2}{6^2} - \dfrac{y^2}{2^2} = 1$
center: $(0, 0)$; $a = 6, b = 2$;
$(6, 0), (-6, 0)$

76. $x^2 - y^2 = -25$ $\dfrac{y^2}{25} - \dfrac{x^2}{25} = 1$
center: $(0, 0)$; $a = 5, b = 5$;
$(0, 5), (0, -5)$

77. $x^2 - y^2 = 1.44$ $\dfrac{x^2}{(1.2)^2} - \dfrac{y^2}{(1.2)^2} = 1$
center: $(0, 0)$; $a = 1.2, b = 1.2$;
$(1.2, 0), (-1.2, 0)$

78. $x^2 - y^2 - 0.04 = 0$ $\dfrac{x^2}{(0.2)^2} - \dfrac{y^2}{(0.2)^2} = 1$
center: $(0, 0)$; $a = 0.2, b = 0.2$;
$(0.2, 0), (-0.2, 0)$

79. $6(x-5)^2 - 24(y-1)^2 = 24$ $\dfrac{(x-5)^2}{2^2} - \dfrac{(y-1)^2}{1^2} = 1$
center: $(5, 1)$; $a = 2, b = 1$;
$(3, 1), (7, 1)$

80. $9(x-6)^2 - (y-2)^2 = 36$ $\dfrac{(x-6)^2}{2^2} - \dfrac{(y-2)^2}{6^2} = 1$
center: $(6, 2)$; $a = 2, b = 6$;
$(4, 2), (8, 2)$

81. $9x^2 - y^2 = 9$ $\dfrac{x^2}{1^2} - \dfrac{y^2}{3^2} = 1$
center: $(0, 0)$; $a = 1, b = 3$;
$(1, 0), (-1, 0)$

82. $49y^2 - x^2 = 49$ $\dfrac{y^2}{1^2} - \dfrac{x^2}{7^2} = 1$
center: $(0, 0)$; $a = 7, b = 1$;
$(0, 1), (0, -1)$

83. $4x^2 - y^2 = -100$ $\dfrac{y^2}{10^2} - \dfrac{x^2}{5^2} = 1$
center: $(0, 0)$; $a = 5, b = 10$;
$(0, 10), (0, -10)$

84. $25(x-3)^2 - (y+12)^2 = 100$
$\dfrac{(x-3)^2}{2^2} - \dfrac{[y-(-12)]^2}{10^2} = 1$; center: $(3, -12)$; $a = 2, b = 10$;
$(1, -12), (5, -12)$

85. $4x^2 - 9y^2 = 36$ $\dfrac{x^2}{3^2} - \dfrac{y^2}{2^2} = 1$
center: $(0, 0)$; $a = 3, b = 2$; $(3, 0), (-3, 0)$

86. $(x+10)^2 - 25(y-2)^2 = 25$
$\dfrac{[x-(-10)]^2}{5^2} - \dfrac{(y-2)^2}{1^2} = 1$; center: $(-10, 2)$; $a = 5, b = 1$;
$(-15, 2), (-5, 2)$

In Exercises 87 through 92, each equation is the equation of a parabola, a circle, an ellipse, or a hyperbola. Complete the square, where necessary, and write each in graphing from. Then identify the conic section described by each equation.

87. $x^2 + y^2 - 10x + 2y + 25 = 0$
$(x-5)^2 + [y-(-1)]^2 = 1$, circle

88. $6x - 3y^2 - 12y + 4 = 0$
$x = \frac{1}{2}(y+2)^2 - \frac{8}{3}$, parabola

89. $x^2 - y^2 - 6x + 8y - 3 = 0$
$\dfrac{(y-4)^2}{4} - \dfrac{(x-3)^2}{4} = 1$, hyperbola

90. $4x^2 + 9y^2 - 8x + 36y + 4 = 0$ **91.** $4x^2 - y^2 + 32x + 6y + 67 = 0$ **92.** $x^2 + y^2 - 6x + 5 = 0$

$\frac{(x-1)^2}{9} + \frac{(y+2)^2}{4} = 1$, ellipse $\frac{(y-3)^2}{12} - \frac{[x-(-4)]^2}{3} = 1$, hyperbola $(x-3)^2 + y^2 = 4$, circle

MIXED PRACTICE

By doing these exercises, you will practice the topics up to this point in the chapter.

93. Write the equation for the ellipse that has its center at $(-13, -11)$, a minor axis of length 28, and a vertical major axis of length 44. $\frac{(x+13)^2}{196} + \frac{(y+11)^2}{484} = 1$

94. Write in graphing form the equation for a circle with its center at $(-2, -1)$ and radius 3. $(x+2)^2 + (y+1)^2 = 9$

95. Find the center and the length of the major and minor axes of the graph of $\frac{(x-6)^2}{225} + \frac{(y-8)^2}{144} = 1$.   center: $(6, 8)$; major axis: 30; minor axis: 24

96. This is the equation of a conic section. Which one?

$$3y + 2x^2 + 8x - 4 = 0 \qquad y = -\frac{2}{3}(x+2)^2 + 4, \text{ parabola}$$

97. Write the equation of the graph and identify the vertices of a hyperbola with center at $(3, 4)$, $a = 2$, $b = 2$, and the x^2-term subtracted. $\frac{(y-4)^2}{4} - \frac{(x-3)^2}{4} = 1$; vertices: $(3, 2), (3, 6)$

98. Find the vertex of the graph of $y = 3x^2 - 2x + 1$ and the direction in which it opens.   vertex: $\left(\frac{1}{3}, \frac{2}{3}\right)$; opens upward

99. Find the length of the major and minor axes, the center, and the x- and y-intercepts for the ellipse whose equation is

$$\frac{x^2}{4} + \frac{y^2}{49} = 1$$

Then graph the equation.   major axis: 14; minor axis: 4; center: $(0, 0)$; $(2, 0)$, $(-2, 0)$; $(0, 7)$, $(0, -7)$

100. This is the equation of a conic section. Which one?

$$8y^2 + 25x^2 - 48y + 50x + 47 = 0 \qquad \frac{[x-(-1)]^2}{2} + \frac{(y-3)^2}{6.25} = 1, \text{ ellipse}$$

EXCURSIONS

Posing Problems

1. Ask and answer four questions. Try to ask questions that other students would want to answer.

Golf

Specifications of ball: Broadened to require that the ball be designed to perform as if it were spherically symmetrical. The weight of the ball shall not be greater than 1.620 ounces avoirdupois, and the size shall not be less than 1.680 inches in diameter.

Velocity of ball: Not greater than 250 feet per second when tested on USGA apparatus, with 2 percent tolerance.

Hole: 4-1/4 inches in diameter and at least 4 inches deep.

Clubs: 14 is the maximum number permitted.

Overall distance standard: A brand of ball shall not exceed a distance of 280 yards plus 6% when tested on USGA apparatus under specified conditions, on an outdoor range at USGA Headquarters.

Source: 1996 Information Please® Almanac (©1995 Houghton Mifflin Co.), p. 1004. All rights reserved. Used with permission by Information Please LLC.

2. Ask and answer four questions. Try to ask questions that other students would want to answer.

Hockey

Size of rink: 200 feet long by 85 feet wide surrounded by a wooden wall not less than 40 inches and not more than 48 inches above level of ice.

Size of goal: 6 feet wide by 4 feet in height.

Puck: 1 inch thick and 3 inches in diameter, made of vulcanized rubber; weight 5-1/2 to 6 ounces.

Length of stick: Not more than 60 inches from heel to end of shaft nor more than 12-1/2 inches from heel to end of blade. Blade should not be more than 3 inches in width but not less than 2 inches—except goal keeper's stick, which shall not exceed 3-1/2 inches in width except at the heel, where it must not exceed 4-1/2 inches, nor shall the goalkeeper's stick exceed 15-1/2 inches from the heel to the end of the blade.

Source: 1996 Information Please® Almanac (©1995 Houghton Mifflin Co.), p. 1004. All rights reserved. Used with permission by Information Please LLC.

3. Ask and answer four questions. Try to ask questions that other students would want to answer.

Football
(NCAA)

Length of field: 120 yards (including 10 yards of end zone at each end).

Width of field: 53-1/3 yards (160 feet).

Height of goal posts: At least 30 feet.

Height of crossbar: 10 feet.

Width of goal posts (above crossbar): 18 feet 6 inches, inside to inside.

Length of ball: 10-7/8 to 11-7/16 inches (long axis).

Circumference of ball: 20-3/4 to 21-1/4 inches (middle): 27-3/4 to 28-1/2 inches (long axis).

Source: 1996 Information Please® Almanac (©1995 Houghton Mifflin Co.), p. 1004. All rights reserved. Used with permission by Information Please LLC.

4. Ask and answer four questions. Try to ask questions that other students would want to answer.

Boxing

Ring: Professional matches take place in an area not less than 18 nor more than 24 feet square including apron. It is enclosed by four covered ropes, each not less than one inch in diameter. The floor has a 2-inch padding of Ensolite (or equivalent) underneath ring cover that extends at least 6 inches beyond the roped area in the case of elevated rings. For USA Boxing or Olympic-style boxing, not less than 16 nor more than 20 feet square within the ropes. The floor must extend beyond the ring ropes not less than 2 feet. The ring posts shall be connected to the four ring ropes with the extension not shorter than 18 inches and must be properly padded.

Gloves: In professional fights, not less than 8-ounce gloves generally are used. USA Boxing, 10 ounces for boxers 106 pounds through 156 pounds; 12 ounces for boxers 165 pounds through 201+ pounds; for international competition, 8 ounces for lighter classes, 10 ounces for heavier divisions.

Headguards: Mandatory in Olympic-style boxing.

9.4 Applications Involving Conic Sections

SECTION LEAD-IN

Regular figures are figures with equal sides and equal angles. Find the number of degrees in each angle of a

1. regular triangle (What is this called?)
2. regular hexagon
3. regular pentagon
4. regular dodecagon (research to find the number of sides)

To find the answers, first you may have to determine the total number of degrees in the figure.

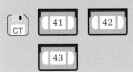

SECTION GOALS

- *To solve verbal problems that involve conic sections*
- *(Optional Topic) To solve nonlinear systems of equations*

Applications in Geometry

A geometric figure is **inscribed** in a second figure if it is drawn within the second figure such that the two figures touch in as many points as possible.

· · ·

EXAMPLE 1

What is the area of the largest square that can be inscribed in a circle of radius 10 inches?

SOLUTION

Every solution involving geometric figures should begin with a sketch. The following figure on the left shows a square with sides x inscribed within a circle. It also shows that a diagonal of the square is a diameter of the circle. Because a diameter is two radii in length, we have

$$\text{Diagonal} = 20 \text{ inches}$$

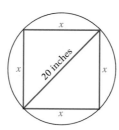 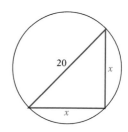

Now look at either of the two triangles cut off by the diagonal—say, the lower one. As is shown in the figure on the right above, it is a triangle with two sides of length x and hypotenuse of length 20.

By the Pythagorean theorem, then,

$$x^2 + x^2 = 20^2 \qquad \text{Substituting in } a^2 + b^2 = c^2$$
$$2x^2 = 400 \qquad \text{Simplifying}$$
$$x^2 = 200 \qquad \text{Multiplying by } \tfrac{1}{2}$$
$$x = \pm 10\sqrt{2} \quad \text{Taking square roots}$$

The square has sides of length $10\sqrt{2}$. Its area is $A = \left(10\sqrt{2}\right)^2 = 200$ square units.

▶ CHECK **Warm-Up 1**

To work the next example, we need two pieces of information: First, a geometric figure is **circumscribed** about a second figure if it is drawn *around* the second figure such that the two figures touch at as many points as possible. Second, the area A of an ellipse can be found as

$$A = \pi ab$$

where a and b are half the lengths of the major and minor axes.

∎∎∎

EXAMPLE 2

A rectangle is circumscribed about an ellipse with the equation

$$\frac{(x+6)^2}{4} + \frac{(y-7)^2}{25} = 1$$

What is the difference between the area of the ellipse and the area of the rectangle?

SOLUTION

The ellipse has a graphing equation of the form

$$\frac{(x-h)^2}{a^2} + \frac{(y-k)^2}{b^2} = 1$$

So here $a = 2$ and $b = 5$. The given ellipse, then, has a horizontal axis of length

$$2a = 2(2) = 4$$

and a vertical axis of length

$$2b = 2(5) = 10$$

The values of *h* and *k* affect only the position of the ellipse on the coordinate plane, not its shape. They can be ignored here. We sketch the ellipse and then circumscribe a rectangle as shown in the accompanying figure.

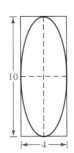

As the sketch shows, the rectangle has a width *W* of 4 units, and a length *L* of 10 units. Its area is $L \cdot W = 40$ square units.

When we approximate π as 3.14, the area of the ellipse is

$$\pi ab = 3.14(2)(5) = 31.4$$

The difference in areas is

$$40 - 31.4 = 8.6 \text{ square units}$$

▶ *CHECK* **Warm-Up 2**

Applications in Engineering

The shape of the cables that support the roadway on a suspension bridge can be approximated by a parabola. This fact is important to the designers and builders of bridges.

▪ ▪ ▪

EXAMPLE 3

Suppose a small suspension bridge has the dimensions shown in the following figure. Find the equation of the (approximate) parabolic shape assumed by either cable. (Place the coordinate axes such that the origin is at the intersection of the roadway and the middle tower).

INSTRUCTOR NOTE

The curve here is actually a catenary.

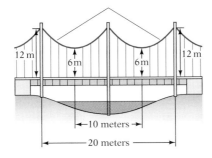

SOLUTION

With the axes located as required, we have the situation shown in the accompanying figure. As a result of the symmetry of a parabola, the vertices of the parabolic cables must be located at $(-5, 6)$ and $(5, 6)$. The parabola passes through the points $(-10, 12)$, $(0, 12)$, and $(10, 12)$, as shown. We shall find the equation of the parabola on the right.

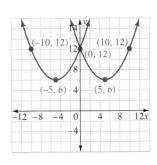

We begin with the graphing form of the equation of a parabola:

$$y = a(x - h)^2 + k$$

Substituting the coordinates of the vertex $(5, 6)$ for (h, k), we get

$$y = a(x - 5)^2 + 6$$

We still need to find the value of a. To do so, we substitute the coordinates of some other point of the parabola for x and y and solve the result for a. Using the point $(10, 12)$, we get

$$y = a(x - 5)^2 + 6 \quad \text{The equation so far}$$
$$12 = a(10 - 5)^2 + 6 \quad \text{Substituting for } x \text{ and } y$$
$$6 = 25a \quad \text{Simplifying}$$
$$\frac{6}{25} = a \quad \text{Multiplying by } \frac{1}{25}$$

The equation of the right cable is then

$$y = \frac{6}{25}(x - 5)^2 + 6$$

One use for this equation is in laying out and checking the cable's shape on design drawings and during construction of the bridge.

▶ *CHECK* **Warm-Up 3**

The orbits of all the planets of our solar system are ellipses, with the sun at one focus. We use that fact in the next example.

■ ■ ■

EXAMPLE 4

The closest Mercury gets to the sun is 46×10^6 kilometers from it, and the farthest is 69.8×10^6 kilometers. How far from the sun is the center of Mercury's elliptical orbit?

SOLUTION

We start with a sketch of Mercury's orbit, placing the sun at one focus. For simplicity, we place the origin at the center of the ellipse, as shown below.

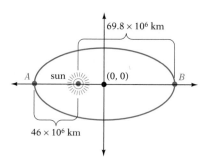

Mercury is closest to the sun when it is at point A. (All other points of the ellipse are farther from the left focus than point A.) The planet is farthest from the sun when it is at point B. So the distances are as marked in the figure.

The distance from A to B is the length $2a$ of the major axis of the ellipse. That distance is equal to the sum of the distances from A to the sun and from the sun to B. So

$$2a = 46 \times 10^6 + 69.8 \times 10^6$$
$$= 115.8 \times 10^6 \text{ km}$$

Then the distance from A to the center is

$$a = \frac{115.8 \times 10^6}{2} = 57.9 \times 10^6 \text{ km}$$

The distance from the sun to the center of the ellipse is then

$$57.9 \times 10^6 - 46 \times 10^6 = 11.9 \times 10^6 \text{ km}$$

▶ *CHECK* **Warm-Up 4**

Optional Topic: Solving Nonlinear Systems of Equations

An equation in two variables is a linear equation if its graph is a straight line. We can recognize a linear equation by its form, $ax + by = c$. An equation in two variables is *nonlinear* if it is not linear. Thus, for example, equations that represent conic sections are nonlinear.

Nonlinear Systems of Equations

If at least one equation in a system of two (or more) equations is nonlinear, then we say the *system* is nonlinear. In this section, we solve nonlinear systems whose equations represent lines or conic sections. These systems may have as many as four solutions when they are not dependent systems.

▪▪▪

EXAMPLE 5

Suppose a nonlinear two-variable system consists of the equations of an ellipse and a parabola. Sketch possible graphs of these equations in which the system has (a) zero, (b) one, and (c) two solutions.

SOLUTION

If a system has two equations, then the graph of any solution of the system is a point where the graphs of the two equations intersect.

a.   If there are no solutions, then the graphs of the equations do not intersect. One possible situation is shown in figure (a).

b.   With one solution, the ellipse and parabola must intersect at one point, perhaps as in figure (b).

c.   If there are two solutions, the graphs intersect at two points. One possibility is shown in figure (c).

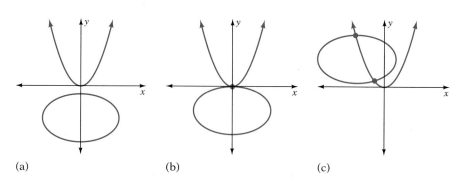

(a)   (b)   (c)

▶ CHECK **Warm-Up 5**

A system of nonlinear equations in two variables can be solved just as we solved systems of linear equations.

Solution by Addition

The addition (elimination) method usually works when both equations are of the same degree—all conic sections have equations of the second degree.

■ ■ ■
EXAMPLE 6

Solve the system of equations: $2x + 3y^2 = 10$
$$4x - 3y^2 = 8$$

SOLUTION

If we add the left and right sides of these equations, the y^2-terms will be eliminated. Then we can solve the resulting equation for x.

$$
\begin{array}{ll}
2x + 3y^2 = 10 & \\
\underline{4x - 3y^2 = \ 8} & \\
6x + 0 \ = 18 & \text{Adding left and right sides} \\
x = 3 & \text{Multiplying by } \tfrac{1}{6}
\end{array}
$$

We can substitute 3 for x in either equation and solve for y.

$$
\begin{array}{ll}
2x + 3y^2 = 10 & \text{First equation} \\
2(3) + 3y^2 = 10 & \text{Substituting 3 for } x \\
3y^2 = 4 & \text{Adding } -6 \text{ to both sides} \\
y^2 = \tfrac{4}{3} & \text{Multiplying by } \tfrac{1}{3} \\
y = \sqrt{\tfrac{4}{3}} \ \text{ and } \ y = -\sqrt{\tfrac{4}{3}} & \text{Taking square roots} \\
y = \tfrac{2}{\sqrt{3}} \ \text{ and } \ y = \tfrac{-2}{\sqrt{3}} & \text{Simplifying}
\end{array}
$$

To rationalize the denominators, we multiply the numerator and denominator of each solution by $\sqrt{3}$, getting

$$y = \frac{2\sqrt{3}}{3} \quad \text{and} \quad y = \frac{-2\sqrt{3}}{3}$$

WRITER'S BLOCK
What is the difference between linear and nonlinear systems of equations?

Thus we obtain the solution $\left(3, \frac{2\sqrt{3}}{3}\right)$ and $\left(3, \frac{-2\sqrt{3}}{3}\right)$. Both must be checked in both equations. For the first of these solutions,

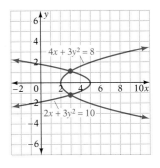

$$2x + 3y^2 = 10$$
$$2(3) + 3\left[\left(\frac{2\sqrt{3}}{3}\right)\right]^2 = 10$$
$$6 + 3\left(\frac{12}{9}\right) = 10$$
$$6 + 4 = 10 \quad \text{True}$$

$$4x - 3y^2 = 8$$
$$4(3) - 3\left[\left(\frac{2\sqrt{3}}{3}\right)\right]^2 = 8$$
$$12 - 3\left(\frac{12}{9}\right) = 8$$
$$12 - 4 = 8 \quad \text{True}$$

You should check for the second solution, $\left(3, \frac{-2\sqrt{3}}{3}\right)$.

The solution set is $\left\{\left(3, \frac{2\sqrt{3}}{3}\right), \left(3 \frac{-2\sqrt{3}}{3}\right)\right\}$. The graphs are sketched in the accompanying figure.

▶ CHECK **Warm-Up 6**

Solution by Substitution

The substitution method usually works best when one of the equations in a system is a linear equation. That linear equation can be easily solved for one variable in terms of the other.

▪▪▪

EXAMPLE 7

Solve: $y - 2x = 4$

$\qquad y^2 - x^2 = -4$

SOLUTION

The first of the given equations is linear. Adding $2x$ to both sides yields

$$y = 2x + 4$$

Thus, we can substitute $2x + 4$ for y in the second equation, simplify as needed, and solve for x.

$$y^2 - x^2 = -4 \quad \text{Original equation}$$
$$(2x + 4)^2 - x^2 = -4 \quad \text{Substituting for } y$$
$$4x^2 + 16x + 16 - x^2 = -4 \quad \text{Squaring}$$
$$3x^2 + 16x + 20 = 0 \quad \text{Simplifying}$$

This quadratic equation can be solved by factoring. We factor on the left to get

$$(3x + 10)(x + 2) = 0$$

and then write

$$3x + 10 = 0 \qquad\qquad x + 2 = 0$$
$$x = \frac{-10}{3} \qquad\qquad\quad x = -2$$

So $x = \frac{-10}{3}$ and $x = -2$.

We now substitute these values for x in one of the original equations and solve for the corresponding values of y. The first equation is simpler. We get

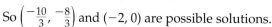

For $x = \frac{-10}{3}$:   For $x = -2$:

$$y = 2x + 4 \qquad\qquad y = 2x + 4$$

$$y = 2\left(\frac{-10}{3}\right) + 4 \qquad y = 2(-2) + 4$$

$$= \frac{-20}{3} + \frac{12}{3} \qquad\qquad = -4 + 4$$

$$= \frac{-8}{3} \qquad\qquad\qquad = 0$$

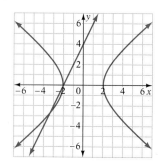

So $\left(\frac{-10}{3}, \frac{-8}{3}\right)$ and $(-2, 0)$ are possible solutions.

To check them, you should substitute them, in turn, in both original equations. You will find that the solution set is $\left\{\left(\frac{-10}{3}, \frac{-8}{3}\right), (-2, 0)\right\}$. The graphs of the equations and their solutions are sketched in the accompanying figure.

▶ CHECK **Warm-Up 7**

Practice what you learned.

SECTION FOLLOW-UP

Each angle in a regular, or equilateral, triangle has 60°. Because a triangle has 180° and 3 angles, $180 \div 3 = 60°$.

What did you find out about the other figures?

9.4 WARM-UPS

Work these problems before you attempt the exercises.

1. The area of a circle is 314 square inches. A square is inscribed inside. What is the difference between the area of the circle and the area of the largest square that could be inscribed in the circle? (Use $\pi \approx 3.14$.)   114 square inches

2. A round tablecloth measures 6.5 feet in diameter. It is folded in half and then in half again so that it looks like a piece of pie. What are the length and width of the smallest drawer that it can fit into without folding it further?   3.25 by 3.25 feet

3. Find the equation of the shape of the cable on the left in Example 3. $y = \frac{6}{25}(x + 5)^2 + 6$

4. The *eccentricity* of an ellipse indicates how close its equation is to that of a circle. We can calculate eccentricity by

$$e = \frac{\sqrt{a^2 - b^2}}{a}$$

Find the eccentricity of Mercury's orbit. (See Example 4.)
2.06×10^{-1}

5. For the system in Example 5 sketch possible graphs in which the system has (a) three and (b) four solutions.   Answers may vary.

6. Solve by addition: $x^2 + y^2 = 4$
$$x^2 - y^2 = 4 \quad \{(2, 0), (-2, 0)\}$$

7. Solve by substitution: $x^2 + y^2 = 25$
$$3x + 4y = 0 \quad (4, -3), (-4, 3)$$

9.4 EXERCISES

Note: Use your graphing calculator to check your results wherever possible.

In Exercises 1 through 20, use 3.14 as an approximation for π when you need it to solve problems.

1. *Mirror, Mirror*   A circular mirror with a 91.06-inch circumference is mounted on a square backing and framed in a square frame that is 1 inch wide. What are the outside dimensions of the frame.   31 inches by 31 inches

2. *Packing Pizza*   A round pizza has a diameter of 18 inches and is $1\frac{1}{2}$ inches thick. What is the interior volume of the smallest square pizza box that will hold this pizza?   486 cubic inches

3. a. What is the area of the largest ellipse that can be cut out of a rectangular sheet of paper with dimensions 17 inches by 24 inches?   320.28 square inches

   b. How much paper will be thrown away?   87.72 square inches

4. *Wrestling Ring*   The circular wrestling area in a wrestling ring has an area of 683.15 square feet. It is enclosed in a square roped-off area that covers an additional 4147 square feet outside of the ring. What is the perimeter of this roped-off area?   278 feet

5. *Inscribed Rectangle*   What is the area of the smallest rectangle that an ellipse can be inscribed in if the ellipse has an area of 27π square inches and a major axis of 18 inches?   108 square inches

6. *Magnet Size*   The world's largest circular magnet (located in Russia) has a diameter of 196.1 feet. A square fence is to be built around this magnet that allows a minimum of 10 feet between fence and magnet. What length of fencing is required?   864.4 feet

7. *Maxima and Minima* The vertex of a parabola $ax^2 + bx + c = y$ is also called a minimum or maximum point. If a is positive, substituting $x = \frac{-b}{2a}$ gives a minimum value for y. If a is negative, substituting $x = \frac{-b}{2a}$ gives a maximum value for y. A ball is shot out of a cannon directly upward from the ground at an initial velocity of 128 feet per second. The equation describing this shot is $y = -16x^2 + 128x$. Find the maximum height the ball reaches by finding the value of y at the vertex of the graph of this equation.   256 feet

8. *Bacterial Growth* The formula $C = 20t^2 - 200t + 640$ describes the concentration C of bacteria after t doses of an antibiotic. Find the number of doses t that gives the minimum concentration of bacteria C.   5 doses of medicine (140 bacteria)

9. *Halley's Comet* Halley's comet has an orbit that is elliptical with the sun as a focus. Its equation is

$$\frac{x^2}{(18.09)^2} + \frac{y^2}{(4.56)^2} = 1$$

with a and b given in astronomical units. Use the fact that $c^2 = a^2 - b^2$ and the foci at $(c, 0)$ and $(-c, 0)$ to find, in astronomical units, the comet's distance from the sun at its closest approach to the sun. (*Hint:* How far is it from one end of the major axis to the focus?)   0.58 astronomical unit

10. *Satellite Path* The equation $x^2 = -16,400(y - 4100)$ models the path of a satellite as it escapes Earth's gravity. What type of conic section is this? Solve the equation for x when $y = 0$.   parabola; $x = 8200$

11. *Eccentricity of an Ellipse* The eccentricity e, of an ellipse is given by

$$e = \frac{\sqrt{a^2 - b^2}}{a}$$

The closer this ratio is to zero, the more the ellipse resembles a circle. The eccentricities of the orbits of the planets are

Mercury:	2.06×10^{-1}	Saturn:	5.1×10^{-2}
Venus:	7×10^{-3}	Uranus:	4.6×10^{-2}
Earth:	1.7×10^{-2}	Neptune:	5×10^{-3}
Mars:	9.3×10^{-2}	Pluto:	2.5×10^{-1}
Jupiter:	4.9×10^{-2}		

Which planet has the most nearly circular orbit?   Neptune has an eccentricity closest to zero and therefore has the most nearly circular orbit.

12. *Geometry: Ellipses* Find the equation of an ellipse that is centered at the origin and has an eccentricity of 0.5 and a horizontal major axis of 10. Use

$$e = \frac{\sqrt{a^2 - b^2}}{a} \qquad \frac{x^2}{5^2} + \frac{y^2}{(4.33)^2} = 1$$

13. *Geometry: Ellipses* There is no simple formula to compute the circumference of an ellipse. However, there are three elementary formulas that can approximate it.

$$C_1 = \pi\left[a + b + \frac{1}{2}\left(\sqrt{a} - \sqrt{b}\right)^2\right] \quad \text{(Suggested by Giuseppe Peano in 1887)}$$

$$C_2 = \pi\left[3(a + b) - \sqrt{(a + 3b)(3a + b)}\right]$$ (Suggested by Srinivasa Ramanujan in 1914)

$$C_3 = \frac{\pi}{2}\left(a + b + \sqrt{2(a^2 + b^2)}\right)$$ (Suggested by R. A. Johnson in 1930)

a. Find the circumference of the ellipse $\frac{x^2}{9} + \frac{y^2}{25} = 1$ using each of these formulas. Round to 3 decimal places. $C_1 = 25.519, C_2 = 25.514, C_3 = 25.507$

b. Suppose the actual circumference, rounded to three decimal places, is 25.527. Compare the values of C_1, C_2, C_3. Which answer is closest to the given answer? C_1 is the closest. All are identical to the nearest tenth.

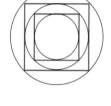

14. *Toy Sizes* A child's toy has a series of "round" and "square" cups that fit inside each other as shown in the figure at the right. Ignore the width of the sides of the cups. If the outside one has a radius of 2 inches, what is the radius of the inside cup? 1 inch

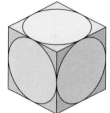

15. *Sculpture* A cubical sculpture was designed and constructed out of brass and steel. Each side is a circle of brass inscribed in a steel square. (See the accompanying figure.) If each square is 10 feet on a side, what is the difference between the surface areas of the brass and steel? 342 square feet more brass (57 square feet more per side)

16. *Sports: Football* The playing field for Australian football is not rectangular but rather is somewhat elliptical in shape. Its maximum length measures 185 meters, and its maximum width measures 155 meters. Write the equation of an ellipse that has its center at the origin and approximates the shape of this playing field. $\frac{x^2}{(92.5)^2} + \frac{y^2}{(77.5)^2} = 1$

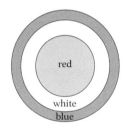

17. *Target Design* A target has three regions—red, white, and blue—and the area of each portion is equal to those of the other portions. (See the accompanying figure.) The diameter is 3 inches. What are the radii of the three concentric circles formed by these regions? radius of red circle $= \sqrt{0.75}$ inch; radius of red and white circle $= \sqrt{1.5}$ inches; radius of big circle $= 1.5$ inches

18. *Castle Moat* A moat 12 feet wide surrounds a square castle "inscribed" on a circular piece of land with a diameter of 314 feet. (See the accompanying figure.) What are the areas of the castle, the land surrounding the castle, and the moat? Use 3.14 for π. castle: 49,298 square feet; land: 28,099.86 square feet; moat: 12,283.68 square feet

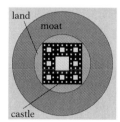

19. *Garden Size* An elliptical garden is planted inside a rectangular fence that has a perimeter of 116 feet and surrounds an area of 805 square feet. What is the area (to the nearest square foot) of the largest such garden? 632 square feet

20. *Clock Box* The circumference of a clock is 31.4 inches, and the clock is 2 inches deep. What are the dimensions of the smallest box it will fit into? (Use $\pi \approx 3.14$.) $10 \times 10 \times 2 = 200$ cubic inches

Suppose a nonlinear, two-variable system consists of the equations of two ellipses. In Exercises 21 through 23, sketch possible graphs of these equations in which the system has the indicated number of solutions. Answers may vary.

21. zero   **22.** one   **23.** two

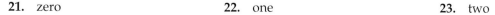

Suppose a nonlinear, two-variable system consists of the equations of a hyperbola and an ellipse. In Exercises 24 through 26, sketch possible graphs of these equations in which the system has the indicated number of solutions.   Answers may vary.

24. zero

25. two

26. four

Suppose a nonlinear, two-variable system consists of the equations of an ellipse and a circle. In Exercises 27 through 29, sketch possible graphs of these equations in which the system has the indicated number of solutions.   Answers may vary.

27. zero

28. two

29. three

In Exercises 30 through 32, sketch the intersection of a parabola and a hyperbola in such a way that their equations form a system with the given number of solutions.   Answers may vary.

30. three solutions

31. four solutions

32. two solutions

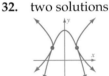

In Exercises 33 through 48, solve the nonlinear system.

33. $y = x + 3$
$y = 9 - x^2$
$\{(-3, 0), (2, 5)\}$

34. $y = x - 2$
$y = 4 - x^2$
$\{(-3, -5), (2, 0)\}$

35. $x^2 + 4y^2 = 4$
$x^2 - 6y^2 = 12$
no real-number solutions

36. $y = x - 7$
$y = 49 - x^2$
$\{(-8, -15), (7, 0)\}$

37. $x = 3y^2 - 5$
$x = y^2 + 3$
$\{(7, 2), (7, -2)\}$

38. $y = 4x^2 - 2$
$y = x^2 + 1$
$\{(1, 2), (-1, 2)\}$

39. $x^2 + y^2 = 1$
$y = x + 1$
$\{(0, 1), (-1, 0)\}$

40. $\dfrac{x^2}{4} + \dfrac{y^2}{9} = 1$
$x^2 + y^2 = 9$
$\{(0, 3), (0, -3)\}$

41. $6x^2 + y^2 = 10$
$2x^2 + 4y^2 = 40$
$\{(0, \sqrt{10}), (0, -\sqrt{10})\}$

42. $x^2 + y^2 = 6$
$y = x^2$
$\{(\sqrt{2}, 2), (-\sqrt{2}, 2)\}$

43. $xy = 1$
$x - 2y = 1$
$\{(2, \tfrac{1}{2}), (-1, -1)\}$

44. $y = x^2 - 3$
$x^2 + y^2 = 9$
$\{(0, -3), (\sqrt{5}, 2), (-\sqrt{5}, 2)\}$

45. $20x^2 + 4y^2 = 4$
$8x^2 - y^2 = 12$
no real-number solutions

46. $x^2 + 2y^2 = 15$
$2x^2 + y^2 = 24$
$\{(\sqrt{11}, \sqrt{2}), (\sqrt{11}, -\sqrt{2}),$
$(-\sqrt{11}, \sqrt{2}), (-\sqrt{11}, -\sqrt{2})\}$

47. $x^2 + y^2 = 36$
$35y = x^2$
$\{(\sqrt{35}, 1), (-\sqrt{35}, 1)\}$

48. $x^2 + y^2 = 9$
$2x^2 - y^2 = -6$
$\{(1, 2\sqrt{2}), (1, -2\sqrt{2}),$
$(-1, 2\sqrt{2}), (-1, -2\sqrt{2})\}$

In Exercises 49 and 52, graph the system of equations to determine the number of solutions.

49. $\dfrac{x^2}{16} + y^2 = 1$
$y = (x + 2)^2 - 4$ three solutions

50. $(x - 4)^2 + (y + 2)^2 = 4$
$\dfrac{x^2}{9} + \dfrac{y^2}{4} = 1$ two solutions

51. $x^2 + y^2 = 1$
$4x^2 + 9y^2 = 36$
zero solutions

52. $\dfrac{x^2}{25} + \dfrac{y^2}{16} = 1$
$y = 2(x - 2)^2 - 1$ two solutions

MIXED PRACTICE

By doing these exercises, you will practice the topics up to this point in the chapter.

53. Write $9x^2 + 6x + 6 = y$ in graphing parabola form, find the vertex of its graph, and determine in which direction the graph of this equation opens.

$y = 9\left(x + \dfrac{1}{3}\right)^2 + 5;$

vertex: $\left(-\dfrac{1}{3}, 5\right)$; opens upward

54. Find the equation of the circle with its center at the point $(1, 0)$ and radius 4. $(x - 1)^2 + y^2 = 16$

55. Find the center and the values of a and b for the ellipse represented by the equation

$$\frac{(x - 12)^2}{100} + \frac{(y + 10)^2}{121} = 1$$

center: $(12, -10)$; $a = 10, b = 11$

56. Write the equation of the hyperbola for which $a = 2$ and $b = 1$, with center at $(2, 4)$, and with vertices on the horizontal axis. $\dfrac{(x - 2)^2}{4} - (y - 4)^2 = 1$

57. Find the center and radius of the circle whose equation is

$$(x - 4)^2 + (y + 2)^2 = 32$$

center: $(4, -2)$; radius: $4\sqrt{2}$

58. Solve: $y = x + 5$
$y = 25 - x^2$

$\{(-5, 0), (4, 9)\}$

59. Find the center, the length of the major and minor axes, and the x- and y-intercepts of the ellipse described by the equation

$$\frac{x^2}{25} + \frac{y^2}{49} = 1$$

center: $(0, 0)$
major axis: 14 (vertical);
minor axis: 10 (horizontal)
x-intercepts: $(5, 0)$, $(-5, 0)$
y-intercepts: $(0, 7)$, $(0, -7)$

60. Find the equation of the ellipse with $a = 2$, $b = 42$, and its center at the point $(3, 5)$. $\dfrac{(x - 3)^2}{4} + \dfrac{(y - 5)^2}{1764} = 1$

61. Graph the hyperbola described by the equation

$$\frac{(x + 2)^2}{9} - \frac{(y - 2)^2}{1} = 1$$

62. Solve: $x^2 + y^2 = 12$
$x = y^2$ $\{(3, \sqrt{3}), (3, -\sqrt{3})\}$

63. Give the center and the lengths of the major and minor axes of the graph of $\dfrac{(x - 7)^2}{81} + \dfrac{(y + 9)^2}{64} = 1$ center: $(7, -9)$; major axis: 18 (horizontal); minor axis: 16 (vertical)

64. Write the equation of a circle that is centered at the origin and has an area of 9.8596 square inches. Use $\pi \approx 3.14$. $x^2 + y^2 = 3.14$

65. Write the equation of the circle whose center is at (2, 3) and whose radius is $2\sqrt{2}$. $(x - 2)^2 + (y - 3)^2 = 8$

66. Write the equation of the ellipse with center (19, 27) and with $a = 60$ and $b = 20$. $\dfrac{(x - 19)^2}{3600} + \dfrac{(y - 27)^2}{400} = 1$

67. Graph the hyperbola whose equation is
$$\frac{(y + 2)^2}{9} - \frac{(x + 1)^2}{1} = 1$$

68. Graph the hyperbola whose equation is
$$\frac{(x + 1)^2}{9} - \frac{(y + 1)^2}{4} = 1$$

68.

69. Graph the parabola whose equation is $y = 0.5(x + 1)^2 - 1$.

70. Rewrite the circle equation $y^2 + 4x + x^2 - 6y = 0$ in graphing form and find the center and radius. $(x + 2)^2 + (y - 3)^2 = 13$; center: $(-2, 3)$, radius: $\sqrt{13}$

EXCURSIONS

Exploring Geometry

1. Find the length of the side of the square that can be inscribed inside the ellipse $4x^2 + 9y^2 = 36$.

2. Here are several designs used in making quilts.

Research   Choose a conic section and research to find examples of its use in design.

Class Act

3. Here is an example of a "mat" that will transform into a box with a top when folded. How many ways can you arrange six squares to do that?

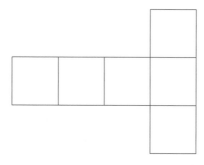

CHAPTER LOOK-BACK

All the figures tessellate except for the pentagon. There is a very good reason why a regular pentagon cannot tessellate, and the other regular figures we have talked about can.

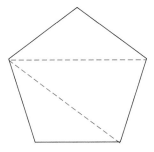

- Total degrees: $3 \times 180° = 540°$
- Degrees in each angle of a regular pentagon: $540 \div 5 = 108°$

No combination of these angles will total 360°:

$$3 \cdot 108° = 324° \quad \text{Too small}$$
$$4 \cdot 108° = 432° \quad \text{Too big}$$

So a regular pentagon will not tessellate.

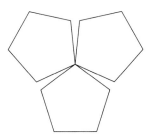

CHAPTER **9**
REVIEW PROBLEMS

The following exercises will give you a good review of the material presented in this chapter.

SECTION 9.1

1. Sketch the graph of $y = -0.5(x + 3)^2 - 1$ by using the graphing form of the equation of a parabola to identify the vertex and the direction in which the graph opens.   vertex: $(-3, -1)$; opens downward

2. Write the equation of the parabola that has vertex $(-2, 3)$ and passes through the point $(0, 0)$. $y = -0.75(x + 2)^2 + 3$

3. Rewrite the parabolic equation $x^2 - 4x + 15 = y$ in graphing form, give the vertex, and tell in which direction the graph opens. $y = (x - 2)^2 + 11$; vertex: $(2, 11)$; opens upward

4. Sketch the graph of $y = -5(x - 4)^2 + 4$ and identify the vertex and two points.   vertex: $(4, 4)$; points: $(3, -1), (5, -1)$

5. Find the vertex and the direction in which a parabola opens if its equation is $y = -6x^2 - 12x + 2$.   vertex: $(-1, 8)$; opens downward

6. Describe the graph of $y = x^2 - 12x + 36$ by giving the vertex, the direction the parabola opens, and the equation written in standard form. vertex: $(6, 0)$; opens upward; $y = (x - 6)^2$

SECTION 9.2

7. Find the center and the radius of the circle that has the equation $x^2 + y^2 - 64 = 0$. Then sketch the circle.   center: $(0, 0)$; radius: 8

8. Write the equation of the circle that has its center at $(0, 0)$ and radius $\sqrt{3}$. $x^2 + y^2 = 3$

9. Graph $\dfrac{x^2}{4} + y^2 = 1$ by writing the equation in graphing form, and then find the center, the length of the major and minor axes, and the x- and y-intercepts of its graph. $\dfrac{x^2}{2^2} + \dfrac{y^2}{1^2} = 1$
   center: $(0, 0)$
   major axis: 4 (horizontal)
   minor axis: 2 (vertical)
   x-intercepts: $(2, 0), (-2, 0)$
   y-intercepts: $(0, 1), (0, -1)$

10. Graph $\dfrac{x^2}{25} + \dfrac{y^2}{4} = 1$ and find the major and minor axes.
    major axis: 10 (horizontal); minor axis: 4 (vertical)

11. Graph: $y^2 - x^2 = 1$ 12. Graph the hyperbola: $\dfrac{x^2}{4} - \dfrac{y^2}{1} = 1$

SECTION 9.3

13. Write the equation of the circle that has its center at $(0, -2)$ and radius 5. $x^2 + (y + 2)^2 = 25$

14. Write the equation of the circle that has its center at $(-3, -5)$ and passes through the point $(0, 0)$. $(x + 3)^2 + (y + 5)^2 = 34$

15. Write the equation of the circle that has its center at $(5, 7)$ and passes through the point $(0, 3)$. $(x - 5)^2 + (y - 7)^2 = 41$

16. Determine the equation of the circle that has its center at $(2, 0)$ and radius 3. $(x - 2)^2 + y^2 = 9$

17. Find the equation of the ellipse for which $a = 16$ and $b = 8$ and that has its center at $(12, 3)$. $\dfrac{(x - 12)^2}{256} + \dfrac{(y - 3)^2}{64} = 1$

18. Write the equation of the ellipse for which $a = 17$ and $b = 5$ and that has its center at $(-15, 6)$. $\dfrac{(x + 15)^2}{289} + \dfrac{(y - 6)^2}{25} = 1$

19. Find a and b for the equation $\dfrac{(x + 18)^2}{16^2} + \dfrac{(y - 22)^2}{169} = 1$, and find the center of its graph. $a = 16, b = 13$; center; $(-18, 22)$

20. Write the equation of the ellipse for which $a = 18$ and $b = 24$ and that has its center at $(16, -13)$. $\dfrac{(x - 16)^2}{324} + \dfrac{(y + 13)^2}{576} = 1$

21. Write the equation $y^2 - x^2 = -25$ in graphing form; then identify the center, a, b, and the vertices of its graph. $\dfrac{x^2}{5^2} - \dfrac{y^2}{5^2} = 1$; center: $(0, 0)$; $a = 5$, $b = 5$; vertices: $(5, 0)$, $(-5, 0)$

22. Write the equation $4x^2 - y^2 = -36$ in graphing form; then identify the center, a, b, and the vertices of its graph. $\dfrac{y^2}{6^2} - \dfrac{x^2}{3^2} = 1$; center $(0, 0)$; $a = 3$, $b = 6$; vertices: $(0, 6)$, $(0, -6)$

23. Write the equation for the hyperbola that has $a = 5$, $b = 1$, a subtracted y^2-term, and its center at $(-20, 30)$. Then find its vertices. $\dfrac{(x + 20)^2}{25} - \dfrac{(y - 30)^2}{1} = 1$; vertices: $(-25, 30)$, $(-15, 30)$

24. Write the equation for the hyperbola that has $a = 30$, $b = 12$, a subtracted x^2-term, and its center at $(-10, 5)$. Then find its vertices.
$\dfrac{(y - 5)^2}{144} - \dfrac{(x + 10)^2}{900} = 1$; vertices: $(-10, -7)$, $(-10, 17)$

SECTION 9.4

In Section 9.4, use $\pi \approx 3.14$.

25. *Bat Box* The widest point of a baseball bat is 2.75 inches in diameter. The bat is shipped in a box that is 4 feet long (outside dimension) and has sides $\frac{1}{2}$ inch thick. What is the outside surface area of the smallest box that can hold this bat?   748.125 square inches

26. *Football Shape* A football drawn from a side view has the shape, approximately, of an ellipse. Its maximum length is 72.4 centimeters. Its maximum width is 54.6 centimeters. Write the equation of an ellipse, centered at the origin, that approximates this shape. $\dfrac{x^2}{(36.2)^2} + \dfrac{y^2}{(27.3)^2} = 1$

27. *Company Logo* The logo of a company consists of an ellipse with three circles inscribed in it (see the accompanying figure). The long axis is $1\frac{3}{8}$ inches, the short axis is $\frac{1}{2}$ inch, and the two identical small circles have a diameter that is $\frac{7}{8}$ that of the large circle. What is the difference between the area of the ellipse and that of the three circles combined?
about 0.04 square inch

28. A circle and an ellipse have equal areas. The long axis of the ellipse is 10 inches. The diameter of the circle is 8 inches. The center of the ellipse is at $(-2, 3)$. Write the equation of the ellipse. $\dfrac{(x + 2)^2}{5^2} + \dfrac{(y - 3)^2}{(3.2)^2} = 1$

29. *Target Design* Four concentric circles are painted alternating red and white from the inside out. The radius of the entire figure is 1 foot. The radius of the inner circle is $\frac{1}{4}$ foot, and each of the three remaining circles is $\frac{1}{4}$ foot wide. What is the difference between the areas of the red and white regions? (See the accompanying figure.) (*Hint:* the area of the entire figure is 3.14 square feet.)   The white area is 0.785 square foot larger than the red area.

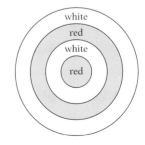

30. A spherical clock with a radius of 7.75 inches is to be packed inside a cubical box. The box is $\frac{1}{2}$ inch wider than the clock $\left(\frac{1}{4}\text{ inch on each side}\right)$. If packing material is to be put inside the box to keep the clock from shifting, what volume of packing material is needed?   about 2147 cubic inches

31. Sketch the graph of a hyperbola and an ellipse in such a way that they share just one solution.

32. Sketch the graph of an ellipse and a straight line in such a way that they share two solutions.

33. Solve this system:　$y = x - 10$
$y = x^2 - 100$　$\{(10, 0), (-9, -19)\}$

34. Solve by addition:　$x^2 - y^2 = 8$
$2x^2 + y^2 = 19$　$\{(3, 1), (3, -1), (-3, 1), (-3, -1)\}$

35. Solve this system by substitution:　$x^2 + y^2 = 17$
$16y = x^2$　　$\{(4, 1), (-4, 1)\}$

36. Solve this system by substitution:　$49y = x^2$
$-x^2 + 50 = y^2$　$\{(7, 1), (-7, 1)\}$

MIXED REVIEW

37. Write in graphing form the equation of the circle that has its center at $(-2, -3)$ and radius $3\sqrt{3}$.　$(x + 2)^2 + (y + 3)^2 = 27$

38. Find the equation of the ellipse whose center is at $(15, -3)$ and for which $a = 12$ and $b = 23$.　$\dfrac{(x - 15)^2}{144} + \dfrac{(y + 3)^2}{529} = 1$

39. Find the center and radius of the circle with equation $(x + 5)^2 + (y - 3)^2 = 48$.　center at $(-5, 3)$; radius: $4\sqrt{3}$

40. Find the vertex of the parabola whose equation is $y = 2x^2 + 5x - 10$, and determine in which direction the parabola opens.　vertex: $\left(-\dfrac{5}{4}, -\dfrac{105}{8}\right)$; opens upward

41. Find the vertex of, and then graph, the parabola whose equation is $y = -0.5(x - 1)^2 + 4$.　vertex: $(1, 4)$

42. Give the center and the length of the major and minor axes of the graph of
$$\dfrac{(x - 15)^2}{400} + \dfrac{(y + 10)^2}{81} = 1$$
center: $(15, -10)$
major axis: 40 (horizontal)
minor axis: 18 (vertical)

43. Find the equation of the circle that has its center at $(1, -1)$ and radius 1.
$(x - 1)^2 + (y + 1)^2 = 1$

44. Write the equation and identify the vertices for the hyperbola with a subtracted y^2-term, its center at $(5, -6)$ and values $a = 2$ and $b = 42$.　$\dfrac{(x - 5)^2}{4} - \dfrac{(y + 6)^2}{1764} = 1$; vertices: $(3, -6), (7, -6)$

45. Find the center of the conic section whose equation is
$$\dfrac{(x + 4)^2}{16} + \dfrac{(y - 33)^2}{121} = 1$$
Also find a and b for the conic section.　center: $(-4, 33)$; $a = 4, b = 11$

46. Write in graphing form the equation for the circle that has its center at the point $(6, -5)$ and passes through the point $(0, 0)$.　$(x - 6)^2 + (y + 5)^2 = 61$

47. Find the vertex of the parabola whose equation is $y = -5x^2 + 10x - 1$, and find the x- and y-intercepts.

47.　vertex: $(1, 4)$
x-intercepts: $\left(\dfrac{5 - 2\sqrt{5}}{5}, 0\right)$, $\left(\dfrac{5 + 2\sqrt{5}}{5}, 0\right)$
y-intercepts: $(0, -1)$

48. Find the center and the length of the major and minor axes for the graph of the equation

$$\frac{(x - 21)^2}{16} + \frac{(y + 10)^2}{144} = 1$$

center: (21, −10)
major axis: 24
minor axis: 8
ellipse

What is this figure?

center: (0, 0)
major axis: 8 (horizontal)
minor axis: 6 (vertical)
x-intercepts: (4, 0), (−4, 0)
y-intercepts: (0, 3), (0, −3)

49. What is the center of $(x - 2)^2 + (y + 2)^2 = 16$? What is the radius of this circle?   center: (2, −2); radius: 4

CHAPTER 9 TEST

This exam tests your knowledge of the material in Chapter 9.

1. a. Find the vertex of the graph of $y = x^2 + 8x$, and determine in which direction the graph opens. Write the equation in graphing form.
(−4, −16); opens upward; $y = [x - (-4)]^2 - 16$

b. Locate the vertex of $y = 3(x - 3)^2 - 1$ and then graph this equation, using a table of values.   vertex: (3, −1)

x	y
2	2
4	2

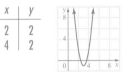

c. Write an equation for the vertical parabola that has its vertex at (−1, −4) and passes through the point (−2, −2). $y = 2(x + 1)^2 - 4$

2. a. Determine the center and radius of the circle with an equation of $x^2 = 400 - y^2$.   center: (0, 0); radius: 20

b. Find the center, major and minor axes, and x- and y-intercepts for the graph of

$$\frac{x^2}{16} + \frac{y^2}{9} = 1$$

center: (0, 0)
major axis: 8 (horizontal)
minor axis: 6 (vertical)
x-intercepts: (4, 0), (−4, 0)
y-intercepts: (0, 3), (0, −3)

c. Find the vertices of, and then graph, the hyperbola defined by $\frac{y^2}{1} - \frac{x^2}{9} = 1$.   vertices: (0, 1), (0, −1)

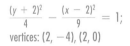

3. a. Write the equation of an ellipse that has a horizontal axis of length 5, a vertical axis of length 1, and its center at the point (−5, −2). $\frac{(x + 5)^2}{6.25} + \frac{(y + 2)^2}{0.25} = 1$

b. Write the equation and identify the vertices for the hyperbola that has its center at (2, −2), values of $a = 3$ and $b = 2$, and y-intercepts. $\frac{(y + 2)^2}{4} - \frac{(x - 2)^2}{9} = 1$; vertices: (2, −4), (2, 0)

c. Write the equation of the circle that has its center at (2, −3) and radius $3\sqrt{2}$. $(x - 2)^2 + (y + 3)^2 = 18$

4. a. An ice hockey rink has a shape that approximates a rectangle with a semicircle (half-circle) on either end. The length is 200 feet at the longest point, and the width is 100 feet. Find the area of the rink. Use $\pi \approx 3.14$.   17,850 square feet

b. A penny has a diameter of 0.748 inches. Two rolls of pennies are put side-by-side into a very small rectangular box that just holds the rolls. What is the interior volume of the box if the rolls are $2\frac{1}{2}$ inches long?
≈ 2.8 cubic inches (= 2.79752 cubic inches)

c. An ellipse has a length of 20 inches and an area of 200 square inches. What is an equation of the ellipse centered at $(6, -5)$? 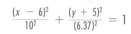 $\dfrac{(x-6)^2}{10^2} + \dfrac{(y+5)^2}{(6.37)^2} = 1$

5. a. Suppose a nonlinear, two-variable system consists of the equations of a circle and an ellipse. Sketch possible graphs of these equations in which the system has four solutions.

b. Find the solution by substitution: $y = x - 8$
$$y + 64 = x^2 \quad \{(8, 0), (-7, -15)\}$$

c. Find the solution by addition: $x^2 + y^2 = 2$
$$x = y^2 \qquad \{(1, 1), (1, -1)\}$$

CUMULATIVE REVIEW

CHAPTERS 1–9

The following exercises will help you maintain the skills you have learned in this and previous chapters.

1. Arrange $|-2.3|, -|-2.3|$, and $[-(-2.8)]$ in order from least to greatest. $\quad -|-2.3| < |-2.3| < [-(-2.8)]$

2. Add: $-3\frac{7}{8} + 5\frac{1}{2} \quad 1\frac{5}{8}$

3. Graph: $3x - 5y < 9$
$\qquad 2x - 4y > 6$

4. Simplify: $\dfrac{-10\frac{1}{2}}{2\frac{1}{3}} \quad -4\frac{1}{2}$

5. Solve: $3x + 4y = 9$
$\qquad 2x + 5y = 12 \quad x = -\frac{3}{7}, y = \frac{18}{7}$

6. Fill in the boxes with numbers that make the statement true. What property is illustrated?
$$(3x + ■)17 = ■ \cdot 17 + 28x \cdot ■$$
$28x; 3x; 17$ distributive property

7. Write $-\left(\dfrac{-7}{9}\right)^{-2}$ without an exponent. $\quad -\frac{81}{49}$

8. Simplify: $5^2 - \frac{2}{4} + 17 \cdot 65^0 \quad 41\frac{1}{2}$

9. Solve: $-3x \le 24 \quad x \ge -8$

10. Define two sets, sets A and B, in such a way that $A \cup B = \{1, 5, 7\} \quad A = \{1, 7\}; B = \{5, 7\}$ Answers may vary.

11. If the circumference of a circle becomes 5 times its original measure, how much larger is the new radius? $\quad$ 5 times larger

12. Divide $x^5 - 1$ by $x - 1$. $\quad x^4 + x^3 + x^2 + x + 1$

13. Evaluate $3x^2 - 5y^2 + 2xy$ when $x = -1$ and $y = -2 \quad -13$

14. Simplify: $8y - 3(x + y) + 4(y - x) \quad -7x + 9y$

15. Solve: $2\sqrt{x} + 3 = 10 \quad x = \frac{49}{4}$

16. Solve: $\dfrac{3x}{2} = \dfrac{15}{18} \quad x = \frac{5}{9}$

17. Solve: $|m - 5| = 10 \quad m = 15$ or $m = -5$

18. Solve for t: $D = rt + h \quad \dfrac{D - h}{r} = t$

19. Solve: $\dfrac{13 + x}{20 + x} = \dfrac{2}{3} \quad x = 1$

20. Solve: $-3y + 2x = 15 \quad \left\{\left(\frac{3}{7}, -\frac{33}{7}\right)\right\}$
$\qquad y = 3x - 6$

EXPONENTIAL AND LOGARITHMIC FUNCTIONS

Logarithms can be used for graphing and describing the growth of bacterial populations. For example, if a species of bacteria has a generation time of 30 minutes, a single bacteria could result in 2^2 bacteria at the end of 60 minutes, and 2^4, or 16, bacteria at the end of two hours. The overall pattern of population growth includes an initial lag time and then a rapid growth period. After the exponential phase, a plateau is reached, followed by a population decline. What do you think might cause a population of bacteria to decline?

■ *What other growth processes (in biological or any other context) might be described as exponential or logarithmic?*

SKILLS CHECK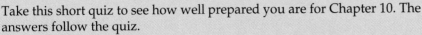

Take this short quiz to see how well prepared you are for Chapter 10. The answers follow the quiz.

1. Solve for y: $x = \dfrac{3 - 2y}{4}$

2. Evaluate $y = 2^{x+3}$ when $x = -1$, when $x = 0$, and when $x = 2$. Give your results as ordered pairs (x, y).

3. Factor 240 into primes.

4. Simplify: $(2^2 x^2 y^4)^{-\frac{1}{2}}$

5. Simplify: $(2x^2 y)^2 (2x^3 y^2 z)^3$

6. Solve: $(x + 3)^2 = -9$

7. Explain why $\{(0, 3), (3, 3)\}$ is or is not a function.

ANSWERS: **1.** $y = \dfrac{3 - 4x}{2}$ [Section 2.3] **2.** $(-1, 4), (0, 8), (2, 32)$ [Section 2.3]
3. $2^4 \cdot 3 \cdot 5$ [Section 1.1] **4.** $\dfrac{1}{2xy^2}$ [Section 7.1] **5.** $32^{13} y^8 z^3$ [Section 5.1] **6.** $x = -3 + 3i$ and $x = -3 - 3i$ [Section 8.2] **7.** This is a function because each first element has just one second element. [Section 3.5]

CHAPTER LEAD-IN

Scientists in certain fields work with very small or very large numbers if their subjects are very small, very large, or very old. Virologists measure viruses in millimicrons—0.000001 mm. Astronomers measure the vast distances of space using light years—the 9,460,000,000,000 kilometers that light travels in one year. And geologists estimate Earth's age at 4,600,000,000 years. The use of exponents and polynomials makes calculations and measurements easier to work with such numbers. Using exponents, the above numbers can also be written, respectively, as 10^{-6}, 9.46×10^{12}, and 4.6×10^9.

Name quantities that would be easier to write using exponents.

10.1 One-to-One and Inverse Functions

SECTION LEAD-IN

A particular bacteria will double in number every 10 hours. How long would it take (in theory) for 1 bacterium to become 1,000,000?

A second organism numbers 738,292. Its doubling time is $1\frac{1}{2}$ hours. How long ago was there just one of these?

Introduction

Recall from Chapter 3 that a function is a rule for assigning exactly one element from a set Y to each element of a set X. The set X is the domain of the function, and the set Y is its range.

There are several ways we can write functions, some of which you have already seen. We can actually list the pairs (x, y) of the function, as in

$$f = \{(1, 2), (2, 3), (4, 0)\}$$

Or we can use functional notation to give a rule for pairing elements, as in

$$f = \{(x, f(x)) \mid x \text{ is real and } f(x) = 2x + 2\}$$

We usually write this last equation more simply as

$$f(x) = 2x + 2 \quad \text{or} \quad y = 2x + 2$$

where it is understood that the function pairs are $(x, f(x))$ or (x, y). Finally, we can draw a diagram or a graph to specify which range element is paired with which domain element.

One-to-One Functions

A set of ordered pairs (x, y) is a function if no domain element x is assigned more than one range element y. The function is a **one-to-one function** if, furthermore, no range element y is assigned to more than one domain element x. More precisely,

---

A function f is one-to-one if

$$f(a) = f(b) \text{ only when } a = b$$

One way to determine whether a function is one-to-one is to examine its ordered pairs.

▪▪▪

EXAMPLE 1

Determine whether each of the following sets of pairs is a one-to-one function.

a. $\{(2, 4), (2, 5), (3, 2)\}$ **b.** $\{(3, 4), (2, 4), (1, 5), (4, 6)\}$ **c.** $\{(3, 5), (4, 7), (6, 8), (2, 9)\}$

SOLUTION

a. This is not a function because the first element 2 is paired with two different second elements (4 and 5).

b. This is a function, but it is not one-to-one because the second element 4 is paired with two different first elements (3 and 2).

c. This is a one-to-one function because it fulfills both requirements.

▶ CHECK **Warm-Up 1**

It is not so easy to determine whether a function is one-to-one if it is described by an equation. However, you can graph the equation and then apply two tests:

Vertical-Line Test

An equation specifies a function if every vertical line intersects its graph at only one point.

Horizontal-Line Test

A function is one-to-one if every horizontal line intersects its graph at only one point.

As an example, the graph in the following figure is the graph of a certain function $y = f(x)$. We can verify that it is a function by noting that any vertical line crosses the graph at only one point. The function is one-to-one because every horizontal line we can draw crosses the graph at only one point.

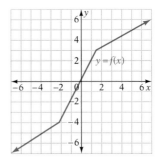

■ ■ ■

EXAMPLE 2

Graph the following functions to determine whether they are one-to-one functions.

a. $y = 2x - 2$ b. $y = \frac{1}{2}x^2 + 1$

SOLUTION

We assume that these equations specify functions because they are referred to as functions.

a. This linear function is graphed in the following figure. Obviously, any horizontal line will intersect the line in only one point, so the function is one-to-one. Actually, *every* linear function is one-to-one.

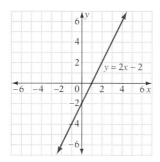

b. This function has, as its graph, a parabola that opens upward and is centered at (0, 1); see the figure below. Any horizontal line more than 1 unit above the *x*-axis intersects the graph at two points, so the function is not one-to-one.

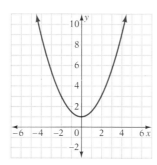

▶ *CHECK* **Warm-Up 2**

Inverse Functions

Every one-to-one function has an *inverse function.*

> **Inverse Function**
>
> For every one-to-one function f, there is an **inverse function** f^{-1} that is obtained by interchanging the first and second elements of the ordered pairs of f.

As an example, let

$$f = \{(1, 3), (2, -3), (-3, -1), (-2, 7)\}$$

Because f is one-to-one, it has an inverse function

$$f^{-1} = \{(3, 1), (-3, 2), (-1, -3), (7, -2)\}$$

Remember that the superscript of -1 on f means "inverse of f," not $\frac{1}{f}$. A function that is not one-to-one may have an inverse, but that inverse is not a function.

STUDY HINT

To find the inverse of a function

1. *Write it in the form* $y =$ *function.*
2. *Exchange x and y in the equation.*
3. *Solve for y.*
4. *Replace y with* f^{-1}.

Always check to see that the inverse of a function is a function.

WRITER'S BLOCK

The term *one-to-one* is used in describing some functions. Explain what it means in your own words.

To find the inverse of a function that is specified by an equation $y = f(x)$, just interchange x and y. Thus if f is given by $y = 3x - 2$, then f^{-1} is given by

$$x = 3y - 2$$

Functions (including inverse functions) are usually solved for y in terms of x. If we solve this last equation for y, we get

$$y = \frac{x + 2}{3}$$

Thus f^{-1} is also given by $y = \frac{x + 2}{3}$. Or we can say more simply that

$$f^{-1}(x) = \frac{x + 2}{3}$$

ERROR ALERT

Identify the error and give a correct answer.

Find $f^{-1}(x)$ when $f(x) = 2 - x$.

Incorrect Solution:

$f^{-1}(x) = \dfrac{1}{2 - x}$

EXAMPLE 3

Let f be given by $f(x) = 5 - 2x$. Find $f^{-1}(x)$.

SOLUTION

We first have to write the function equation as $y = 5 - 2x$.

Now we interchange x and y.

$$x = 5 - 2y$$

Solving for y yields

$$
\begin{aligned}
x - 5 &= -2y & &\text{Adding } -5 \\
\frac{-(x - 5)}{2} &= y & &\text{Multiplying by } \frac{-1}{2} \\
\frac{5 - x}{2} &= y & &\text{Simplifying}
\end{aligned}
$$

WRITER'S BLOCK

Does the inverse of a function always have an inverse? Explain.

So f^{-1} is given by $y = \frac{5 - x}{2}$, which we show by writing

$$f^{-1}(x) = \frac{5 - x}{2}$$

▶ CHECK **Warm-Up 3**

Composition of a Function and Its Inverse Function

If a function f and its inverse function f^{-1} are given in functional notation, then for all x in the domain of f,

$$(f^{-1} \circ f)(x) = x \quad \text{and} \quad (f \circ f^{-1})(x) = x$$

This **composition property** can be used to check that two functions are indeed inverses of each other.

EXAMPLE 4

Use the composition property to check the result of Example 3.

SOLUTION

In Example 3 we have

$$f(x) = 5 - 2x \quad \text{and} \quad f^{-1}(x) = \frac{5-x}{2}$$

Their compositions are

$$(f^{-1} \circ f)(x) = f^{-1}(f(x)) = f^{-1}(5 - 2x)$$

$$= \frac{5 - (5 - 2x)}{2}$$

$$= \frac{2x}{2} = x$$

$$(f \circ f^{-1})(x) = f(f^{-1}(x)) = f\left[\frac{5-x}{2}\right]$$

$$= 5 - 2\left(\frac{5-x}{2}\right)$$

$$= 5 - 5 + x = x$$

So f and f^{-1} are inverse functions.

> ▶ CHECK **Warm-Up 4**

> ■■■
>
> **WRITER'S BLOCK**
>
> Is the inverse of a function always a function? Explain.

Graphs of Inverse Functions

If an ordered pair (a, b) is part of a function f, then the ordered pair (b, a) is part of the inverse function f^{-1}. This means that (a, b) is a point on the graph of f and that (b, a) is a point on the graph of f^{-1}.

■ ■ ■

EXAMPLE 5

Graph the function f given by $f(x) = 1 + 2x$, and then graph its inverse.

SOLUTION

Because the function is linear, it is one-to-one and its inverse function exists. The inverse is

$$f^{-1}(x) = \frac{x}{2} - \frac{1}{2}$$

We use the x values 1, 0, and -1 to make a table of values for the function $y = 1 + 2x$. Then we simply reverse the pairs to find points for the inverse function $y = \frac{x}{2} - \frac{1}{2}$.

$f: y = 1 + 2x$

x	y
-1	-1
0	1
1	3

Reversed pairs

$f^{-1}: y = \frac{x}{2} - \frac{1}{2}$

x	y
-1	-1
1	0
3	1

The two functions are graphed in the following figure.

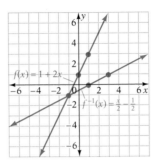

▶ *CHECK* **Warm-Up 5**

The next figure shows the graphs of f and f^{-1} from the previous figure, with the line $y = x$ added as a dashed line. For every point on the graph of f, there is a "mirror image" point on the graph of f^{-1}. Each point on one graph is reflected through the line $y = x$ to the other graph. Why? Because the domain of the function is the range of its inverse, and the range of the function is the domain of its inverse. This is a property of all graphs of inverses. We can use this property to graph inverse functions without even making a table of values.

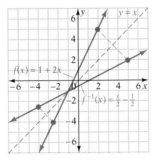

▪ ▪ ▪

EXAMPLE 6

Two functions are graphed in figure (a) and figure (b). Sketch the graphs of their inverses.

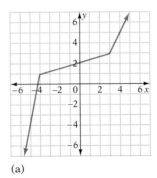

(a)

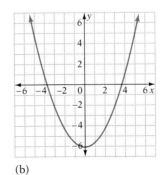

(b)

SOLUTION

a. We first graph the line $y = x$ [figure (a)]. We then reflect several points of the original graph through $y = x$. Each reflected (inverse) point is as distant from $y = x$ as the original point from which it is reflected; and each dashed line in figure (b) that connects a reflected point with an original point is perpendicular to $y = x$.

We graph enough new points to see the shape of the inverse graph. We then sketch the inverse graph through those points, as in figure (c).

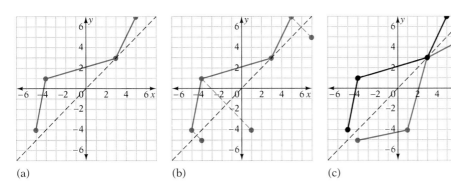

(a)   (b)   (c)

b. Again, we use the graph of the line $y = x$, and we identify a few points and sketch the graph of the inverse of this quadratic function, a parabola.

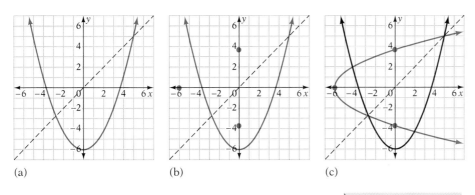

(a)   (b)   (c)

▶ *CHECK* **Warm-Up 6**

Calculator Corner

Most graphing calculators have a **DRAW INVERSE** feature that enables you to draw the inverse of a function. Care must be taken, however, when using the **DRAW INVERSE** feature of many graphing calculators. Try the following example. The screens shown are for the *TI-82/83* graphing calculator. Consult your calculator's manual to see how to graph the inverse of a function on your particular model.

Graph the function $y = x^3 + 2x - 1$ and the line $y = x$ on the same screen. (It is easier to see the inverse of a function if you choose dimensions for your viewing rectangle that are fairly *square*.) Use the horizontal-line test to check if the cubic function has an inverse. If it does, **DRAW** the **INVERSE** on the same viewing rectangle.

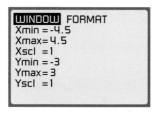
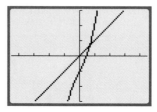

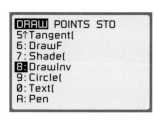
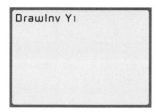
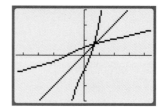

You can see that to draw the inverse of the original function you reflect the original function about the line $y = x$. The two graphs are *mirror images* of each other. Furthermore, by investigating the graphs of the two functions, you can see that the point (1, 2) is on the original graph $y = x^3 + 2x - 1$, and the point (2, 1) appears to be on the inverse graph. (Unfortunately, on most graphing calculators you cannot **TRACE** on the inverse function when using **DRAW INVERSE**.) In general, if the point (a, b) is a point in the function f, then the point (b, a) is a point in the inverse of f denoted by f^{-1}.

Now try some examples on your own. First use the paper-and-pencil technique illustrated in the previous examples. Then use the **DRAW INVERSE** utility of your graphing calculator to draw the function's inverse. Check yourself by finding some points (a, b) on the original function and see if the points (b, a) are on the graph of the inverse.

a. $y = \dfrac{3}{x - 3}$ **b.** $y = \dfrac{2}{x}$

c. $y = \sqrt{2x + 5}$ **d.** $y = x^3$

Again, we want to point out how a function and its inverse are related:

> The domain of a function is the range of its inverse.

and

> The range of a function is the domain of its inverse.

Practice what you learned.

SECTION FOLLOW-UP

To find the number of bacteria, we start by using 2^n, where n changes from 0 to any positive integer.

$$1 = 2^0$$
$$2 \times 1 = 2^1 \cdot 2^0 = 2^1 = 2$$
$$2 \times 2 \times 1 = 2^2 \cdot 2^0 = 2^2 = 4$$

It turns out that

$$2^{19} = 524{,}288$$

and

$$2^{20} = 1{,}048{,}576$$

It also turns out that the time needed to go from 1 bacterium to 2 bacteria is 10 hours. And it takes 10 hours to go from 2 bacteria to 4.

$$2^0 \longrightarrow 2^1 = 10 \text{ hours}$$
$$2^0 \longrightarrow 2^2 = 2 \times 10 = 20 \text{ hours}$$
$$2^0 \longrightarrow 2^3 = 3 \times 10 = 30 \text{ hours}$$

So from

$$2^0 \longrightarrow 2^{19} = \quad 19 \quad \times \quad \underline{10} \quad = 190 \text{ hours}$$
$$2^0 \longrightarrow 2^{20} = \underline{\ 20\ } \times \quad 10 \quad = \underline{\ 200\ } \text{ hours}$$

So it would take from 190 to 200 hours. How many days is this? minutes? seconds? about 8 days   11,400–12,000 minutes   684,000–720,000 seconds

The organism that numbers 738,292 started doubling about 30 hours ago.

$$2^{19} = 524{,}288$$
$$2^{20} = 1{,}048{,}576$$

So this organism had doubled about 20 times: $20 \times 1.5 = 30$ hours.

What assumptions do we have to make to solve this problem?

10.1 WARM-UPS

Work these problems before you attempt the exercises.

1. Determine whether $f(x) = \{(2, 3), (3, 5), (6, 7)\}$ is a one-to-one function. one-to-one

2. Graph $f(x) = x^2$ to determine whether it is a one-to-one function.   not one-to-one

3. Let $f(x) = 2x - 7$. Find $f^{-1}(x)$. $\frac{x + 7}{2}$

4. Verify that $f(x) = 2 - x$ and $f^{-1}(x) = 2 - x$ are inverses by computing $f \circ f^{-1}(x)$ and $f^{-1} \circ f(x)$. $f \circ f^{-1}(x) = 2 - (2 - x) = 2 - 2 + x = x, f^{-1} \circ f(x) = 2 - (2 - x) = 2 - 2 + x = x$

5. Graph the function $f(x) = 2x - 3$ and then find and graph its inverse. $f^{-1}(x) = \frac{x + 3}{2}$

6. Graph the inverse of the function $f(x) = -2x + 3$ without a table of values.

10.1 EXERCISES

Note: Use your graphing calculator to check your results whenever possible.

In Exercises 1 through 8, determine whether those that are functions are one-to-one.

1. $\{(1, 2), (1, 5), (2, 3), (4, 8)\}$
   not a function

2. $\{(6, 1), (8, 2), (9, 3), (5, 4), (3, 1)\}$
   function; not one-to-one

3. $\{(2, 6), (6, 2), (5, 3), (3, 5)\}$
   one-to-one function

4. $\{(1, 1), (2, 2), (3, 3), (4, 4)\}$
   one-to-one function

5. $\{(3, 1), (4, 1), (5, 1), (6, 1)\}$
   function; not one-to-one

6. $\{(4, 9), (2, 8), (1, 1), (0, 0)\}$
   one-to-one function

7. $\{(3, 18), (18, 3), (39, 93), (39, 93)\}$
   one-to-one function

8. $\{(6, 25), (8, 20), (5, 22), (1, 20)\}$
   function; not one-to-one

In Exercises 9 through 16, graph each equation and determine whether each specifies a function. If it does, determine whether the function is one-to-one.

9. $x = y^2$
   not a function

10. $y = x - 2$
    one-to-one function

11. $y = |x|$
    function; not one-to-one

12. $y = x^2 - 5$
    function; not one-to-one

13. $x = \frac{y}{2}$
    one-to-one function

14. $x = y^2 - 1$
    not a function

15. $y = \frac{x - 2}{2}$
    one-to-one function

16. $y = \sqrt{x}$
    one-to-one function

In Exercises 17 through 24, find the inverse of the given function.

17. $f(x) = x - 3$
    $x + 3$

18. $f(x) = 2 - 2x$
    $1 - \frac{x}{2}$

19. $f(x) = \frac{2x + 1}{2}$
    $\frac{2x - 1}{2}$

20. $f(x) = x^3$
    $\sqrt[3]{x}$

21. $f(x) = 3x + 1$
$\frac{x - 1}{3}$

22. $f(x) = 2x - 2$
$\frac{x + 2}{2}$

23. $f(x) = 10x - 1$
$\frac{x + 1}{10}$

24. $f(x) = \frac{x - 3}{5}$
$5x + 3$

In Exercises 25 through 32, verify the inverses that you found in Exercises 17 through 24 by finding $(f \circ f^{-1})(x)$ and $(f^{-1} \circ f)(x)$ for the given functions.

25. $f(x) = x - 3$ $(x + 3) - 3 = x; (x - 3) + 3 = x$

26. $f(x) = 2 - 2x$ $2 - 2\left(1 - \frac{x}{2}\right) = x; 1 - \frac{(2 - 2x)}{2} = x$

27. $f(x) = \frac{2x + 1}{2}$ $\dfrac{2\left(\frac{2x - 1}{2}\right) + 1}{2} = x;$ $\dfrac{2\left(\frac{2x + 1}{2}\right) - 1}{2} = x$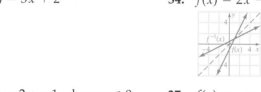

28. $f(x) = x^3$ $(\sqrt[3]{x})^3 = x; \sqrt[3]{x^3} = x$

29. $f(x) = 3x + 1$ $3\left(\frac{x - 1}{3}\right) + 1 = x; \frac{(3x + 1) - 1}{3} = x$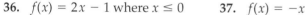

30. $f(x) = 2x - 2$ $2\left(\frac{x + 2}{2}\right) - 2 = x; \frac{(2x - 2) + 2}{2} = x$

31. $f(x) = 10x - 1$ $10\left(\frac{x + 1}{10}\right) - 1 = x; \frac{(10x - 1) + 1}{10} = x$

32. $f(x) = \frac{x - 3}{5}$ $\frac{(5x + 3) - 3}{5} = x; 5\left(\frac{x - 3}{5}\right) + 3 = x$

In Exercises 33 through 40, graph the given function $f(x)$ or $f^{-1}(x)$ by using a table of values. Then reverse those values and use them to graph the inverse. Label $f(x)$ and $f^{-1}(x)$

33. $f(x) = 3x + 2$

34. $f(x) = 2x - 1$

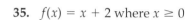

35. $f(x) = x + 2$ where $x \geq 0$

36. $f(x) = 2x - 1$ where $x \leq 0$

37. $f(x) = -x$

38. $f(x) = \dfrac{-1}{x}$

39. $f^{-1}(x) = \frac{1}{x} + 1$ where $0 < x \leq 1$

40. $f^{-1}(x) = 3x - 1$

In Exercises 41 through 48, use a vertical-line test to decide whether each graph is the graph of a function. If it is a function, is it one-to-one? (Use a horizontal-line test.) Assume that each graph is defined only where it is shown.

41.

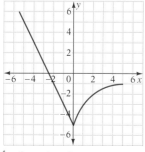

function; not one-to-one

42.

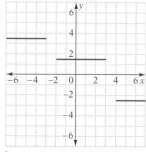

function; not one-to-one

43.

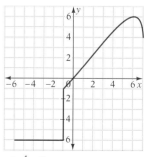

not a function

44.

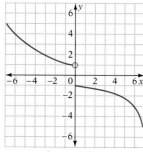

one-to-one function

45.

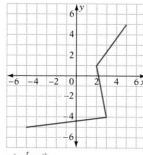

not a function

46.

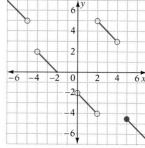

one-to-one function

47.

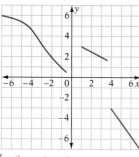

function; not one-to-one

48.

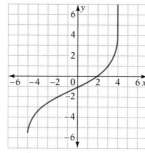

not a function

In Exercises 49 through 54, sketch the inverse by reflecting through the line $y = x$.

49.

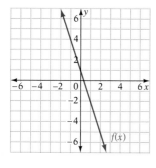

50.

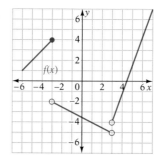

51.

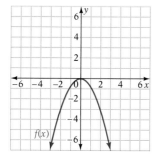

52.

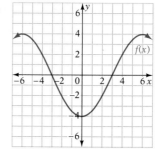

53.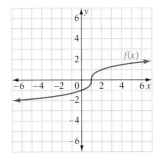

54.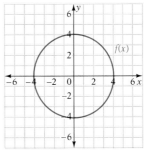

The function and its inverse are identical.

EXCURSIONS

Posing Problems

1. Have an adventure! Write a paragraph about something that takes place in one of these cities. Use some of the data here in your story. How much did your adventure cost?

International Price Comparisons ($)

Item	Madrid	Moscow	Paris	Rome	São Paolo	Tokyo	Toronto	Vienna
Chocolate candy bar (150 g)	1.18	2.18	1.66	1.24	1.21	4.06	1.53	1.66
Carbonated soft drink (6 pack)	4.20	3.73	2.48	2.81	3.39	7.62	2.84	2.85
Bottled mineral water (1 liter)	0.33	0.96	0.35	0.33	0.44	1.73	0.83	0.89
Wine (750 ml)	7.75	20.63	5.00	9.12	19.33	16.20	7.13	9.35
Dry cleaning (man's suit—1 piece)	14.41	27.98	23.69	9.14	19.63	26.87	7.75	15.81
Woman's haircut—wash/dry	34.10	35.15	60.89	43.37	28.94	94.37	33.12	50.31
Toothpaste (100 ml)	2.51	3.87	2.40	2.51	1.35	7.40	1.31	2.52
Deodorant (155 ml)	2.71	7.83	2.49	4.51	4.50	15.35	1.86	4.21
Aspirin (20 units)	8.62	19.53	10.46	14.68	6.13	25.83	3.77	12.04
Blank videotape (1 unit)	4.67	14.21	8.31	4.98	9.45	8.02	5.11	13.32
Camera film (36 exposures)	20.71	26.36	30.77	14.02	33.69	14.76	16.25	17.30
Paperback book (1 unit)	9.96	10.51	12.23	9.60	8.88	15.62	7.39	14.16
Movie ticket (1 unit)	4.99	7.57	8.43	6.20	7.70	23.08	5.93	6.74
Taxi ride (2 km)	2.61	NA	3.98	6.01	0.96	6.02	3.86	4.42
Business lunch (for two)	36.33	72.00	80.79	59.83	75.83	45.45	36.75	91.58
Hotel (daily rate)	185.30	365.00	277.06	167.24	287.91	454.54	166.78	269.18

Note: Prices recorded between October 1994 and February 1995.
Source: Organization Resources Counselors as cited in the *1996 Information Please® Business Almanac* (©1995 Houghton Mifflin Co.), p. 408. All rights reserved. Used with permission by Information Please LLC.

2. Ask and answer four questions using this data.

What They've $hown for the Money

In the heat of summer, NBA teams spent more than $1.5 billion to sign the picks of the litter of the free-agent class of '96. *USA TODAY* writer Greg Boeck takes a midwinter's look at who's delivering and who's not.

Underpaid

Player, team	Years	Contract	Comment
Michael Jordan, Bulls	1	$30.1 million	Best bargain in hoops
Tim Hardaway, Heat	4	$13.0 million	On a mission
Buck Williams, Knicks	3	$4.3 million	Big bank for Buck
Mark Davis, 76ers	1	$247,500	Nice pickup
Derek Strong, Magic	1	$247,500	Big year off bench
Dominique Wilkins, Spurs	1	$247,500	Playing like 27 instead of 37
Gerald Wilkins, Magic	1	$247,500	Like brother, like brother
Walt Williams, Raptors	1	$247,500	Eye on '97-'98
Rex Chapman, Suns	1	$247,500	Shooting for makeup contract

Earning their keep

Player, team	Years	Contract	Comment
Shaquille O'Neal, Lakers	7	$121.0 million	MVP candidate
Alonzo Mourning, Heat	7	$105.0 million	Hardest-working center in NBA
Gary Payton, Sonics	7	$88.0 million	Rolls-Royce of point guards
Dikembe Mutombo, Hawks	5	$57.0 million	His offense is a bonus
Kenny Anderson, Blazers	7	$45.0 million	Loves West Coast
Steve Smith, Hawks	7	$45.0 million	Triple-double threat
Antonio Davis, Pacers	7	$38.0 million	Steady season
P. J. Brown, Heat	7	$36.0 million	Unsung defender
Reggie Miller, Pacers	4	$36.0 million	Good but not All-Star season
Latrell Sprewell, Warriors	4	$28.0 million	Great scorer, lousy team
Chris Childs, Knicks	6	$24.0 million	Making his point
Chris Gatling, Mavericks	5	$22.0 million	Deserves more minutes
Bryant Stith, Nuggets	5	$22.0 million	Typical season
Robert Pack, Nets	5	$18.0 million	Nets miss him
John Stockton, Jazz	3	$15.0 million	Money in the bank
Don MacLean, 76ers	4	$12.0 million	Perked to life lately
Dan Majerle, Heat	3	$8.5 million	Back derails comeback
Kevin Willis, Rockets	4	$8.0 million	Nice fit
Vernon Maxwell, Spurs	1	$1.0 million	Quietly doing his job

Overpaid

Player, team	Years	Contract	Comment
Juwan Howard, Bullets	7	$105.0 million	Off All-Star form
Allan Houston, Knicks	7	$56.0 million	Team's second-best shooter
Horace Grant, Magic	5	$50.0 million	Injury-prone
Elden Campbell, Lakers	7	$49.0 million	Cruising, not bruising
Dale Davis, Pacers	7	$42.0 million	Subpar season
Jim McIlvaine, Sonics	7	$35.0 million	One-dimensional
Tracy Murray, Bullets	7	$19.0 million	Inconsistent
Hersey Hawkins, Sonics	5	$18.0 million	Off season
Brent Price, Rockets	7	$17.0 million	Price isn't right
Craig Ehlo, Sonics	3	$13.0 million	Booed at Key Arena
Lucious Harris, 76ers	7	$13.0 million	Hasn't delivered
Mark Price, Warriors	3	$12.0 million	Like brother, like brother
Dennis Rodman, Bulls	1	$9.0 million	Act has staled
Oliver Miller, Mavericks	1	$247,500	Overweight, too

Source: USA Today, 7 February 1997.

3. Ask and answer four questions that use this data.

Fiscal Year 1995 Federal Tax Burden per Capita and
Fiscal Year 1994 Federal Expenditures per Dollar of Taxes

	1995 Federal tax burden per capita	Rank	1994 Federal expenditures per dollar of taxes[1]	Rank
United States	**$4,996**	—	**$1.00**	—
Alabama	3,922	41	1.37	6
Alaska	5,797	6	1.32	11
Arizona	3,981	40	1.17	19
Arkansas	3,751	46	1.26	14
California	5,130	17	0.97	34
Colorado	5,182	15	1.01	30
Connecticut	7,769	1	0.66	50
Delaware	5,969	5	0.71	48
Florida	4,974	20	1.03	28
Georgia	4,414	31	1.04	27
Hawaii	5,370	12	1.21	18
Idaho	3,986	39	1.11	22
Illinois	5,739	8	0.75	47
Indiana	4,518	28	0.87	40
Iowa	4,287	33	1.09	23
Kansas	4,773	24	1.04	25
Kentucky	3,836	44	1.21	17
Louisiana	3,848	43	1.33	10
Maine	4,094	36	1.34	8
Maryland	5,777	7	1.28	13
Massachusetts	6,113	4	0.97	33
Michigan	5,072	19	0.82	41
Minnesota	5,220	14	0.80	44
Mississippi	3,170	50	1.69	2
Missouri	4,585	25	1.33	9
Montana	4,060	37	1.36	7
Nebraska	4,583	27	1.02	29
Nevada	5,401	10	0.77	45
New Hampshire	5,441	9	0.76	46
New Jersey	6,889	2	0.70	49
New Mexico	3,654	47	1.88	1
New York	6,185	3	0.82	43
North Carolina	4,220	34	0.98	32
North Dakota	4,026	38	1.55	3
Ohio	4,836	22	0.91	39
Oklahoma	3,908	42	1.26	15
Oregon	4,585	26	0.93	38
Pennsylvania	5,173	16	1.00	31
Rhode Island	5,279	13	1.05	24
South Carolina	3,788	45	1.25	16
South Dakota	4,181	35	1.29	12
Tennessee	4,391	32	1.12	21
Texas	4,501	29	0.97	35
Utah	3,574	48	1.12	20

Vermont	4,493	30	0.94	36
Virginia	5,090	18	1.39	5
Washington	5,374	11	0.93	37
West Virginia	3,547	49	1.51	4
Wisconsin	4,783	23	0.82	42
Wyoming	4,844	21	1.04	26
Washington, D.C.	7,396	—	5.27	—

1. When calculating this ratio, expenditures by state were adjusted downward to account for deficit spending. As might be expected, states with either low federal tax burdens and/or high federal expenditures tend to benefit most, on net from federal fiscal operations. The most obvious example of this is New Mexico, the state with the highest federal expenditure/tax ratio of 1.88 (or $1.88 federal spending for every $1 of federal taxes.) *Source:* Tax Foundation; Census Bureau, as cited in the *1996 Information Please® Almanac* (©1995 Houghton Mifflin Co.), p. 45. All rights reserved. Used with permission by Information Please LLC.

4. Study the following data and then answer the questions.

The Public Debt

Year	Gross debt Amount (in millions)	Per capita	Year	Gross debt Amount (in millions)	Per capita
1800 (Jan. 1)	$ 83	$ 15.87	1950	$ 256,087[1]	$ 1,688.30
1860 (June 30)	65	2.06	1955	272,807[1]	1,650.63
1865	2,678	75.01	1960	284,093[1]	1,572.31
1900	1,263	16.60	1965	313,819[1]	1,612.70
1920	24,299	228.23	1970	370,094[1]	1,807.09
1925	20,516	177.12	1975	533,189	2,496.90
1930	16,185	131.51	1980	907,701	3,969.55
1935	28,701	225.55	1985	1,823,103	7,598.51
1940	42,968	325.23	1990	3,233,313	12,823.28
1945	258,682	1,848.60	1994	4,643,711	17,805.64

1. Adjusted to exclude issues to the International Monetary Fund and other international lending institutions to conform to the budget presentation. *Source:* Department of the Treasury, Financial Management Service, as cited in the *1996 Information Please® Almanac* (©1995 Houghton Mifflin Co.), p. 55. All rights reserved. Used with permission by Information Please LLC.

a. How much do you think we will owe per person in 2000?

b. Ask and answer four questions using this data.

Data Analysis

5. Study the following data and then answer the questions.

The Six Best-Paid NFL Players at Each Position, 1996 (salaries in millions)

Offense		Defense	
Name, team	**Salary**	**Name, team**	**Salary**
Quarterback		**Cornerback**	
Troy Aikman, Dallas	$5.371	Troy Vincent, Phila.	$4.6
Dan Marino, Mia.	$5.327	Aeneas Williams, Ariz	$3.4
Steve Young, S.F.	$4.975	Cris Dishman, Hou.	$3.06
Scott Mitchell, Det.	$4.317	Rod Woodson, Pitts.	$3
John Elway, Den.	$4.266	Eric Davis, Carol.	$2.75
Brett Favre, G.B.	$4.175	Eric Allen, N.O.	$2.7
Running Back		**Safety**	
Barry Sanders, Det.	$4.347	Eric Turner, Balt.	$3.317
Emmitt Smith, Dallas	$4.001	Carnell Lake, Pitts.	$2.4
Chris Warren, Sea.	$2.700	Henry Jones, Buff.	$2.2
Marshall Faulk, Ind.	$2.208	Steve Atwater, Den.	$2.16
Thurman Thomas, Buff.	$2.175	Stanley Richard, Wash.	$2.025
David Meggett, N.E.	$2.1	Tim McDonald, S.F.	$1.972
Receivers		**Defensive End**	
Carl Pickens, Cin.	$3	Reggie White, G.B.	$3.625
Rob Moore, Ariz.	$3	Neil Smith, K.C.	$3.25
Alvin Harper, T.B.	$2.85	Leslie O'Neal, St. Louis	$3.167
Jerry Rice, S.F.	$2.521	Bruce Smith, Buff.	$2.967
Michael Haynes, N.O.	$2.5	Alonzo Spellman, Chi.	$2.9
Cris Carter, Minn.	$2.475	Renaldo Turnbull, N.O.	$2.862
Tight Ends		**Defensive Tackle**	
Jackie Harris, T.B.	$1.625	Eric Swann, Ariz.	$3.5
Ben Coates, N.E.	$1.5	Dana Stubblefield, S.F.	$2.942
Mark Chmura, G.B.	$1.5	John Randle, Minn.	$2.625
Tony McGee, Cin.	$1.5	Cortez Kennedy, Sea.	$2.492
Shannon Sharpe, Den.	$1.45	Michael Dean Perry, Den.	$2.4
Aaron Pierce, N.Y.G.	$1.283	Henry Thomas, Det.	$2.4
Offensive Line		**Linebacker**	
Leon Searcy, Jack.	$3.3	Junior Seau, S.D.	$3.99
Bruce Matthews, Hou.	$3.26	Hardy Nickerson, T.B.	$3.6
Lomas Brown, Ariz.	$3	Quentin Coryatt, Ind.	$3.5
Richmond Webb, Mia.	$2.9	Derrick Thomas, K.C.	$3.1
Randall McDaniel, Minn.	$2.725	Cornelius Bennett, Atl.	$3.05
Bruce Armstrong, N.E.	$2.675	Lamar Lathon, Carol.	$3
Punters/Kickers			
John Carney, S.D.	$1.05		
John Kasay, Carol.	$900,000**		
Chris Mohr, Buff.	$825,000**		
Morten Andersen, Atl.	$800,000**		
Pete Stoyanovich, K.C.	$770,000**		
Matt Stover, Balt.	$739,000**		

*Includes base pay, prorated signing bonus, and other bonuses paid in contract year.
**In thousands.
Source: National Football League.

a. Which team has the most high-priced players?

b. What is the mean price paid to each position (of the best-paid NFL players)?

c. What is the mean price paid by each team for these players?

d. Ask and answer another question.

SECTION GOALS

▪ *To graph exponential functions*

▪ *To translate between exponential and logarithmic notations*

▪ *To graph logarithmic functions*

▪ *To graph natural exponential and logarithmic functions*

10.2 Exponential and Logarithmic Functions

SECTION LEAD-IN

Many biological systems grow at a rate that is exponential. This growth shows a pattern. What other things have a pattern that describes change?

Length of side n	Area n^2	Difference 1	Difference 2
1	1		
2	4	3	2
3	9	5	2
4	16	7	2
5	25	9	2
6	36	11	2
7	49	13	2
8	64	15	

▪ Find some other patterns that involve exponents.

Introduction

In the remainder of this chapter, we discuss two functions that are inverses of each other, *exponential* and *logarithmic functions.*

Exponential Function

An **exponential function** has the form

$$f(x) = b^x$$

where the base b is a positive real number not equal to 1.

Exponential functions describe growth and decay in many real systems, from populations of organisms to radioactive elements to interest payments on savings accounts.

The inverses of exponential functions are called logarithmic functions.

Logarithmic Function

If x and b are positive real numbers and b is not equal to 1, then the function
$$f(x) = \log_b x$$
is called the **logarithmic function** with base b.

Graphing the Functions

Graphs of Exponential Functions

We graph exponential functions the same way we graph any function.

▪▪▪

EXAMPLE 1

Evaluate $f(x) = 2^x$ for integers from -3 to 3, and graph the results.

SOLUTION

We substitute each value for x and evaluate the function. We get the following table of values:

x	$f(x)$
-3	$2^{-3} = \frac{1}{8}$
-2	$2^{-2} = \frac{1}{4}$
-1	$2^{-1} = \frac{1}{2}$
0	$2^0 = 1$
1	$2^1 = 2$
2	$2^2 = 4$
3	$2^3 = 8$

We graph the points $(x, f(x))$ and connect them with a smooth curve to obtain the graph in the following figure.

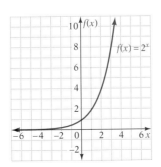

▶ CHECK **Warm-Up 1**

The graph in the figure in Example 1 is a typical exponential graph with $b > 1$. The graph passes both the vertical-line test and the horizontal-line test, so it is the graph of a one-to-one function. The domain of the function is all real numbers.

Also, $f(x) = 2^x$ is never negative; as you can see in the figure in Example 1, it has the x-axis as an asymptote. Its range is all real numbers greater than zero.

Finally, note that the values of $f(x)$ generally increase as x increases.

Calculator Corner

a. Plot the following function for each value of x on a piece of graph paper.

$y = b^x$ for $b = 1, 2, 3$, and 4 (that is, graph $y = 1^x$, $y = 2^x$, and so on)

What conjecture can you make about this family of curves?

Without using your graphing calculator, how do you anticipate the graph of $y = 5^x$ will look? Sketch your answer on a piece of graph paper and support with your graphing calculator.

b. Plot the following function for each value of x on a piece of graph paper.

$y = -b^x$ for $b = 1, 2, 3$, and 4

What conjecture can you make about this family of curves?

Without using your graphing calculator, how do you anticipate the graph of $y = -5^x$ will look? Sketch your answer on a piece of graph paper and support with your graphing calculator.

▪ ▪ ▪

EXAMPLE 2

Graph: $f(x) = 2^{x-2}$

SOLUTION

We substitute some integer values for x and evaluate the function. We get the following table of values.

x	$f(x)$
-1	$2^{-1-2} = 2^{-3} = \frac{1}{8}$
0	$2^{0-2} = 2^{-2} = \frac{1}{4}$
1	$2^{1-2} = 2^{-1} = \frac{1}{2}$
2	$2^{2-2} = 2^{0} = 1$
3	$2^{3-2} = 2^{1} = 2$
4	$2^{4-2} = 2^{2} = 4$
5	$2^{5-2} = 2^{3} = 8$
6	$2^{6-2} = 2^{4} = 16$

We graph the points $(x, f(x))$ and connect them with a smooth curve to obtain the graph shown in the following figure.

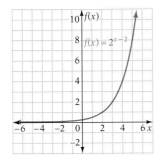

▶ *CHECK* **Warm-Up 2**

Note that the graph in the figure in Example 2 has the same shape as that for $f(x) = 2^x$ in the figure in Example 1. The graph is simply shifted to the right 2 units.

Calculator Corner

Use your graphing calculator to graph $f(x) = 2^{x-4}$, $f(x) = 2^{x-6}$, and $f(x) = 2^{x-8}$. The graph should shift to the right each time.

■ ■ ■

EXAMPLE 3

Graph: $f(x) = \left(\frac{1}{2}\right)^x$

SOLUTION

We substitute integer values from -3 to 3 and evaluate the function. We get the following table of values:

x	$f(x)$
-3	$\left(\frac{1}{2}\right)^{-3} = 8$
-2	$\left(\frac{1}{2}\right)^{-2} = 4$
-1	$\left(\frac{1}{2}\right)^{-1} = 2$
0	$\left(\frac{1}{2}\right)^{0} = 1$
1	$\left(\frac{1}{2}\right)^{1} = \frac{1}{2}$
2	$\left(\frac{1}{2}\right)^{2} = \frac{1}{4}$
3	$\left(\frac{1}{2}\right)^{3} = \frac{1}{8}$

We graph the points $(x, fx))$ and connect them with a smooth curve to obtain the graph shown in the following figure.

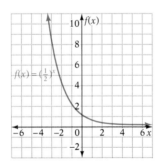

▶ CHECK **Warm-Up 3**

Compare the figure in Example 3 with the figure in Example 1. Note that when $b < 1$, the graph is "turned around"; that is, the values of $f(x)$ generally decrease as x increases.

Calculator Corner

Use your graphing calculator to graph $f(x) = \left(\frac{1}{4}\right)^x$, $f(x) = \left(\frac{1}{5}\right)^x$, and $f(x) = \left(\frac{1}{8}\right)^x$. Describe how the graph changes as the base decreases.

Exponential and Logarithmic Notation

We start with an exponential function f given by

$$f(x) = 4^x$$

We first write the function as

$$y = 4^x$$

Then we interchange x and y, obtaining

$$x = 4^y$$

Now we must solve for y, but we have not yet discussed a way to do that. The best we can do is write, in words,

$$y = \text{the power of 4 that yields } x$$

The inverse function f^{-1} is given by this "equation."

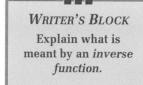

WRITER'S BLOCK

Explain what is meant by an *inverse function*.

Note that f^{-1} is a perfectly good function: The foregoing "equation" is a clearly stated rule for assigning one y-value to each value of x. In fact, f and f^{-1} have been graphed in the following figure; as you can see, they are mirror images about the line $y = x$.

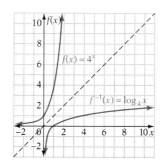

Mathematicians have given the name *logarithmic functions* to functions like f^{-1}, and they have produced a shorthand notation for them. With this shorthand, we write the function equation for f^{-1} as

$$y = \log_4 x$$

We read the right side as "the logarithm of x to the base 4." It means "the exponent on 4 that yields x."

As an example,

$$\log_5 x = 2 \quad \text{means} \quad x = 5^2$$

▪▪▪

EXAMPLE 4

Write (a) $2^3 = 8$ and (b) $x = 10^5$ in logarithm form.

SOLUTION

a. The base is 2 and the exponent is 3, so

$$2^3 = 8 \quad \text{means} \quad \log_2 8 = 3$$

b. The base is 10 and the exponent is 5, so

$$x = 10^5 \quad \text{means} \quad \log_{10} x = 5$$

▶ CHECK **Warm-Up 4**

▪ ▪ ▪

EXAMPLE 5

Write (a) $\log_2 32 = 5$ and (b) $\log_3 x = \frac{2}{5}$ in exponential form.

SOLUTION

a. The base is 2 and the logarithm (exponent) is 5, so

$$\log_2 32 = 5 \quad \text{means} \quad 2^5 = 32$$

b. The base is 3 and the logarithm is $\frac{2}{5}$ so

$$\log_3 x = \frac{2}{5} \quad \text{means} \quad 3^{\frac{2}{5}} = x$$

▶ CHECK **Warm-Up 5**

> ▪▪▪
> **WRITER'S BLOCK**
>
> The log function $y = \log x$ is not defined if x is not positive. Explain why in your own words. (*Hint:* Rewrite the function in exponential form.)

Graphs of Logarithmic Functions

The easiest way to graph a logarithmic function is to translate its equation to exponential form and then graph that form. We do so in the next example.

▪ ▪ ▪

EXAMPLE 6

a. Graph the function f given by $f(x) = \log_2 x$.

b. Find and graph f^{-1}.

SOLUTION

a. Because we want to translate the function equation to exponential form, we first write it as

$$y = \log_2 x$$

Then, with base 2 and logarithm y, this is equivalent to

$$2^y = x$$

We shall graph this function. To do so, we assign values to y and solve for x, obtaining the following pairs:

y	3	2	1	0	-1	-2	-3
x	8	4	2	1	$\frac{1}{2}$	$\frac{1}{4}$	$\frac{1}{8}$

These pairs are graphed and connected with a smooth curve in the following figure. Note that the domain (x-values) is $\{x \mid x$ is real *and* $x > 0\}$, and the range is all real numbers.

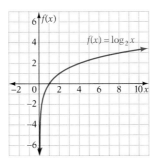

b. The inverse function f^{-1} is found by starting with the equation for f,

$$y = \log_2 x$$

interchanging x and y to get

$$x = \log_2 y$$

and solving for y. We solve for y by translating to exponential form, obtaining

$$y = 2^x$$

Thus f^{-1} is the function given by $y = 2^x$.

To obtain a table of values, we assign values to x and find the corresponding values of y. We get

x	4	3	2	1	0	-1	-2	-3
y	16	8	4	2	1	$\frac{1}{2}$	$\frac{1}{4}$	$\frac{1}{8}$

Then we graph these points and draw a smooth curve through them. The result is shown in the following figure, along with the graph of $y = \log_2 x$.

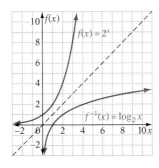

▶ CHECK **Warm-Up 6**

Note that the two curves in the figure in Example 6 are reflections of each other through the line $y = x$. If you also compare the two tables of values, you will find that for every pair (a, b) in one function, there is a pair (b, a) in the inverse function.

The Number *e*

The mathematician Leonhard Euler (1707–1783) identified a number that is the result of evaluating

$$\left(1 + \frac{1}{x}\right)^x$$

as x takes on increasing positive values.

$$\text{When} \qquad x = 10, \left(1 + \frac{1}{x}\right)^x = 2.59374246$$

$$100, \left(1 + \frac{1}{x}\right)^x = 2.704813829$$

$$1000, \left(1 + \frac{1}{x}\right)^x = 2.716923932$$

$$10{,}000, \left(1 + \frac{1}{x}\right)^x = 2.718145927$$

$$100{,}000, \left(1 + \frac{1}{x}\right)^x = 2.718268237$$

$$1{,}000{,}000, \left(1 + \frac{1}{x}\right)^x = 2.718280469$$

$$10{,}000{,}000, \left(1 + \frac{1}{x}\right)^x = 2.718281693$$

$$100{,}000{,}000, \left(1 + \frac{1}{x}\right)^x = 2.718281815$$

$$1{,}000{,}000{,}000, \left(1 + \frac{1}{x}\right)^x = 2.718281827$$

As x gets greater and greater, the result gets closer and closer to the irrational number

$$2.718281828\ldots$$

This number occurs in so many natural processes that it has been given the special symbol e. You will see it often in equations in physics, biology, and other sciences, usually as the base for an exponential function.

▪▪▪

EXAMPLE 7

Graph the function $f(x) = e^x$.

SOLUTION

The simplest way to find a table of values for this graph is to use a calculator. To the nearest tenth, we obtain the values

x	e^x
-2	0.1
-1	0.4
0	1
1	2.7
2	7
3	20

and the graph in the following figure.

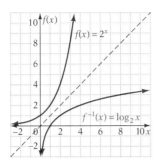

▶ *CHECK* **Warm-Up 7**

The function we just graphed is called the **natural exponential function.**

Calculator Corner

The number e and its powers are so useful that there is an e^x key on all scientific calculators. On our graphing calculator, we use this key by entering 2nd e^x, then the value of x, and then ENTER .

Natural Logarithms

The logarithmic function $f(x) = \log_e x$ with the base $e = 2.71828...$ is called the **natural logarithmic function.** Values of $f(x)$ for specific values of x are called natural logarithms of numbers. For example, one of the pairs of $f(x) = \log_e x$ is (2, 0.6931), so 0.6931 is the natural logarithm of 2.

A special symbol, ln, has been assigned to the natural logarithm function, so we write it as $f(x) = \ln x$. In other words,

$$f(x) = \ln x \quad \text{means} \quad f(x) = \log_e x$$

and

$$\ln 2 = 0.6931 \quad \text{means} \quad \log_e 2 = 0.6931$$

We will generally approximate logarithms to four decimal places, as is done here.

! ! !

ERROR ALERT

Identify the error and give a correct answer.

Write $\ln 7.39 = 2$ in exponential form.

Incorrect Solution:

$e^{7.39} = 2$

▪▪▪

EXAMPLE 8

Graph: $f(x) = \ln x$

SOLUTION

There is no need to translate to the exponential form $x = e^y$ to graph this function, because all scientific calculators have a natural logarithm key. The table of values looks like this:

x	$\ln x$
10	2.3026
4	1.3863
3	1.0986
2	0.6931
1	0
0.5	−0.6931
0.25	−1.3863

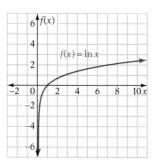

We graph these points as closely as possible and sketch a smooth curve through them, obtaining the graph shown above.

▶ *CHECK* **Warm-Up 8**

Calculator Corner

a. Plot each of the following functions on a piece of graph paper.

$$y = \ln ax \quad \text{for } a = 1, 2, 3, \text{ and } 4$$

What conjecture can you make about this family of curves?

Without using your graphing calculator, how do you anticipate the graph of $y = \ln 5x$ will look? Sketch your answer on a piece of graph paper and support with your graphing calculator.

b. Sketch the following three equations on a piece of graph paper: $y = x$, $y = \ln x$, and $y = e^x$. Then put all that you have observed in problems (a) and (b) together with these sketches and make a conjecture about these two families of functions.

c. Use your **DRAW INVERSE** feature to first draw the inverse of $y = \ln x$. What do you notice? Sketch your results on a piece of graph paper.

Now **DRAW** the **INVERSE** of $y = e^x$. What do you notice?

Sketch your results on a piece of graph paper.

Practice what you learned.

SECTION FOLLOW-UP

Many patterns exist in nature and in mathematics. You could have used a pattern to determine the units digit of

$$2^{17}$$
$$2^{23}$$
$$2^{40}$$
$$2^{100}$$

Power of 2	Units digit	Power of 2	Units digit
2^1	2**	2^7	8***
2^2	4	2^8	6*
2^3	8***	2^9	2**
2^4	6*	2^{10}	4
2^5	2**	2^{11}	8***
2^6	4	2^{12}	6*

Observations:

   *The power of 2 is divisible by 4 and the units digit is 6. So, the units digit of 2^{40} and of 2^{100} is 6.
  **The power of 2 is in the form $4n + 1$ and the units digit is 2. So, the units digit of $2^{17} = 2$ because $17 = 4 \cdot 4 + 1$.
 ***The power of 2 is in the form $4n + 3$ and the units digit is 8. So, the units digit of $2^{23} = 8$ because $23 = 4 \cdot 5 + 3$.

10.2 WARM-UPS

Work these problems before you attempt the exercises.

1. Graph: $f(x) = 2^{2x}$

2. Graph: $f(x) = 2^{x+2}$

3. Graph: $f(x) = 0.8^x$

4. Write $125 = 5^3$ and $x = 7^5$ in logarithmic form.
$\log_5 125 = 3; \log_7 x = 5$

5. Write $\log_3 81 = 4$ and $\log_{10} 0.001 = -3$ in exponential form.
$3^4 = 81; 10^{-3} = 0.001$

6. Find and graph $f^{-1}(x)$ when $f(x) = \log_3 x$. $f^{-1}(x) = 3^x$

7. Graph: $f(x) = e^{0.25x}$

8. Graph: $f(x) = \frac{1}{5} \ln x$

10.2 EXERCISES

Note: Use your graphing calculator to check your results whenever possible.

In Exercises 1 through 12, graph the given function.

1. $f(x) = 3^x$

2. $f(x) = 1.5^x$

3. $f(x) = 2.5^x$

4. $f(x) = 5^x$

5. $f(x) = 1.1^x$

6. $f(x) = 1.2^x$

7. $f(x) = 1.3^x$

8. $f(x) = 2^{2x-1}$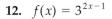

9. $f(x) = 2^{2x+1}$

10. $f(x) = 3^{x-2}$

11. $f(x) = 3^{x+2}$

12. $f(x) = 3^{2x-1}$

In Exercises 13 through 20, write each statement in logarithm form.

13. $3^4 = 81$

$\log_3 81 = 4$

14. $6^3 = 216$

$\log_6 216 = 3$

15. $9^x = 27$

$\log_9 27 = x$

16. $3^{-x} = \frac{1}{9}$

$\log_3 \frac{1}{9} = -x$

17. $\left(\frac{1}{2}\right)^{-2} = x$

$\log_{\frac{1}{2}} x = -2$

18. $\left(\frac{3}{5}\right)^2 = x$

$\log_{\frac{3}{5}} x = 2$

19. $4^{-3} = \frac{1}{64}$

$\log_4 \frac{1}{64} = -3$

20. $\left(\frac{2}{3}\right)^{-3} = \frac{27}{8}$

$\log_{\frac{2}{3}} \frac{27}{8} = -3$

In Exercises 21 through 28, write each statement in exponential form.

21. $\log_3 9 = 2$

$3^2 = 9$

22. $\log_2 32 = 5$

$2^5 = 32$

23. $\log_8 4 = \frac{2}{3}$

$8^{\frac{2}{3}} = 4$

24. $\log_{\frac{1}{5}} 25 = -2$

$\left(\frac{1}{5}\right)^{-2} = 25$

25. $\log_{10} 100{,}000 = 5$

$10^5 = 100{,}000$

26. $\log_6 \left(\frac{1}{1296}\right) = -4$

$6^{-4} = \frac{1}{1296}$

27. $\log_{\frac{2}{3}} \frac{27}{8} = -3$

$\left(\frac{2}{3}\right)^{-3} = \frac{27}{8}$

28. $\log_4 \frac{1}{64} = -3$

$4^{-3} = \frac{1}{64}$

In Exercises 29 through 32, (a) complete the table of ordered pairs, and (b) graph and label $f(x)$ and $f^{-1}(x)$.

29. $f(x) = \log_3 x$

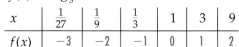

x	$\frac{1}{27}$	$\frac{1}{9}$	$\frac{1}{3}$	1	3	9
$f(x)$	-3	-2	-1	0	1	2

30. $f(x) = \log_5 x$

x	$\frac{1}{25}$	$\frac{1}{5}$	1	5	25
$f(x)$	-2	-1	0	1	2

31. $f(x) = \log_{\frac{1}{2}} x$

x	$\frac{1}{8}$	$\frac{1}{4}$	$\frac{1}{2}$	1	2	4	8	16
$f(x)$	3	2	1	0	-1	-2	-3	-4

32. $f(x) = \log_{\frac{1}{3}} x$

x	$\frac{1}{27}$	$\frac{1}{9}$	$\frac{1}{3}$	1	3	9
$f(x)$	3	2	1	0	-1	-2

In Exercises 33 through 36, graph the exponential functions.

33. $f(x) = 10e^{0.5x}$ **34.** $f(x) = -0.1e^{0.5x}$ **35.** $f(x) = 0.1e^{0.5x}$ **36.** $f(x) = -10e^{0.5x}$

In Exercises 37 through 40, graph the given function.

37. Graph $f(x) = \ln 2x$

38. Graph $f(x) = \ln \frac{1}{2}x$

39. Find and graph $f^{-1}(x)$ when $f(x) = \ln 2x$.
$f^{-1}(x) = \frac{1}{2}e^x$

40. Find and graph $f^{-1}(x)$ when $f(x) = \ln \frac{1}{2}x$. $f^{-1}(x) = 2e^x$

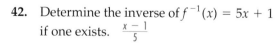

MIXED PRACTICE

By doing these exercises, you will practice the topics up to this point in the chapter.

41. Find $f(0)$, $f(a)$, and $f(2x)$ when $f(x) = 2^{x-3}$.
$\frac{1}{8}; 2^{a-3}; 2^{2x-3}$

42. Determine the inverse of $f^{-1}(x) = 5x + 1$ if one exists. $\frac{x-1}{5}$

43. Is $f(x) = 2x^2 - 6$ a one-to-one function? Explain your answer.   No, because, for example, $f(1) = 2(1)^2 - 6 = 2 - 6 = -4$, and $f(-1) = 2(-1)^2 - 6 = 2 - 6 = -4$.

44. Graph: $f(x) = 3^{x-2}$

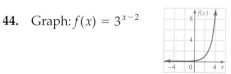

45. Find $f^{-1}(7)$ if $f(x) = 2x + 3$.   2

46. Evaluate $2e^{0.5x}$ when $x = 0$, when $x = 2$, and when $x = 4$.   2; 5.44; 14.778

EXCURSIONS

Exploring Numbers

1. Find the replacements that make these two statements true at the same time. Use only the digits 1, 2, 3, 4, and 6. Each letter stands for a unique digit.

$$a^b + b^b + c^b = dee$$
$$d^b + e^b + e^b = abc$$

2. **a.** Assign the first three non-negative odd integers to the letters a, b, and c in some order so that the following statement is true. Treat abc as a three-digit number.

$$abc = a^a + b^2 + c^b$$

   b. In words, tell what this statement says.

3. Replace a, b, c, d, and n with the digits 1, 2, 3, 6, and 9 to make these two statements true. Treat abc and cba as three-digit numbers; treat ad and da as two-digit numbers.

$$abc = (ad)^n$$
$$cba = (da)^n$$

4. The number 1201 can be expressed in the form

$$x^2 + ny^2$$

for all values of n from 1 to 10. Find the values that x and y take on for each n.

5. 204^2 is the sum of three consecutive cubes. Find them.

10.3 Properties of Logarithms

Section Goals

▪ *To apply properties of logarithms*

▪ *To solve exponential equations by using the powering rule*

▪ *To solve equations that involve logarithms*

Section Lead-In

There is a constantly changing sign in Times Square in New York City that shows the National Debt. What is meant by "National Debt"? Have you thought about just how big one trillion is? If you write it, it is 1,000,000,000,000.

a. If you started counting from one to one trillion and counted one number per second, do you know how many years it would take? ten years? fifty? one thousand?

b. If we could halve a debt of one trillion dollars every year, how many years would be necessary to reduce the debt to less than $1?

Introduction

The properties discussed in this section are used to simplify expressions and equations that contain logarithms. Most of them arise directly from the properties of exponents, along with the definition of logarithms.

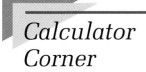

Calculator Corner

a. Graph each of the following functions on the same piece of graph paper. How many graphs do you see when you graph the four functions? Which equations are equal?

```
Y₁ = ln 3X
Y₂ = (ln 3)(ln X)
Y₃ = ln 3+ln X
Y₄ = 3ln X
Y₅ =
Y₆ =
Y₇ =
Y₈ =
```

b. Now repeat part (a) with four new equations. Which equations are equal?

```
Y₁ = ln .5X
Y₂ = (ln .5)(ln X)
Y₃ = ln .5+ln X
Y₄ = .5ln X
Y₅ =
Y₆ =
Y₇ =
Y₈ =
```

c. Now try other combinations of equations on your own. Do you obtain the same results? Does it make a difference if the 0.5 is changed to −0.5? Can you make a conjecture about a rule of logs?

d. Graph each of the following functions on the same piece of graph paper. How many graphs do you see when you graph the five functions? Which equations are equal?

```
Y₁ = ln (X/3)
Y₂ = (ln 3)(ln X)
Y₃ = ln 3+ln X
Y₄ = ln 3−ln X
Y₅ = ln X−ln 3
Y₆ =
Y₇ =
Y₈ =
```

e. Graph each of the following functions on the same piece of graph paper. How many graphs do you see when you graph the five functions? Which equations are equal?

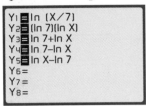

```
Y₁ = ln (X/7)
Y₂ = (ln 7)(ln X)
Y₃ = ln 7+ln X
Y₄ = ln 7-ln X
Y₅ = ln X-ln 7
Y₆ =
Y₇ =
Y₈ =
```

f. Try some other combinations of equations on your own. Do you obtain the same results? Can you make a conjecture about a rule of logs?

g. Graph each of the following functions on the same piece of graph paper. How many graphs do you see when you graph the five functions? Which equations are equal?

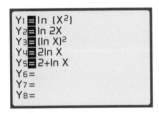

```
Y₁ = ln (X²)
Y₂ = ln 2X
Y₃ = (ln X)²
Y₄ = 2ln X
Y₅ = 2+ln X
Y₆ =
Y₇ =
Y₈ =
```

h. Does it make a difference whether the exponents are positive or negative? Which equations are equal for positive exponents?

i. Graph each of the following functions on the same piece of graph paper. How many graphs do you see when you graph the five functions? Which equations are equal?

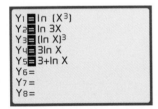

```
Y₁ = ln (X³)
Y₂ = ln 3X
Y₃ = (ln X)³
Y₄ = 3ln X
Y₅ = 3+ln X
Y₆ =
Y₇ =
Y₈ =
```

j. Try some other combinations of equations on your own. Do you obtain the same results? Can you make a conjecture about a rule of logs?

We will learn more about these rules in this chapter.

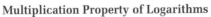

Multiplication Property of Logarithms

For positive real numbers x, y, and b, with b not equal to 1,

$$\log_b (x \cdot y) = \log_b x + \log_b y$$

For natural logarithms, the multiplication property states

$$\ln (x \cdot y) = \ln x + \ln y$$

In all the examples and exercises for this section, we shall assume that variables represent positive real numbers.

▪▪▪

EXAMPLE 1

Write as a single logarithm:

a. $\log_5 17 + \log_5 9$ **b.** $\log_b 2 + \log_b x$

Write as a sum of logarithms:

c. $\ln y\sqrt{5}$ **d.** $\log_5 (3xy)$

SOLUTION

We use the multiplication property of logarithms.

a. $\log_5 17 + \log_5 9 = \log_5 (17 \cdot 9) = \log_5 163$

b. $\log_b 2 + \log_b x = \log_b (2 \cdot x) = \log_b (2x)$

c. $\ln y\sqrt{5} = \ln y + \ln \sqrt{5}$

Note that $\ln a$ is nothing more than $\log_e a$, so all the properties of logarithms hold for natural logarithms. Hence, we can write $\ln y\sqrt{5} = \ln y + \ln \sqrt{5}$.

d. $\log_5 (3xy) = \log_5 3 + \log_5 x + \log_5 y$

▶ *CHECK* **Warm-Up 1**

> **Division Property of Logarithms**
>
> For positive real numbers x, y, and b, with b not equal to 1,
> $$\log_b \left(\frac{x}{y}\right) = \log_b x - \log_b y$$

For natural logarithms, the division property states

$$\ln \left(\frac{x}{y}\right) = \ln x - \ln y$$

▪▪▪

EXAMPLE 2

a. Write $\log_b 5 - \log_b y$ as a single term.

Write as a difference of logarithms:

b. $\log_2 \left(\frac{x}{8}\right)$

c. $\log_n \left(5 \cdot \frac{ab}{c}\right)$

! ! !
ERROR ALERT

Identify the error and give a correct answer.

Rewrite: $\log (8 \cdot 9)$

Incorrect Solution:
$\log (8 \cdot 9) = (\log 8)(\log 9)$

■■■
WRITER'S BLOCK

The rules for exponents, such as $x^m \cdot x^n = x^{m+n}$, are similar to those for logarithms, such as $\log_a (m \cdot n) = \log_a m + \log_a n$. In your own words, explain why that is so.

SOLUTION

We use the division property of logarithms.

a. $\log_b 5 - \log_b y = \log_b \dfrac{5}{y}$

b. $\log_2 \dfrac{x}{8} = \log_2 x - \log_2 8$

c. $\log_n \left(5 \cdot \dfrac{ab}{c}\right) = \log_n 5ab - \log_n c$

$\qquad\qquad = (\log_n 5 + \log_n a + \log_n b) - \log_n c$ Multiplication property

▶ *CHECK* **Warm-Up 2**

ERROR ALERT

Identify the error and give a correct answer.

Rewrite: $2 \log 6 - \log 5$

Incorrect Solution:

$2 \log 6 - \log 5$
$= \log 6^2 - \log 5$
$= \dfrac{\log 6^2}{\log 5}$

Power Property of Logarithms

For positive real numbers x and b, with b not equal to 1, and for a real number n

$$\log_b x^n = n \log_b x$$

For natural logarithms, the power property states

$$\ln x^n = n \ln x$$

■■■
EXAMPLE 3

Express as the sum, difference, or product of logarithms:

a. $\ln x^2$ b. $\log_2 9a^2$ c. $\log_5 \sqrt[4]{\dfrac{x^3}{y^4}}$

Rewrite as a single logarithm:

d. $2 \log_3 a - 3 \log_3 b$ e. $4 \log_3 x + \dfrac{1}{3} \log_3 x$

SOLUTION

ERROR ALERT

Identify the error and give a correct answer.

Rewrite:
$\log 3 + \log 5$

Incorrect Solution:

$\log 3 + \log 5 = \log 8$

a. $\ln x^2 = 2 \ln x$

b. $\log_2 9a^2 = \log_2 9 + \log_2 a^2$ Multiplication property
$\qquad\quad\;\; = \log_2 9 + 2 \log_2 a$ Power property

c. $\log_5 \sqrt[4]{\dfrac{x^3}{y^4}} = \log_5 \left(\dfrac{x^3}{y^4}\right)^{\frac{1}{4}}$ Rewriting radical with fractional exponent

$\qquad\qquad = \dfrac{1}{4} \log_5 \left(\dfrac{x^3}{y^4}\right)$ Power property

$\qquad\qquad = \dfrac{1}{4}(\log_5 x^3 - \log_5 y^4)$ Division property

$\qquad\qquad = \dfrac{1}{4}(3 \log_5 x - 4 \log_5 y)$ Power property

$\qquad\qquad = \dfrac{3}{4} \log_5 x - \log_5 y$ Distributive property

d. $2 \log_3 a - 3 \log_3 b$

$\qquad = \log_3 a^2 - \log_3 b^3$ Power property

$\qquad = \log_3 \left(\dfrac{a^2}{b^3}\right)$ Division property

e. $4 \log_3 x + \frac{1}{3} \log_3 y$

$$= \log_3 x^4 + \log_3 y^{\frac{1}{3}} \quad \text{Power property}$$
$$= \log_3 x^4 \sqrt[3]{y} \quad \text{Multiplication property}$$

▶ *CHECK* **Warm-Up 3**

Two Special Properties

For a positive real number b not equal to 1, and for real n,

$$\log_b b^n = n$$
$$\log_b 1 = 0$$

! ! !

ERROR ALERT

Identify the error and give a correct answer.

Rewrite: $\log_b 2a^3$

Incorrect Solution:

$\log_b 2a^3$
$= 3 \log_b 2a$
$= 3(\log_b 2 + \log_b a)$
$= 3 \log_b 2 + 3 \log_b a$

For natural logarithms, these properties become

$$\ln e^n = n \quad \text{and} \quad \ln 1 = 0$$

▪▪▪

EXAMPLE 4

Simplify:

a. $\log_b b$ **b.** $\ln e^{2x+1}$ **c.** $\log_\pi 1$

SOLUTION

We use the special properties.

a. $\log_b b = \log_b b^1 = 1$

b. $\ln e^{2x+1} = 2x + 1$

c. $\log_\pi 1 = 0$ because the logarithm of 1 is zero no matter what the base.

▶ *CHECK* **Warm-Up 4**

Solving Exponential and Logarithmic Equations

Solving Exponential Equations

Because exponential functions are so common in real situations, they appear often in equations. We can use the **powering rule** to solve them:

Powering Rule

For x and y real numbers, and b and c positive, real, and not equal to 1,

$$\text{if } b^x = b^y, \text{ then } x = y$$
$$\text{if } b^x = c^x \text{ and } x \text{ is not equal to zero, then } b = c$$

▪ ▪ ▪

EXAMPLE 5

Solve: $3^{x+2} = 243$

SOLUTION

The left side of the equation has the base 3, and we can rewrite the right side with that same base. Because 3^5 is 243, we obtain

$$3^{x+2} = 3^5$$

Now both sides have the same base, and the powering rule tells us that

$$x + 2 = 5$$

Solving this equation for x, we get

$$x = 3$$

We check by substituting in the original equation.

$$3^{x+2} = 243$$
$$3^{3+2} = 243$$
$$3^5 = 243 \quad \text{True}$$

▶ CHECK **Warm-Up 5**

▪ ▪ ▪

EXAMPLE 6

Solve: $3^{2x+3} = (x + 2)^{2x}(x + 2)^3$

SOLUTION

Again, we must begin by rewriting the given equation so that both sides have the same base or the same exponent. We rewrite the right side.

$$(x + 2)^{2x}(x + 2)^3 = (x + 2)^{2x+3}$$

Then the given equation becomes

$$3^{2x+3} = (x + 2)^{2x+3}$$

By the powering rule,

$$3 = x + 2$$

Solving for x, we get

$$x = 1$$

You should check this solution by substituting in the original equation.

▶ CHECK **Warm-Up 6**

In this next example, we must use the power rules $x^a \cdot x^b = x^{a+b}$ and $(x^a)^b = x^{ab}$.

▪ ▪ ▪
EXAMPLE 7

Solve: $4^x(64^{x+2}) = 16^x$

SOLUTION

As always, we must begin by rewriting both sides with the same base. All the numbers in the given equation can be written as powers of 2.

$$4^x = (2^2)^x = 2^{2x}$$
$$64^{x+2} = (2^6)^{x+2} = 2^{6(x+2)} = 2^{6x+12}$$

and

$$16^x = (2^4)^x = 2^{4x}$$

We substitute these powers of 2, obtaining

$$2^{2x}(2^{6x+12}) = 2^{4x} \quad \text{Rewriting bases as powers of 2}$$
$$2^{2x+6x+12} = 2^{4x} \quad \text{Simplifying the left side}$$
$$2^{8x+12} = 2^{4x} \quad \text{Simplifying the left exponent}$$

Now by using the powering rule, we get

$$8x + 12 = 4x$$
$$4x = -12$$
$$x = -3$$

You should check this solution by substituting in the original equation.

▶ *CHECK* **Warm-Up 7**

Solving Equations that Contain Logarithms

The ability to translate between exponential form and logarithmic form enables us to solve equations that contain logarithms.

▪ ▪ ▪
EXAMPLE 8

Solve: $\log_5 x = 3$

SOLUTION

The base is 5 and the logarithm is 3, so the exponential form of this equation is

$$5^3 = x$$

By multiplying out on the left, we obtain $x = 125$.

▶ *CHECK* **Warm-Up 8**

■ ■ ■

EXAMPLE 9

Find $\log_{10} \frac{1}{10}$.

SOLUTION

This is not in equation form, but we can write it as

$$\log_{10}\left(\frac{1}{10}\right) = x$$

The base is 10 and the logarithm is x, so we can write the equation in exponential form as

$$10^x = \frac{1}{10} = 10^{-1}$$

Then, by the powering rule,

$$x = -1$$

So $\log_{10} \frac{1}{10} = -1$.

▶ CHECK **Warm-Up 9**

■ ■ ■

EXAMPLE 10

Solve: $x^2 - 1 = \log_3 81$

SOLUTION

Here, the base is 3 and the logarithm is $x^2 - 1$, so the exponential form of this equation is

$$3^{x^2-1} = 81 = 3^4$$

The powering rule gives us

$$x^2 - 1 = 4$$

from which we find that

$$x = \sqrt{5} \quad \text{and} \quad x = -\sqrt{5}$$

You should check both solutions.

▶ CHECK **Warm-Up 10**

■ ■ ■

EXAMPLE 11

Solve: $\log_2 128 = x^3 - 1$

SOLUTION

Here the base is 2 and the logarithm is $x^3 - 1$. The exponential form of this equation is

$$2^{x^3-1} = 128$$

Since $128 = 2^7$, we have

$$2^{x^3-1} = 2^7$$

The powering rule gives

$$x^3 - 1 = 7$$

So,

$$x^3 = 8$$
$$x = 2$$

▶ *CHECK* **Warm-Up 11**

Another rule that helps us solve more complex logarithmic equations is

For positive real numbers x, y, and b, with b not equal to 1,

$$\text{if } \log_b x = \log_b y, \text{ then } x = y$$

For natural logarithms, this property becomes

$$\text{if } \ln x = \ln y, \text{ then } x = y$$

■ ■ ■

EXAMPLE 12

Solve: **a.** $\log_a x = 2 \log_a 9 - \log_a 3$ **b.** $2 \log_4 (x + 1) = \log_4 1$

SOLUTION

a. The right side of this equation can be rewritten. We get

$$
\begin{aligned}
\log_a x &= 2 \log_a 9 - \log_a 3 && \text{Original equation} \\
&= \log_a 9^2 - \log_a 3 && \text{Power property} \\
&= \log_a 81 - \log_a 3 && \text{Simplifying} \\
&= \log_a \tfrac{81}{3} && \text{Division property}
\end{aligned}
$$

So the original equation becomes

$$\log_a x = \log_a 27$$

Because the logarithms (to the same base a) are equal, the quantities must be equal, so

$$x = 27$$

We check by substituting $x = 27$ in the original equation.

$$
\begin{aligned}
\log_a x &= 2 \log_a 9 - \log_a 3 \\
\log_a 27 &= \log_a 9^2 - \log_a 3 \\
&= \log_a \tfrac{81}{3} \\
&= \log_a 27 && \text{True}
\end{aligned}
$$

b. $2 \log_4 (x + 1) = \log_4 1$ Original equation
$\log_4 (x + 1)^2 = \log_4 1$ Power property
$(x + 1)^2 = 1$ "Same-base" rule

This equation gives us

$$x + 1 = 1 \quad \text{and} \quad x + 1 = -1$$
$$x = 0 \quad \Big| \quad x = -2$$

We check both solutions, obtaining

$2 \log_4 (x + 1) = \log_4 1$	$2 \log_4 (x + 1) = \log_4 1$
$2 \log_4 (0 + 1) = \log_4 1$	$2 \log_4 (-2 + 1) = \log_4 1$
$2 \log_4 (1) = \log_4 1$	$2 \log_4 (-1) = \log_4 1$
$2(0) = 0$ True	

We stop the check of the solution -2 at this point, because $\log_4 (-1)$ is not defined.

Only $x = 0$ can be a solution of the given equation.

▶ *CHECK* **Warm-Up 12**

Calculator Corner

Solve each of the following equations using your graphing calculator and the multi-graph method of solving equations. Sketch your results on a piece of graph paper.

a. $\ln x^2 - \ln 2x = 1$ **b.** $\ln x = 4 \ln 2 - 2 \ln 3 - \ln 5$

Practice what you learned.

SECTION FOLLOW-UP

a. It would take more than 31,000 years.

Name at least four situations you can think of where numbers larger than one million are used.

b. We will be paying off the U.S. national debt for a lot of years:

$$\$1,000,000,000,000 \div 2^{10} = 976,562,500$$
$$976,562,500 \div 2^{10} = 953,674$$
$$953,674 \div 2^{10} = 931$$
$$931 \div 2^{10} = \$0.90$$

It will take 40 years to reduce the debt if we halve it every year.

Suppose we pay \$199 a month? How many years will it take us if no additional interest is added to the debt?   418,760,469 years

Research   Redo the problem using the actual U.S. national debt.

10.3 WARM-UPS

Work these problems before you attempt the exercises.

1. Write $\log_d 8 + \log_d x$ as a single logarithm. $\log_d(8x)$

2. Write $\log_9 \left(\frac{3a}{6}\right)$ as a sum and difference of logarithms.
    $\log_9 3 + \log_9 a - \log_9 6$

3. Rewrite $\log_b (2x)^2 (3a)^{-5}$ by using the power and multiplication properties. $2\log_b 2 + 2\log_b x - 5\log_b 3 - 5\log_b a$

4. Simplify: $10 \ln e + \ln e^0$ 10   5. Solve: $2^{x-5} = 64$ $x = 11$

6. Solve: $27^{2x-3} = 9\left(\frac{1}{81}\right)$ $x = \frac{7}{6}$ 7. Solve: $(25^x)(125^{x-3}) = 5^x$ $x = \frac{9}{4}$

8. Solve: $\log_x 64 = 3$ $x = 4$ 9. Find $\log_3 81$.   4

10. Solve: $\log_2 0.25 = x - 2$ 11. Solve: $\log_7 343 = 3x + 1$ $x = \frac{2}{3}$
     $x = 0$

12. Solve: $\log_3 x + 2\log_3 5 = \log_3 15$ $x = \frac{3}{5}$

10.3 EXERCISES

Note: Use your graphing calculator to check your results whenever possible.

In Exercises 1 through 12, write the number that each expression represents.

1. $\log_8 1$ 0

2. $\log_9 1$ 0

3. $3 \log_{10} 1$ 0

4. $5 \log_9 1$ 0

5. $\log_5 5$ 1

6. $\log_4 4$ 1

7. $2 \log_8 8$ 2

8. $8 \log_3 3$ 8

9. $\ln e^6$ 6

10. $\ln e^4$ 4

11. $\ln e^{x+2}$ $x + 2$

12. $\ln e^{x-5}$ $x - 5$

In Exercises 13 through 36, rewrite each expression as a sum or difference of logarithms.

13. $\log_2 3a$
     $\log_2 3 + \log_2 a$

14. $\log_3 2x$
     $\log_3 2 + \log_3 x$

15. $\log_5 (6 \cdot 7)$
     $\log_5 6 + \log_5 7$

16. $\log_4 (5 \cdot 3)$
     $\log_4 5 + \log_4 3$

17. $\log_5 \frac{3}{4}$
$\log_5 3 - \log_5 4$

18. $\log_2 \frac{5}{6}$
$\log_2 5 - \log_2 6$

19. $\log_2 x^3 y$
$3\log_2 x + \log_2 y$

20. $\log_3 x^5 y^2$
$5\log_3 x + 2\log_3 y$

21. $\log_5 2x^3$
$\log_5 2 + 3\log_5 x$

22. $\log_6 3x^5$
$\log_6 3 + 5\log_6 x$

23. $\log_7 \left(\frac{2x^3}{5}\right)$
$\log_7 2 + 3\log_7 x - \log_7 5$

24. $\log_2 \left(\frac{3x^5}{12}\right)$
$\log_2 3 + 5\log_2 x - \log_2 12$

25. $\ln\left(\frac{5\sqrt{2}}{c^2}\right)$ $\quad \ln 5 + \frac{1}{2}\ln 2 - 2\ln c$

26. $\ln 3 \sqrt{\frac{x}{y}}$ $\quad \ln 3 + \frac{1}{2}\ln x - \frac{1}{2}\ln y$

27. $\log \sqrt{\frac{r^3}{ts}}$ $\quad \frac{3}{2}\log r - \frac{1}{2}\log t - \frac{1}{2}\log s$

28. $\log \frac{\sqrt{3a}}{\sqrt{2c}}$
$\frac{1}{2}\log 3 + \frac{1}{2}\log a - \frac{1}{2}\log 2 - \frac{1}{2}\log c$

29. $\log_2 \left(\frac{3ab}{c^2}\right)$
$\log_2 3 + \log_2 a + \log_2 b - 2\log_2 c$

30. $\log_3 \left(\frac{2a}{b^2}\right)$
$\log_3 2 + \log_3 a - 2\log_3 b$

31. $\log_8 \left(\frac{10x}{5y}\right)^3$
$3\log_8 10 + 3\log_8 x - 3\log_8 5 - 3\log_8 y$

32. $\log_3 \left(\frac{5x}{2y}\right)^5$
$5\log_3 5 + 5\log_3 x - 5\log_3 2 - 5\log_3 y$

33. $\log_8 \sqrt{\frac{3a}{5}}$
$\frac{1}{2}\log_8 3 + \frac{1}{2}\log_8 a - \frac{1}{2}\log_8 5$

34. $\log_3 \sqrt{\frac{5x^2}{y^2}}$
$\frac{1}{2}\log_3 5 + \log_3 x - \log_3 y$

35. $\log_3 6 \sqrt[3]{\frac{8x^3}{2y^6}}$
$\log_3 6 + \log_3 x + \frac{2}{3}\log_3 2 - 2\log_3 y$

36. $\log_4 3 \sqrt[6]{\frac{(2x)^3}{5y^6}}$
$\log_4 3 + \frac{1}{2}\log_4 2 + \frac{1}{2}\log_4 x - \frac{1}{6}\log_4 5 - \log_4 y$

In Exercises 37 through 48, rewrite each expression as a single logarithm.

37. $\log_4 x + \log_4 t$ $\quad \log_4 xt$

38. $\log_5 a + \log_5 b$ $\quad \log_5 ab$

39. $\log_8 6 - \log_8 c$ $\quad \log_8 \left(\frac{6}{c}\right)$

40. $\log_2 5 - \log_2 x$ $\quad \log_2 \left(\frac{5}{x}\right)$

41. $\log_3 7 - 2\log_3 x$ $\quad \log_3 \left(\frac{7}{x^2}\right)$

42. $2\log_5 8 - \log_5 z$ $\quad \log_5 \left(\frac{8^2}{z}\right)$

43. $\frac{1}{2}\log_2 7 - \frac{1}{2}\log_2 x$ $\quad \log_2 \sqrt{\frac{7}{x}}$

44. $\frac{1}{3}\log_5 8 - \frac{1}{3}\log_5 c$ $\quad \log_5 \sqrt[3]{\frac{8}{c}}$

45. $\log_4 6 + 3\log_4 x - 2\log_4 z$
$\log_4 \left(\frac{6x^3}{z^2}\right)$

46. $\log_2 5 + 2\log_2 x - 3\log_2 y$
$\log_2 \left(\frac{5x^2}{y^3}\right)$

47. $\log_5 9 + \log_5 x - 2\log_5 y$
$\log_5 \left(\frac{9x}{y^2}\right)$

48. $\log_4 x - 2\log_4 y - \log_4 z$
$\log_4 \left(\frac{x}{y^2 z}\right)$

In Exercises 49 through 60, solve the equation by using the powering rule.

49. $9 = 3^x$ $\quad x = 2$

50. $8^x = 4$ $\quad x = \frac{2}{3}$

51. $2^{x+1} = 32$ $\quad x = 4$

52. $3^{5-x} = 81$ $\quad x = 1$

53. $\left(\frac{1}{4}\right)^x = 64$ $\quad x = -3$

54. $2^{-x} = \left(\frac{1}{32}\right)$ $\quad x = 5$

55. $9^{2x} = 81\left(\frac{1}{9}\right)$ $\quad x = \frac{1}{2}$

56. $\left(\frac{1}{4}\right)(8^x) = 2^{6-x}$
$x = 2$

57. $\frac{1}{9} = \frac{1}{81}(9^{2-x})$ $\quad x = 1$

58. $27^{x-1} = \frac{(9x)^x}{9x}$ $\quad x = 3$

59. $2x = 2^{3x}(4^2)$ $\quad x = -2$

60. $\frac{1}{25} = 125\left(\frac{1}{5^{x-1}}\right)$
$x = 6$

In Exercises 61 through 68, solve for x.

61. $\log_4 x = 3$ $\quad x = 64$

62. $\log_5 x = 2$ $\quad x = 25$

63. $\log_2 x = -3$ $\quad x = \frac{1}{8}$

64. $\log_5 x = -2$ $\quad x = \frac{1}{25}$

65. $\log_x \left(\frac{1}{9}\right) = -2$ $\quad x = 3$

66. $\log_x \left(\frac{1}{8}\right) = -3$ $\quad x = 2$

67. $\log_x 8 = -3$ $\quad x = \frac{1}{2}$

68. $\log_x 27 = -3$ $\quad x = \frac{1}{3}$

In Exercises 69 through 76, evaluate each expression.

69. $\log_{10} 100$ $\quad 2$

70. $\log_8 64$ $\quad 2$

71. $\log_2 \left(\frac{1}{8}\right)$ $\quad -3$

72. $\log_{10} 0.0001$ $\quad -4$

73. $\log_{16} 64$ $\quad \frac{3}{2}$

74. $\log_{25} 125$ $\quad \frac{3}{2}$

75. $\log_{\frac{1}{2}} 8$ $\quad -3$

76. $\log_{\frac{3}{4}} \left(\frac{16}{9}\right)$ $\quad -2$

In Exercises 77 through 82, solve for x.

77. $\log_2 8 = x^2 - 1$
$x = 2$ and $x = -2$
$x = 2$ and

78. $\log_3 27 = x^2 + 2$
$x = 1$ and $x = -1$
$x = 1$ and

79. $\log_5 125 = 2x^2 + 1$
$x = 1$ and $x = -1$
$x = 1$ and

80. $\log_2 32 = x^2 + 1$ $\quad x = -2$

81. $\log_2 \left(\frac{1}{8}\right) = x^2 - 4$ $\quad x = -1$

82. $\log_{\frac{1}{2}} 4 = x^2 - 3$ $\quad x = -1$

In Exercises 83 through 94, solve the equation.

83. $\log_4 (x - 1) = 2$ $x = 17$

84. $\log_5 (x + 1) = 3$ $x = 124$

85. $\log_3 (2x - 1) = 4$ $x = 41$

86. $\log_2 (3x + 2) = 4$ $x = 4\frac{2}{3}$

87. $\log x + \log (3x - 7) - 1 = 0$ $x = \frac{10}{3}$

88. $\log x + \log (x - 3) = 1$ $x = 5$

89. $\log_2 x + \log_2 (x - 1) = \log_2 6$ $x = 3$

90. $\log_4 x + \log_4 (x - 3) = \log_4 18$ $x = 6$

91. $\log_5 8x - \log_5 3 = \log_5 (x^2 - 1)$ $x = 3$

92. $\log_5 2 + \log_5 (x^2 - 1) = \log_5 3x$ $x = 2$

93. $\log_3 x - \log_3 2 = -\log_3 (3x - 5)$ $x = 2$

94. $\log_3 (6x - 7) - \log_3 5 = -\log_3 x$ $x = \frac{5}{3}$

MIXED PRACTICE

By doing these exercises, you will practice the topics up to this point in the chapter.

95. Find $f(x)$ when $f^{-1}(x) = \frac{x + 3}{2}$. $2x - 3$

96. Solve: $9^{5-x} = 81$ $x = 3$

97. Solve: $\log_2 4 - \log_2 8 = x$ $x = -1$

98. Rewrite $\log_{10} \left(\frac{5x^2}{6}\right)^4$ as a sum or difference of logs. $4\log_{10} 5 + 8\log_{10} x - 4\log_{10} 6$

99. Find $f^{-1}(x)$ when $f(x) = 3x - 2$. $\frac{x + 2}{3}$

100. Solve: $8^x = \left(\frac{1}{4}\right)^{(8x + 1)}$ $x = -\frac{2}{19}$

EXCURSIONS

Posing Problems

1. Ask and answer four questions using this data.

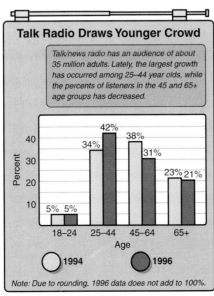

Source: Interep Research.

2. A lot of money is spent on entertainment. Ask and answer four questions using this information.

Entertainment and the U.S. Economy
The entertainment industry revenues listed below equaled $127.47 billion. Figures are based on 1994 statistics unless otherwise noted.

Industry	Sales
Film	$5.4 billion in box-office receipts
Television	
Network, national spot and local spot	$29.05 billion based on 1994 advertising revenue (network: $11.08 billion; national spot: $8.74 billion; local spot; $9.24 billion)
Cable	$24.08 billion based on 1994s pay-per-view and premium subscription fees and advertising revenue (basic services: $15.6 billion; pay-per-view: $2.97 billion; premium: $2.58 billion; advertising revenue: $2.93 billion)
Radio	$10.30 billion based on network, national spot, and local spot advertising revenue
Recorded Music	$12.07 billion based on manufacturers recommended retail price
Publishing	
Book	$18.70 billion based on end-user spending
Magazine	$8.5 billion based on advertising revenue
Performing Arts	
Theater	$129.53 million in ticket sales based on a survey of 231 theaters nationwide, assuming ticket sales account for 46.7% of annual earned income
Opera	$176.7 million based on preliminary data
Dance	No comprehensive data available

Source: 1996 Information Please® Business Almanac (©1995 Houghton Mifflin Co.), p. 6. All rights reserved. Used with permission by Information Please LLC.

3. What interesting data for mathematics students! Ask and answer four questions using this data.

The U.S. median income for men age 30 or older with bachelor's degrees and earning wages or salaries was $43,856[1]. Here are the five fields of study with the highest median income.

Major	Median Income
Engineering	$52,998
Mathematics	$52,316
Physics	$51,819
Pharmacy	$50,805
Economics	$50,360

1. Excludes self-employed, through 1993 (latest year available).
Source: Bureau of Labor Statistics, *Occupational Outlook Quarterly,* (Summer 1996); in *USA Today,* 15 January 1997.

4. Ask and answer four questions. Try to ask questions that other students would want to answer.

Median Weekly Earnings of Full-Time Workers by Occupation and Sex

Occupation	Men		Women		Total	
	Number of workers (in thousands)	Median weekly earnings	Number of workers (in thousands)	Median weekly earnings	Number of workers (in thousands)	Median weekly earnings
Managerial and prof. specialty	13,021	$803	12,187	$592	25,208	$683
Executive admin., and managerial	6,785	797	5,548	541	12,333	658
Professional specialty	6,236	809	6,639	623	12,875	705
Technical, sales, and admin. support	9,764	548	15,954	376	25,718	420
Technicians and related support	1,638	622	1,536	466	3,174	534
Sales occupations	4,836	575	3,633	324	8,470	450
Administrative support, incl. clerical	3,289	482	10,785	374	14,074	392
Service occupations	4,784	350	4,702	257	9,486	294
Private household	14	(¹)	311	166	324	179
Protective service	1,674	538	277	430	1,951	517
Service, except private household and protective	3,096	293	4,115	256	7,211	271
Precision production, craft, and repair	9,824	515	970	370	10,795	504
Mechanics and repairers	3,593	519	160	520	3,753	519
Construction trades	3,407	492	52	408	3,460	491
Operators, fabricators, and laborers	11,333	406	3,412	293	14,745	373
Machine operators, assemblers, and inspectors	4,469	415	2,563	292	7,032	361
Transportation and material moving occupations	3,854	469	242	361	4,096	461
Handlers, equipment cleaners, helpers, and laborers	3,010	319	608	279	3,617	311
Farming, forestry, and fishing	1,265	290	161	234	1,426	282

1. Data not shown where base is less than 100,000. *Note:* Figures are for the year 1994. *Source:* U.S. Department of Labor, Bureau of Labor Statistics, "Employment and Earnings," January 1995, as cited in the *1996 Information Please® Almanac* (©1995 Houghton Mifflin Co.), p. 54. All rights reserved. Used with permission by Information Please LLC.

5. Ask and answer four questions using this data. Try to ask questions that other students would want to answer.

**Comparison of Median Earnings of Year-Round
Full-Time Workers 15 Years and Over, by Sex, 1960 to 1993**

Year	Median earnings		Earnings gap in current dollars	Women's earnings as a percent of men's	Percent men's earnings exceeded women's	Earnings gap in constant 1993 CPI-U-xl adjusted dollars
	Women	Men				
1960	$ 3,257	$ 5,368	$2,111	60.7	64.8	$9,473
1970	5,323	8,966	3,643	59.4	68.4	12,746
1980	11,197	18,612	7,415	60.2	66.2	13,019
1984	14,780	23,218	8,438	63.7	57.1	11,735
1985	15,624	24,195	8,571	64.6	54.9	11,510
1986	16,232	25,256	9,024	64.3	55.6	11,898
1987	16,911	25,946	9,035	65.2	53.4	11,493
1988	17,606	26,656	9,050	66.0	51.4	11,054
1989	18,769	27,331	8,562	68.7	45.6	9,977
1990	19,822	27,678	7,856	71.6	39.6	8,685
1991	20,553	29,421	8,868	69.9	43.1	9,408
1992r	21,375	30,197	8,822	70.8	41.3	9,086
1993	21,747	30,407	8,660	71.5	39.8	8,660

r = based on implementation of 1990 census population controls. *Source*: Department of Commerce, Bureau of the Census; as cited in the *1996 Information Please*® *Almanac* (©1995 Houghton Mifflin Co.), p. 67. All rights reserved. Used with permission by Information Please LLC.

Data Analysis

6. Study the following table and then answer the questions.

Profile of Daily Listeners of Radio Stations by Format

Format	% Male	% Female	Median Age	Median Income ($)	% 1+ Yrs. of College
Total	47.97	52.03	41.88	35,696	40.83
Adult contemporary	42.08	57.92	37.39	42,293	48.57
All news	52.87	47.13	48.02	43,940	49.78
Album-oriented rock	63.88	36.12	30.47	44,248	50.59
Black/R&B	41.27	58.73	34.01	30,356	38.06
Classic rock	60.70	39.30	32.31	38,281	47.83
Classical	56.55	43.45	48.27	42,085	62.49
Country	47.86	52.14	39.96	36,016	36.97
Easy listening	44.16	55.84	53.59	39,799	46.78
Educational	78.09	21.91	42.07*	42,409*	78.94*

Ethnic	43.37	56.63	37.45	54,446	53.22
Golden oldies	52.88	47.12	40.89	43,278	47.86
Jazz	55.82	44.18	40.29	40,908	47.97
MOR/nostalgia	44.37	55.63	58.90	34,107	39.22
New age	67.94	32.06	32.56	45,820	70.30
News/talk	55.59	44.41	50.64	43,784	49.43
Religious	37.18	62.82	41.80	35,345	42.71
Soft contemporary	43.88	56.12	41.88	40,423	46.14
Urban contemporary	46.32	53.68	34.34	25,500	30.56
Variety	46.58	53.42	41.65	43,218	54.29

*Number of cases too small for reliability.
Source: Simmons, Study of Media Markets, 1994; as cited in the *1996 Information Please® Business Almanac* (©1995 Houghton Mifflin Co.), p. 486. All rights reserved. Used with permission by Information Please LLC.

    a. Organize this data so you can say something about the preferences of

       i.   men

      ii.   women

     iii.   older listeners

     iv.   younger listeners

      v.   people with higher incomes

     vi.   people with more education

    b. What other questions would you like to ask about this data? Answer your questions.

10.4 Applications of Logarithmic and Exponential Functions

SECTION LEAD-IN

You start with $10 in a bank account and have the option to let your account (D) double every two years or (T) triple every three years.

    a. Which account will have the most money 5 years after you start? 10 years?

    b. You can remove your money during any year that is divisible by 6 after it has been deposited for 20 years. If you first invest your money in 1992, which account would have the most money at the earliest time you can withdraw?

    c. At some point after 30 years, one account becomes consistently greater than the other. Find the last year that the accounts "switch" status. Which account becomes the larger at that time and stays that way?

Computing with Logarithms

Logarithms with base 10 have been given a special name, **common logarithms,** and a special symbol, log (without a base subscript). In other words,

$$\log x \quad \text{means} \quad \log_{10} x$$

Also, if $\log x = y$, then $10^y = x$. You can find the common logarithm of a positive real number with the $\boxed{\text{LOG}}$ key on a scientific calculator.

If $\log x = y$, we call x the **antilogarithm** of y.

The common antilogarithm of a number y is abbreviated antilog y. That is,

$$\text{antilog } y = x \quad \text{means} \quad \log x = y$$

You can find common antilogarithms of numbers by using the $\boxed{10^x}$ key on a calculator.

▪▪▪
Example 1

Evaluate:

a. log 800   **b.** log 45.6   **c.** antilog 0.06   **d.** antilog 1.06

Solution

a. $\boxed{\text{LOG}}$ 800 $\boxed{\text{ENTER}}$ = 2.9031

b. $\boxed{\text{LOG}}$ 45.6 $\boxed{\text{ENTER}}$ = 1.6590

c. Antilog 0.06 = y means $10^{0.06} = y$, so we key in $\boxed{10^x}$ .06 $\boxed{\text{ENTER}}$ and read that

$$10^{0.06} = 1.1482$$

d. We key in $\boxed{10^x}$ 1.06 $\boxed{\text{ENTER}}$ and read that

$$10^{1.06} = 11.4815$$

▶ *CHECK* **Warm-Up 1**

Change-of-Base Rule

For positive real numbers x, a, and b, with a and b not equal to 1,

$$\log_b x = \frac{\log_a x}{\log_a b}$$

▪▪▪
Example 2

Evaluate $\log_2 30$ in two ways, by rewriting it with (a) base 10 and (b) base e.

SOLUTION

a. Using the change-of-base rule with 10 as the base, we rewrite

$$\log_2 30 = \frac{\log 30}{\log 2}$$

and use the calculator to get

$$\frac{\log 30}{\log 2} = \frac{1.477121255}{0.3010299957} = 4.9069$$

We can check by verifying that $2^{4.9069} = 30$.

b. Using the change-of-base rule with e as the base, we rewrite

$$\log_2 30 = \frac{\ln 30}{\ln 2}$$

and use the calculator to get

$$\frac{\ln 30}{\ln 2} = \frac{3.401197382}{0.6931471806} = 4.9069$$

▶ *CHECK* **Warm-Up 2**

Natural Logarithms

Recall that "natural logarithm of" is abbreviated ln and that

$$\ln x = y \quad \text{means} \quad e^y = x$$

The **natural antilogarithm** of a number y, which is abbreviated antiln y, is the number x whose natural logarithm is y. Thus

$$\text{antiln } y = x \quad \text{also means} \quad e^y = x$$

In words, x is the natural antilogarithm of y if y is the natural logarithm of x. You can find the natural antilogarithms of numbers with the $\boxed{e^x}$ key on a calculator.

▪▪▪

EXAMPLE 3

Evaluate (a) antiln 16 and (b) antiln 4.

SOLUTION

a. If antiln $16 = y$, then $e^{16} = y$. Using the graphing calculator, we key in $\boxed{e^x}$ 16 $\boxed{\text{ENTER}}$ and obtain

$$e^{16} = 8.88610521\text{E}6 = 8.886110521 \times 10^6$$
$$= 8{,}886{,}110.521$$

b. antiln 4, so $e^4 = x$. Using the calculator and rounding to four decimal places, we find that

$$e^4 = 54.5982$$

▶ *CHECK* **Warm-Up 3**

Solving Exponential Equations by Using Logarithms

An equation with a variable exponent is called an exponential equation. In Section 10.3, we solved simple exponential equations by applying the powering rule—but that rule works only for the "right" bases. Now, by using logarithms, we can solve many more types of exponential equations. For some, we use this rule, which should look familiar:

> For positive real numbers x, y, and b, with b not equal to 1,
> $$\text{if } x = y, \text{ then } \log_b x = \log_b y$$

▪ ▪ ▪

EXAMPLE 4

Solve: $e^{x-1} = 5$

SOLUTION

We take the natural logarithms of both sides of the equation.

$$\ln e^{x-1} = \ln 5$$
$$x - 1 = 1.609437912$$
$$x = 2.609437912$$
$$x = 2.6094 \qquad \text{Rounded}$$

▶ *CHECK* **Warm-Up 4**

▪ ▪ ▪

EXAMPLE 5

Solve: $5^x = 7$

SOLUTION

Taking the logarithms (to the base 10) of both sides gives us

$$\log 5^x = \log 7 \qquad \text{Taking logarithms}$$
$$x \log 5 = \log 7 \qquad \text{Power property}$$
$$x = \frac{\log 7}{\log 5} \qquad \text{Multiplying by } \frac{1}{\log 5}$$
$$= \frac{0.84509804}{0.6989700043} \qquad \text{Finding logarithms}$$
$$= 1.2091 \qquad \text{Rounded}$$

We can check by verifying that

$$5^{1.2091} = 7$$

▶ *CHECK* **Warm-Up 5**

▪ ▪ ▪

EXAMPLE 6

Solve: $\left(\frac{1}{2}\right)^{2x+3} = 3^x$

SOLUTION

Taking the logarithms (again to the base 10) of both sides gives

$$\log \left(\tfrac{1}{2}\right)^{2x+3} = \log 3^x \quad \text{Taking logarithms}$$

$$(2x + 3) \log \tfrac{1}{2} = x \log 3 \quad \text{Power property}$$

Then we use a calculator to find $\log \tfrac{1}{2}$ and $\log 3$ to four decimal places, substitute, and get

$$(2x + 3)(-0.3010) = x(0.4771)$$

The rest of the solution should follow easily:

$$-0.6020x - 0.9030 = 0.4771x \quad \text{Multiplying out}$$

$$-0.9030 = 1.0791x \quad \text{Adding } 0.6020x \text{ to both sides}$$

$$-0.8368 = x \quad \text{Multiplying by } \tfrac{1}{1.0791}$$

You should check this result by substituting in the original equation.

▶ *CHECK* **Warm-Up 6**

Applications of Logarithms and Exponential Expressions

There are many useful applications of logarithms or exponential expressions.

▪ ▪ ▪
EXAMPLE 7

Chemists determine the acidity of a solution by measuring its pH, for which they use the formula

$$\text{pH} = -\log [H_3O^+]$$

The symbol $[H_3O^+]$ stands for the concentration of hydronium ions (in moles per liter) in the solution. An *acidic* solution has $\text{pH} < 7$; a *basic* solution has $\text{pH} > 7$.

a. Find the pH of a vinegar solution that has a hydronium ion concentration of 7.1×10^{-6} moles per liter.

b. Find the concentration of hydronium ions in a solution with $\text{pH} = 9.6$.

SOLUTION

a. We substitute in the original equation.

$$\text{pH} = -\log [H_3O^+]$$

$$= -\log (7.1 \times 10^{-6}) \quad \text{Substituting the } [H_3O^+]$$

$$= -(\log 7.1 + \log 10^{-6})$$

$$= -0.8513 + 6 \quad \text{Evaluating logarithms}$$

$$\text{pH} = 5.1487 \approx 5.1$$

b. $pH = -\log[H_3O^+]$

$$9.6 = -\log[H_3O^+] \quad \text{Substituting the pH}$$
$$-9.6 = \log[H_3O^+] \quad \text{Multiplying by } -1$$
$$10^{-9.6} = [H_3O^+] \quad \text{Definition of common logarithms}$$
$$2.51 \times 10^{-10} = [H_3O^+] \quad \text{Simplifying by calculator}$$

▶ *CHECK* **Warm-Up 7**

∎∎∎
EXAMPLE 8

Under laboratory conditions, a certain sample of bacteria grows according to the equation

$$N = 1.24e^{0.1t}$$

where N is the number of bacteria present in millions, and t is the time that has passed in hours. Find the number of bacteria present at times (a) $t = 0$ and (b) $t = 24$.

SOLUTION

a. Substituting 0 for t, we have

$$N = 1.24e^{0.1(0)}$$
$$= 1.24e^0$$
$$= 1.24$$

There are approximately 1.24 million bacteria at the start of the experiment.

b. Substituting 24 for t, we have

$$N = 1.24e^{0.1(24)} = 1.24e^{2.4}$$
$$= 1.24\,(11.02317638)$$
$$= 13.66873871$$

After 24 hours, there are approximately 13.7 million bacteria.

▶ *CHECK* **Warm-Up 8**

∎∎∎
EXAMPLE 9

Carbon dating is used to determine the ages of objects ranging from paintings to ancient bones. The nuclear activity A of radioactive carbon 14 in the object is measured and substituted in the equation.

$$A = 0.23e^{-0.00012t}$$

The equation is then solved for t, which is the age of the object in years.

When the Dead Sea Scrolls were found in 1947, their carbon 14 activity was measured to be 0.18. How old were they, to the nearest ten years?

Solution

Substituting 0.18 for A in the given equation yields

$$A = 0.23e^{-0.00012t}$$
$$0.18 = 0.23e^{-0.00012t}$$

Multiplying both sides by $\frac{1}{0.23}$ to isolate the power of e on the right, we get

$$0.7826 = e^{-0.00012t}$$

The base is e and the logarithm (exponent) is $-0.00012t$, so we translate this equation to logarithm form as

$$\ln 0.7826 = -0.00012t$$

We find the natural logarithm on the left with a calculator, obtaining

$$-0.2451 = -0.00012t$$

Finally, multiplying by $\frac{-1}{0.00012}$ to isolate t yields

$$2042.5 = t$$

Rounding to the nearest ten years, we get 2040 years. The Dead Sea Scrolls were about 2040 years old in 1947.

▶ *CHECK* **Warm-Up 9**

▪▪▪
EXAMPLE 10

Compound interest is interest that is paid on a deposited amount plus all previously paid interest (if no money is withdrawn from the account). If P dollars are deposited, and compound interest is paid each year at interest rate I (expressed as a decimal number), then at the end of t years, the initial deposit will have grown to the amount

$$A = P(1 + I)^t$$

a. How much do you have to deposit now to have $1000 exactly 10 years from now, if compound interest is paid at 8% per year?

b. How long does it take for $500 to grow to $1000 if compound interest is paid at $3\frac{1}{2}\%$ per year?

Solution

a. We substitute $t = 10$, $I = 0.08$, and $A = 1000$ into the formula and solve for P.

$$A = P(1 + I)^t$$
$$1000 = P(1 + 0.08)^{10} \qquad \text{Substituting and simplifying}$$
$$1000 = P(1.08)^{10} \qquad \text{in parentheses}$$
$$\log 1000 = \log P(1.08)^{10} \qquad \text{Taking common logarithms}$$
$$3 = \log P + 10 \log 1.08$$
$$3 = \log P + 10(0.0334)$$
$$2.666 = \log P$$

Thus

$$10^{2.666} = P \quad \text{Definition of common logarithms}$$
$$463.447 = P \quad \text{Simplifying by calculator}$$

You should deposit $463.45.

b. We substitute $I = 0.035$, $A = 1000$, and $P = 500$ into the formula and solve for t.

$$A = P(1 + I)^t$$
$$1000 = 500(1 + 0.035)^t \quad \text{Substituting}$$
$$2 = 1.035^t \quad \text{Multiplying by } \tfrac{1}{500}$$
$$\log 2 = t \log 1.035 \quad \text{Taking common logarithms}$$
$$\frac{\log 2}{\log 1.035} = t \quad \text{Multiplying by } \tfrac{1}{\log 1.035}$$
$$\frac{0.3010}{0.01494} = t \quad \text{Simplifying by calculator}$$
$$20.15 = t \quad \text{Simplifying}$$

Under these conditions, it takes slightly more than 20 years for $500 to grow to $1000.

▶ CHECK **Warm-Up 10**

Practice what you learned.

SECTION FOLLOW-UP

a. D = $40 after 5 years
    D = $320 after 10 years

b. T = $65,610; D = $40,960

c. T, after 32 years

10.4 WARM-UPS

Work these problems before you attempt the exercises.

1. Evaluate: antilog 2.5
   316.2278

2. Evaluate $\log_{31} 18$ by using 10 as the base.   0.8417

3. Find the number x such that antiln $x = 1.3$. $x \approx 0.2624$

4. Solve: $10^{2x+1} = 120$
   $x \approx 0.5396$

5. Solve: $7^x = 350$ $x \approx 3.0104$

6. Solve: $\dfrac{2^{3x}}{2^5} = 9^x$ $x = -29.4247$

7. What is $[H_3O^+]$ for a neutral solution—that is, a solution of pH = 7? 1.0×10^{-7}

8. Using the formula in Example 8, find N after 48 hours.   150.7 million bacteria

9. Find, to the nearest hundred years, the age of a bone with a carbon 14 activity of 0.11.   6100 years

10. How long would it take for $2000 to triple at 5% interest compounded annually?   about 22.52 years

10.4 EXERCISES

Note: Use your graphing calculator to check your results whenever possible.

In Exercises 1 through 8, evaluate each of the following by using a calculator. Write each answer to the nearest ten thousandth.

1. log 520
   2.7160

2. antilog 0.1608
   1.4481

3. $10^{3.15} = x$
   $x = 1412.5375$

4. log 0.032
   -1.4949

5. antilog -5
   0.00001

6. log 16.3
   1.2122

7. log 5200
   3.7160

8. $10^{-2.8} = x$
   $x = 0.0016$

In Exercises 9 through 16, use a change of base to evaluate each expression in two ways—using 10 as a base and using e as a base. Write each answer to four decimal places.

9. $\log_2 8$ $\frac{\ln 8}{\ln 2} = 3.0000$; $\frac{\log 8}{\log 2} = 3.0000$

10. $\log_3 81$ $\frac{\log 81}{\log 3} = 4.0000$; $\frac{\ln 81}{\ln 3} = 4.0000$

11. $\log_5 1000$ $\frac{\ln 1000}{\ln 5} = 4.2920$; $\frac{\log 1000}{\log 5} = 4.2920$

12. $\log_6 256$ $\frac{\log 256}{\log 6} = 3.0948$; $\frac{\ln 256}{\ln 6} = 3.0948$

13. $\log_4 132$ $\frac{\ln 132}{\ln 4} = 3.5222$; $\frac{\log 132}{\log 4} = 3.5222$

14. $\log_7 15.6$ $\frac{\ln 15.6}{\ln 7} = 1.4118$; $\frac{\log 15.6}{\log 7} = 1.4118$

15. $\log_{12} 0.06$ $\frac{\ln 0.06}{\ln 12} = -1.1322$; $\frac{\log 0.06}{\log 12} = -1.1322$

16. $\log_8 0.0016$ $\frac{\ln 0.0016}{\ln 8} = -3.0959$; $\frac{\log 0.0016}{\log 8} = -3.0959$

In Exercises 17 through 24, evaluate each of the following by using a calculator. Write each answer to the nearest ten thousandth.

17. ln 12
    2.4849

18. antiln 18
    65,659,969.14

19. $e^{2.5} = x$
    $x = 12.1825$

20. ln 230
    5.4381

21. antiln 0.321
    1.3785

22. ln 5.7
    1.7405

23. ln 82.6
    4.4140

24. $e^{-1.2} = x$
    $x = 0.3012$

In Exercises 25 through 32, use properties of logs to solve the exponential equations. Round each answer to the nearest ten thousandth.

25. $e^{3x} = 9$
    $x = 0.7324$

26. $e^{2x} = 14$
    $x = 1.3195$

27. $e^{x-2} = 210$
    $x = 7.3471$

28. $e^{2x+1} = 35$
    $x = 1.2777$

29. $10^x = 15$
    $x = 1.1761$

30. $10^{2x} = 250$
    $x = 1.1990$

31. $10^{x-1} = 7$
    $x = 1.8451$

32. $10^{x+3} = 1500$
    $x = 0.1761$

In Exercises 33 through 40, solve for x. Round each answer to the nearest ten thousandth.

33. $5^x = 1300$
$x = 4.4550$

34. $6^x = 2.7$
$x = 0.5543$

35. $4^x = 0.002$
$x = -4.4829$

36. $2^x = 100$
$x = 6.6439$

37. $3^x = 5$
$x = 1.4650$

38. $7^x = 4500$
$x = 4.3228$

39. $2^x = 5002$
$x = 12.2883$

40. $5^x = 0.15$
$x = -1.1787$

In Exercises 41 through 48, solve for x. Round each answer to the nearest ten thousandth.

41. $2^{x-5} = 5^x$
$x = -3.7824$

42. $2^x = 8^{2x+2}$
$x = -1.2$

43. $9^x = 3^{x+3}$
$x = 3$

44. $7^{3x-1} = 3^x$
$x = 0.4106$

45. $\dfrac{4^x}{4^5} = 3^{x+2}$
$x = 31.7319$

46. $5^{2x} = \dfrac{9^6}{9^x}$
$x = 2.4341$

47. $7^x = \dfrac{8^{3x}}{8^4}$
$x = 1.9378$

48. $\dfrac{6^5}{6^x} = 2^{x-2}$
$x = 4.1632$

Work the following word problems.

49. *Population Growth* $f(x) = 4500(2)^{0.4x}$ is a model for the population of a midwestern town x years after 1976.

   a. What was the 1976 population? 4500

   b. What was the population in 1980? 13,641

   c. Suppose this function is also valid for years before 1976. What was the population in 1970? about 853

   d. What population is predicted for 2001? 4,608,000

50. *Fantasy Banking* Suppose a certain bank account (fictitious, unfortunately) allows you to invest one cent for a month (30 days). At the end of each day, the bank doubles the amount of money in your account.

   a. The function $f(t) = (0.01)2^t$ gives the value of your account at the end of the tth day, in dollars. How much is in your account at the end of a week? $1.28

   b. After 30 days, the bank offers you $1 million for the money in your account. Should you accept the offer? Why? No, because it's only one tenth the money in the account.

   c. The function given in part (a) is called a **doubling function.** The function $f(t) = (0.01)3^t$ gives the value of your account at the end of the tth day if your money is *tripling*. How much more would you have at the end of one week if this function were used than you would have if the doubling function were used? $20.59

   d. Suppose you started with 5 million dollars. If your money were halved each day, how long would it take until you had $1.00 or less remaining? (To answer, you first have to write a "halving function.") 23 days

51. *Buying Manhattan* A model for determining the value of an account is given by

$$A = Pe^{rx}$$

This account has its interest calculated continuously. A is the accumulated value when an amount or principal P is invested at an interest rate r for a number of years x.

a. Find the accumulated value when $1000 is invested for 5 years at 4% compound interest. $1221.40

b. Which is worth more after 5 years, $750 invested at 10% interest or $900 invested at 8% compound interest? $900 invested at 8% interest

c. In 1626, Peter Minuit "stole" Manhattan from the Man-a-hat-a Indians for a few beads valued at $24. If the $24 had been invested at 5% compound interest, what would it be worth in 2001?   over $3 billion: $3,336,051,738

d. What percent interest is needed to quadruple an investment of $100 in 10 years?   14%

52. *Used Cars*   The function $f(t) = 2^{4-0.2t}$ is a model for the value of a certain car (in thousands of dollars) at the end of t years.

a. How much did the car cost originally? $16,000

b. What is the value of the car after 2 years? $12,126

c. How much does the car depreciate (drop in value) between the end of its fifth year and the end of its sixth year? $1036

d. When is the car worth $4000?   after 10 years

53. *Deer Population*   When a number of deer were introduced into a new area, their population p at times t months after their introduction was
$$p = 180 \log_{10} (t + 10)$$

a. How many deer were originally introduced into the area? (*Hint*: Find p when $t = 0$.)   180 deer

b. How many deer were there a year later?   242

54. *Advertising*   Ms. Hamby, an advertising executive, has found the total new sales for a business, in thousands of dollars, after a big advertising campaign are $S = 1000 \log_2 (x + 8)$, where x is the time in weeks after the campaign was introduced.

a. What were the sales before the campaign started? (*Hint:* Find S when $x = 0$.) $3,000,000

b. What were the sales at the end of the second month ($x = 8$ weeks)? $4,000,000

55. *Bacteria Growth*   A culture of bacteria grows according to $C(t) = 8000e^{0.4t}$, where $C(t)$ is the number of bacteria present at time t hours.

a. What is the number of bacteria present at time $t = 0$?   8000 bacteria

b. After how many hours will the culture contain 32,000 bacteria?   after $3\frac{1}{2}$ hours

56. *Controlling Bacteria*   When an antibiotic is introduced into a colony of bacteria, the number of bacteria $B(t)$ present t hours after treatment is $B(t) = 50,000e^{-0.01t}$.

a. Find the number of bacteria present at time $t = 0$.   *50,000 bacteria*

b. In how many days will fewer than 1,000 bacteria be present?   *16.3 days*

57. ***Flu Outbreak***   The function $f(x) = \dfrac{50,000}{1 + 20e^{-1.5x}}$ describes the number of people who became ill with the flu x months after its initial outbreak.

   a. How many became ill 4 months after the initial outbreak?   *47,638*

   b. How many became ill after 1 week $\left(\frac{1}{4} \text{ month}\right)$?   *3391*

   c. About how many weeks after the initial outbreak did 10,000 people become ill?   *about 4 weeks*

58. ***Packing Boxes***   The function $f(x) = 800 - 700e^{-0.2x}$ describes the number of boxes a worker can pack per hour after x months of training.

   a. How many boxes can a worker pack per hour with no training?   *100 boxes*

   b. How many boxes can a worker pack per hour with 2 months of training?   *331 boxes*

   c. When will a worker be able to pack 500 boxes per hour?   *after about $4\frac{1}{4}$ months of training*

59. ***Population Growth***   Kenya has a population of 2.3 million people. It has been estimated that the population will double in 19 years. Answer the questions that follow by using the formula

   $$P = P_0 2^{\frac{t}{d}}$$

   where

   $P_0 = $ original population
   $P = $ the result of "doubling"
   $d = $ doubling time in years
   $t = $ time that has passed

   a. How long will it take for the population to grow to 3.68 million?   *about 13 years*

   b. How many years ago was the population 460,000? (Write this number as 0.46 million.)   *44 years ago, if population growth has been constant*

60. ***Measuring Earthquakes***   The Richter scale is used to measure the intensity of an earthquake. The formula for the Richter scale number (M) is $M = \frac{2}{3} \log\left(\frac{E}{E_0}\right)$, where E_0 is $10^{4.40}$ joules and E is the energy released by the earthquake in joules.

   a. The San Francisco earthquake of 1906 registered 8.25 on the Richter scale. How much energy was released by this earthquake?   *$E = 5.96 \times 10^{16}$ joules*

   b. On October 1, 1989, a 15-second earthquake hit San Francisco. It released energy that was measured at 5.62×10^{14} joules. What was the rating of this earthquake on the Richter scale?   *6.9*

MIXED PRACTICE

By doing these exercises, you will practice the topics up to this point in the chapter.

61. Determine whether $f(x) = \{(2, 3), (3, 5), (6, 7), (9, 8), (8, 9)\}$ is a one-to-one function. If it is, find its inverse.   yes; {(3, 2), (5, 3), (7, 6), (8, 9), (9, 8)}

62. Solve: $\log_5 x = 0$ x = 1

63. Evaluate $\log_4 8$. Use $\log 8 = 0.9031$ and $\log 4 = 0.6021$ 1.4999

64. Rewrite $-6 \log_2 5 + 2 \log_2 x$ as a single logarithm. $\log_2 \frac{x^2}{5^6}$

65. Evaluate $\log_7 49$.   x = 2

66. Evaluate $\log (4 \cdot 8)$. Use $\log 8 = 0.9031$ and $\log 4 = 0.6021$.   1.5052

67. Evaluate $\log_3 5$. Use $\log 3 = 0.4771$ and $\log 5 = 0.6990$.   1.4651

68. Solve: $27^{x-1} = 81$ $x = \frac{7}{3}$

69. Rewrite $\log_b 5\sqrt{\frac{2}{y^2}}$ as sums and differences of logarithms. $\log_b 5 + \frac{1}{2} \log_b 2 - \log_b y$

70. Evaluate log 2 by using $\log 8 = 0.9031$ and $\log 4 = 0.6021$. $x \approx 0.301$

71. Write $\log_{\frac{1}{2}} 4 = -2$ in exponential form, and verify the truth of the statement.
   $\left(\frac{1}{2}\right)^{-2} = 4$ $(2^{-1})^{-2} = 4$ $2^2 = 4$ $4 = 4$

72. Determine whether $\{(2, 3), (3, 5), (2, 4)\}$ is a function and, if so, whether it is a one-to-one function.   not a function

73. Solve: $\log_b 27 = -3$ $b = \frac{1}{3}$

74. Solve: $16^{x-2} = \frac{1}{64}$ $x = \frac{1}{2}$

75. Solve: $\log_6 x = 3$ x = 216

76. Find $f^{-1}(x)$ when $f(x) = 4x - 5$. $\frac{x + 5}{4}$

77. Solve: $\left(\frac{1}{243}\right)^x = 27$ $x = -\frac{3}{5}$

78. Evaluate:
   $\log_{10} 100 + \log_{10} 0.01 - 3 \log_{10} 1000$ -9

EXCURSIONS

Class Act

1. In an article in a scientific journal, we are told that "A 12-inch test span can support 650 pounds."

   a. Criticize this statement. What do we need to know? What might the article mean?

   b. What if they had said "A piece of this material (of a standard thickness) 3 inches by 2 inches can support 2100 pounds." How many pounds would be supported by one square inch? one square centimeter?

2. In 1996, archaeologists from the University of Capetown in South Africa discovered cave paintings that were 500 (plus or minus 150) years old.

   a. What is the percent error here? That is, what percent of the number 500 is 150?

   b. If the same percent of error persisted, by how many years might paintings from 19,000 years ago be misdated?

   c. A researcher claims that cave paintings 19,000 years old might really have been painted 26,000 years ago. Is this possible? By what percent of error might this be true?

3. Too many people have problems dealing with credit cards. To see how your purchases can cost more (much more) when you charge, do the following exercise.

   Tracy's Department Store advertises an 18% interest rate on purchases with no interest if you pay off your bill within 30 days of purchase.

   You spend $100 and decide to pay the minimum of $20. Therefore, during the first month you must pay interest. You have

   $$\$100 - 20 = \$80$$

   left to pay on your debt, plus interest on the $80.

   a. Use d_R to stand for "debt remaining." Write a function (a rule) for calculating each month's remaining debt.

   b. A rule that can be used to calculate each month's remaining debt is

   $$d_R + 0.18d_R - 20 = \text{remaining debt}$$

   Apply the rule repeatedly until the remaining debt is zero.

   i. How much interest would you pay?

   ii. How many months will it take to pay the debt?

   iii. How much would your $100 charge really cost?

   iv. ✏ Would the rule

   $$\left(1 + \frac{0.18}{12}\right)d_R - 20 = \text{remaining debt}$$

   work also? Explain why.

   v. ✏ Explain how you can use the ⎡ ANS ⎤ button on your calculator to keep from re-entering the previous month's numbers.

Exploring Problem Solving

4. Students work with exponents a lot. The topic is tested often on exams used to determine a student's ability in mathematics. In 1997, Colin Rizzio, a high school student from Contoocook Valley Regional High School in Peterborough, New Hampshire, got an apology and additional points from the SAT testing agency when he got the right answer and it was marked wrong. Test your wits against theirs.

Solve this problem when a is positive. Solve this problem when a is negative.

The Question that Tripped Up the SAT

Consider the following sequence:

- $1, a, a$ squared, a to the third power . . . a to the nth power. (Each succeeding term is the product of a and the preceding term.)
- Now here are two quantities based on that sequence: **Column A:** The median of the sequence if n is a positive even integer; **Column B:** a to the power of one-half of n.
- Compare the quantities and pick: **A** if the quantity in Column A is greater; **B** if the quantity in Column B is greater; **C** if the two are equal; **D** if the relationship cannot be determined from the information given.
- **Explanation:** The test writers intended for you to assume that a is a positive number, in which case the correct answer is C. But if you assume that a could be either positive or negative, the correct answer is D.

Source: USA Today, 7 February 1997.

Exploring with Calculators

5. **a.** Use your graphing calculator to explore the following functions on the same viewing rectangle.

$$y_1 = e^x \quad y_2 = 1 + x \quad y_1 = \left(1 + \frac{x}{2}\right)^2 \quad y_4 = \left(1 + \frac{x}{4}\right)^4 \quad y_5 = \left(1 + \frac{x}{8}\right)^8$$

Sketch each function on a piece of graph paper. What appears to be happening to the graphs of y_2 through y_5 as the denominator of the coefficient of the variable x and the exponent of the function get larger?

b. Now plot $y_1 = e^x$ and $y_2 = \left(1 + \frac{x}{100}\right)^{100}$ on the same piece of graph paper. Then make a conjecture about the value of e^x as x gets bigger and bigger.

c. Graph $y = \ln(1 + e^x)$ and $y = x$ on the same piece of graph paper. What do you see happening as x gets larger and larger?

CHAPTER LOOK-BACK

Throughout the text, you have had the opportunity to see how numbers, equations, and other elements of mathematics are used.

As you read your next magazine or newspaper, see if you can find additional examples that are interesting to you.

We are always looking for interesting examples. If you would like to share number facts about your school or hometown that can be used in word problems, send them to the authors at Hunter College, 695 Park Avenue, New York, New York 10021.

CHAPTER **10**
REVIEW PROBLEMS

The following exercises will give you a good review of the material presented in this chapter.

SECTION 10.1

1. Determine whether $\{(2, 3), (4, 5), (6, 8)\}$ is a one-to-one function. If so, determine its inverse.
   yes; $\{(3, 2), (5, 4), (8, 6)\}$

2. Find the inverse of $f^{-1}(x) = \dfrac{2 - 3x}{4}$. $\quad y = \dfrac{2 - 4x}{3}$

3. Find $(f \circ f^{-1})(x)$ for $f^{-1}(x) = 3x - 16$ and its inverse. $\quad \dfrac{(3x - 16) + 16}{3} = x$

4. Find $(f \circ f^{-1})(x)$ for $f(x) = \dfrac{3 - 5x}{2}$ and its inverse. $\quad \dfrac{3 - 5\left(\frac{3 - 2x}{5}\right)}{2} = x$

5. Graph: $f(x) = -2x + 1$.

6. Graph the inverse of $f(x) = -2x + 1$.

SECTION 10.2

7. Graph: e^{-x}

8. Graph: $f(x) = e^{x+1}$

9. Evaluate: $f(x) = 2^{2x}$ when $x = -1$, when $x = 0$, and when $x = 2$. $\quad \frac{1}{4}; 1; 16$

10. Simplify: $\ln e^{6x-2}$ $\quad 6x - 2$

11. Simplify: $6 \log_3 3$ $\quad \log_3 3^6 = 6$

SECTION 10.3

12. Solve: $8^x = \dfrac{1}{2}$ $\quad x = \dfrac{-1}{3}$

13. Solve: $5^{2x-3} = 125$ $\quad x = 3$

14. Solve: $16^x \left(\dfrac{1}{32}\right)^3 = 2^{x-3}$ $\quad x = 4$

15. The function $P(t) = 5300(e^{0.07t})$ yields an estimate of the population of a Chilean city t years after 1986. What was its population in 1986? in 1992? $\quad 5300; 8066$

16. Write $3^8 = 6561$ in logarithmic form. $\quad \log_3 6561 = 8$

17. Write $\log_5 625 = 4$ in exponential form. $\quad 5^4 = 625$

18. Solve: $\log_3 x = 4$ $\quad x = 81$

19. Solve: $\log_{10} 10{,}000 = x$ $\quad x = 4$

20. Solve: $x^2 + 2 = \log_2 16$ $\quad x = \sqrt{2} \text{ or } x = -\sqrt{2}$

21. Solve: $16^x = \dfrac{1}{2}$ $\quad x = \dfrac{-1}{4}$

22. Write $\frac{1}{2} \log_2 3 - 3 \log_2 5$ as a single logarithm with a coefficient of 1. $\quad \log_2\left(\dfrac{\sqrt{3}}{5^3}\right)$

23. Write $2 \log_a 6 + (3 \log_a 5 - \log_a 9)$ as a single logarithm with a coefficient of 1. $\quad \log_a\left(\dfrac{6^2 \cdot 5^3}{9}\right)$

24. Solve for x: $\log_a x = 3 \log_a 16 - \log_a 4$ $\quad x = 1024$

25. Solve: $\ln 10 = 2 \ln 100 - \ln x$ $\quad x = 1000$

SECTION 10.4

26. *Measuring Sound* A scale of sound intensity has units called decibels, named after Alexander Graham Bell. A formula for determining the decibel level D of a sound is

$$D = 10 \log\left(\frac{I}{I_0}\right)$$

where I is the intensity of the sound, in watts per square meter, and I_0 is the intensity of the least audible sound that an average person can hear, which is 10^{-12} watt/square meter.

a. What is the decibel level of heavy traffic if $I = 8.5 \times 10^{-4}$? $D = 89.2942$ decibels

b. What is the intensity of a whisper if its decibel level is 27? $I = 5 \times 10^{-10}$ watt/square meter

c. What is the decibel level of normal conversation if I is 3.2×10^{-6}? $D = 65.0515$ decibels

In Exercises 27 through 32, evaluate each expression.

27. $\ln 6$ 1.7918

28. antiln 5   148.4132

29. $\log 1600$ 3.2041

30. antilog -6.2 6.3096×10^{-7}

31. $\log_5 18$ 1.7959

32. Solve: $e^{2x} = 6$ $0.8959 = x$

MIXED REVIEW

33. Solve: $\left(\frac{1}{3}\right)^{2x+1} = 8^x$ $x = -0.2569$

34. Determine whether $f(x) = 2x + 10$ is a one-to-one function, and then find its inverse if it has one.   yes; $\dfrac{x - 10}{2}$

35. What is the inverse of $f(x) = \log_3 x$? 3^x

36. Write $\frac{1}{3}(\log_3 5 + \log_3 7) - \log_3 8$ as a single logarithm with a coefficient of 1. $\log_3\left(\dfrac{\sqrt[3]{35}}{8}\right)$

37. Solve: $e^x = 2$ $x = 0.6931$

38. Determine whether $f(x) = x^2 + 2$ is a one-to-one function, and then find its inverse if it has one.   no, because $\pm\sqrt{y - 2} = x$ so there are two x-values for each value of y

39. Solve: $5^{x-3} = 30^x$ $x \approx -2.6947$

40. Write $5\log_3 16 - 6\log_3 1 + 1$ as a single logarithm with a coefficient of 1. $\log_3 (3 \cdot 16^5)$, or $\log_3 3{,}145{,}728$

41. Evaluate: $\log_{\frac{1}{2}} 16$ -4

42. Solve: $18^x = 36$ $x = 1.2398$

43. *Radioactive Decay*   The function $f(t) = 500e^{-0.8t}$ describes the mass in grams of a radioactive substance after t minutes.

a. What is the mass of the substance at time $t = 0$?   500 grams

b. How many grams remain after 10 minutes?   0.17 gram

c. When is half of the original amount left? This time is the half-life of the radioactive substance.   52 seconds

d. The half-life of carbon 14 is approximately 5570 years, and a general formula for the radioactive decay of carbon 14 is

$$A = A_0 e^{-0.0001244t}$$

Substitute $A_0 = 1000$ and $t = 5570$ years in the formula to verify that it gives the correct amount A. (*Hint:* What *should* A be?)

$A = A_0 e^{-0.0001244t}$
$500 = 1000e^{-0.0001244(5570)}$
$500 = 1000e^{-0.693}$
$500 = 1000(0.5)$
$500 = 500$ True

44. Graph: $f(x) = 3^{2x}$

CHAPTER 10 TEST

This exam tests your knowledge of the material in Chapter 10.

1. **a.** Is {(3, 2), (4, 8), (5, 8) (7, 2)} a function? Is it one-to-one? Justify your answer.   function; not one-to-one. See the Solutions Manual for a complete justification.

   b. Find $(f \circ f^{-1})(x)$ and $(f^{-1} \circ f)(x)$ for $f(x) = 2x$. $2\left(\frac{x}{2}\right) = x; \frac{(2x)}{2} = x$

   c. Graph $y = x - 2$ and its inverse.

   $y = x + 2; y = x - 2$

2. **a.** Graph: $f(x) = 2^{x+1}$

   b. Write $\log_a 6 = 15$ in exponential form. $a^{15} = 6$

   c. Write $e^5 = x$ in logarithmic form. $\ln x = 5$

3. **a.** Solve: $2^{5x+2} = \left(\frac{1}{32}\right)^2$ $x = -2.4$

   b. Solve: $4x^2 - 1 = \log_3 81^2$ $x = \frac{3}{2}$ or $x = \frac{-3}{2}$

   c. Solve: $\log_x 125 = -3$ $x = \frac{1}{5}$

4. **a.** Write $3 \log_a 2 - 4 \log_a 6 + \log_a 3$ as a single logarithm with a coefficient of 1. $\log_a\left(\frac{3 \cdot 2^3}{6^4}\right)$

   b. Write $\log_3\left(\frac{5x}{3y}\right)^6$ as sums and differences of logarithms. $6 \log_3 5 + 6 \log_3 x - 6 \log_3 y - 6$

   c. Simplify: $\ln e^{3x-2}$ $3x - 2$

5. Use $\log 2 = 0.3010$ and $\log 3 = 0.4771$ to evaluate:

   a. $\log_2 3$ 1.5850

   b. $\log \frac{2}{3}$ -0.1761

   c. $\log 2^{20}$ 6.02

6. *Thin Air*   The function $f(x) = 14.7e^{-0.2x}$ gives the average atmospheric pressure (in pounds per square inch) at any location that is x miles above sea level.

   a. What is the average atmospheric pressure in Denver, the "mile-high city"?   12.03 pounds per square inch

   b. Mt. McKinley is about 3.85 miles high. Mt. Everest is about 5.5 miles high. What is the difference between the atmospheric pressure at the top of Mt. Everest and that at the top of Mt. McKinley?   1.9 pounds per square inch

   c. Marie Byrd Land in Antarctica is about 1.5 miles below sea level. What is the average atmospheric pressure in Marie Byrd Land?   19.8 pounds per square inch

CUMULATIVE REVIEW

CHAPTERS 1–10

The following exercises will help you maintain the skills you have learned in this and previous chapters.

1. Solve: $\dfrac{4}{x^2 + 3x - 10} + \dfrac{1}{x^2 + 9x + 20} = \dfrac{2}{x^2 + 2x - 8}$ $x = \dfrac{-4}{3}$

2. Simplify: $3\sqrt{24} + 2\sqrt{48} + \sqrt{36} - \sqrt{-27}$ $6\sqrt{6} + 8\sqrt{3} + 6 - 3i\sqrt{3}$

3. *Optics*   A formula used in the study of optics is

$$f = \dfrac{1}{\dfrac{1}{p} + \dfrac{1}{q}}$$

   where

   p = distance from the object to the lens
   q = distance from the image to the lens
   f = focal length of the lens

   Determine p when q is 15 centimeters and f is 10 centimeters   30 centimeters

4. Factor: $2x^2 + 3x - 9$ $(2x - 3)(x + 3)$

5. A fraction has a numerator that is $\dfrac{1}{3}$ the denominator. If 5 is added to the numerator and subtracted from the denominator, the new fraction is equivalent to 1. What was the original fraction? $\dfrac{5}{15}$

6. Rewrite using only positive exponents: $\left(\dfrac{3x^3y^{-2}z^{-5}}{x^{-5}y^3z^{-2}}\right)^{-2}$ $\dfrac{y^{10}z^6}{9x^{16}}$

7. If $f(x) = 2x^2 + 4x - 9$, and $g(x) = -3x^2 - 4x + 12$, find $\left(\dfrac{f}{g}\right)x$. $\dfrac{2x^2 + 4x - 9}{-(3x^2 + 4x - 12)}$

8. Solve: $-3(x - 4) + 5(x + 9) = -6(x - 3)$ $x = \dfrac{-39}{8}$

9. When the object is 60 centimeters from the lens and q is 12 centimeters, what is the focal length? (Use the formula from Problem 3.)   The focal length is 10 centimeters.

10. Graph: $-3x + 2y = 9$

11. Simplify and write using one radical: $10^{\frac{2}{3}} \cdot 100^{\frac{1}{2}} \cdot 1000^{\frac{3}{4}} \cdot 8^{\frac{2}{3}} \cdot 25^{\frac{2}{3}}$ $2^5 \cdot 5^5 \sqrt[12]{2^{11}5^3}$

12. Simplify: $\sqrt{20x} \cdot \sqrt{45x}$ $30x$

13. Use the quadratic formula to solve $4x^2 - 3x + 5 = 0$. $x = \dfrac{3 + i\sqrt{71}}{8}$ and $x = \dfrac{3 - i\sqrt{71}}{8}$

14. Find the solution set: $\dfrac{(-5x)}{8} \geq \dfrac{1}{64}$ $\left\{x \mid x \leq \dfrac{-1}{40}\right\}$

15. Find the slope of $8x - 2y + 5 = 0$.   slope: 4

16. Write the equation of the line that passes through the points (1, 2) and (4, 3). $y = \frac{1}{3}x + 1\frac{2}{3}$

17. **a.** When the height of a trapezoid is reduced by 2 and the sum of the two bases is tripled, the new area is 2.25 times the original area. If the sum of the original bases is 44, find the original height. 8 units

    b. Find the difference between the areas. 220 square units

18. Evaluate the determinant: $\begin{vmatrix} 15 & 8 \\ 3 & 0 \end{vmatrix}$ -24

19. Simplify using scientific notation: $\dfrac{(27{,}000{,}000)(3{,}000)}{(81{,}000{,}000{,}000)}$ 1

20. Let $f(x) = 400x^2 - 49x + 3$. Find $f(0)$ and $f(-1)$. $f(0) = 3; f(-1) = 452$

ANSWERS to Warm-Ups

SECTION 1.1 *(page 13)*

1. [number line with points at $-5.111\ldots$, $-\frac{2}{3}$, and 2.17; marks at -3, 0, 2.17]   **2.** $2\frac{1}{2} > -3.927 > -195.62 > -503$ **3.** $-17; 29$ **4.** $170; 174.0; 200; 173.985$
5. 3   **6.** 7.054×10^7 **7.** 0.000000719

SECTION 1.2 *(page 30)*

1. a. $-180\frac{1}{2}$ **b.** -157.972 **2. a.** $-14\frac{5}{6}$ **b.** 12.92 **3.** 3.73 **4. a.** 0.7828 **b.** $\frac{2}{5}$
5. a. 0.08 **b.** -0.48 **6.** $2\frac{4}{11}$ **7.** $2\frac{4}{5}$

SECTION 1.3 *(pages 45–46)*

1. a. true; associative property of multiplication   **b.** true; commutative property of addition
2. multiplication property of -1; distributive property; multiplication property of -1 **3.** $-\frac{1}{8}; -16$ **4.** $-\frac{1}{2^2}; \frac{1}{3^3}$
5. $\left(\frac{3}{2}\right)^1; \left(\frac{2}{3}\right)^2$ **6.** $4; -\frac{8}{27}$ **7.** $256; 9$ **8.** 92 **9.** 74.75 **10.** $-10\frac{5}{6}$ **11. a.** 1080 **b.** 0.000009

SECTION 1.4 *(page 61)*

1. $\{x \mid x$ is an integer greater than -2 and less than $5\}; \{-1, 0, 1, 2, 3, 4\}$ **2. a.** $\{1, 2, 3, 4, 5, 7\}$
b. $\{1, 2, 3, 4, 5, 7\}$ **3. a.** $A' = \{6, 7, 8, 9, 10\}$ **b.** $B' = \{2, 4, 6, 7, 8, 9, 10\}$ **c.** $(A \cap B)' = B'$
4. 16 times   **5.** about 3 minutes   **6.** 30%   **7.** 117 people

SECTION 2.1 *(page 85)*

1. -222 **2.** -114 **3.** $-102\frac{1}{9}$ **4.** $33 + 12t$ **5.** $-3r + 2r^2$ **6.** $3xy^2 + 3x^2y$ **7.** $5x^2 - 5$

SECTION 2.2 *(page 99)*

1. $n = -13$ **2.** $n = -9$ **3.** $n = -459$ **4.** $x = -100$ **5.** $y = -1\frac{1}{4}$ **6.** $t = 20$ **7.** $x = 4\frac{1}{6}$
8. $n = -4$ **9. a.** all real numbers   **b.** no solution, or $\varnothing$ **10.** no solution   **11.** $w = 6; w = 4$

SECTION 2.3 *(page 112)*

1. $3n + 62$ **2.** number of dimes in stack $= d; 10d$ cents   **3.** Answers may vary. The sum of c and 9, multiplied
by twenty-seven   **4.** $r = \frac{14n}{3}$ **5.** $y = \frac{5}{4}x - \frac{5}{2}$ **6.** approximately 272 trees   **7.** 77 miles per hour

SECTION 2.4 *(pages 134–135)*

1. $\{y \mid y \le -1\}$; Answers may vary. Sample:
   $y = -2$ $y = 0$
   $0.3 \ge -0.7$ True $0.3 \ge 1.3$ False
2. $\{r \mid r \ge 9\}$ **3.** $\left\{r \mid r > -20\frac{4}{7}\right\}$ **4.** $r > 1.45$ **5.** $\{m \mid m < 2\}$ **6.** $\{x \mid x < 3\}$ **7.** $\{x \mid -11 < x \le 14\}$
[number line: $-12\ -8\ -4\ \ 0\ \ 4\ \ 8\ \ 12\ 16$]   **8.** $\{x \mid x \ge 10\}$ [number line: $0\ \ 5\ \ 10\ \ 15$]   **9.** $\{n \mid 1 \le n \le 11\}$ [number line: $1\ \ 3\ \ 5\ \ 7\ \ 9\ \ 11$]
10. $\left\{x \mid x < -4 \cup x > 1\frac{1}{2}\right\}$ [number line: $-4\ \ -2\ \ 0\ \ 2$]

SECTION 2.5 *(page 145)*

1. 14 accounts   **2.** The boyfriend is 59; Anita is 49.   **3.** $19, 21, 23$ **4.** $6\frac{2}{3}$ hours   **5.** from 460 to 1300
beats per 10 minutes   **6.** from \$11.97 to \$14.22

SECTION 3.1 *(page 163)*

1. yes   **2.** $\frac{1}{2}$ **3.**   **4.** $(2, 4) = C$, quadrant I; $(-6, -5) = D$, quadrant III; $(2, -7) = B$,

quadrant IV; $(-5, 3) = A$, quadrant II   **5.** Tables may vary.

x	y
0	-3
1	-1
2	1

SECTION 3.2 *(page 181)*

1. Tables may vary.

x	y
1	-4
0	-6
3	0

    2.   **3. a.** x- and y-intercepts: $(0, 0)$

b. y-intercept: $(0, 2)$; x-intercept: $\left(\frac{-2}{5}, 0\right)$ **c.** y-intercept: $\left(0, -1\frac{1}{2}\right)$; x-intercept: $(-3, 0)$ **4.** $\frac{-2}{9}$ **5. a.** $m = 0$

b. The x-intercept is every point on the x-axis. The y-intercept is the origin, the point $(0, 0)$.   **6.** slope of $\overline{AB} = 0$;

slope of $\overline{BC}$ is undefined; slope of $\overline{CD} = 0$; slope of $\overline{AD}$ is undefined; x-intercepts: $(2, 0)$ and $(-4, 0)$; y-intercepts: $(0, -5)$

and $(0, 4)$ **7.**

x	y
21	48
3	30
13	40

SECTION 3.3 *(page 199)*

1. $y = -3x + 6$; $m = -3$; y-intercept: $(0, 6)$ **2.** (a) $y = 2x - 2$ (b) $y = 2x + 6$ (c) $y = \frac{-1}{2}x + 5$ (d) $y = \frac{1}{2}x - 2$.
The graphs of equations (a) and (b) are parallel. The graphs of equations (a) and (c) and those of equations (b) and (c) are perpendicular and intersect. The graphs of equations (a) and (d), (b) and (d), and (c) and (d) intersect only.
3. $y = \frac{-3}{8}x$ **4.** $y = 2.6x + 8.7$ **5.** $y = \frac{-1}{2}x - 4$ **6.** $y = -2x - 3$ **7.** $y = 2x$ **8.** $126.10

SECTION 3.4 *(page 215)*

1.   **2.**   **3.**   **4.**   **5.**

SECTION 3.5 *(page 225)*

1. $R = \{(1, 3), (2, 4), (3, 3)\}$; Domain: $\{1, 2, 3\} = A$; Range: $\{3, 4\} = B$ **2.** $\left\{\left(\frac{1}{2}, 1\right), (1, 2), \left(1\frac{1}{2}, 3\right)\right\}$; Range: $\{1, 2, 3\}$
3. $f(0) = -2$; $f(2) = 10$; $f(a) = 3a^2 - 2$ **4.** Both graphs show functions.

SECTION 4.1 *(page 245)*

1. $\{(3, 1)\}$ **2.** $\{(-3, -4)\}$ **3.** $\{(-3, 2)\}$ **4.** $\left\{\left(5\frac{2}{9}, 2\frac{1}{6}\right)\right\}$

5. $\{(-1.1, 1.4)\}$ **6.** $\{(2, -1, 1)\}$ **7.** $\{(1, 2)\}$ **8.** $\{(15, -6, 2)\}$

SECTION 4.2 *(page 263)*

1. 12.5 mph   **2.** 28 days   **3.** adult tickets: 125; student tickets: 150   **4.** Pointers Nuts: 7.5 ounces; Nestors Nuts: 12.5 ounces   **5.** At 11 miles the charge will be the same for both car services.   **6.** $4000 at 5%; $3500 at 6%; $500 at 4%   **7.** Blake weighs 64 pounds. Catherine weighs 55 pounds. Bobby weighs 78 pounds.; yes

SECTION 4.3 *(page 280)*

1.   **2.**   **3.**   **4.**

SECTION 4.4 *(page 296)*

1. -22 **2.** -4 **3.** 79   **4.** -36; calculations may vary.   **5.** $\{(0, 1)\}$ **6.** $\{(0, -2, -7)\}$

SECTION 5.1 *(page 317)*

1. $-6n^6 t$ **2.** $73.6x^{11}y^8$ **3.** $72r^2 t^2$ **4.** $4t^8 r^2$ **5.** $\dfrac{1}{6x^2 y^2 z}$ **6.** $81n^4 y^8$ **7.** $\dfrac{1}{64m^3 n^6}$ **8.** $\dfrac{x^8}{256r^8 t^{24}}$
9. $\dfrac{r^{16}t^{16}}{81}$ **10. a.** 1.5834×10^{10} **b.** 4.0×10^{-12}

SECTION 5.2 *(pages 333–334)*

1. degree 4   **2.** $12m^2 + 2m + 5$ **3.** $3y^2 + 16y - 20$ **4.** $8n^2 + 3$ **5.** $x^2 y - 7xy - 9y + 5$
6. $7t^2 - 17t$ **7.** 4   **8.** $(f - g)(x) = -x^2 + x + 2$; domain: {all real numbers}   **9.** $2x^4 - 2x^3 + 6x^2$
10. $\dfrac{14x^2}{3} - 2x + \dfrac{2}{3x^2}$ **11.** $-3nx^2 + \dfrac{3}{5}$ **12.** They are equal: $\left(-\dfrac{11}{9}\right)$.

SECTION 5.3 *(page 352)*

1. $6y^3 - 11y^2 - 43y + 84$ **2.** $72t^3 - 96t^2 - 138t + 84$ **3.** $28n^2 - 18n - 16$ **4.** $t^2 - 16$
5. $25x^2 - 10x + 1$ **6.** $216x^3 - 216x^2 + 72x - 8$ **7.** $81x^4 - 72x^2 y^2 + 16y^4$ **8.** $x - 4$
9. $x - 1 + \dfrac{1}{3x - 2}$ **10.** $t^3 - 3$ **11.** $x + 4$ **12.** $n^3 + 2$

SECTION 5.4 *(pages 368–369)*

1. $4xy^4$ **2.** $12x^2 y^3(4y^2 + 3x^2)$ **3.** $24r^4 y^3(y^3 - 2r)$ **4.** $2(y - 3)(y - 1)$ **5.** $(2t + 3)(2t + 5)$
6. $(17 + n)(17 - n)$ **7.** $(11t + 2)(11t - 2)$ **8.** yes   **9.** $(r - 10)^2$ **10.** $3(t - 11y)^2$
11. $27(3y - x)(9y^2 + 3xy + x^2)$ **12. a.** $y(y^2 - 9y + 27)$ **b.** $17(2x - 1)$

SECTION 5.5 *(page 390)*

1. $(n + 5)(n + 1)$ **2.** $(n - 10)(n - 2)$ **3.** $4(x - 6)(x + 3)$ **4.** $2(x - 3y)(x - 12y)$ **5.** $(3x + 2)(x + 1)$
6. $(35x + 1)(x - 7)$ **7.** $(g \circ f)(x) = 14x^2 - 1$; $(g \circ f)(-2) = 55$ **8.** $(f \circ h \circ g)(x) = 4x^2 + 28x + 48$;
$(g \circ f \circ h)(x) = 4x^2 - 4x + 4$ **9.** $\{1, 2\}$ **10.** $\left\{-\dfrac{1}{3}, 0, 4\right\}$ **11.** $\left\{\dfrac{3}{7}, -\dfrac{9}{5}\right\}$ **12.** 27 and 31

Section 6.1   *(page 410)*

1. $\{y \mid y$ is real and $y \neq 2\}$ 2. $\frac{y}{n^2}$ 3. $\frac{m+5}{m+2}$ 4. $\frac{7}{v-15}$ 5. $x-7$ 6. 9 7. $\frac{2n}{n+6}$

8. $h(x) = \frac{x-3}{x+4}; h(-2) = \frac{-5}{2}; h(a-2) = \frac{a-5}{a+2}$ 9. $f(x) = 0$ when $x = -16$ or $x = 16; f(x)$ is undefined when

$x = -2$ 10. $\left(\frac{f}{g}\right)(y) = \frac{y-6}{y+4}$; domain of $f(y)$: $\{y \mid y$ is real and $y \neq 7, -4\}$; domain of $g(y)$: $\{y \mid y$ is real and $y \neq 7, 6\}$;

domain of $\left(\frac{f}{g}\right)(y)$: $\{y \mid y$ is real and $y \neq -4, 2, 4, 6, 7\}$

Section 6.2   *(page 421)*

1. $\frac{7y}{8}$ 2. $24(x-1)^2$, or $24x^2 - 48x + 24$ 3. $\frac{3x-4}{(x+2)(x-2)}$ and $\frac{2x^2+9x+10}{(x+2)(x-2)}$ 4. $\frac{2t+3}{3(t-5)}$ 5. $\frac{x+6}{3(x-3)}$

6. $\frac{3x^2}{2(x+5)(x-5)}$ 7. $\frac{y^2-11y+12}{3(y-2)}$

Section 6.3   *(page 441)*

1. 20 2. $\frac{8(y+21)}{7(y-16)}$ 3. $\frac{4}{r^2}$ 4. $\frac{3x^2-12x+21}{-x^2-17x-25}$ 5. $x=3$ 6. $n=4\frac{3}{5}$ 7. $\frac{P}{v^3} = K; K \approx 0.0148$

Section 6.4   *(page 455)*

1. 6 and 48   2. 110 tourists   3. The wind speed was 1.6, or $1\frac{3}{5}$ kilometers per hour. It took the skater $5\frac{5}{24}$
hours to complete the skate-a-thon.   4. 24 minutes   5. Hose 1 fills the pool in 32 minutes; hose 2 fills the pool
in 96 minutes.   6. $t = kw$ 7. $f = \frac{k}{t}$ 8. $F = ksh$ 9. 75

Section 7.1   *(page 481)*

1. $9; \frac{5}{6}x^2$ 2. $|y^2-3|$; no real root   3. $20; 2x^2\sqrt{65y}; 3y^2\sqrt[3]{y}$ 4. $20\sqrt{5}; 15$ 5. $\frac{2}{x}; \frac{y^2}{x}\sqrt[5]{\frac{5}{6}}$
6. $\sqrt[3]{x^3y^4} = xy\sqrt[3]{y}; \sqrt[4]{3x^2}$ 7. $74^{\frac{1}{2}}x^{\frac{1}{2}}; 3^{-\frac{1}{3}}x^{\frac{1}{3}}$ 8. $x^{\frac{2}{5}}y^{\frac{3}{5}}$ 9. 8

Section 7.2   *(page 500)*

1. $8\sqrt{5}$ 2. $4xy\sqrt[3]{2y} + 2xy\sqrt{2y}$ 3. $-4\sqrt{6y} + 5\sqrt{3y}$ 4. $102 - 12\sqrt{30}$ 5. $32\sqrt{2} + 1$
6. $\sqrt{6} - \sqrt{3}$ 7. We cannot simplify through division because the radicals do not have the same index.
8. $\frac{\sqrt{6}}{9}$ 9. $\sqrt[5]{(6x)^2}$ 10. $-15 + 3\sqrt{30} + 25\sqrt{3} - 15\sqrt{10}$ 11. $18\sqrt{2}$ 12. 16

Section 7.3   *(pages 516–517)*

1. $r = 25$ 2. $r = 9\frac{1}{5}$ 3. $r = -1$ or $r = -3$ 4. $t = \frac{121}{25}$ 5. $x = \sqrt{69}$ or $x = -\sqrt{69}$ 6. $x = 10$
7. $x = 64$ 8. $a = 18.04$ inches   9. $V = 215.11$ feet per second

Section 7.4   *(page 534)*

1. $-1; -i$ 2. $5i; 4i\sqrt{2}$ 3. $27\sqrt{-1}; -\sqrt{26}$ 4. $-60i; \frac{-7i}{3}$ 5. $-26 + 18i$
6. $20i$ 7. $48 - 140i$ 8. 234 9. $\frac{15-9i}{34}$
10. $|-2+i| = \sqrt{5}; |0 - 3i| = 3; |\sqrt{-25}| = 5$

SECTION 8.1 *(pages 554–555)*

1. $\left\{-\frac{1}{2}, 5\right\}$ **2.** $x = -3 - 3i$ and $x = -3 + 3i$ **3.** Answers may vary. Examples: $x^2 - 6x + 13 = 0$; $2x^2 - 12x + 26 = 0$; $-5x^2 + 30x - 65 = 0$ **4.** $x = \pm 8i$ **5.** $x = \pm 2\sqrt{2}$ **6.** $x = \frac{2 \pm 7i}{3}$ **7.** $\left\{\frac{4}{7}, -\frac{5}{7}\right\}$ **8.** $x = -3 \pm i\sqrt{3}$ **9.** $x = \frac{1}{4} + 3i$ and $x = \frac{1}{4} - 3i$

SECTION 8.2 *(page 510)*

1. $\left\{\frac{5}{4}, -\frac{2}{3}\right\}$ **2.** $\left\{5 + \sqrt{34}, 5 - \sqrt{34}\right\}$ **3.** two complex solutions; $\left\{-1 + i\sqrt{2}, -1 - i\sqrt{2}\right\}$ **4.** $\{9\}$ **5.** $\left\{\pm\sqrt{2}, \pm\sqrt{3}\right\}$ **6.** $\{256\}$ **7.** $\left\{\sqrt[6]{4}, \sqrt[6]{-3}\right\}$ **8.** $\left\{-2\frac{1}{2}, -3\frac{1}{3}\right\}$ **9.** $\{25, 1\}$ **10.** $\{9\}$

SECTION 8.3 *(page 578)*

1. 12.24 feet tall **2.** $144\frac{2}{3}$ feet **3.** $y = 4.16$ and $y = -2.16$

SECTION 8.4 *(page 595)*

1.   **2.** Estimates may vary but should be close to $x = 7.5$ and $x = 0.5$.

3. $\{y \mid y \le -10 \text{ or } y \ge 19\}$ 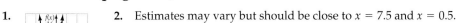 **4.** $\{x \mid -3 < x < 0 \text{ or } x > 3\}$

5. $\{x \mid x < -\frac{1}{3} \text{ or } x \ge 0\}$

SECTION 9.1 *(page 620)*

1. 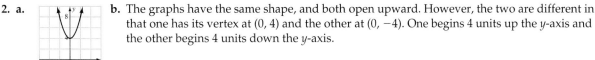 The graphs are different in that one opens upward and the other opens downward. The graphs have the same vertex and the same shape.

2. a. 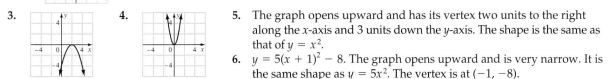 **b.** The graphs have the same shape, and both open upward. However, the two are different in that one has its vertex at $(0, 4)$ and the other at $(0, -4)$. One begins 4 units up the y-axis and the other begins 4 units down the y-axis.

3. 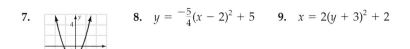 **4.** **5.** The graph opens upward and has its vertex two units to the right along the x-axis and 3 units down the y-axis. The shape is the same as that of $y = x^2$.

6. $y = 5(x + 1)^2 - 8$. The graph opens upward and is very narrow. It is the same shape as $y = 5x^2$. The vertex is at $(-1, -8)$.

7. **8.** $y = \frac{-5}{4}(x - 2)^2 + 5$ **9.** $x = 2(y + 3)^2 + 2$

SECTION 9.2   *(page 635)*

1. a. $x^2 + y^2 = 49$ **b.**   **2.** major axis: 12; minor axis: 10   **3.**

SECTION 9.3   *(page 653)*

1. $(x - 12)^2 + (y + 3)^2 = 4$ **2.** $(x - 10)^2 + y^2 = 82$ **3.** center: $(2, -5)$; radius: $2\sqrt{7}$
4. ellipse with center at $(6, 18)$; major axis: $2 \cdot 9 = 18$ (horizontal); minor axis: $2 \cdot 7 = 14$

5.   **6.**   **7.** $\dfrac{(y - 2)^2}{4} - \dfrac{(x - 5)^2}{144} = 1$ **8.** ellipse

SECTION 9.4   *(pages 666–667)*

1. 114 square inches   **2.** 3.25 by 3.25 feet   **3.** $y = \frac{6}{25}(x + 5)^2 + 6$ **4.** 2.06×10^{-1}
5. Answers may vary. Examples: **6.** $\{(2, 0), (-2, 0)\}$ **7.** $\{(4, -3), (-4, 3)\}$

(a) three solutions   (b) four solutions

SECTION 10.1   *(pages 689–690)*

1. one-to-one   **2.**   not one-to-one   **3.** $\dfrac{x + 7}{2}$

4. $f \circ f^{-1}(x) = 2 - (2 - x) = 2 - 2 + x = x$; $f^{-1} \circ f(x) = 2 - (2 - x) = 2 - 2 + x = x$
5. $f^{-1}(x) = \dfrac{x + 3}{2}$ **6.**

SECTION 10.2   *(page 709)*

1.   **2.**   **3.**   **4.** $\log_5 125 = 3$; $\log_7 x = 5$

5. $3^4 = 81$; $10^{-3} = 0.001$ **6.**   **7.**   **8.**

SECTION 10.3 *(page 723)*

1. $\log_d (8x)$ 2. $\log_9 3 + \log_9 a - \log_9 6$ 3. $2 \log_b 2 + 2 \log_b x - 5 \log_b 3 - 5 \log_b a$ 4. 10 5. $x = 11$
6. $x = \dfrac{7}{6}$ 7. $x = \dfrac{9}{4}$ 8. $x = 4$ 9. 4 10. $x = 0$ 11. $x = \dfrac{2}{3}$ 12. $x = \dfrac{3}{5}$

SECTION 10.4 *(pages 736–737)*

1. 316.2278 2. 0.8417 3. $x = 0.2624$ 4. $x = 0.5396$ 5. $x = 3.0104$ 6. $x = -29.4247$
7. 1.0×10^{-7} 8. 150.7 million bacteria   9. 6100 years   10. about 22.52 years

ANSWERS to Chapter 1 Odd-Numbered Exercises

SECTION 1.1 *(pages 14–19)*

1. $2\frac{1}{2}$ 3. -1 5. 22.1 7. -69 9. -38 11. $-x$ if $x \geq 0$, or x if $x < 0$ 13.

15.   17. $-|-2981| < -|-|264|| < -|7|$ 19. $-58 < -\left|48\frac{1}{5}\right| < 6.4 < 28$ 21. $<$
23. $=$ 25. $>$ 27. $<$ 29. $246; 245.9; 245.908$ 31. $8; 7.7; 7.725$ 33. $2; 1.6; 1.618$ 35. $0; 0.2;$
0.174 37. greater than $\frac{1}{2}$ 39. less than $\frac{1}{2}$ 41. 15 43. 32 45. rubber, petroleum, alcohol,
beechwood   47. $0.08333333 > 0.07692307 > 0.07142857 > 0.06666666$ 49. **a.** $6.85; 6.94999; 6.9049$ **b.** $7.23;$
7.2349875 **c.** $1.99; 1.957867$ **d.** $3.0399876; 3.04499$ **e.** $2.1949; 2.18997; 2.1854$ **f.** $3.61107; 3.610997$
51. $10{,}000{,}000$ vehicles   53. $20{,}000$ graduates   55. **a.** violet **b.** yellow **c.** red **d.** violet **e.** blue
f. orange **g.** yellow **h.** red   57. 1.4×10^1 grams per cubic centimeter   59. 1.0×10^0 gram per cubic
centimeter   61. 8.0×10^8 calls   63. 2.6×10^{14} ergs   65. 0.0000000667 centimeter per second per second
67. $43{,}560$ square feet   69. 0.0000068 mm   71. $\$5{,}200{,}000{,}000$ 73. 3.65×10^9 pounds   75. 2.92×10^{11}
calls   77. 3.0×10^9 years   79. 6.0×10^9 ergs   81. $5{,}800{,}000$ pounds   83. $450{,}000{,}000{,}000$ gallons
85. $2{,}678{,}400$ seconds   87. Answers may vary.

SECTION 1.2 *(pages 30–32)*

1. -487 3. 43.8 5. $-43\frac{1}{4}$ 7. 1.02 9. -63 11. -57.6 13. 39.7 15. -2.22 17. $16\frac{7}{8}$
19. $-10\frac{1}{4}$ 21. $-2\frac{9}{10}$ 23. $-21\frac{2}{3}$ 25. $1\frac{5}{21}$ 27. 37.1 29. greater by $44\frac{1}{24}$ 31. less by 2.26
33. $-35\frac{2}{9}$ 35. $-24\frac{1}{6}$ 37. $\frac{-1}{6}$ 39. 0 41. -54 43. $30\frac{1}{3}$ 45. 0.0084 47. 70
49. -1960 51. $-1\frac{1}{4}$ 53. 0.8 55. $-18\frac{1}{2}$ 57. 0.04 59. $4\frac{2}{3}$ 61. $\frac{-1}{2}$ 63. $\frac{11}{2}$
65. -11.55 67. $-1\frac{1}{4}$ 69. $\frac{-6}{25}$ 71. -0.02 73. $1\frac{1}{8}$ 75. 24.986 77. 848 79. $\frac{3}{10}$
81. 1.45 83. -47.4 85. $-1\frac{3}{8}$ 87. $\left|-4\frac{3}{7}\right|$ 89. $-1\frac{19}{85}$ 91. $-9\frac{5}{8}$ 93. -3.6 95. -328
97. $2\frac{1}{4}$ 99. $1\frac{83}{203}$

SECTION 1.3 *(pages 46–49)*

1. commutative property of addition
$$336 + [24 + (-5)] = [24 + (-5)] + 336$$
$$336 + 19 = 19 + 336$$
$$355 = 355$$

3. distributive property
$$6(5 - 7) - 9 = (30 - 42) - 9$$
$$(6 \cdot 5 - 6 \cdot 7) - 9 = -12 - 9$$
$$(30 - 42) - 9 = -21$$
$$-12 - 9 = -21$$
$$-21 = -21$$

5. multiplication property of -1
$$25 = (-1)(-25)$$
$$25 = 25$$

7. associative property of addition
$$5 + (3 + 2) = (5 + 3) + 2$$
$$5 + 5 = 8 + 2$$
$$10 = 10$$

9. associative property of multiplication
$$\left(5\frac{1}{3}\right)\left[\left(-2\frac{1}{2}\right)(-4)\right] = \left[\left(5\frac{1}{3}\right)\left(-2\frac{1}{2}\right)\right](-4)$$
$$\left(5\frac{1}{3}\right)\left[\left(-\frac{5}{2}\right)(-4)\right] = \left[\left(\frac{16}{3}\right)\left(-\frac{5}{2}\right)\right](-4)$$
$$\left(5\frac{1}{3}\right)[(-5)(-2)] = \left[\frac{(8)(-5)}{3}\right](-4)$$
$$\left(\frac{16}{3}\right)[10] = \left[\frac{-40}{3}\right](-4)$$
$$\frac{160}{3} = \frac{160}{3}$$

11. associative property of multiplication
$$(2 \cdot 5) \cdot 9 = 2 \cdot (5 \cdot 9)$$
$$10 \cdot 9 = 2 \cdot 45$$
$$90 = 90$$

13. distributive property
$$8(6 + 16) = (8)(6) + (8)(16)$$
$$8(22) = 48 + 128$$
$$176 = 176$$

15. 5; 715; commutative property of addition   **17.** $-2\frac{1}{2}$; 53; associative property of addition
19. 16; 18^2; commutative property of multiplication   **21.** 27; $6a$; commutative property of addition
23. 2; 8; 5; 3; commutative property of multiplication   **25.** 321; 5; 3; commutative property of addition
27. 256; $5\frac{1}{2}$; associative property of multiplication   **29.** distributive property; multiplication and multiplication property of -1 **31.** distributive property; multiplication and multiplication property of -1 **33.** commutative property of addition; associative property of addition   **35.** associative property of multiplication; commutative property of multiplication; associative property of multiplication   **37.** false   **39.** false   **41.** false
43. true   **45.** false   **47.** false   **49.** false   **51.** true   **53.** 1   **55.** -1 **57.** $\left(\frac{6}{5}\right)^3$ **59.** $\left(\frac{-3}{4}\right)^2$
61. 3   **63.** 1000   **65.** -22 **67.** -64 **69.** 3.4   **71.** 29   **73.** 0.995   **75.** 9   **77.** -1
79. -38 **81.** 24,404   **83.** $-6,000,000,000,000,000$ **85.** 10   **87.** 1,440,000,000   **89.** -27
91. 0.004   **93.** 4,800,000,000   **95.** $-7\frac{2}{5} < 7\frac{1}{4} < 7.3$ **97.** $1\frac{1}{4}$ **99.** commutative property of addition

SECTION 1.4   *(pages 61–69)*

1. Not empty. 0 is an element.   **3.** Empty. No number is both even and odd.   **5.** $\{0, 2, 4, 6, \dots\}$
7. $\{2, 3, 5, 7, 11, 13, 17, 19, 23, 29\}$ **9.** $\{1, 2, 3, 4, 6, 12\}$ **11.** $\{-4, -3, -2, -1, 0, 1, 2, 3, 4, 5, 6, 7, 8, 9\}$
13. true   **15.** true   **17.** $\{0, 2, 3, 4, 5, 9\}$ **19.** $\{0\}$ **21.** $A \cup B = \{$integers$\}$ or $A \cup B = B$; $A \cap B = \{$whole numbers$\}$ or $A \cap B = A$ **23.** $A \cup B = \{n \mid n$ is a whole number$\}$; $A \cap B = \varnothing$ **25.** $2\frac{3}{4}$ inches
27. $181\frac{2}{5}$ miles   **29.** 25 steps   **31.** 666,000 bills   **33.** 37 touchdowns   **35.** 13 feet $2\frac{1}{16}$ inches
37. 36 times longer   **39.** 12.35%   **41.** 100 bpm   **43.** 125°F   **45.** $9\frac{5}{8}$ **47.** 24 less than at the start
49. 5.5 meters   **51.** 9,269,000 members   **53.** $6.87   **55.** 24%   **57.** 22.8 ounces   **59.** 75% ocean;
25% land   **61.** 19,775 feet $\frac{59}{144}$ inches

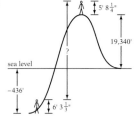

63. a. $66.08   **b.** Answers may vary.

65. a. $\frac{1}{5}; \frac{1}{7}; \frac{2}{7}$; $\frac{1}{7}, \frac{1}{5}, \frac{5}{7}$ **b.** Answers may vary.   **67.** $62.10   **69.** 1337 cubic centimeters
71. Answers may vary.   **73.** Answers may vary.   **75.** 15 semesters   **77.** 6 trips   **79.** $533
81. 2290 feet higher   **83.** 550   **85.** $-7\frac{11}{24}$ **87.** $1\frac{9}{16}$ **89.** $\varnothing$ **91.** distributive property
93. 64.0; 63.99; 63.985   **95.** $\frac{1}{9}$ **97.** 58,433,000 square miles

CHAPTER 1 REVIEW PROBLEMS   *(pages 71–73)*

1. true   **2.** -6 **3.** $-8.932 < -8.75 < -8.7$ **4.** $2.4 > \left\| \frac{-9}{5} \right\| > -\left| \frac{11}{3} \right|$ **5.** 86.3; 86.29; 86.294   **6.** 20
7. 40,000   **8.** 7000   **9.** $-20\frac{3}{20}$ **10.** $2\frac{5}{6}$ **11.** $-1\frac{5}{56}$ **12.** -2.389 **13.** -6.3 **14.** 54
15. 204.6   **16.** 18   **17.** $-3\frac{1}{3}$ **18.** -0.24 **19.** $\frac{-1}{2}$ **20.** $\frac{-2}{3}$ **21.** 5; $3x$; commutative property of multiplication   **22.** 7; -63; 16; commutative property of addition   **23.** 27; 17; associative property of addition
24. 15; $2x$; 9; distributive property   **25.** 144   **26.** 9   **27.** 40   **28.** $-8\frac{17}{20}$ **29.** 1.96   **30.** $\frac{5}{8}$
31. -5 **32.** 676   **33.** $\{5, 6\}$ **34.** $\{5, 6\}$ **35.** $\{0, 5, 6, 13\}$ **36.** $2000   **37.** $\{x \mid x$ is an even integer and $-6 \le x \le 4\}$; $\{-6, -4, -2, 0, 2, 4\}$ **38. a.** $\{0, 1, 2, 3, 4\}$ **b.** $\{0, 1, 2\}$ **39.** 1800 people
40. 40 mannequins   **41.** 27   **42.** $-\left| \frac{-23}{5} \right|$ is larger   **43.** 957.0; 956.98; 956.981   **44.** 76.5   **45.** -40
46. 3969

ANSWERS to Chapter 2 Odd-Numbered Exercises

SECTION 2.1 *(pages 85–87)*

1. terms: $-11r^2t, 5xt^3, -8rt^3, 7rx$; coefficients: $-11, 5, -8, 7$ **3.** terms: $xtz, -3xz^2, tz$; coefficients: $1, -3, 1$
5. -3 **7.** -512 **9.** -768 **11.** $\frac{1}{16}$ **13.** -760 **15.** $28\frac{1}{2}$ **17.** -108 **19.** -7 **21.** -20
23. -16 **25.** $-\frac{2}{3}$ **27.** -57 **29.** -656 **31.** 1560 **33.** -54 **35.** $\frac{9}{32}$ **37.** $-8\frac{11}{18}$
39. $-14\frac{2}{3}$ **41.** $-450 - 4n$ **43.** $283\frac{2}{3} - 2t$ **45.** variable parts: ny^2, ny^3, ny, n^2y;
$-9ny^2 - ny^3 - 12ny + 8n^2y - 3$ **47.** variable parts: ry, rt^2, rt; $ry + 14rt^2 + 4rt - 3$ **49.** $2n^3 + 3n^2 - n + 6$
51. $12t^2 - 3t$ **53.** $18t - 7$ **55.** $-12kt + 56$ **57.** $7k^3n + 61$ **59.** $7k + 51$ **61.** $-m - 8y - 33$
63. $8r^2t - 48rt + 21$ **65.** $5n - 9x + 32$ **67.** $3.2y - 44.4$ **69.** $-1.1n + 3.4x$
71. $-180n^2y - 15ny^2 - 9$

SECTION 2.2 *(pages 100–101)*

1. $x = 7$ **3.** $y = -50$ **5.** $y = \frac{-5}{8}$ **7.** $n = -0.7$ **9.** $x = 4$ **11.** $n = -40$ **13.** $n = -3$
15. $t = -0.06$ **17.** $x = -0.475$ **19.** $n = -0.15$ **21.** $r = 0.4$ **23.** $n = -90$ **25.** $x = -12\frac{3}{5}$
27. $n = 56$ **29.** $x = 44$ **31.** $n = 32$ **33.** $x = -280$ **35.** $r = 6\frac{2}{3}$ **37.** $x = -4\frac{7}{10}$ **39.** $r = 120$
41. $x = -2\frac{1}{2}$ **43.** $y = 2\frac{5}{6}$ **45.** $n = 9$ **47.** conditional **49.** contradiction **51.** $v = 3$ **53.** $y = 2$
55. $y = 8\frac{1}{2}$ **57.** $n = -6$ **59.** $x = \frac{-3}{32}$ **61.** $x = 22$ **63.** $x = \frac{-6}{13}$ **65.** $r = 2\frac{4}{13}$ **67.** $x = -13\frac{1}{3}$
69. $x = \frac{-9}{34}$ **71.** $t = 2$ **73.** $x = \frac{-1}{41}$ **75.** $x = 4\frac{19}{22}$ **77.** identity; all real numbers **79.** $t = \frac{-23}{27}$
81. $p = 4; p = -4$ **83.** no solution **85.** $x = 6\frac{3}{4}; x = -6\frac{1}{4}$ **87.** $x = 1; x = \frac{-1}{3}$ **89.** $x = \frac{1}{2}; x = -5\frac{1}{2}$
91. no solution **93.** $x = -3; x = \frac{-1}{3}$ **95.** $x = 3; x = 13$ **97.** no solution **99.** $t = -2$
101. $x = 112$ **103.** variable part: a^3; exponent: 3; coefficient: -7 **105.** $t = -10\frac{1}{2}$ **107.** $y = \frac{3}{4}$
109. $y = -14$ **111.** $-44m - 10$ **113.** $x = 4\frac{2}{3}; x = -1\frac{1}{3}$

SECTION 2.3 *(pages 112–118)*

1. $n - 4$ **3.** t^2 **5.** $528 \cdot n$ **7.** $0.85 + n$ **9.** price of butter $= B; B - \$1.94$ **11.** cost of
leather $= C; 1\frac{1}{4}C$ **13.** annual insurance bill $= I; \frac{1}{12} \cdot I$ **15.** cost of one item $= n; 15n$ **17.** difference of 8
and $x = 8 - x; 2(8 - x)$ **19.** number of dimes $= n$, number of nickels $= y; n + y$ **21.** Catherine's age
now $= C; 2(C + 3)$ **23.** product of 8 and $x = 8x; 8x + 6$ **25.** a number $= n; n + \left(n - \frac{1}{5}\right)$
27. a number $= n; n^2 + (n + 2)$ **29.** a number $= n; (4n)^2 \div 23$ **31.** a number $= n; n - (n + 6)$
33. a number $= n; \left(n - \frac{4}{7}\right) + n$ **35.** a number $= n; n - (n + 0.3)$ **37.** the difference of 15 and y, or y less
than 15 **39.** the quotient of 19.6 and x **41.** the sum of the product of 3.4 and k, and 6 **43.** the sum of 6
and the square of x **45.** the product of 2.5 and the difference of y and 3 **47.** the product of 5.2 and x divided
by the product of 18 and x **49.** the sum of n and 6 times c **51.** the quotient of n and 7 plus 2 **53.** the
product of 6 and n divided by c **55.** the difference of n and 10 plus the product of d and c **57.** $155r$ dollars
59. $d - 4.15$ dollars **61.** $50n + 6p$ cents **63.** nh hamburgers **65.** c stamps **67.** np people
69. $6n + 80d$ dollars **71.** $(n + 2) + (n - d + 2)$ years **73.** $f = e - y$ **75.** $w = -r + j$ **77.** $w = jr$
79. $x = \frac{he}{t}$ **81.** $j = \frac{1}{4}(r - w)$, or $\frac{r}{4} - \frac{w}{4}$ **83.** $m = r - \frac{x}{y}$ **85.** $x = -yz - tz$ **87.** $m = -xr + 4xy$
89. $x = \frac{4}{5}y + \frac{12}{5}$ **91.** $y = \frac{z}{6(t + x)}$ **93.** $P = \frac{I}{RT}$ **95. a.** $F = \frac{1}{3.1}(S + 23.9)$ **b.** about 10.6 inches
97. a. $W = \frac{P}{2} - L$ **b.** $L = \frac{P}{2} - W$ **99.** $P = \frac{MS}{m}$ **101.** $V = \frac{D - r}{Q}$ **103.** $D_a = \frac{L}{V} + D_h$ **105.** $T = 3$
107. about 3.25 feet **109.** about 8 minutes and 21 seconds, or 8.35 minutes, per mile **111.** $13x + 21y$
113. 62 **115.** no solution, or $\varnothing$ **117.** $c = h(y + w)$ **119.** $x = 2\frac{2}{5}$ **121.** $w = \frac{P}{2} - l$

Section 2.4 *(pages 135–137)*

1. no **3.** no **5.** no **7.** yes **9.** $\{t \mid t > 33\}$ **11.** $\{y \mid y \geq 6\}$ **13.** $\{y \mid y < 11\}$

15. $\{t \mid t > -5\}$ **17.** $\left\{m \mid m \leq \frac{1}{3}\right\}$ **19.** $\{y \mid y \geq -1\}$

21. $\left\{y \mid y < -1\frac{5}{9}\right\}$ **23.** $\left\{y \mid y \geq -45\right\}$

25. $\left\{m \mid m > \frac{3}{8}\right\}$ **27.** $\{y \mid y \leq 3\}$

29. $\left\{y \mid y \geq \frac{7}{4}\right\}$ **31.** $\left\{y \mid y < \frac{2}{3}\right\}$

33. $\left\{y \mid y > \frac{2}{9}\right\}$ **35.** $\{m \mid m \geq 96\}$

37. $\{y \mid y \geq -5\}$ **39.** $\{m \mid m < -4.6\}$ **41.** All real numbers

43. No solution **45.** No solution **47.** **49.**

51. **53.** $\{x \mid 9 < x < 15\}$

55. $\left\{x \mid -9\frac{1}{2} < x < 3\frac{1}{2}\right\}$ **57.** $\{x \mid -11 < x < -8\}$

59. $\{x \mid -7 < x \leq -3\}$ **61.** $\{x \mid x < 1 \cup x > 4\}$

63. $\{x \mid x > 2\}$ **65.** $\varnothing$; no real solutions No solution

67. $\{x \mid -1 \leq x \leq 4\}$ **69.** $\{y \mid y < 2 \cup y > 10\}$

71. $\{x \mid x < 1\}$ **73.** $\{m \mid -3 < m < 3\}$

75. $\{x \mid x \geq 8 \cup x \leq -10\}$ **77.** $\left\{y \mid y > 2 \text{ or } y < \frac{-14}{3}\right\}$

79. $\{r \mid r \geq 6 \cup r \leq -2\}$ **81.** $\left\{y \mid y \leq \frac{-29}{15}\right\}$

83. $\{y \mid y \geq -2.3\}$ **85.** $\{y \mid y > -5\}$

87. $\left\{y \mid y > \frac{49}{75}\right\}$ **89.** $x = 216$ **91.** $\{x \mid x \leq -7\}$ **93.** $\left\{x \mid x > -\frac{1}{10}\right\}$ **95.** $x = \frac{f - e}{r}$

97. a number $= n$; product of 4 and $n = 4n$; $\frac{1}{3}(4n)$ **99.** $W = \frac{V}{LH}$

Section 2.5 *(pages 145–149)*

1. 572 units **3.** 1200 millimeters **5.** -323 and -275 **7.** The price for towing 25 miles in Chicago is $50; in New Jersey, $41; in Atlanta, $30. **9.** Joella is 61; her husband is 57. **11.** 20 years old **13.** -321, -319, and -317 **15.** $-2, -1, 0,$ and 1 **17.** 8891 **19.** Thompson's time was 14.33 seconds; O'Brien's time was 13.98 seconds. **21.** after 3.5 hours **23.** 18 miles per hour **25.** 13 days **27. a.** 44 ounces **b.** Answers may vary. **29.** Marisol is 38; Sansi is 22. **31.** 72,000,000 bags were exported. Each bag weighed 132 pounds. **33.** from $207 to $945 **35.** least: 1620 ounces; most: 1738.8 ounces **37.** from 1260 to 2800 calories **39.** smallest: 4325 cubic inches; largest: 4362 cubic inches **41.** $\{x \mid -13 < x < -10\}$ **43.** $x = \frac{y - b}{a}$ **45.** $x = 0.04$ **47.** $\{x \mid x \geq -0.025\}$ **49.** -96 **51.** $n + (n \div 10)$ **53.** the difference between the product of 4 and a number and the product of 7 and that number

Chapter 2 Review Problems *(pages 151–152)*

1. variable part: x^4; coefficient: -6 **2.** $-2, 6x, -5x^2$ **3.** -2 **4.** 2 **5.** -30 **6.** -4 **7.** 0 **8.** $4rt^2 - 6rt + 2r^2t$ **9.** $ab + 6a + 11b$ **10.** $a + 18$ **11.** $16 - 4x$ **12.** $8y - 9x$ **13.** $x = -58$ **14.** $t = -94\frac{1}{2}$ **15.** $x = 88$ **16.** $x = -16$ **17.** $y = \frac{11}{3}$ **18.** $r = -2184$ **19.** $x = 1$ **20.** $y = 20; y = 4$ **21.** $x = \frac{1}{8}$ **22.** $x = 5\frac{1}{2}$ **23.** $y = -206; y = 206$ **24.** $m = 1\frac{1}{6}$ **25.** $e = \frac{11}{5}f$ **26.** $x = a(c - f)$ **27.** $s = \frac{11r + 4w}{t}$ **28.** $w = \frac{xt - mr}{m}$ **29.** $\{x \mid x \geq -8.2\}$ **30.** $\left\{x \mid x < 5\frac{1}{6}\right\}$ **31.** no solution **32.** $\left\{c \mid c < 21\frac{1}{3}\right\}$ **33.** $\{x \mid x > -6\}$ **34.** $\{x \mid x > 20 \cup x < -10\}$ **35.** $\left\{x \mid x < \frac{4}{7} \text{ or } x \geq 5\right\}$ **36.** $\{x \mid x \leq -1\}$

37. the difference between x and 7   **38.** size of a walnut: w; $2\frac{2}{3}w$ **39.** $-x$ **40.** x **41.** 13   **42.** 17

43. 15.6 feet and 5.2 feet   **44.** $n = 102$ **45.** $x = \frac{22}{17}$ **46.** $y = \frac{-54}{29}$ **47.** 640   **48.** $7r^2t - rt^2 - 9$

49. $\left\{y \mid y > -\frac{18}{7}\right\}$

$\overset{-20}{7}\ \overset{-19}{7}\ \overset{-18}{7}\ \overset{-17}{7}\ \overset{-16}{7}$

50. $\frac{rw + e - 3}{e}$ **51.** $-\frac{5}{961}$ **52.** $\left\{y \mid \frac{7}{8} < y < \frac{9}{8}\right\}$

$\overset{7}{8}\ \ 1\ \ \overset{9}{8}$

53. $\{y \mid y \text{ is a real number}\}$

All real numbers

54. $\{x \mid -1 \leq x \leq 18\}$

$\overset{-10}{\bullet}\qquad\qquad\overset{18}{\bullet}$

CUMULATIVE REVIEW *(pages 153–154)*

1. $-33\frac{5}{8}$ **2.** $x = -25$ **3.** $A \cup B = \{0, 1, 2, 3, 8\}$ **4.** $\left\{x \mid -\frac{4}{3} < x < 3\right\}$ **5.** $y = \frac{8}{3} - \frac{2}{3}x$ **6.** 80; 83; 82.98; 82.977   **7.** 30   **8.** $m = \frac{y - b}{x}$ **9.** $\left|-\frac{18}{3}\right| > \left|-\frac{14}{6}\right| > -\left|-\frac{12}{5}\right|$ **10.** 23,010   **11.** $\frac{47}{4}$ **12.** 4.303   **13.** $t \leq -1$ **14.** 304   **15.** $x = 18$ **16.** $(x + 6) \cdot y$ **17.** 0.042   **18.** $-\frac{4}{3}$ **19.** $14x - 4$ **20.** \$18.48

Answers to Chapter 3 Odd-Numbered Exercises

SECTION 3.1 *(pages 164–167)*

1. nonlinear   **3.** linear   **5.** nonlinear   **7.** nonlinear   **9.** $x = \frac{-25}{91}y - \frac{5}{91}$; $y = \frac{-91}{25}x - \frac{1}{5}$

11. $x = -36y - 9$; $y = -\frac{1}{36}x - \frac{1}{4}$ **13.** $x = \frac{-17}{78}y + \frac{19}{78}$; $y = \frac{-78}{17}x + \frac{19}{17}$ **15.** $x = \frac{4}{27}y + \frac{2}{9}$; $y = \frac{27}{4}x - \frac{3}{2}$

17. To find y: Multiply 17 times -1, obtaining -17. Then subtract 21 (or add negative 21), obtaining -38. So y is equal to -38.   **19.** To find x: Multiply 0.3 times 50, obtaining 15. Then compute 15 plus -17, obtaining -2. So x is equal to -2.   **21.** yes; yes   **23.** no; yes   **25.** yes; yes   **27.** no; yes   **29. a.** (5, 17) **b.** $(-3, -23)$ **c.** $\left(2\frac{2}{5}, 4\right)$ **d.** $\left(\frac{2}{5}, -6\right)$ **31. a.** (5, 46) **b.** $(-3, -10)$ **c.** $(-1, 4)$ **d.** $\left(-2\frac{3}{7}, -6\right)$ **33.** $\frac{3}{5}$ **35.** 1   **37.** $\frac{6}{7}$ **39.** $-1\frac{1}{2}$

41.

x	y
6	10
2	0
-2	-10

43.

x	y
2	10
-2	-10
6	30

45. $A = (1, 1)$, quadrant I; $B = (-6, 5)$, quadrant II; $C = (-5, -5)$, quadrant III; $D = (7, -7)$, quadrant IV

47. $A = (15, -15)$, quadrant IV; $B = (-30, -30)$, quadrant III; $C = (-20, 5)$, quadrant II; $D = (20, 25)$, quadrant I

49. Tables may vary.   **51.** Tables may vary.   **53.** Tables may vary.   **55.** Tables may vary.

x	y
0	6
-2	8
-3	9

x	y
-10	0
0	-40
-9	-4

x	y
2	18
-2	-2
0	8

x	y
0	0
2	-4
-2	4

57. Tables may vary.   **59.** Tables may vary.   **61.** Tables may vary.

x	y
1	0
0	-2
-2	-6

x	y
8	0
0	-8
16	8

x	y
0	10
-3	1
-5	-5

63. Tables may vary.

x	y
-4	1
-14	2
6	0

65. Tables may vary.

x	y
0	0
2	-1
4	-2

67. Tables may vary.

x	y
0	$\frac{1}{2}$
2	$6\frac{1}{2}$
-2	$-5\frac{1}{2}$

69.

71.

SECTION 3.2 *(pages 182–186)*

1. **3.** **5.** **7.** **9.**
   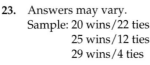

11. **13.** **15.** **17.** **19.**

21. $102.95

23. Answers may vary.
Sample: 20 wins/22 ties
25 wins/12 ties
29 wins/4 ties

25. a. $8.95
b. 25 pounds
c. 75 pounds

27. a. 80 miles
b. 50 miles
c. The point (30, 35) means that when the bus has traveled 30 miles, the car has traveled 35 miles.

29. x-intercept: $(-2, 0)$
y-intercept: $(0, 2)$

31. x-intercept: $(-3, 0)$
y-intercept: $(0, -3)$

33. x-intercept: $\left(\frac{2}{3}, 0\right)$
y-intercept: $(0, -2)$

35. x-intercept: $\left(\frac{2}{5}, 0\right)$
y-intercept: $(0, -2)$

37. x-intercept: $\left(2\frac{2}{3}, 0\right)$
y-intercept: $(0, 4)$

39. x-intercept: $\left(\frac{2}{3}, 0\right)$
y-intercept: $\left(0, \frac{-2}{5}\right)$

41. x-intercept: $\left(-\frac{3}{2}, 0\right)$
y-intercept: $(0, 3)$

43. x-intercept: $\left(-\frac{3}{2}, 0\right)$
y-intercept: $(0, -3)$

45. x-intercept: $\left(-\frac{3}{2}, 0\right)$
y-intercept: $(0, 3)$

47. no x-intercept
y-intercept: $(0, 2)$

49. x-intercept: $(6, 0)$
y-intercept: $(0, -2)$

51. $m = \frac{2}{5}$ **53.** $m = 3$ **55.** $m = 5$ **57.** $m = \frac{-3}{13}$ **59.** $m = 0$ **61.** $m = \frac{-4}{5}$ **63.** $m = \frac{-2}{3}$

65. $m = 1; m = -1; m = 1; m = -1$ **67.** $m = \frac{2}{7}; m = 8; m = \frac{2}{7}; m = 8$ **69.** Answers may vary. Sample: $(1, 3)$

71. Answers may vary. Sample: $(7, -5)$ **73.** undefined **75.** $m = -\frac{1}{3}$ **77.** $x = \frac{-2}{3}$ **79.** no

81. yes **83.** yes **85.** $x = 8y + 9; y = \frac{1}{8}x - \frac{9}{8}$ **87.**

89. x-intercept: $\left(\frac{-7}{2}, 0\right)$; y-intercept: $(0, 7)$ **91.** Tables may vary.

x	y
0	$\frac{1}{2}$
1	0
-1	1

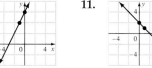

93. yes

SECTION 3.3 *(pages 199–204)*

1. $m = -2$; y-intercept: $(0, 4)$ **3.** $m = 3$; y-intercept: $(0, -5)$ **5.** $m = \frac{1}{2}$; y-intercept: $\left(0, \frac{3}{2}\right)$

7. $m = 2$; y-intercept: $(0, 3)$ **9.** **11.** **13.**

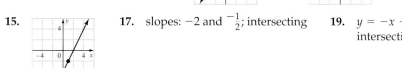

15. **17.** slopes: -2 and $\frac{-1}{2}$; intersecting **19.** $y = -x + 7$; slopes: 1 and -1; perpendicular and intersecting

21. $y = x + 3$; slopes: -1 and 1; perpendicular and intersecting **23.** $y = \frac{1}{2}x + 1; y = \frac{1}{2}x - 1$; slopes: $\frac{1}{2}$ and $\frac{1}{2}$; parallel **25.** $y = \frac{1}{8}x + 7$; slopes: 8 and $\frac{1}{8}$; intersecting **27.** $y = 8x - 6$; slopes: 8 and 8; parallel

29. $y = \frac{-1}{2}x + \frac{2}{3}; y = -2x + \frac{5}{3}$; slopes: $\frac{-1}{2}$ and -2; intersecting **31.** $y = \frac{1}{3}x - 1; y = 3x - \frac{1}{2}$; slopes: $\frac{1}{3}$ and 3; intersecting **33.** $y = \frac{-1}{2}x + 11$; slopes: -2 and $\frac{-1}{2}$; intersecting **35.** $y = -4x + 7$; slopes: $\frac{1}{4}$ and -4; perpendicular and intersecting **37.** **39.** **41.**

43. **45.** $y = -2x + \frac{1}{3}$ **47.** $y = \frac{-1}{2}x + 8$ **49.** $y = \frac{3}{5}x - 6$ **51.** $y = -4x - 38$

53. $y = 2x + 9$ **55.** $y = \frac{1}{2}x - 4\frac{1}{2}$ **57.** $y = \frac{-1}{4}x - \frac{3}{4}$ **59.** $y = 2\frac{2}{3}x + 3\frac{2}{3}$ **61.** $y = -3\frac{1}{2}x + 9$

63. $y = x + 7$ **65.** $y = \frac{-1}{5}x$ **67.** $y = -2x$ **69.** $y = -3x - 8$ **71.** $y = 5x - 22$ **73.** $y = \frac{1}{3}x - 3\frac{1}{3}$

75. $y = \frac{3}{4}x + 1\frac{3}{4}$ **77.** $y = -2$ **79.** $y = -x - 1$ **81.** $y = \frac{1}{14}x + 2\frac{2}{7}$ **83.** $y = 2x + 3$

85. $y = 3x + 25$ **87.** $y = \frac{2}{3}x - 5$ **89.** $y = \frac{1}{3}x + 7$ **91.** $y = 4x + 8$ **93.** $y = \frac{1}{3}x - 3\frac{2}{3}$

95. When the temperature is 0°C, the equivalent temperature is 32°F. For every 5° that the temperature changes on the Celsius scale, it changes 9°F. **97. a.** slope: -125 **b.** $y = -125T + 16{,}250$ **c.** T-intercept: (130, 0); V-intercept: (0, 16,250). The T-intercept has no real-life meaning because 130 is not a reasonable temperature. The V-intercept means that when it is 0 degrees, 16,250 people will vote. **99. a.** $y = 29x + \$2.00$ **b.** No. Because x represents distance, and distance is *not* negative. **c.** No. (0, \$2) would mean that you rode no miles (no tenths of a mile) and paid \$2.00. Hardly likely. The y-intercept would require x to be negative. That is not possible here. **101.** no

103. no **105.** $x = -y - 7; y = -x - 7$ **107.** x-intercept: $\left(-\frac{3}{7}, 0\right)$; y-intercept: $\left(0, -\frac{3}{2}\right)$

109. slope: 2; y-intercept: $\left(0, -\frac{9}{2}\right)$

SECTION 3.4 *(pages 216–219)*

1. **3.** **5.** **7.**  **9.**

11. **13.** **15.** **17.** **19.**

21. **23.** **25.** 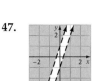 **27.** **29.**

31. **33.** **35.** A **37.** B **39.** B **41.** D

43. **45.** **47.** **49.** **51.**

53. **55.** Answers may vary. Sample: (0, 9), (0, 10), (0, 11) **57.** $y = \frac{14}{9}x - \frac{1}{3}$ **59.** $y = -6x - 4$

61.

SECTION 3.5 *(pages 225–227)*

1. Domain: {even whole numbers} Range: {odd whole numbers}

3. Domain: {3, 4} Range: {all real numbers}

5. Domain: {whole numbers}; Range: {5}

7. Answers may vary. Sample: $\{(0, 0), (1, 3), (2, 6), (3, 9)\}$ **9.** Answers may vary. Sample: $\{(0, -3), (1, -2), (2, -1),$ $(3, 0)\}$ **11.** Answers may vary. Sample: $\{(0, -4), (1, -2), (2, 0), (3, 2)\}$ **13.** Answers may vary. Sample: $\{(0, 0),$ $(1, 0), (2, 2), (3, 6)\}$ **15.** $f(0) = 7; f(2) = 11; f(-1) = 5$ **17.** $f(0) = -2; f(4) = 14; f(-3) = 7$
19. $f(0) = 1; f(a) = -2a + 1; f(x + 2) = -2x - 3$ **21.** $f(0) = 3; f(a) = 3 - 2a; f(x - 1) = 5 - 2x$ **23.** function
25. function **27.** not a function **29.** **31.** -8

33. Answers may vary. Sample: $(0, 1), (0, 0), (0, 2)$ **35.** Range: $\{-3, 2, 3\}$; Domain: $\{0, 5, 6\}$ **37.** $\frac{-6}{5}$
39. $D - 1.0825T$

CHAPTER 3 REVIEW PROBLEMS *(pages 229–230)*

1. $y = 20\frac{2}{3}$ **2.** no **3.** $A = -6\frac{1}{2}$ **4.** $A = (-5, 1)$, quadrant II; $B = (3, -5)$, quadrant IV; $C = (-1, -2)$,
quadrant III; $D = (2, 4)$, quadrant I **5.** Answers may vary. Sample: $\left(0, \frac{3}{2}\right), \left(1, \frac{5}{2}\right), \left(-1, \frac{1}{2}\right)$ **6.** $\left(\frac{13}{2}, 1\right)$
7. **8.** **9.** **10.**

11. a. $100 \stackrel{?}{=} 8(12 - 0) + 4$ **b.** A score of 8 correct (4 missed) gives a 68 on the test. **12.** 84 **13.** $m = \frac{5}{4}$
$100 \stackrel{?}{=} 8(12) + 4$
$100 \stackrel{?}{=} 96 + 4$
$100 = 100$ True
14. y-intercept: $(0, -22)$; x-intercept: $\left(-9\frac{3}{7}, 0\right)$ **15.** $m = \frac{-4}{11}$ **16.** $m = \frac{2}{17}$ **17.** $y = x - 8$ **18.** parallel
19. $y = \frac{4}{3}x + \frac{3}{2}$ **20.** intersecting only **21.** $y = x - 1$ **22.** intersecting only **23.** $y = -2x - 22$
24. $y = \frac{1}{2}x - \frac{1}{2}$ **25.** $y = -6x - 3$ **26.** $y = 5x - 12$ **27.** $y = -2x - 1$ **28.** $y = -x + 6$
29. **30.** Answers may vary. Sample: $(0, 0)$ and $(1, 0)$ **31.** **32.**

33. $f(0) = -3; f(-1) = -5$ **34.** $f(0) = 5; f(-2) = -1$ **35.** Domain: $\{0, 1, 4, 8\}$; Range: $\{3, 5, 6\}$
36. Answers may vary. Sample: $(2, 1), (6, 3), (4, 2), (10, 5)$ **37.** $y = \frac{1}{6}x - \frac{5}{6}$ **38.**
39. $y = -x + 3$ **40.** no

CUMULATIVE REVIEW *(page 232)*

1. 3.3009 **2.** $\frac{-1}{25}$ **3.** $2\frac{4}{5}$ **4.** $x = a(c - b)$ or $x = ac - ab$ **5.** She has $25(n + 6)$ cents. **6.** no solution
7. 10 **8.** 90; 89.4; 89.39, 89.395 **9.** 7; 3; 3; distributive property **10. a.** $A \cap B = \{1, 3, 5\}$

b. $A \cup B = \{0, 1, 2, 3, 5\} = B$ **11.** $x \le -0.9$ **12.**   **13.** $y = \frac{7}{12}x + 1\frac{1}{2}$ **14.** $2\frac{5}{6}$

15. $x = 3\frac{1}{3}$ **16.** -56 **17.**   **18.** 4   **19.** $y < \frac{1}{3}$

20. x-intercept: $\left(5\frac{2}{3}, 0\right)$; y-intercept: $\left(0, -8\frac{1}{2}\right)$

ANSWERS to Chapter 4 Odd-Numbered Exercises

SECTION 4.1   (pages 245–247)

1. yes   **3.** yes   **5.** yes   **7.** consistent   **9.** consistent   **11.** dependent and consistent
13. 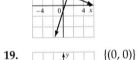 $(1, 2)$ **15.** $(4, 4)$ **17.** no point of intersection

19. $\{(0, 0)\}$ **21.** $\{(0, 1)\}$ **23.** $\{(4, 1)\}$ **25.** $\{(3, 5)\}$

27. $\{(1, -1)\}$ **29.** $\{(3, -4)\}$ **31.** $\{(3, 2)\}$

33. $\{(-3, 0)\}$ **35.** $\{(-8, -2)\}$ **37.** $\{(-5, 7)\}$ **39.** $\left\{\left(3, 1\frac{3}{4}\right)\right\}$

41. The system is inconsistent. There is no solution.   **43.** $\{(5, 4)\}$ **45.** $\{(-9, 9)\}$ **47.** $\{(-3, 12)\}$
49. $\left\{\left(1\frac{1}{2}, 5\frac{1}{2}\right)\right\}$ **51.** $\{(4.6, -3.4)\}$ **53.** $\{(-3.25, 1)\}$ **55.** $\left\{\left(-\frac{1}{2}, -\frac{2}{3}\right)\right\}$ **57.** $\{(9, 2)\}$ **59.** $\{(-3, -2)\}$
61. $\{(11, 2)\}$ **63.** $\{(0.8, -0.5)\}$ **65.** $\{(0.8, -0.5)\}$ **67.** $\{(-2, 3)\}$ **69.** $\left\{\left(\frac{1}{2}, -\frac{2}{5}\right)\right\}$ **71.** $\{(3, 0, 1)\}$
73. $\{(7, -3, 2)\}$ **75.** $\{(-3, -13, -5)\}$ **77.** $\{(-1, -2, -1)\}$ **79.** $\{(3, -12, 1)\}$ **81.** $\{(3, 7, -5)\}$
83. $\{(3, 1, 4)\}$ **85.** $\{(8, -5, -10)\}$ **87.** $\{(-2, 11, 5)\}$

SECTION 4.2   (pages 264–268)

1. 2-point baskets: 5; 3-point baskets: 3   **3.** wind: 50 km/hr; plane: 550 km/hr   **5.** 9.4 pounds
7. cabinet maker: $21 per hour; apprentice: $11 per hour   **9.** A: $654; B: $446   **11.** 194,716,200 cubic feet
13. 1932; 2000 letters   **15.** 55%   **17.** 60 stories; 22 mph   **19.** first cauldron: 40 gallons;
second cauldron: 60 gallons   **21.** 5723 square feet   **23.** rollerblader: 6.75 mph; race walker: 2.25 mph

25. day 1: 26 quarts; day 2: 18 quarts; day 3: 20 quarts   **27.** $5000 at 5%; $4000 at 7%; $6000 at 9%
29. 297   **31.** Jackson: 44; Lewis: 35; Gwyn: 19   **33.** 78 and 57   **35.** $\{(21, -13)\}$
37. 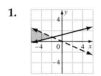 $\{(4, -3)\}$ **39.** $\{(0.5625, -0.1875)\}$ or $\left\{\left(\frac{9}{16}, -\frac{3}{16}\right)\right\}$ **41.** $\{(1, -1)\}$

SECTION 4.3   *(pages 280–282)*

1. 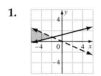   **3.**   **5.**   **7.**   **9.**

11.   **13.**   **15.**   **17.**   **19.**

21.   **23.**   **25.** $\geq$; $>$; $<$; $\leq$ **27.** $>$; $\geq$ **29.** $>$; $\leq$; $\leq$

31. $>$; $<$; $\geq$ **33.**   **35.**   **37.** $\{(-2, 2.5)\}$ **39.** consistent

41. $\{(1.5, -8)\}$ **43.** $\{(-5, 8, -2)\}$

SECTION 4.4   *(pages 297–299)*

1. 2   **3.** 2   **5.** 8   **7.** -80 **9.** 194   **11.** 78   **13.** 0; 24   **15.** $-30; 4$ **17.** $-14; -2$
19. $10; -6$ **21.** -403 **23.** 2097   **25.** 311   **27.** -2 **29.** $\left\{\left(-1, \frac{1}{2}\right)\right\}$ **31.** $\left\{\left(1\frac{20}{23}, 2\frac{2}{23}\right)\right\}$
33. $\{(0.2, -2)\}$ **35.** $\{(-3, 5)\}$ **37.** $\{(8, -4)\}$ **39.** $\{(9, 6)\}$ **41.** $\{(-3.6, -0.4)\}$
43. solution cannot be found by Cramer's rule because $D = 0$ **45.** $\{(1, 4, -1)\}$ **47.** $\{(1, 1, 3)\}$
49. $\left\{\left(-2\frac{1}{3}, 3\frac{1}{3}, 5\right)\right\}$ **51.** $\{(3.5, 4.5, 1)\}$ **53.**  $\{(-2, 2)\}$ **55.** 8   **57.** $\{(5, -3, 2)\}$ **59.** -74

61. plane: 125 mph; wind: 25 mph   **63.** I am 50, Bill is 60, and Barbara is 41.

CHAPTER 4 REVIEW PROBLEMS   *(pages 303–304)*

1. $\{(4, 2)\}$ **2.** $\{(-4, -5)\}$ **3.** $\{(-4, 5)\}$

4. $\{(2, -3)\}$ **5.** $\{(4, 3)\}$ **6.** $\{(-3, 2)\}$ **7.** $\{(7, -3)\}$ **8.** $\{(10, 1)\}$ **9.** $\{(3, 5, -2)\}$

10. $\{(-1, 2, 1)\}$ **11.** 13 quarters; 7 dimes **12.** 2:30 P.M.; 70 miles
13. 5.5 pounds of potato salad; 4.5 pounds of fruit salad **14.** 45 ounces **15.** $1.35
16. Ruth ran 9 miles; Anita ran 11 miles; Vincent ran 6 miles.
17.   **18.**   **19.**   **20.**

21. $\{(18, -7)\}$ **22.** $\{(-5, 2)\}$ **23.** $\{(-4, 8, 6)\}$ **24.** $\{(6, 3)\}$ **25.**

26. $\{(10. 5. -2)\}$

CUMULATIVE REVIEW *(pages 305–306)*

1. $-1 < x < 2$ **2.** $2(x - y)$ **3.** perpendicular and intersecting **4.** 3; 3; -7; distributive property
5. $x = 4\frac{5}{6}$ **6.** x-intercept: $(-5, 0)$; y-intercept: $(0, -5)$ **7.** 16 **8.** $324\frac{1}{2}$ **9.** -6
10. $15x^2y - 8xy^2 + 11x^3 + 12y^2$ **11.**   **12.** $4\frac{1}{9}$ **13.** -180 **14.** 22

15.   **16.**

x	y
-1	1
$\frac{2}{3}$	0
4	-2

17. $\{(-2, -1)\}$ **18.** $y = \frac{1}{3}x - 3$

19. $-3x^2 + 2x - 7$ **20.** 1

ANSWERS to Chapter 5 Odd-Numbered Exercises

SECTION 5.1 *(pages 318–319)*

1. $-12x^5y^3$ **3.** $\frac{-5x^5y^7}{2}$ **5.** $\frac{x^9y^2}{5}$ **7.** $-9.03x^5y^7$ **9.** $\frac{1}{m^4}$ **11.** $\frac{1}{p^2}$ **13.** $\frac{3}{x^2y}$ **15.** $\frac{-6y^3}{x^5}$
17. $\frac{-60}{x^8}$ **19.** $12xy$ **21.** $\frac{-84m^5}{t^5}$ **23.** $\frac{34wx^2}{y}$ **25.** $121x^4y^{10}$ **27.** $-27x^3y^6$ **29.** $\frac{32w^{15}x^{20}}{243y^{20}}$
31. $\frac{625y^8}{x^{12}}$ **33.** $\frac{4y^6}{x^{10}}$ **35.** $\frac{64x^6y^8}{81}$ **37.** $\frac{1}{w^6x^3y^3}$ **39.** $\frac{y^{15}}{32x^5}$ **41.** $-\frac{1}{64w^6x^9}$ **43.** $\frac{49x^{14}y^{12}}{25w^8}$ **45.** $\frac{w}{x^2}$
47. $\frac{27w^{13}}{x^2}$ **49.** $\frac{-w^3x^2}{2}$ **51.** $\frac{64x^{18}}{729w^6y^6}$ **53.** $-81w^{29}x^9y^5$ **55.** $\frac{9w^2x^{14}}{4y^{42}}$ **57.** 6.0×10^{-4} **59.** 3.0×10^{-1}
61. 3.0×10^{10} **63.** 5.0×10^3 **65.** 2.0×10^{-1} **67.** 2.0×10^7 **69.** 8.0×10^{-11} **71.** 4.0×10^{-7}
73. 5.0×10^1 **75.** 5.0×10^9

SECTION 5.2 *(pages 334–337)*

1. degree 3   **3.** degree 3   **5.** $13x^2y - 3x - 4y$ **7.** $11t + 27$ **9.** $-r^2 - 2r + 8$
11. $-12y^3 - 18y^2 - 13y + 1$ **13.** $3w^3 - 18w^2 + 18w - 50$ **15.** $2t^2 + 7t - 4$ **17.** $-6t - 8t^3 + 5 + 9t^2r$
19. $12y^2 + 12$ **21.** $3w^2$ **23.** $-9y^2 - 15y - 13$ **25.** $w^2 - 23w + 34$ **27.** $26y^2 - 39y + 8$
29. $-42w^2 - 12w + 26$ **31.** $-4x^2 - 20x + 2$ **33.** $13w^3 - 33w^2 - 11w - 1$
35. $h(x) = 5x^2; h(0) = 0; f(0) = +1; g(0) = -1$ **37.** $h(x) = 5x^2 + 3x + 3; h(0) = 3; f(0) = 0; g(0) = 3$
39. $h(x) = -x^2 - 2x + 1; h(0) = 1; f(0) = 0; g(0) = 1$ **41.** $h(x) = 4; h(-1) = 4; f(-1) = 1; g(-1) = -3$
43. $h(x) = 10x^2 - 12; h(-1) = -2; f(-1) = -1; g(-1) = 1$ **45.** $h(x) = -3x^2 - 1; h(-1) = -4; f(-1) = 1; g(-1) = 5$
47. $(f + g)(x) = x^2 + x + 4$; domain: all real numbers   **49.** $(f - g)(x) = 5x + 5$; domain: all real numbers
51. $(f + g)(x) = -1$; domain: all real numbers   **53.** $7x^2 + 28x$ **55.** $8x^2 + 72x$ **57.** $\frac{16}{9k^2} + 1$
59. $21x^3 + 35x^2 - 56x$ **61.** $1 + \frac{3}{2t}$ **63.** $45x^3 + 90x^2$ **65.** $\frac{9}{2} + x^5$ **67.** $n^2 - \frac{3}{8n^2} + \frac{1}{6}$
69. $\frac{2xy}{3} + \frac{1}{9x^2} - \frac{1}{y}$ **71.** $\frac{7}{t} + \frac{2}{x} - \frac{7}{2}$ **73.** $\frac{4}{r^2u^3} + \frac{2}{3u^4} - \frac{7}{3ru^3}$ **75.** $66y^3 + 88y^2$ **77.** $1 + \frac{1}{9x^{11}y^5}$
79. $-48y^3 + 24y^2 + 16y$ **81.** $6n^2 - \frac{3}{4n^2} + \frac{1}{n}$ **83.** $2xy + \frac{1}{3x^2} - \frac{1}{y}$ **85.** $-6x^3 - 8x^2 + 6x$
87. $\left(\frac{f}{g}\right)(x) = \frac{2}{x}$; domain: x is a real number and $x \neq 0$ **89.** $(fg)(x) = 24x^3 - 12x$; domain: x is a real number
91. $(fg)(x) = 9x^2 - 18x^3$; domain: x is a real number   **93.** 441   **95.** $\frac{-x^2y^4}{3}$ **97.** $45x^3$ **99.** 0.012

SECTION 5.3 *(pages 352–354)*

1. $54y^3 - 21y^2 - 62y + 35$ **3.** $-120x^3 + 66x^2 - 42x - 12$ **5.** $-84y^3 + 143y^2 - 105y + 25$
7. $x^2 + 5x + 6$ **9.** $x^2 - 10x + 24$ **11.** $4x^2 + 13x - 35$ **13.** $5x^2 - 41x - 36$ **15.** $6x^2 + 7x + 2$
17. $21x^2 - x - 36$ **19.** $x^2 - 1$ **21.** $4x^2 - 1$ **23.** $t^2 + 4t + 4$ **25.** $4y^2 - 20y + 25$
27. $49x^2 - 16t^2$ **29.** $25x^2 + 2xy + 0.04y^2$ **31.** 399   **33.** 6396   **35.** 809,900   **37.** 805,100
39. 2601   **41.** 6084   **43.** 3025   **45.** 12,100   **47.** $x^3 + 6x^2 + 12x + 8$ **49.** $x^4 + 4x^3 + 6x^2 + 4x + 1$
51. $x^6 - 3x^4 + 3x^2 - 1$ **53.** $16x^4 - 8x^2 + 1$ **55.** $x - 3$ **57.** $x + 2$ **59.** $x^2 + x - 12$
61. $2x^2 + 15x - 8$ **63.** $6x^2 - 7x - 3$ **65.** $x^2 + 5x - 1$ **67.** $n - 10 + \frac{25}{n + 2}$ **69.** $x - 4$
71. $2x + 1$ **73.** $x^2 - 2x - 4$ **75.** $2x^2 + 2x - 5$ **77.** $y^2 + y + 1$ **79.** $y^2 - 3y + 9$
81. $3x^3 + x + 1$ **83.** $x^3 + x^2 + x + 1$ **85.** $x^2 + 3x + 9$ **87.** $x^7 - x^6 + x^5 - x^4 + x^3 - x^2 + x - 1$
89. $x^3 - 2$ **91.** $-x^2 - 11x - 8$ **93.** $6x^3y^2 - 8x^2y^2 + 4xy$ **95.** 3

SECTION 5.4 *(pages 369–371)*

1. $5x$ **3.** $9t^2$ **5.** $18xy$ **7.** $20t^2u$ **9.** xy **11.** m **13.** $4x^4(2 - x^2)$ **15.** $7x(6 + x^5)$
17. $5xy^6(5 + 9y)$ **19.** $8t^4(8t^2 - 3x^3)$ **21.** $2xy(4xy^5 - 2x^3y^5 + 1)$ **23.** $8xy(6x^8y^5 - 3x^3y^3 - 1)$
25. $5tx^4(5t^5x^2 + 6 - t^8x^2)$ **27.** $2t(12t^5 - 4x^4 + 9xt^5)$ **29.** $5x^6y(13y^3 - 5 + x^3y^5)$
31. $3tx(3x - 17t^3 + 27x^2)$ **33.** $(2x - 1)(5x - 3)$ **35.** $2x - 9$ **37.** $(x + 4)(x + 2)$ **39.** $(r + 3)(r + 2)$
41. $2(3t - 5)(t - 2)$ **43.** $(r - 2)(r - 1)$ **45.** $(7x - 8)(4x - 3)$ **47.** $(n - 1)(16n - 1)$ **49.** $(t + 5)(t - 5)$
51. $(13 + r)(13 - r)$ **53.** $(17u + 1)(17u - 1)$ **55.** $(7t^4 + 3)(7t^4 - 3)$ **57.** $(x + 2y^{18})(x - 2y^{18})$
59. $(5x^8 + 11)(5x^8 - 11)$ **61.** $(y - 5)^2$ **63.** $(y + 2)^2$ **65.** $(6x - 1)^2$ **67.** $4(3x + 1)^2$ **69.** $3(y + 5)^2$
71. $6(y - 3)^2$ **73.** $(5x + 2)^2$ **75.** $(7x - 3)^2$ **77.** $(6y - 7)^2$ **79.** $(x - 5)(x^2 + 5x + 25)$
81. $(5r - 4)(25r^2 + 20r + 16)$ **83.** $12(y - 1)(y^2 + y + 1)$ **85.** $(x^2 + 4)(x + 2)(x - 2)$
87. $(x^2 + 1)(x + 1)(x - 1)$ **89.** $(9x^2 + 1)(3x + 1)(3x - 1)$ **91.** 72,900 or 7.29×10^4 **93.** $3x^2y - 2$
95. $5x^2y^2z(3yz^3 - 4x + 5x^3y^4)$ **97.** $-3x^2 + 2x - 1$ **99.** $5x^2 + x + 15$

SECTION 5.5 *(pages 390–393)*

1. $(y + 2)(y + 9)$ **3.** $(x + 4)[x + (-3)]$ **5.** $[y + (-5)](y + 8)$ **7.** $[t + (-1)](t - 2)$ **9.** $-; +$
11. $+; -$ **13.** $-; +$ **15.** $-; +$ **17.** $(x + 6)(x + 1)$ **19.** $(n + 5)(n - 7)$ **21.** $(y + 9)(y + 9)$
23. $(15x + 2)(3x - 1)$ **25.** $3(x + 5)(x + 6)$ **27.** $8(x - 8t)(x + 2t)$ **29.** $5(x - 6)(x - 7)$

31. $6(x - 8y)(x + 9y)$ **33.** $2(2v - t)(3v + 4t)$ **35.** $9(-t + 3x)(t + 2x)$ **37.** $5(t + 6)(t - 5)$
39. $7(6v + 1)(7v - 8)$ **41.** $-9(8v - 5)(2v - 9)$ **43.** $3(t - 6)(t - 9)$ **45.** $4(9v + 5)(7v - 9)$
47. $3x^4 + 12x^3 - 5x - 20$ **49.** $7x^3 - 49x^2 - 3x + 21$ **51.** $-9x^4 + 63x^3 - 6x + 42$ **53.** $\left(\frac{f}{g}\right)(x) = 3$
55. $(fg)(x) = 2x - 20x^2 + 50x^3$ **57.** $(f \circ g)(x) = 2x^2 - 7; (f \circ g)(0) = -7$
59. $(f \circ h)(x) = -6x - 5; (f \circ h)(0) = -5$ **61.** $(f \circ g \circ h)(x) = 18x^2 - 7; (f \circ g \circ h)(0) = -7$
63. $(h \circ g \circ f)(x) = -12x^2 + 60x - 72; (h \circ g \circ f)(0) = -72$ **65.** equal **67.** not equal **69.** not equal
71. $\{2, 5\}$ **73.** $\left\{-2, \frac{1}{2}, 3\right\}$ **75.** $\{-7\}$ **77.** $\left\{\frac{-1}{5}\right\}$ **79.** $\left\{-2\frac{1}{3}, 2\right\}$ **81.** $\left\{0, \frac{-3}{4}, \frac{1}{2}\right\}$ **83.** 6 units
85. base: 15 inches; height: 14 inches **87.** 32 **89.** 17 sides **91.** $(6x + 7)(6x - 7)$ **93.** $x^2 + 1$
95. $48x^2y^2 + 32xy^2 - 3x - 2$ **97.** $\frac{1}{5} - \frac{x}{3y} + \frac{1}{xy}$ **99.** $(x + 2)(8x - 5)$

Chapter 5 Review Problems *(pages 394–396)*

1. $\frac{3m^5}{x^2}$ **2.** $9x^7y^5$ **3.** $\frac{49z^6}{36x^{14}}$ **4.** 4 **5.** 65,000 **6.** 2000 **7.** 0.0053 **8.** 0.25
9. $3x^2 - 6x + 13$ **10.** $11xy^2 - 4x^2 - 5y^2$ **11.** $-3xy + 3xy^2$ **12.** $xy + 8y^2 + y - 5$
13. $6x^3y^2 - 9x^2y + 15xy$ **14.** $\frac{x}{2} - \frac{1}{6}$ **15.** $\frac{5x^2}{2} - 2x + \frac{1}{x}$ **16.** $\frac{72x^3}{y^2} - 36x^2y$ **17.** $-x$ **18.** 9
19. $15x^2 - 6xy + 19x - 4y + 6$ **20.** $-6x^2 + 22x - 12$ **21.** $x^3 - 3x^2 + 3x - 1$ **22.** $18x^2 - 2y^2$
23. $x^2 + x + 1$ **24.** $16x^2 + 3x + 13 + \frac{5}{x - 1}$ **25.** $(x - 7)(x + 9)$ **26.** $(x + 10)(x + 11)$
27. $(x + 7)(x + 7)$ **28.** $4(x - 20)(x + 10)$ **29.** $(3x - 2)(5x + 3)$ **30.** $-5(2x - 7)(3x - 1)$
31. $3(x + 5)(x - 5)$ **32.** $(5x - 7y)^2$ **33.** $x = 5; x = -3$ **34.** $y = 2\frac{1}{2}; y = 2\frac{2}{3}$ **35.** $\{-9, 9\}$
36. $\left\{-3\frac{1}{2}, -2, 1\frac{2}{3}\right\}$ **37.** $8x^4 - 16x^3 + 2x^2 + 6x$ **38.** $8x^4 - 24x^3 + 14x^2 + 6x$ **39.** $(5x - 7)(2x + 1)$
40. $-3(2x + 5)(4x - 7)$ **41.** $2(x + 5)^2$ **42.** $(10x + 9y)(10x - 9y)$ **43.** $\frac{1}{64x^6y^{14}t^6}$
44. $(x - 2)(x - 1)(x + 1)$ **45.** $r = 6; r = -2$ or $\{6, -2\}$ **46.** $3(t - 6)(t - 7)$ **47.** $mn^2(3mn - 9m^2 + 11)$
48. $r = \frac{-2}{5}; r = -1$ **49.** 3 **50.** $48y^2 - 7y - 28$ **51.** $(7y + 12x)(7y - 12x)$ **52.** $(8x - 3)(7x + 1)$

Cumulative Review *(pages 397–398)*

1. $r = 13; r = -5$ **2.** $\frac{x^8}{81y^4}$ **3.** the product of 4 and x minus the product of 3 and the difference between 2 and x
4. $x - 3$ **5.** $11xy^3 - 7.3xy^2 - 24xy + 3.2$ **6.** 8.2656 **7.** $t = \frac{-4}{3}$
8. length: 35 inches; width: 12 inches **9.** $\{x \mid 1 < x < 2\}$ ⟨number line⟩ **10.** $-13\frac{7}{15}$
11. $8y - 12x$ **12.** $y = \frac{-2}{3}x$ **13.** $f(0) = -8; f(-3) = -14$ **14.** perpendicular **15.** $\frac{y - b}{x} = m$
16. 6.15 **17.** $6x^2 + 2$ **18.** 270 **19.** $\{t \mid t \geq 0.2\}$ **20.** $x = -4; y = 12$

ANSWERS to Chapter 6 Odd-numbered Exercises

Section 6.1 *(pages 410–412)*

1. $\{x \mid x \text{ is real and } x \neq 0\}$ **3.** $\{t \mid t \text{ is real and } t \neq 2\}$ **5.** $\{y \mid y \text{ is real and } y \neq 1, 2\}$ **7.** $\{n \mid n \text{ is real and}$

$n \neq -3, 3\}$ **9.** x^2 **11.** $\frac{n^2}{x^2}$ **13.** $\frac{9n^2}{16m^2}$ **15.** $\frac{21}{20xy}$ **17.** $n + \frac{5}{n^5}$ **19.** $\frac{1}{x^2} + \frac{8}{x^3}$ **21.** $\frac{1}{2y} - \frac{3y}{x}$

23. $\frac{2}{3x} - \frac{4}{3tx} + 2tx$ **25.** $\frac{2}{3w} - \frac{5}{6x} + \frac{10}{9w^2x^2}$ **27.** $\frac{5}{3}$ **29.** $\frac{2}{5}$ **31.** $3y(x - 2y)$ **33.** $\frac{1}{y + 4}$ **35.** $\frac{2}{y + 2}$

37. $n + 2$ **39.** $\frac{6y}{y - 1}$ **41.** $\frac{n + 2}{n - 2}$ **43.** $\frac{y + 6}{y + 5}$ **45.** $\frac{3(3 + x)}{9 + 3x + x^2}$ **47.** $2n^3x^3$ **49.** $\frac{r^8x^3}{2}$ **51.** $\frac{1}{4}$

53. $\frac{3(x - 4)}{2}$ **55.** $\frac{(x - 2)(x - 50)}{(x - 4)(x + 4)}$ **57.** $\frac{x + 8}{2(x + 7)}$ **59.** $\frac{t}{x}$ **61.** $\frac{8}{tx}$ **63.** $\frac{x + 6}{6x}$ **65.** $\frac{x + 5}{x}$

67. $\frac{3(x - 3)}{2(x + 14)}$ **69.** $\frac{(x - 4)(x - 8)}{(x - 11)(x + 10)}$ **71.** $\frac{x - 1}{x + 3}$ **73.** $\frac{-a}{b^2}$ **75.** $\frac{(y + 2)(y + 12)}{(y + 1)(y + 7)}$ **77.** $(r + 10)(r + 15)$

79. $\frac{xy(x - y)}{x - 1}$ **81.** $\frac{(r - m)^2}{(r^2 + rm + m^2)(r + m)}$ **83.** $2(x - y)$ **85.** $\frac{1}{x - 4}$ **87.** $\frac{(2x + 5)^2}{(2x + 3)^2}$ **89.** $\frac{z - v}{z + v}$

91. x **93.** not defined for $x = 0$ **95.** 3 **97.** 0 **99.** -3

SECTION 6.2 *(pages 421–424)*

1. $f(x) + g(x) = 2x; f(x) - g(x) = \frac{x}{2}$ 3. $f(x) + g(x) = 2(x + 2); f(x) - g(x) = \frac{2(x + 2)}{5}$

5. $f(x) + g(x) = \frac{2(x - 1)}{x}; f(x) - g(x) = \frac{6x - 38}{5x}$ 7. $f(x) + g(x) = \frac{2(x + 1)}{x}; f(x) - g(x) = \frac{2(x - 7)}{5x}$ 9. $24xy^2$

11. $360x^2 y$ 13. $150xyz^2$ 15. $84xy^2 z^2$ 17. $\frac{48y^2}{30xy}, \frac{80x^2}{30xy}$ 19. $\frac{48}{288x^2 y}, \frac{9xy + 9x}{288x^2 y}$

21. $\frac{2x^2 + 2x}{(x + 1)(x - 1)}, \frac{x^2 - x}{(x + 1)(x - 1)}$ 23. $\frac{5x - 10}{(x - 6)(x - 2)}, \frac{x - 5}{(x - 6)(x - 2)}$ 25. $\frac{9r + 11rx}{99x}$ 27. $\frac{47}{18t}$ 29. $\frac{x(8y - 11)}{88y}$

31. $\frac{10 - 3y}{6y}$ 33. $\frac{8t^2 - 35t + 6}{12t}$ 35. $\frac{-4r^3 + 18r^2 - 35}{14r^3}$ 37. $\frac{2r^2 - 34}{(r - 3)(r + 5)}$ 39. $\frac{-11}{(x - 5)(x + 6)}$

41. $\frac{8r - 31}{(r + 3)(r - 8)}$ 43. $\frac{t + 35}{(t - 5)(t + 5)}$ 45. $\frac{9r^2 + 33r + 26}{(r + 2)(r - 2)}$ 47. $\frac{t^2 + 20t - 4}{(t + 2)(t - 3)}$ 49. $\frac{23r + 5}{2(2r - 3)}$

51. $\frac{15t - 29}{6(t + 2)}$ 53. $\frac{-3y^2 + 4y + 6}{(y + 2)(y - 2)}$ 55. $\frac{-9t^2 + 10t + 28}{(t + 5)(t - 5)}$ 57. $\frac{-(t^2 - 9t - 20)}{(t - 4)^2 (t + 4)}$ 59. $\frac{r^2 - 25r + 132}{(r + 12)^2 (r - 12)}$

61. $\frac{1}{(t - 3)(t - 5)}$ 63. $\frac{1}{(t - 2)(t + 2)}$ 65. $\frac{9y^2 - 25y + 30}{(y + 3)^2 (y - 3)}$ 67. $\frac{-(x^2 + x + 8)}{(x - 2)(x + 1)}$

69. $f(x) + g(x) = \frac{x^2 - 5x + 12}{3(x - 2)}; g(x) - f(x) = \frac{x^2 - 11x + 12}{3(x - 2)}$ 71. $f(x) + g(x) = \frac{5 - 4x}{3x^2}; f(x) - g(x) = \frac{5 - 8x}{3x^2}$

73. $f(x) + g(x) = \frac{x - 2}{2(x - 5)}; g(x) - f(x) = \frac{x - 6}{2(x - 5)}$ 75. $f(x) + g(x) = \frac{4x^2 - x + 2}{(x + 2)(x - 2)^2}; g(x) - f(x) = \frac{2x^2 + 9x - 10}{(x + 2)(x - 2)^2}$

77. $f(x) + g(x) - h(x) = \frac{x(x + 20)}{2(x + 5)(x - 5)}; f(x) - g(x) - h(x) = \frac{-x(x - 20)}{2(x + 5)(x - 5)}; g(x) - f(x) + h(x) = \frac{x(x - 20)}{2(x + 5)(x - 5)}$

79. $f(y) - g(y) + h(y) = \frac{12 - y}{(y + 3)^2}; f(y) + g(y) - h(y) = \frac{y + 4}{(y + 3)^2}; f(y) + g(y) + h(y) = \frac{y + 18}{(y + 3)^2}$ 81. $\frac{1}{x - 4}$

83. $\frac{5x - 19}{x^2 - 16}$ 85. $\frac{1}{(x + 5)^2}$ 87. $f(0) = \frac{3}{4}$ 89. $\frac{5}{x - 5}$

SECTION 6.3 *(pages 442–444)*

1. $\frac{n}{r^2}$ 3. $\frac{1}{k}$ 5. $\frac{2r}{5}$ 7. $\frac{3n + 54}{-72 + 2n}$ 9. $\frac{189 + 7k}{3k + 168}$ 11. $\frac{8r}{9r - 30}$ 13. $\frac{-r^2 + 20r - 96}{8r^2}$ 15. 2

17. $\frac{7k}{4(k + 3)^2}$ 19. $\frac{1}{2(4 - y)}$ 21. $\frac{30}{47}$ 23. $\frac{x^3 + 5x}{x^3 - 2}$ 25. $\frac{t - 5}{t + 3}$ 27. $\frac{1}{t + 1}$ 29. $\frac{k - 1}{k + 1}$ 31. $2x^2 - 1$

33. $y = 9$ 35. $x = 4$ 37. $y = -3.5$ 39. $y = 5$ 41. $y = 7$ 43. $x = 1$ 45. $x = -3$

47. no solution   49. $x = -1.5$ 51. $y = \frac{1}{6}$ 53. $y = \frac{-2}{5}$ 55. $x = -5$ 57. $x = -4$

59. $t = 0; t = 7$ 61. $y = 8; y = 1$ 63. $t = -6; t = 3$ 65. $x = 8; x = -1$ 67. $t = 1$

69. no solution   71. $x = -56$ 73. $t = 4.5$ 75. no solution   77. $t = \frac{24}{11}$ 79. $t = -2$

81. $E = 0.315$ 83. $t = 9.38d^2 + 2.25d$ 85. $r_1 = \frac{kfr_2}{r_2 + kf}$ 87. $\frac{x^2 + 4x + 3}{3}$ 89. $\frac{x + 4}{x}$ 91. $\frac{3}{16}$

93. $n^2 - n$ 95. $x = 5$ 97. $\frac{-v - 14}{(v + 2)(v - 2)}$ 99. t^2

SECTION 6.4 *(pages 456–462)*

1. 18 centimeters   3. $n = 3$ 5. 5 mph   7. $\frac{15}{19}$ 9. 35 inches   11. Players C and D will win in 2 hours and 24 minutes.   13. 2 and 10   15. **a.** The cost to remove 90% of the impurities is $36,000. The cost to remove 95% of the impurities is $76,000.   **b.** As x gets larger (and closer to 100), C also gets larger (at an increasing rate).   **c.** domain: $\{x \mid 0 \le x < 100\}$ 17. Plant A requires 12 days. Plant B requires 24 days.   19. 60 cookies

21.
$$p + \left(\frac{p-1}{2}\right)^2 = \left(\frac{p+1}{2}\right)^2$$

$$p + \frac{p^2 - 2p + 1}{4} = \frac{p^2 + 2p + 1}{4}$$

$$\frac{4p + p^2 - 2p + 1}{4} = \frac{p^2 + 2p + 1}{4}$$

$$\frac{p^2 + 2p + 1}{4} = \frac{p^2 + 2p + 1}{4}$$

23. $3\frac{1}{3}$ hours **25.** $\frac{12.4}{9.4}$ **27.** 1 hour and 8 minutes **29.** $s = kd$ **31.** $F = mk$ **33.** $t = \frac{k}{T}$

35. $n = \frac{kP_1P_2}{d}$ **37.** $I = \frac{k}{d^2}$: When d doubles, I is $\frac{1}{4}$ its original value. **39.** V is doubled. **41.** If a basketball were shrunk to the size of a volleyball, it would be much heavier. For a basketball, $k = 8\frac{1}{3}$ g/cm, while for the volleyball, $k \approx 4.2$ g/cm. The difference in weight is probably due to the weight of materials in the balls. **43.** 70 miles

45. 3460 feet **47.** 119.6 pounds **49.** 9 square feet **51.** $\frac{35}{12x}$ **53.** $t = 3$ **55.** $x = 2$

57. 187.5 inches **59.** about 1.28 hours **61.** $p = k\sqrt{L}$; p doubles **63.** 0 **65.** It is meaningless when $x + 8 = 0$, and $x = -8$. **67.** $\frac{21x + 28}{14 + x^2}$ **69.** $\frac{42x + 6y}{x + y}$ **71.** $\frac{c^2 - 5c}{b} - 8 = a$

CHAPTER 6 REVIEW PROBLEMS   *(pages 465–468)*

1. $\frac{8y^2z^3}{9x}$ **2.** nx^2 **3.** $x - 1$ **4.** $\frac{x - 6}{x + 6}$ **5.** $x - 1$; $h(-5) = -6$; $h(x + 1) = x$

6. $-1(x + 7)$; $h(-3) = -4$; $h(x - 7) = -x$ **7.** $\frac{1}{3(x + 3)}$ **8.** $\frac{3(x + 2y)}{16(x - 2y)}$ **9.** $\frac{128}{x}$ **10.** $1701x^{12}y^{17}$

11. $\frac{9}{49y^3}$ **12.** $\frac{x^2}{y}$ **13.** $\frac{-x^2 - 6x - 5}{2(x + 3)}$ **14.** $\frac{15t^2 - 69t + 26}{(3t - 2)(t - 4)}$ **15.** $\frac{12}{x - 3}$ **16.** $\frac{41}{60x}$ **17.** $\frac{6x - 14}{(x - 3)(x - 1)}$

18. $\frac{x^2 + 6x + 13}{2(x + 3)}$ **19.** $-\frac{1}{2x + 1}$ **20.** $\frac{m - 1}{m + 1}$ **21.** $\frac{x + 1}{x - 1}$ **22.** $\frac{3}{n}$ **23.** $\frac{-x}{y}$ **24.** $\frac{31}{19}$ **25.** $m = 6$

26. $t = 6$ **27.** $x = 4$ **28.** $t = -4$ **29.** $m = 1$ **30.** $m = 4$ **31.** 10 miles per hour; 10.5 hours
32. The new copier would take 4 days. The old copiers would take 8 days each . **33.** 10 women **34.** 30 slices
35. 36 kilometers per hour; $11\frac{1}{9}$ hours **36.** 5.87 hours; $\frac{1}{189}$ job per minute **37.** $9\frac{7}{9}$ seconds
38. The golf ball would weigh 8.2 ounces, which is much less than the bowling ball weighs. The difference is due to the difference in the weights of the materials they are made from. **39.** L is 4 times greater. **40.** R doubles
41. $T = kh^3$ **42.** $L = \frac{kwh^2}{d}$ **43.** $m = 3$ **44.** $\frac{12(x - 1)}{(x - 4)(x + 2)}$ **45.** $\frac{2x(x + 5)}{(x + 6)(x - 5)}$ **46.** $\frac{1}{m - 4}$

47. $\frac{5}{16xy^3}$ **48.** $\frac{1}{n + 2}$ **49.** $\frac{9x - 15}{x^2 - 25}$ **50.** $x = 2$ **51.** Jokie would take 36 hours, Max would take 45 hours.

52. 30 burlap belts **53.** $\frac{1}{2}$ article per hour **54. a.** $5\frac{5}{13}$ miles per hour **b.** The same distance would be

covered in half the time. **55.** $A = kd_1d_2$ **56.** 81 **57.** $3\frac{1}{3}$ minutes **58.** $s = kt^2$ **59.** 10 doctors

60. $\frac{5x + 3}{6}$ **61.** $\frac{1}{m^2}$ **62.** $t = -5$; $t = 8$ **63.** $\frac{1 + x}{xy - x}$ **64.** $\frac{n}{x}$ **65.** $\frac{3}{8}$ **66.** $\frac{y + 3}{y + 5}$

CUMULATIVE REVIEW   *(page 470)*

1. -0.000009 **2.** $-1\frac{5}{6}$ **3.** $y = -6$ **4.** $\frac{x^4y}{2z^2}$ **5.** $2(x + 5)^2$ **6.** 0.209 **7.** -13

8. $t = 6$; $t = -6$ **9.** Answers may vary. The two sets should be disjoint except for the element 5, which must be in both sets. **10.** $y = -x + 3$ **11.** $\frac{9}{4}$ **12.** $\left\{\left(\frac{5}{7}, 1\frac{4}{7}\right)\right\}$ **13.** 2.44 **14.** $3y - \frac{3y^2}{2x} + 1$ **15.** -3
16. $8(x + y)(x - y)$ **17.** 2.125 **18.** $x = 132$ **19.** $56 and $44.80 **20.** 42 adults

ANSWERS to Chapter 7 Odd-Numbered Exercises

SECTION 7.1   *(pages 482–438)*

1. $4x^2$ **3.** $7y^4$ **5.** $16|xy|$ **7.** $20t^2|v|$ **9.** $17|s|t^2$ **11.** $40v^2x^2$ **13.** $11|x|y^2z^2$

15. $12w^2y^2|z|$ **17.** $\frac{2}{3}$ **19.** $\frac{11}{13}$ **21.** $\frac{4}{25}x^2y^4$ **23.** $0.01t^2z^2$ **25.** $0.9|x|y^2z^2$ **27.** $0.2w^2y^2|z|$

29. $\frac{5}{19}|x^3y^3z^3|$ **31.** $0.17|xz^3|y^2$ **33.** $5x^2$ **35.** $6x^2$ **37.** $-7x^4y^5$ **39.** $9x^4y^5$ **41.** $14x^2y^2z^2$

43. $-11xyz^2$ **45.** 70 **47.** 90 **49.** 20 **51.** 60 **53.** $4x\sqrt{15}$ **55.** $10xy\sqrt{10}$ **57.** $2y\sqrt{89xy}$

59. $36\sqrt{xy}$ **61.** $2xy\sqrt{114}$ **63.** $12xy\sqrt{10}$ **65.** $3x\sqrt{109}$ **67.** $4xy\sqrt{330}$ **69.** $26x\sqrt{10}$

71. $14x^2\sqrt{21}$ **73.** $63x^2$ **75.** $120xy$ **77.** $35xy\sqrt{3}$ **79.** $19xy\sqrt{3x}$ **81.** $30xy\sqrt{6x}$ **83.** $125x\sqrt{2y}$

85. $\frac{20}{9}$ **87.** $\sqrt{14}$ **89.** $\frac{25xy}{12}$ **91.** $\frac{2\sqrt{5}}{y}$ **93.** $\frac{8x\sqrt{2}}{3y}$ **95.** $\frac{4x\sqrt{5}}{9y}$ **97.** $\frac{\sqrt{3}}{2x}$ **99.** $\frac{6x\sqrt{3}}{7y}$

SECTION 7.2   *(pages 500–502)*

1. $3\sqrt{2}+3\sqrt{3}$ **3.** $23\sqrt{3}-8\sqrt{7}$ **5.** $\sqrt{13}-8\sqrt{11}$ **7.** $8\sqrt{2}-2\sqrt[3]{2}$ **9.** $5\sqrt[4]{8}-3\sqrt{2}+4\sqrt[3]{4}$

11. $14\sqrt{7}-19\sqrt[3]{7}$ **13.** $4\sqrt{5}$ **15.** $10\sqrt{3}$ **17.** $2\sqrt{2}$ **19.** $29\sqrt{2}$ **21.** $-7x\sqrt{3}$ **23.** $8\sqrt{5x}$

25. $75\sqrt{2}+30\sqrt{5}$ **27.** $-10xy\sqrt{6y}$ **29.** $42\sqrt{5}-18\sqrt{2}$ **31.** $48\sqrt{5}-12\sqrt{10}$ **33.** $-17+\sqrt{3}$

35. 1 **37.** 10 **39.** $112-24\sqrt{3}$ **41.** $6-2\sqrt{2}$ **43.** -4 **45.** 29 **47.** $\frac{4}{xy^2}-2+3x^2$

49. $\frac{\sqrt{2}}{2}$ **51.** $\frac{3\sqrt{5}}{5}$ **53.** $\frac{\sqrt{3}}{6}$ **55.** $\frac{\sqrt{42}}{3t}$ **57.** $\frac{\sqrt{39}}{3}+\sqrt{2}$ **59.** $\sqrt{7}-\frac{\sqrt{22}}{2}$ **61.** $\sqrt{3}-\frac{2\sqrt{10}}{5}$

63. $\frac{6\sqrt{35}}{7}+\frac{3\sqrt{5}}{2}$ **65.** $-\sqrt{2}-\sqrt{3}$ **67.** $\frac{\sqrt{6}}{4}-\frac{\sqrt{2}}{4}$ **69.** $3\sqrt{2}-2\sqrt{3}$ **71.** $\frac{6\sqrt{3}}{7}+\frac{3\sqrt{5}}{7}$ **73.** $-17\frac{1}{3}$

75. $\frac{-34}{243}$ **77.** $-1+\sqrt{2}+\sqrt{3}-\sqrt{6}$ **79.** -1 **81.** $3+2\sqrt{2}$ **83.** $\sqrt[3]{2}$ **85.** $\frac{\sqrt[3]{25}}{5}$ **87.** $\frac{\sqrt[4]{24}}{2}$

89. $2\sqrt[3]{2}$ **91.** $5xy\sqrt{10y}$ **93.** $\sqrt{2}$ **95.** $-x^2y^2\sqrt[3]{xy}$ **97.** $-2xy$ **99.** $7-4\sqrt{3}$

SECTION 7.3   *(pages 517–519)*

1. $x=36$ **3.** $x=9$ **5.** $x=8$ **7.** $x=-125$ **9.** $x=1$ **11.** $x=-1$ **13.** $x=25$

15. $x=1$ **17.** $n=2$ **19.** $t=3$ **21.** $x=16$ **23.** no real solution   **25.** no real solution

27. $m=7$ **29.** $x=9$ **31.** no real solution   **33.** $y=\frac{9}{16}$ **35.** no real solution   **37.** $x=2$

39. $x=5$ **41.** $x=5$ **43.** $x=\frac{15}{2}$ **45.** $x=5$ **47.** $x=1$ **49.** $x=3$ **51.** $x=0$

53. $x=-\frac{1}{2}$ **55.** $x=7$ **57.** $x=3$ **59.** no real solution   **61.** $x=64$ **63.** $x=-1$ **65.** $x=36$

67. $x=144$ **69.** $x=\frac{1}{9}$ **71.** $x=4$ **73.** 36 months   **75.** 36 weeks   **77.** $\frac{x\sqrt{10y}}{2}$ **79.** $\frac{2\sqrt{5xy}}{5y}$

81. $7xy$ **83.** $x=9$ **85.** $40\sqrt{5}$

SECTION 7.4   *(pages 534–536)*

1. i **3.** 1 **5.** -1 **7.** -1 **9.** -1 **11.** i **13.** i **15.** -1 **17.** 1 **19.** i **21.** $7i$

23. $8i$ **25.** $-10i$ **27.** $-15i$ **29.** $-2i\sqrt{10}$ **31.** $2i\sqrt{3}$ **33.** -16 **35.** $15\sqrt{-1}$ **37.** $-30\sqrt{-1}$

39. -19 **41.** $16\sqrt{-1}$ **43.** 44 **45.** $-3+4i$ **47.** $-12+2i$ **49.** -100 **51.** $-24i$

53. $1+19i$ **55.** $-24+3i$ **57.** $18-60i$ **59.** $-28-24i$ **61.** $5-12i$ **63.** 10 **65.** $\frac{3}{2}-i$

67. $\frac{2}{13}-\frac{3}{13}i$ **69.** $\frac{3}{2}-\frac{i\sqrt{7}}{2}$ **71.** $\frac{1}{17}-\frac{13i}{17}$ **73.** $5\sqrt{2}$ **75.** $2\sqrt{29}$ **77.** $6+5i$ **79.** $2+2i$

81. 153   **83.** 481   **85.** 24,400   **87.** $\frac{89}{1600}$ **89.** $(5 + 6i)\left(\frac{5}{61} - \frac{6i}{61}\right) = \frac{25}{61} - \frac{30i}{61} + \frac{30i}{61} - \frac{36(-1)}{61} = \frac{25 - (-36)}{61} =$
$\frac{61}{61} = 1$; Because the product is 1, these factors are multiplicative inverses.   **91.** $-\frac{5}{29} - \frac{2}{29}i$ **93.** $14 + 4\sqrt{10}$

95. 68   **97.** $3 - 2\sqrt{2}$ **99.** $x^3 y^3 (36xy^2)^{\frac{1}{3}} - x^2 y^2 (28y)^{\frac{1}{3}}$

CHAPTER 7 REVIEW PROBLEMS   *(pages 539–540)*

1. $5xy$ **2.** $21xy^2$ **3.** $-10xy^3$ **4.** $4x^2y^3$ **5.** $(x + 4)^2$ **6.** $(x - 3)^4$ **7.** $1250x^4 y\sqrt{3x}$
8. $x\sqrt{5xy}$ **9.** $10xy\sqrt{2xy}$ **10.** $13\sqrt{6}$ **11.** $2yx\sqrt[4]{x^3 yz^2}$ **12.** $2xy^2 z^2 \sqrt[3]{3xz^2}$ **13.** $29\sqrt{2}$
14. $135\sqrt[3]{3}$ **15.** 38   **16.** 8   **17.** $x - 12\sqrt{x} + \sqrt{3x} - 12\sqrt{3}$ **18.** $41 + 12\sqrt{5}$ **19.** $\frac{\sqrt{3xy}}{2y}$
20. $\frac{\sqrt{6}}{4}$ **21.** $\frac{5\sqrt{x} + 15}{x - 9}$ **22.** $\frac{16 + 5\sqrt{10}}{6}$ **23.** $16x^2 y^7 \sqrt{x}$ **24.** $8x^3 y^{10}\sqrt{15y}$ **25.** $\{4\}$ **26.** $\{6\}$
27. $\{6\}$ **28.** $\{13\}$ **29.** $x = 10,000$ **30.** $x = 32$ **31.** i **32.** $3\sqrt{5}$ **33.** $-28 - 4i$
34. $-16 - 2i$ **35.** -52 **36.** -153 **37.** $6|xy|$ **38.** $5xy^4 \sqrt[3]{x^2 y^2}$ **39.** $2x\sqrt{30xy}$ **40.** $\frac{2\sqrt{15}}{5}$
41. $x = 27$ **42.** $5\sqrt{5}$ **43.** 1   **44.** $9y^2$ **45.** $4\sqrt{2}$ **46.** $18x - 13\sqrt{3x} - 5$ **47.** $30\sqrt{2}$
48. $x = 4$ **49.** $3\sqrt{2} - 3$ **50.** 90   **51.** $4x^7 y^4\left(6^{\frac{1}{2}}\right)$ **52.** $\frac{\sqrt{10}}{2}$ **53.** $20x^2$ **54.** $\frac{\sqrt{15y}}{y}$
55. $4x^2 \sqrt[3]{x^2 y^2}$ **56.** -1 **57.** $3x^3 y^4 (2x^2 y)^{\frac{1}{3}}$ **58.** $-i$ **59.** $\frac{2\sqrt{5}}{xy^2}$ **60.** $3xy^3 (10y)^{\frac{1}{2}}$

CUMULATIVE REVIEW   *(pages 541–542)*

1. $1\frac{3}{4}$ **2.** Answers may vary, but neither set may have elements other than 2 or 3.
Examples: $A = \{2\}$, $B = \{3\}$; $A = \{2, 3\}$, $B = \{2\}$; $A = \{2, 3\}$, $B = \{2, 3\}$; $A = \{\ \}$, $B = \{2, 3\}$ **3.** $A = 8$ years
4. $0.00099 < 0.00100 < 0.0019$ **5.** $2(10^2)$ or 200   **6.** $\frac{yz^7}{3x}$ **7.** $x = 7, y = 3,$ or $\{(7, 3)\}$ **8.** $\frac{(x + 1)^2}{3}$
9. 2.0965   **10.** 1650%   **11.** $\{x \mid x < -2 \text{ and } x > 6\}$ ⟵━┼━┼━●━┼━┼━●━┼━⟶   **12.** $y = \frac{1}{5}x + 2$
                                                                             $-2\ \ 0\ \ 2\ \ 4\ \ 6\ \ 8$
13. $r = \frac{I}{pt}$ **14.** slope $= 2$ **15.** $34x + 10$ **16.** $\{-2, 16\}$ **17.** $-x^3 y - 11x^3 + 15x^2 y$
18. 81 three-cent stamps and 62 one-cent stamps   **19.** $c = 1.25h + 0.75d$ **20.** 1500 miles

ANSWERS to Chapter 8 Odd-Numbered Exercises

SECTION 8.1   *(pages 555–556)*

1. $\{-1, -6\}$ **3.** $\{-9, 4\}$ **5.** $\left\{\frac{5}{3}, -\frac{3}{2}\right\}$ **7.** $\left\{-\frac{7}{8}, -\frac{1}{3}\right\}$ **9.** $\left\{-\frac{3}{4}, 5\right\}$ **11.** $\left\{-\frac{7}{3}, \frac{1}{2}\right\}$ **13.** $\left\{\frac{2}{9}, -\frac{1}{3}\right\}$
15. $\left\{-\frac{4}{7}, \frac{8}{9}\right\}$ **17.** $\left\{-\frac{8}{9}, \frac{4}{3}\right\}$ **19.** $x^2 - 9x + 18 = 0$ **21.** $x^2 + 10x + 24 = 0$ **23.** $3x^2 + 7x - 6 = 0$
25. $x^2 + 2.8x + 0.75 = 0$ **27.** $x^2 - 6x + 7 = 0$ **29.** $x^2 - 2x - 11 = 0$ **31.** $x^2 - 10x + 26 = 0$
33. $x^2 - 6x + 13 = 0$ **35.** $x = 17$ and -17 **37.** $x = \frac{6}{5}$ and $\frac{-6}{5}$ **39.** $x = \frac{7}{8}$ and $\frac{-7}{8}$
41. $x = -\sqrt{35}$ and $\sqrt{35}$ **43.** $x = \pm\sqrt{6}$ **45.** $x = \pm 5i$ **47.** $x = -1 + \sqrt{6}$ and $-1 - \sqrt{6}$
49. $x = \frac{-3 \pm \sqrt{5}}{2}$ **51.** $x = 6 \pm 5i$ **53.** $x^2 - 16x = -180; \left(\frac{b}{2}\right)^2 = 64$ **55.** $x^2 - \frac{1}{3}x = 6; \left(\frac{b}{2}\right)^2 = \frac{1}{36}$
57. $\left(x + \frac{3}{2}\right)^2 = 0$ **59.** $(x - 10)^2 = 0$ **61.** $\{29, -9\}$ **63.** $\{4, -24\}$ **65.** $\{3, -29\}$ **67.** $\left\{\frac{3}{5}, -\frac{1}{2}\right\}$
69. $\left\{1\frac{1}{3}, \frac{2}{3}\right\}$ **71.** $\left\{1\frac{1}{2}, -1\frac{1}{3}\right\}$ **73.** $\{1.3, -0.3\}$ **75.** $\{0.4, 20\}$ **77.** $\left\{-2 + 2\sqrt{3}, -2 - 2\sqrt{3}\right\}$
79. $\{-5 + 3i, -5 - 3i\}$ **81.** $\{3 + 4i, 3 - 4i\}$ **83.** $\left\{-\frac{3}{2} + i, -\frac{3}{2} - i\right\}$

SECTION 8.2 *(pages 570–572)*

1. $\{5, 3\}$ **3.** $\{3\}$; double root   **5.** $\{-8, -15\}$ **7.** $\left\{-\frac{2}{3}, -\frac{3}{4}\right\}$ **9.** $\left\{-\frac{1}{5}, -\frac{1}{2}\right\}$ **11.** $\left\{\frac{3}{4}, \frac{2}{3}\right\}$

13. $\left\{\sqrt{10}, -\sqrt{10}\right\}$ **15.** $\left\{3\sqrt{5}, -3\sqrt{5}\right\}$ **17.** $\left\{2 + \sqrt{3}, 2 - \sqrt{3}\right\}$ **19.** $\{4 + i, 4 - i\}$ **21.** $\{6i, -6i\}$

23. $\left\{-2 + 2\sqrt{3}, -2 - 2\sqrt{3}\right\}$ **25.** two real, irrational roots   **27.** one real, rational root (double root)

29. two real, irrational roots   **31.** two real, irrational roots   **33.** two complex roots

35. two real, rational roots   **37.** $x^2 - 5x - 50 = 0$; $\{-5, 10\}$ **39.** $4x^2 - 17x - 15 = 0$; $\left\{5, -\frac{3}{4}\right\}$

41. $x^2 + 6x - 315 = 0$; $\{15, -21\}$ **43.** $3u^2 - 2u + 8 = 0$ **45.** $2u^2 + 2u - 7 = 0$ **47.** $3u^2 + 2u - 7 = 0$

49. $\{1, -1, 3, -3\}$ **51.** $\{256, 81\}$ **53.** $\left\{\frac{1}{3}, -\frac{1}{3}, i, -i\right\}$ **55.** $\{-2, 2, -5, 5\}$ **57.** $\{6561, 1\}$ **59.** $\left\{-\frac{1}{2}, 1\right\}$

61. $\{1\}$ **63.** $\{8\}$ **65.** $\{-2, -1\}$ **67.** $\{5\}$ **69.** $\{3, -3, 2, -2\}$ **71.** $\{81\}$ **73.** $\{-5, -6\}$

75. $\left\{-\frac{3}{8}, -\frac{3}{4}\right\}$ **77.** $\{41, 30\}$ **79.** $\left\{\frac{17}{2}\right\}$ **81.** $\left\{\frac{4}{3}, -1, -\frac{5}{3}, 2\right\}$ **83.** $\left\{\frac{1 + i}{4}, \frac{1 - i}{4}\right\}$ **85.** $\{9998, 14{,}639\}$

87. $\{5 \pm 4i\}$ **89.** $\{5, 1\}$ **91.** $\left\{\frac{-5 \pm \sqrt{57}}{2}\right\}$ **93.** $\{-2, -9\}$ **95.** $\left\{1, -\frac{5}{2}\right\}$ **97.** $\left\{\frac{1 \pm \sqrt{19}}{3}\right\}$

99. $x^2 - 3x - 10 = 0$; $\{5, -2\}$

SECTION 8.3 *(pages 578–582)*

1. 29 and 31   **3.** The ball that was simply dropped lands first (after 9 seconds), then 5 seconds later the "thrown" ball lands (14 seconds from time "zero").   **5.** $5 + i\sqrt{15}$ and $5 - i\sqrt{15}$ **7.** about 1970   **9.** 11.4 hours

11. The last car reaches the bottom 5 seconds after it starts. Its speed at that time is 40.6 feet per second.

13. after 10 seconds   **15.** $n = 64$ or $n = 164$ **17.** 3 seconds   **19.** 20 days   **21.** 18 people; 100 square feet

23. 2.5 seconds   **25.** 1 second: 4.9 meters; 2 seconds: 19.6 meters; 3 seconds: 44.1 meters; 4 seconds: 78.4 meters

27. 3.1 seconds; the other answer is zero.   **29.** $x = \pm 6.4$ **31.** $x = \pm 1.2$ **33.** $x = \pm 0.9$ **35.** $x = \pm 5.9$

37. $x = \pm 2.4$ **39.** $x = 1.4$ or $x = -3.4$ **41.** $x = -0.4$ or $x = -2.6$ **43.** $x = 11.1$ or $x = 0.9$

45. $x = 8.6$ or $x = -13.6$ **47.** $\{6, 5\}$ **49.** $\left\{\sqrt{2}, -\sqrt{2}, \frac{i\sqrt{7}}{7}, \frac{-i\sqrt{7}}{7}\right\}$ **51.** $\{-15 + 7i$ and $-15 - 7i\}$

53. $9x^2 - 18x + 13 = 0$ **55.** $x = 8.5$ or $x = -10.5$ **57.** 15 compact discs

SECTION 8.4 *(pages 596–599)*

1. $(0, 1), (-1, 6), (-2, 15), (1, 0), (2, 3), (3, 10)$ **3.** $(0, 1), (-1, -3), (-2, -9), (1, 3), (2, 3), (3, 1)$

5. $(0, -2), (-1, 11), (-2, 40), (1, 1), (2, 20), (3, 55)$ **7.**   **9.**   **11.**

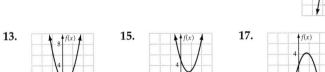

13.   **15.**   **17.**

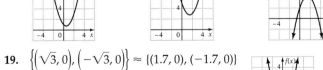

19. $\left\{\left(\sqrt{3}, 0\right), \left(-\sqrt{3}, 0\right)\right\} \approx \{(1.7, 0), (-1.7, 0)\}$

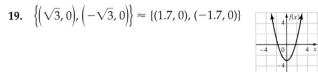

21. $\left\{\left(1 + \sqrt{3}, 0\right), \left(1 - \sqrt{3}, 0\right)\right\} \approx \{(2.7, 0), (-0.7, 0)\}$

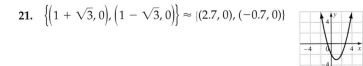

23. $\left\{\left(-2 + \sqrt{5}, 0\right), \left(-2 - \sqrt{5}, 0\right)\right\} = \{(0.2, 0), (-4.2, 0)\}$

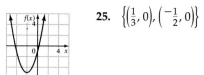

25. $\left\{\left(\frac{1}{3}, 0\right), \left(-\frac{1}{2}, 0\right)\right\}$

27. $\left\{(1, 0), \left(-\frac{1}{2}, 0\right)\right\}$

29. $\left\{\left(\frac{-3 + \sqrt{17}}{4}, 0\right), \left(\frac{-3 - \sqrt{17}}{4}, 0\right)\right\} = \{(0.3, 0), (-1.8, 0)\}$

31.    **33.**   **35.**   **37.**

39.   **41.**   **43.** $-3 \le x \le 6$ **45.** $x < -28$ or $x > 17$

47. $18 \le x \le 40$ **49.** $x \le -29$ or $x > 15$ **51.** $1 < x \le 6$ **53.** $x \le -18$ or $x > 100$

55. $\{x \mid x < -7 \text{ or } x > 7\}$ **57.** $\{x \mid 2 < x < 4\}$

59. $\{x \mid x \le -5 \text{ or } x \ge 1\}$ **61.** $\{x \mid x \le 1 \text{ or } x \ge 2\}$

63. $\{x \mid x < -1 \text{ or } x > -1\}$ **65.** $\{x \mid x < 7 \text{ or } x > 11\}$

67. $\{x \mid -2 < x \le 8\}$ **69.** $\{x \mid -9 < x \le 5\}$

71. $\{x \mid -2 < x < 1\}$ **73.** $\{x \mid x < 5 \text{ or } x \ge 7\}$

75. $\{x \mid -3 \le x \le 2 \text{ or } x > 5\}$ **77.** $\left\{x \mid -2 \le x \le \frac{3}{2} \text{ or } x > 5\right\}$

79. $\left\{-\frac{6}{5}, -2\right\}$ **81.** $8x^2 - 8x - 7 = 0$ **83.** $\{x \mid -7 \le x \le 1\}$

85. Answers may vary. Examples: $(0, -6), (-1, -5), (-2, -2), (1, -5), (2, -2), (3, 3)$ **87.** $\left\{\frac{-3 + \sqrt{10}}{2}, \frac{-3 - \sqrt{10}}{2}\right\}$

89. $\left\{\left(\frac{-5 + \sqrt{73}}{6}, 0\right), \left(\frac{-5 - \sqrt{73}}{6}, 0\right)\right\} = \{(-2.26, 0), (0.59, 0)\}$

91. $\left\{\frac{-1 + i\sqrt{7}}{2}, \frac{-1 - i\sqrt{7}}{2}\right\}$

93. $\{-5 \pm 10i\}$ **95.** two real roots   **97.** 47 and 49 or -47 and -49 **99.** 12.65 inches by 12.65 inches

CHAPTER 8 REVIEW PROBLEMS *(pages 602–603)*

1. $\{2\sqrt{3}, -2\sqrt{3}\}$ **2.** $\{-5 -7\}$ **3.** $\{7, 3\}$ **4.** $x = 23$ and $x = 1$ **5.** $\{14, -5\}$ **6.** $\left\{2\frac{1}{2}, -1\frac{1}{2}\right\}$ **7.** $\{1, 1\}$

8. $\left\{\frac{4}{7}, \frac{-4}{9}\right\}$ **9.** $\left\{1\frac{1}{4}, \frac{-2}{3}\right\}$ **10.** $x^2 - 10x + 29 = 0$ **11.** $\{27\}$ **12.** $\{10\}$ **13.** $\{-0.5, -1.7\}$

14. $x = 0.4096$ and $x = 0.0081$ **15.** $\left\{\frac{1}{27}, \frac{-1}{8}\right\}$ **16.** $\left\{\frac{5 + i\sqrt{23}}{4}, \frac{5 - i\sqrt{23}}{4}\right\}$ **17.** 19 yards by 30 yards

18. Thursday   **19.** 0.0102 centimeter   **20.** $10 + 5i$ and $10 - 5i$ **21.** $(0, 0), (-1, 6), (-2, 16), (1, -2), (2, 0), (3, 6)$

22. $(0, -2), (-1, 7), (-2, 52), (1, 25), (2, 88), (3, 187)$ **23.** $\{(0.625, 0), (0, 0)\}$

24. $\left\{\frac{1}{2}, \frac{-2}{3}\right\}$

25. $\{x \mid x \le -6 \text{ or } x \ge 12\}$ **26.** $\{x \mid x = 0\}$

27. $\{x \mid x < -6 \text{ or } x > 1\}$ **28.** $\{x \mid \sqrt{2} < x \le 5\}$ **29.**

30. $\left\{x \mid x \le \frac{3}{4} \text{ or } x > 1\right\}$ **31.** $\{-7 + 9i, -7 - 9i\}$ **32.** $\left\{\frac{-16 + \sqrt{6}}{4}, \frac{-16 - \sqrt{6}}{4}\right\}$

33. $\left\{3\frac{1}{3}, \frac{2}{3}\right\}$ **34.** $x = 30.5$ and $x = 363$

35. Estimates may vary. Sample: $\left\{\left(\frac{-1 + \sqrt{3}}{2}, 0\right), \left(\frac{-1 - \sqrt{3}}{2}, 0\right)\right\}$ or $\{-1.37, 0.37\}$

36. $\{x \mid -1 \le x \le 5\}$ **37.** $\left\{4\frac{2}{3}, -3\frac{1}{3}\right\}$ **38.** $\{0.3 + 1.1i, 0.3 - 1.1i\}$

39. two real, rational roots   **40.** $\{x \mid -12 < x < -11\}$ **41.** $x^2 - 6x + 4 = 0$

42. $\left\{2\sqrt{3}, -2\sqrt{3}\right\}$ **43.**   **44.** $\{27, -12\}$

CUMULATIVE REVIEW *(page 605)*

1. base $= 10$ feet; height $= 6.5$ feet   **2.** $20.7x^4y^5$ **3.** $y = -4.4$ **4.** $\frac{2}{x + 5}$ **5.** $0.06a + b$

6. $x^2 + 4x + 2$ **7.** $11\sqrt{2}$ **8.** $t = 5\frac{47}{93}$ **9.** $\left\{9, \frac{1}{5}\right\}$ **10.** $t = 208$ **11.** $8y - 12x$

12. $\frac{3r}{(r - 1)^2(r - 4)}$ **13.** $x = 7$ **14.** $\frac{14\sqrt[3]{2x^2}}{x^3}$ **15.** $r = \frac{4s}{3t - 1}$ **16.** 10 **17.** $y = -3x + 5$

18. $14t\sqrt{2}$ **19.** $\frac{2}{9x}$ **20.** $\frac{x^2 + 1}{x^2 - 1}$

ANSWERS to Chapter 9 Odd-Numbered Exercises

SECTION 9.1 *(pages 620–622)*

1. Graphs (a) and (b) are similar in that they are the same shape; both have their vertex at the origin. They are different in that graph (a) opens up, and graph (b) opens down. Graphs (c) and (d) are similar in that they are the same shape; both have their vertex at the origin. They are different in that graph (c) opens up, and graph (d) opens down. Graphs (a) and (c) and (b) and (d) are similar in that they both have their vertex at the origin. They are different in that one graph in each pair [(a) and (b) is narrow and the other [(c) and (d)] is wide. Also, (a) and (c) open upward, and (b) and (d) open downward. All four graphs are similar in that their vertex is at the origin, and they share the axis $x = 0$.

3. Graphs (a) and (b) are similar in they are the same shape; both have their vertex at the origin. They are different in that graph (b) opens up, and graph (a) opens down. Graphs (c) and (d) are similar in that they are the same shape; both have their vertex at the origin. They are different in that graph (d) opens up, and graph (c) opens down. Graphs (a) and (c) and (b) and (d) are similar in that they both have their vertex at the origin. They are different in that one graph in each pair [(c) and (d)] is narrow and the other [(a) and (b)] is wide. Also, (b) and (d) open upward, and (a) and (c) open downward. All four graphs are similar in that their vertex is at the origin, and they share the axis $x = 0$.

5.   **7.**   **9.**   **11.**   **13.**

15.   **17.** $y = \frac{2}{9}(x - 1)^2 + 3$ **19.** $y = \frac{7}{4}(x + 1)^2 - 2$ **21.** $y = \frac{-5}{9}(x - 1)^2 - 3$

23. $y = -3x^2$ **25.** $y = 2(x - 0)^2 + 18$; $(0, 18)$; upward   **27.** $y = 2\left[x - \left(\frac{-1}{4}\right)\right]^2 + 7\frac{7}{8}$; $\left(\frac{-1}{4}, 7\frac{7}{8}\right)$; upward

29. $y = \frac{-1}{6}(x - 0)^2 + \frac{1}{36}$; $\left(0, \frac{1}{36}\right)$; downward   **31.** $y = 4[x - (-2)]^2 - 71$; $(-2, -71)$; upward

33. $(-1, -15)$; upward   **35.** $(-2, -3)$; upward   **37.** $\left(\frac{-1}{2}, -12\frac{1}{2}\right)$; upward   **39.** $(1.5, -0.5)$; downward

41. $\left(\frac{5}{6}, \frac{-23}{12}\right)$; downward   **43.** $(0, 12)$; downward   **45.** $x = \frac{8}{25}(y - 3)^2 - 3$ **47.** $x = \frac{-1}{3}(y + 1)^2 - 2$

49.   **51.**   **53.**

SECTION 9.2 *(pages 635–637)*

1. $x^2 + y^2 = 25$ **3.** $x^2 + y^2 = 18$ **5.** $x^2 + y^2 = 50$ **7.** $x^2 + y^2 = 500$

9. center: $(0, 0)$; radius: 4   **11.** center: $(0, 0)$; radius: 1

13. major axis: 8 (horizontal);
minor axis: 4 (vertical);
x-intercepts: $(4, 0)$, $(-4, 0)$;
y-intercepts: $(0, 2)$, $(0, -2)$

15. major axis: 10 (horizontal);
minor axis: 4 (vertical);
x-intercepts: $(5, 0)$, $(-5, 0)$;
y-intercepts: $(0, 2)$, $(0, -2)$

17. major axis: 6 (horizontal);
minor axis: 2 (vertical);
x-intercepts: $(3, 0)$, $(-3, 0)$;
y-intercepts: $(0, 1)$, $(0, -1)$

19. major axis: 10 (vertical);
minor axis: 6 (horizontal);
x-intercepts: $(3, 0)$, $(-3, 0)$;
y-intercepts: $(0, 5)$, $(0, -5)$

21.   **23.**   **25.**   **27.**   **29.**

31.   **33.** $y = (x + 7)^2$; vertex: $(-7, 0)$; opens upward   **35.**

SECTION 9.3 *(pages 653–657)*

1. $(x - 4)^2 + (y + 1)^2 = 25$ **3.** $(x + 1)^2 + (y - 2)^2 = 27$ **5.** $x^2 + (y - 2)^2 = 0.16$

7. $(x + 1)^2 + (y + 2)^2 = \frac{25}{16}$ **9.** center: $(0, -1)$; radius: 3

11. center: $(-4, 0)$; radius: 2   **13.** center: $(-2, -2)$; radius: $2\sqrt{3}$

15. center: $(1, 1)$; radius: 2   **17.** center: $(1, 0)$; radius: 5

19. center: (1, 2); radius: 4   **21.** $r = 5; x^2 + y^2 = 25$ **23.** $r = \sqrt{34}; (x - 3)^2 + (y + 5)^2 = 34$

25. $r = 13; (x - 2)^2 + (y - 3)^2 = 169$ **27.** $r = 3\sqrt{2}; (x - 2)^2 + (y + 3)^2 = 18$

29. $\frac{x^2}{1^2} + \frac{y^2}{3^2} = 1$; major axis: 6 (vertical); minor axis: 2 (horizontal); x-intercepts: $(1, 0), (-1, 0)$; y-intercepts: $(0, 3), (0, -3)$

31. $\frac{x^2}{3^2} + \frac{y^2}{6^2} = 1$; major axis: 12 (vertical); minor axis: 6 (horizontal); x-intercepts: $(3, 0), (-3, 0)$; y-intercepts: $(0, 6), (0, -6)$

33. $\frac{x^2}{2^2} + \frac{y^2}{5^2} = 1$; major axis: 10 (vertical); minor axis: 4 (horizontal); x-intercepts: $(2, 0), (-2, 0)$; y-intercepts: $(0, 5), (0, -5)$

35. $\frac{x^2}{2^2} + \frac{y^2}{6^2} = 1$; major axis: 12 (vertical); minor axis: 4 (horizontal); x-intercepts: $(2, 0), (-2, 0)$; y-intercepts: $(0, 6), (0, -6)$

37. center: (1, 3); horizontal axis: 4; vertical axis: 14   **39.** center: $(-3, 6)$; horizontal axis: 8; vertical axis: 6

41. $\frac{(x - 3)^2}{4} + \frac{(y - 4)^2}{1} = 1$ **43.** $\frac{(x + 1)^2}{4} + \frac{(y - 4)^2}{9} = 1$ **45.** $\frac{(x - 3)^2}{36} + \frac{(y - 4)^2}{1} = 1$ **47.** $\frac{(x - 2)^2}{144} + \frac{(y - 4)^2}{625} = 1$

49. $\frac{x^2}{3^2} + \frac{y^2}{12^2} = 1$; center: (0, 0); major axis: 24 (vertical); minor axis: 6 (horizontal)

51. $\frac{[x - (-3)]^2}{3^2} + \frac{y^2}{2^2} = 1$; center: $(-3, 0)$; major axis: 6 (horizontal); minor axis: 4 (vertical)

53. $\frac{(x - 6)^2}{2^2} + \frac{[y - (-2)]^2}{6^2} = 1$; center: $(6, -2)$; major axis: 12 (vertical); minor axis: 4 (horizontal)

55. $\frac{[x - (-1)]^2}{1^2} + \frac{(y - 2)^2}{3^2} = 1$; $a = 1, b = 3$; center at $(-1, 2)$

57. $\frac{(x - 1)^2}{2^2} + \frac{[y - (-2)]^2}{3^2} = 1$; $a = 2, b = 3$; center at $(1, -2)$

59. $\frac{(x - 1)^2}{2^2} + \frac{y^2}{3^2} = 1$; $a = 2, b = 3$; center at $(1, 0)$ **61.**   **63.**

65.   **67.** $\frac{(x - 10)^2}{484} - \frac{(y + 11)^2}{2809} = 1$; vertices: $(-12, -11), (32, -11)$

69. $\frac{(y - 3)^2}{2500} - \frac{(x - 2)^2}{576} = 1$; vertices: $(2, -47), (2, 53)$ **71.** $\frac{(y - 5)^2}{9} - \frac{(x - 3)^2}{4} = 1$; vertices: $(3, 2), (3, 8)$

73. $\frac{(x + 5)^2}{9} - \frac{(y + 2)^2}{16} = 1$; vertices: $(-8, -2), (-2, -2)$

75. $\frac{x^2}{6^2} - \frac{y^2}{2^2} = 1$; center: (0, 0); $a = 6, b = 2$; vertices: $(6, 0), (-6, 0)$

77. $\frac{x^2}{(1.2)^2} - \frac{y^2}{(1.2)^2} = 1$; center: (0, 0); $a = 1.2, b = 1.2$; vertices: $(1.2, 0), (-1.2, 0)$

79. $\frac{(x - 5)^2}{2^2} - \frac{(y - 1)^2}{1^2} = 1$; center: (5, 1); $a = 2, b = 1$; vertices: $(3, 1), (7, 1)$

81. $\frac{x^2}{1^2} - \frac{y^2}{3^2} = 1$; center: (0, 0); $a = 1, b = 3$; vertices: $(1, 0), (-1, 0)$

83. $\frac{y^2}{10^2} - \frac{x^2}{5^2} = 1$; center: (0, 0); $a = 5, b = 10$; vertices: $(0, 10), (0, -10)$

85. $\frac{x^2}{3^2} - \frac{y^2}{2^2} = 1$; center: (0, 0); $a = 3, b = 2$; vertices: $(3, 0), (-3, 0)$ **87.** $(x - 5)^2 + [y - (-1)]^2 = 1$, circle

89. $\frac{(y - 4)^2}{4} - \frac{(x - 3)^2}{4} = 1$, hyperbola   **91.** $\frac{(y - 3)^2}{12} - \frac{[x - (-4)]^2}{3} = 1$, hyperbola   **93.** $\frac{(x + 13)^2}{196} + \frac{(y + 11)^2}{484} = 1$

95. center: (6, 8); major axis: 30; minor axis: 24   **97.** $\frac{(y - 4)^2}{4} - \frac{(x - 3)^2}{4} = 1$; vertices: $(3, 2), (3, 6)$

99. major axis: 14;
minor axis: 4; center: (0, 0);
x-intercepts: (2, 0), (−2, 0);
y-intercepts: (0, 7), (0, −7)

SECTION 9.4 *(pages 667–672)*

1. 31 inches by 31 inches   **3. a.** 320.28 square inches   **b.** 87.72 square inches   **5.** 108 square inches
7. 256 feet   **9.** 0.58 astronomical unit   **11.** Neptune has an eccentricity closest to zero and therefore has the most nearly circular orbit.   **13. a.** $C_1 = 25.519, C_2 = 25.514, C_3 = 25.507$ **b.** C_1 is the closest. All are identical to the nearest tenth.   **15.** 342 square feet more brass (57 square feet more per side)   **17.** radius of red circle $= \sqrt{0.75}$ inch; radius of red and white circle $= \sqrt{1.5}$ inches; radius of big circle $= 1.5$ inches   **19.** 632 square feet
21. Answers may vary. Example: **23.** Answers may vary. Example:

25. Answers may vary. Example: **27.** Answers may vary. Example:

29. Answers may vary. Example: **31.** Answers may vary. Example:

33. $\{(-3, 0), (2, 5)\}$ **35.** no real-number solutions   **37.** $\{(7, 2), (7, -2)\}$ **39.** $\{(0, 1), (-1, 0)\}$
41. $\{(0, \sqrt{10}), (0, -\sqrt{10})\}$ **43.** $\left\{\left(2, \tfrac{1}{2}\right), (-1, -1)\right\}$ **45.** no real-number solutions   **47.** $\left\{\left(\sqrt{35}, 1\right), \left(-\sqrt{35}, 1\right)\right\}$
49.   **51.**   **53.** $y = 9\left(x + \tfrac{1}{3}\right)^2 + 5$; vertex: $\left(-\tfrac{1}{3}, 5\right)$; opens upward

55. center: $(12, -10)$; $a = 10, b = 11$ **57.** center: $(4, -2)$; radius: $4\sqrt{2}$
59. center: $(0, 0)$; major axis: 14 (vertical); minor axis: 10 (horizontal); x-intercepts: $(5, 0), (-5, 0)$; y-intercepts: $(0, 7), (0, -7)$
61.   **63.** center: $(7, -9)$; major axis: 18 (horizontal); minor axis: 16 (vertical)

65. $(x - 2)^2 + (y - 3)^2 = 8$ **67.**   **69.**

CHAPTER 9 REVIEW PROBLEMS *(pages 673–677)*

1. vertex: $(-3, -1)$; opens downward   **2.** $y = -0.75(x + 2)^2 + 3$

3. $y = (x - 2)^2 + 11$; vertex: $(2, 11)$; opens upward **4.** vertex: $(4, 4)$; points: $(3, -1)$, $(5, -1)$

5. vertex: $(-1, 8)$; opens downward **6.** vertex: $(6, 0)$; opens upward; $y = (x - 6)^2$
7. center: $(0, 0)$; radius: 8 **8.** $x^2 + y^2 = 3$

9. $\frac{x^2}{2^2} + \frac{y^2}{1^2} = 1$; center: $(0, 0)$; **10.** major axis: 10 (horizontal); minor axis: 4 (vertical)

major axis: 4 (horizontal);
minor axis: 2 (vertical);
x-intercepts: $(2, 0)$, $(-2, 0)$;
y-intercepts: $(0, 1)$, $(0, -1)$

11. **12.** **13.** $x^2 + (y + 2)^2 = 25$ **14.** $(x + 3)^2 + (y + 5)^2 = 34$

15. $(x - 5)^2 + (y - 7)^2 = 41$ **16.** $(x - 2)^2 + y^2 = 9$ **17.** $\frac{(x - 12)^2}{256} + \frac{(y - 3)^2}{64} = 1$

18. $\frac{(x + 15)^2}{289} + \frac{(y - 6)^2}{25} = 1$ **19.** $a = 16$, $b = 13$; center: $(-18, 22)$ **20.** $\frac{(x - 16)^2}{324} + \frac{(y + 13)^2}{576} = 1$

21. $\frac{x^2}{5^2} - \frac{y^2}{5^2} = 1$; center: $(0, 0)$; $a = 5$, $b = 5$; vertices: $(5, 0)$, $(-5, 0)$

22. $\frac{y^2}{6^2} - \frac{x^2}{3^2} = 1$; center: $(0, 0)$; $a = 3$, $b = 6$; vertices: $(0, 6)$, $(0, -6)$

23. $\frac{(x + 20)^2}{25} - \frac{(y - 30)^2}{1} = 1$; vertices: $(-25, 30)$, $(-15, 30)$ **24.** $\frac{(y - 5)^2}{144} - \frac{(x + 10)^2}{900} = 1$; vertices: $(-10, -7)$, $(-10, 17)$

25. 748.125 square inches **26.** $\frac{x^2}{(36.2)^2} + \frac{y^2}{(27.3)^2} = 1$ **27.** about 0.04 square inch **28.** $\frac{(x + 2)^2}{5^2} + \frac{(y - 3)^2}{(3.2)^2} = 1$

29. The white area is 0.785 square foot larger than the red area. **30.** 2147 cubic inches
31. Answers may vary. Example: 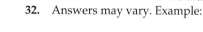 **32.** Answers may vary. Example:

33. $\{(10, 0), (-9, -19)\}$ **34.** $\{(3, 1), (3, -1), (-3, 1), (-3, -1)\}$ **35.** $\{(4, 1), (-4, 1)\}$ **36.** $\{(7, 1), (-7, 1)\}$

37. $(x + 2)^2 + (y + 3)^2 = 27$ **38.** $\frac{(x - 15)^2}{144} + \frac{(y + 3)^2}{529} = 1$ **39.** center: $(-5, 3)$; radius: $4\sqrt{3}$

40. vertex: $\left(-\frac{5}{4}, -\frac{105}{8}\right)$; opens upward **41.** vertex: $(1, 4)$

42. center: $(15, -10)$; major axis: 40 (horizontal); minor axis: 18 (vertical) **43.** $(x - 1)^2 + (y + 1)^2 = 1$

44. $\frac{(x - 5)^2}{4} - \frac{(y + 6)^2}{1764} = 1$; vertices: $(3, -6)$, $(7, -6)$ **45.** center: $(-4, 33)$; $a = 4$, $b = 11$

46. $(x - 6)^2 + (y + 5)^2 = 61$ **47.** vertex: $(1, 4)$; x-intercepts: $\left(\frac{5 - 2\sqrt{5}}{5}, 0\right)$, $\left(\frac{5 + 2\sqrt{5}}{5}, 0\right)$; y-intercept: $(0, -1)$

48. center: $(21, -10)$; major axis: 24; minor axis: 8; ellipse **49.** center: $(2, -2)$; radius: 4

CUMULATIVE REVIEW   *(page 678)*

1. $-|-2.3| < |-2.3| < [-(-2.8)]$ **2.** $1\frac{5}{8}$ **3.**

4. $-4\frac{1}{2}$ **5.** $x = \frac{-3}{7}; y = \frac{18}{7}$

6. $28x; 3x; 17$; distributive property   **7.** $\frac{-81}{49}$ **8.** $41\frac{1}{2}$ **9.** $x \geq -8$

10. Answers may vary. Example: $A = \{1, 7\}; B = \{5, 7\}$ **11.** 5 times larger   **12.** $x^4 + x^3 + x^2 + x + 1$

13. -13 **14.** $-7x + 9y$ **15.** $x = \frac{49}{4}$ **16.** $x = \frac{5}{9}$ **17.** $m = 15$ and $m = -5$ **18.** $\frac{D - h}{r} = t$

19. $x = 1$ **20.** $\left\{\left(\frac{3}{7}, \frac{-33}{7}\right)\right\}$

ANSWERS to Chapter 10 Odd-Numbered Exercises

SECTION 10.1   *(pages 690–693)*

1. not a function   **3.** one-to-one function   **5.** function; not one-to-one   **7.** one-to-one function

9. not a function

11. function; not one-to-one

13. one-to-one function

15. one-to-one function

17. $x + 3$ **19.** $\frac{2x - 1}{2}$

21. $\frac{x - 1}{3}$ **23.** $\frac{x + 1}{10}$ **25.** $(x + 3) - 3 = x; (x - 3) + 3 = x$

27. $\dfrac{2\left(\frac{2x - 1}{2}\right) + 1}{2} = \dfrac{2x - 1 + 1}{2} = \dfrac{2x}{2} = x; \dfrac{2\left(\frac{2x - 1}{2}\right) - 1}{2} = \dfrac{2x + 1 - 1}{2} = \dfrac{2x}{2} = x$

29. $3\left(\frac{x - 1}{3}\right) + 1 = x - 1 + 1 = x; \dfrac{(3x + 1) - 1}{3} = \dfrac{3x}{3} = x$

31. $10\left(\frac{x + 1}{10}\right) - 1 = x + 1 - 1 = x; \dfrac{(10x - 1) + 1}{10} = \dfrac{10x}{10} = x$ **33.**

35.

37.

39.

41. function; not one-to-one   **43.** not a function   **45.** not a function

47. function; not one-to-one   **49.**

51.

53.

SECTION 10.2 *(pages 710–711)*

1.   **3.**   **5.**   **7.**   **9.**

11.   **13.** $\log_3 81 = 4$ **15.** $\log_9 27 = x$ **17.** $\log_{\frac{1}{2}} x = -2$ **19.** $\log_4 \frac{1}{64} = -3$ **21.** $3^2 = 9$

23. $8^{\frac{2}{3}} = 4$ **25.** $10^5 = 100{,}000$ **27.** $\left(\frac{2}{3}\right)^{-3} = \frac{27}{8}$ **29.** $-3; -2; -1; 0; 1; 2$

31. $3; 2; 1; 0; -1; -2; -3; -4$ **33.**   **35.**   **37.**

39.   **41.** $\frac{1}{8}; 2^{a-3}; 2^{2x-3}$

43. No, because, for example, $f(1) = 2(1)^2 - 6 = 2 - 6 = -4$, and $f(-1) = 2(-1)^2 - 6 = 2 - 6 = -4$ **45.** 2

SECTION 10.3 *(pages 723–725)*

1. 0   **3.** 0   **5.** 1   **7.** 2   **9.** 6   **11.** $x + 2$ **13.** $\log_2 3 + \log_2 a$ **15.** $\log_5 6 + \log_5 7$
17. $\log_5 3 - \log_5 4$ **19.** $3 \log_2 x + \log_2 y$ **21.** $\log_5 2 + 3 \log_5 x$ **23.** $\log_7 2 + 3 \log_7 x - \log_7 5$

25. $\ln 5 + \frac{1}{2} \ln 2 - 2 \ln c$ **27.** $\frac{3}{2} \log r - \frac{1}{2} \log t - \frac{1}{2} \log s$ **29.** $\log_2 3 + \log_2 a + \log_2 b - 2 \log_2 c$

31. $3 \log_8 10 + 3 \log_8 x - 3 \log_8 5 - 3 \log_8 y$ **33.** $\frac{1}{2} \log_8 3 + \frac{1}{2} \log_8 a - \frac{1}{2} \log_8 5$

35. $\log_3 6 + \frac{2}{3} \log_3 2 + \log_3 x + 2 \log_3 y$ **37.** $\log_4 xt$ **39.** $\log_8 \left(\frac{6}{c}\right)$ **41.** $\log_3 \frac{7}{x^2}$ **43.** $\log_2 \sqrt{\frac{7}{x}}$

45. $\log_4 \left(\frac{6x^3}{z^2}\right)$ **47.** $\log_5 \left(\frac{9x}{y^2}\right)$ **49.** $x = 2$ **51.** $x = 4$ **53.** $x = -3$ **55.** $x = \frac{1}{2}$ **57.** $x = 1$

59. $x = -2$ **61.** $x = 64$ **63.** $x = \frac{1}{8}$ **65.** $x = 3$ **67.** $x = \frac{1}{2}$ **69.** 2   **71.** -3 **73.** $\frac{3}{2}$

75. -3 **77.** $x = 2$ and $x = -2$ **79.** $x = 1$ and $x = -1$ **81.** $x = 1$ and $x = -1$ **83.** $x = 17$
85. $x = 41$ **87.** $x = \frac{10}{3}$ **89.** $x = 3$ **91.** $x = 3$ **93.** $x = 2$ **95.** $2x - 3$ **97.** $x = -1$

99. $\frac{x + 2}{3}$

SECTION 10.4 *(pages 737–741)*

1. 2.7160   **3.** $x = 1412.5375$ **5.** 0.00001   **7.** 3.7160   **9.** $\frac{\ln 8}{\ln 2} = 3.0000; \frac{\log 8}{\log 2} = 3.0000$

11. $\frac{\ln 1000}{\ln 5} = 4.2920; \frac{\log 1000}{\log 5} = 4.2920$ **13.** $\frac{\ln 132}{\ln 4} = 3.5222; \frac{\log 132}{\log 4} = 3.5222$

15. $\frac{\ln 0.06}{\ln 12} = -1.1322; \frac{\log 0.06}{\log 12} = -1.1322$ **17.** 2.4849 **19.** $x = 12.1825$ **21.** 1.3785 **23.** 4.4140

25. $x = 0.7324$ **27.** $x = 7.3471$ **29.** $x = 1.1761$ **31.** $x = 1.8451$ **33.** $x = 4.4550$ **35.** $x = -4.4829$

37. $x = 1.4650$ **39.** $x = 12.2883$ **41.** $x = -3.7824$ **43.** $x = 3$ **45.** $x = 31.7319$ **47.** $x = 1.9378$

49. a. 4500 **b.** 13,641 **c.** about 853 **d.** 4,608,000 **51. a.** \$1221.40 **b.** \$900 invested at 8% interest

c. \$3,336,051,738 **d.** 14% **53. a.** 180 deer **b.** 242 **55. a.** 8000 bacteria **b.** after $3\frac{1}{2}$ hours

57. a. 47,638 **b.** 3391 **c.** about 4 weeks **59. a.** about 13 years **b.** 44 years ago, if population growth

has been constant **61.** yes; {(3, 2), (5, 3), (7, 6), (8, 9), (9, 8)} **63.** 1.4999 **65.** $x = 2$ **67.** 1.4651

69. $\log_b 5 + \frac{1}{2}\log_b 2 - \log_b y$

71. $\left(\frac{1}{2}\right)^{-2} = 4$ **73.** $b = \frac{1}{3}$ **75.** $x = 216$ **77.** $x = \frac{-3}{5}$

$\left(2^{-1}\right)^{-2} = 4$
$2^2 = 4$
$4 = 4$

CHAPTER 10 REVIEW PROBLEMS *(pages 744–745)*

1. yes; {(3, 2), (5, 4), (8, 6)} **2.** $y = \frac{2 - 4x}{3}$ **3.** $\frac{(3x - 16) + 16}{3} = x$ **4.** $\frac{3 - 5\left(\frac{3 - 2x}{5}\right)}{2} = x$

5. **6.** **7.** **8.** **9.** $\frac{1}{4}$; 1; 16

10. $6x - 2$ **11.** $\log_3 3^6 = 6$ **12.** $x = \frac{-1}{3}$ **13.** $x = 3$ **14.** $x = 4$ **15.** 5300; 8066

16. $\log_3 6561 = 8$ **17.** $5^4 = 625$ **18.** $x = 81$ **19.** $x = 4$ **20.** $x = \sqrt{2}$ or $x = -\sqrt{2}$ **21.** $x = \frac{-1}{4}$

22. $\log_2\left(\frac{\sqrt{3}}{5^3}\right)$ **23.** $\log_a\left(\frac{6^2 \cdot 5^3}{9}\right)$ **24.** $x = 1024$ **25.** $x = 1000$ **26. a.** $D = 89.2942$ decibels

b. $I = 5 \times 10^{-10}$ watt/square meter **c.** $D = 65.0515$ decibels **27.** 1.7918 **28.** 148.4132 **29.** 3.2041

30. 6.3096×10^{-7} **31.** 1.7959 **32.** $x = 0.8959$ **33.** $x = -0.2569$ **34.** yes; $\frac{x - 10}{2}$ **35.** 3^x

36. $\log_3\left(\frac{\sqrt[3]{35}}{8}\right)$ **37.** $x = 0.6931$ **38.** no, because $\pm\sqrt{y - 2} = x$, there are two x values for each value of y.

39. $x = -2.6947$ **40.** $\log_3(3 \cdot 16^5)$, or $\log_3 3,145,728$ **41.** -4 **42.** $x = 1.2398$ **43. a.** 500 grams

b. 0.17 gram **c.** after about 1 minute (52 seconds) **d.** $A = A_0 e^{-0.0001244t}$ **44.**

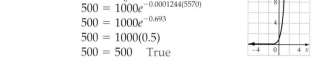

$500 = 1000e^{-0.0001244(5570)}$
$500 = 1000e^{-0.693}$
$500 = 1000(0.5)$
$500 = 500$ True

CUMULATIVE REVIEW *(pages 747–748)*

1. $x = \frac{-4}{3}$ **2.** $6\sqrt{6} + 8\sqrt{3} + 6 - 3i\sqrt{3}$ **3.** 30 centimeters **4.** $(2x - 3)(x + 3)$ **5.** $\frac{5}{15}$ **6.** $\frac{y^{10}z^6}{9x^{16}}$

7. $\frac{2x^2 + 4x - 9}{-(3x^2 + 4x - 12)}$ **8.** $x = \frac{-39}{8}$ **9.** The focal length is 10 centimeters. **10.**

11. $2^5 \cdot 5^5 \sqrt[12]{2^{11}5^3}$ **12.** $30x$ **13.** $x = \frac{3 + i\sqrt{71}}{8}$ and $x = \frac{3 - i\sqrt{71}}{8}$ **14.** $\left\{x \mid x \le \frac{-1}{40}\right\}$ **15.** slope: 4

16. $y = \frac{1}{3}x + 1\frac{2}{3}$ **17. a.** 8 units **b.** 220 square units **18.** -24 **19.** 1, or 1×10^0

20. $f(0) = 3; f(-1) = 452$

INDEX